Topics in Mining, Metallurgy and Materials Engineering

Series editor

Carlos P. Bergmann, Porto Alegre, Brazil

More information about this series at http://www.springer.com/series/11054

Ling Bing Kong · Yizhong Huang
Wenxiu Que · Tianshu Zhang
Sean Li · Jian Zhang · Zhili Dong
Dingyuan Tang

Transparent Ceramics

Springer

Ling Bing Kong
School of Materials Science and Engineering
Nanyang Technological University
Singapore
Singapore

Yizhong Huang
School of Materials Science and Engineering
Nanyang Technological University
Singapore
Singapore

Wenxiu Que
School of Electronic and Information Engineering, Electronic Materials Research Laboratory
Xi'an Jiaotong University
Xi'an, Shaanxi
China

Tianshu Zhang
Anhui Target Advanced Ceramics Technology
Hefei, Anhui
China

Sean Li
School of Materials Science and Engineering
The University of New South Wales
Sydney
Australia

Jian Zhang
Jiangsu Key Laboratory of Advanced Laser Materials and Devices
Jiangsu Normal University
Xuzhou, Jiangsu
China

Zhili Dong
School of Materials Science and Engineering
Nanyang Technological University
Singapore
Singapore

Dingyuan Tang
School of Electrical and Electronic Engineering
Nanyang Technological University
Singapore
Singapore

ISSN 2364-3293 ISSN 2364-3307 (electronic)
Topics in Mining, Metallurgy and Materials Engineering
ISBN 978-3-319-18955-0 ISBN 978-3-319-18956-7 (eBook)
DOI 10.1007/978-3-319-18956-7

Library of Congress Control Number: 2015938749

Springer Cham Heidelberg New York Dordrecht London

Printed on acid-free paper

Springer International Publishing AG Switzerland is part of Springer Science+Business Media (www.springer.com)

Preface

Transparent polycrystalline ceramics have been found various applications as laser hosts, infrared (IR) windows/domes, lamp envelopes, and transparent armors, instead of their glass and single-crystal counterparts, due to their processing flexibility in fabricating items with large sizes and complex shapes and more importantly cost-effectiveness. High optical transparent ceramics require high purity and high density. To achieve high-purity final products, it is necessary to use high-purity precursor powders, which requires the use of various advanced synthetic methods, such as chemical coprecipitation, sol-gel, thermal spray and spray pyrolysis. To get full density, various sintering technologies have been employed, such as high-pressure (HP) sintering, hot isostatic pressure (HIP) sintering, vacuum sintering and spark plasma sintering (SPS). It is important to understand the interrelationship among powder synthesis, characteristics, compaction, sintering, microstructure, grain size and performances in solid-state lasers and other applications.

Although various books on solid-state lasers and ceramic processing have been available in the open literature, very less information is present to cover both the solid-state lasers and processing of transparent ceramics, due to the multi-disciplinary or interdisciplinary characteristics. The purpose of this book is to bridge these two very important fields. It consists of 10 chapters, covering the topics of transparent ceramic materials, ceramic powder synthesize, powder characterization and compaction, fundamentals of sintering, new sintering techniques, microstructural development of transparent ceramics, densification of transparent ceramics by using different methods, solid-state ceramic lasers and other applications of transparent ceramics. The book can be used as a reference for senior undergraduate students, postgraduate students, researchers and engineers, in materials science and engineering, applied physics, solid-state lasers, solid-state physics, and so on.

The financial supports from the start-up grant (SUG/2012) from Nanyang Technological University and AcRF Tier 1 project (RG44/12) from Ministry of Education, Singapore, are acknowledged.

Ling Bing Kong
Yizhong Huang
Wenxiu Que
Tianshu Zhang
Sean Li
Jian Zhang
Zhili Dong
Dingyuan Tang

Contents

Chapter 1
Introduction

1.1 Transparency

Transparent materials form a special group of materials that have a wide range of applications in various aspects of our daily life. Transparency, also known as pellucidity or diaphaneity, is a unique physical property of materials, which measures their ability to allow light to pass through them without the presence of scattering. At macroscopic scale, the behavior of the photons follows Snell's law, because this dimension is much larger than the wavelength of photons. Translucency is another property closely related to transparency, which is also known as translucence or translucidity. In this case, the materials allow light to pass through, but Snell's law may not be followed at macroscopic scale. In other words, the photons could be scattered either internally or at the interfaces where the index of refraction is varied. The opposite property of translucency and transparency is opacity. Transparent materials usually have a clear appearance, either in one color or any combination leading to a spectrum of multiple colors. When light shines on a material, they could have several interactions, which are dependent on both the wavelength of the light and the nature of the material. Photons interact with an object in the forms of reflection, absorption, and transmission [1].

1.2 Transparent Materials

Conventional materials that are optically transparent mainly include glasses, polymers, and alkali hydrides, which have various applications in industries and daily life. However, these materials have relatively poor mechanical strength and sometimes insufficient chemical and physical stabilities. More importantly, conventional transparent materials have strong absorption in the IR range, making them not suitable for applications in this spectral range. Also, they usually possess

L.B. Kong et al., *Transparent Ceramics*, Topics in Mining,
Metallurgy and Materials Engineering, DOI 10.1007/978-3-319-18956-7_1

relatively low melting temperatures (<600 °C), so that they cannot be used for applications at high temperatures. As a result, it is desirable to develop new transparent materials that can be used for applications operating in harsh and extreme environments.

With the advancement in technologies in material growth, single crystals of some compounds appeared as new transparent materials. Compared with conventional transparent materials, single crystals, such as sapphire (Al_2O_3) for IR windows, YAG (yttrium garnet, $Y_3Al_5O_{12}$) for lasers, and PZN-PT (lead zinc niobate-lead titanate, $PbZn_{1/3}Nb_{2/3}O_3$-$PbTiO_3$) for electro-optics, addressed most of the problems encountered by conventional transparent materials. However, growth of single crystals requires sophisticated facilities and is time-consuming, thus leading to expensive products that are only used in the case cost is not necessarily considered. Furthermore, as-grown single crystals have shapes determined more by their lattice structures and less by processing conditions. Therefore, machining of single crystals to meet the requirement of specific applications is a difficult task. Other problems of single crystals include difficulty of large-scale production and mechanical brittleness of some materials. In this respect, transparent ceramics become more and more important. Transparent ceramics have various advantages over single crystals, such as cost-effectiveness, large-scale production, feasibility of shape controlling, and better mechanical properties.

Different from single crystals, ceramics have various sites to scatter light, including residual pores within grains and at grain boundaries, grain boundaries, second phases (impurities) at the grain boundaries, and double refraction from birefringent materials, as shown schematically in Fig. 1.1 [2].

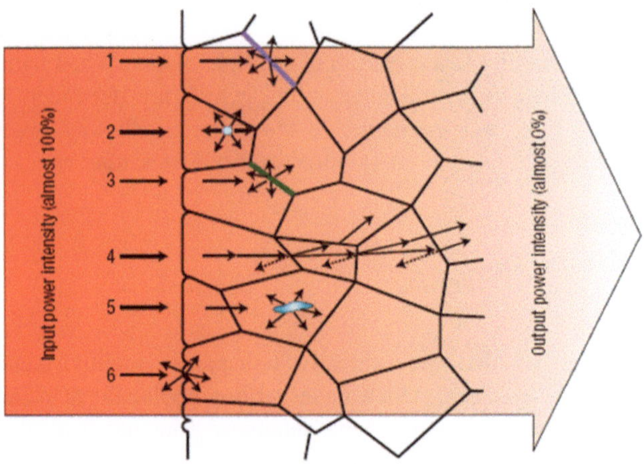

Fig. 1.1 Schematic microstructure of a ceramic with various light-scattering sources: (*1*) a grain boundary, (*2*) residual pores, (*3*) secondary phase, (*4*) double refraction, (*5*) inclusions and (*6*) surface roughness in ceramics prohibits applications in optics. Reproduced with permission from [3]. Copyright © 2008, Nature Publishing Group

The most significant factor for transparency of ceramics is porosity. The surface of a pore is a boundary between phases with sharply different optical characteristics, which therefore intensely reflects and refracts light. The presence of a large number of pores makes ceramics opaque. Pores could be interbrain or intragrain. The elimination of intragrain pores, even if they are submicron in size, is a difficult and longer process than the elimination of closed interbrain pores. Intergrain pores are located at grain boundaries, acting as sinks of vacancies, which can be easily removed compared with those at other locations.

Ceramics consist of grains and grain boundaries. If there is a difference in properties, e.g., composition, between grains and grain boundaries, the interfaces between them will become scattering sites of light. To be transparent, the difference in optical behaviors between grains and grain boundaries should be minimized. Transparent ceramics have less clear grain boundaries, while opaque ceramics have clear grain boundaries. The presence of a second phase at the grain boundaries is generally the most common reason. Therefore, to fabricate transparent ceramics, it is necessary to use raw material of high purity and avoid any possible contamination during processing. Besides, care should be taken when selecting additives. For example, the amount of additives must be as low as possible, so that they completely dissolve in the solid solution with the main phase, without the presence of second phases.

Crystal structure plays an important role in determining whether ceramics can be optically transparent or not. In ceramics of optically anisotropic crystals, additional scattering of light arises at the boundaries when the light travels from one grain to another. This is the reason why transparent ceramics generally have a cubic lattice structure, which is isotropic, such as MgO, Y_2O_3, YAG ($Y_3Al_5O_{12}$), and $MgAl_2O_4$ (spinel) [4–6]. However, several materials that are not cubic crystal have also been made to be transparent, such as mullite, Al_2O_3 and some ferroelectrics. This means that the principles governing transparency of ceramics should be further studied. In fact, since the explanation was first proposed, it has never been challenged. For example, the first point is grain boundary. It is well known that ceramics always have grain boundaries. Therefore, this point should be revised to be quality of grain boundary, which means that grain boundaries cannot be avoided in ceramics, but the quality of grain boundary affects transparency.

Besides the internal factors, there are also external factors that affect transparency of a ceramic sample, including thickness and surface finish. A rough surface means a significant diffuse scattering, so the sample should be as smooth as possible. Generally, transparency decreases with increasing thickness. Thickness-independent transparency is only possible when the material reaches its theoretical maximum of in-line transmission.

In summary, the key strategy to develop transparent ceramics is to eliminate all possible scattering sites of light, including least porosity (>99.9 % of theoretical density), absence of pores at grain boundaries or pores with size smaller than the wavelength of light, absence of second phase (impurity or glass phase) at grain boundaries (negligible difference in optical property between grains and grain boundaries), high quality grain boundaries, small grain size (as compared to the wavelength of visible light) with uniform size distribution, isotropic lattice structure, and high surface finish.

1.3 Important Issues on Transparency Important

Various transparent ceramics have been developed using different sintering techniques, combined with the use of ultrafine precursor powders with high sinterability. Although significant progress has been made on the one hand, there are certain problems in research in this area, of which a detailed discussion is presented in Ref. [7].

It is necessary to emphasize that a transparent component means that when it is used as a window a clear image of subjects far away from it can be formed. The prerequisite of such a clear image is the absence of any scattering when the light travels through the transparent window. Qualitative evidence by observation is used to show the transparency of transparent ceramics. The right way to do this is to maintain a sufficiently large distance between the objects and the transparent samples, as shown in Fig. 1.2a. In this way, it is clearly demonstrated that the Al_2O_3

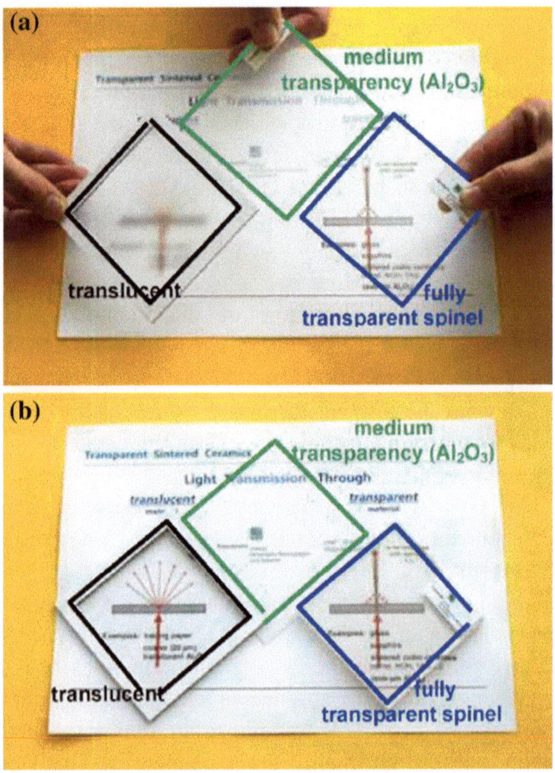

Fig. 1.2 Photographs of a thin translucent organic material of 0.06 mm, a sintered window of sub-μm Al_2O_3 (thickness 0.8 mm, real in-line transmission ~60 %) and 6 mm thick almost fully transparent spinel for comparison. The characteristic difference in transmission among the three samples is clearly demonstrated in **a**, while there is no clear difference due to the wrong positioning in **b**. Reproduced with permission from [7]. Copyright © 2009, Elsevier

sheet is translucent, while the spinel and organic samples are transparent. However, more generally, samples have been laid on background with a direct contact, as shown in Fig. 1.2b. In this case, all samples look like transparent. Another concern with accurate presentation of transparency is measurement. The apertures of most commonly used photospectrometers are in the range of 3°–5°, with which the measured transparencies are usually overestimated by including a fairly large amount of the scattered portion. For example, as the opening angle is increased from 0.5° to 5°, the recording area is increased by 100 times. Therefore, it is urgent to establish an international standard for the measurement of transparency of transparent ceramics.

Another important aspect is the so-called thickness effect if transparency of a material is below the theoretical maximum. As shown in Fig. 1.3a, b, the sample has 60 % of its theoretical transparency. As the sample thickness is increased from (b) to (a), the amount of scattered light is greatly increased. In other words, the real in-line transparency of the sample is reduced with increasing thickness. To exclude the contribution of scattered light, a narrow aperture should be used. Otherwise, the measured transparency will be overestimated. Only when a material reaches its theoretical maximum transparency, its transparency is not affected by thickness, as shown in Fig. 1.3c. In this respect, it is always desired to fabricate transparent ceramics with maximum transparency without any scattering or absorption. This is especially important for applications, such as armors, where a relative large thickness is necessary to ensure sufficiently high ballistic strengths.

As discussed earlier, various scattering losses are the main contribution to the decrease in in-line transmission, with respect to the theoretical maximum, of transparent ceramics, as summarized in Fig. 1.4 [7, 8]. Without the presence of these losses, the theoretical maximum of transmission is 100 % minus the reflection

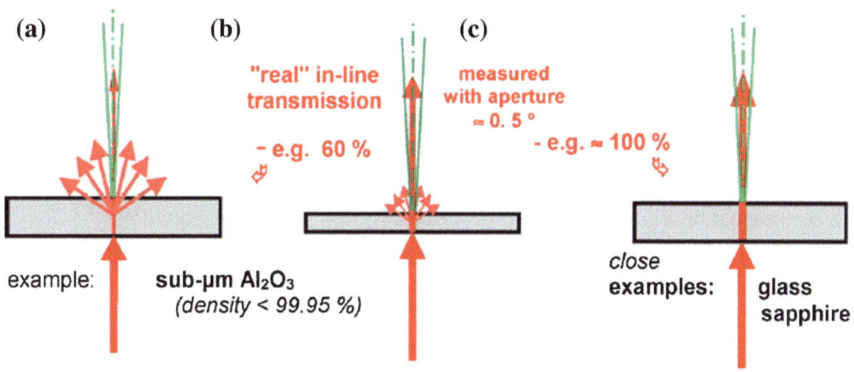

Fig. 1.3 Thickness effect (**b** → **a**) on in-line transmission of transparent of the transparent ceramics with an in-line transmission of <100 %, due to the scattering and/or absorption. "Real" in-line transmission is measured by using narrow measuring aperture (**b**). This thickness effect is absent as the transmission approaches the theoretical maximum (**c**). Reproduced with permission from [7]. Copyright © 2009, Elsevier

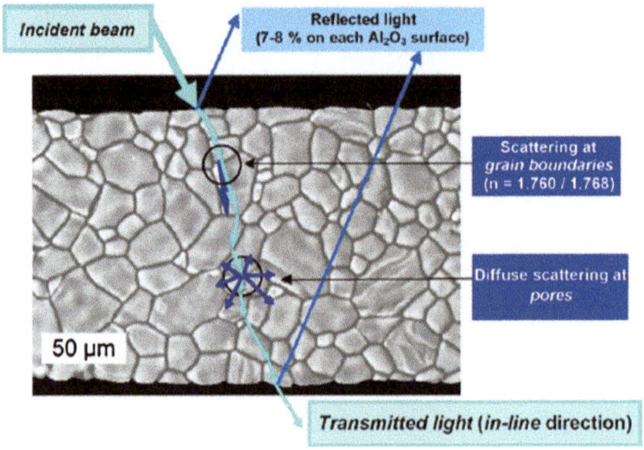

Fig. 1.4 Light transmission through polycrystalline translucent ceramics (e.g. coarse Al_2O_3). Loss in intensity of the incident light due to absorption, reflection on the two surfaces, diffuse scattering at pores and scattering caused by the birefringent splitting (specifically for noncubic materials). Reproduced with permission from [7]. Copyright © 2009, Elsevier

on both surfaces of a piece of the material. At normal incidence, the reflection R_1 on one surface is related to the material refractive index n, given as:

$$R_1 = \left(\frac{n-1}{n+1}\right)^2, \tag{1.1}$$

and the total reflection loss, including multiple reflection, is

$$R_2 = \frac{2R_1}{1+R_1}. \tag{1.2}$$

Therefore, the maximum transmission is as follows:

$$T_{\text{th}} = 1 - R_2 = \frac{2n}{n^2+1}. \tag{1.3}$$

When all the factors are present, the real in-line transmission (RIT) of a fully dense transparent ceramics can be described by the following equation:

$$\text{RIT} = T_{\text{th}} \exp\left[-3\pi^2\left(\frac{\Delta n}{n}\right)^2 \frac{rd}{\lambda^2}\right], \tag{1.4}$$

where $\Delta n/n$ is the ratio of refractive index difference between polarization perpendicular and parallel to c-axis to the average index n, $2r$ is grain size of the ceramics, λ is the wavelength of incident light, and d is thickness of the sample. This equation implies that, at a given thickness, the RIT is closely related to the $\Delta n/n$ and r. The smaller the values of $\Delta n/n$ and r, the high the RIT will be. If $\Delta n/n$ is an intrinsic property of the materials, r is an extrinsic parameter that can be controlled through material processing. Therefore, for the given materials, the grain size of ceramics should be sufficiently small to achieve high RIT. For example, the grain size of a highly dense sintered Al_2O_3 has to be about 0.5 μm for RIT of 60–65 % at $\lambda = 640$ nm and $d = 1$ mm [7].

Besides, according to the definition of T_{th}, the transmission of transparent ceramics can never be 100 %. It is also necessary to mention that T_{th} in Eqs. (1.3) and (1.4) is calculated by assuming that the reflection occurs on a theoretically perfect surface. Because theoretically perfect surface is an ideal assumption, real ceramics inevitably have a certain value of roughness. Therefore, technical roughness gives rise to additional loss—another mechanism which decreases the real in-line transmission of products depending on the quality of surface preparation [7].

Traditional high-tech ceramics (e.g., for hip implants or cutting tools) are regarded as dense when their relative density reaches, e.g., 99.7 %. However, the starting point of increase in real in-line transmission is usually at residual porosities of <0.1–0.2 %. The key technology to achieve highly transparent ceramics is the use of high sintering temperatures (>1700 °C), even in cases when sintering facilities (HP, HIP or SPS) are used. As a result, transparent ceramics often have extremely large grain sizes (50–300 μm), which is detrimental to mechanical properties. In this light, development of transparent ceramics with fine-grain microstructure remains a challenge to the research community. Systematic knowledge of the relationship among the properties of precursor powders, green body preparation, as well as the combined use of various technologies, will be useful to address this problem, which is currently still lacking and thus deserves to be a subject of future research in this area.

1.4 History of Solid-State Lasers

1.4.1 Lasers

Lasers have been the main component in the optical range of quantum electronics, science of generation and amplification, control of properties, such as wavelength, temporal regime and power range, of radiation by the processes of interaction of electromagnetic radiation with quantum systems. The word LASER is an acronym, for light amplification by stimulated emission of radiation. The driving mechanism that determines the amplification of radiation in a laser is the stimulated emission in an ensemble of identical quantum systems, i.e., the active medium, with inversion of populations between a high energy level and a lower energy level, which act as emitting and terminal energy levels for the emission, in a resonator that provides the necessary conditions for amplification and extraction of the amplified radiation beam.

To discuss solid-state lasers, it is necessary to start with lasers in general. Laser was invented in the 1960s and has been growing quickly, with a wide range of applications in industry, defense, and daily life [2, 9–11]. According the properties of the gain medium used, lasers can be classified into four types: (i) semiconductor lasers, (ii) gas lasers, (iii) liquid (dye) lasers, and (iv) solid-state lasers.

Semiconductor lasers are mainly used as optical sources in optical communications and information storages. Gas lasers are optical sources usually used for measurements (He–Ne lasers), etching or lithography (Excimer) and industrial cutting, welding, or consolidating (CO_2 lasers). Dye lasers are highly tunable and can produce pulses of very short duration (of the order of femtosecond). Solid-state lasers represent a new class of lasers, including glass lasers and crystal lasers. Glass lasers are used in nuclear fusion reactors, while single crystal or ceramic (e.g., YAG) lasers have found applications in mechanical processing and medical equipment.

Various materials have been developed for solid-state lasers [12]. Crystalline solid-state laser materials include garnets, such as $Y_3Al_5O_{12}$ (YAG), $Gd_3Ga_5O_{12}$ (GGG), $Gd_3Sc_2Ga_3O_{12}$ (GSGG), and $Y_3Sc_2Ga_3O_{12}$ (YSGG) [13–16], fluorides such as CaF_2 and MgF_2 [17, 18], alexandrites [19–21] and other oxide-based materials [22–24], which can be either single crystals or polycrystalline ceramics. Noncrystalline materials are mainly silicate and phosphate glasses, containing Er, Yb, or Ce [25–28]. Physical properties of materials for solid-state lasers include high thermal conductivity, high chemical stability, and high mechanical strength. In terms of optical properties, they must have a large cross-section of stimulated emission (σ), long fluorescence lifetime (τ), low possible laser thresholds, and stable laser oscillation in continuous wave (CW). The best candidate is YAG, because it meets almost all the requirements.

Single crystals appeared earlier than ceramics for solid-state lasers. Melt-growth processes are used for single crystals. The Verneuil process is among the early single crystal growth techniques, also known as flame fusion process. Single crystals prepared using this method encountered quality problems in terms of laser performances. Currently, the most widely used single crystal growth technique is the Czochralski (Cz) process. It is well known that single crystal growth is very time-consuming. Usually, only limited portions of a single crystal have optical homogeneity for laser oscillations, because of the intrinsic core and facets that are determined by the growth characteristics. As a result, fabricated solid-state single crystal lasers are costly for large-scale applications.

Comparatively, the use of transparent ceramics could address these technical problems. However, due to the presence of grain boundaries, ceramic solid-state lasers have a common problem, i.e., the efficient amplification of the coherence beam in the materials could be disturbed. This problem was serious during the early state of ceramic solid-state lasers, such as Dy:CaF_2 and Nd:ThO_2–Y_2O_3. Although laser oscillation was observed by using these ceramics, the resulting beam quality and lasing efficiency were not sufficient for practical applications. However, the success of Nd:YAG ceramic laser revived ceramic solid-state laser. More importantly, ceramic materials can now exhibit equivalent laser performance of single crystals. Therefore, the demonstration of Nd:YAG ceramic laser is a great milestone in ceramic solid-state lasers. This is because ceramic processing is significantly simpler than single crystals. It is believed that transparent ceramics will be dominant candidates for solid-state lasers in the future.

1.4.2 Single Crystals

In general, the performance of melt-growth technologies, such as homogeneity, uniformity, crystal size, doping concentration, productivity, and so on, are largely determined by thermophysical properties of the material, like melting temperature and thermal conductivity, thermochemical or thermostructural properties, and compositional limitation. On the other hand, growth equipment, including size of the crucible, thermal field distribution, mechanical stability, and so on, also plays an important role in determining the quality of the grown crystals.

The dominant single crystal for solid-state lasers is YAG, which is produced using the Cz melt-growth process [29]. Transition metal elements or lanthanide rare earth elements are used as laser active ions that are doped in YAG host material. Due to its narrow spectral width and high quantum efficiency, Nd ion, a four-level laser system has been acknowledged to be the most popular active ion.

Nd^{3+} ion has three fluorescent emissions: (i) 0.9 μm ($^4F_{3/2} \rightarrow {}^4I_{9/2}$), (ii) 1.06 μm ($^4F_{3/2} \rightarrow {}^4I_{11/2}$), and (iii) 1.3 μm ($^4F_{3/2} \rightarrow {}^4I_{13/2}$), with radiative transition possibilities of 0.25, 0.60, and 0.15, respectively [30–32]. Energy efficiency emission (ii) is the highest, so that Nd:YAG crystals are used for 1.06 μm generation. To achieve laser action in Nd:YAG, the electrons of Nd ions at the ground state are first excited with a xenon (Xe) or krypton (Kr) lamp, i.e., white light source from a discharge lamp, so that the electrons are pumped up to energy levels higher than $^4F_{3/2}$. Semiconductor lasers (LD, laser diode) of a certain wavelength can also be used to pump ground state electrons to the $^4F_{5/2}$ band. The excited electrons at the upper levels would decay rapidly to the $^4F_{3/2}$ level, which is a non-radiative transition. When these electrons are transferred to $^4F_{13/2}$, $^4F_{11/2}$, and $^4F_{9/2}$, three fluorescent emissions, radiative transitions with λ = 0.9, 1.06, and 1.3 μm are obtained. As fluorescent emission is amplified by using a set of mirrors as resonator, an intensified light with a single wavelength, i.e., monochromatic light, is finally emitted.

When exciting with lamp source, electron transitions from the upper levels to the $^4F_{3/2}$ level are non-radiative and energy is released in the form of heat. As a result, effective laser oscillation cannot be obtained by using a system of lamp excitation. However, lamp excitation systems can offer high output powers, because lamps can supply very high input powers. If LD excitations are used, laser diodes of wavelengths 808 and 885 nm would pump the ground state electrons directly to upper levels, $^4F_{5/2}$ and $^4H_{3/2}$ levels, respectively. Because there are no excited electrons for non-radiative transitions in this case, efficient laser oscillation can be obtained. The only problem is that it was difficult to have high power LDs.

As stated earlier, YAG single crystals are grown using the Cz method. High purity starting materials, i.e., over 4N or 99.99 %, Y_2O_3, Al_2O_3 and Nd_2O_3 powders, have to be used to grow high purity single crystals. The powders were first mixed according to the stoichiometric composition of Nd:YAG. They were then pressed into pellets and sintered at certain temperatures. The sintered Nd:YAG pellets were put in an iridium (Ir) crucible and then melted through high frequency induction heating to a temperature of over 1950 °C. On the top surface of the YAG melt, a YAG seed crystal was positioned, which was then continuously pulled at very slow rates, e.g., 0.2 mm h^{-1} and rotated at speeds of 10–30 rpm. Because of large surface energy, the YAG seed crystals used have an orientation of $\langle 111 \rangle$. Of course, seeds with $\langle 110 \rangle$ and $\langle 100 \rangle$ orientations can also be used.

In YAG crystals, Nd^{3+} substitutes for Y^{3+}. Since the ionic radius of Nd^{3+} is larger than that of Y^{3+}, the segregation coefficient of Nd^{3+} in YAG, defined as the ratio of Nd concentration in the crystal to that in the melt, is as low as $\sim$0.2 [11]. To grow an Nd:YAG crystal with desired concentration of Nd, it is necessary to use a precursor melt with 2–3 times higher Nd concentrations. In other words, 2–3 at.% Nd:YAG pellets have to be used to grow a 1 at.% Nd:YAG crystal. It has been found that inhomogeneous distribution of Nd is a bid problem of Nd:YAG single crystals. This is the reason why only Nd:YAG single crystals with Nd concentrations of ≤1 at.% are available for laser applications. Although at such low concentration, inhomogeneity is occasionally observed in parts, like the core and the (211) facets, which should be avoided when the crystals are used to make laser devices [11]. It is also observed that the as-grown Nd:YAG crystals often have a slightly gradient concentration of Nd in the longitudinal direction, simply because the concentration of Nd in the melt is increased gradually when the crystal becomes bigger and bigger, which forms another type of inhomogeneity.

In addition, the facilities as well as the Ir crucibles used for the growth of YAG crystals are extremely expensive. Also, due to slow growth rate, it usually takes months for one round of growth. Therefore, the growth process is energy-consuming. Furthermore, the yield percentage of high quality crystal is very low. All these disadvantages are intrinsic to melt-growth technologies [1, 11, 33]. To address these problems, it is necessary to use new processing, i.e., ceramic process.

1.4.3 Transparent Ceramics

The development of transparent ceramics is not so straightforward. Prior to the presence of transparent ceramics, translucent ceramics (Al_2O_3) were first reported [34]. However, these translucent ceramics could not be used for laser applications that require high optical quality, whereas they only found applications in the forms of thin layers, such as discharge tubes in high-pressure sodium vapor lamps. It was once believed that polycrystalline ceramics should be always opaque, i.e., ceramics could never be transparent. In this regard, the discovery of translucent Al_2O_3 ceramics was an encouraging event, because it demonstrated that light can transmit through ceramics, as the bulk density of a ceramic is very close to its theoretical density.

Another exciting achievement is the development of Nd doped 10 % ThO_2–Y_2O_3 transparent ceramics, which enabled the production of laser oscillation at room temperature. That was the first polycrystalline ceramic gain medium for laser applications. Nd:ThO_2–Y_2O_3 transparent ceramics were prepared by sintering Y_2O_3 powder at high temperature of 2200 °C for a prolonged duration of about 100 h, with 10 % ThO_2 used as sintering aid [35–37]. Although pulse laser oscillation could be obtained by flashlamp excitation at room temperature, the performance was not sufficient for practical applications, e.g., the slope efficiency was only 0.1 %. The research on transparent ceramics has flourished since the discovery of transparent Nd:YAG ceramics.

For laser applications, because light amplification takes place in the gain medium and the amplified light travels through the gain medium repeatedly, even an optical loss would have a significant effect on laser oscillation. Optical loss is also called optical attenuation coefficient, which is used to quantitatively characterize the quality of transparent materials. The optical loss of high quality commercial single crystals is in the range of 0.3–0.2 % cm^{-1}, which can offer efficient laser generation. Comparatively, transparent ceramics have higher optical losses. This means that the optical quality of transparent ceramics should be further improved.

Till date, various transparent ceramics with high purity and high density have been produced, including simple oxides such as MgO, Y_2O_3, and ZnO, composite or complex oxides such as ZrO_2–Y_2O_3, YAG, spinel ($MgAl_2O_4$), ferroelectric PLZT ($(PbLa)(ZrTi)O_3$), SBN ($Sr_xBa_{1-x}Nb_2O_6$), Gd_2O_2S:Pr, and $(YGd)O_3$:Eu and even non-oxides such as AlON and AlN, for applications in solid-state laser, electro-optical devices, X-ray scintillators, and thermally conductive components, have been developed and/or commercialized [38–42]. Figure 1.5 shows a photograph of a high quality 1.0 at% Nd:YAG ceramic disk fabricated with optimized processing conditions [43].

The main strategies to ensure high optical quality of transparent ceramics include controlling the purity, particle size, and homogeneity of the precursor materials, exploring and understanding the sintering mechanisms of different materials, and the use of advanced sintering techniques.

Ceramic processing typically consists of three main steps: (i) synthesis or preparation of precursor powders, (ii) consolidation or packing of the powders into green bodies, and (iii) sintering [44]. Every step has a significant effect on the microstructure and optical performance of the final transparent ceramics. There are a number of parameters relevant to the quality of the powders, facility, and way of consolidation and techniques of sintering, which can be used to optimize the fabrication process as a whole.

Precursor powders of ceramics can be prepared using solids, such as oxides, hydroxides, and carbonates, as starting materials [45–50]. In this case, it is called the solid-state reaction process. The precursor powders can also be synthesized by wet-chemical methods, such as chemical precipitation or co-precipitation [51–59], sol–gel [60, 61], gel combustion [62–65] and hydrothermal synthesis [66–69]. The solid-state powders react at a certain temperature to form compounds with desired compositions together with dopants, which is called calcination or pre-sintering. If the reaction is difficult, this step can be repeated. This step cannot be skipped even though no reaction is involved, e.g., the fabrication of simple oxide transparent ceramics. If a relative low calcination temperature is used, the reaction of the starting components cannot be triggered. In this case, the reaction will take place during the sintering process, so that the process is called reactive sintering. The products of chemical processes also need to have thermal treatment. For example, chemical precipitation usually yields carbides or hydroxides, which are decomposed into oxides by releasing water molecules or carbon dioxide. All the remaining steps are the same for the solid-state reaction process and the wet-chemical process.

These two methods have their own advantages and disadvantages. Solid-state reaction method is simple and can be easily scaled-up and thus is suitable for large-scale industrial applications. However, it has encountered various problems, including poor homogeneity in distribution of dopants and sintering aids, high sintering temperature, long sintering time duration, as well as poor microstructure and optical performance. Chemical procedures could have high homogeneity, fine powders, and thus low temperatures and short time duration. Their problems include complicated process, expensive chemicals, less quantity, and so on. Therefore, these methods are mainly used for fundamental studies of low quantity and small sizes in laboratories.

High doping of active components is an important requirement of transparent ceramics for laser applications. Homogeneous distribution of these elements in the final ceramics has a significant effect of laser performances. Also, when sintering aids are used to promote the sintering, their quantity should be restrictedly controlled, which must be as low as possible, in order to minimize their genitive effect on the optical properties of transparent ceramics. In this case, a combination of solid-state reaction and chemical processing can be used [70]. The main bodies of the materials are solids, while the additives (dopants and sintering aids) are introduced through wet-chemical routes. This is because chemical routes allow more homogeneous mixing.

It has been proved that the application of cold isostatic compression (CIP) before sintering is a crucial step to further increase the density of pre-sintered items, even though the increase in density is just in the range of 5–10 % [2, 9]. Special sintering techniques that are used to obtain high quality transparent ceramics include vacuum sintering, high-pressing (HP) sintering, hot isostatic pressure (HIP) sintering, as well as newly emerged spark plasma sintering (SPS), and microwave sintering. Specifically, SPS and HIP can be used to sinter transparent ceramics, with restrained grain growth [71–77]. This is especially important when mechanical strength is required for some applications.

1.5 Performance of Solid-State Lasers

A set of parameters, including efficiency, extension of wavelength range, diversification of temporal regime, power or energy scaling, and so on, are used to evaluate the performance of solid-state lasers [11]. The flow of excitation inside pumped laser materials should be optimized, so that there is a concurrent response of all the three major parts of a laser, i.e., the laser material, the pumping system, and the laser resonator. Good thermal management should be maintained, which determines the scaling to higher power or energy. This effect is caused by heat generation due to non-radiative de-excitation. The laser emission should be maximized while the heat generation should be minimized. Detailed discussion can be found in Ref. [11], a brief description will be given to each subject as follows.

1.5.1 Laser Materials

The core components of solid-state lasers are laser materials that allow for the inversion of population and amplification of radiation through stimulated emission. The properties of the laser materials determine the ways to design pumping system and laser resonator of a solid-state laser. Because the characteristics of laser active centers are determined by the physical processes related to the laser materials, while there are various possible interactions between the active centers and the electromagnetic radiations, the interrelationship among the composition, structure, properties, and functionality of laser materials is very complicated, leading the research in this field to be unlimited.

Similar to most advanced engineering materials, transparent ceramic laser materials also have various requirements for practical applications. The fabrication and processing process should be reliable, consistent, feasible, reproducible, energy saving, cost-effective, and scalable. More importantly, the compositions, both the main components and the dopants, must be under good control.

Optical properties of transparent ceramics are the first and foremost requirement for laser applications. For instance, they must be highly optically transparent in the spectral regions in order for pumping process and laser emission. Other optical properties include high optical homogeneity, minimized second-order refractive index, and high temperature stability in refractive index. Chemical stability and mechanical rigidity of the transparent ceramics are required to ensure the stability and lifetime of solid-state lasers. Thermophysical properties are also concerned in terms of stable laser performances, such as low thermal expansion coefficient, high thermal conductivity, strong thermomechanical shock-resistant capability, and acceptable temperature dependence of material properties.

The materials should meet specific requirements in compositions at microscopic levels. The substitution sites of low symmetry should be specific and well-defined in monochromatic lasers, while inhomogeneous broadening of lines are preferred in tunable or ultrashort pulse lasers. In addition, the doping ions should preferentially have the same valences as the substituted ions, which is called charge balance requirement. Otherwise, charge compensation should be considered during processing of the materials.

The performances of laser materials have a close relation to their spectroscopic properties [11]. The combination of doping ions and host materials must provide electronic energy levels and transition probabilities to support the efficient laser emission schemes in the desired temporal regime: (i) strong absorption of the pump radiation and the possibility of controlling the pump rate distribution at pump wavelengths that allow a high quantum defect (Stokes) ratio $\eta_{qd}^{(l)}$, (ii) efficient de-excitation of the pump level to the emitting level, i.e., $\eta_p \approx 1$, (iii) long radiative lifetime τ_{rad} of the emitting level, (iv) negligible electron–phonon de-excitation W_{nr} of the emitting level, but very efficient de-excitation of the pump level to the emitting level and the terminal level to the ground state, (v) reduced self-quenching by downconversion or upconversion energy transfer and high emission quantum efficiency η_{qe}, to give long τ_{eff}; (vi) emission cross-section σ_e adapted to the temporal regime, (vii) reduced parasitic de-excitation by excited state absorption from the emitting level of the pump or laser emission, and (viii) reduced reabsorption of laser emission.

1.5.2 Pumping Systems and Laser Resonators

Regarding pumping systems, the emission spectrum should have promising superposition of absorption for laser materials, in order to minimize the content of quantum defects. Also, the pumping systems should have desired temporal regime for the continuous wave or pulsed wave, high power output, long life span and stable emission characteristic, high electrical–optical efficiency, high transfer efficiency, and so on.

Laser resonators should have appropriate laser transition, good control of outcoupling, high quality factor to ensure the desired temporal regime of emission, control of the mode structure and beam size, control of the thermal field (dissipation of heat, external cooling). Technical requirements include mechanical robustness, volumetric compact, lowest maintenance, and simple replacement of parts/components.

1.6 Selection of Laser Material

Laser performances are determined by the properties of laser materials. Therefore, the selection of laser materials is an important step in building a high performance solid-state laser. The absorption and emission properties of the doping ions should be appropriate for lasing requirements, whereas the host materials should have required optical, mechanical, and thermal properties.

1.6.1 Laser Active Ions

Laser active ions are those that have energy level structures and transition probabilities, thus limiting to those with ground configuration of $3d^n$ or $4f^n$. In other words, laser active ions are from transitional elements.

1.6.1.1 3d Ions

The 3d ions for laser applications include Ti^{3+}, Cr^{n+} ($n = 2, 3, 4$) V^{3+}, and Co^{2+}. Ti^{3+}-doped Al_2O_3 crystal (sapphire) belongs to the family of vibronic laser materials, which are the most promising candidates for tunable or short pulse laser emissions. The Ti^{3+}-doped Al_2O_3 system offers tunable emission at about 800 nm, covering a range of about 230 nm, with femtosecond pulses. Lasers with peak powers of petawatts have been available in chirped pulse amplification (CPA) systems based on Ti:sapphire systems. Due to the difficulties in pumping and making large size sapphire crystals, this type of laser has limited applications.

With promising absorption and emission properties, Cr^{n+} ions with different valences ($n = 2, 3, 4$) can be used as either laser active ions or sensitizers for weakly absorbing 4f ions. Cr^{4+} ion tends to occupy sites with tetrahedral coordination, which has strong absorption in the range of wavelength from visible to 1 μm, with σ_a of 3.5–7.0 × 10^{-18} cm^2. The ion has infrared emission over 1350–1580 nm, with a high intensity σ_e of 3–5 × 10^{-19} cm^2, short lifetime of microseconds, and a broad range of >200 nm, depending on host materials. Cr^{3+} ion prefers to take six-fold coordinated sites, which can have broad vibronic emission lines near infrared when the excited level $^4T_2$ is below the ^2E level, because there is a strong spin-allowed emission to the ground $^4A_2$ level. Also, since it can be sufficiently thermalized, a relatively long effective emission lifetime is available, which is 60–100 μs. Additionally, due to strong vibronic absorption lines in visible region, Cr^{3+} ion can be used for direct lamp or diode laser pumping. At tetrahedral coordination sites, Cr^{2+} ion has a strong broad absorption centered at about 1750 nm and a short strong broad emission centered at about 2450 nm, which are corresponding to the vibronic transitions between $^5T_2$ and ^5E.

V^{3+} ion at octahedral coordination sites exhibit strong absorption over 1.05–1.5 μm, with an absorption cross-section of ~1.3 × 10^{-18} cm^2, which can be used for *Q*-switching of 1.3 μm lasers. Similarly, Co^{2+} ion at weak field tetrahedral coordination sites can also be used for *Q*-switching of 1.3 μm lasers.

1.6.1.2 4f Ions

Due to their wide range of sharp fluorescent transitions, which cover almost the whole region from the visible and near-infrared portions of the electromagnetic spectrum, all the rare earth (RE) ions are promising candidates as active ions in solid-state lasers. More significantly, it is found that all these lines are sufficiently sharp for strong lasing applications, even though strong local fields of crystals are present, due to the shielding effects of the outer electrons.

Rare earth atoms are characterized with a ground-state electronic configuration, which consists of a core that is identical to that of xenon (Xe), while the remaining electrons occupy higher orbital levels. The shells of Xe atom, with quantum numbers $n = 1$, 2, and 3 are completely filled, while in the shell with $n = 4$, s, p, and d subshells are fully filled, whereas the 4f subshell capable of accommodating up to 14 electrons is empty. However, in the shell of $n = 5$, the 5s and 5p orbits are fully filled with eight electrons.

The elements after Xe have this electronic configuration, plus electrons in the orbits of 4f, 5d, 6s, and so on. There are three elements, Cesium (Cs), barium (Ba), and lanthanum (La), between Xe and the RE elements. Cs has one 6s electron and Ba has two 6s electrons, while La has two 6s and one 5d electrons. When it comes to RE elements, the electrons start to fill the inner vacant 4f orbits. For instance, the first RE element Ce has only one electron in the f orbit, so that its configuration is $[Xe]6s^24f^15d^1$, while neodymium (Nd) has four electrons in the f orbit, i.e., its electron structure is $[Xe]6s^24f^4$.

RE ions are usually trivalent or occasionally divalent at a certain special environment. For RE^{3+} ions, outermost 6s electrons are all given up and the 5d electron is lost, if there is. If there is no 5d electron, one of the 4f electrons will be lost. For example, Ce^{3+} ion has an electronic configuration of $[Xe]4f^1$, while the electronic configuration of Nd^{3+} ion is $[Xe]4f^3$. Therefore, after becoming ions, the electronic configurations of the RE elements become simpler. As listed in Table 1.1, the RE ions are only different in the number of electrons in the 4f shell [78]. IF the RE elements are present as divalent ion, only the two outermost 6s electrons are lost.

Owing to the presence of potential electronic transitions between different levels of the partially filled 4f shell, RE ions offer a wealth of fluorescence spectra. Electrons in the 4f shell can be excited into the unoccupied 4f levels when absorbing light radiation. Because the 4f states are well shielded by the outer filled shells of 5s and 5p, the emission lines are usually narrow and the energy level structure varies only slightly for different host materials. According to crystal field theories, the effect of the crystal field is considered to be a perturbation on the energy levels of free ions. Comparatively, this perturbation is smaller than the spin-orbital and electrostatic interactions among the 4f electrons. The variation in the energy levels is a splitting of each of the free-ion levels into many closely spaced levels, which is known as the Stark effect of the crystal field. In this case, the levels of free ions are referred to as manifolds in crystals. All the RE^{3+} ions, except Yb^{3+}, exhibit a pretty complex energy level structure, so that there are laser emissions on several inter-manifold transitions [11]. As an example, Fig. 1.6 shows the splitting

Table 1.1 Electronic configurations of trivalent rare earth (RE) elements [78]

Element No.	RE element	Trivalent ions	4f electrons	Ground state
58	Cerium	Ce^{3+}	1	$^2F_{5/2}$
58	Praseodymium	Pr^{3+}	2	$^3H_4$
60	Neodymium	Nd^{3+}	3	$^4I_{9/2}$
61	Promethium	Pm^{3+}	4	$^5I_4$
62	Samarium	Sm^{3+}	5	$^4H_{5/2}$
63	Europium	Eu^{3+}	6	$^7F_0$
64	Gadolinium	Gd^{3+}	7	$^8S_{7/2}$
65	Terbium	Tb^{3+}	8	$^7F_6$
66	Dysprosium	Dy^{3+}	9	$^6H_{15/2}$
67	Holmium	Ho^{3+}	10	$^5I_8$
68	Erbium	Er^{3+}	11	$^4I_{15/2}$
69	Thulium	Tm^{3+}	12	$^3H_6$
70	Ytterbium	Yb^{3+}	13	$^2F_{7/2}$
71	Lutetium	Lu^{3+}	14	$^1S_0$

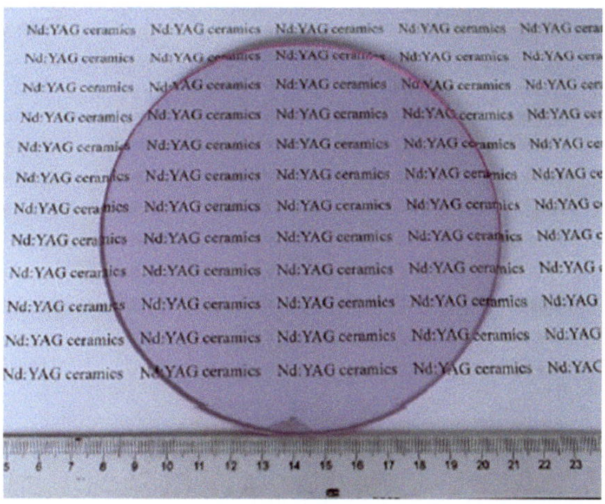

Fig. 1.5 Photograph of a high quality 1.0 at% Nd:YAG ceramic disk with a dimension of Ø130 mm × 6 mm. Reproduced with permission from [43]. Copyright © 2012, Elsevier

behavior of the Nd^{3+} manifolds into various sublevels, under the perturbation of the YAG crystal field [78].

Pr^{3+} Ion Due to the manifold $^3P_0$ located in the 20,500 cm^{-1} region, Pr^{3+} ion has been demonstrated to exhibit laser transitions in the visible, from blue, green, orange to red, as well as near-infrared region. The manifold $^3P_0$ has a large emission cross-section, especially for the red transition of $^3P_0 \rightarrow {}^3F_2$. Therefore, Pr^{3+} has been extensively studied as a laser active ion. However, Pr^{3+} ion has a relatively

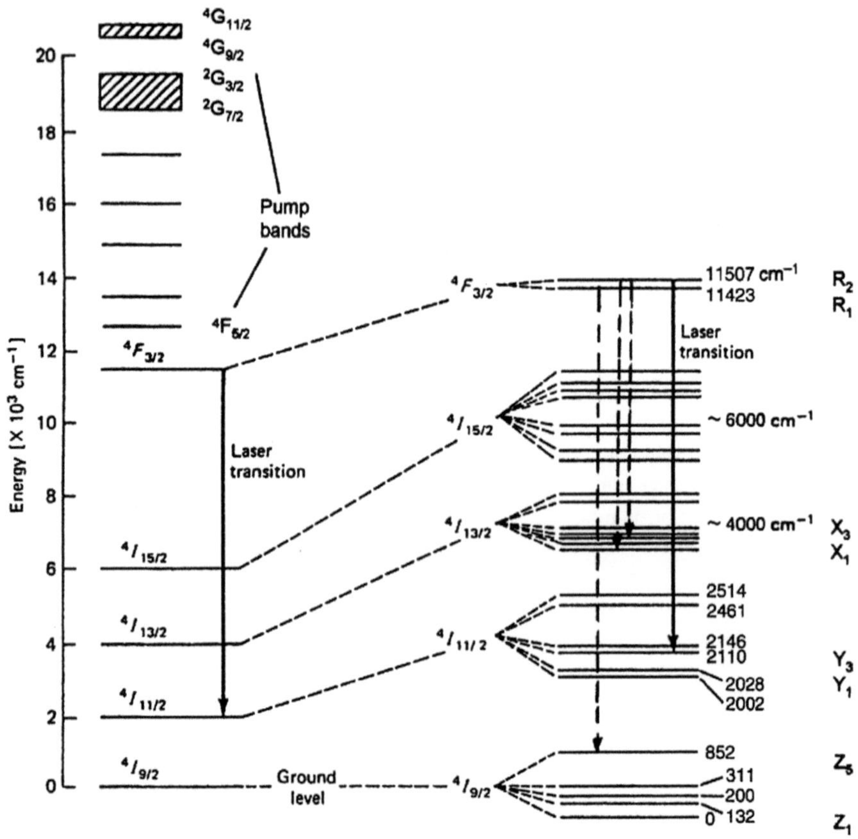

Fig. 1.6 Energy level diagram of Nd:YAG, in which the *solid line* represents the major transition at 1064 nm, while the dashed lines are transitions at 1319, 1338, and 946 nm. Reproduced with permission from [78]. Copyright © 2006, Springer

short lifetime, which is about tens of microseconds. The main problems of Pr^{3+} ion is the lack of suitable pump sources and the weak absorption.

Nd^{3+} Ion With a ground configuration of $4f^3$, Nd^{3+} ion has a complicated structure of energy manifolds. Among these manifolds, only that of $^4F_{3/2}$, at about 11,500 cm^{-1}, is in a metastable state. Depending on the host material that the ion is in, it has radiative lifetimes in the range of 80–500 μs. Below the metastable manifold, four manifolds are observed, which are ^4I_J, with $J = 9/2$, 11/2, 13/2, and 15/2. These manifolds stem from the same spectral term ^4I. The gap between $^4F_{3/2}$ and the one nearest it, $^4I_{15/2}$, is at about 5500 cm^{-1}. This gap has p of 7–10, which can be bridged only when very high order electron–phonon processes are present. If the concentrations of Nd are sufficiently low, the measured luminescence lifetime at temperatures can be used estimate the radiative lifetime. At low symmetry crystal field, each of these energy manifolds will be split into $(2J + 1)/2$ Kramers doublets, which are labeled as R_i with $i = 1, 2$ for $^4F_{3/2}$, X_j with $j = 1$ to 7 for $^4I_{13/2}$, Y_j with $j = 1$–6 for $^4I_{11/2,}$ and Z_j with $j = 1$–5 for $^4I_{9/2}$. The global crystal field splitting behaviors of the manifolds in different host materials can be greatly different, which are in the ranges of 0–250 cm^{-1} for $^4F_{3/2}$, 400–1000 cm^{-1} for $^4I_{15/2}$, 240–690 cm^{-1} for $^4I_{13/2}$, 190–525 cm^{-1} for $^4I_{11/2,}$ and 280–880 cm^{-1} for $^4I_{9/2}$ [11].

Because laser emission of Nd^{3+} ion is attributed to optical transitions between the Stark levels of the emitting and terminal manifolds, it has $2j$ possible optical transitions between the two Stark levels of the emitting manifold $^4F_{3/2}$ and the Stark levels of lower manifolds. The effective lifetime τ_{eff} of all these transitions is the same, whereas their laser quantum defect ratio $\eta_{\mathrm{qd}}^{(l)}$ and emission cross-section σ_{e} are different. Since temperature has an effect on the fractional thermal population coefficients $f_{1,2}$ of the levels $R_{1,2}$ of $^4F_{3/2}$, the effective emission cross-sections σ_{eff} and laser threshold are also dependent on temperature. The slope efficiency $\eta_{\mathrm{sl}}^{(a)}$ of all the laser transitions is only determined by the particular quantum defect, which is also independent on temperature.

Many of the transitions in the range of near 900 nm to the ground manifold $^4I_{9/2}$ possess laser emission, while some of them have cross-sections of $>10^{-20}$ cm^2 at room temperature. The Judd–Ofelt parameters have been used to calculate the branching ratios β_{ij} of the emission transitions due to the excited manifold. For emission of $^4F_{3/2}$, the quality factor $X = \Omega_4/\Omega_6$ is particularly important. For very small values of X, the branching ratio $\beta_{3/2,11/2}$ is dominant, which is about 0.66 for $X = 0$ and is found to decrease with increasing X. In contrast, $\beta_{3/2,9/2}$ is relatively low, which is only about 0.17 for $X = 0$, but it increases with increasing X. As a result, there is a crossing point for these two branching ratios, which is about $X \approx 1.15$. The branching ratio $\beta_{3/2,13/2}$ decreases from 0.17 at $X = 0$ to less than 0.1 for $X = 2$, while $\beta_{3/2,15/2}$ is very small, which is usually negligible.

Due to the high density of the excited levels above the emitting level, excitations with a broad range of wavelengths in the visible and near-infrared region have been observed. Because all these excitations are relaxed to $^4F_{3/2}$ through a series of very fast processes, the pumping of Nd lasers can be achieved using various optical sources, e.g., lamps, solar, diode lasers, and so on.

Nd^{3+} ion has a disadvantage, which is called self-quenching of emission, due to its complex electronic structure. For downconversion, an excited Nd^{3+} ion in the $^4F_{3/2}$ level could transfer part of the excitation to an unexcited Nd^{3+} ion. At room temperature, the process is dominated by the cross-relaxation process, i.e., ($^4F_{3/2}$, $^4I_{9/2}$) $\rightarrow$ ($^4I_{15/2}$, $^4I_{15/2}$). In most Nd-doped laser materials, the final levels of the donor and acceptor are further de-excited to the ground state through efficient low order electron–phonon processes, so that the initial excitation of the donor ion is completely wasted as the form of heat. The interaction between two excited Nd^{3+} ions can induce upconversion of excitation to an upper level, through the cross-relaxation processes, ($^4F_{3/2}$, $^4F_{3/2}$) $\rightarrow$ one of ($^4I_{15/2}$, $^4G_{5/2}$), ($^4I_{13/2}$, $^4G_{7/2}$), ($^4I_{11/2}$, $^4G_{9/2}$) and ($^4I_{9/2}$, $^2P_{1/2}$). Due to the high density of the levels above $^4F_{3/2}$, all the excitation transferred to the acceptor will be lost through the electron–phonon interaction, so that the excited acceptor returns to its initial excited level $^4F_{3/2}$. Therefore, downconversion is more wasteful than upconversion. Nevertheless, Nd^{3+} ion has been acknowledged to be one of the most popular laser active ions, due to its wide range of possibilities.

Ho^{3+} Ion Ho^{3+} has a ground electronic configuration of $4f^{10}$, with several well separated levels for emissions in the visible or infrared range. Among these emissions, the 2.1 μm one, i.e., $^5I_7 \rightarrow {}^5I_8$, is the most important, which has a cross-section of about 10^{-20} cm^2 and lifetime of about 7 ms. Because the visible and near-infrared absorption lines of Ho^{3+} are relatively weak, it is usually sensitized using either Tm^{3+} or Cr^{3+}–Tm^{3+}. It has also been demonstrated that efficient 2 μm laser emission with low quantum defect can be realized by pumping directly into the emitting level $^5I_7$ with a Tm^{3+} laser or diode lasers. Also, excitation of visible emission can be obtained through various upconversion schemes.

Er^{3+} Ion The ground configuration of Er^{3+} is $4f^{11}$, which offers a very rich electronic level structure. Several levels in the visible and near infrared are well separated to give emission. It exhibits accidental coincidence of the gaps between various pairs of energy levels, which facilitates a large variety of ET downconversion or upconversion processes. Some of the low energy levels, such as $^4I_{11/2}$ located at about 10,000 cm^{-1} and $^4I_{13/2}$ at about 6500 cm^{-1} have long radiative lifetimes of 6–9 ms. Efficient emission can be obtained on the four-level transition, $^4I_{11/2} \rightarrow {}^4I_{13/2}$, in the range of 2.7–2.95 μm. However, the luminescence lifetime of the terminal level is longer than that of the emitting level, which makes the laser emission to be rapidly self-saturated. This problem can be addressed by using high Er doping, because the migration-assisted ET upconversion, $({}^4I_{13/2}, {}^4I_{13/2}) \rightarrow ({}^4I_{9/2}, {}^4I_{15/2})$, can efficiently depopulate the level, so as to recirculate part of the excitation to the $^4I_{9/2}$ level, which could be relaxed back to the emitting level $^4I_{11/2}$. Another upconversion from the level $^4I_{11/2}$, through $({}^4I_{11/2}, {}^4I_{11/2}) \rightarrow ({}^4S_{3/2}, {}^4I_{15/2})$ could reduce the effect of that upconversion. Therefore, these two upconversion processes must be well balanced.

$^4I_{13/2} \rightarrow {}^4I_{15/2}$ is a quasi-three-level transition, with emission cross-section of about 5×10^{-21} cm^2, which can be used for laser emission in the range of 1.55–1.65 μm. This laser emission is eye-safe radiation, which can be used to replace the eye-dangerous 1 μm lasers. It also has potential application in telecommunications, due to matching with the range of minimal absorption of glass fibers. There are also transitions from high energy levels for Er^{3+} laser emissions in the visible region, especially green or red emission from $^4S_{3/2}$.

Weak pump absorption is a disadvantage of Er^{3+}, but this problem has now been addressed through co-doping with sensitizer ions. For instance, sensitization with Cr^{3+} has been used in lamp pumping. It is possible to excite the $^4I_{11/2}$ level using 960 nm diode lasers, whose low efficiency due to the low absorption has been addressed by using co-doping with Yb^{3+}, because Yb^{3+} has larger absorption and can transfer the excitation efficiently to Er^{3+}. It has been found that $^4I_{11/2}$ can be rapidly de-excited to $^4I_{13/2}$ by using the ET to other ions, such as Ce^{3+}, because the absorption of Ce^{3+} is resonant with this energy gap. This can be used to improve the efficiency of excitation of the 1.6 μm $^4I_{13/2}$ emission with Er^{3+} through sensitized pumping. The upper energy levels for visible emission can be excited by ET upconversion and/or excited state absorption of the infrared pump radiation, while the energy transfer from Yb^{3+} can be used for sensitization of similar upconversion processes.

Tm^{3+} Ion Tm^{3+} has a ground configuration of $4f^{12}$, with well-separated levels that facilitate emissions in the visible and infrared range. Among them, the quasi-three-level emission in the range of 2 μm due to the transition from the first excited manifold $^3F_4$, located near 5000 cm^{-1}, to the ground level $^3H_6$, is the most important one. Tm^{3+} has strong electron–phonon interaction, thus leading to significant homogeneous broadening of the lines. The emission cross-section in this transition is relatively small, of the order of 10^{-21} cm^2, which is compensated by the long lifetime of level $^3F_4$, i.e., 8–10 ms. It is also found that the $^3F_4$ level can be efficiently populated through quantum splitting of the excitation of level $^3H_4$, located at about 12,500 cm^{-1}, via cross-relaxation, ($^3H_4$, $^3H_6$) → ($^3F_4$, $^3F_4$). Weak absorption is a major problem for pumping Tm^{3+} in the visible and infrared region, which can be sensitized by using 3d ions such as Cr^{3+} or Fe^{3+}. The visible emission from Tm^{3+} can be excited by upconversion processes, especially the Yb-sensitized upconversion.

Yb^{3+} Ion The ground electronic configuration of Yb^{3+} is $4f^{13}$, which has only two manifolds, $^2F_{7/2}$ at ground and $^2F_{5/2}$ in the region of 10,000 cm^{-1}. Such a configuration precludes emission self-quenching through cross-relaxation or excited state absorption. Because it is at the end of the lanthanide series, Yb^{3+} has strongest electron–phonon interaction and crystal field effects. As a result, Yb-doped materials have broad emission lines and strong vibronic satellites. The global crystal field splitting of $^2F_{5/2}$ with three Stark levels and $^2F_{7/2}$ with four Stark levels can reach 600 and 1200 cm^{-1}, respectively. The emission transitions are in the range 960–1060 nm. The lower Stark levels of $^2F_{7/2}$ can experience thermal population, which could result in strong reabsorption, thus limiting the useful range for laser emission to 1025–1060 nm. Compared with the emitting level $^4F_{3/2}$ of Nd^{3+}, the level $^2F_{5/2}$ of Yb^{3+} has longer lifetime, but lower absorption and smaller emission cross-sections.

The $^2F_{7/2}(1) \rightarrow {}^2F_{5/2}(1)$ absorption line in the range of 960–980 nm, with the highest peak absorption cross-section, is very sharp and exhibits strong homogeneous broadening. Therefore, diode pumping of this line requires restrictive control of the peak wavelength and FWHM of the pump. The broader $^2F_{7/2}(1) \rightarrow {}^2F_{5/2}(2)$ absorption line at about 940 nm is used for diode laser pumping, because its relatively small cross-section can be addressed by various techniques, such as fiber or thin-disk multi-pass laser configurations. Bulk Yb lasers require high doping concentrations. Reduction in thermal conductivity could be a problem of Yb^{3+} lasers, especially when the mass difference between the substituted ion and Yb^{3+} is significantly large. In most Yb laser materials, the $^2F_{5/2}(1) \rightarrow {}^2F_{7/2}(3)$ emission line has the highest peak cross-section, but with reabsorption.

Yb laser materials have shown to be promising candidates for highly efficient and low-heat diode pumped laser emission in various regimes, due to their unique properties, such as large product $\sigma_{eff}\tau_{eff}$ to enable efficient CW emission, long τ_{eff} to ensure efficient storage of inversion for Q-switched emission, broad emission lines suitable for mode-locked short pulse emission, and so on.

1.6.2 Host Materials

Due to the special characteristics of the laser emission process and the parasitic non-radiative de-excitation, it is necessary to carefully select the laser materials, including both the active ions and host materials. In addition, the characteristics of dopants and the states of doping have also played a crucial role in determining the performances of laser materials and thus the solid-state lasers. The efficiency and effectiveness of doping is mainly determined by the degree of matching in ionic radii between the dopant ions and substituted cations. The Shannon ionic radii of the RE^{3+} ions in condensed state with anionic coordination number of 6 and 8 are r_6 = 0.103–0.115 nm and r_8 = 0.113–0.128 nm, respectively. In both cases, the radius decreases with increasing atomic number [79]. These ions can substitute for host cations with similar ionic radius, such as Ca^{2+}, Sr^{2+}, La^{3+}, Gd^{3+}, Y^{3+}, Lu^{3+}, Sc^{3+}, and so on. The efficiency of the substitution is closely related to the conditions of fabrication and processing of the laser materials, such as properties of starting materials, synthesis methods, processing parameters, and so on.

It has been found that doping with laser active ions can greatly reduce the thermal conductivity of the host materials. Therefore, doping leads to an additional criterion for the selection of the host material, i.e., the atomic mass of the substituted cation should be as close as possible to that of the doping ion. The ionic radii of both the laser active ions and host cations have been well documented as database in the open literature. The doping efficiency is also related to valences of the doping and substituted ions. The difference in valence requires charge compensation, which is realized by creating lattice defects, such as vacancies or interstitials. Occasionally, heavy doping of ions with a significant valence difference could result in severe crystalline lattice restructuring or phase transformation.

There are also other requirements for the host materials, such as mechanical strength, chemical stability, fabrication processibility, and so on, which should be taken into account during the selection of host materials. All raw materials and procedures involved should be cost-effective, especially for large-scale production, otherwise no industry will adopt.

1.7 Other Applications of Transparent Ceramics

Besides the main application in solid-state lasers, transparent ceramics have also found various other applications, including lighting, scintillators, armors, optical devices, electro-optical devices, and biomaterials. While all these applications require high optical transmission, in some cases, high mechanical strengths are also important, for example, in armor applications. As stated earlier, mechanical strength is closely related to the grain size and size distribution of transparent ceramics. Reducing grain size is the key strategy to increase the mechanical performance of ceramic materials.

1.8 Motivation and Objectives of the Book

Due to their important applications in solid-state lasers and many other areas, transparent ceramics have attracted the attention of researchers from laser physics, applied physics, chemistry and materials science and engineering. However, ceramic lasers are a highly multidisciplinary or interdisciplinary subject, involving laser physics and materials processing at the same time. There are numerous books available in either solid-state layers [80–83] or ceramic processing [44, 84–86], but very few that cover evenly both fields [11, 87]. Therefore, the aim of this book is to bridge the gap between solid-state transparent ceramic lasers and transparence ceramic processing. At the same time, transparent ceramics for other applications will also be elaborated.

1.9 Outline of the Book

This book consists of eight chapters. Chapter 1 provides a brief introduction to transparent materials, transparent ceramics, issues on transparency of transparent materials, and solid-state lasers (including single crystals and ceramics). Requirements and availabilities of materials for solid-state lasers are also evaluated and discussed in detail.

In Chap. 2, materials that can be processed into ceramics for solid-state laser and related applications are elaborated, including brief descriptions on crystal structure, physical properties, phase formation, and processing requirements. The dominant transparent ceramics are still for solid-state laser applications. Among the laser ceramics, YAG-based ones are the most widely studied and described. With the development of new sintering technologies, especially SPS, various new transparent ceramics, which cannot be obtained using conventional processing techniques, have been fabricated. New transparent ceramics derived from glass precursors will also be touched upon in this chapter.

Chapter 3 covers synthesis of precursor powders of transparent ceramics. These synthetic methods are also widely used for other ceramics or other materials. For each method, a brief description, together with examples of transparent ceramics derived from the powder synthesized by using the method, will be presented. There is no attempt to comment on whether one method is superior to another. In fact, every method has its own advantages and disadvantages. Different methods could be suitable for different transparent ceramic materials.

Chapter 4 consists of two parts: (i) powder characterization and (ii) consolidation of ceramic powders. In the powder characterization part, efforts have been made to include as many techniques as possible, which have been used or are potentially useful to characterize the precursor powders of transparent ceramics. In the powder consolidation part, various packing techniques are discussed and demonstrated with examples of transparent ceramics.

Various conventional sintering techniques, such as vacuum sintering, high-pressure sintering, and high-isostatic pressure sintering, are discussed in Chap. 5, in terms of producing high quality advanced ceramics in general and transparent ceramics in particular. Fundamental issues regarding conventional sintering are re-emphasized, including driving forces of sintering, progress of sintering, and liquid-phase sintering.

New sintering technologies, mainly electrical current aided sintering (ECAS) and microwave sintering, are presented in Chap. 6, where ECAS is also known as SPS. Theoretical considerations, simulation, and applications of the two new sintering technologies are discussed first in a more general way, which is not restricted to transparent ceramics. For the theoretical and modeling part of SPS, not only the achievements on ceramics are described, but also some related contents of metallic materials are included as well for the purpose of completeness.

In Chap. 8, grain growth and microstructure development of ceramics as a result of sintering are presented. Besides general theoretical and experimental studies, some topics are specifically emphasized, such as abnormal grain growth (AGG) and two-step sintering. AGG is mainly used to convert polycrystalline ceramics into single crystals, which is also known as solid-state conversion of single crystal. Two-step sintering is employed to control grain size, which is of special importance for the transparent ceramics with the requirement of high mechanical strength.

Applications of transparent ceramics are covered in the last two chapters, with Chap. 9 focusing on solid-state lasers with transparent ceramics and Chap. 10 on all other applications of transparent ceramics. In Chap. 9, besides traditional transparent laser ceramics, advanced ceramic laser technologies, including composite ceramics and crystal fibers (not ceramics), are also included, in order to demonstrate new research and development direction of solid-state lasers. In Chap. 10, other applications, such as lighting, scintillation, armor, potential biomaterials, and so on, are summarized and discussed.

Although each chapter covers only a specific topic, all the chapters are compiled in such a way that each one is self-consistent, so as to be read as an independent monograph.

Acknowledgments One of the authors (LBK) would like to acknowledge the financial supports from the start-up grant (SUG/2012) from Nanyang Technological University and AcRF Tier 1 project (RG44/12) from Ministry of Education, Singapore.

References

1. Wang SF, Zhang J, Luo DW, Gu F, Tang DY, Dong ZL et al (2013) Transparent ceramics: processing, materials and applications. Prog Solid State Chem 41:20–54
2. Ikesue A, Kinoshita T, Kamata K, Yoshida K (1995) Fabrication and optical properties of high-performance polycrystalline Nd-YAG ceramics for solid-state lasers. J Am Ceram Soc 78:1033–1040
3. Ikesue A, Aung YL (2008) Ceramic laser materials. Nat Photonics 2:721–727
4. Sanghera J, Bayya S, Villalobos G, Kim W, Frantz J, Shaw B et al (2011) Transparent ceramics for high-energy laser systems. Opt Mater 33:511–518
5. Sanghera J, Kim W, Villalobos G, Shaw B, Baker C, Frantz J et al (2012) Ceramic laser materials. Materials 5:258–277
6. Sanghera J, Kim W, Villalobos G, Shaw B, Baker C, Frantz J et al (2013) Ceramic laser materials: past and present. Opt Mater 35:693–699
7. Krell A, Hutzler T, Klimke J (2009) Transmission physics and consequences for materials selection, manufacturing, and applications. J Eur Ceram Soc 29:207–221
8. Apetz R, van Bruggen MPB (2003) Transparent alumina: a light-scattering model. J Am Ceram Soc 86:480–486
9. Ikesue A, Kamata K, Yoshida K (1996) Synthesis of transparent Nd-doped HfO_2-Y_2O_3 ceramics using HIP. J Am Ceram Soc 79:359–364
10. Ikesue A, Kamata K (1996) Microstructure and optical properties of hot isostatically pressed Nd:YAC ceramics. J Am Ceram Soc 79:1927–1933
11. Ikesue A, Aung YL, Lupei V (2014) Ceramic lasers. Cambridge University Press, Cambridge
12. Maiman TH (1960) Optical and microwave-optical experiments in ruby. Phys Rev Lett 4:564–566
13. Drube J, Struve B, Huber G (1984) Tunable room-temperature CW laser action in Cr^{3+}-GdScAl-garnet. Opt Commun 50:45–48
14. Struve B, Huber G (1985) The effect of the crystal-field strength on the optical-spectra of Cr^{3+} in gallium garnet laser crystals. Appl Phys B Photophysics Laser Chem 36:195–201
15. Moulton PF, Manni JG, Rines GA (1988) Spectroscopic and laser characteristics of Er, Cr-YSGG. IEEE J Quantum Electron 24:960–973
16. Caird JA, Shinn MD, Kirchoff TA, Smith LK, Wilder RE (1986) Measurements of losses and lasing efficiency in GSGG-Cr, Nd and YAG-Nd laser rods. Appl Opt 25:4294–4305
17. Harrison J, Welford D, Moulton PF (1989) Threshold analysis of pulsed lasers with application to a room-temperature Co-MgF_2 laser. IEEE J Quantum Electron 25:1708–1711
18. Welford D, Moulton PF (1988) Room-temperature operation of a Co-MgF_2 laser. Opt Lett 13:975–977
19. Walling JC, Jenssen HP, Morris RC, Odell EW, Peterson OG (1979) Broad-band tuning of solid-state Alexandrite laser. J Opt Soc Am 69:373
20. Walling JC, Peterson OG (1980) High-gain laser performance in Alexandrite. IEEE J Quantum Electron 16:119–120
21. Walling JC, Peterson OG, Jenssen HP, Morris RC, Odell EW (1980) Tunable alexandrite lasers. IEEE J Quantum Electron 16:1302–1315
22. Moulton PF (1986) Spectroscopic and laser characteristics of Ti-Al_2O_3. J Opt Soc Am B Opt Phys 3:125–133

23. Tsuiki H, Kitazawa K, Masumoto T, Shiroki K, Fueki K (1980) Single-crystal growth of pure and Nd-doped Y_2O_3 by floating zone method with Xe arc lamp imaging furnace. J Cryst Growth 49:71–76
24. Tsuiki H, Masumoto T, Kitazawa K, Fueki K (1982) Effect of point-defects on laser oscillation properties of Nd-doped Y_2O_3. Jpn J Appl Phys Part 1 Regul Papers Short Notes Rev Papers 21:1017–1021
25. Ebendorffheidepriem H, Seeber W, Ehrt D (1993) Dehydration of phosphate-glasses. J Non-Cryst Solids 163:74–80
26. Heumann E, Ledig M, Ehrt D, Seeber W, Duczynski EW, Vanderheide HJ et al (1988) CW laser action of Er^{3+} in double sensitized fluoroaluminate glass at room-temperature. Appl Phys Lett 52:255–256
27. Ledig M, Heumann E, Ehrt D, Seeber W (1990) Spectroscopic and laser properties of Cr^{3+}, Yb^{3+}, Er^{3+} fluoride phosphate-glass. Opt Quant Electron 22:S107–S122
28. Seeber W, Ehrt D, Ebendorffheidepriem H (1994) Spectroscopic and laser properties of Ce^{3+}-Cr^{3+}-Nd^{3+} co-doped fluoride phosphate and phosphate-glasses. J Non-Cryst Solids 171:94–104
29. Geusic JE, Marcos HM, Vanuitert LG (1964) Laser oscillation in Nd-doped yttrium aluminium, yttrium gallium and gadolinium garnets. Appl Phys Lett 4:182–184
30. Deshazer LG, Komai LG (1965) Fluorescence conversion efficiency of neodymium glass. J Opt Soc Am 55:940
31. Kushida T, Geusic JE, Marcos HM (1968) Optical properties of YAIG-Nd^{3+}. IEEE J Quantum Electron 4:316
32. Kushida T, Marcos HM, Geusic JE (1968) Laser transition cress section and fluorescence branching ratio for Nd^{3+} in yttrium aluminum garnet. Phys Rev 167:289
33. Pan YB, Li J, Jiang XB (2013) Advanced optical functional transparent ceramics. Science Press, Beijing
34. Silvfast WT (2004) Laser fundamentals. Cambridge University Press, Cambridge
35. Greskovich C, Chernoch JP (1974) Improved polycrysalline ceramic lasers. J Appl Phys 45:4495–4502
36. Greskovich C, Woods KN (1973) Fabrication of transparent ThO_2-doped Y_2O_3. Am Ceram Soc Bull 52:473–478
37. Greskovich C, Woods KN (1972) Recent advances in processing of thorium-doped yttrium oxide ceramics. Am Ceram Soc Bull 51:326
38. Miles GD, Sambell RAJ, Rutherfo J, Stephens GW (1967) Fabrication of fully dense transparent polycrystalline magnesia. Trans Br Ceramic Soc 66:319
39. Rice RW (1971) Hot-pressing of MgO. J Am Ceram Soc 54:205–207
40. Greskovich C, Curran MJ, Oclair CR (1972) Preparation of transparent Y_2O_3-doped ThO_2. J Am Ceram Soc 55:324–325
41. Greskovich C, Brewer JA (2001) Solubility of magnesia in polycrystalline alumina at high temperatures. J Am Ceram Soc 84:420–425
42. Scott C, Kaliszewski M, Greskovich C, Levinson L (2002) Conversion of polycrystalline Al_2O_3 into single-crystal sapphire by abnormal grain growth. J Am Ceram Soc 85:1275–1280
43. Liu WB, Li J, Jiang BX, Zhang D, Pan YB (2012) 2.44 KW laser output of Nd:YAG ceramic slab fabricated by a solid-state reactive sintering. J Alloy Compd 538:258–261
44. Rahaman MN (2003) Ceramic processing and sintering, 2nd edn. CRC Press, New York
45. Ikesue A, Furusato I, Kamata K (1995) Fabrication of polycrystalline transparent YAG cramics by solid-state reaction method. J Am Ceram Soc 78:225–228
46. Li HL, Liu XJ, Huang LP (2005) Fabrication of transparent cerium-doped lutetium aluminum garnet (LuAG:Ce) ceramics by a solid-state reaction method. J Am Ceram Soc 88:3226–3228
47. Li HL, Liu XJ, Huang LP (2006) Fabrication of transparent Ce:LuAG ceramics by a solid-state reaction method. J Inorg Mater 21:1161–1166
48. Li J, Chen Q, Yang LL, Feng GY, Wu WJ, Zheng FS et al (2011) High transmittance of Nd-doped YAG transparent ceramics prepared by solid-state reaction method. Ferroelectrics 411:62–68

49. Wang NL, Zhang XY, Jiang HT, Dong TT, Yang D (2012) Fabrication of Er^{3+}/Yb^{3+} co-doped Y_2O_3 transparent ceramics by solid-state reaction method and its up-conversion luminescence. Mater Chem Phys 135:709–713
50. Wu YS, Li J, Qiu FG, Pan YB, Liu Q, Guo JK (2006) Fabrication of transparent Yb, Cr:YAG ceramics by a solid-state reaction method. Ceram Int 32:785–788
51. Lu J, Prabhu M, Song J, Li C, Xu J, Ueda K et al (2000) Optical properties and highly efficient laser oscillation of Nd:YAG ceramics. Appl Phy B Lasers Opt 71:469–473
52. Arabgari S, Malekfar R, Motamedi K (2011) Parameters effects on the surface morphology and structure of Nd:YAG nanopowders synthesized by co-precipitation method. J Nanopart Res 13:597–611
53. Chen JY, Shi Y, Shi JL (2004) Synthesis of (Y, Gd)2O3: Eu nanopowder by a novel co-precipitation processing. J Mater Res 19:3586–3591
54. Huang YH, Jiang DL, Zhang JX, Lin QL (2009) Precipitation synthesis and sintering of lanthanum doped yttria transparent ceramics. Opt Mater 31:1448–1453
55. Huang ZG, Sun XD, Xiu ZM, Chen SW, Tsai CT (2004) Precipitation synthesis and sintering of yttria nanopowders. Mater Lett 58:2137–2142
56. Ji XB, Deng JG, Kang B, Huang H, Wang X, Jing W et al (2013) Thermal decomposition of $Y_3Al_5O_{12}$ precursor synthesized by urea homogeneous co-precipitation. J Anal Appl Pyrol 104:361–365
57. Li HL, Liu XJ, Xie RJ, Zeng Y, Huang LP (2006) Fabrication of transparent cerium-doped lutetium aluminum garnet ceramics by co-precipitation routes. J Am Ceram Soc 89:2356–2358
58. Ikegami T, Li JG, Mori T, Moriyoshi Y (2002) Fabrication of transparent yttria ceramics by the low-temperature synthesis of yttrium hydroxide. J Am Ceram Soc 85:1725–1729
59. Li JG, Ikegami T, Mori T (2003) Fabrication of transparent Sc_2O_3 ceramics with powders thermally pyrolyzed from sulfate. J Mater Res 18:1816–1822
60. Guo K, Chen HH, Guo XG, Yang XX, Xu FF, Zhao JT (2010) Morphology investigation of yttrium aluminum garnet nano-powders prepared by a sol–gel combustion method. J Alloy Compd 500:34–38
61. Li DY, Hui Y, Lian JS, Xie TT (2005) Progress in research on nanometer Al_2O_3 fabricated by sol–gel method. J Rare Earths 23:600–605
62. Biswas A, Prabhakaran K, Gokhale NM, Sharma SC (2007) Synthesis of nanocrystalline yttria doped ceria powder by urea-formaldehyde polymer gel auto-combustion process. Mater Res Bull 42:609–617
63. Prabhakaran K, Patil DS, Dayal R, Gokhale NM, Sharma SC (2009) Synthesis of nanocrystalline magnesium aluminate ($MgAl_2O_4$) spinel powder by the urea-formaldehyde polymer gel combustion route. Mater Res Bull 44:613–618
64. Su J, Miao J-h, Xu L-h, Ji Y-q, Wang C-q (2012) Synthesis and characterization of nanocrystalline Nd^{3+}-doped gadolinium scandium aluminum garnet powders by a gel-combustion method. Mater Res Bull 47:1709–1712
65. Wang N, Zhang X, Bai Z, Sun H, Liu Q, Lu L et al (2011) Synthesis of nanocrystalline ytterbium-doped yttria by citrate-gel combustion method and fabrication of ceramic materials. Ceram Int 37:3133–3138
66. Huang BT, Ma YQ, Qian SB, Zou D, Zheng GH, Dai ZX (2014) Luminescent properties of low-temperature-hydrothermally-synthesized and post-treated YAG: Ce (5 %) phosphors. Opt Mater 36:1561–1565
67. Moore CA, McMillen CD, Kolis JW (2013) Hydrothermal growth of single crystals of $Lu_3Al_5O_{12}$ (LuAG) and its doped analogues. Cryst Growth Des 13:2298–2306
68. Qian S, Ma Y, Zan F, Zou D, Dai Z, Zheng G et al (2013) Fine YAG:Ce^{3+} nanoparticles synthesised by supercritical hydrothermal reaction. Micro Nano Lett 8:201–205
69. Mancic L, Lojpur V, Marinkovic BA, Dramicanin MD, Milosevic O (2013) Hydrothermal synthesis of nanostructured Y_2O_3 and $(Y_{0.75}Gd_{0.25})_2O_3$ based phosphors. Opt Mater 35:1817–1823

70. Sang YH, Qin HM, Liu H, Zhao LL, Wang YN, Jiang HD et al (2013) Partial wet route for YAG powders synthesis leading to transparent ceramic: a core-shell solid-state reaction process. J Eur Ceram Soc 33:2617–2623
71. Ballato J, Serivalsatit K (2011) Sub-micron grained highly transparent sesquioxide ceramics: synthesis, processing, and properties. In: Laser Technology for Defense and Security Vii. 8039
72. Kokuoz BY, Serivalsatit K, Kokuoz B, Geiculescu O, McCormick E, Ballato J (2009) Er-doped Y_2O_3 nanoparticles: a comparison of different synthesis methods. J Am Ceram Soc 92:2247–2253
73. Serivalsatit K, Ballato J (2010) Submicrometer grain-sized transparent erbium-doped scandia ceramics. J Am Ceram Soc 93:3657–3662
74. Serivalsatit K, Kokuoz B, Yazgan-Kokuoz B, Kennedy M, Ballato J (2010) Synthesis, processing, and properties of submicrometer-grained highly transparent yttria ceramics. J Am Ceram Soc 93:1320–1325
75. Serivalsatit K, Kokuoz BY, Kokuoz B, Ballato J (2009) Nanograined highly transparent yttria ceramics. Opt Lett 34:1033–1035
76. Fedyk R, Hreniak D, Lojkowski W, Strek W, Matysiak H, Grzanka E et al (2007) Method of preparation and structural properties of transparent YAG nanoceramics. Opt Mater 29:1252–1257
77. Pazik R, Gluchowski P, Hreniak D, Strek W, Ros M, Fedyk R et al (2008) Fabrication and luminescence studies of Ce:$Y_3Al_5O_{12}$ transparent nanoceramic. Opt Mater 30:714–718
78. Koechner W (2006) Solid-state laser engineering, 6th edn. Springer, Berlin
79. Shannon RD (1976) Revised effective ionic radii and systematic studies of interatomic distances in halides and chalcogenides. Acta Crystallogr Sect A 32:751–767
80. Siegman AE (1986) Lasers. University Science Books, Sausalito
81. Powell RC (1988) Physics of solid state laser materials. Springer, New York
82. Henderson B, Bartram RH (2000) Crystal field engineering of solid-state laser materials. Cambridge University Press, Cambridge
83. Svelto O (2010) Principles of lasers, 5th edn. Springer, New York
84. Carter CB, Norton MG (2007) Ceramics materials: science and engineering. Springer, Berlin
85. Aldinger F, Claussen N, Kaneno M, Koumoto K, Sōmiya S, Spriggs RM et al (2003) Handbook of advanced ceramics: volume II processing and their applications. Elsevier, Amsterdam
86. Barsoum MW (2003) Fundamentals of ceramics. Institute of Physics Publishing (IoP), UK
87. Pan YB, Li J, Jiang BX (2013) Advanced optical functional transparent ceramics. Science Press, Beijing

Chapter 2
Transparent Ceramic Materials

2.1 Introduction

As stressed in previous chapter, to be transparent ceramics, materials should have an isotropic lattice structure, i.e., cubic structure. Actually, there have been various transparent ceramics with such a lattice structure, including simple oxides [1–9], sesquioxides [10–19], complex oxides [20–26], and aluminum oxynitride [27–30]. However, various transparent ceramics that have noncubic lattice structures, such as tetrahedral ferroelectric PLZT [31–33], orthorhombic mullite [34, 35], and hexagonal Al_2O_3 [36–51], emerged in recent years. To make these transparent ceramics, various sintering techniques, including vacuum sintering, high pressure (HP) and hot isostatic pressure (HIP), spark plasma sintering (SPS), and microwave sintering, have been used for different materials. In order to improve the sintering behaviors of the precursor powders, various synthesis methods, especially wet-chemical routes, such as chemical co-precipitation, sol–gel and gel combustion, have been employed to produce powders with ultrafine particle/grain sizes. In some cases, special packing technologies, like tape casting, slip casting, or gel casting, have been used to form green bodies of the transparent ceramics. For example, tape casting is able to produce thin sheets with large sizes, while slip or gel casting enables to form rod-like or specially shaped items. This chapter is aimed to cover the ceramic materials that could be made to be transparent, which have been reported in the open literature.

2.2 Simple Oxides

2.2.1 Alumina (Al_2O_3)

The only thermodynamically stable crystallographic modification of alumina (Al_2O_3) is α-Al_2O_3, or known as corundum. Corundum has a hexagonal crystal

L.B. Kong et al., *Transparent Ceramics*, Topics in Mining,
Metallurgy and Materials Engineering, DOI 10.1007/978-3-319-18956-7_2

lattice with cell parameters $a = 4.754$ Å and $c = 12.99$ Å. The O^{2-} ions are arranged in close hexagonal arrangement, with the cation Al^{3+} occupying two-thirds of the octahedral interstitial positions [52]. It has a density of 3.95 g cm^{-3}, with melting point of 2072 °C and boiling point of 2977 °C.

Alumina exhibits many interesting properties, such as high strength, high hardness, and excellent corrosive resistance. This makes transparent Al_2O_3 ceramics a promising candidate for applications as electromagnetic windows, transparent armor, and envelopes of HP metal halide lamps [53]. Sintered Al_2O_3 ceramics with submicron grain size are the hardest materials among all transparent armors, even including sapphire. Therefore, transparent polycrystalline alumina is believed to be a promising alternative for sapphire. Photographs of representative Al_2O_3 transparent ceramics, which were processed by using SPS, are shown in Fig. 2.1 [53].

Traditional transparent Al_2O_3 ceramics were prepared by sintering in hydrogen at temperatures of generally above 1700 °C [54, 55]. The high sintering temperature caused extensive grain growth, which in turn seriously affected mechanical strength and hardness of the materials. Also, the large grain sizes of up to 410 μm led to significant light scattering caused by the birefringence of the coarse Al_2O_3 grains [50]. Therefore, its in-line transmission was typically below 10 %. The low strength and in-line transmission posed an almost insurmountable obstacle for their applications in fields where high transparence and good mechanical properties are required.

Optical and mechanical properties of transparent Al_2O_3 ceramics are highly dependent on their grain size and residual porosity. Various strategies have been employed to control the grain sizes and minimize the residual porosity. For this purpose, fine-grained transparent Al_2O_3 ceramics have recently attracted much

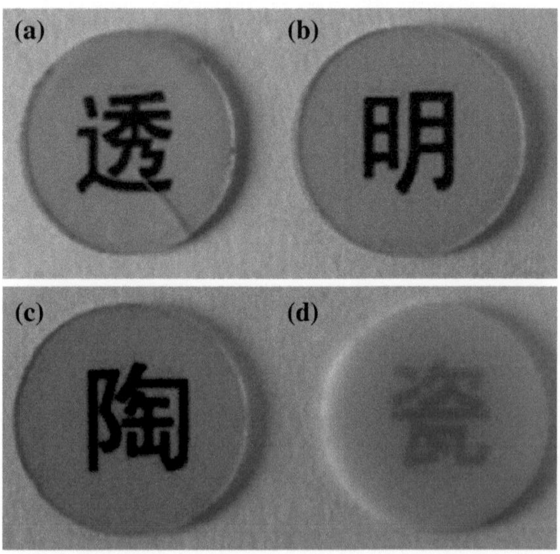

Fig. 2.1 Photographs of representative alumina (Al_2O_3) ceramics that were spark plasma sintered at **a** 1250 °C, **b** 1300 °C, and **c–d** 1350 °C, using the pretreated (**a–c**) and the untreated Al_2O_3 powder (**d**). Reproduced with permission from [53]. Copyright © 2010, John Wiley & Sons

attention [45–51]. The fine-grained ceramics demonstrated a significant improvement in mechanical strength and optical transparency. It has been reported that typical fine-grained transparent Al_2O_3 ceramics had strength of up to 600–800 MPa and high in-line transmission of up to 60 % [47, 49]. Various strategies and technologies have been used to fabricate fine-grained transparent Al_2O_3 ceramics.

A multiple-step pressureless sintering has been used to obtain high quality transparent Al_2O_3 ceramics [56]. Nanosized Al_2O_3 ceramics with an average grain size of 70 nm and a relative density of 95 % were obtained by using a two-step sintering. In this case, α-Al_2O_3 powders, with a mean particle size of about 10 nm and a weak agglomeration, were synthesized by using an α-Al_2O_3 seeded poly-acrylamide gel method. The densification and the suppression of the grain growth are realized by exploiting the difference in kinetics between grain boundary diffusion and grain boundary migration. The densification was promoted by the slower grain boundary diffusion without promoting grain growth in second-step sintering. Unfortunately, the nanosized Al_2O_3 ceramics were not optically transparent.

By float packing and sintering in air, transparent alumina with submicrometer grains were obtained [57]. Commercial α-alumina powder with high purity, submicrometer particle size, and narrow particle-size distribution was used as starting material. The powder was dispersed and stabilized in a water-based suspension. Controlled consolidation and drying by float packing led to homogeneous green compacts, which could be densified without additives by sintering in air at 1275 °C to transparency, while the mean grain size remained to be 0.4 μm. The in-line transmittance at wavelengths of 300–450 nm was comparable to commercial polycrystalline alumina tubes for lighting technologies, whose grain sizes are larger by a factor of 40.

A method of combining injection molding and pressureless sintering in hydrogen was reported to manufacture transparent alumina ceramics [58]. Transparent alumina components with small size and high precision were fabricated in such a stable, efficient, and low-cost route. Excellent rheological properties of the feedstocks for injection molding were obtained through a method of powder pretreatment with stearic acid induced by ball milling. The average grain size of the sintered body was 30–50 mm, with no significant pores and abnormal grain growth observed. The real in-line transmission was higher and more stable than those via other forming technologies and pressureless sintering reported in the literatures.

A similar powder injection molding (PIM) was employed to produce translucent alumina brackets [59]. For PIM, alumina precursor powders were mixed with a binder consisting of a mixture of paraffin wax and polyethylene. After injection molding, debinding was performed using the wicking method. Samples were sintered in a vacuum at 1700 °C to achieve high density. Sintering aids, including MgO, La_2O_3, and Y_2O_3, were used to promote the densification and decrease the porosity of the final ceramics. An improvement in translucency by promoting grain growth during the pressureless sintering was observed. Vickers hardness, bending strength, density, and transmittance of the fabricated items were found to be comparable with those of the commercially available dental brackets.

Solid loading is a critical factor to the fabrication of ceramic compacts with high densities by using ceramic injection molding. Different from the traditional approaches in which stearic acid (SA) was introduced just in the powder blending process, a new strategy has been developed to prepare feedstock with a much higher solid loading, which was up to 64 vol% by using a prior ball milling treatment of the ceramic powders with a small amount of SA [60]. In this way, SA could be coated homogeneously on the powder surfaces, due to the chemical reaction induced by the ball milling treatment. Highly translucent Al_2O_3 ceramics were fabricated from the precursors.

Doping has been found to be an effective technique in fabricating transparent ceramics. MgO doping, combined with pre-sintering heat treatment, was used to accomplish a high translucency in polycrystalline Al_2O_3 ceramics sintered at 1700 °C in vacuum [61]. The concentrations of MgO were 140, 500, and 2500 ppm. A pre-sintering heat treatment was 800 °C for 50 h in air. It was found that the pre-sintering heat treatment improved the sample transmittance in the visible range (400–700 nm) significantly. This enhanced transmittance was explained in terms of the removal of residual pores and the homogenization of the microstructure, due to the lowering of the boundary mobility, as a result of the MgO addition and the suppression of local densification through the pre-coarsening step.

Wet-chemical approaches have been used to fabricate transparent alumina ceramics, which had higher homogeneous dopant distribution [62]. In this approach, alumina powder was first mixed with metal nitrates and dispersed by PEG-2000 as dispersant, and then the pH value was controlled by introducing the $NH_3 \cdot H_2O$ into the suspension to favor the precipitation of Mg^{2+} and Y^{3+} onto the Al_2O_3 matrix. It is found that the dopants were smaller in size and more homogeneously dispersed through the chemical precipitation in the translucent alumina sample as compared with those derived from conventional ball milling. The samples prepared by doping through chemical precipitation had much higher transparency.

Translucent alumina ceramics were also obtained through incorporating $MgO/Y_2O_3/La_2O_3$ additives using infiltration and gelling technique, combined with sintering in H_2 atmosphere [63]. The improved microstructural homogeneity, finer grain size, and enhanced transmission properties of infiltration processed samples over those processed by conventional ball milling method were corroborated by experimental results. Triple doping via infiltration appears to be significantly beneficial for achieving enhanced transmission. The sample with a thickness of 0.75 mm exhibited a high transmittance of 36.3 % at wavelength of 800 nm.

Advanced powder compaction technologies have been widely adopted in fabrication of transparent alumina ceramics. For example, a slip casting method was reported to create transparent alumina ceramics, with the addition of oligosaccharide alcohol to the alumina slurry [64]. Transmittance of the Al_2O_3 ceramics produced by slip casting using the slurry with both NH_4-PMA and oligosaccharide alcohol was higher than that of the ceramics produced by using slip casting form the slurry with NH_4-PMA alone. The increased optical property of the ceramics was attributed to the low viscosity of the slurry, due to the addition of oligosaccharide alcohol, at a high solid content.

Sintering techniques are equally important in reducing sintering temperature and enhancing transparency of transparent ceramics. For instance, it was reported that by HIP, the sintering temperature of Al_2O_3 could be reduced to 1200–1300 °C, at which the porosity could be easily decreased to less than 0.05 % and the grain size was suppressed to be less than 1 μm [47, 50]. Similarly, an effective method was developed to fabricate Al_2O_3 ceramics with almost 100 % relative density by using HIP, which was combined with colloidal dispersion casting and a natural pre-sintering step [65]. The Al_2O_3 ceramics prepared in this way had 60 % of relative transparency for a 1 mm thick sample, which was enough to observe well-resolved pictures at several kilometers through such windows. The transparent ceramics had an average grain size of about 600 nm. The properties of transparent Al_2O_3 ceramics are also related to the characteristics of the starting Al_2O_3 powders [64].

SPS has become a popular new technique to sinter alumina ceramics. A self-doping approach, combined with SPS, was reported to prepare transparent alumina ceramics with controlled grain growth [66]. Highly transparent Al_2O_3 ceramics were obtained by using SPS from both pure and aluminum ethoxide-doped powders. It was found that the porosity was reduced in the self-doped samples, when the doping content is below 1 %. In this case, the improvement in in-line transmittance of the ceramics could be explained by considering the grain and pore size distributions.

A new trend in transparent ceramics is the presence of nanocomposite transparent ceramics. For example, transparent alumina/ceria nanocomposites have been developed by using SPS [67]. It was found that the ceria nanoparticles strongly enhanced the transparency of the SPS sintered compacts, which was attributed to at least two factors. On the one hand, the ceria nanoparticles served as powder lubricant, increasing the initial density of the powder in the SPS die by about 15 %. On the other hand, the ceria nanoparticles, have a very low solid solubility in the alumina grains, so that they were located at grain boundaries, thus hindering alumina grain growth through pinning effect. Therefore, densification could be accomplished at 1430 °C and 80 MPa for 2 min. However, this effect was found to be effective only at vacuum conditions.

Effects of the nature of the dopants, thermal pretreatment, and sintering temperature on SPS of transparent alumina have been systematically investigated [68]. A slurry of α-Al_2O_3 was doped with Mg, Zr, and La nitrates or chlorides, with concentrations of 150–500 weight ppm and then freeze-dried to produce nanosized doped powders ($\sim$150 nm). The powders were sintered by SPS to yield transparent polycrystalline alumina ceramics. Transparency of the nanosized Al_2O_3 ceramics was shown to depend mainly on the way the powder was prepared, as well as the nature of the dopants. RIT values at 640 nm of the samples doped with ZrO_2, MgO, and La_2O_3 were 40.1, 44.1, and 48.1 %, as compared to 30.5 % for pure alumina.

A two-step pressing method was reported to be able to significantly improve the optical properties of alumina ceramics with SPS at high heating rates [69]. In this case, commercial alumina powder could be consolidated at 1150 °C at a heating rate of 100 °C min^{-1}. The effects of the pressure application mode were examined, in terms of microstructure, porosity, and optical transparency. Two ways of application of pressure were compared. In the first case, a constant pressure of 80 MPa was applied for the entire duration of the sintering process, which was called constant pressure. In the second case, an initial pressure of 35 MPa was applied, which was subsequently increased to a higher level, in 3 min after the beginning of dwelling time, i.e., two-step pressure. The application of two-step pressure allowed for homogeneously densified translucent alumina ceramics at a high heating rate.

The influences of SPS sintering pressure and temperature, as well as Mg, Y, and La single or co-doping with concentrations of 75–450 ppm, on the RIT and grain size of Al_2O_3 ceramics, were systematically studied [70]. Using optimized sintering parameters, RITs of >50 % could be obtained in the visible wavelength (640 nm) for 0.8 mm thick samples for almost all the different methods. The optimized doping was the triple dopant, showing a RIT of 57 % at a total dopant level of 450 ppm, which was the highest value among those processed by using SPS in the open literature. It means that a combination of doping and SPS could produce transparent Al_2O_3 ceramics with improved performances.

Microwave sintering is another new sintering technique that has been applied to transparent Al_2O_3 ceramics [71]. With no holding at the sintering temperature, microwave-sintered samples could reach 95 % density at 1350 °C, which was much lower than the 1600 °C for conventionally heated samples. The microwave-sintered sample was almost fully densified after sintering at 1400 °C, while the density of the sample sintered through the conventional sintering was only about 50 %. It was also found that grain growth behavior of alumina ceramics was not affected by the way of sintering. Therefore, the enhancement in densification of Al_2O_3 by the microwave sintering was not attributed to the fast heating rate. However, the microwave densification had an apparent activation energy of 85 ± 10 kJ mol^{-1}, which was much lower than the value of 520 ± 14 kJ mol^{-1} for the conventional sintering, implying that the densification mechanisms of the two sintering techniques are different.

Mechanical strength of transparent Al_2O_3 ceramics is closely related to their grain size and size distribution [47, 51, 72–74]. Fine-grained transparent Al_2O_3 ceramics, with high optical and mechanical properties, can be developed by using various methods and strategies, as discussed above. Mechanical strength is crucial for certain applications, such as IR windows and armors.

Besides the applications for lighting and domes, transparent nanostructured γ-Al_2O_3 ceramics have various other applications, such as a humidity sensor of γ-Al_2O_3 [75]. Nanostructured γ-Al_2O_3 ceramics were prepared from Al-Sec-Butoxide ($C_{12}H_{27}AlO_3$) by using a sol–gel process. The sensors had a long term stability of up to two years.

2.2.2 Zirconia (ZrO_2)

Zirconia (ZrO_2), with a density of 5.68 g cm^{-3}, melting point of 2715 °C, and boiling point of 4300 °C, has been widely used because of its exceptional mechanical and functional properties, such as high toughness, high oxygen diffusivity, and low thermal conductivity. Based on these properties, transparent ZrO_2 ceramics, especially yttria-stabilized zirconia or YSZ, have been extensively and widely investigated. The introduction of yttria can stabilize the cubic or tetragonal structure of zirconia with the presence oxygen vacancies, as shown in phase diagram in Fig. 2.2 [76]. This improves ionic conductivity and makes stabilized ZrO_2 appropriate for use as an electrolyte in solid oxide fuel cells.

Recently, polycrystalline transparent cubic yttria-stabilized zirconia (ZrO_2-8 mol% Y_2O_3, c-YSZ) ceramics have attracted increasing attention because of their unique combination of mechanical and optical properties [77–87]. Cubic zirconia has a refractive index of 2.2, which has never been attained in optical glasses and is higher than other oxides. Therefore, high transparencies, together with large refractive indices and high dielectric constants, make YSZ ceramics to be a new group of transparent ceramics [78, 88, 89]. Figure 2.3 shows photographs of representative YZS transparent

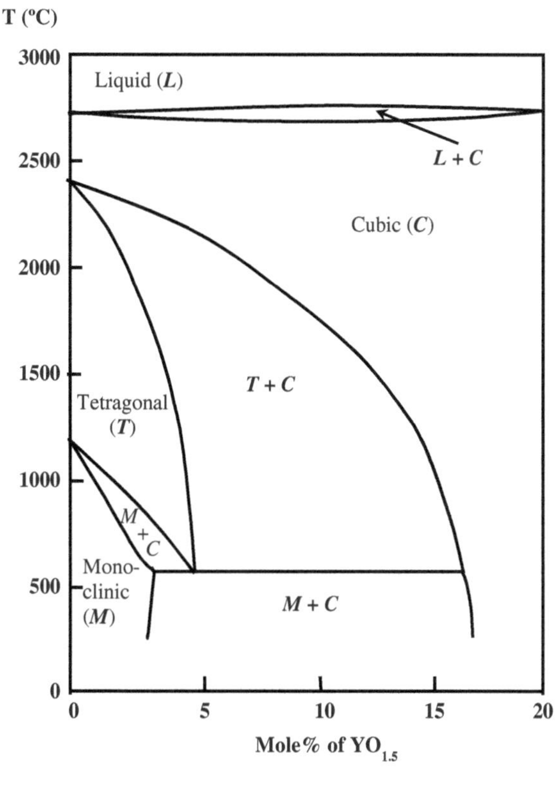

Fig. 2.2 Portional binary phase diagram of ZrO_2–Y_2O_3

Fig. 2.3 Photographs of representative YZS transparent ceramics, which were made from commercial powder with hot isostatic pressing sintering method. Reproduced with permission from [88]. Copyright © 2008, John Wiley & Sons

ceramics, which were made from commercial powder by hot isostatic pressing sintering method [88].

Translucent cubic zirconia (15 mol% Y_2O_3–ZrO_2) was prepared by using a HP hot pressing technique [80]. In this report, submicron particles of monoclinic ZrO_2, with and without stabilizing oxides of CaO or Y_2O_3, were subjected to pressures ranging from 0.5 to 30 kbars and sintered at temperatures ranging from 1300 to 1750 °C, at heating rates of 1000 °C h^{-1} and cooling rates of 500 °C h^{-1}, so as to translucent monoclinic and cubic stabilized ZrO_2. Samples with near theoretical density, including monoclinic ZrO_2 = 5.61 g cm^{-3}, 15 mol% CaO-stabilized cubic ZrO_2 = 5.54 g cm^{-3}, and 15 mol% Y_2O_3-stabilized cubic ZrO_2 = g cm^{-3}, could be achieved by using HP hot pressing. The samples had an average in-line light transmittance of 12 %. An increase in grain size from 1 to 5 μ was observed from the center of the samples to their outer edge, due to the pressure gradients.

Translucent cubic zirconia (6 mol% Y_2O_3–ZrO_2) was fabricated by using the conventional sintering technique [81]. The precursor powder of high surface activity with a grain size of less than 1 μm was synthesized through simultaneous decomposition of yttrium and zirconium alkoxides. The mixed oxide powder was consolidated by calcining at 1000 °C for 30 min. High density fully stabilized 6 mol% Y_2O_3–ZrO_2 ceramics were obtained by sintering at 1450 °C. Translucent tetragonal zirconia (3 mol% Er_2O_3–ZrO_2) by the conventional sintering technique, using powders synthesized through alkoxides hydrolysis method [82]. These studies indicated that sintering process and sinterability of the starting powders are critical parameters for translucency.

Transparent polycrystalline c-YSZ could be prepared by using hot isostatic pressing (HIP) [79, 85]. It was found that TiO_2 was a good additive for developing transparent c-YSZ by using HIP, as demonstrated by the in-line transmittance curves of the ZrO_2 (10 % Y_2O_3) ceramics doped with TiO_2 [79]. The transparent ceramics were made from cubic stabilized ZrO_2 (c-ZrO_2) powder by sintering at vacuum of 1×10^{-3} Pa at 1650 °C for 3 h, followed by HIP treatment at 1750 °C for 1 h at a pressure of 196 MPa. Post-thermal annealing at 1000 °C was conducted to decolorize the as-sintered samples. The role of TiO_2 was regarded as a grain growth promoter and a pore scavenger. The optical properties of these zirconia ceramics were comparable to those of their single crystal counterparts. In addition, transmission performances of the ceramics were sensitive to the microstructure of the pre-sintered compacts. Microstructural features, such as fine grains and small intergranular pores, were critical to achieve high transparency. Therefore, high sintering temperatures of HIP led to samples with large grains, which is not desirable in terms of mechanical properties [79, 85].

SPS has been shown to be better than HIP in fabricating transparent c-YSZ ceramics. For example, SPS has been used to process a nanosized cubic yttria-stabilized zirconia (ZrO_2-8 mol% Y_2O_3) powder synthesized by using a glycine–nitrate process combined with high-energy ball milling [76]. The powder could be consolidated by using SPS at 1200–1350 °C for 5 min, while transparent ceramics could be fabricated at 1300 °C. Another example was the use of HP SPS to fabricate translucent cubic zirconia ceramics at 1000–1200 °C [89]. The total forward transmittance and in-line transmittance of the SPS samples were mainly affected by the color centers, with a limited contribution from residual pores, whereas the changes in reflectance were only related to the porosity. Because the number of color centers and the density of the residual pores increased with increasing sintering temperature, the total forward and in-line transmittances of the as-sintered zirconia ceramics were decreased.

A cost-effective technology has been reported for the development of larger quantities of transparent cubic ZrO_2 ceramics [90]. The key strategy was the application of a multistep sintering process. Yttrium-stabilized ZrO_2 powder was treated by dry pressing, together with subsequent cold isostatic pressing. The obtained ZrO_2 ceramics were reported to be comparable to perovskite ceramics, in terms of optical performance. The in-line transmittance was related to thickness of the samples. The real in-line transmissions of the unscattered transmitted light at a wavelength of 640 nm were 57 % for 4 mm thickness, 59 % for 3 mm thickness, and 65 % for 2 mm thickness, respectively. Due to the high refractive index of ZrO_2, the reflection losses are 12 % on both surfaces. Therefore, depending on the sample thickness, the transmission of the 4 mm plate was 0.75 of the theoretical maximum, while that of the 2 mm plate was 0.85. Hardness HV10 measured at a testing load of 10 kg was up to 13 GPa, approaching the upper limit of ZrO_2 ceramics. Average grain size of the ZrO_2 ceramics was about 5 μm.

It was found that cubic ZrO_2 ceramic has relatively low mechanical strength, which is <600 MPa. In contrast, tetragonal ZrO_2 ceramics have a strength of >900 MPa. Also, nearly dense YSZ ceramics with a grain size of <100 nm can exhibit a fracture toughness of about 8 MPa $m^{1/2}$, which is higher than a typical cubic material with a toughness of 2.8 MPa $m^{1/2}$. Therefore, tetragonal ZrO_2 ceramics deserve to be further developed, especially mechanical strength is an important requirement [86, 91].

The influences of grain size and birefringence on transparency of partially Y_2O_3-stabilized tetragonal ZrO_2 (YSZ) ceramics have been systematically studied [92]. It was confirmed that optical birefringence was the main factor that limited the transparency of tetragonal pore-free ZrO_2 ceramics. The conditions to obtain transparent tetragonal ZrO_2 ceramics (partially stabilized by yttria) have been predicted by comparing the experimental data for sintering of nanopowder with scattering calculations. The simulations were validated well by experimental data for tetragonal ZrO_2 (3 mol% Y_2O_3) ceramics made from a powder with an initial particle size of ~10 nm. The sintering was carried out in air and with HIP. A maximum in-line transmission of about 77 % was observed at IR wavelengths of 3–5 μm in the samples.

It was found that tetragonal ZrO_2 ceramics with grain sizes of 50–55 nm could be obtained by using SPS, which suppressed effectively the grain growth of the nanosized powders [86]. These works further confirmed that dense tetragonal ZrO_2 ceramics are less transparent than their cubic counterparts.

Munir et al. prepared transparent tetragonal (3 mol% yttria) and cubic (8 mol% yttria) YSZ ceramics derived from nanosized powders using a pulsed electric current sintering process [78]. Both the tetragonal and cubic ceramics had an average grain size of about 50 nm. Consolidation pressure was found to show a positive effect on the transparency of both samples. Transmittances of the samples with 1 mm thickness in the near-infrared region were >60 % for the cubic 8 % YSZ and >50 % for the tetragonal 3 % YSZ ceramics, representing between 70 and 80 % of the theoretical values of the two materials. The samples had a yellowish-brown color, which was attributed to the presence of color centers. Annealing in oxygen improved the transmittance initially, but prolonged annealing resulted in translucent samples.

By using powders with very fine particles, nanocrystalline 3YSZ ceramics with grain sizes of 22–45 nm have been developed by using HIP [93]. Amorphous to nanocrystalline ZrO_2-3 wt% Y_2O_3 powders were synthesized by using a chemical precipitation method from mixed nitrate salt solutions. The powders were cold pressed and pre-sintered in air for 2–6 h at temperatures of 1100–1300 °C. HIP was carried out at 1150–1350 °C for 2–3 h at 150 MPa in argon. Fully dense pellets with grain size of 22–45 nm were formed by application of low pre-sintering temperatures. However, these samples had a significant amount of monoclinic phase.

2.2.3 Sesquioxides

2.2.3.1 Yttria (Y_2O_3)

Y_2O_3, as a promising optical material, has excellent physical and chemical properties, such as high melting point (2430 °C), broad range of transparency (0.2–8 μm), and high corrosion resistance, with density of 5.01 g cm^{-3} and boiling point of 4300 °C. Typical physical properties of Y_2O_3 include thermal conductivity of 13.6 W m^{-1} K^{-1} at 300 K, refractive index of ~1.935, density of 5.04 g cm^{-3}, and effective atomic number of 36.7 [94]. Y_2O_3 has a cubic crystal structure with group of Ia3, as shown in Figs. 2.4 and 2.5.

Due to its high effective atomic number and high density, i.e., 35 and 4.56 g cm^{-3}, respectively, Y_2O_3 could be a more effective scintillator than Yttrium

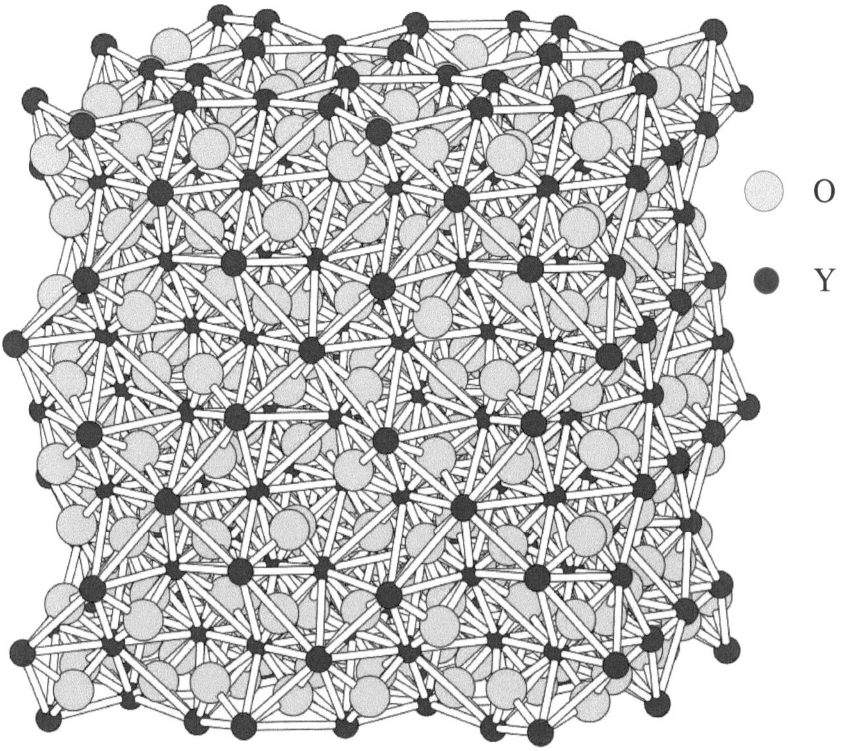

Fig. 2.4 Crystal structure of Y_2O_3, which was created by using Electron Micsrocopy Software (EMS) (version 3.3111U2008)

Fig. 2.5 Crystal structure of R_2O_3 structure with the two crystallographic cationic sites C_2 and C_{3i}

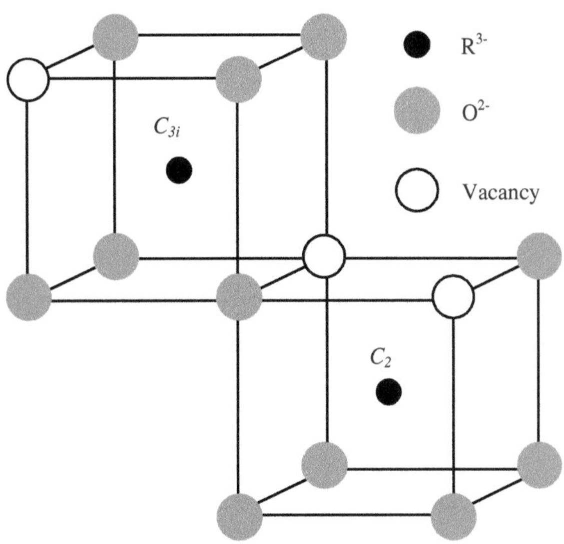

aluminum garnet (YAG) [95]. Y_2O_3 possesses a higher thermal conductivity and a lower thermal expansion coefficient than the widely used YAG, which is critical for thermal management, because more heat will be generated, as laser powers continue to increase. Other potential applications of Y_2O_3 transparent ceramics include infrared domes, nozzles, refractories, components of semiconductor devices, and efficient NIR-visible upconverter [96–98]. Due to its refractory characteristics, Y_2O_3 has been mainly developed in transparent ceramics rather than single crystals. The first transparent yttria ceramic was reported about 50 years ago [99].

There have been quite a number of reports on Y_2O_3 transparent ceramics by using pressureless sintering in vacuum or in H_2 atmosphere [100–107]. One example is the preparation of transparent ZrO_2-doped Y_2O_3 ceramics through a combined processing of vacuum sintering to fabricate from commercial power using aqueous slip casting technique [103]. Figure 2.6 shows photographs of the ZrO_2-doped Y_2O_3 ceramics with different contents of ZrO_2, after sintering at 1860 °C for 8 h. The samples with ZrO_2 of above 2 mol% were all highly transparent. The starting Y_2O_3 power had an average particle size of 2 μm, which was reduced to about 0.34 μm after ball milling in ethanol for 12 h. The powder with reduced particle size was beneficial for densification. It was shown that 5 mol% ZrO_2 was the optimum concentration in terms of in-line transmittance of the final ceramics. The optical properties of the ceramics were attributed to the enhanced densification and suppressed grain growth behaviors due to the presence of ZrO_2.

Advanced sintering techniques, including HP [108, 109], HIP [110, 111], and SPS [112–116], have all been employed to obtain transparent Y_2O_3 ceramics. For example, transparent Y_2O_3 ceramics were fabricated by using HP at 1800 °C for 20 h [109]. Another example is the application of a uniaxial hot pressing to develop Y_2O_3 ceramics, at a maximum pressure of 40 MPa and a maximum temperature of 1580 °C [108]. Optical transparency could be optimized through a stepwise application of pressure. The use of SPS could reduce the processing temperature of Y_2O_3 ceramics [116]. It was reported that the samples sintered at 1300 °C and annealed at 1050 °C had a transmittance of 81.7 % at wavelength of 2000 nm. Due to the low temperature, SPS is able to control the grain growth of yttria ceramics [114]. The Y_2O_3 specimen sintered at 1050 °C for 1 h exhibited an in-line transmittance of 6–46 % in a wavelength range of 400–800 nm. Heating rate and sintering temperature could be optimized to improve the optical properties of SPS-processed Y_2O_3 ceramics [117].

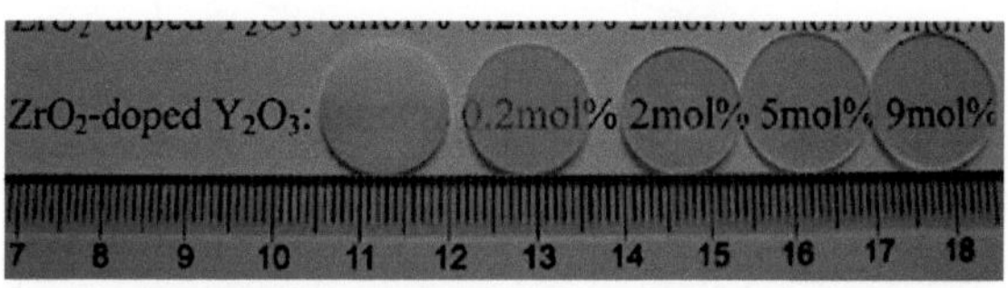

Fig. 2.6 Photographs of representative Y_2O_3 ceramics doped with different levels of ZrO_2, processed through the combination of slip casting and vacuum sintering. Reproduced with permission from [103]. Copyright © 2010, Elsevier

Various strategies, such as use of nanosized precursor powder, introduction of sintering aids, and adoption of multiple-step sintering process, have been employed to obtain Y_2O_3 ceramics with relatively high density, small grain size, and high transparency. Wet-chemical methods, such as chemical precipitation and co-precipitation, have been widely used to synthesize pure and doped Y_2O_3 nanopowders, in order to achieve high sinterability [101, 118, 119].

An yttria nanopowder with an average particle size of 60 nm was synthesized by using a chemical precipitation process [118]. Hydroxide precursor of Y_2O_3 with an approximate composition of $Y_2(OH)_5NO_3{\cdot}H_2O$ was obtained by using ammonia water as precipitant and yttrium nitrate as the starting salt. By employing appropriate striking method and optimum synthetic conditions, yttrium hydroxide with a card-house or spherical structure was formed. Normal striking-derived powders had higher sinterability than the reverse striking-derived powders. The addition of a small amount of ammonia sulfate in the yttrium nitrate solution could decrease the agglomeration and particle size of the yttria powder. Nanocrystalline yttria powders were obtained by calcining the precursor at 1100 °C for 4 h. Transparent yttria ceramics were fabricated from the nano Y_2O_3 powder by vacuum sintering at 1700 °C for 4 h, with optimized in-line transmittance of 52 % at wavelength of 1000 nm.

Various sintering aids, such as ThO_2 [105], La_2O_3 [106], HfO_2 [120], LiF [121], and ZrO_2 [122], as well as their combinations [123], have been used to fabricate transparent Y_2O_3 ceramics. The use of sintering aids led to Y_2O_3 ceramics with small grain sizes due to the reduced sintering temperatures. Doping effect could be enhanced, when combining with compaction and sintering techniques, such as slip casting, vacuum sintering [103], and sintering in oxygen atmosphere [124]. With the addition of zirconia and other additives to control the grain growth, pores could be eliminated at a temperature of <1650 °C, deriving Y_2O_3 ceramics with an average grain size of 1 μm [124].

Optical properties of Y_2O_3 ceramics, doped with rare-earth ions, are of special interest for some applications, such as solid-state lasers and scintillators [95]. Luminescence properties, such as upconversions, are also of special significance. For instance, RE/Yb co-doped Y_2O_3 transparent ceramics (RE = Er, Ho, Pr and Tm), exhibited very efficient NIR-visible upconverters [96].

Through the combination of two-step sintering and vacuum sintering, transparent lanthanum-doped yttrium oxide ceramics have been fabricated [100]. The samples were first sintered at an intermediate temperature T_1 (1450 °C) in air. Then, the temperature was decreased quickly to a lower point T_2 (900–1100 °C) in air and held for more than 20 h in order to facilitate densification with limited grain growth. Nanocrystalline yttria ceramics with a high relative density were then obtained, which could be further densified at higher temperature T_3 (1700 °C) in vacuum. The final transparent yttria ceramics doped with 10 % lanthanum, with an average grain size of about 25 μm, had a transmittance of 77 % at 580 nm. Microstructural analysis results indicated that the mean particle size of the samples after the second step was just slightly larger than that of those after the first step processing, but the relative density was over 90 %. However, because the significant shrinkage was mainly observed in this step, pores were isolated by grains. Therefore, grain boundary diffusion was thought to be the main matter transport mechanism.

A similar two-step sintering combined with HIP was reported to fabricate Y_2O_3 ceramics with submicron grain sizes [125]. Microhardness and fracture toughness of the transparent yttria ceramics with an average grain size of 0.3 μm made by using the modified two-step sintering were higher than those of the ones fabricated by using conventional sintering, by 25 and 70 %, respectively. The conventional processing led to ceramics with an average grain size of ~300 μm. The submicron transparent yttria ceramics exhibited a transparency of single crystals in the near-infrared spectral region.

2.2.3.2 Scandia (Sc_2O_3)

Sc_2O_3 (scandia) belongs to the cubic space group $IA\bar{3}$ [126]. Its unit cell contains 16 f.u., with 32 cations that form 24 sites of C_2 symmetry and 8 sites of C_{3i} symmetry. The C_2 site is an eightfold cubic structure with two oxygen vacancies on a face diagonal, while the C_{3i} site corresponds to a cube with two vacancies on a body diagonal. The RE^{3+} dopants are assumed to occupy randomly both sites, but the induced electric dipole transitions are allowed only for C_2 centers. The cationic density (3.338×10^{22} cm^{-3}) is rather high compared to other oxides. Sc_2O_3 has the smallest lattice parameter of 0.986 nm, high melting point, high bulk refractive index of $n_H = 2.0$ at $\lambda = 300$ nm, high band gap of 5.7 eV (corresponding to an ultraviolet cutoff of 215 nm), high thermal conductivity of 17 W m^{-1} K^{-1}, and high chemical stability in alkali metal steam.

The combination of these excellent optical and thermal properties makes Sc_2O_3 attractive for applications as damage-resistant and high reflection materials in light-emitting diodes and high power pulsed ultraviolet lasers. It is also very useful as infrared transmitting devices, host materials for high power and ultrashort pulse solid-state lasers, and heat-resistant optical windows [127–133]. Due to its high melting point (2430 °C), it is difficult to fabricate Sc_2O_3 single crystal. Therefore, transparent polycrystalline Sc_2O_3 ceramics are expected to be more and more important.

Vacuum sintering is the most widely used method to prepare transparent Sc_2O_3 ceramics, with nanosized precursor powders to provide additional benefit to ensure high quality of the final products [127, 128, 131–135]. Various chemical routes have been used to synthesize nanosized Sc_2O_3 powders. For example, a homogeneous precipitation method was used to synthesize nanocrystalline Sc_2O_3 powder [135]. A highly reactive Sc_2O_3 powder was made through pyrolyzing a basic sulfate precursor, ($Sc(OH)_{2.6}(SO_4)_{0.2}{\cdot}H_2O$), at 1100 °C, which was precipitated from scandium sulfate solution with $(CH_2)_6N_4$ as the precipitant [127]. Similar reactive Sc_2O_3 powders were obtained through thermal pyrolysis of a scandium sulfate salt, $Sc_2(SO_4)_3{\cdot}7.8H_2O$ [131]. The Sc_2O_3 ceramics obtained by using vacuum sintering have relatively large grain sizes, due to the significant grain growth during the final stage of sintering [127, 128, 131–135].

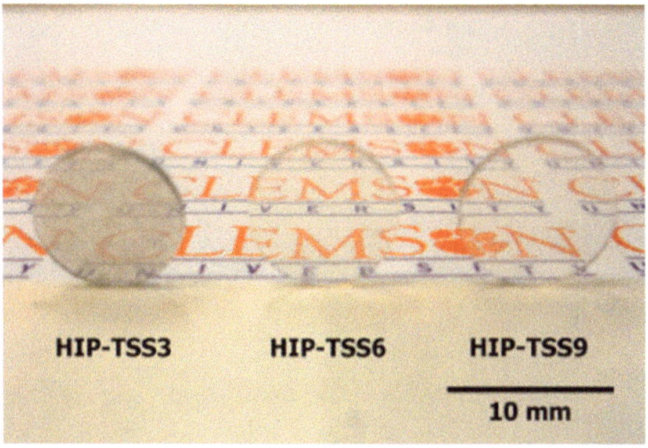

Fig. 2.7 Photographs of representative Er-doped Sc_2O_3 ceramics hot isostatic pressing two-step sintered (HIP-TSS). Reproduced with permission from [136]. Copyright © 2010, John Wiley & Sons

Pressure-assisted sintering techniques have been used to develop transparent Sc_2O_3 ceramics with smaller grain sizes. For example, HIP was used to fabricate erbium-doped transparent Sc_2O_3 ceramics with an average grain size of 0.3 μm, from solution-derived nanoparticles, using a two-step sintering process [136]. The two-step sintered ceramics showed a remarkable decrease in grain size with relative density of over 98 % before HIP. After HIP, the Sc_2O_3 ceramics became highly transparent with negligible grain growth. The samples exhibited an optical transmittance of about 80 %. Photographs of representative Sc_2O_3 ceramics are shown in Fig. 2.7.

Spectroscopic properties of Sc_2O_3 ceramics doped with various ions are attractive for different applications. For instance, spectral characteristics of Ho^{3+}: Sc_2O_3 transparent ceramics have been analyzed in detail [126]. Spectroscopic parameters, including JO intensity parameters, oscillator strengths, radiative transitions probabilities radiative lifetimes and branching ratios, were obtained. Intensity parameters of Tm^{3+} doped transparent Sc_2O_3 ceramics as laser materials have also been reported. They are potential candidates for applications in diode-pumped solid-state laser, emitting at about 2 μm [17]. Crystal field analysis of Er^{3+} in Sc_2O_3 transparent ceramics has been made, which was compared with rare-earth ions in Y_2O_3 [137].

2.2.3.3 Lutetia (Lu_2O_3)

Lu_2O_3 (lutetia) has a cubic crystal structure, with a representative structure shown in Fig. 2.8. It is an attractive host material for various activators because of its wide band gap and favorable properties, such as phase stability, low thermal expansion, and chemical stability. In particular, Lu_2O_3 is the one with the highest doping concentration of Yb^{3+}, while maintaining high heat conductivity, due to the

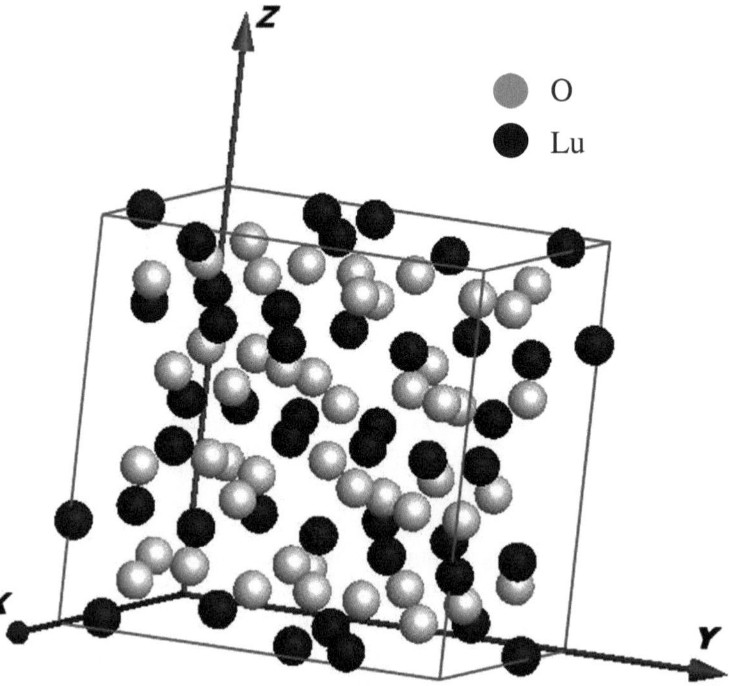

Fig. 2.8 Crystal structure of Lu_2O_3, which was created by using Electron Micsrocopy Software (EMS) (version 3.3111U2008)

comparable mass and ionic radius between Lu^{3+} and Yb^{3+} [138]. The superiority of Lu_2O_3 to Y_2O_3 is its high density of ~9.42 g cm^{-3}, as well as the high *Z* number of Lu (71), which makes Lu_2O_3 to have an exceptionally high stopping power for ionizing radiation. Lu_2O_3 ceramics doped with Nd or Yb are promising high power ceramic laser materials. Europium-doped Lu_2O_3 ceramic scintillator can find applications in digital X-ray imaging technology, due to its high density resulting in high stopping power, efficient X-ray to visible light conversion of up to 90,000 photons MeV^{-1} [16], and visible emission at ~600 nm coupling well with the silicon CCD spectral response [16, 139]. The use of an optically transparent scintillator screen requires high transparent Lu_2O_3: Eu to avoid excess background noise, loss of resolution, and scatter-induced ghosting.

A co-precipitation method was used to synthesize Lu_2O_3 ultrafine powder, which could be used to fabricate transparent ceramics [12]. Eu-doped Lu_2O_3 powder was prepared by using ammonium hydroxide ($NH_3{\cdot}H_2O$) and ammonium hydrogen carbonate (NH_4HCO_3) as mixed precipitant, from solutions of nitrate salts. The precipitates were calcined at 1000 °C for 2 h, which was the optimized condition to

Fig. 2.9 Representative photographs of mirror-polished 5 % Eu^{3+}-doped Lu_2O_3 ceramics sintered at 1850 °C for 6 h, with thickness of 1 mm. Reproduced with permission from [12]. Copyright © 2006, John Wiley & Sons

obtain precursor powder with high dispersity and desired sinterability. Highly transparent polycrystalline Eu-doped Lu_2O_3 ceramics with a relative density of ~99.9 % were developed by using pressureless sintering in flowing H_2 at 1850 °C for 6 h without any additives. The final ceramics with grain sizes of 50–60 μm exhibited optical transmittance in the visible light region of 600–800 nm with a thickness of 1 mm reached 80.3 %. Figure 2.9 shows photographs of representative Eu-doped Lu_2O_3 ceramics.

HIP has been combined with vacuum sintering to fabricate Eu:Lu_2O_3 ceramics with high performances [15]. Nanosized precursor powder, synthesized by using a co-precipitation method, was first uniaxially pressed and sintered under high vacuum at temperatures of 1575–1850 °C to reach relative densities of 94–99 %. The sintered compacts were then subjected to reach full density by using HIP, which resulted in highly transparent ceramics without the need of further air annealing. The Eu:Lu_2O_3 ceramics showed about four times higher light yield than commercially used scintillating glass, indicating that transparent ceramics could be used for high-energy radiography devices with improved performances. Vacuum sintering above 1650 °C led to rapid grain growth prior to densification, so that the pores were immobile. Sintering between 1600 and 1650 °C resulted in closed porosity. Because the grains were fine, the pores were mobile during the subsequent HIP step. As a result, fully dense Eu:Lu_2O_3 transparent ceramics could be prepared without the requirement of subsequent air annealing.

Transparent Lu_2O_3 ceramics have also been produced by using SPS from commercial powders [138, 140]. Sintering conditions, such as sintering temperature, pressure, and holding time, all have effects on the density and optical properties of the final ceramics. It was found that transparency of the Lu_2O_3 ceramics could be further improved by using a two-step pressure profile, combined with a low heating rate [138]. The effects of preload pressures from 10 to 100 MPa and heating rates from 0.03 to 1.67 K s^{-1} on microstructures and optical properties of the materials were investigated. The samples obtained at a preload pressure of 30 MPa at a heating rate of 0.17 K s^{-1} had transmittances of 60 % at 550 nm and 79 % at 2000 nm, showing a significant improvement in optical properties.

2.2.4 Magnesia (MgO)

MgO is has cubic crystal structure, as shown in Fig. 2.10. MgO ceramics have excellent thermal and mechanical properties with a high melting point of 2850 °C and low density of 3.58 g cm^{-3}. It has an isotropic cubic crystal structure, which meets one of the requirements to be transparent. Polycrystalline infrared-transparent MgO is a potential substitute for sapphire IR windows and protectors for sensors. Due to the high sintering temperature required for full densification of MgO, fabrication of transparent MgO ceramics is still a challenge. Almost all techniques discussed above have been used to prepare transparent MgO ceramics [3].

LiF is a popular sintering aid for the fabrication of transparent MgO ceramics. One example is the development of transparent MgO ceramics from commercial powder [4]. The commercial MgO powder had very high purity, high surface area, fine crystallites, and low aggregation. LiF suspended in 2-propanol was added to the nano-MgO powder at content levels of 2 and 4 wt%. After compaction, the pellets were burnout binder and then were loaded in a graphite die for hot pressing carried out in Ar at 45 MPa, at heating rates of 3 or 10 °C min^{-1}. Figure 2.11 shows photographs of representative samples of the transparent MgO ceramics.

The addition of LiF as sintering aid could reduce the sintering temperature of MgO transparent ceramics [141, 142]. It has been acknowledged that densification

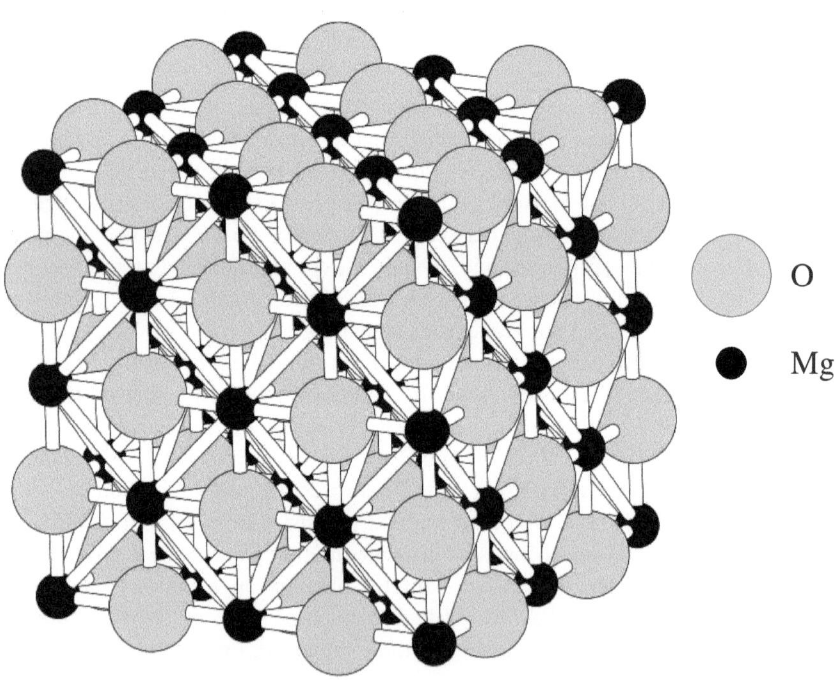

Fig. 2.10 Crystal structure of MgO, which was created by using Electron Micsrocopy Software (EMS) (version 3.3111U2008)

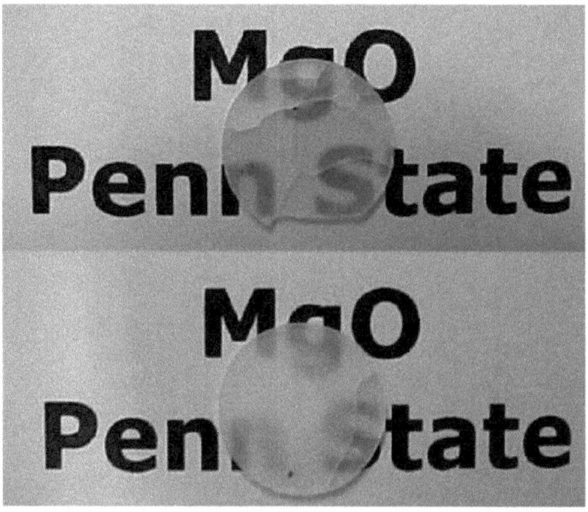

Fig. 2.11 Photographs of representative as-annealed MgO ceramics fabricated through hot pressing of the nano-MgO powder containing 4 % LiF at 1100 °C for 1 h in Ar, with sizes of 12.7 mm in diameter and 0.5 mm in thickness. Reproduced with permission from [4]. Copyright © 2004, Elsevier

of MgO with LiF was enhanced due to the formation of a liquid phase, which acted initially as a lubricant for the rearrangement of the MgO particles and later as a material transport medium to support pressure-enhanced liquid phase sintering [143].

Although sintering aids can be used to reduce the sintering temperature of MgO, they could have negative effects on the intrinsic properties of MgO. It was demonstrated that transparent MgO ceramics could be at 1600 °C by a HIP technique, without using sintering aids [144]. Before HIP, MgO compacts were obtained by using pressureless sintering. A nonagglomerated MgO powder with an average primary particle size of 57 nm was used as precursor. The relative density and average grain size of MgO compact pressureless sintered at 1600 °C for 5 h were 96.7 % and 10.7 μm, respectively.

Transparent MgO ceramics have also been fabricated by using SPS [1]. As-received commercial powder was put into graphite die with an inner diameter of 12 mm, and prepressed to 150 MPa before heating. The pressure was then reduced to 10 MPa during the heating procedure. The pressure was increased to 100 MPa within 20 s upon reaching the final sintering temperature. The sintering was conducted at that temperature for 5 min hold. Lower sintering temperatures were used, when a high pressure of 150 MPa was applied for 3 and 5 min during the sintering. With SPS, fully dense optically transparent MgO ceramics from nanocrystalline powders could be produced at 800 °C and 150 MPa for 5 min.

A simple pressureless sintering was applied to MgO transparent ceramics [3]. The powder used in this study was synthesized by using a chemical precipitation method. Magnesium nitrate hexahydrate ($Mg(NO_3)_2 \cdot 6H_2O$, 98 % purity), ammonium hydrogen carbonate (NH_4HCO_3), and ammonium hydroxide (NH_4OH, 28.0 %) were used the precipitation. The precipitate was aged for 24 h, before centrifugation, washing, and drying. The dried powder was then calcined at various temperatures for 2 h. The powder calcined at 700 °C for 2 h was used to obtain

MgO ceramics. The MgO green bodies were sintered at 1400 °C for 2 h in ambient atmosphere. The as-sintered MgO ceramics had a relative density of 98.1 % with an average grain size of 6 μm and an average hardness of 6.8 GPa [3].

2.3 Complex Oxides

2.3.1 Garnet

Yttrium aluminum garnet ($Y_3Al_5O_{12}$ or YAG) is a crystalline material of the garnet group. It is also one of the three phases of yttria–aluminum compounds, while the other two are yttrium aluminum monoclinic (YAM) and yttrium aluminum perovskite (YAP), as illustrated in the binary phase diagram of Y_2O_3 and Al_2O_3 in Fig. 2.12. YAG has a cubic crystal structure and belongs to the isometric crystal system [145]. Crystal structure of YAG with the three cationic sites, i.e., dodecahedral, octahedral, and tetrahedral, is schematically shown in Fig. 2.13, while a more comprehensive crystal structure is illustrated in Fig. 2.14.

Due to the centrosymmetric cubic crystalline structure, its optical properties are isotropic. Owing to its high thermal stability, stable chemical properties, and unique homogenous optical properties, transparent YAG ceramics are not only an important high-temperature structural material, but also an excellent host material for

Fig. 2.12 Phase diagram of Y_2O_3 and Al_2O_3 with the formation of YAG phase

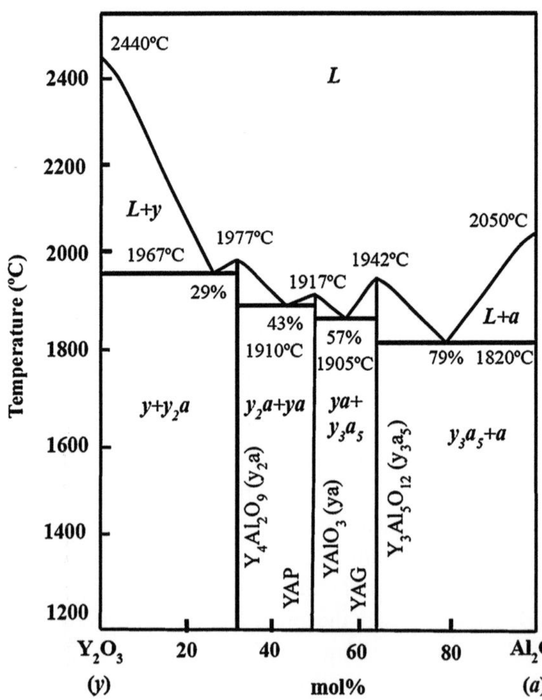

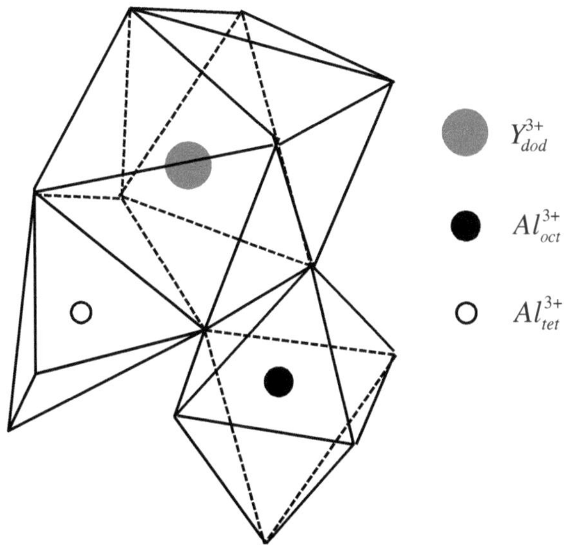

Fig. 2.13 Crystal structure of YAG with the three cationic sites: dodecahedral, octahedral, and tetrahedral

fluorescence application and high power solid-state lasers [146]. Nd- and Er-doped YAG ceramics are typical laser materials. Ce-doped YAG (Ce:YAG) is used as phosphor in cathode ray tubes and white light-emitting diodes and as scintillator. Since the first report on transparent polycrystalline Nd:YAG ceramics for laser media by Ikesue et al. [147] in 1995, extensive attentions have been paid to the development and fabrication of YAG ceramics [148].

Solid-state reaction (SSR) has been widely used to fabricate transparent polycrystalline YAG ceramics. The most cost-effective SSR is the use of commercial oxide powders. For example, mixture of commercial α-Al_2O_3 and Y_2O_3 powders, together with tetraethoxysilane (TEOS) and MgO as sintering aids, was used as precursors to prepare transparent YAG ceramics through SSR [149]. The mixture was made by using ball milling. The milled mixture was dried and compacted. Pressed samples were sintered from 1500 to 1750 °C in vacuum. Figure 2.15 shows XRD patterns of the green body (YAG composition, Y:Al = 3:5) after vacuum sintering at different temperatures for 10 h. As stated earlier, the Al_2O_3–Y_2O_3 system experienced three phases during the reaction or reactive sintering, i.e., monoclinic phase (YAM, $Y_4Al_2O_9$), perovskite phase (YAP, $YAlO_3$), and cubic phase (YAG, $Y_3Al_5O_{12}$). Phase compositions and formation temperature ranges for each composition are described as follows:

$$2Y_2O_3 + Al_2O_3 \rightarrow Y_4Al_2O_9(\text{YAM}) \quad 900{-}1100^\circ\text{C} \tag{2.1}$$

$$Y_4Al_2O_9 + Al_2O_3 \rightarrow 4YAlO_3(\text{YAP}) \quad 1100{-}1300^\circ\text{C} \tag{2.2}$$

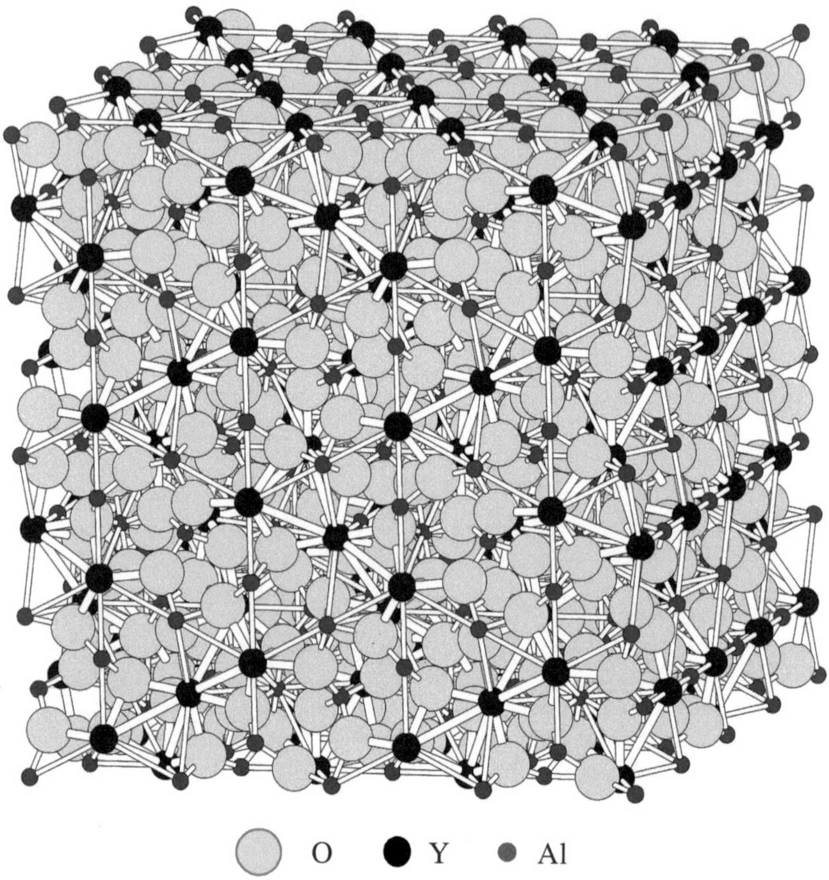

Fig. 2.14 A schematic of crystal structure of YAG, which was created by using Electron Micsrocopy Software (EMS) (version 3.3111U2008)

$$3YAlO_3 + Al_2O_3 \rightarrow Y_3Al_5O_{12}(YAG) \quad 1400{-}1600\,^{\circ}C \tag{2.3}$$

Undoped YAG transparent ceramics are usually prepared by using vacuum sintering, HP, and SPS. A YAG precursor powder was synthesized through a co-precipitation method using ammonium hydrogen carbonate as precipitant, which could be used to develop transparent YAG ceramics by using vacuum sintering at 1700 °C for 1 h [150]. With the mixture of Y_2O_3 and Al_2O_3 commercial powders, YAG transparent ceramics could be obtained by using HP at 1750 °C for 4 h at a pressure of about 300 atm [151]. SPS was applied to fabricate YAG transparent ceramics at 1400 °C for 3 min starting from nanocrystalline YAG powders [23]. In this case, fully dense and transparent polycrystalline cubic YAG ceramics with micrometer grain sizes were obtained. Nanosized YAG transparent ceramics have also been developed by using HP sintering process [152]. The nanostructured YAG

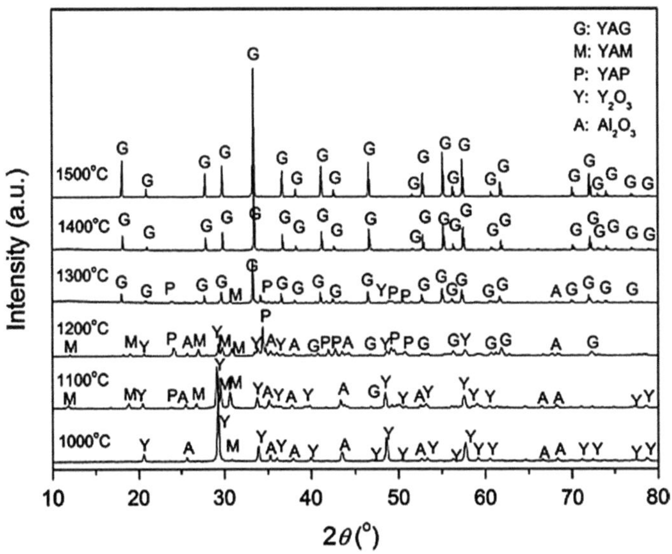

Fig. 2.15 XRD patterns of the samples for the preparation of YAG sintered at temperatures of 1000–1500 °C for 10 h. Reproduced with permission from [149]. Copyright © 2014, Elsevier

compacts were prepared at high pressure and modest temperature, or HPMT, 2.0–5.0 GPa and 300–500 °C. Prior to sintering, pellets were also heat treated in vacuum. After the samples were treated at 2.0–5.0 GPa and 300–500 °C for 30 min, the pressure was reduced to ambient pressure, and then the samples were quenched to room temperature at a cooling rate of ~15 °C min^{-1}. The temperature was found to be sufficiently low to prohibit the grain growth of the YAG ceramics, which is beneficial for improvement in mechanical properties.

For practical laser applications, YAG ceramics should be doped with different elements. For example, highly transparent polycrystalline Er^{3+}:$Y_3Al_5O_{12}$ (Er:YAG) ceramics with concentrations of Er^{3+} ion from 1 to 90 % were prepared by the SSR and the vacuum sintering technique [153]. In-line transmittances of mirror-polished Er:YAG ceramics were all up to 84 % at 1100 nm wavelength, which were attributed to high density of the samples.

A two-step sintering process was developed to synthesize Nd:YAG ceramics [154]. In this two-step sintering process, the furnace temperature was first arisen to a higher temperature T_1 (1800 °C), at a heating rate of 10 °C min^{-1}, and then the temperature was rapidly decreased to a certain sintering temperature T_2 (1550–1750 °C) for 1–8 h in a vacuum graphite tube furnace at 1.0×10^{-3} Pa during the holding process. The transparent ceramics prepared by using this method exhibited pore-free and homogeneous microstructure, with transparency of up to 84.98 % at the visible and near-infrared band and 87 % in the mid-infrared wavelength range. The absorption cross section of a 0.3 % Nd:YAG sample at 808 nm was 5.47×10^{-20} cm^2, while the emission cross section at 1064 nm was 4.66×10^{-19} cm^2.

To achieve full density and highest transparency, YAG ceramics have also been prepared by using sintering aids [155–159]. MgO is one of the most effective sintering aids to produce Nd:YAG ceramics [160]. The addition of small amount of MgO could reduce the degree of agglomeration and particle size of the Nd:YAG precursor powders. The MgO doped Nd:YAG powders possessed better dispersion characteristics as compared with the undoped powders. The sample with 0.01 wt% MgO were of spherical particles of 100 nm in diameter. The well-dispersed precursor powders resulted in ceramics with homogenous microstructure. Optimized ceramic samples exhibited transmission of 82.6 % at the wavelength of 1064 nm, which was comparable to that of Nd:YAG single crystals. La_2O_3 is another effective sintering aid for Nd:YAG ceramics [156]. The addition of 0.8 wt% La_2O_3 was at optimized concentration, which led to Nd:YAG transparent ceramics having the highest transmittance in the wavelength region from 400 to 1100 nm.

Co-doping with MgO + SiO_2 has been found to be a promising technique to improve the sinterability of YAGs [161, 162]. By optimizing the concentrations of the dopants, transparent YAG ceramics with desired optical properties can be readily achieved. For example, the Nd:YAG ceramic samples with 0.4 wt% TEOS as the precursor of SiO_2 combined with 0.08 wt% MgO possessed the highest transparency at 1064 nm [162]. It was also found that the function of SiO_2 was to eliminate pores and remove secondary phases, while MgO contributed to homogeneous microstructure of the ceramics, which has been confirmed by scanning electron microscopy (SEM) analysis results [157].

It is well known that the sinterability of a ceramic precursor powder is highly dependent on various characteristics, such as particle size, size distribution, morphology, and degree of agglomeration. This is also applicable to YAG transparent ceramics [163, 164]. The effect of powder properties on the densification and performance of YAG transparent ceramics has been systematically studied [165]. It was observed that the YAG ceramics derived from nanosized powder exhibited higher optical transparency, due to their smaller grain size and less residual pores. The properties of precursor powders could be also modified by adjusting pH values when using co-precipitation method [164]. pH values could affect the performance of the final Nd:YAG ceramics by influencing the morphology as well as the (Nd + Y):Al mole ratio of the powders. Optimized pH values were found in the range of 7.9–8.2.

If the (Nd + Y):Al molar ratio was much lower than 0.6, superfluous Al_2O_3 was present, while it was much higher than 0.6, superfluous Y_2O_3 was present which led to intermediate phases (YAP and YAM). The powders prepared within the range of pH value of 7.9–8.2 had (Nd + Y):Al molar ratio very close to 0.6 (0.598 and 0.603). In this case, no secondary phases were observed and fully dense Nd:YAG ceramics with homogeneous microstructures were obtained. The conclusion was further confirmed by the cross-sectional SEM analysis. The fracture styles of samples with (Nd + Y):Al molar ratios of 0.576 and 0.648 are both intracrystalline, because secondary phases and pores in them enhanced strength of the grain boundaries. In contrast, the samples from the powders with ratios of 0.598 and

0.603 possessed pore-free microstructure and clean grain boundaries, thus having intergranular fracture behavior.

Other technologies used to further improve the quality of precursor powers of transparent YAG ceramics include spray drying [166, 167] and freeze-drying [168, 169]. High purity Al_2O_3, Y_2O_3, and Nd_2O_3 powder mixture was processed using spray drying technology. Polyvinyl butyral (PVB) was used for granulation of the powders and inhibition of the compositions segregation. The effect of the addition of PVB (0, 1, 2, and 3 wt%) on the morphologies and compositions of the spray-dried powders was evaluated. The powder with 2 wt% PVB after calcination at 1000 °C for 2 h led to Nd:YAG transparent ceramics. Figure 2.16 shows photographs of representative samples of the Nd:YAG transparent ceramics [167].

It has been demonstrated that freeze-drying could play an important role in the synthesis of high quality Nd:YAG nanosized powders [169]. Submicronic neodymium-doped YAG ($Y_3Al_5O_{12}$) powder was synthesized from freeze-dried precursors. The powder calcined at 1200 °C, with small crystallite size and the lowest amount of organic residues, showed the highest sinterability. With these powder, although color centers were detected by transmission, transparent ceramics were obtained after a vacuum sintering at 1700 °C for 3 h and completed with a HIP treatment at 1700 °C, 160 MPa of Ar for 90 min.

Optical properties of rare-earth doped YAG ceramics are important requirements for practical applications, which has been a subject of numerous studied in the literature [170–176]. For example, thermal annealing had a strong effect on optical properties of Cr-doped YAG ceramics [170]. After annealing at high temperatures, the absorption intensity was increased and the two peak wavelengths shifted from 430 to 600 nm to approximately 465 and 612 nm, respectively, which could be attributed to valence changing from Cr^{3+} to Cr^{4+}. The presence of Cr^{4+} ion was confirmed by emission measurement.

The application of transparent Ce:YAG ceramic phosphors for white LED has been reported widely. For instance, the Ce:YAG ceramics had a broad emission band peaked at 530 nm due to the 5d-4f transition of Ce^{3+} [172]. The transmittances of the samples were 70–87 % at 800 nm. The absorption coefficient and emission intensity of Ce^{3+} were increased with increasing sample thickness. Under 465 nm LED excitation, the color coordinates of the Ce:YAG ceramics shifted from blue

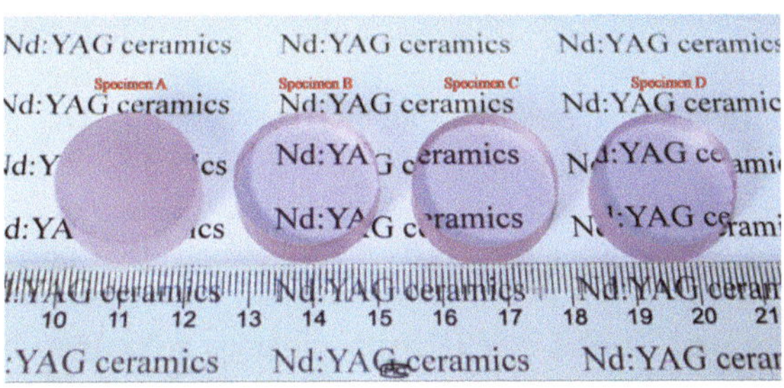

Fig. 2.16 Photographs of representative samples of the Nd:YAG transparent ceramics. Reproduced with permission from [167]. Copyright © 2012, Elsevier

region to yellow region with increasing sample thickness, passing nearby the theoretical white point in the chromaticity diagram. The highest value of luminous efficacy of the ceramic white LED was 73.5 lm W^{-1}. The LED's current dependence of luminous efficacy and luminous intensity for a 0.632 mm thick sample under 465 nm LED excitation has been studied. The luminous efficacy was decreased from 74 to 44 lm W^{-1}, while the luminous intensity was increased from 100 to 1550 mCd with increasing current.

When studying optical properties of transparent Eu^{3+}:$Y_3Al_{5(1-x)}Sc_{5x}O_{12}$ ceramics, a reduced peak splitting of Eu^{3+} for $^5D_0$–$^7F_1$ and $^5D_0$–$^7F_2$ was observed when 10 at.% Al_{31} was substituted by Sc_{31} [173]. The enhanced symmetry of the Eu sites in YAG lattice, which resulted from the expanded YASG lattice due to the Sc_{31} doping, was the main reason for the reduced peak splitting.

Other garnets, such as Lu_3Al5O_{12} (LuAG) and Tm_3Al5O_{12} (TmAG), have emerged as new members of transparent ceramics of garnet family [177–186]. Yb: LuAG laser ceramics with different Yb^{3+} doping concentrations were fabricated by using a solid-state reactive sintering method [182]. SEM results demonstrate that the samples had a dense and pore-free microstructure. According to the spectroscopic studies, the ceramics exhibited a large emission cross section of 2.7×10^{-20} cm^2 at 1030 nm emission peak. CW laser operation of the samples generated 7.2 W output power with 65 % slope efficiency.

2.3.2 *Spinel* ($MgAl_2O_4$)

The crystal structure of magnesium aluminate "spinel" ($MgAl_2O_4$) is based on an FCC close-packed oxygen sublattice, in which a fraction of the octahedral and tetrahedral sites is filled. It is characterized by magnesium ions tetrahedrally coordinated with oxygen and aluminum octahedrally coordinated with oxygen. A unit cell contains 32 oxygen ions, 16 octahedral cations, and 8 tetrahedral cations, as shown schematically in Fig. 2.17. The polycrystalline structure of the magnesia spinel is optically isotropic. Magnesia spinel undergoes no polymorphic transformations and hence is devoid of any thermally induced phase changes.

Spinel is transparent to electromagnetic radiation from the ultraviolet through the mid-infrared (0.2–5.5 μm) [187–189]. Spinel has a distinct transmission advantage over sapphire and AlON in the range of 4.5–5.5 μm, a region of particular importance for seeker and electro-optic imaging systems. The combination of unique optical and mechanical properties makes spinel one of the most outstanding optically transparent ceramics. Figure 2.18 shows photographs of two spinel ceramics without and with sintering aid of LiF, which were processed by using SPS [190].

Given its high strength and excellent transmittance properties up to the mid-IR wavelength, spinel is a very attractive material to manufacture high performance optical components, such as lenses, IR windows, and domes. Very good UV-VIS and medium wavelength IR transparency coupled with high hardness and high resistance

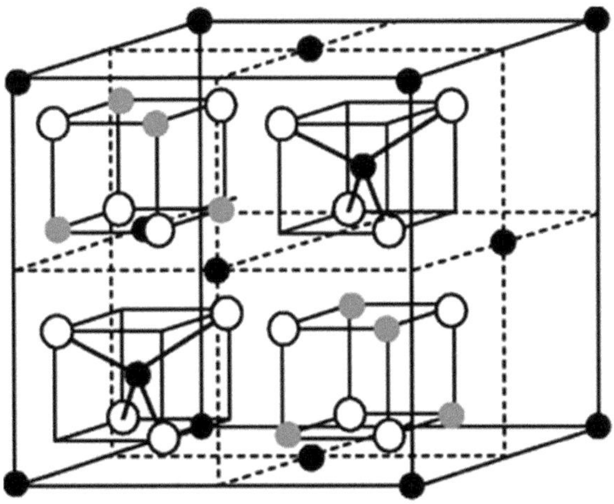

Fig. 2.17 Crystal structure of $MgAl_2O_4$ spinel, *filled black circles*—Mg site, *filled gray circles*—Al site, and *open circles*—O site

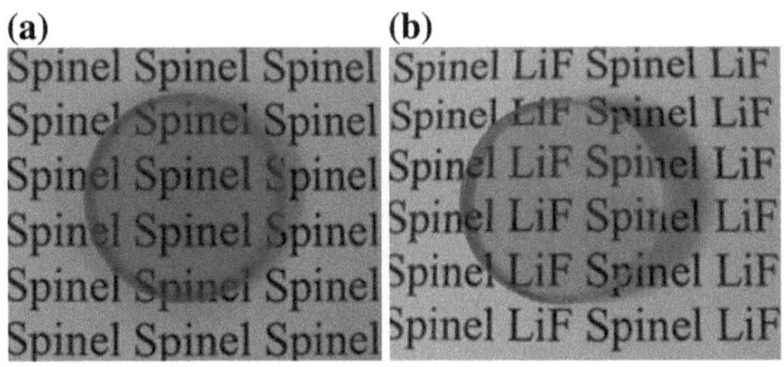

Fig. 2.18 SPS processed spinel ceramics: **a** undoped sample fabricated at sintering temperature of 1300 °C, at heating rate of 2 °C min^{-1} for 30 min, applied pressure of 63 MPa, and thickness of 2 mm, **b** 1.0 wt% LiF-doped sample fabricated at sintering temperature of 1400 °C, with heating rate of 15 °C min^{-1}, for 120 min, and thickness of 2 mm

to erosion make spinel an excellent material for protective windows for aircraft, ship, and submarine IR sensors, as well as missile domes. Spinel is also very promising for applications as windows or windshields for military vehicles such as large and bulletproof windows, due to its mechanical and ballistic properties [191].

Since it was shown to be transparent in early 1960s, spinel has received considerable worldwide attention [192, 193]. Transparent spinel is difficult to fabricate directly from high purity precursor powders by using the conventional pressureless sintering techniques [192–194]. Therefore, it is necessary to use HP, HIP, or SPS [193, 195–202]. By the way, fine precursor powders, together with sintering aids, such as B_2O_3 [203], CaO [192], and LiF [204], have been shown to be very helpful in developing transparent spinel ceramics.

It was reported that by using a combination process of pre-sintering and HIP, fine-grained highly transparent spinel ceramics could be obtained from powders with a wide range of properties [205]. Clear transparent components with a thickness of up to 20 mm and lateral dimensions up to 240 mm were manufactured equally successfully from both the spinel powders with lower specific surface areas and the nanosized powders. Recently, a similar process was developed to fabricate spinel ceramics with large-scale production [206].

A similar two-step process, i.e., HP followed by HIP, has been used to prepare optically transparent polycrystalline spinel ceramics [193, 195]. The effect of HP temperature on optical properties of the final ceramics HIP sintered at 1900 °C for 1 h under a pressure of 189 MPa was studied. Three HP temperatures, 1400, 1450, and 1500 °C in vacuum of 10^{-2} Pa for 1 h at 50 MPa, were examined. The sample HP pre-sintered at 1400 °C had the maximum transmittance of over 60 and 70 % in the UV-visible and near-IR wavelength regions, respectively. The light transmittance was well correlated with the microcracked grain boundary surface area per unit volume of the samples.

Very recently, SPS has been proved to be an alternative technique for the fabrication of fine-grained transparent spinel. Several studies have been conducted for enhancing the transparency by using additives [199], controlling heating rates [197, 202], or applying two-step pressure profile during sintering [196]. Although the transparency of the spinel fabricated by SPS is lower than that by HIP at present, an improvement for higher transmission can be made in SPS by controlling the sintering parameters such as pressure, temperature, heating rate, atmosphere, and powder preparation.

A commercial magnesium aluminate powder could be made into transparent ceramics by using SPS, at the low heating rate of 10 °C min^{-1}, with LiF as sintering aid, [199]. The presence of 1 wt% LiF dopant affected dramatically the spinel morphology, which caused the formation of large well-defined grains and thus eliminated residual boundary phases. Therefore, relatively high levels of light transmission could be achieved in this way.

In a separate study, the densification of a fine-grained high purity spinel was realized for only 20 min at a low temperature of 1300 °C without any sintering aids. For heating rates <10 °C·min^{-1}, the spinel exhibits an in-line transmission of 47 % for a visible wavelength of 550 nm and a fracture strength of >500 MPa. The high in-line transmission can be explained by lower residual porosity (<0.5 %) [197].

Preheating the precursor powder is a useful strategy to enhance the transparency of spine ceramics processed by using SPS [201]. The commercial $MgAl_2O_4$ spinel powder had a purity of >99.97 % and an average particle size of 0.36 μm. The as-received powder was heated at 800–1100 °C for 3 h in air and before subject to SPS processing. Preheating at 1100 °C had almost no influence on final density of the ceramics, while the particle growth during the preheating led only slight increase in grain size, because of the large particle size of the starting powder. Preheating at lower temperatures of ≤1000 °C had less effect on the microstructure, and preheating at higher temperatures of ≥1200 °C made it impossible to obtain transparent spinel with large grain sizes. Therefore, 1100 °C was the optimized preheating temperature.

Both the pressure sintering and SPS have the problem of contamination [197, 199]. It has been shown that graphite contamination has a negative effect on the optical properties of transparent spinel obtained by SPS [207]. They found that residual porosities and second-phase graphite particles had a strong negative influence on in-line transmittance of the ceramics. The carbon contamination was proposed to be arisen from a CO-containing residual atmosphere during the SPS experiments, because of the graphite configuration of the equipment. During SPS, a rotary pump was used to generate vacuum in which the residual gases gave rise to an atmosphere containing CO_2. Such an atmosphere infiltrated the porous body at the beginning of sintering and CO_2 was possibly reduced to CO at high pressure within the shrinking pores, which was finally carbonized into particles. The presence of such carbon particles made the in-line transmittance to be less than theoretical value of single crystalline $MgAl_2O_4$. Therefore, special carefulness should be taken when using pressure-aided techniques to fabricate transparent ceramics and more systematic studies should be carried out on this important issue.

Because the performance of spinel windows is also very sensitive to the quality of the precursor materials (both purity and morphology), there have been researches that can be found in the open literature. To produce better spinel precursor powders, various methods have been developed. For example, a new method has been used to synthesize spinel powders from boehmite, in which Mg^{2+} ion was metal exchanged onto the surface of boehmite particles [208]. In this method, particle size, size distribution, purity, and stoichiometry of the Mg-doped boehmite powders could be well controlled. Such powders exhibited good sinterability and resulted in spinel ceramics with desired optical properties.

Fine precursor powder has been synthesized by using flame spray pyrolysis to prepare, in order to prepare high quality spinel transparent ceramics [209]. The powder possessed excellent sinterability, allowing the attainment of very high densification levels after the pre-sintering stage at 1400 °C. HIPing is necessary to attain high transparency for thickness of >2 mm. The minimal HIPing conditions (for a reasonable transparency are for specimens sintered at 1400 °C for 80 h). A low-temperature HIP processing, 1500 °C, 3 h and 200 MPa, led to 2 mm thick samples with a real in-line transmission of 63 % at 635 nm. The samples had an average grain size of 2.2 μm and a hardness of 13.2 GPa.

An alternative method, by using $Mg(OH)_2$, γ-Al_2O_3 and AlOOH, toward the direct production of transparent $MgAl_2O_4$ ceramics, was reported [210]. The samples derived from the mixture of $Mg(OH)_2$ and γ-Al_2O_3 under optimized processing conditions showed in-line transmittances of 84.2 ± 1.0 % at 550 nm, with thicknesses of 3.5–4 mm. Typical mechanical properties observed in the ceramics included Knoop hardness of ~12 GPa, elastic modulus of ~280 GPa and biaxial flexural strength of 85–136 MPa. There was a slight difference between γ-Al_2O_3 and AlOOH as the precursor components.

A simple reactive method named self-heat-sustained (SHS) technique employing metallic aluminum as one of the reactants to produce very high phase-purity magnesium aluminate powder under rather mild experimental conditions was also developed more recently [211]. Powders with small particle size and narrow particle size distribution were obtained, which could be used to fabricate spinel ceramics with high optical performances.

Another strategy for developing transparent spinel ceramics is the use of sintering aids. It was found that the addition of a small amount of boron oxide (0.05–0.5 mass%) was very effective in promoting the sintering behavior of a fine $MgAl_2O_4$ powder processed by using HIP [203]. Experimental results indicated that the presence of B_2O_3 not only reduced the sintering temperature of the precursor powder, but also suppressed the grain growth of the ceramics which was important to ensure high transparency. The optimal concentration of B_2O_3 was 0.15 wt%, with which the samples could achieve a high optical transmittance of 80 % at 1 mm thickness.

Besides B_2O_3, LiF has also been widely used as a sintering aid for spinel ceramics [212–216]. By using 1 wt% LiF, $MgAl_2O_4$ transparent ceramics could be obtained by using HP process at a temperature as low as 1550 °C for 2 h [213]. The samples had an excellent visible transmittance of ~85 %. The effect of LiF on sintering kinetics of $MgAl_2O_4$ was also investigated [215]. Powders with 0, 0.5, and 1.0 wt% LiF were densified by using a vacuum HP under a range of unaxial pressures. According to the sintering mechanisms derived from the experimental studies, the vacuum HP processing schedule could be further optimized.

According to MgO–Al_2O_3 phase diagram, spinel is the only compound, as shown in Fig. 2.19. However, the spine phase can act as a parent structure type of solid solutions whose compositional range increases with increasing temperature. The compositions can be expressed by the formulation of MgO·nAl_2O_3, with n = 1.0 corresponding to the stoichiometry of $MgAl_2O_4$. At appropriate temperatures, n can range from 0.8 (MgO-rich) to 3.5 (Al_2O_3-rich). Conceptually, as all solid solution compositions share 1.0-spinel's cubic crystal structure, they should

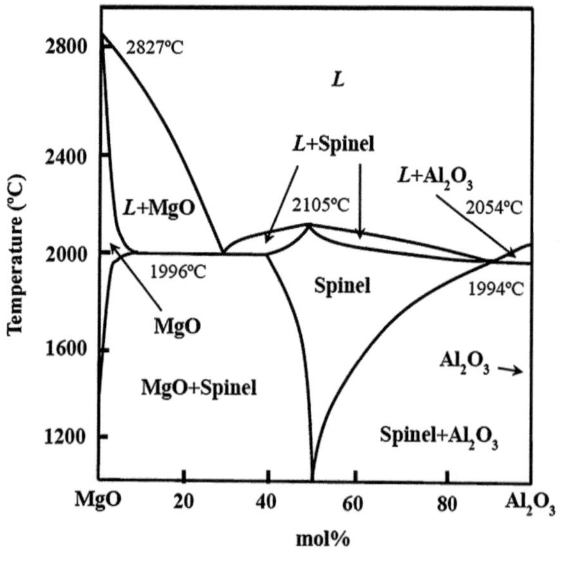

Fig. 2.19 Binary phase diagram of MgO and Al_2O_3 with the formation of spinel $MgAl_2O_4$ and solid solutions at high temperatures

all have the potential to be prepared as transparent, single-phase ceramics; however, exsolution of excess alumina frequently undermines the formation of single-phase $MgO{\cdot}nAl_2O_3$ solid solution polycrystalline ceramics. It has been reported that $MgO{\cdot}nAl_2O_3$ polycrystalline ceramics with Al-rich compositions ($n > 1$) possess varied levels of light transmission at infrared wavelengths, although generally the transmission in the visible spectrum is low. As a consequence, there has been increasing interest in $MgO{\cdot}nAl_2O_3$ ceramics with $n > 1$ [193, 217, 218].

Transparent spinel ceramics with a composition of $MgO{\cdot}1.2Al_2O_3$ have been developed by using HP and HIP [218]. The processing consisted of HP at 1600 °C for 5 h under vacuum and 20 MPa uniaxial load for pre-sintering and HIP at 1850 °C for 5 h under 200 MPa in argon. The samples obtained exhibited an average in-line transmission of 84.8 ± 2.7 % at 550 nm and >82 % throughout the visible spectrum. Final grain sizes of the ceramics ranged between 300 and 1000 μm, with flexural strength of 176.8 ± 46.2 MPa, hardness of 12.3 ± 0.2 GPa, and elastic modulus of 292.9 ± 7.5 GPa.

A detailed study on the relationship between composition and mechanical properties of spinel transparent ceramics has been reported [195]. In this study, HP and HIP were also used, but with commercial MgO and Al_2O_3 powders as the precursor. Three compositions of $MgO{\cdot}nAl_2O_3$ were investigated, with n = 1:0, 1.5, and 2.0. The spinel ceramic with MgO to Al_2O_3 molar ratio of 1:1 demonstrated the highest scattering coefficient due to its highest amount of grain boundary microcracking. Al_2O_3 rich compositions possessed a specular transmission of 40–60 % in the same wavelength range. The spinel ceramic with n = 2.0 exhibited the highest fracture toughness with a mean value of ~2.02 MPa $m^{1/2}$. Based on their optical and mechanical properties, the Al_2O_3-rich non-stoichiometric polycrystalline spinel ceramics would be potential candidate for engineering applications requiring high optical transparency and fracture toughness.

Other studies on spinel transparent ceramics include the effect of grain boundary microcracking on optical performance [193], nanoporosity effect, and elimination of precursor powders [219]. Light transmittance losses occurred when the sintered $MgAl_2O_4$ ceramics had microcracked grain boundaries which facilitated light scattering [193]. It was recognized that the nanoscale pores of <100 nm could be readily eliminated by using pressure-assisted reactive sintering of $MgO–Al_2O_3$ mixtures, starting with coarser powders with median particle sizes of 100–200 nm instead of sintering pre-calcined raw spinel powders <100 nm [219].

Due to the special requirement in mechanical properties for some specific applications, such as armor [220], a new direction of research on spinel transparent ceramics appeared recently, which is nanocomposite transparent ceramics. This concept was well demonstrated in a recent publication [221]. In this case, the nanocomposites were $MgAl_2O_4$ spinel matrix dispersed with Si_3N_4 nanoparticles. The as-processed nanocomposites were found to have a high transparency of >70 % in the critical infrared wavelength range of 3–4.5 μm and an acceptable transparency in the visible region. The optical properties of the nanocomposites remained almost unchanged upon heat treatment at 1000 °C for 4 h in air, which, however, resulted in a 29 % increase in the average strength, accompanied by almost doubling of the Weibull modulus, and an 85 % increment in the indentation toughness. The improvements in mechanical properties after the heat treatment in these nanocomposites were attributed to surface-oxidation-induced surface compression and flaw healing.

TEM observation indicated that no residual porosity could be found in this as-processed spinel/Si_3N_4 nanocomposite. The spinel grain size is 300 nm and the size of is about 100 nm. The nanosized Si_3N_4 particles were located both at grain boundaries and within the spinel grains. The as-processed spinel/Si_3N_4 nanocomposite ceramics with a thickness of 2 mm showed good transparency in the visible range, allowing to view distant objects through it. Therefore, this type of transparent ceramics could be used for possible armor applications [221].

More recently, a nanocrystalline spinel ceramics with 50 % increase in hardness was reported [222]. It is well known that mechanical strengthening by grain refinement is a method, i.e., the strength and hardness of a material can be increased by decreasing its grain size, as described by the empirical Hall–Petch relationship, which provided a guidance to produce stronger materials. Base on this principle, an integrated approach was developed by using nanosized spinel powder, combined with HP and low-temperature sintering to fabricate fully dense and high purity nanocrystalline ceramics with nanometersized grains. The Hall–Petch relationship was confirmed by the hardness of the transparent spinel ceramics with grain sizes down to 28 nm. As a result, the nanosized spinel ceramics exhibited a 50 % increase in hardness without a decline in fracture resistance. More importantly, the nanocrystalline spinel ceramics had near theoretical optical transparency. However, this technique has a difficulty in making large-sized items.

2.4 Electro-optic Ferroelectric Ceramics

Opto-ceramics is a family of special transparent oxide materials. Their crystallography structure is perovskite type with formulation of ABO_3, with unit cell shown schematically in Fig. 2.20, which is tetragonal with $c_T > a_T = b_T$. Due to its small size, the center ion (B) could be displaced either upward or downward to form spontaneous polarization, which is also the reason why the polarization can be switched by the application of external electric fields. This behavior is known as ferroelectricity.

The typical representatives of this family include $Pb_{1-x}La_x(Zr_yTi_{1-y})_{1-x/4}O_3$ (PLZT), $Pb(Mg_{1/3}Nb_{2/3})O_3$-$PbTiO_3$ (PMN-PT), and $Pb(Zn_{1/3}Nb_{2/3})O_3$-$PbTiO_3$ (PZN-PT). The PLZT formula assumes that La substitutes for Pb^{2+} in the A-site and the B-site vacancies are created for electrical balance. To achieve highest transparency and electro-optic coefficient, some elements, such as Ba and/or La, are usually introduced into the solid solutions of PMN-PT and PZN-PT. Material synthesis and properties of transparent electro-optic ceramics, including PLZT and PMN-PT, have well been summarized and documented [223].

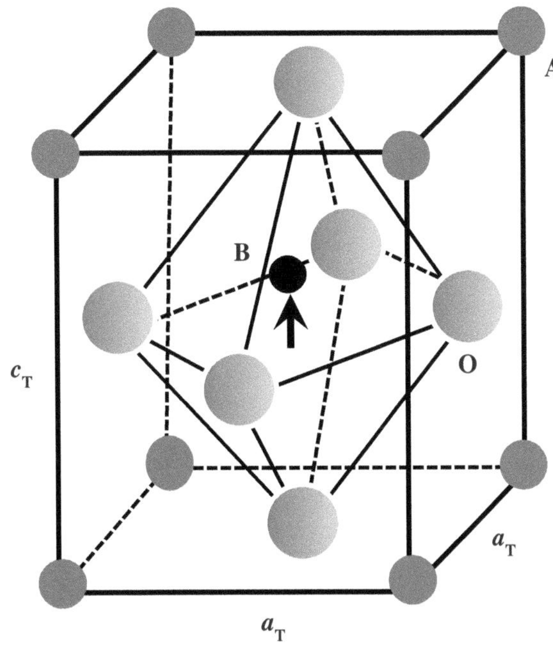

Fig. 2.20 Unit cell of perovskite ABO_3, in which the center ion (*B*) is displaced upward to form a spontaneous polarization. The center ion can also be displaced downward

2.4.1 PLZT Ceramics

PLZT ceramics possess much higher electro-optic coefficient than $LiNbO_3$ single crystals. PLZT ceramics also have higher transparency than other ferroelectric materials. Specifically, $Pb_{(1-x)}La_x(Zr_yTi_z)_{1-x/4}O_3$, i.e., PLZT9/65/35, exhibits outstanding electro-optic coefficients with little hysteresis and no remnant polarization. Therefore, PLZT9/65/35 ceramics are potential candidates for applications in digital light modulators and large aperture light shutters [31, 33].

Transparent PLZT ferroelectric ceramics were reported by Haertling in 1971 [32]. The quality of transparent PLZT ceramics is mainly determined by the properties of the precursors. High quality powders should have small particle size, narrow size distribution, and no hard agglomeration. Several wet chemistry-based methods, such as chemical co-precipitation, sol–gel processing, and hydrothermal processing, as well as dry method of high-energy ball milling, have been developed to prepare PLZT powders. Moreover, special sintering techniques, such HP/HIP and atmosphere-controlled sintering, are also very useful in fabricating high quality transparent PLZT ceramics [32].

The sintering atmosphere is demonstrated to be important in determining the final microstructure of the PLZT ceramic because of the volatility of lead oxide (PbO). A controlled oxygen partial pressure sintering method was established to obtain transparent (Pb, La)(Zr, Ti)O_3 ceramics [224]. PLZT(7/60/40) ceramics with large piezoelectric coefficients were prepared by using a two-step sintering process with controlled oxygen partial pressure. At the first step, the samples were sintered under a low oxygen pressure (vacuum) at low temperature of 1150 °C, while they were then processed under higher oxygen pressure of 0.22 MPa at higher temperature (1250 °C), at the second step. Fully densified transparent PLZT ceramics with an average grain size of 3 μm could be readily fabricated in this way. The samples possessed k_p and d_{33} of as high as 0.71 and 850 × 10^{-12} C/N, respectively. The piezoceramics were transparent, with an optical transmittance of 15 % (λ = 610 nm), for a sample thickness of 1 mm.

Pressure sintering techniques have also been used to fabricate transparent lanthanum-doped lead zirconate titanate (PLZT) ceramics in air and in oxygen-gas atmosphere [31]. The microstructure of the sintered body was not uniform; it was completely dense near the surface, but it was porous at the center. The thickness of the dense layer increased with sintering time and oxygen-gas pressure in the sintering atmosphere. Vaporization of PbO from the specimen surface and resultant formation of lattice vacancies were attributed to this microstructural evolution. Diffusion of the gas trapped in the pores was also important in determining the thickness of the dense layer. When the PLZT specimen was sintered in air at 1200 °C for 8 h, the thickness of the dense layer was 0.25 mm. Therefore, if the specimen thickness was 0.5 mm, the whole specimen was dense and transparent. When the specimen was sintered in an oxygen-gas atmosphere under the same conditions, the specimen thickness increased markedly.

When a specimen with the same composition was sintered in argon, the thickness of the dense layer decreased. Furthermore, it is apparent that the outer region was not completely pore-free as was the case of the specimen sintered in air. On the other hand, when the specimen was sintered in a 33 % oxygen-gas atmosphere, the dense layer became thicker than the case of sintering in air. As the oxygen partial pressure of the sintering atmosphere was further increased, the dense layer continued to grow until the whole specimen became dense in a pure oxygen atmosphere.

A three-stage-atmosphere-sintering technique was used to obtain PLZT transparent ceramics [225]. This technique consists of three stages: (i) sintering in an oxygen atmosphere, (ii) elimination of pores in a carbon dioxide atmosphere, and (iii) elimination of oxygen vacancies in an oxygen atmosphere. The experimental results reveal that use of a carbon dioxide atmosphere effectively decreases residual pores and improves optical transmittance. From commercially available raw powders, an optical transmittance of 51 % at wavelength of 550 nm can be achieved for 0.7 mm thick polished PLZT9/65/35 ceramics using a carbon dioxide atmosphere, whereas the value is only 34 % without a carbon dioxide atmosphere. The advantage of this technique is that PLZT ceramics having high optical quality can be obtained using conventional sintering tools.

A simple and feasible repeated annealing process was available to obtain transparent PLZT(9/65/35) ceramics [226]. The experiment was started from nanosized powder of PLZT, which was synthesized directly from commercially available constituent oxides by using a high-energy ball milling at near room temperature. Transparent PLZT ceramics were obtained by annealing the samples pre-sintered at 1125 °C for 6 h, which was repeated for 4 times without using hot pressing or oxygen flow. The annealing temperature was 1200 °C. The optical properties together with the electrical characteristics of the PLZT ceramics produced in this way were comparable with the literature results. The most significant aspect of this study is the low-cost large-scale fabrication of the PLZT nanosized powders, which is not available by most of the wet-chemical synthesis routes.

SPS has been confirmed to be a viable technique to synthesize PLZT transparent ceramics. An example is the preparation of PLZT(8/65/35) ceramics [227]. The PLZT powder was synthesized by using a wet–dry combination method. Due to the contamination of carbon from the graphite mold, the samples after the SPS sintering should be annealed in air or oxygen. Optical properties of the samples were affected by the SPS temperature, as well as the post-annealing temperature. It was found that 700 °C was the optimal sintering temperature when transmittance was considered.

2.4.2 *Other Ferroelectric Ceramics*

$Pb(Mg_{1/3}Nb_{2/3})O_3$ (PMN) with perovskite structure is known as one of the best representative relaxor ferroelectrics [228]. However, the Curie point (T_C) of PMN is below room temperature and the synthesis of single PMN is always a challenge. Another perovskite structure ferroelectric $PbTiO_3$ (PT) has a high T_C of about 490 °C and very phase stability. Incorporation of PT with PMN has been widely used to form new solid solution PMN-PT, which has T_C of above room temperature and is easy to be synthesized [229–234]. More importantly, PMN-PT has an isotropic minimum energy stable structure and a higher electro-optic coefficient than PLZT (about 10–30 times). To the same index change (or phase shift), the applied voltage onto PMN-PT could be much lower than that required when using PLZT. Moreover, the electric hysteresis of PMN-PT can be readily minimized [229]. In addition, PMN-PT is the most promising nonlinear optical materials because of its extremely high electro-optic (EO) coefficient and strong photorefractive effect [235]. Its EO effect is about 2–5 times higher than that of PLZT and nearly 100 times than that of $LiNbO_3$ at room temperature. The remarkably good transparency over a wide wavelength range of 500–7000 nm makes PMN-PT suitable for almost all the visible to mid-IR optical applications.

Detailed investigation on the optical characteristics, including the electro-optic phase modulation, electric hysteresis property, and thermo-optic coefficient, of transparent PMN-PT electro-optic ceramics have been conducted [229]. A polarization independent PMNT electro-optic switch by using π-shifted fiber Sagnac interferometer structure was constructed and analyzed experimentally. Some switch performances, including thermal characteristic and different switching frequency response, were also realized.

Preparation and characterization of transparent ceramics of La-doped PMN-0.25PT by a two-stage sintering method from conventional raw materials were reported in Ref. [236]. Optical characteristics of Er^{3+}-doped PMN-PT transparent ceramics have been investigated [237], with doping concentrations of 0.5, 1.0, 2.0, and 5.0 mol%. Figure 2.21 shows photographs of the PMN-PT ceramics.

The presence of Er^{3+} over 0.5–2.0 mol% could enhance the optical transmittance of the samples and modify the optical absorption near the band gap energy of the materials. Transmittance of the undoped PMN-0.25PT in the visible region was estimated to be 43.0 %. The transparent ceramics doped with Er^{3+} levels of 0.5, 1.0, and 2.0 mol% were 50, 53, and 57 %, respectively. However, the transmittance was decreased to 28.0 % in the sample with Er^{3+} of 5.0 mol%, which was considered to be closely related to the formation of a trace of pyrochlore phase. By the way, the absorption edge shifted to higher energy as the doping concentrations were <2.0 mol% and a redshift was observed in the samples with higher doping concentration (5.0 mol%). The observations were explained by considering the combined effects of ion polarization and localized defect energy level.

$Pb(Zn_{1/3}Nb_{2/3})O_3$ (PZN) single crystal exhibits outstanding dielectric and electrostrictive properties and also shows a very large electro-optic coefficient [238–241]. However, single-phase PZN ceramics cannot be prepared due to the thermodynamics inhibition. Therefore, it becomes a natural idea to combine PMN and PLZT, aiming at creating new compounds with advantages of both materials, such as excellent piezoelectric and dielectric properties, high optical transmittance in a wide range of wavelength, and strong electro-optical effect. Transparent PZN-PLZT ferroelectric ceramics have been prepared by using HP sintering in oxygen atmosphere [242]. Uniform cylindrical samples with diameter as large as 50 mm could be readily obtained. The product showed a pure perovskite phase, thin grain boundaries, and homogeneous dense microstructures. The ceramics exhibited very good transparency from near ultraviolet to infrared regions.

Besides ferroelectric transparent ceramics, there have also been reports on antiferroelectric ceramics. For example, lead zirconate stannate titanate ($Pb_{0.97}La_{0.02}(Zr_{0.77}Sn_{0.14}Ti_{0.09})O_3$, PLZSnT) ceramics by using SPS at 900–950 °C [243]. The samples had nearly full density and fine grain size of about 300 nm. Compared with samples made by using conventional sintering (CS), the SPS sintered samples showed higher permittivity and lower dielectric loss tangent, as well as special strain hysteresis loops similar to those of ferroelectric relaxors.

Fig. 2.21 Photographs of the Er^{3+}-doped PMN-0.25PT transparent ceramics with a thickness of 1.5 mm. The numbers in the photographs are the Er^{3+} doping levels in mol%. Reproduced with permission from [237]. Copyright © 2012, Elsevier

Lead-free ferroelectric and some microwave dielectric ceramics have been made to be transparent [244–248]. Bismuth-modified potassium sodium niobate, $K_{0.48}Na_{0.52}Nb_{1-x}Bi_xO_3$ ($x = 0.05$–0.15), transparent ceramics were fabricated by a hot pressing method [244]. Both hot pressing conditions, including pressure, temperature, and duration, and Bi content had strong influences on densification, optical transparency, electro-optical coefficient, and electrical properties of the ceramics. Under the hot pressing conditions of 1060 °C for 5 h at 5.5 MPa, the optimized Bi content was 7.5 at.%. These ceramics display good optical transmission in the visible and especially in the infrared region. In visible region, the highest transmission of a polished 0.5-mm thick plate was about 60 % at a wavelength of 800 nm, while in infrared region, the transmission value could be 80 % at 4000 nm. The effective linear electro-optic coefficient of the ceramics was 42 pV m^{-1}. The dielectric permittivity and loss tangent of the KNNB ($x = 0.075$) ceramics were 716 and 0.031 at room temperature. Representative photographs of the ceramics are shown in Fig. 2.22.

Another group of lead-free transparent electro-optic ceramics, $(K_{0.5}Na_{0.5})_{1-x}Li_xNb_{1-x}Bi_xO_3$, has also been fabricated by using the similar process [245]. The co-modification with Li and Bi was able to transfer the ceramics into a nearly cubic structure with minimal optical anisotropy. At the same time, a diffuse phase transformation was induced, making the ceramics to be more relaxor-like and contain more polar nanoregions. These would reduce the light scattering by the grains, at the grain boundaries and at the domain walls, thus leading to ceramics with high optical transparency. For the ceramic modified with 5 mol% Li^+ and Bi^{5+}, the optical transmittance reached as high as 60 % in the near-IR region. The ceramics also exhibited a strong linear EO response, offering a large effective linear EO coefficient of 120–200 ppm V^{-1}. Figure 2.23 shows photographs of representative samples of the transparent ceramics.

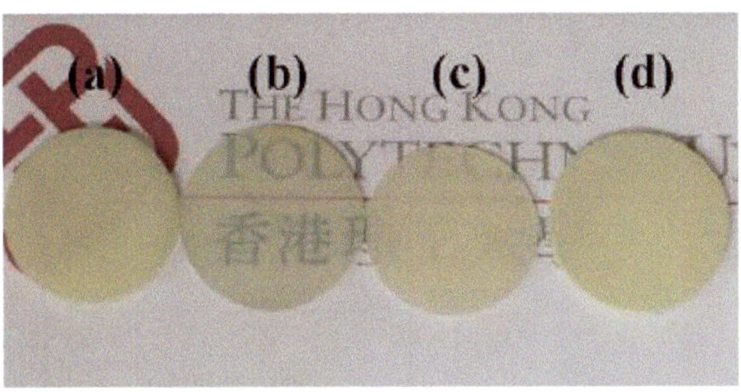

Fig. 2.22 Photographs of the $K_{0.48}Na_{0.52}Nb_{1-x}Bi_xO_3$ ceramics with different compositions: **a** $x = 0.05$, **b** $x = 0.075$, **c** $x = 0.10$, and **d** $x = 0.15$, which were hot pressed at 1060 °C for 5 h at a pressure of 6.1 MPa. Reproduced with permission from [244]. Copyright © 2011, Elsevier

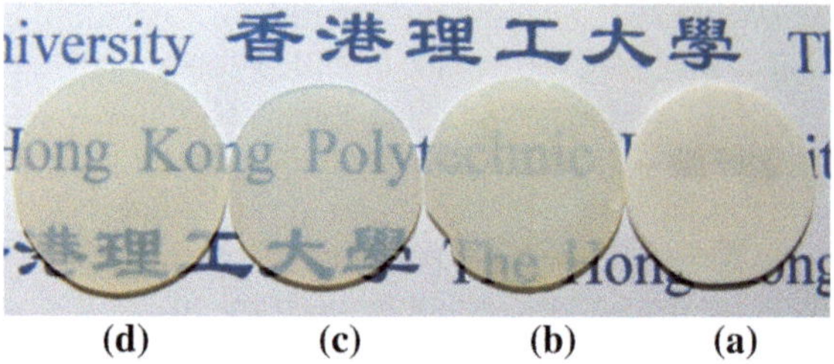

Fig. 2.23 Photographs of the KNNLB-x transparent ceramics: **a** $x = 0.02$, **b** $x = 0.03$, **c** $x = 0.05$, and **d** $x = 0.07$. Reproduced with permission from [245]. Copyright © 2013, Elsevier

2.5 Mullite

Mullite, a crystalline aluminum silicate phase, with a stoichiometric composition of $3Al_2O_3{\cdot}2SiO_2$, is a very interesting mineral constituting silicate ceramics. With an orthorhombic crystal structure, its lattice parameters are a = 7.5785(6) Å, b = 7.6817(7) Å and c = 2.8864(3) Å. Depending on the real composition, mullite ceramics have densities in the range of 3.11–3.26 g cm^{-3}, with a melting point of 1840 °C. Figure 2.24 shows a binary phase diagram of SiO_2 and Al_2O_3.

It is one of the most widely used ceramic materials because of its unique properties, like high-temperature strength, thermal shock resistance, and low dielectric constant [249]. It is also a good refractory material, due to its high melting

Fig. 2.24 Binary phase diagram of SiO_2 and Al_2O_3 with the formation of mullite

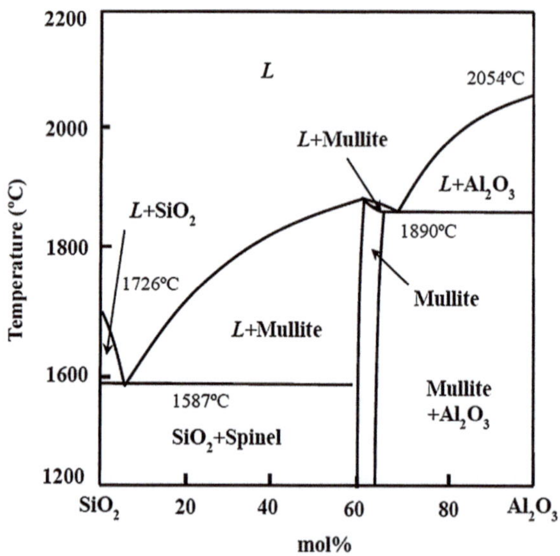

point [250], low thermal expansion coefficient, low thermal conductivity, high strength, and good creep resistance. Due to its low dielectric constant, $\varepsilon = 6.7$, which is lower than 9.8 of Al_2O_3, it can be used as a promising candidate for substrates in electronic packaging [251]. Because mullite has a coefficient of thermal expansion (CTE) that is the closest to that of silicon (Si), it has been acknowledged to be the most promising candidate as the substrates of Si thin film solar cell [252–255]. Moreover, mullite can be employed as window materials, especially at high temperatures, because it has a good transmission at high temperatures in the middle infrared band. Low dielectric constant and optical transmittance of fine-grained polycrystalline mullite could be a good host material of solid-state laser activators [34].

Pore-free translucent mullite was reported in 1972 by using HP sintering at 1500–1650 °C and 30–50 MPa, and vacuum HP, with high purity submicron-mixed oxide precursor powders derived from metal alkoxides [256]. Infrared-transparent mullite ceramics with 72 wt% Al_2O_3 were obtained by using HP sintering at 1630 °C and with 76 wt% Al_2O_3 by using HIP pressing at 1650 °C [257]. Visible spectrum translucent and infrared-transparent mullite ceramics were prepared by using pressureless sintering at 1750 °C [258]. Transparent mullite ceramics could also be developed through a combination of HIP pressing and pressureless sintering using commercial fused mullite precursor [259]. Both processing methods led to mullite ceramics with fine-grained microstructures (mean grain size of 5–10 μm) with equiaxed grains and a narrow grain size distribution. The pressureless sintered mullite had a porosity of 5.5 % and a transmittance of 20 % in the VIS and NIR ranges, while the HIP sintered mullite showed a porosity of 1 % and a transmittance of 40 % in the VIS range and up to 80 % in NIR range. This optically translucent mullite ceramics are suitable for optical windows in the VIS and NIR ranges [259].

Mullite ceramics, transparent in both mid-infrared and visible light regions, have been reported by using SPS from monophasic precursors [35]. In this method, the effects of calcination temperature, properties of precursor powders and sintering parameters on microstructure, and optical performances of the mullite ceramics were systematically investigated. The samples with a relative density of above 97.5 % exhibited an infrared transmittance of 75–82 % in wavelength of 2.5–4.3 μm. It was found that, as the precursor powders were calcined at temperatures of below 1100 °C, the residual OH, H_2O, and organic compound could not be completely eliminated, which in turn had negative effects on optical properties of the final ceramics. However, if the calcination temperature was too high, the precursors could not fully densified, due to the absence of viscous flow of amorphous phase. At a given calcination temperature, optical transmittance of the mullite ceramics was decreased with increasing sintering temperature above 1450 °C owing to the elongated grain growth. In addition, microwave sintering provided a rapid heating, accelerated mullitization, enhanced densification, and limited grain growth for developing transparent mullite ceramics with diphasic gels [34].

2.6 Other Oxide Ceramics

2.6.1 Newly Emerged Oxide Ceramics

Various transparent ceramics of complex oxides with pyrochlore structure, including $Lu_2Ti_2O_7$ [260–262], $Lu_2Zr_2O_7$ [263, 264], Lu_3NbO_7 [265, 266], $LaGdZr_2O_7$ [267], have been fabricated by using SPS. Nonmagnesium spinel transparent ceramics, e.g., zinc spinel or $ZnAl_2O_4$, have been reported [268–270]. Lanthanum silicate oxyapatite, with a formula of $La_{9.33}Si_6O_{26}$, ion-conducting materials presenting a strong aversion against densification, has been prepared in the form of dense transparent ceramics, by combining the beneficial use of freeze-drying and SPS methods [271]. Another example is $Sr_2Y_8(SiO_4)_6O_2$ oxyapatite translucent ceramics, which were derived from pure powder by using SPS combined with ball milling [272]. The microstructures and transmittances in the completely and incompletely densified ceramics were investigated. The ceramic sintered at 1500 °C is translucent with 0.7 % porosity and its total forward light transmittance can reach about 52 % in the infrared region. The transmittance was interpreted based on the microstructures of the ceramic samples and the Mie scattering theory. It was accepted previously that these ceramics cannot be processed to be transparent by using the conventional sintering techniques.

A simple combustion method was used to synthesize $LaGdZr_2O_7$ powder, which was then used to produce $LaGdZr_2O_7$ transparent ceramics by using vacuum sintering at 1850 °C for 6 h [267]. The final transparent ceramics, with a density of 6.46 g cm^{-3}, had an in-line transmittance of 70.7 % at 1000 nm and a refractive index of 2.08 at 632.8 nm. Figure 2.25 shows photographs of representative samples of the transparent ceramics.

The starting materials were $Zr(NO_3)_4{\cdot}3H_2O$, $La(NO_3)_3{\cdot}6H_2O$, and glycine. Gadolinium nitrate solution was prepared by dissolving Gd_2O_3 (99.99 %) in excess nitric acid solution. Stoichiometric amounts of the reactants were mixed thoroughly to form clear solutions. After that, ammonia solution was added into the solutions, so that their pH value was about 4. The solutions were then heated to facilitate gelation and finally trigger the combustion. The as-synthesized precursors were calcined at 800 °C for 2 h to obtain oxide powders. The powders were ball-milled with zirconia balls as grinding media 20 h. After drying, the powders were cold

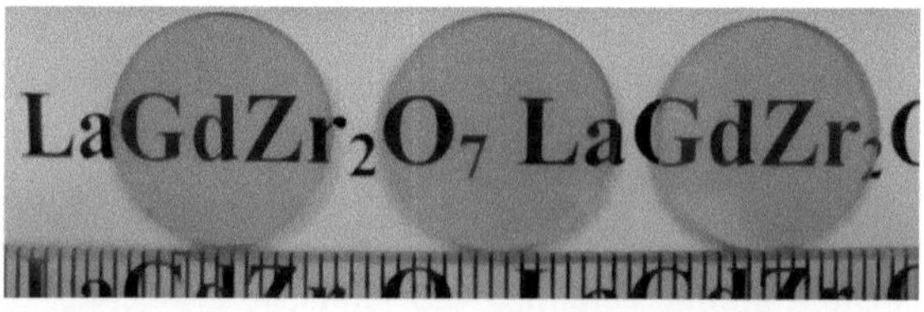

Fig. 2.25 Photographs of the $LaGdZr_2O_7$ ceramics vacuum sintered at 1850 °C for 6 h and annealed at 1500 °C for 5 h in air, with a thickness of 1 mm. Reproduced with permission from [267]. Copyright © 2013, Elsevier

isostatically pressed into pellets (Ø20 mm × 2.5 mm) at 200 MPa for 2 min. The green pellets were sintered at 1850 °C for 6 h in vacuum of 10^{-3} to 10^{-4} Pa and then annealed at 1500 °C for 5 h in air.

2.6.2 Transparent Ceramics Derived from Glasses

More recently, a new method to fabricate transparent ceramics was demonstrated, in which glasses with certain compositions were prepared first and then converted to transparent ceramics by using simple annealing at very low temperatures in air. Such transparent ceramics include $BaAl_4O_7$ [273], $Sr_3Al_2O_6$ [274], and $Sr_{1+x/2}Al_{2+x}Si_{2-x}O_8$ ($0 \leq x \leq 0.4$) [275], which will be discussed in the following part.

The first example is barium aluminate, $BaAl_4O_7$, which had two orthorhombic polymorphs with micrometer grain size and both were optically transparent in the visible region [273]. Starting precursor materials of $BaCO_3$ (>99.8 % purity) and Al_2O_3 (>99.997 % purity) were weighed, mixed in a 1:2 molar ratio, and pressed into pellets. Glasses with that composition were first prepared by using an aerodynamic levitator with a conical nozzle equipped with two CO_2 lasers. A single glass bead was obtained by melting the mixed precursor pellet of ~0.1 g at 2100 °C, followed by rapid quenching. The glassy beads were then crystallized into transparent ceramics by annealing in air. Crystallization temperatures were selected according to the differential scanning calorimetry (DSC) measurements. Figure 2.26 shows photographs of the glass and the two ceramic samples, together with the two crystal structures and thermal analysis result of the glass sample.

The method was also applied to the conversion of $Sr_3Al_2O_6$ and $Sr_3Ga_2O_6$ glasses [274]. The $Sr_3Al_2O_6$ and $Sr_3Ga_2O_6$ glasses were first synthesized from high purity $SrCO_3$ (99.99 %), Al_2O_3 (99.98 %), and Ga_2O_3 (99.99 %) using the method as discussed above. For doped samples, Eu_2O_3, Er_2O_3, Ho_2O_3, and CeO_2 oxides were used. The starting materials were made into pellets and then melted to form glasses. Full glasses were finally annealed at 840 °C for 5 h in air to be converted into transparent ceramics.

Figure 2.27 shows a photograph of the $Sr_3Al_2O_6$ polycrystalline ceramics obtained by full crystallization from the glasses, together with transmission curve. It was highly transparent in the visible region, covering a wide range from the visible to infrared ranges, i.e., 300 nm–6 μm. The absorption below 300 nm was due to the optical band gap absorption of the strontium aluminate, whereas the absorption observed around 3 μm corresponds to moisture in the sample, i.e., O–H bonds. The difference between the measured transparency and the theoretical maximum value of 87.4 ± 0.5 % for an average refractive index of 1.70 ± 0.02 was very small in the visible range and was negligible in the IR range, which implied that there were only limited number of scattering centers within the polycrystalline ceramics. Furthermore, luminescence of the $Sr_3Al_2O_6:Eu^{3+}$ and $Sr_3Al_2O_6:Er^{3+}$ ceramics, with strong emissions in the visible and infrared ranges, was observed. The $Sr_3Al_2O_6:Ce^{3+}$ ceramics exhibited scintillation properties under X-ray excitation.

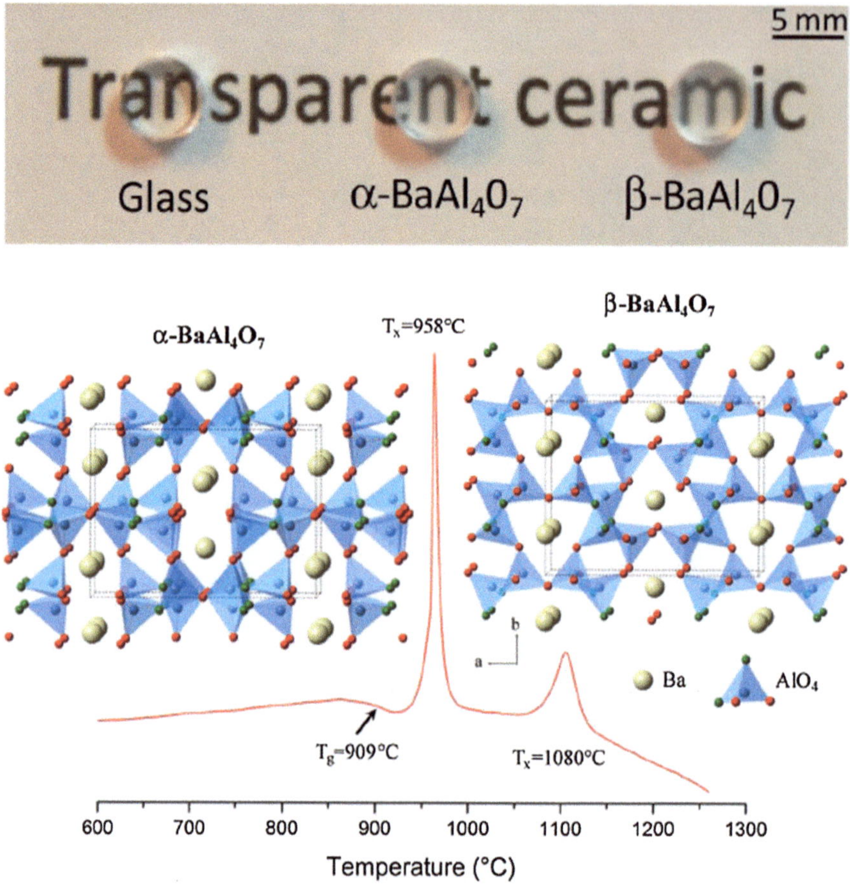

Fig. 2.26 Top panel: dense ~1.5 mm thick disks of the $BaAl_4O_7$ parent glass sample, α-polymorph and β-polymorph transparent ceramics. *Middle panel* Crystal structures of the α- and β-$BaAl_4O_7$ polymorphs. The corner-sharing AlO_4 tetrahedra form channels occupied by barium. The two polymorph frameworks are not structurally related, but both of them exhibit structural anisotropy. The tricluster oxygen ions appear in *green*. *Bottom panel* Differential scanning calorimetry measurement of the $BaAl_4O_7$ glass showing the glass transition at 909 ± 1 °C and a strong exothermic peak corresponding to crystallization of the α phase at 958 ± 1 °C followed by a broader second exothermic peak corresponding to the α- to β-$BaAl_4O_7$ transition at 1080 ± 1 °C. Reproduced with permission from [273]. Copyright © 2012, John Wiley & Son

The latest development was to use normal glass technique to large size and flat samples [275]. The five $Sr_{1+x/2}Al_{2+x}Si_{2-x}O_8$ (x = 0, 0.1, 0.2, 0.3, and 0.4) polycrystalline ceramics were first prepared as glass materials from high purity $SrCO_3$ (99.9 %), Al_2O_3 (99.98 %), and SiO_2 (99.8 %) precursors. Stoichiometric ratios for each composition were thoroughly mixed. The mixed powders were then placed in rhodium/platinum crucibles and calcined at 1700 °C for 10 min in a preheated

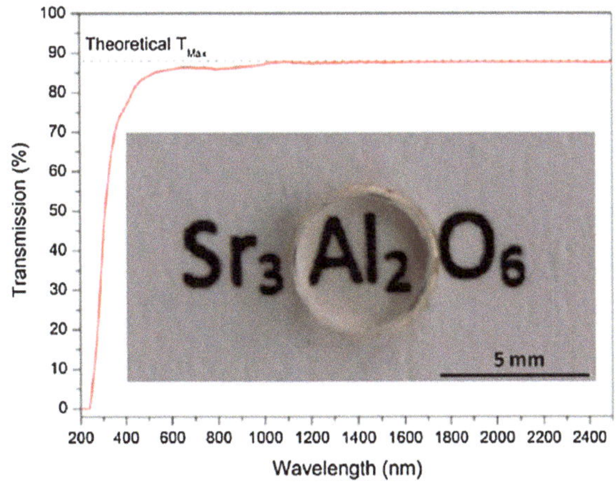

Fig. 2.27 Transmission spectra of the $Sr_3Al_2O_6$ transparent polycrystalline ceramic (obtained from a 1.5 mm thick sample). The *dotted black line* corresponds to the theoretical maximum transmission calculated as 87.4 % for an average refractive index of 1.70. A photograph of the $Sr_3Al_2O_6$ material is also shown. Reproduced with permission from [274]. Copyright © 2013, American Chemical Society

electric furnace first, and then heated to 1750 °C for 10 min and back to 1700 °C for 30 min. The samples were then cooled in air. The glasses were annealed at 1020 °C for 3 h in air to form transparent ceramics. Representative samples and optical properties of the transparent ceramics are shown in Fig. 2.28.

2.6.3 Multiphase Transparent Ceramics

An open question is whether binary or multicomponent systems that are from different transparent ceramics are still transparent. Recent studies provided with positive answers to the question [276, 277]. $MgAl_2O_4$/Ce:YAG transparent ceramics have been fabricated by using hot pressing sintering and hot isostatic pressing sintering techniques, with high purity yttrium aluminum garnet (YAG:Ce) powder and $MgAl_2O_4$ powder [276]. The $MgAl_2O_4$ powder was synthesized by using a metal alkoxide method with magnesium, aluminum, and isopropanol as precursors. The transparent ceramics contained both magnesium-aluminum spinel ($MgAl_2O_4$) and YAG, in which the YAG phase was dispersed uniformly in the matrix of $MgAl_2O_4$. The grain size of YAG was about 8 μm. The ceramics of 2 mm in thickness had an in-line transmittance of >40 % at 530 nm. The main absorption peak centered at 460 nm, indicates that the $MgAl_2O_4$/Ce:YAG transparent ceramics could be promising fluorescent materials and have potential for application in white LED.

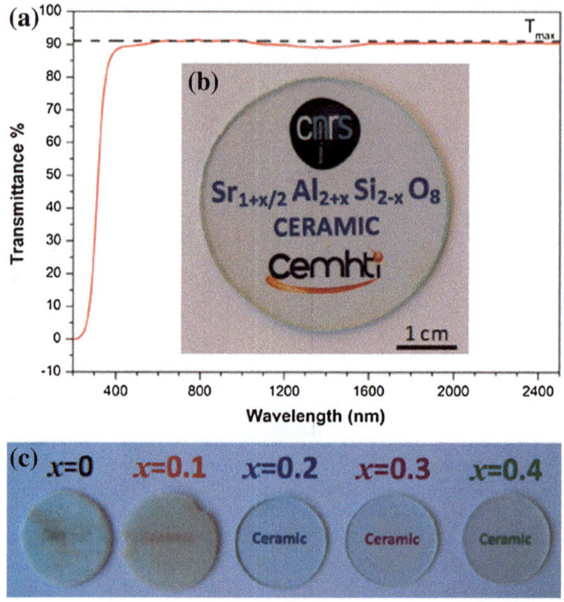

Fig. 2.28 Transparency of the $Sr_{1+x/2}Al_{2+x}Si_{2-x}O_8$ ($0 \leq x \leq 0.4$) polycrystalline ceramics. **a** Transmission spectrum measured on the sample with $x = 0.2$ ($Sr_{1.1}Al_{2.2}Si_{1.8}O_8$) at a thickness of 1.5 mm. The $n = 1.59$ refractive index implies a 90.1 % maximum theoretical transparency (*dashed line*). **b** Photograph of the same material illustrating the almost perfect 90 % transparency. **c** Photograph of the $Sr_{1+x/2}Al_{2+x}Si_{2-x}O_8$ ($x = 0$, 0.1, 0.2, 0.3, and 0.4) polycrystalline ceramics showing transparency evolution along the reported solid solution. Reproduced with permission from [275]. Copyright © 2015, American Chemical Society

2.7 Nonoxide Ceramics

2.7.1 AlON

Aluminum oxynitride (AlON) has a defect cubic spinel crystal structure with chemical formula of $Al_{(64+x)/3}O_{32-x}N_x$ ($2 \leq x \leq 5$). The addition of a small amount of nitrogen converts the rhombohedral crystal structure of alumina into the cubic spinel structure of AlON. Nitrogen stabilizes the cubic spinel crystal structure over a wide range of compositions. AlON is optically transparent (≥80 %) over the region from near-ultra violet to visible and near-infrared wavelength of the electromagnetic spectrum. It is 4 times harder than fused silica glass, 85 % as hard as sapphire, and nearly 15 % harder than $MgAl_2O_4$ spinel. Furthermore, the cubic structure of AlON exhibits isotropic optical properties. Combined with its good optical and mechanical properties, AlON can be used in defence and commercial applications, such as military aircraft and missile domes, transparent armors, IR windows, hyper-hemispherical domes, laser windows, military aircraft lenses, semiconductor processing applications, and scanner windows. The history of development of AlON transparent ceramics has been well documented in the open literature [278].

Over the years, several processing routes have been used to produce fully dense transparent polycrystalline AlON ceramics. Initial work was reported by McCauley et al. [279] in 1979, using reactive sintering of Al_2O_3-AlN mixtures. Then the reactive sintering technique was widely used by other researchers [280].

The conventional method for the fabrication of transparent AlON ceramics involves several steps, including synthesis of precursor AlON powder, forming green body with the powder, and then sintering in a nitrogen atmosphere at high temperatures (>1850 °C) for extended period (20–100 h). AlON precursor powders can be synthesized by several methods, such as reaction of Al_2O_3 and AlN, carbothermal reduction (CR) of Al_2O_3, SSR, plasma arc synthesis, and self-propagating high-temperature synthesis (SHS) [281]. Among these methods, CR and SSR have been the most widely used ones, because CR is very cost effective while SSR is extremely simple. However, CR method has too many influencing factors, making it difficult to synthesize pure AlON powders without residual AlN or Al_2O_3 [282]. SSR method requires high purity and fine AlN powders to ensure high sinterability. In this respect, a combinational method of CR and SSR synthesis of γ-AlON powders was developed [283]. Ultrafine, phase pure, and low-cost γ-AlON powders were easily synthesized by using organic sucrose instead of carbon black as a reducing agent.

Highly transparent AlON ceramics have been made with powders synthesized by using direct nitridation method [284, 285]. As an example, the work of [285] is presented here. Commercial available submicron Al_2O_3 (99.99 % purity) and Al (99.95 % purity) were used as starting materials, while MgO, Y_2O_3, and La_2O_3 (99.99 % purity) were used as sintering additives. The reactants were dispersed in alcohol and then vacuum dried at 80 C for 3 h. The homogeneous mixture was heated a BN crucible to 1750 °C gradually for 3 h in flowing N_2. The as-synthesized AlON powder was mixed homogeneously with 0.10 wt% MgO, 0.08 wt% Y_2O_3, and 0.025 wt% La_2O_3. The resulting powder was made into pellets at low pressure and then cold isostatically pressed at 200 MPa. The green bodies were pressurelessly sintered in a graphite furnace at 1950 °C for 12 h in nitrogen atmosphere. Figure 2.29 shows a photograph of the AlON transparent ceramics sintered at 1950 °C for 12 h.

Similar to other transparent ceramics, AlON ceramics have also been processed by using various sintering techniques, such as pressureless sintering [286], microwave sintering [27, 287], HP [288], HIP [286, 289, 290], and SPS [291]. For example, a simple one-step pressureless sintering was proposed, using α-Al_2O_3 and aluminum powder as starting material [286]. The starting material was first nitridised into high activity AlN, from which $Al_{23}O_{27}N_5$ was then synthesized in situ by reacting with α-Al_2O_3. The optimal one-step pressureless sintering process was observed to occur at 1750 °C for 2 h, which resulted in samples with flexural strength of as high as 321 MPa.

A reactive SPS method has been developed to prepare AlON transparent ceramics [291]. Al_2O_3 and AlN powder mixtures were used as precursors. Reactive SPS was conducted at temperatures between 1400 °C and 1650 °C for 15–45 min at 40 MPa

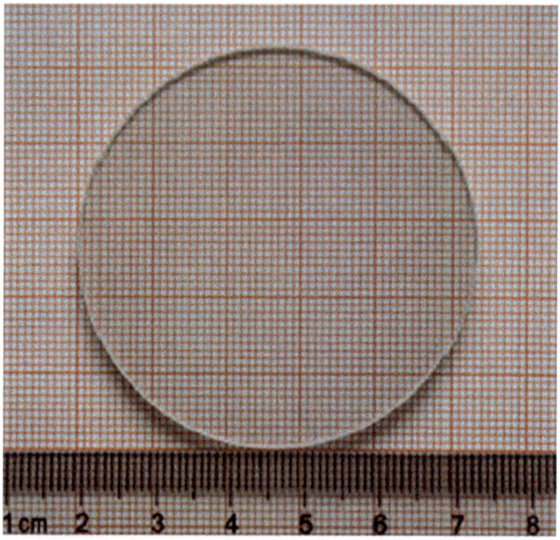

Fig. 2.29 Photograph of the AlON transparent ceramics sintered at 1950 °C for 12 h. Reproduced with permission from [285]. Copyright © 2015, Elsevier

under N_2 gas flow. AlON phase formation was initiated in the samples sintered above 1430 °C, according to the XRD analysis. The complete transformation from the precursor phases (Al_2O_3 and AlN) to AlON was observed at 1650 °C for 30 min at 40 MPa. High temperature and low heating rate were favorable to the formation of AlON. The highest value of hardness was recorded to be 16.7 GPa. The sample with highest content of AlON possessed a fracture toughness of 3.95 MPa $m^{1/2}$.

A two-step method to prepare AlON ceramics was developed, with nanosized Al_2O_3 and AlN as precursors [292]. Single-phase AlON powder was obtained at 1750 °C for 4 h in flowing nitrogen atmosphere through direct reaction of the nanosized Al_2O_3 and AlN. After ball milling, the powder had an average size of several microns and showed quite high sintering activity. The high reactive AlON powder was used to fabricate transparent ceramics at 1880 °C.

By using MgO and Y_2O_3 as sintering aids, transparent AlON ceramics were successfully prepared by solid-state reactive sintering [293]. The effects of sintering aids and holding time on densification of AlON ceramics were investigated. Co-doping of MgO and Y_2O_3 was found to be more effective in enhancing the densification of AlON than the doping of either MgO or Y_2O_3. It was observed that, at a given doping level of Y_2O_3, e.g., 0.08 wt%, transmission coefficient of the ceramics increased with increasing concentration of MgO. A 1 mm thick sample, co-doped with 0.08 wt% Y_2O_3 and 1 wt% MgO, sintered at 1950 °C for 12 h in N_2 atmosphere, had an in-line transmittance of 60 % at 600 nm.

The single-phase AlON powder used in this experiment was obtained at relatively low temperature by the solid-state reaction method with nanosized AlN powder and nanosized alumina powder [292]. The obtained powder was ground by ball milling and doped with Y_2O_3. Then, it was shaped into pellets. Transparent ceramics sintering was carried out at 2153 K (1880 °C) for 5, 10, and 20 h. Obtained samples for three sintering times were all transparent, while the light transmittance of samples increased with increasing holding time at the sintering temperature. The transmittance of the sample sintered for 20 h is 55 pct at near 5 μm. X-ray diffraction (XRD) measurement indicated that all samples were single-phase AlON. The fracture surfaces of samples from SEM analysis showed that, with increasing the holding time, the grain size grew slightly, while the pore size and porosity decreased rapidly, so that the transmittance increased. The pore size and the porosity were the main factors affecting the light transmitting properties of samples.

2.7.2 Aluminum Nitride (AlN)

An effective processing method was studied to fabricate the translucent AlN ceramics [294]. AlN ceramics with 80 mm diameter have been prepared, which exhibited promising optical performances, e.g., transparent over wavelengths of 0.2–2.2 μm.

Transient liquid phase (TLP) sintering of CaF_2 additive on the densification behaviors and microstructural development of AlN ceramics are investigated [295]. It is found that 1 wt% CaF_2 can effectively promote densification process. Increasing content of CaF_2 results in finer grain size and slower densification during intermediate sintering stage. XRD results show that grain boundary phase of $CaAl_4O_7$ is formed at 1150 °C from reactions of AlN–CaF_2–Al_2O_3. With further temperature increasing, the grain boundary phases of CA2 and $CaAl_{12}O_{18}$, which were formed from the reaction between CaF_2 and oxide layers, experienced transformations first into $CaAl_4O_7$ above 1600 °C and into $CaAl_2O_4$ at higher temperatures. SEM and TEM results show that formed grain boundary phases can evaporate from sintering bodies during further soaking, leaving clean grain boundaries. The efficiency of TLP sintering mechanism is further manifested by the preparation of transparent AlN ceramics with good combination properties.

Optical spectroscopy of Er^{3+} doped into bulk AlN ceramics has been reported [296]. The material was prepared by using hot press sintering of AlN with Er_2O_3 and $(NH_4)(ErF_4)$, which yielded fully dense, translucent, hexagonal AlN. The Er^{3+} concentration was a small fraction of a percent, and resided in multiple sites, with one type of center dominant. A number of the energy levels of Er^{3+} were identified for this center. The temperature dependent fluorescence lifetime was probably radiative, with which the stimulated emission and absorption cross section spectra were derived for the $^4I_{13/2} \rightarrow {}^4I_{15/2}$ transitions.

2.7.3 Sialon

SiAlONs are ceramics based on the elements of silicon (Si), aluminum (Al), oxygen (O), and nitrogen (N). They are solid solutions based on silicon nitride (Si_3N_4), with three basic forms. Each form is isostructural with one of the two common forms of silicon nitride, i.e., β and α, and silicon oxynitride. Sialon ceramics are a special class of high-temperature refractory materials, with high strength at both room temperature and high temperatures, good thermal shock resistance, and exceptional resistance to wetting or corrosion by molten nonferrous metals, compared to other refractory materials such as, for example, alumina. Recently, with the advancement in sintering techniques, translucent or transparent Sialon ceramics have emerged as a new group of engineering ceramics with special optical properties.

A report is available on microstructure and properties of translucent Mg-sialon ceramics prepared by SPS [297]. The translucent Mg-sialon ceramics were prepared using SPS α-Si_3N_4 powders with 9 wt% AlN and 3 wt% MgO as sintering additives. Microstructural observations indicate that the optical and mechanical properties of Mg-sialon ceramics are affected by the density and α′:β′-phase ratio in sintered bodies, which are tailored by controlling the content of formed liquid phase and optimizing the parameters of SPS in present study. The material is toughened by the existence of a small amount of β′-sialon. The reason that the two-phase composite does not greatly compromise optical property could be attributed to the fine equiaxed microstructures and low content of β′-phase. Translucent Mg-sialon ceramics achieving 66.4 % of maximum transmittance, 21.4 ± 0.3 GPa hardness, and 6.1 ± 0.1 MPa $m^{1/2}$ fracture toughness were prepared by SPS at 1850 °C for 5 min. Figure 2.30 shows photographs of representative samples of the Mg-sialon ceramics.

Another example is (Mg, Y)-α/β-Sialon translucent ceramics processed by using hot pressing [298–302]. Although MgO is a suitable sintering aid for the development of translucent Mg-α/β-Sialon ceramics by using HP sintering, it is difficult to achieve complete densification, because MgO has high viscosity and high melting point. It has been acknowledged that Y_2O_3 is one of the promising sintering aids to facilitate complete densification and provide good mechanical properties of ceramics based on silicon nitride. Therefore, a combination of Y_2O_3 and MgO was proposed to prepare Mg-α/β-Sialon ceramics [298]. It was found that the amount of α-Sialon phase was increased with increasing content of Y_2O_3 up to 1 wt%. Too high content, e.g., 2 wt% Y_2O_3, β-Sialon phase was observed. Thermal

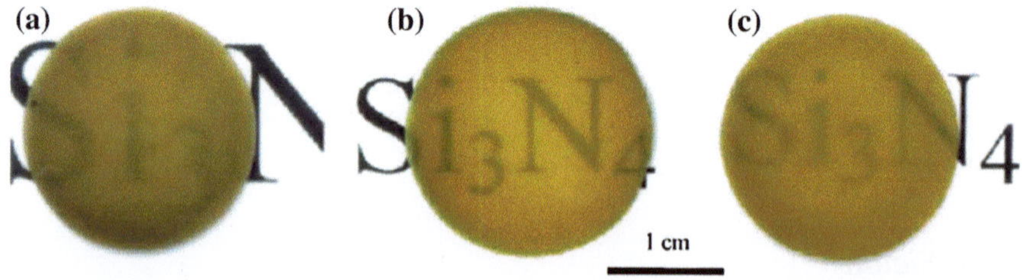

Fig. 2.30 Photographs of the translucent Mg-sialon ceramic slices of 0.5 mm in thickness sintered at different temperatures: **a** 1800 °C, **b** 1850 °C and, **c** 1870 °C. Reproduced with permission from [285]. Copyright © 2015, Elsevier

conductivity and CTE values were increased with increasing amount of Y_2O_3. However, the presence of Y_2O_3 had a negative effect on transparency of the final ceramics.

Very few works have been report on the processing and characterization of Si_3N_4 transparent ceramics, because in principle they cannot be transparent. More recently, a study indicated that Si_3N_4 translucent ceramics could be obtained by using SPS [303]. Sintering aids, including Y_2O_3, MgO and Al_2O_3, were used to promote the densification of Si_3N_4. The density, crystal phase, microstructure, transmittance, hardness, and fracture toughness of sintered Si_3N_4 were investigated. Highly densified sintered bodies were obtained in single Y_2O_3 and MgO systems, but not in single Al_2O_3 system. The XRD results indicated that sintered bodies were composed of β-Si_3N_4. The SEM images showed that all the sintered bodies had a fine-grained microstructure with an average diameter of 0.29–0.37 lm. The thickness of grain boundary was changed with additive content. The transmittance (T) and the wavelength (λ) followed the relationship of $T \propto \exp(-\lambda^{-2.3})$ due to the light scattering. The transmittance was mainly influenced by the refractive index of additives and the thickness of grain boundary phase. The hardness and fracture toughness of sintered ceramics were 12.6–15.5 GPa and 6.2–7.2 MPa $m^{1/2}$, respectively.

2.8 Summary

Various transparent ceramics have been developed by using various sintering technologies. Cubic crystal structure is one of the requirements for ceramics to be transparent, which should be modified, because noncubic crystalline transparent ceramics have been available. Furthermore, with the advancement of new sintering technologies, especially SPS, some ceramics that cannot be made to be transparent by using the conventional sintering methods, have emerged more recently. Although transparent ceramics for solid-state laser applications still form the main stream of these special materials, the number of transparent ceramics for other applications is rapidly increasing. It is highly suggested to pay attention to the new development of transparent ceramics, such as transparent ceramics derived from glasses and multicomponent transparent ceramics, which are different from the traditional transparent ceramics.

References

1. Chaim R, Shen ZJ, Nygren M (2004) Transparent nanocrystalline MgO by rapid and low-temperature spark plasma sintering. J Mater Res 19:2527–2531
2. Chaim R, Shen ZJ, Nygren M (2005) Transparent nanocrystalline MgO by low temperature spark plasma sintering. In: Lu S, Hu MZ, Gogotsi Y (eds) Ceramic nanomaterials and nanotechnology III. American Ceramic Society, Westerville, pp 21–30

3. Chen DY, Jordan EH, Gell M (2008) Pressureless sintering of translucent MgO ceramics. Scripta Mater 59:757–759
4. Fang Y, Agrawal D, Skandan G, Jain M (2004) Fabrication of translucent MgO ceramics using nanopowders. Mater Lett 58:551–554
5. Ikegami T, Matsuda SI, Suzuki H (1974) Effect of halide dopants on fabrication of transparent polycrystalline MgO. J Am Ceram Soc 57:507
6. Sanamyan T, Cooper C, Gilde G, Sutorik AC, Dubinskii M (2014) Fabrication and spectroscopic properties of transparent Nd^{3+}:MgO and Er^{3+}:MgO ceramics. Laser Phys Lett 11:065801
7. Miles GD, Sambell RAJ, Rutherfo J, Stephens Gw (1967) Fabrication of fully dense transparent polycrystalline magnesia. Trans Br Ceram Soc 66:319
8. Misawa T, Moriyoshi Y, Yajima Y, Takenouchi S, Ikegami T (1999) Effect of silica and boron oxide on transparency of magnesia ceramics. J Ceram Soc Jpn 107:343–348
9. Suzuki M, Ikegami T, Yokoyama M, Komatsu T, Fukahori A (2005) Effects of chloride ion on densification transparency magnesia ceramics. J Ceram Soc Jpn 113:149–153
10. An LQ, Ito A, Zhang J, Tang DY, Goto T (2014) Highly transparent Nd^{3+}:Lu_2O_3 produced by spark plasma sintering and its laser oscillation. Opt Mater Express 4:1420–1426
11. Antipov OL, Golovkin SY, Gorshkov ON, Zakharov NG, Zinovev AP, Kasatkin AP et al (2011) Structural, optical, and spectroscopic properties and efficient two-micron lasing of new Tm^{3+}:Lu_2O_3 ceramics. Quant Electron 41:863–868
12. Chen QW, Shi Y, An LQ, Chen JY, Shi JL (2006) Fabrication and photoluminescence characteristics of Eu^{3+}-Doped Lu_2O_3 transparent ceramics. J Am Ceram Soc 89:2038–2042
13. Kim W, Baker C, Bowman S, Florea C, Villalobos G, Shaw B, et al (2013) Laser oscillation from Ho^{3+} doped Lu_2O_3 ceramics. Opt Mater Express 3:176WL
14. Lu J, Takaichi K, Uematsu T, Shirakawa A, Musha M, Ueda K et al (2002) Promising ceramic laser material: highly transparent Nd^{3+}:Lu_2O_3 ceramic. Appl Phys Lett 81:4324–4326
15. Seeley ZM, Kuntz JD, Cherepy NJ, Payne SA (2011) Transparent Lu_2O_3: Eu ceramics by sinter and HIP optimization. Opt Mater 33:1721–1726
16. Shi Y, Chen QW, Shi JL (2009) Processing and scintillation properties of Eu^{3+} doped Lu_2O_3 transparent ceramics. Opt Mater 31:729–733
17. Gheorghe C, Lupei A, Lupei V, Ikesue A, Enculescu M (2011) Intensity parameters of Tm^{3+} doped Sc_2O_3 transparent ceramic laser material. Opt Mater 33:501–505
18. Gheorghe C, Lupei A, Voicu F, Enculescu M (2012) Sm^{3+}-doped Sc_2O_3 polycrystalline ceramics: spectroscopic investigation. J Alloy Compd 535:78–82
19. Lupei V, Lupei A, Ikesue A (2005) Transparent Nd and (Nd, Yb)-doped Sc_2O_3 ceramics as potential new laser materials. Appl Phys Lett 86
20. Alderighi D, Pirri A, Toci G, Vannini M, Esposito L, Costa AL et al (2010) Characterization of Yb:YAG ceramics as laser media. Opt Mater 33:205–210
21. Ba XW, Li J, Pan YB, Zeng YP, Kou HM, Liu WB et al (2013) Comparison of aqueous- and non-aqueous-based tape casting for preparing YAG transparent ceramics. J Alloy Compd 577:228–231
22. Bagayev SN, Osipov VV, Solomonov VI, Shitov VA, Maksimov RN, Lukyashin KE et al (2012) Fabrication of Nd^{3+}:YAG laser ceramics with various approaches. Opt Mater 34:1482–1487
23. Chaim R, Kalina M, Shen JZ (2007) Transparent yttrium aluminum garnet (YAG) ceramics by spark plasma sintering. J Eur Ceram Soc 27:3331–3337
24. Chen ZH, Li JT, Hu ZG, Xu JJ (2008) Fabrication of YAG transparent ceramics by two step sintering process. J Inorg Mater 23:130–134
25. Frage N, Kalabukhov S, Sverdlov N, Ezersky V, Dariel MP (2010) Densification of transparent yttrium aluminum garnet (YAG) by SPS processing. J Eur Ceram Soc 30:3331–3337

26. Guo W, Lu TC, Tong SH (2007) Effect of phase of YAG powder synthesized by co-precipitation on transparent ceramic sintering. In: Pan W, Gong JH (eds) High-performance ceramics IV, Pts 1-3. Trans Tech Publications Ltd, Stafa-Zurich, pp 2054–2057
27. Agrawal D, Cheng JP, Roy R (2002) Microwave reactive sintering to fully transparent aluminum oxynitride (AlON) ceramics. In: McCauley JW, Crowson A, Gooch WA, Rajendran AM, Bless SJ, Logan KV et al (eds) Ceramic armor materials by design. American Ceramic Society, Westerville, pp 587–593
28. Hartnett TM, Bernstein SD, Maguire EA, Tustison RW (1998) Optical properties of AlON (aluminum oxynitride). Infrared Phys Technol 39:203–211
29. Jiang HW, Du HB, Tian TY, Wu H (2010) Influence of Y_2O_3 additive on transparent of AlON ceramics. In: Pan W, Gong JH (eds) Chinese ceramics communications, pp 580–581
30. Wang J, Zhang F, Chen F, Zhang J, Zhang HL, Tian R et al (2015) Effect of Y_2O_3 and La_2O_3 on the sinterability of gamma-AlON transparent ceramics. J Eur Ceram Soc 35:23–28
31. Choi JJ, Ryu J, Kim HE (2001) Microstructural evolution of transparent PLZT ceramics sintered in air and oxygen atmospheres. J Am Ceram Soc 84:1465–1469
32. Hertling GH (1971) Improved hot-pressed electrooptic ceramics in (Pb, La)(Zr, Ti)O_3 system. J Am Ceram Soc 54:303–309
33. Haertling GH (1999) Ferroelectric ceramics: history and technology. J Am Ceram Soc 82:797–818
34. Fang Y, Roy R, Agrawal DK, Roy DM (1996) Transparent mullite ceramics from diphasic aerogels by microwave and conventional processings. Mater Lett 28:11–15
35. Zhang GM, Wang YC, Fu ZY, Wang H, Wang WM, Zhang JY et al (2009) Transparent mullite ceramic from single-phase gel by spark plasma sintering. J Eur Ceram Soc 29:2705–2711
36. Dang KQ, Takei S, Kawahara M, Nanko M (2011) Pulsed electric current sintering of transparent Cr-doped Al_2O_3. Ceram Int 37:957–963
37. Jiang D, Hulbert DM, Anselmi-Tamburini U, Ng T, Land D, Mukherjee AK (2008) Optically transparent polycrystalline Al_2O_3 produced by spark plasma sintering. J Am Ceram Soc 91:151–154
38. Krell A, Klimke J, Hutzler T (2009) Advanced spinel and sub-μm Al_2O_3 for transparent armour applications. J Eur Ceram Soc 29:275–281
39. Lu SZ, Yang QH (2009) Fluorescencec characteristics of Al_2O_3 transparent ceramics. Chin J Inorg Chem 25:1642–1645
40. Nanko M, Dang KQ (2014) Two-step pulsed electric current sintering of transparent Al_2O_3 ceramics. Adv Appl Ceram 113:80–84
41. Penilla EH, Kodera Y, Garay JE (2013) Blue-green emission in terbium-doped alumina (Tb: Al_2O_3) transparent ceramics. Adv Funct Mater 23:6036–6043
42. Roussel N, Lallemant L, Chane-Ching JY, Guillemet-Fristch S, Durand B, Garnier V et al (2013) Highly tense transparent-Al_2O_3 ceramics from ultrafine nanoparticles via a standard SPS sintering. J Am Ceram Soc 96:1039–1042
43. Yang QH, Zeng ZJ, Xu J, Ding J, Su LB (2006) Spectroscopic characteristics of Cr^{4+} in transparent polycrystalline Al_2O_3. Acta Phys Sin 55:4166–4169
44. Zhang XJ, Qiao GJ, Zhang X (2011) Effect of Mg doping on the microstructure and properties of α-Al_2O_3 transparent ceramics. In: Tan HH (ed) Mechanical, materials and manufacturing engineering, Pts 1–3. Trans Tech Publications Ltd, Stafa-Zurich, pp 1264–1269
45. Kim BN, Hiraga K, Morita K, Yoshida H (2007) Spark plasma sintering of transparent alumina. Scripta Mater 57:607–610
46. Kim BN, Hiraga K, Morita K, Yoshida H, Miyazaki T, Kagawa Y (2009) Microstructure and optical properties of transparent alumina. Acta Mater 57:1319–1326
47. Krell A, Blank P, Ma HW, Hutzler T, van Bruggen MPB, Apetz R (2003) Transparent sintered corundum with high hardness and strength. J Am Ceram Soc 86:12–18
48. Kim BN, Hiraga K, Morita K, Yoshida H (2009) Effects of heating rate on microstructure and transparency of spark-plasma-sintered alumina. J Eur Ceram Soc 29:323–327

49. Mizuta H, Oda K, Shibasaki Y, Maeda M, Machida M, Ohshima K (1992) Preparation of high strength and translucent alumina by hot isostatic pressing. J Am Ceram Soc 75:469–473
50. Apetz R, van Bruggen MPB (2003) Transparent alumina: a light-scattering model. J Am Ceram Soc 86:480–486
51. Krell A, Klimke J (2006) Effects of the homogeneity of particle coordination on solid-state sintering of transparent alumina. J Am Ceram Soc 89:1985–1992
52. Pauling L, Hendricks SB (1925) The crystal structures of hematite and corundum. J Am Chem Soc 47:781–790
53. Jin XH, Gao L, Sun J (2010) Highly transparent alumina spark plasma sintered from common-grade commercial powder: the effect of powder treatment. J Am Ceram Soc 93:1232–1236
54. Wei GC, Rhodes WH (2000) Sintering of translucent alumina in a nitrogen-hydrogen gas atmosphere. J Am Ceram Soc 83:1641–1648
55. Mao XJ, Wang SW, Shimai S, Guo JK (2008) Transparent polycrystalline alumina ceramics with orientated optical axes. J Am Ceram Soc 91:3431–3433
56. Li JG, Ye YP (2006) Densification and grain growth of Al_2O_3 nanoceramics during pressureless sintering. J Am Ceram Soc 89:139–143
57. Godlinski D, Kuntz M, Grathwohl G (2002) Transparent alumina with submicrometer grains by float packing and sintering. J Am Ceram Soc 85:2449–2456
58. Liu W, Bo TZ, Xie ZP, Wu Y, Yang XF (2011) Fabrication of injection moulded translucent alumina ceramics via pressureless sintering. Adv Appl Ceram 110:251–254
59. Roh JY, Kwon J, Lee CS, Choi JS (2011) Novel fabrication of pressure-less sintering of translucent powder injection molded (PIM) alumina blocks. Ceram Int 37:321–326
60. Liu W, Xie ZP, Bo TZ, Yang XF (2011) Injection molding of surface modified powders with high solid loadings: a case for fabrication of translucent alumina ceramics. J Eur Ceram Soc 31:1611–1617
61. Kim DS, Lee JH, Sung RJ, Kim SW, Kim HS, Park JS (2007) Improvement of translucency in Al_2O_3 ceramics by two-step sintering technique. J Eur Ceram Soc 27:3629–3632
62. Liu W, Xie ZP, Liu GW, Yang XF (2011) Novel preparation of translucent alumina ceramics induced by doping additives via chemical precipitation method. J Am Ceram Soc 94:3211–3215
63. Liu GW, Xie ZP, Liu W, Chen LX, Wu Y (2012) Fabrication of translucent alumina ceramics from pre-sintered bodies infiltrated with sintering additive precursor solutions. J Eur Ceram Soc 32:711–715
64. Hotta Y, Tsugoshi T, Nagaoka T, Yasuoka M, Nakamura K, Watari K (2003) Effect of oligosaccharide alcohol addition to alumina slurry and translucent alumina produced by slip casting. J Am Ceram Soc 86:755–760
65. Petit J, Dethare P, Sergent A, Marino R, Ritti MH, Landais S et al (2011) Sintering of alpha-alumina for highly transparent ceramic applications. J Eur Ceram Soc 31:1957–1963
66. Suarez M, Fernandez A, Menendez JL, Torrecillas R (2009) Grain growth control and transparency in spark plasma sintered self-doped alumina materials. Scripta Mater 61:931–934
67. Alvarez-Clemares I, Mata-Osoro G, Fernandez A, Lopez-Esteban S, Pecharroman C, Palomares J et al (2010) Transparent alumina/ceria nanocomposites by spark plasma sintering. Adv Eng Mater 12:1154–1160
68. Roussel N, Lallemant L, Durand B, Guillemet S, Ching JYC, Fantozzi G et al (2011) Effects of the nature of the doping salt and of the thermal pre-treatment and sintering temperature on spark plasma sintering of transparent alumina. Ceram Int 37:3565–3573
69. Grasso S, Hu CF, Maizza G, Kim BN, Sakka Y (2011) Effects of pressure application method on transparency of spark plasma sintered alumina. J Am Ceram Soc 94:1405–1409
70. Stuer M, Zhao Z, Aschauer U, Bowen P (2010) Transparent polycrystalline alumina using spark plasma sintering: effect of Mg, Y and La doping. J Eur Ceram Soc 30:1335–1343
71. Brosnan KH, Messing GL, Agrawal DK (2003) Microwave sintering of alumina at 2.45 GHz. J Am Ceram Soc 86:1307–1312

72. Watzig K, Hutzler T, Krell A (2009) Transparent spinel by reactive sintering of different alumina modifications with MgO. CFI-Ceram Forum Int 86:E47–E49
73. Krell A, Baur G, Dahne C (2003) Transparent sintered sub-μm Al_2O_3 with IR transmissivity equal to sapphire. In: Tustison RW (ed) Window and dome technologies VIII. SPIE-Int Soc Optical Engineering, Bellingham, pp 199–207
74. Krell A (1998) The effects of load, grain size, and grain boundaries on the hardness of alumina. In: Bray D (ed) 22nd annual conference on composites, advanced ceramics, materials, and structures: B, pp 159–68
75. Saha D, Mistry KK, Giri R, Guha A, Sensgupta K (2005) Dependence of moisture absorption property on sol–gel process of transparent nano-structured γ-Al_2O_3 ceramics. Sens Actuators B-Chem 109:363–366
76. Lei LW, Fu ZY, Wang H, Lee SW, Niihara K (2012) Transparent yttria stabilized zirconia from glycine-nitrate process by spark plasma sintering. Ceram Int 38:23–28
77. Zhang HB, Kim BN, Morita K, Yoshida H, Lim JH, Hiraga K (2010) Optimization of high-pressure sintering of transparent zirconia with nano-sized grains. J Alloy Compd 508:196–199
78. Anselmi-Tamburini U, Woolman JN, Munir ZA (2007) Transparent nanometric cubic and tetragonal zirconia obtained by high-pressure pulsed electric current sintering. Adv Funct Mater 17:3267–3273
79. Peuchert U, Okano Y, Menke Y, Reichel S, Ikesue A (2009) Transparent cubic-ZrO_2 ceramics for application as optical lenses. J Eur Ceram Soc 29:283–291
80. Vahldiek FW (1967) Translucent ZrO_2 prepared at high pressures. J Less-Common Metals 13:530–540
81. Mazdiyas KS, Lynch CT, Smith JS (1967) Cubic phase stabilization of translucent yttria-zirconia at very low temperatures. J Am Ceram Soc 50:532–537
82. Duran P, Recio P, Jurado JR, Pascual C, Moure C (1989) Preparation, sintering and properties of translucent Er_2O_3-doped tetragonal zirconia. J Am Ceram Soc 72:2088–2093
83. Srdic VV, Winterer M, Hahn H (2000) Sintering behavior of nanocrystalline zirconia prepared by chemical vapor synthesis. J Am Ceram Soc 83:729–736
84. Srdic VV, Winterer M, Hahn H (2000) Sintering behavior of nanocrystalline zirconia doped with alumina prepared by chemical vapor synthesis. J Am Ceram Soc 83:1853–1860
85. Tsukuma K (1986) Transparent titania yttria zirconia ceramics. J Mater Sci Lett 5:1143–1144
86. Casolco SR, Xu J, Garay JE (2008) Transparent/Translucent polycrystalline nanostructured yttria stabilized zirconia with varying colors. Scripta Mater 58:516–519
87. Alaniz JE, Perez-Gutierrez FG, Aguilar G, Garay JE (2009) Optical properties of transparent nanocrystalline yttria stabilized zirconia. Opt Mater 32:62–68
88. Tsukuma K, Yamashita I, Kusunose T (2008) Transparent 8 mol% Y_2O_3–ZrO_2 (8Y) ceramics. J Am Ceram Soc 91:813–818
89. Zhang H, Kim B-N, Morita K, Hiraga HYK, Sakka Y (2011) Effect of sintering temperature on optical properties and microstructure of translucent zirconia prepared by high-pressure spark plasma sintering. Sci Technol Adv Mater 12:055003
90. Klimke J, Krell A (2005) Polycrystalline ZrO_2-transparent ceramics with high refractive index. Fraunhofer Institute for Ceramic Technologies and Sintered Materials (IKTS), p 23
91. Trunec M, Chlup Z (2009) Higher fracture toughness of tetragonal zirconia ceramics through nanocrystalline structure. Scripta Mater 61:56–59
92. Klimke J, Trunec M, Krell A (2011) Transparent tetragonal yttria-stabilized zirconia ceramics: influence of scattering caused by birefringence. J Am Ceram Soc 94:1850–1858
93. Chaim R, Hefetz M (1998) Fabrication of dense nanocrystalline ZrO_2-3 wt.% Y_2O_3 by hot-isostatic pressing. J Mater Res 13:1875–1880
94. Nigara Y (1968) Measurement of optical constants of yttrium oxide. Jpn J Appl Phys 7:404–408
95. Fukabori A, Yanagida T, Pejchal J, Maeo S, Yokota Y, Yoshikawa A et al (2010) Optical and scintillation characteristics of Y_2O_3 transparent ceramic. J Appl Phys 107:073501

96. Hou XR, Zhou SM, Jia TT, Lin H, Teng H (2011) Investigation of up-conversion luminescence properties of RE/Yb co-doped Y_2O_3 transparent ceramic (RE = Er, Ho, Pr, and Tm). Phys B-Condens Matter 406:3931–3937
97. Micheli AL, Dungan DF, Mantese JV (1992) High-density yttria for practical ceramic applications. J Am Ceram Soc 75:709–711
98. Iwasawa J, Nishimizu R, Tokita M, Kiyohara M, Uematsu K (2007) Plasma-resistant dense yttrium oxide film prepared by aerosol deposition process. J Am Ceram Soc 90:2327–2332
99. Lefever RA, Matsko J (1967) Transparent yttrium oxide ceramics. Mater Res Bull 2:865–869
100. Huang YH, Jiang DL, Zhang JX, Lin QL (2009) Fabrication of transparent lanthanum-doped yttria ceramics by combination of two-step sintering and vacuum sintering. J Am Ceram Soc 92:2883–2887
101. Ikegami T, Li JG, Mori T, Moriyoshi Y (2002) Fabrication of transparent yttria ceramics by the low-temperature synthesis of yttrium hydroxide. J Am Ceram Soc 85:1725–1729
102. Ikegami T, Mori T, Yajima Y, Takenouchi S, Misawa T, Moriyoshi Y (1999) Fabrication of transparent yttria ceramics through the synthesis of yttrium hydroxide at low temperature and doping by sulfate ions. J Ceram Soc Jpn 107:297–299
103. Jin LL, Zhou GH, Shimai S, Zhang J, Wang SW (2010) ZrO_2-doped Y_2O_3 transparent ceramics via slip casting and vacuum sintering. J Eur Ceram Soc 30:2139–2143
104. Zhang J, An LQ, Liu M, Shimai S, Wang SW (2009) Sintering of Yb^{3+}:Y_2O_3 transparent ceramics in hydrogen atmosphere. J Eur Ceram Soc 29:305–309
105. Greskovich C, Chernoch JP (1973) Polycrystalline ceramic lasers. J Appl Phys 44:4599–4606
106. Rhodes WH (1981) Controlled transient solid 2nd-phase sintering of yttria. J Am Ceram Soc 64:13–19
107. Saito N, Matsuda S, Ikegami T (1998) Fabrication of transparent yttria ceramics at low temperature using carbonate-derived powder. J Am Ceram Soc 81:2023–2028
108. Podowitz SR, Gaume R, Feigelson RS (2010) Effect of europium concentration on densification of transparent Eu:Y_2O_3 scintillator ceramics using hot pressing. J Am Ceram Soc 93:82–88
109. Hou X, Zhou S, Jia T, Lin H, Teng H (2011) Effect of Nd concentration on structural and optical properties of Nd:Y_2O_3 transparent ceramic. J Lumin 131:1953–1958
110. Mouzon J, Maitre A, Frisk L, Lehto N, Oden M (2009) Fabrication of transparent yttria by HIP and the glass-encapsulation method. J Eur Ceram Soc 29:311–316
111. Eilers H (2007) Fabrication, optical transmittance, and hardness of IR-transparent ceramics made from nanophase yttria. J Eur Ceram Soc 27:4711–4717
112. Bezdorozhev O, Borodianska H, Sakka Y, Vasylkiv O (2011) Tough yttria-stabilized zirconia ceramic by low-temperature spark plasma sintering of long-term stored nanopowders. J Nanosci Nanotechnol 11:7901–7909
113. Yoshida H, Morita K, Kim B-N, Hiraga K, Kodo M, Soga K et al (2008) Densification of nanocrystalline yttria by low temperature spark plasma sintering. J Am Ceram Soc 91:1707–1710
114. Yoshida H, Morita K, Kim B-N, Hiraga K, Yamanaka K, Soga K et al (2011) Low-temperature spark plasma sintering of yttria ceramics with ultrafine grain size. J Am Ceram Soc 94:3301–3307
115. Chaim R, Shlayer A, Estournes C (2009) Densification of nanocrystalline Y_2O_3 ceramic powder by spark plasma sintering. J Eur Ceram Soc 29:91–98
116. An L, Ito A, Goto T (2012) Transparent yttria produced by spark plasma sintering at moderate temperature and pressure profiles. J Eur Ceram Soc 32:1035–1040
117. Zhang HB, Kim BN, Morita K, Yoshida H, Hiraga K, Sakka Y (2011) Fabrication of transparent yttria by high-pressure spark plasma sintering. J Am Ceram Soc 94:3206–3210
118. Wen L, Sun XD, Lu Q, Xu GX, Hu XZ (2006) Synthesis of yttria nanopowders for transparent yttria ceramics. Opt Mater 29:239–245
119. Huang YH, Jiang DL, Zhang JX, Lin QL, Huang ZR (2011) Synthesis of mono-dispersed spherical Nd:Y_2O_3 powder for transparent ceramics. Ceram Int 37:3523–3529

120. Ikesue A, Kamata K, Yoshida K (1996) Synthesis of transparent Nd-doped HfO_2–Y_2O_3 ceramics using HIP. J Am Ceram Soc 79:359–364
121. Majima K, Niimi N, Watanabe M, Katsuyama S, Nagai H (1993) Effect of LiF addition on the preparation of transparent Y_2O_3 by vacuum hot-pressing method. J Jpn Inst Met 57:1221–1226
122. Hou XR, Zhou SM, Li YK, Li WJ (2010) Effect of ZrO_2 on the sinterability and spectral properties of $(Yb_{0.05}Y_{0.95})_2O_3$ transparent ceramic. Opt Mater 32:920–923
123. Yi Q, Zhou S, Teng H, Lin H, Hou X, Jia T (2012) Structural and optical properties of Tm: Y_2O_3 transparent ceramic with La_2O_3, ZrO_2 as composite sintering aid. J Eur Ceram Soc 32:381–388
124. Huang YH, Jiang DL, Zhang JX, Lin QL, Huang ZG (2010) Sintering of transparent yttria ceramics in oxygen atmosphere. J Am Ceram Soc 93:2964–2967
125. Serivalsatit K, Kokuoz B, Yazgan-Kokuoz B, Kennedy M, Ballato J (2010) Synthesis, processing, and properties of submicrometer-grained highly transparent yttria ceramics. J Am Ceram Soc 93:1320–1325
126. Gheorghe C, Lupei A, Lupei V, Gheorghe L, Ikesue A (2009) Spectroscopic properties of Ho^{3+} doped Sc_2O_3 transparent ceramic for laser materials. J Appl Phys 105:123110
127. Li JG, Ikegami T, Mori T (2005) Fabrication of transparent, sintered Sc_2O_3 ceramics. J Am Ceram Soc 88:817–821
128. Li JG, Ikegami T, Mori T, Yajima Y (2004) Sc_2O_3 nanopowders via hydroxyl precipitation: effects of sulfate ions on powder properties. J Am Ceram Soc 87:1008–1013
129. Kuck S, Fornasiero L, Mix E, Huber G (2000) Spectroscopic properties of Cr-doped Sc_2O_3. J Lumin 87–9:1122–1125
130. Fornasiero L, Mix E, Peters V, Petermann K, Huber G (1999) New oxide crystals for solid state lasers. Cryst Res Technol 34:255–260
131. Li JG, Ikegami T, Mori T (2003) Fabrication of transparent Sc_2O_3 ceramics with powders thermally pyrolyzed from sulfate. J Mater Res 18:1816–1822
132. Lupei A, Lupei V, Gheorghe C, Ikesue A (2008) Excited states dynamics of Er^{3+} in Sc_2O_3 ceramic. J Lumin 128:918–920
133. Longuet L, Bravo AC, Autissier D, Vissie P, Longuet JL, Lambert S (2009) Preparation of Yb-doped Sc_2O_3 transparent ceramics for laser applications. In: Dierolf V, Fujiwara Y, Hommerich U, Ruterana P, Zavada JM (eds) Rare-earth doping of advanced materials for photonic applications. Materials Research Society, Warrendale, pp 161–166
134. Li JG, Ikegami T, Mori T (2004) Solution-based processing of Sc_2O_3 nanopowders yielding transparent ceramics. J Mater Res 19:733–736
135. Wang Y, Lu B, Sun X, Sun T, Xu H (2011) Synthesis of nanocrystalline Sc_2O_3 powder and fabrication of transparent Sc_2O_3 ceramics. Adv Appl Ceram 110:95–98
136. Serivalsatit K, Ballato J (2010) Submicrometer grain-sized transparent erbium-doped scandia ceramics. J Am Ceram Soc 93:3657–3662
137. Trabelsi I, Maalej R, Dammak M, Lupei A, Kamoun M (2010) Crystal field analysis of Er^{3+} in Sc_2O_3 transparent ceramics. J Lumin 130:927–931
138. An LQ, Ito A, Goto T (2011) Two-step pressure sintering of transparent lutetium oxide by spark plasma sintering. J Eur Ceram Soc 31:1597–1602
139. Wang ZF, Zhang WP, Lin L, You BG, Fu YB, Yin M (2008) Preparation and spectroscopic characterization of Lu_2O_3: Eu^{3+} nanopowders and ceramics. Opt Mater 30:1484–1488
140. An LQ, Ito A, Goto T (2011) Fabrication of transparent lutetium oxide by spark plasma sintering. J Am Ceram Soc 94:695–698
141. Kan A, Moriyama T, Takahashi S, Ogawa H (2011) Low-temperature sintering and microwave dielectric properties of MgO ceramic with LiF addition. Jpn J Appl Phys 50:09NF2
142. Rhodes WH, Sellers DJ (1967) Mechanism of pressure sintering MgO with LiF additions. Am Ceram Soc Bull 46:469
143. Hart PE, Pask JA (1971) Effect of LiF on creep of MgO. J Am Ceram Soc 54:315–316

144. Itatani K, Tsujimoto T, Kishimoto A (2006) Thermal and optical properties of transparent magnesium oxide ceramics fabricated by post hot-isostatic pressing. J Eur Ceram Soc 26:639–645
145. Xu GG, Zhang XD, He W, Liu H, Li H, Boughton RI (2006) Preparation of highly dispersed YAG nano-sized powder by co-precipitation method. Mater Lett 60:962–965
146. Ikesue A, Furusato I, Kamata K (1995) Fabrication of polycrystalline transparent YAG ceramics by a solid-state reaction method. J Am Ceram Soc 78:225–228
147. Ikesue A, Kinoshita T, Kamata K, Yoshida K (1995) Fabrication and optical properties of high performance polycrystalline Nd-YAG ceramics for solid-state lasers. J Am Ceram Soc 78:1033–1040
148. Zhang XD, Liu H, He W, Wang JY, Li X, Boughton RI (2005) Novel synthesis of YAG by solvothermal method. J Cryst Growth 275:E1913–E1917
149. Liu Q, Liu J, Li J, Ivanov M, Medvedev A, Zeng Y et al (2014) Solid-state reactive sintering of YAG transparent ceramics for optical applications. J Alloy Compd 616:81–88
150. Li JG, Ikegami T, Lee JH, Mori T (2000) Low-temperature fabrication of transparent yttrium aluminum garnet (YAG) ceramics without additives. J Am Ceram Soc 83:961–963
151. Zych E, Brecher C (2000) Temperature dependence of host-associated luminescence from YAG transparent ceramic material. J Lumin 90:89–99
152. Liu J, Liu K, Wang HS, Gao F, Liao R (2012) Preparation of silicon nitride porous ceramics. In: Pan W, Gong JH (eds) High-performance ceramics IIi, Parts 1 and 2. Trans Tech Publications Ltd, Stafa-Zurich, pp 824–827
153. Zhou J, Zhang WX, Wang LA, Shen YQ, Li J, Liu WB et al (2011) Fabrication, microstructure and optical properties of polycrystalline Er^{3+}:$Y_3Al_5O_{12}$ ceramics. Ceram Int 37:119–125
154. Li J, Chen Q, Feng GY, Wu WJ, Xiao DQ, Zhu JG (2012) Optical properties of the polycrystalline transparent Nd:YAG ceramics prepared by two-step sintering. Ceram Int 38: S649–S652
155. Dewith G, Vandijk HJA (1984) Transparent $Y_3Al_5O_{12}$ ceramics. Mater Res Bull 19:1669–1674
156. Liu W, Li J, Jiang B, Zhang D, Pan Y (2012) Effect of La_2O_3 on microstructures and laser properties of Nd:YAG ceramics. J Alloy Compd 512:1–4
157. Yang H, Qin X, Zhang J, Ma J, Tang D, Wang S et al (2012) The effect of MgO and SiO_2 codoping on the properties of Nd:YAG transparent ceramic. Opt Mater 34:940–943
158. Appiagyei KA, Messing GL, Dumm JQ (2008) Aqueous slip casting of transparent yttrium aluminum garnet (YAG) ceramics. Ceram Int 34:1309–1313
159. Stevenson AJ, Li X, Martinez MA, Anderson JM, Suchy DL, Kupp ER et al (2011) Effect of SiO_2 on densification and microstructure development in Nd:YAG transparent ceramics. J Am Ceram Soc 94:1380–1387
160. Liu WB, Zhang WX, Li J, Kou HM, Zhang D, Pan YB (2011) Synthesis of Nd:YAG powders leading to transparent ceramics: the effect of MgO dopant. J Eur Ceram Soc 31:653–657
161. Chen C, Zhou SM, Lin H, Yi Q (2012) Selection of different sintering aids and heat-treatment of Y_2O_3 raw powders for Yb^{3+}:$Y_3Al_5O_{12}$ transparent ceramics. In: Shao J, Sugioka K, Stolz CJ (eds) Pacific rim laser damage 2011: optical materials for high power lasers, p 820620
162. Li YK, Zhou SM, Lin H, Hou XR, Li WJ, Teng H et al (2010) Fabrication of Nd:YAG transparent ceramics with TEOS, MgO and compound additives as sintering aids. J Alloy Compd 502:225–230
163. Chen PL, Chen IW (1996) Sintering of fine oxide powders: I, Microstructural evolution. J Am Ceram Soc 79:3129–3141
164. Liu WB, Zhang WX, Li J, Kou HM, Shen YH, Wang L et al (2010) Influence of pH values on (Nd + Y): Al molar ratio of Nd:YAG nanopowders and preparation of transparent ceramics. J Alloy Compd 503:525–528
165. Esposito L, Piancastelli A (2009) Role of powder properties and shaping techniques on the formation of pore-free YAG materials. J Eur Ceram Soc 29:317–322

166. Serantoni M, Piancastelli A, Costa AL, Esposito L (2012) Improvements in the production of Yb:YAG transparent ceramic materials: spray drying optimisation. Opt Mater 34:995–1001
167. Liu WB, Zhang WX, Li J, Zhang D, Pan YB (2012) Preparation of spray-dried powders leading to Nd:YAG ceramics: the effect of PVB adhesive. Ceram Int 38:259–264
168. Gong H, Zhang J, Tang DY, Xie GQ, Huang H, Ma J (2011) Fabrication and laser performance of highly transparent Nd:YAG ceramics from well-dispersed Nd:Y_2O_3 nanopowders by freeze-drying. J Nanopart Res 13:3853–3860
169. Rabinovitch Y, Bogicevic C, Karolak F, Tetard D, Dammak H (2008) Freeze-dried nanometric neodymium-doped YAG powders for transparent ceramics. J Mater Process Technol 199:314–320
170. Yagi H, Takaichi K, Ueda K, Yanagitani T, Karninskii AA (2006) Influence of annealing conditions on the optical properties of chromium-doped ceramic $Y_3Al_5O_{12}$. Opt Mater 29:392–396
171. Li J, Wu Y, Pan Y, Guo J (2006) Fabrication of Cr^{4+}, Nd^{3+}:YAG transparent ceramics for self-Q-switched laser. J Non-Cryst Solids 352:2404–2407
172. Nishiura S, Tanabe S, Fujioka K, Fujimoto Y (2011) Properties of transparent Ce:YAG ceramic phosphors for white LED. Opt Mater 33:688–691
173. Feng T, Shi JL, Jiang DY (2006) Preparation and optical properties of transparent Eu^{3+}: $Y_3Al_{5(1-x)}Sc_{5x}O_{12}$ ceramics. J Am Ceram Soc 89:1590–1593
174. Fu Y, Li J, Liu Y, Liu L, Zhao H, Pan Y (2014) Effect of air annealing on the optical properties and laser performance of Nd:YAG transparent ceramics. Opt Mater Express 4:2108–2115
175. Zhang W, Lu T, Wei N, Ma B, Li F, Lu Z et al (2012) Effect of annealing on the optical properties of Nd:YAG transparent ceramics. Opt Mater 34:685–690
176. Yanagida T, Kamada K, Fujimoto Y, Yokota Y, Yoshikawa A, Yagi H et al (2011) Scintillation properties of transparent ceramic and single crystalline Nd:YAG scintillators. Nucl Instr Meth Phys Res Sect A 631:54–57
177. Jiang BX, Lu X, Zeng YP, Liu SP, Li J, Liu WB et al (2013) Synthesis and properties of Yb: LuAG transparent ceramics. Phys Status Solidi C 10(6):958–961
178. Li HL, Liu XJ, Huang LP (2005) Fabrication of transparent cerium-doped lutetium aluminum garnet (LuAG:Ce) ceramics by a solid-state reaction method. J Am Ceram Soc 88:3226–3228
179. Li HL, Liu XJ, Huang LP (2006) Fabrication of transparent Ce:LuAG ceramics by a solid-state reaction method. J Inorg Mater 21:1161–1166
180. Liao YK, Jiang DY, Feng T, Zhang N (2007) Preparation, spectroscopic properties and enhanced luminescence of Tb3+-doped LuAG phosphors and transparent ceramics by introduction of Sc3+. J Mater Sci 42:5406–5410
181. Luo D, Zhang J, Xu C, Lin H, Yang H, Zhu H et al (2013) Mode-locked Yb:LuAG ceramics laser. Phys Status Solidi C 10(6):967–968
182. Luo D, Zhang J, Xu C, Yang H, Lin H, Zhu H et al (2012) Yb:LuAG laser ceramics: a promising high power laser gain medium. Opt Mater Express 2:1425–1431
183. Pirri A, Vannini M, Babin V, Nikl M, Toci G (2013) CW and quasi-CW laser performance of 10 at.% Yb3+:LuAG ceramic. Laser Phys 23
184. Shen YQ, Feng XQ, Babin V, Nikl M, Vedda A, Moretti F et al (2013) Fabrication and scintillation properties of highly transparent Pr:LuAG ceramics using Sc, La-based isovalent sintering aids. Ceram Int 39:5985–5990
185. Wagner N, Herden B, Dierkes T, Plewa J, Justel T (2012) Towards the preparation of transparent LuAG:Nd^{3+} ceramics. J Eur Ceram Soc 32:3085–3089
186. Zhang WX, Li J, Liu WB, Pan YB, Guo JK (2009) Fabrication and properties of highly transparent $Tm_3Al_5O_{12}$ (TmAG) ceramics. Ceram Int 35:2927–2931
187. Goldstein A (2012) Correlation between $MgAl_2O_4$-spinel structure, processing factors and functional properties of transparent parts (progress review). J Eur Ceram Soc 32:2869–2886

188. du Merac MR, Kleebe HJ, Muller MM, Reimanis IE (2013) Fifty years of research and development coming to fruition; unraveling the complex interactions during processing of transparent magnesium aluminate ($MgAl_2O_4$) spinel. J Am Ceram Soc 96:3341–3365
189. Ganesh I (2013) A review on magnesium aluminate ($MgAl_2O_4$) spinel: synthesis, processing and applications. Int Mater Rev 58:63–112
190. Rothman A, Kalabukhov S, Sverdlov N, Dariel MP, Frage N (2014) The effect of grain size on the mechanical and optical properties of spark plasma sintering-processed magnesium aluminate spinel $MgAl_2O_4$. Int J Appl Ceram Technol 11:146–153
191. Sepulveda JL, Loutfy RO, Chang SY, Ibrahim S (2011) High performance spinel ceramics for IR windows and domes. In: Tustison RW (ed) Window and dome technologies and materials, vol XII, p 801604
192. Bratton RJ (1974) Translucent sintered $MgAl_2O_4$. J Am Ceram Soc 57:283–286
193. Dericioglu AF, Kagawa Y (2003) Effect of grain boundary microcracking on the light transmittance of sintered transparent $MgAl_2O_4$. J Eur Ceram Soc 23:951–959
194. Li JG, Ikegami T, Lee JH, Mori T (2000) Fabrication of translucent magnesium aluminum spinel ceramics. J Am Ceram Soc 83:2866–2868
195. Dericioglu AF, Boccaccini AR, Dlouhy I, Kagawa Y (2005) Effect of chemical composition on the optical properties and fracture toughness of transparent magnesium aluminate spinel ceramics. Mater Trans 46:996–1003
196. Wang C, Zhao Z (2009) Transparent $MgAl_2O_4$ ceramic produced by spark plasma sintering. Scripta Mater 61:193–196
197. Morita K, Kim BN, Hiraga K, Yoshida H (2008) Fabrication of transparent $MgAl_2O_4$ spinel polycrystal by spark plasma sintering processing. Scripta Mater 58:1114–1117
198. Meir S, Kalabukhov S, Froumin N, Dariel MP, Frage N (2009) Synthesis and densification of transparent magnesium aluminate spinel by SPS processing. J Am Ceram Soc 92:358–364
199. Frage N, Cohen S, Meir S, Kalabukhov S, Dariel MP (2007) Spark plasma sintering (SPS) of transparent magnesium-aluminate spinel. J Mater Sci 42:3273–3275
200. Sokol M, Kalabukhov S, Dariel MP, Frage N (2014) High-pressure spark plasma sintering (SPS) of transparent polycrystalline magnesium aluminate spinel (PMAS). J Eur Ceram Soc 34:4305–4310
201. Kim BN, Morita K, Lim JH, Hiraga K, Yoshida H (2010) Effects of preheating of powder before spark plasma sintering of transparent $MgAl_2O_4$ spinel. J Am Ceram Soc 93:2158–2160
202. Morita K, Kim BN, Yoshida H, Hiraga K (2009) Spark-plasma-sintering condition optimization for producing transparent $MgAl_2O_4$ spinel polycrystal. J Am Ceram Soc 92:1208–1216
203. Tsukuma K (2006) Transparent $MgAl_2O_4$ spinel ceramics produced by HIP post-sintering. J Ceram Soc Jpn 114:802–806
204. Villalobos GR, Sanghera JS, Aggarwal ID (2005) Degradation of magnesium aluminum spinel by lithium fluoride sintering aid. J Am Ceram Soc 88:1321–1322
205. Krell A, Hutzler T, Klimke J, Potthoff A (2010) Fine-grained transparent spinel windows by the processing of different nanopowders. J Am Ceram Soc 93:2656–2666
206. DiGiovanni AA, Fehrenbacher L, Roy DW (2005) Hard transparent domes and windows from magnesium aluminate spinel. In: Tustison RW (ed) Window and dome technologies and materials Ix, pp 56–63
207. Bernard-Granger G, Benameur N, Guizard C, Nygren M (2009) Influence of graphite contamination on the optical properties of transparent spinel obtained by spark plasma sintering. Scripta Mater 60:164–167
208. Kleebe HJ, Reimanis IE, Cook RL (2005) Processing and microstructure characterization of transparent spinel monoliths. In: DiAntonio CB (ed) Characterization and modeling to control sintered ceramic microstructures and properties, pp 61–68
209. Goldstein A, Goldenberg A, Yeshurun Y, Hefetz M (2008) Transparent $MgAl_2O_4$ spinel from a powder prepared by flame spray prolysis. J Am Ceram Soc 91:4141–4144

210. Sutorik AC, Gilde G, Swab JJ, Cooper C, Gamble R, Shanholtz E (2012) The production of transparent $MgAl_2O_4$ ceramic using calcined powder mixtures of $Mg(OH)_2$ and γ-Al_2O_3 or AlOOH. Int J Appl Ceram Technol 9:575–587
211. Ping LR, Azad AM, Dung TW (2001) Magnesium aluminate ($MgAl_2O_4$) spinel produced via self-heat-sustained (SHS) technique. Mater Res Bull 36:1417–1430
212. du Merac MR, Reimanis IE, Smith C, Kleebe H-J, Mueller MM (2013) Effect of impurities and LiF additive in hot-pressed transparent magnesium aluminate spinel. Int J Appl Ceram Technol 10:E33–E48
213. Reimanis IE, Kleebe H-J (2007) Reactions in the sintering of $MgAl_2O_4$ spinel doped with LiF. Int J Mater Res 98:1273–1278
214. Reimanis IE, Rozenburg K, Kleebe HJ, Cook RL (2005) Fabrication of transparent spinel: the role of impurities. In: Tustison RW (ed) Window and dome technologies and materials IX, pp 48–55
215. Rozenburg K, Reimanis IE, Kleebe HJ, Cook RL (2007) Chemical interaction between LiF and $MgAl_2O_4$ spinel during sintering. J Am Ceram Soc 90:2038–2042
216. Rozenburg K, Reimanis IE, Kleebe HJ, Cook RL (2008) Sintering kinetics of a $MgAl_2O_4$ spinel doped with LiF. J Am Ceram Soc 91:444–450
217. Sutorik AC, Gilde G, Cooper C, Wright J, Hilton C (2012) The effect of varied amounts of LiF sintering aid on the transparency of alumina rich spinel ceramic with the composition MgO-1.5 Al_2O_3. J Am Ceram Soc 95:1807–1810
218. Sutorik AC, Gilde G, Swab JJ, Cooper C, Gamble R, Shanholtz E (2012) Transparent solid solution magnesium aluminate spinel polycrystalline ceramic with the alumina-rich composition MgO-1.2 Al_2O_3. J Am Ceram Soc 95:636–643
219. Krell A, Waetzig K, Klimke J (2011) Effects and elimination of nanoporosity in transparent sintered spinel ($MgAl_2O_4$). In: Window and dome technologies and materials XII, vol 8016, p 801602
220. Harris DC (2005) History of development of polycrystalline optical spinel in the U.S. In: Tustison RW (ed) Window and dome technologies and materials IX. Spie-Int Soc Optical Engineering, Bellingham, pp 1–22
221. Gledhill AD, Li DS, Mroz T, Goldman LM, Padture NP (2012) Strengthening of transparent spinel/Si_3N_4 nanocomposites. Acta Mater 60:1570–1575
222. Wollmershauser JA, Feigelson BN, Gorzkowski EP, Ellis CT, Goswami R, Qadri SB et al (2014) An extended hardness limit in bulk nanoceramics. Acta Mater 69:9–16
223. Jiang H, Zou YK, Chen Q, Li KK, Zhang R, Wang Y et al (2005) Transparent electro-optic ceramics and devices. In: Ming H, Zhang XP, Chen MY (eds) Optoelectronic devices and integration, Pts 1 and 2. Spie-Int Soc Optical Engineering, Bellingham, pp 380–394
224. Sun P, Xu CN, Akiyama M, Watanabe T (1999) Controlled oxygen partial pressure sintering of (Pb, La)(Zr, Ti)O3 ceramics. J Am Ceram Soc 82:1447–1450
225. Abe Y, Kakegawa K (2002) Fabrication of optically transparent lead lanthanum zirconate titanate ((Pb, La)(Zr, Ti)O3) ceramics by a three-stage-atmosphere-sintering technique. J Am Ceram Soc 85:473–475
226. Kong LB, Ma J, Zhang TS, Zhang RF (2002) Transparent lead lanthanum zirconate titanate ceramics derived from oxide mixture via a repeated annealing process. J Mater Res 17:929–932
227. Wu YJ, Li J, Kimura R, Uekawa N, Kakegawa K (2005) Effects of preparation conditions on the structural and optical properties of spark plasma-sintered PLZT (8/65/35) ceramics. J Am Ceram Soc 88:3327–3331
228. Colla EV, Koroleva EY, Okuneva NM, Vakhrushev SB (1995) Long-time relaxation of the dielectric response in lead magnoniobate. Phys Rev Lett 74:1681–1684
229. Qiao L, Ye Q, Gan JL, Cai HW, Qu RH (2011) Optical characteristics of transparent PMNT ceramic and its application at high speed electro-optic switch. Opt Commun 284:3886–3890
230. Shvartsman VV, Kholkin AL, Orlova A, Kiselev D, Bogomolov AA, Sternberg A (2005) Polar nanodomains and local ferroelectric phenomena in relaxor lead lanthanum zirconate titanate ceramics. Appl Phys Lett 86:202937

231. Shvartsman VV, Kholkin AL, Verdier C, Lupascu DC (2005) Fatigue-induced evolution of domain structure in ferroelectric lead zirconate titanate ceramics investigated by piezoresponse force microscopy. J Appl Phys 98:094109
232. Park SE, Shrout TR (1997) Ultrahigh strain and piezoelectric behavior in relaxor based ferroelectric single crystals. J Appl Phys 82:1804–1811
233. Kamzina LS, Ruan W, Li GR, Zeng JT (2012) Transparent ferroelectric ceramics $PbMg_{1/3}Nb_{2/3}O_3$-$xPbZr_{0.53}Ti_{0.47}O_3$: dielectric and electro-optical properties. Phys Solid State 54:2024–2029
234. Kamzina LS, Wei R, Zeng JT, Li GR (2011) Effect of the La concentration on the dielectric and optical properties of the transparent ferroelectric ceramics $75PbMg_{1/3}Nb_{2/3}O_3$-$25PbTiO_3$. Phys Solid State 53:1608–1613
235. Tong XL, Lin K, Lv DJ, Yang MH, Liu ZX, Zhang DS (2009) Optical properties of PMN-PT thin films prepared using pulsed laser deposition. Appl Surf Sci 255:7995–7998
236. Ruan W, Li GR, Zeng JT, Bian JJ, Kamzina LS, Zeng HR et al (2010) Large electro-optic effect in La-doped $0.75Pb(Mg_{1/3}Nb_{2/3})O_3$-$0.25PbTiO_3$ transparent ceramic by two-stage sintering. J Am Ceram Soc 93:2128–2131
237. Wei ZH, Huang YL, Tsuboi T, Nakai Y, Zeng JT, Li GR (2012) Optical characteristics of Er^{3+}-doped PMN-PT transparent ceramics. Ceram Int 38:3397–3402
238. Ganesamoorthy S, Singh G, Bhaumik I, Karnal AK, Tiwari VS, Kitamura K et al (2005) Growth of relaxor ferroelectric single crystals $PbZn_{1/3}Nb_{2/3}O_3$ (PZN) by high temperature solution growth. Ferroelectrics 326:19–23
239. Wan S, Lynch CS (2001) Characterization of PZN single crystals. In: Streiffer SK, Gibbons BJ, Tsurumi T (eds) Proceedings of the 2001 12th IEEE international symposium on applications of ferroelectrics, vols I and II, pp 347–349
240. Kamzina LS, Krainik NN (1998) Electric-field-induced phase transition in single-crystal lead zinc niobate. Phys Solid State 40:485–488
241. Mulvihill ML, Cross LE, Cao WW, Uchino K (1997) Domain-related phase transitionlike behavior in lead zinc niobate relaxor ferroelectric single crystals. J Am Ceram Soc 80:1462–1468
242. Yin QR, Ding AL, Zheng XS, Qiu PS, Shen MR, Cao WW (2004) Preparation and characterization of transparent PZN-PLZT ceramics. J Mater Res 19:729–732
243. Zhou LJ, Zhao Z, Zimmermann A, Aldinger F, Nygren M (2004) Preparation and properties of lead zirconate stannate titanate sintered by spark plasma sintering. J Am Ceram Soc 87:606–611
244. Li K, Li FL, Wang Y, Kwok KW, Chan HLW (2011) Hot-pressed $K_{0.48}Na_{0.52}Nb_{1-x}Bi_xO_3$ (x = 0.05-0.15) lead-free ceramics for electro-optic applications. Mater Chem Phys 131:320–324
245. Li FL, Kwok KW (2013) Fabrication of transparent electro-optic $(K_{0.5}Na_{0.5})_{(1-x)}Li_xNb_{1-x}Bi_xO_3$ lead-free ceramics. J Eur Ceram Soc 33:123–130
246. Shi YX, Shen J, Zhou J, Xu J, Chen W, Qi YY et al (2015) Structure and optical properties of Sn^{4+} doped $Ba(Mg_{1/3}Nb_{2/3})O_3$ transparent ceramics. Ceram Int 41:253–257
247. Huang YH, Jiang DL, Zhang JX, Lin QL (2010) Fabrication of Sn^{4+} doped $Ba(Mg_{1/3}Ta_{2/3})O_3$ transparent ceramics by a solid state reaction method. Ceram Int 36:1615–1619
248. Kintaka Y, Kuretake S, Tanaka N, Kageyama K, Takagi H (2010) Crystal structures and optical properties of transparent ceramics based on complex perovskite Ba(M4+, B12+, B25+)O3 (M4+=Ti, Sn, Zr, Hf; B12+=Mg, Zn; B25+=Ta, Nb). J Am Ceram Soc 93:1114–1119
249. Schneider H, Schreuer J, Hildmann B (2008) Structure and properties of mullite—a review. J Eur Ceram Soc 28:329–344
250. Aramaki S, Roy R (1962) Revised phase diagram for the systhem Al_2O_3–SiO_2. J Am Ceram Soc 45:229–242
251. Aksay IA, Dabbs DM, Sarikaya M (1991) Mullite for structural, electronic and optical applications. J Am Ceram Soc 74:2343–2358
252. Ohashi M, Iida Y, Wada S (2000) Mullite-based substrates for polycrystalline silicon thin-film solar cells. J Ceram Soc Jpn 108:105–107

253. Bourdais S, Mazel F, Fantozzi G, Slaoui A (1999) Silicon deposition on mullite ceramic substrates for thin-film solar cells. Prog Photovoltaics 7:437–447
254. Slaoui A, Pihan E, Focsa A (2006) Thin-film silicon solar cells on mullite substrates. Sol Energy Mater Sol Cells 90:1542–1552
255. Focsa A, Gordon I, Auger JM, Slaoui A, Beaucarne G, Poortmans J et al (2008) Thin film polycrystalline silicon solar cells on mullite ceramics. Renew Energy 33:267–272
256. Mazdiyas KS, Brown LM (1972) Synthesis and mechanical properties of stoichiometric aluminum silicate (mullite). J Am Ceram Soc 55:548–552
257. Prochazka S, Klug FJ (1983) Infrared-transparent mullite ceramic. J Am Ceram Soc 66:874–880
258. Ohashi M, Tabata H, Abe O, Kanzaki S, Mitachi S, Kumazawa T (1987) Preparation of tranlucent mullite ceramics. J Mater Sci Lett 6:528–530
259. Schneider H, Schmucker M, Ikeda K, Kaysser WA (1993) Optically translucent mullite ceramics. J Am Ceram Soc 76:2912–2914
260. An L, Ito A, Goto T (2012) Effect of calcination temperature on the fabrication of transparent lutetium titanate by spark plasma sintering. Ceram Int 38:4973–4977
261. An L, Ito A, Goto T (2011) Highly transparent lutetium titanium oxide produced by spark plasma sintenng. J Eur Ceram Soc 31:237–240
262. An LQ, Ito A, Goto T (2011) Effects of sintering and annealing temperature on fabrication of transparent $Lu_2Ti_2O_7$ by spark plasma sintering. J Am Ceram Soc 94:3851–3855
263. An LQ, Ito A, Goto T (2011) Fabrication of transparent $La_2Zr_2O_7$ by reactive spark plasma sintering. In: Goto T, Akatsu T (eds) Advanced engineering ceramics and composites. Trans Tech Publications Ltd, Stafa-Zurich, pp 135–138
264. An LQ, Ito A, Goto T (2013) Fabrication of transparent $Lu_2Hf_2O_7$ by reactive spark plasma sintering. Opt Mater 35:817–819
265. An LQ, Ito A, Goto T (2013) Transparent Lu_3NbO_7 bodies prepared by reactive spark plasma sintering and their optical and mechanical properties. Ceram Int 39:383–387
266. An L, Ito A, Goto T (2011) Fabrication of transparent Lu_3NbO_7 by spark plasma sintering. Mater Lett 65:3167–3169
267. Wang Z, Zhou G, Qin X, Yang Y, Zhang G, Menke Y et al (2013) Fabrication of $LaGdZr_2O_7$ transparent ceramic. J Eur Ceram Soc 33:643–646
268. Goldstein A, Yeshurun Y, Vulfson M, Kravits H (2012) Fabrication of transparent polycrystalline $ZnAl_2O_4$-A new optical bulk ceramic. J Am Ceram Soc 95:879–882
269. Kim BN, Hiraga K, Jeong A, Hu C, Suzuki TS, Yun JD et al (2014) Transparent $ZnAl_2O_4$ ceramics fabricated by spark plasma sintering. J Ceram Soc Jpn 122:784–787
270. Xu Y, Fu P, Zhang BH, Gao J, Zhang L, Wang XH (2014) Optical properties of transparent $ZnAl_2O_4$ ceramics: a new transparent material prepared by spark plasma sintering. Mater Lett 123:142–144
271. Chesnaud A, Bogicevic C, Karolak F, Estournes C, Dezanneau G (2007) Preparation of transparent oxyapatite ceramics by combined use of freeze-drying and spark-plasma sintering. Chem Commun 1550–1552
272. Shen Y, Xu J, Tok A, Tang D, Khor KA, Dong Z (2010) Development of translucent oxyapatite ceramics by spark plasma sintering. J Am Ceram Soc 93:3060–3063
273. Allix M, Alahrache S, Fayon F, Suchomel M, Porcher F, Cardinal T et al (2012) Highly transparent $BaAl_4O_7$ polycrystalline ceramic obtained by full crystallization from glass. Adv Mater 24:5570–5575
274. Alahrache S, Al Saghir K, Chenu S, Veron E, Meneses DDS, Becerro AI et al (2013) Perfectly transparent $Sr_3Al_2O_6$ polycrystalline ceramic elaborated from glass crystallization. Chem Mater 25:4017–4024
275. Al Saghir K, Chenu S, Veron E, Fayon F, Suchomel M, Genevois C et al (2015) Transparency through structural disorder: a new concept for innovative transparent ceramics. Chem Mater 27:508–514

276. He LF, Fan GH, Lei MY, Lou ZL, Zheng SW, Su CY et al (2013) Preparation of $MgAl_2O_4$/Ce:YAG transparent ceramics by hot-pressed sintering and its microstructure. Rare Metal Mater Eng 42:463–466
277. He LF, Fan GH, Lei MY, Lou ZL, Chen ZW, Xiao Y et al (2013) Preparation and optical properties of $MgAl_2O_4$/Ce:YAG transparent ceramics. Spectrosc Spectral Anal 33:1175–1179
278. McCauley JW, Patel P, Chen MW, Gilde G, Strassburger E, Paliwal B et al (2009) AlON: a brief history of its emergence and evolution. J Eur Ceram Soc 29:223–236
279. McCauley JW, Corbin ND (1979) Phase-relations and reaction sintering of transparent cubic aluminium oxynitride spinel (AlON). J Am Ceram Soc 62:476–479
280. Kim YW, Park BH, Park HC, Lee YB, Oh KD, Riley FL (1998) Sintering, microstructure, and mechanical properties of AlON-AlN particulate composites. Br Ceram Trans 97:97–104
281. Zientara D, Bucko MM, Lis J (2007) Alon-based materials prepared by SHS technique. J Eur Ceram Soc 27:775–779
282. Rafaniello W, Cutler IB (1981) Preparation of sinterable cubic aluminum oxynitride by the carbothermal intridation of lauminum-oxide. J Am Ceram Soc 64:C128–C
283. Yuan XY, Liu XJ, Zhang F, Wang SW (2010) Synthesis of γ-AlON powders by a combinational method of carbothermal reduction and solid-state reaction. J Am Ceram Soc 93:22–24
284. Jin XH, Gao L, Sun J, Liu YQ, Gui LH (2012) Highly transparent AlON pressurelessly sintered from powder synthesized by a novel carbothermal nitridation method. J Am Ceram Soc 95:2801–2807
285. Su M, Zhou Y, Wang K, Yang Z, Cao Y, Hong M (2015) Highly transparent AlON sintered from powder synthesized by direct nitridation. J Eur Ceram Soc 35:1173–1178
286. Zhang N, Liang B, Wang XY, Kan HM, Zhu KW, Zhao XJ (2011) The pressureless sintering and mechanical properties of AlON ceramic. Mater Sci Eng A-Struct Mater Prop Microstruct Process 528:6259–6262
287. Cheng JP, Agrawal D, Zhang YJ, Roy R (2001) Microwave reactive sintering to fully transparent aluminum oxynitride (AlON) ceramics. J Mater Sci Lett 20:77–79
288. Zientara D, Bucko MM, Lis J (2007) Dielectric properties of aluminium nitride-γ-AlO materials. J Eur Ceram Soc 27:4051–4054
289. Kumar RS, Rajeswari K, Praveen B, Hareesh UNS, Johnson R (2010) Processing of aluminum oxynitride through aqueous colloidal forming techniques. J Am Ceram Soc 93:429–435
290. Clay D, Poslusny D, Flinders M, Jacobs SD, Cutler RA (2006) Effect of $LiAl_5O_8$ additions on the sintering and optical transparency of LiAlON. J Eur Ceram Soc 26:1351–1362
291. Sahin FC, Kanbur HE, Apak B (2012) Preparation of AlON ceramics via reactive spark plasma sintering. J Eur Ceram Soc 32:925–929
292. Qi JQ, Wang YZ, Lu TC, Yu Y, Pan L, Wei NA et al (2011) Preparation and light transmission properties of AlON ceramics by the two-step method with nanosized Al_2O_3 and AlN. Metall Mater Trans A 42A:4075–4079
293. Yuan XY, Zhang F, Liu XJ, Zhang Z, Wang SW (2011) Fabrication of transparent AlON ceramics by solid-state reaction sintering. J Inorgan Mater 26:499–502
294. Zhou YP, Wang DF, Zhuang HR, Wen SL, Guo JK (1998) Study of translucent AlN ceramics. J Inorgan Mater 13:256
295. Xiong Y, Wang H, Fu Z (2013) Transient liquid-phase sintering of AlN ceramics with CaF_2 additive. J Eur Ceram Soc 33:2199–2205
296. Merkle LD, Sutorik AC, Sanamyan T, Hussey LK, Gilde G, Cooper C et al (2012) Fluorescence of Er^{3+}:AlN polycrystalline ceramic. Opt Mater Express 2:78–91
297. Xiong Y, Fu ZY, Wang H, Wang YC, Zhang JY, Zhang QJ (2008) Microstructure and properties of translucent Mg-sialon ceramics prepared by spark plasma sintering. Mater Sci Eng A 488:475–481

298. Joshi B, Gyawali G, Wang H, Sekino T, Lee SW (2013) Thermal and mechanical properties of hot pressed translucent Y_2O_3 doped Mg-α/β-Sialon ceramics. J Alloy Compd 557:112–119
299. Joshi B, Lee HH, Kim YH, Fu Z, Niihara K, Lee SW (2012) Hot pressed translucent (Mg, Y)-alpha/beta-Sialon ceramics. Mater Lett 80:178–180
300. Shan Y, Wang G, Liu G, Sun X, Xu J, Li J (2014) Hot-pressing of translucent Y-α-Sialon ceramics using ultrafine mixed powders prepared by planetary ball mill. Ceram Int 40:11743–11749
301. Su XL, Wang PL, Chen WW, Zhu B, Cheng YB, Yan DS (2004) Translucent α-Sialon ceramics by hot pressing. J Am Ceram Soc 87:730–732
302. Joshi B, Li B, Kshetri YK, Wang H, Lee SW (2014) IR transparent hot pressed Mg-α/β-Sialon:Ba^{2+} ceramics. Ceram Int 40:13041–13047
303. Yang W, Hojo J, Enomoto N, Tanaka Y, Inada M (2013) Influence of sintering aid on the tranlucency of spark plasma-sintered silicon nitride ceramics. J Am Ceram Soc 96:2556–2561

Chapter 3
Ceramic Powder Synthesis

3.1 Introduction

The processing of any ceramics is started from their precursor powders, which are then consolidated into green bodies, followed by sintering at high temperature [1–5]. Of course, there are various other steps that also play significant roles in determining the performance of the final ceramic products. As a result, properties of the powder pose a direct effect on microstructure and thus performance of the final ceramics. The properties of a powder are closely related to the way the powder is synthesized. According to properties and applications, ceramics can be classified into traditional ceramics and advanced ceramics. The functionalities of traditional ceramics are less sensitive to chemical compositions, which usually consist of multiple components, with clays as the main component and various additives [6]. In contrast, the properties and performances of advanced ceramics are very sensitive to their compositions and microstructures. Therefore, they must meet very specific requirements in terms of properties, compositions, and microstructures. Transparent ceramics belong to the category of advanced ceramics. As stated previously, special carefulness is necessary in processing of transparent ceramics, because optical properties are even more sensitive than other properties.

For good control the chemical composition and microstructure of transparent ceramics, the availability of high quality starting powders is a pre-requirement. For starting powders, important characteristics include grain/particle size, size distribution, shape, morphology, degree and state of agglomeration, chemical composition, phase composition, and purity. Physical and chemical properties of surface of the particles also play an important role. Both the powder consolidation and the microstructure development are influenced by the size, size distribution, shape, and state of agglomeration. The size of particle/grain has a specific effect sintering. It is well known that the densification rate increases with decreasing particle/grain size.

According to sizes and size distributions, powders can be classified into two types: (i) polydispersed and (ii) monodispersed powder [7–15]. Polydisperse means

L.B. Kong et al., *Transparent Ceramics*, Topics in Mining,
Metallurgy and Materials Engineering, DOI 10.1007/978-3-319-18956-7_3

that the powder has a wide particle size distribution, while a monodispersed powder has one particle size, but most likely, it means a very narrow size distribution. Polydispersed powders could offer high packing density of green bodies, while there is a negative effect on densification behavior and microstructure development during the sintering. This is because large grains grow much faster by consuming smaller grains surrounding them. In this respect, homogeneous packing of powders with a narrow size distribution is preferred, to from the control of microstructure point of view. Additionally, powders consisting of spherical or near-spherical particles are desirable to have a uniform packing. However, until now, synthesis of monodispersed fine powders is still a challenge of ceramic processing [8, 16–23].

Another challenge of powder processing is the addressing of agglomeration [24–26]. The direct consequence of the presence of agglomeration is the heterogeneity of the packed green bodies, which will result in a phenomenon known as differential sintering [27, 28]. Differential sintering means that different regions of the bodies have different densification behaviors, due to the inhomogeneity. In this case, the final ceramics will likely to have large pores and crack-like voids. Therefore, agglomeration must be avoided in order to fabricate ceramics with high density and homogeneous microstructure. Two types of agglomerations have been observed in ceramic powders: (i) soft agglomerates and (ii) hard agglomerates. Soft agglomerates are made of the particles due to the weak van der Waals forces, while hard agglomerates are formed by chemical bonding. As a result, the negative effect of soft agglomeration is much less than that of hard agglomeration.

Other factors that have significant effects on sintering of a powder include surface characteristics, incomplete chemical reactions, polymorphic phase transformations, and so on. Surface properties affect the dispersion behavior of a powder, thus influencing its sintering efficiency. The unreacted components will react during the sintering, which usually leads to inhomogeneity. Phase transformation could have serious negative effect on microstructure.

In summary, an ideal ceramic starting powder should a small size, e.g., <1 μm or at the scale of nm, a narrow particle or grain size distribution, i.e., monodisperse or near monodisperse, a spherical or near-spherical particle/grain shape, no agglomeration or only soft agglomeration, completed chemical reaction, single phase and high purity. To achieve this, it is important to select the synthesis methods.

Various methods have been developed to synthesize ceramic powders, including physical and chemical methods [29, 30]. The simplest physical method is mechanical milling, in which mechanical action is used to refine the powders [31–33]. Other physical methods include vacuum vapor condensation (VVC) and physical vapor deposition (PVD), and so on. Milling is a general step in the preparation of ceramic powders for the purpose of mixing, if refining is not expected. Physical condensation or deposition is not suitable to produce powders with large quantities. In comparison, chemical methods, such as solid-state reaction and various wet-chemical solution routes are the most widely employed to fabricate advanced ceramics in general and transparent ceramics in particular [34, 35]. In solid-state reaction methods, the use of high-energy milling in recent years has shown certain advantages in reducing the phase formation temperatures of ceramic materials, because transparent ceramics usually have relatively high reaction temperatures [36, 37].

3.2 Synthesis of Precursor Powders

3.2.1 Solid-State Reaction Methods

There are three types of solid-state reaction: (i) chemical decomposition, (ii) chemical reaction between solids, and (iii) chemical reduction. The first two reactions will be introduced, while the third one is commonly used for nonoxides which thus are not included.

Chemical decomposition is usually observed in solid reactions, such as carbonate, hydroxides, nitrate, acetate, oxalates, alkoxides and so on, when they are heated at a certain temperature. The decomposition leads to the formation of a new solid product, together with one or more gaseous phases, which is usually used to produce powders of simple oxides in most cases and complex oxides sometimes. Although this method has not been widely reported for the synthesis of transparent ceramic powders, it could be a potential technique for such a purpose, due to its various advantages, such as simple processing, inexpensive raw materials, and capability of large scale production. In fact, the calcination step involved in most wet-chemical processing routes, especially chemical precipitation or co-precipitation, is chemical decomposition, either from carbonates or hydroxides, as discussed later.

A representative example of chemical decomposition is the formation of magnesia (MgO) from magnesium carbonate ($MgCO_3$). When it is heated, magnesium carbonate decomposes into magnesium oxide and carbon dioxide gas, with reaction equation as follows:

$$MgCO_3\,(s) \rightarrow MgO\,(s) + CO_2\,(g). \tag{3.1}$$

Chemical reactions between solids are used to synthesize mixed powders or complex oxides. The reactants are usually simple oxides, carbonates, nitrates, sulfates, oxalates or acetates, which are mixed according to a target compound with a given stoichiometric composition. An example is the reaction between magnesia and alumina to form magnesium aluminate or spinel, with the following reaction equation:

$$MgO\,(s) + Al_2O_3\,(s) \rightarrow MgAl_2O_4\,(s). \tag{3.2}$$

These methods, involving decomposition of solids and then chemical reaction between solids, are referred to as calcination or pre-sintering in ceramic processing.

3.2.1.1 Decomposition Methods

The principles, kinetics and chemistry of decomposition reactions, such as carbonates and hydroxides, have been extensively studied in the open literature [38–42]. Basic thermodynamics, reaction kinetics and mechanism, as well as closely related process parameters regarding the production of powders, will be elaborated in this subsection.

According to thermodynamics, the decomposition of $MgCO_3$ that is defined by Eq. (3.1), is a strongly endothermic reaction, since ΔH_R^0 is positive, which is similar to most decomposition reactions [43–45]. Therefore, the decomposition is only sustained when the reactant is constantly heated. In thermodynamics, the Gibbs free energy change of a chemical reaction is given by:

$$\Delta G_R = \Delta G_R^0 + RT \ln K, \tag{3.3}$$

where ΔG_R^0 is the free energy change for the reaction when the reactants are in standard state, R is the gas constant, T is the absolute temperature, and K is the equilibrium constant of the reaction. For the reaction defined by Eq. (3.1), there is:

$$K = \frac{a_{\mathrm{MgO}} a_{\mathrm{CO_2}}}{a_{\mathrm{MgCO_3}}} = p_{\mathrm{CO_2}}, \tag{3.4}$$

where a_{MgO} and $a_{\mathrm{MgCO_3}}$ are the activities of pure solids, MgO and $MgCO_3$, respectively, which can be treated as unity, while $a_{\mathrm{CO_2}}$ is the activity of CO_2, which is actually the partial pressure of the gas. Therefore, at equilibrium state, $\Delta G_R = 0$, by combining Eqs. (3.3) and (3.4), there is:

$$\Delta G_R^0 = -RT \ln p_{CO_2}. \tag{3.5}$$

The standard Gibbs free energies for the decomposition of various carbonates and hydroxides, as a function of temperature, as well as the equilibrium partial pressure of the gas for each of the reactions, have been well documented as a database, which can be found in the open literature [46].

In thermodynamics, as the partial pressure of the gaseous product above the solid of a compound is equal to the partial pressure of the gas in the surrounding atmosphere, the compound will become unstable, when it is heated in air. This unstable temperature can be estimated. Thermodynamics predict that the partial pressure of the gas products of many salts, such as acetates, sulfates, oxalates and nitrates, is almost zero, which means that they are unstable at room temperature. However, the fact is that these compounds are stable up to quite high temperatures. Therefore, the decomposition of these compounds is controlled by kinetics, instead of thermodynamics [43–45].

Kinetic studies have been conducted to elucidate the reaction mechanisms and identify the influences of process parameters on the decomposition reactions, such as reaction temperature, particle size of the compound, quantity of the reactant and surrounding environment. These studies can be carried out either isothermally or at a fixed heating rate. Under isothermal conditions, it is impossible to maintain a constant temperature, because heating the sample to a given temperature requires an infinite time. In spite of this, it is easier to analyze the kinetics of isothermal decompositions.

Progress of the reaction is usually measured by monitoring the weight loss, which is plotted as the fraction of the reactant that has decomposed, as a function of time t, defined as:

$$\alpha = \frac{\Delta W}{\Delta W_{\max}}, \tag{3.6}$$

where W and W_{max} are the weight loss at time t and the maximum weight loss after the decomposition reaction is entirely completed, respectively.

Although no general theory has been available to describe the decomposition reactions, a general trend can often be observed. At the initial stage of reaction, decomposition of impurities or unstable superficial items most likely occurs. It then enters an induction period, which is then quickly terminated, due to the development of stable nuclei. After that, the nuclei starts to grow at an accelerated rate, together with further nucleation in some cases. After that, the reaction reaches the maximum rate. With the progress of the reaction, the expansion of nuclei gradually slows down and eventually stops, indicating the starting of a decay period, which continues until the reaction is completed.

Because the molar volume of the solid product is usually smaller than that of the reactant, the product is present as a porous layer covering the nonporous core of the reactant, as shown schematically for the decomposition of $CaCO_3$ in Fig. 3.1 [47]. Therefore, it is a heterogeneous reaction occurring at a sharply defined interface. In this case, there are three processes: (i) reaction at the interface between the reactant and the solid product, (ii) heat transfer to the reaction surface from outside of the particle and (iii) gas diffusion out or permeation from the reaction surface through the porous layer of the product. Any of them could be the controlling factor of kinetics of the decomposition reaction. Generally, it is assumed that the interface moves inward at a constant rate. Therefore, for a spherical reactant with an initial radius r_0, the radius of the unreacted core at time t is given by:

$$r = r_0 - Kt, \tag{3.7}$$

where K is a constant.

One of the parameters that characterize the kinetics of chemical reactions is known as reaction orders. In the simplest case, a reactant A decomposes to a product (P), the reaction equation is written as:

$$A \rightarrow P. \tag{3.8}$$

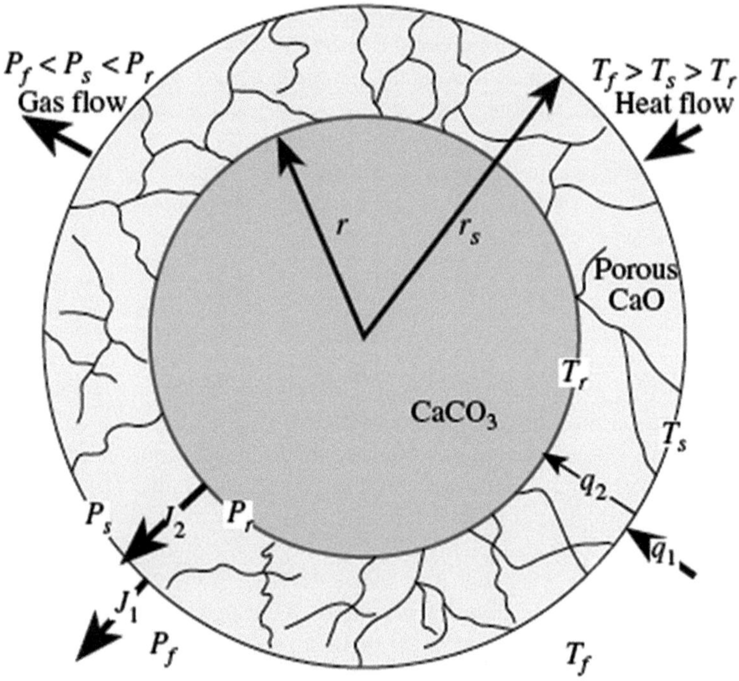

Fig. 3.1 Schematic of the decomposition of calcium carbonate. Reproduced with permission from [47]. Copyright © 2007, Springer

The rate of the reaction can be written as:

$$-\frac{\mathrm{d}C}{\mathrm{d}t} = KC^{\beta}, \tag{3.9}$$

where C is the concentration of the reactant A at time t, K is called the reaction rate constant, and β is an exponent that defines the order of the reaction. The reaction is characterized as first order if $\beta = 1$, second order if $\beta = 2$, and third order if $\beta = 3$. In practice, very high orders of reactions are rarely observed. The orders can be zero or factions.

Several parameters have their influences on the rate of decomposition and the characteristics of the powders produced by the decomposition reaction. These parameters include chemical properties of the reactants, initial particle size and size distribution of the reactants, atmospheric conditions, reaction temperature, and time duration. According to the Arrhenius relation, the rate constant K in the kinetic equation is given by:

$$K = A\exp\left[\frac{-Q}{RT}\right], \tag{3.10}$$

where *A* is a constant which is known as the pre-exponential coefficient or frequency factor, *Q* is the activation energy, *R* is the gas constant, and *T* is the absolute temperature.

Microstructure and morphology of the solid product particles are also affected by the decomposition conditions. Therefore, by controlling the decomposition conditions, it is possible to obtain product powders with fine particles and narrow particle size distribution [48]. The principles of the decomposition method are applicable, when precipitation or co-precipitation methods are used to synthesize oxide powders, where hydroxides or carbonates are formed as precipitates. Decomposition of the hydroxides and carbonates results in oxide powders.

3.2.1.2 Chemical Reaction Methods

The simplest reaction occurs between two solid phases, A and B, to form a solid solution C. A and B are usually polycrystalline compounds. Once the reaction is started, A and B will be separated by the product C. Further reaction is facilitated by the transport of atoms, ions or molecules, which are governed by several possible mechanisms through the phase boundaries and the products of the reaction. Single crystals are generally used to study such reaction mechanisms, because of the simple geometry and the clearly defined boundaries, as shown in Fig. 3.2 [47].

The formation reaction of spinel, $AO + B_2O_3 = AB_2O_4$, is used as an example. Figure 3.3 shows some of the most possible mechanisms involved in the reaction [47]. The first type of mechanism is shown in Fig. 3.3a, b, in which O_2 molecules are transported through the gaseous phase, while the electroneutrality is maintained by electron transport through the layer of the product. In mechanism (ii), i.e., Figure 3.3c, counterdiffusion of the cations with oxygen ions takes place, which has remained stationary. Mechanisms (iii) are shown in Fig. 3.3d, e, in which O_2 ions diffuse through the product layer. Due to their intrinsic properties, the diffusion coefficients of different ions could be significantly different. In spinels, the diffusion of large-sized O_2 ions is much slower than that of cations. As a result, the mechanisms in Fig. 3.3d, e are not dominant. Futhermore, if the phase boundaries have ideal contacts, the transport of O_2 molecules is very slow, so that it is not necessary to consider the mechanisms in Fig. 3.3a, b.

Fig. 3.2 Schematic of solid-state reaction between two single crystals (*A* and *B*). Reproduced with permission from [47]. Copyright © 2007, Springer

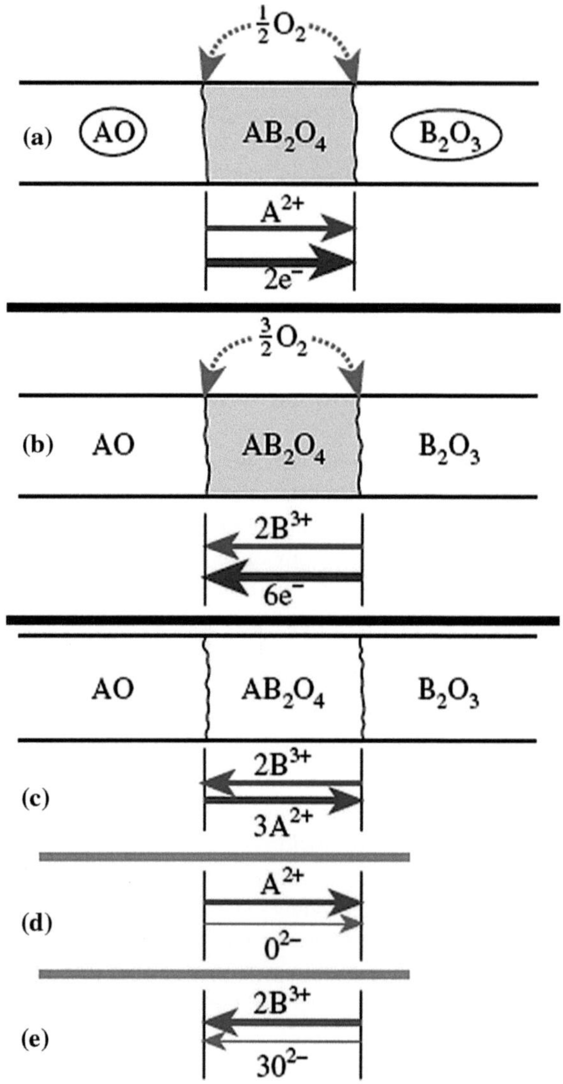

Fig. 3.3 Reaction mechanisms and corresponding net phase boundary reactions for the spinel formation reaction AO $B_2O_3 = AB_2O_4$. Reproduced with permission from [47]. Copyright © 2007, Springer

In this case, the most likely mechanism is the counterdiffusion of cations, i.e., the mechanism of Fig. 3.3c, where the electroneutrality is maintained by the coupling of flux of the cations. When the formation rate of the product is controlled by diffusion through the layer of the product, the thickness of the product layer will follow a parabolic growth law, which is given by:

$$x^2 = Kt, \tag{3.11}$$

where K is a rate constant that can be described by the Arrhenius equation.

Because there are several parameters that should be taken into account to describe the reaction kinetics of reactions in the form of powder mixture, the analysis could be quite complicated. As a result, various assumptions have been made to simplify the analysis models, in order to derive appropriate kinetic equations. For isothermal reactions, it is generally assumed that the particles of reactant A are equal-sized spheres, which are embedded in a quasi-continuous medium of reactant B, so that the reaction product is formed coherently and uniformly on the surface of the A particles [49, 50]. In this case, the volume of the unreacted components at time t is given by:

$$V = \frac{4}{3}\pi(r - y)^3, \tag{3.12}$$

where r is the initial radius of the spherical particles of reactant A and y is the thickness of the product layer. The volume of the unreacted phases can also be expressed as:

$$V = \frac{4}{3}\pi r^3(1 - \alpha), \tag{3.13}$$

where α is the fraction of the volume of the reactants that has already consumed due to the reaction. Combining Eqs. (3.12) and (3.13) yields:

$$y = r[1 - (1 - \alpha)]^{1/3}. \tag{3.14}$$

If it is assumed that y follows the parabolic relationship given by Eq. (3.11), the rate of the reaction becomes:

$$[1 - (1 - \alpha)^{1/3}]^2 = \frac{Kt}{r^2}. \tag{3.15}$$

This is known as Jander equation, which has two oversimplifications that limit its applicability and the capability to predict the rates of most chemical reactions. Firstly, the parabolic growth law for the thickness of the product layer is only valid for one-dimensional reaction across a planar boundary, but not sufficient for the reactions involving spherical particles. In other words, this assumption is only valid for the initial stage of the reaction when $y \ll r$. Second, the changes in molar volumes of the reactants and the products are not taken into account. To address this problem, a more comprehensive equation should be used, which is [51, 52]:

$$[1 + (Z - 1)\alpha]^{2/3} + (Z - 1)(1 - \alpha)^{2/3} = Z + (1 - Z)\frac{Kt}{r^2}, \tag{3.16}$$

where Z is the volume of the product formed from unit volume of the reactant A. This equation is known as the Carter equation, which is applicable to predict the formation of complex oxides.

In practice, the reaction rate decreases with increasing particle size of the reactants due to the increase in the average diffusion distance. According to the Arrhenius relation, the reaction rate increases with increasing temperature. Also, the homogeneity of mixing is an important parameter that determines reaction rate of a reaction involving powder mixtures. This is because it influences the diffusion distance between the reactants and the number of contacts between the reactant particles. It also plays a significant role in ensuring the homogeneity and single-phase product powders. The most effective strategy that can be used to increase the homogeneity of a powder mixture is mechanical milling, which will be discussed in the following subsection.

Solid-state reaction method has been used to fabricate various transparent ceramics, including YAG, LuAG, $MgAl_2O_4$, doped simple oxides and AlON [53–69]. Usually, commercial powders, most likely oxides and sometimes carbonates or hydroxides, are directly used without any special treatment. In some studies, calcination or pre-sintering step is used to form the compounds of interest. Sometimes, this pre-sintering step can be skipped. In this case, it is also called as reactive sintering, simply because the reaction will take place during the sintering process.

The first YAG transparent ceramics for laser applications was made by using solid-state reaction [53]. However, in this report, Al_2O_3 and Y_2O_3 were pre-synthesized through alkoxide precipitation and pyrolysis of $Y_2(OH)Cl_5{\cdot}nH_2O$, respectively. The two powders, with a particle size of <2 μm, were mixed according to the composition of YAG, through the following reaction:

$$3Y_2O_3 + 5Al_2O_3 \rightarrow 2Y_3Al_5O_{12}. \tag{3.17}$$

A similar approach has been reported to synthesize YAG by using pre-synthesized Y_2O_3 [54]. Yttrium nitrate was used as the source of yttrium and aqueous ammonia was used as precipitant. Fine and dendritic precipitate was obtained by adding 0.5 % ammonium sulfate into the precipitation solution. High-purity and low-agglomerated Y_2O_3 powder with spherical particles of about 60 nm in diameter was prepared by calcinating the precipitate at 1100 °C. The highly reactive Y_2O_3 powder was mixed with Al_2O_3 powder to form YAG powder after calination at relatively low temperature. The YAG powder could be sintered to transparent ceramics in vacuum at 1700 °C.

A solid-state reaction method by using commercial powers has been employed to fabricate ytterbium and chromium co-doped YAG (Yb,Cr:YAG) transparent ceramics [58]. CaO and TEOS were used as charge counter element and sintering aid, respectively. High purity Y_2O_3, Al_2O_3, Cr_2O_3, and Yb_2O_3 commercial powders were used as starting materials. CaO provided Ca^{2+} for charge balance when Cr^{3+} was oxidized to Cr^{4+}. The starting powders with a composition of 5 at% Yb, 0.025 at% Cr:YAG were mixed by using ball milling with Ø6 mm ZrO_2 balls in anhydrous alcohol for 10 h. The content of sintering aid was 0.5 wt% tetraethyl orthosilicate (TEOS). Cold isostatic pressure of 200 MPa was applied before sintering. Highly transparent Yb,Cr:YAG ceramics were developed after sintering at 1770 °C for 10 h in vacuum, followed by annealing at 1450 °C for 20 h in air. No calcination or pre-sintering step was involved in this study, so that is reactive sintering, as mentioned earlier.

It has been found that pretreatment could play a significant role in promoting the efficiency of sintering for the fabrication of transparent ceramics. One example has been demonstrated when Y_2O_3 was used to react with Al_2O_3 to form YAG [70]. Commercially available high-purity α-Al_2O_3 (99.98 %), Y_2O_3 (99.99 %) and Nd_2O_3 powders (99.99 %) were used as starting materials. All the raw powders were baked at 200 °C for 24 h and stored in a dry box to prevent water absorption. In addition, the commercial Y_2O_3 powder was calcined in oxygen atmosphere at temperatures of 800, 1000, and 1200 °C for 2 h.

All the untreated and the pretreated Y_2O_3 powders were used to synthesize $Nd_{0.03}Y_{2.97}Al_5O_{12}$. The powder mixtures were ball milled for 12 h with high purity Al_2O_3 balls of 10 mm in diameter in ethanol. 0.1 wt% MgO (99.998 %) and 0.5 wt% tetraethyl orthosilicate (TEOS, 99.99 %) were used as sintering aids. The milled slurries were dried at 80 °C for 2 h. The dried powder mixtures were calcined at 600 °C for 4 h to remove the organic components. Then, the calcined mixtures were dry pressed under 100 MPa into Φ35 mm pellets, followed by the application of cold isostatic pressing (CIP) at pressure of 250 MPa. All the samples were finally sintered at 1745 °C for 50 h in vacuum of 10^{-3} Pa. The sintered ceramics were annealed at 1450 °C for 10 h in air to eliminate the oxygen vacancies.

Figure 3.4 shows SEM images of the untreated commercial Y_2O_3 and α-Al_2O_3 powders. The Y_2O_3 powder had agglomerations, consisting of primary particles with a diameter of about 50 nm. The nanoparticles were aggregated to form secondary particles with sizes in the range of 3–10 μm, as shown in Fig. 3.4a. In addition, the Y_2O_3 powder exhibited a microstructure that was similar to that of nanoceramics. Natural surface of the Y_2O_3 powder, as shown in Fig. 3.4b, had a dense feature formed by the primary particles due to hard aggregates. On the fractured surface, some internal pores were observed within the secondary particles, as illustrated in Fig. 3.4c. The sizes of the primary and secondary particles, as well as the internal porosity, determined particle size distribution and specific surface of the powder, which thus influenced the sintering behavior of the powder and optical properties of the final ceramics. Comparatively, the α-Al_2O_3 powder was not aggregated, with an average particle size of ~260 nm, as shown in Fig. 3.4d.

Figure 3.5 shows SEM images of the untreated Y_2O_3 powder and the powders pretreated at different temperatures in oxygen for 2 h. By comparing Fig. 3.5a, b, it is found that the size of the primary particle was increased from 53 to 62 nm, after treating at 800 °C, and then to 126 nm after calcining at 1000 °C (Fig. 3.5c). 1200 °C was too high for the powder, because a significant grain growth was observed and natural surface of the powder became more rigid, as demonstrated in Fig. 3.5d.

Figure 3.6 shows a schematic illustration of the green body derived from the Y_2O_3 and Al_2O_3 powder mixture after ball milling. It has been accepted that green bodies formed by dry pressing and cold isostatic pressing have a microstructure with two types of pores, i.e., (i) interparticles pores which coexist with (ii) intra-agglomerate pores within the agglomerated Y_2O_3 powders. Because the sintering driving force and shrinkage inside the Y_2O_3 agglomerates were different from those

Fig. 3.4 SEM images of the raw powders: **a** untreated commercial Y_2O_3 powder, **b** natural surface of the untreated Y_2O_3 powder, **c** fractured surface of the untreated Y_2O_3 powder and **d** α-Al_2O_3 powder. Reproduced with permission from [70]. Copyright © 2014, Elsevier

in between the agglomerates and the Al_2O_3 particles, the sintering property of the powder was sensitive to the degree of agglomerates of the Y_2O_3 powders. During the sintering, the elimination of the interparticles pores needed high temperature (>1700 °C). However, due to the high sinterability of the small Y_2O_3 particles, intra-agglomerate pores within the Y_2O_3 powders were easily closed and difficult to be eliminated in the subsequent densification process at such a high temperature (>1700 °C). Therefore, eliminating the intra-agglomerate pores within the Y_2O_3 powders became important. The pretreatment of the Y_2O_3 powders at a low temperature (<1200 °C) in oxygen was beneficial to the elimination of the intra-agglomerate pores. The optimum pretreatment temperature was 1000 °C, because at this temperature there was a synergistic effect of sinterability of the powders and the reduced porosity in green body. This result could be used as a reference, because different materials should have different optimized treatment temperatures.

Besides oxides, no-oxides have also been synthesized by using solid-state reaction method. For example, phase-pure and fine γ-AlON powders have been synthesized by using a combinational method of carbothermal reduction and solid-state reaction [69]. The combined method addressed the disadvantages of the respective individual method. Sucrose was used as the reducer instead of carbon

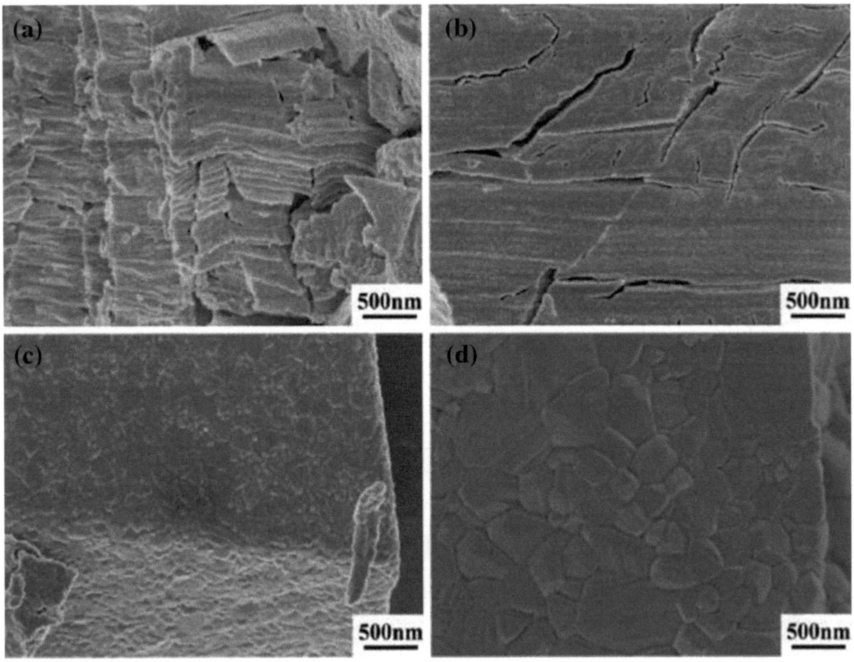

Fig. 3.5 SEM images of the Y_2O_3 powders pretreated at different temperatures for 2 h: **a** without pretreatment, **b** 800 °C, **c** 1000 °C, and **d** 1200 °C. Reproduced with permission from [70]. Copyright © 2014, Elsevier

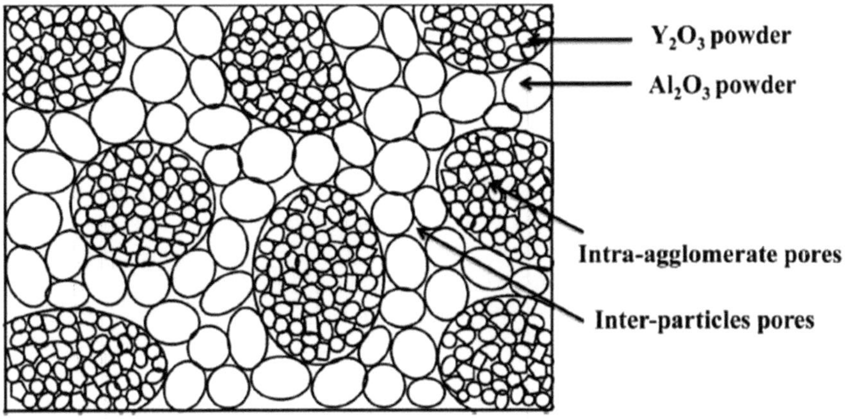

Fig. 3.6 Schematic illustration of the green body derived from the Y_2O_3 and Al_2O_3 powder mixture after ball milling. Reproduced with permission from [70]. Copyright © 2014, Elsevier

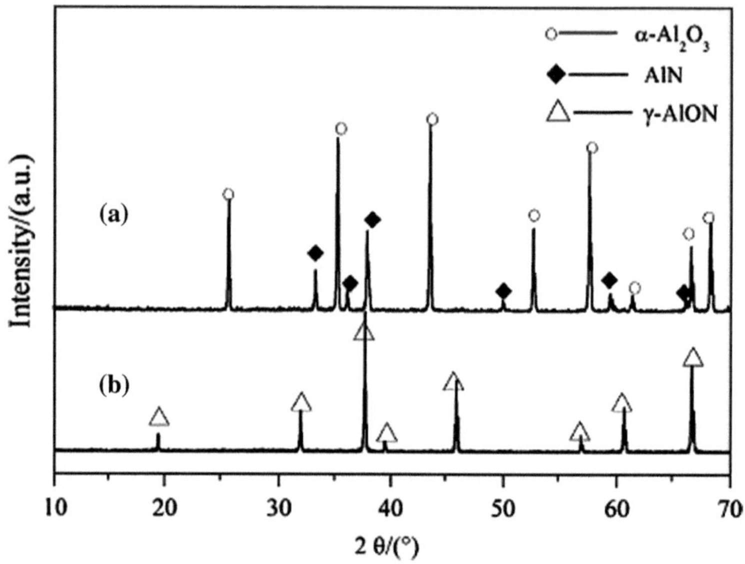

Fig. 3.7 XRD patterns of the **a** nitrided powders and **b** powders heated at 1750 °C for 4 h. Reproduced with permission from [69]. Copyright © 2010, John Wiley & Sons

black. Comparatively, sucrose was more effective in nitridation and hindering the agglomeration of powders.

Figure 3.7 shows XRD patterns of the samples after initial and final treatments [69]. After being heated at 1500 °C for 4 h, with sucrose as the reducer, partial γ-Al_2O_3 powder was nitridized into AlN and the rest was converted from γ to α phase, as shown in Fig. 3.7a. After the residual carbon has been removed, the resultant powder was gray in color, similar to that of commercial AlN powder, In contrast, if carbon black is used as the reducer, the resultant powder was brown owing to the serious carbon infiltration. Schematic diagrams for possible explanation are shown in Fig. 3.8. When sucrose was used as the reducer, Al_2O_3 powder was covered by a layer of carbon after being heated at 900 °C, as shown in Fig. 3.8a. The area for nitridation was maximized, so that nitridation was sufficient. However, if carbon black was used, the efficiency of nitridation was greatly reduced due to the poor contact between the carbon black particles and those of Al_2O_3 powder, as shown in Fig. 3.8b. Therefore, the nitridation was incomplete, thus leading to carbon infiltration.

3.2.1.3 Mechanochemical Synthesis

Mechanochemical synthesis means that chemical reactions that are usually triggered by heating or firing can be facilitated by applying mechanical actions. The most general mechanical action is milling. Various mills have been used to reduce

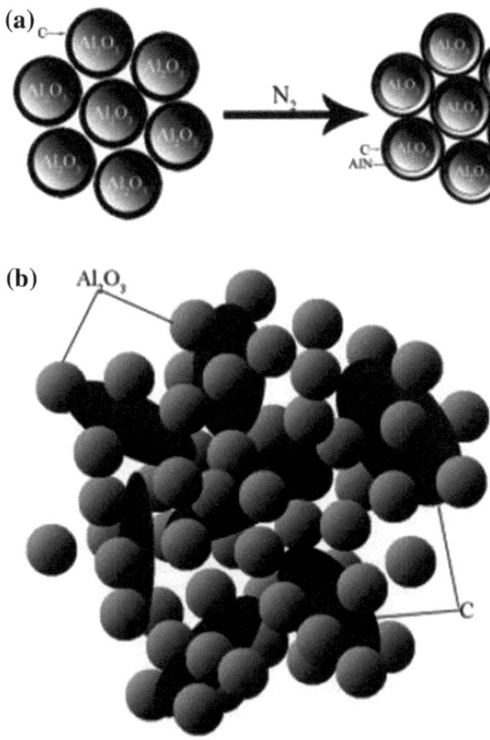

Fig. 3.8 Schematic diagrams for **a** the nitridation with sucrose as reducer and **b** the mixture of γ-Al_2O_3 and carbon black, respectively. Reproduced with permission from [69]. Copyright © 2010, John Wiley & Sons

the particle size and increase the homogeneity of the mixture of reactants for solid-state reactions, including high-compression roller mills, jet mills (also known as fluid energy mills), and ball mills [71, 72]. There are several types of ball mills, which are classified according to the method used to impart motion to the balls, such tumbling, vibration, and agitation.

The energy utilization of a mechanical milling method is defined as the ratio of the new surface area created to the total mechanical energy that has been used. The rate of milling is defined as the amount of new surface area created per unit mass of particles per unit time. Therefore, these two terms can be connected each other. The energy utilization of a mechanical milling method increases with increasing rate of milling. Higher rate of milling means that shorter time is required to achieve a desired particle size. For every method, it is important to understand the relationship between the rate of milling and various experimental parameters. It is also worth mentioning that too high rate of milling could produce high temperature, leading to damage of the milling facilities.

During a milling process, the particles experience mechanical stresses at their contact points due to compression, impact, or shear with the mill medium or with other particles. The mechanical stresses lead to elastic and inelastic deformation of the particles. Once the stress exceeds the ultimate strength of the particles, they will be fractured. Due to the application of the mechanical energy, new surfaces are creased on the particles. At the same time, there are changes in other physical properties of the particles, such as inelastic deformation, increase in localized temperature, lattice rearrangements within the particle and creation of various defects. Changes in chemical properties, especially on the surfaces, could also occur, if a prolonged milling is applied or the milling conditions are very vigorous, such high-energy milling, as discussed later. Generally, the energy utilization of the process is not very high, ranging from <20 % for milling by compression forces to <5 % for milling by impact.

Planetary ball milling is the most widely used technique in both laboratory research and industrial production. The efficiency of milling depends on a number of factors, including the milling parameters, i.e., speed, ball size and ball-to-powder weight ratio, and so on, the properties and types of the milling media, e.g., density and hardness, and the properties of the particles to be milled. For a given size of milling medium, because the mass is proportional to the density, the milling medium should be made of materials with as high density as possible. The size of the milling medium is an important consideration. For a given volume, the number of balls increases inversely as the cube of the radius. As stated earlier, the rate of milling is dependent on the number of contact points between the balls and the powder. The number of contact points is proportional to the surface area of the balls, so that the rate of milling increases inversely as the radius of the balls. However, too small balls cannot supply sufficient mechanical energy to fracture the particles.

The rate of milling also depends on milling time. In the beginning of milling, the decrease in size of the particles is faster. Once the particles are refined to range of about 1 μm to a few micrometers, the rate decreases and further reduction in size becomes more and more difficult. A milling limit is approached, which is determined by several factors. One important factor is the increased tendency of the particles to agglomerate with decreasing particle size. A physical equilibrium is therefore established between the agglomeration and refining processes. Another factor is the decreased probability for the occurrence of a refining event with decreasing particle size.

A disadvantage of ball milling is the wear of the milling medium, which causes potential contamination. For transparent ceramics, the presence of any impurities in the starting powders could be a serious problem. The best solution is to use balls with the materials and composition that are same as the powders to be milled. However, this is only possible for some specific compounds. Therefore, use a milling medium that is chemically inert at the firing temperature of the body (e.g., ZrO_2 balls) or can be removed from the powder by washing (e.g., steel balls). If high-energy milling is required, it is necessary to use WC or at least stainless steel media. A number of commercially available milling balls and their density are listed in Table 3.1.

Tumbling ball mills are the simplest ball mills, consisting of a relatively slow rotating horizontal cylinder that is partly filled with milling balls and the particles to be ground. Besides the factors discussed above, the speed of rotation of the mill is an important parameter, because it determines the trajectory of the balls and the

Table 3.1 Commercially available milling media for ball milling

Milling media	Density (g cm^{-3})
Si_3N_3	2.3
SiC	3.1
Al_2O_3 (<95 % purity)	3.4–3.6
Al_2O_3 (>99 % purity)	3.9
MgO-stabilized ZrO_2	5.5
Y_2O_3-stabilized ZrO_2	6.0
Stainless steel	7.7
Tungsten carbide (WC)	14.5

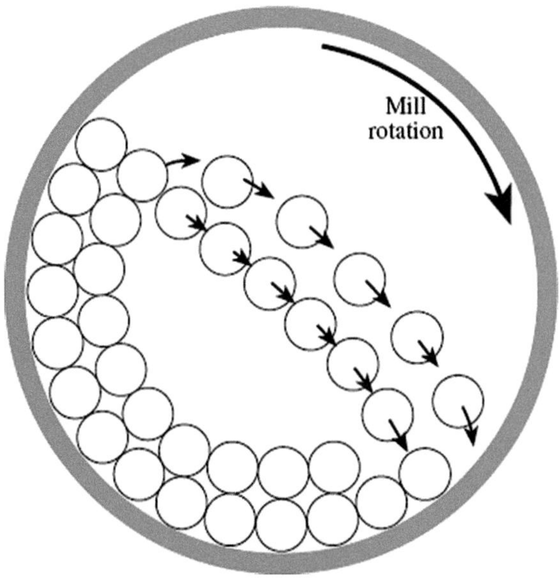

Fig. 3.9 Schematic of a ball mill in cataracting motion showing the movement of the media as the mill rotates about its axis. Reproduced with permission from [47]. Copyright © 2007, Springer

mechanical energy applied to the powders. Identifying the critical speed of rotation as the speed required to take the balls to the apex of revolution, i.e., to the top of the mill, where the centrifugal force just balances the force of gravity. It is found that the critical speed, in revolutions per unit time, is equal to $(g/a)^{1/2}/(2\pi)$, where a is the radius of the mill and g is the acceleration due to gravity. In practice, ball mills are operated at about 75 % of the critical speed so that the balls do not reach the top of the mill, as shown in Fig. 3.9 [47].

An empirical relationship for rate of milling (r_{mill}) is given by:

$$r_{\text{mill}} \approx A a_m^{1/2} \frac{\rho d}{r}, \tag{3.18}$$

where A is a numerical constant that is specific to the mill that is used and the powder that is milled, a is the radius of the mill, ρ is the density of the balls, d is the particle size of the powder, and r is the radius of the balls. According to this relationship, the rate decreases with decreasing particle size. However, there is a practical milling limit that is reached after a certain time of milling. The variation of the rate of milling with the radius of the balls is not always valid, because too small balls cannot supply sufficient energy to fracture the particles.

The objective of the milling is to refine the powders. Therefore, the balls are expected to fall onto the particles at the bottom of the mill rather than onto the mill liner itself. To operate a mill at 75 % of its critical speed, for dry milling, the balls should be 50 % of the mill volume, while the powder filling should be about 25 % of the mill volume. For wet milling, the values should be about 50 % for balls and 40 % for the slurry. At the same time, the solid content in the slurry should be 25–40 %. Comparatively, the energy utilization of wet ball milling is higher that of dry milling by 10–20 %. Furthermore, wet milling produces powders with higher fraction of finer particles. However, wet milling has increased wear of the milling media, needs additional process to dry the powder after milling, and thus easily introduces contamination.

Another type of ball mill is vibrational ball mills or vibro-mills, which consist of a drum, filled with a well-packed arrangement of milling media and the charge of particles. The conventional vibrational mills are operated at frequencies of 10–20 Hz in three dimensions. The milling medium, usually cylindrical in shape, occupies more than 90 % of the mill volume. The amplitude of the vibrations is controlled so as not to disrupt the well-packed arrangement of the milling media. The three-dimensional motion helps the distribution of the particles. In the case of wet milling, segregation of the particles in the slurry can be minimized. The rapid vibratory motion produces an impact energy that is much greater than the energy supplied to the particles in the tumbling ball mills. Vibratory ball mills therefore provide a much more rapid refining process comparatively. They are also more energy efficient than tumbling ball mills. Currently, the vibration frequencies have been significantly increased.

Agitated ball mills, also known as attrition mills or stirred media mills, are also widely used for ceramic processing. One of the characteristics of attrition mills is that the milling chambers are not rotated during the milling. Therefore, the stock of particles and the milling medium are stirred vigorously with a stirrer rotating continuously at frequencies of 1–10 Hz. The milling chamber can be arranged either vertically, as shown in Fig. 3.10, or horizontally with the stirrer located at the center of the chamber [47]. The milling media consist of small spheres, 0.2–10 mm in diameter, which fill 60–90 % of the available volume of the mill. They can be used for either dry milling or wet milling. Agitated ball milling can be conducted in a continuous way, where the slurry of particles to be milled is fed in at one end and the milled product is released at the other end. For intensive milling, a large amount of heat will be produced, which should be taken away by cooling the milling chamber.

Agitated ball mills have a distinct advantage over tumbling ball mills and vibratory ball mills, due to their higher energy utilization. They also allow to deal with slurries with relatively higher contents of solids. The milling efficiency can be

Fig. 3.10 Schematic of an attrition mill system with vertical configuration. Reproduced with permission from [47]. Copyright © 2007, Springer

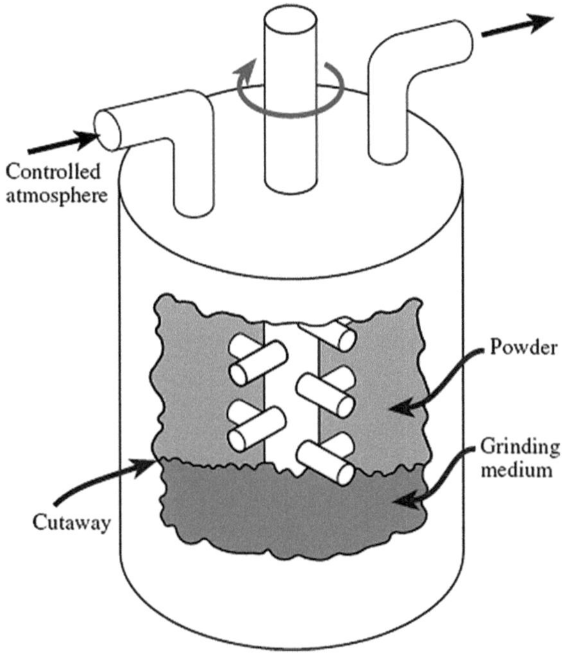

further enhanced by using fine milling media. In this respect, the contamination of the milled powder is less serious, when compared with the tumbling ball mills and vibrational ball mills. Contamination in agitated ball milling can be further reduced by lining the mill chamber with ceramic materials, combined with the use of ceramic stirrers and milling media.

3.2.1.4 Mechanochemical Synthesis and Activation

The initial purpose of mechanical milling is to modify physical characteristics of ceramic powders, such as reducing the particle size and narrowing the size distribution. However, the exploitation of chemical changes during milling for the preparation of powders has received some interest in recent years. Milling enhances the chemical reactivity of powders. Rupture of the bonds during particle fracture results in surfaces with unsatisfied valences. This, combined with the high surface area favors reaction between mixed particles or between the particles and their surroundings.

Powder preparation by high-energy ball milling of elemental mixtures is referred to by various terms, including mechanochemical synthesis, mechanosynthesis, mechanical driven synthesis, mechanical alloying, and high-energy milling. While no term has received widespread acceptance, the term mechanochemical synthesis will be used in this book, unless otherwise specifically stated. The method was initially developed for the production of powders of metals and alloys [73–76]. While less attention has been paid to inorganic systems, the method has been investigated for the preparation of a variety of nonmetallic powders, either directly synthesized or highly activated [36, 71, 77–87].

Mechanochemical synthesis can be carried out in small mills, such as the Spex mill, for synthesizing a few grams of powder or in attrition mills for larger quantities. In the Spex mill, a cylindrical vial containing the milling balls and the charge of particles undergoes large amplitude vibrations in three dimensions at frequencies of up to 1 kHz. The charge occupies about 20 % of the volume of the vial, and the amount of milling media (in the form of balls 5–10 mm in diameter) makes up 2–10 times the mass of the charge. The milling is normally carried out for a few tens of hours for the set of conditions indicated here. The method therefore involves high-intensity vibratory milling for very prolonged time periods.

One of the most distinctive advantages of mechanochemical synthesis is the ease of preparation of powders that can otherwise be difficult to produce, such as those of YAG and spinel. One of the most serious disadvantages is the introduction of impurities from the mill and milling medium into the powders.

The mechanism of mechanochemical synthesis is still not clarified. First, it could be due to the occurrence of the reaction by a solid-state diffusion mechanism. Since diffusion is thermally activated, it would require a significant lowering of the activation energy, a considerable increase in temperature during the milling, or combination of the two. Second, although considerable heat is created, the temperature is significantly lower than that required for a solid-state reaction. Therefore, localized high temperature could trigger the reaction to occur, which then propagandas throughout the whole powder. Additionally, it is also possible that self-propagation reaction could be occurring. However, this is only possible for those highly exothermic reactions, such as the formation of silicides or carbides, not very applicable to oxides like transparent ceramics.

The fabrication of transparent ferroelectric ceramics, involving high-energy ball milling, has been demonstrated for PLZT(x/65/35) (x = 6–9), PMN, and PMN-PT [88–90]. Here, PLZT8/65/35 is used as an example [90]. Commercially available PbO, La_2O_3, ZrO_2, and TiO_2 powders were used as the starting materials with the nominal composition of $(Pb_{0.92}La_{0.08})(Zr_{0.65}Ti_{0.35})_{0.98}O_3$ (PLZT8/65/35), with 10 wt% excess PbO, used to compensate the lead evaporation during the sintering process. The milling operation was carried out by using a planetary high-energy ball milling system in air at room temperature for 36 h. A 250 ml tungsten carbide vial and 100 tungsten carbide balls with diameter of 10 mm were used as a milling medium, with a ball-to-powder weight ratio of 40:1. The milling speed was set at 200 rpm, while the milling was stopped for 5 min for every 25 min of milling to avoid the potential overheating of the system.

The milled powder was then pressed uniaxially into 10 mm diameter pellets. The green pellets of about 70 % of the theoretical density were sintered in air for 4 h at temperatures from 900 to 1000 °C, with both heating and cooling rate being 10 °C min^{-1}. Transparent PLZT ceramics could be obtained by annealing the sintered samples at 1125 °C for 6 h. Such annealing was repeated four times, resulting in a total annealing time of 24 h.

Figure 3.11 shows XRD pattern of the oxide mixture with a composition of PLZT8/65/35, which was high-energy ball milled for 36 h. Perovskite phase of PLZT is the predominant phase, while a peak of PbO is still observed, which can be attributed to the excessive and unreacted PbO. This observation means that the high-energy ball milling can trigger the reaction of the oxide components to form the desired compound. As mentioned earlier, although localized high temperature is possible, the overall temperature of the milling system (milling media plus the milled powder) is very close the room temperature. Therefore, the phase formation mechanism triggered by the high-energy ball milling should be different from that caused by high temperature reaction [91–93]. Figure 3.12 is the SEM image of the

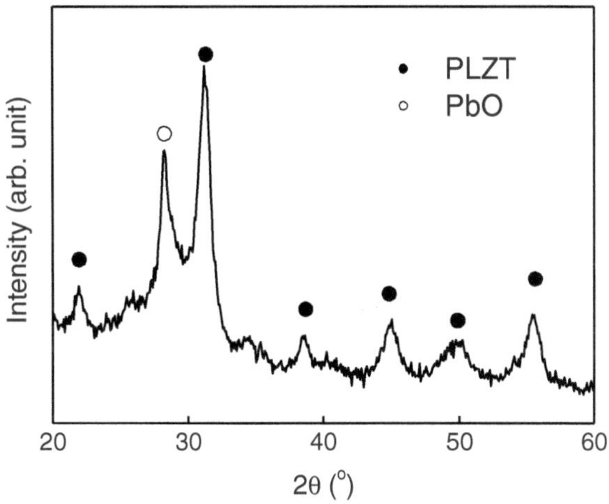

Fig. 3.11 XRD pattern of the oxide mixture with a nominal composition of PLZT8/65/35 milled for 36 h in air at room temperature. Reproduced with permission from [90]. Copyright © 2001, Elsevier

Fig. 3.12 SEM image of the milled oxide mixture. Reproduced with permission from [90]. Copyright © 2001, Elsevier

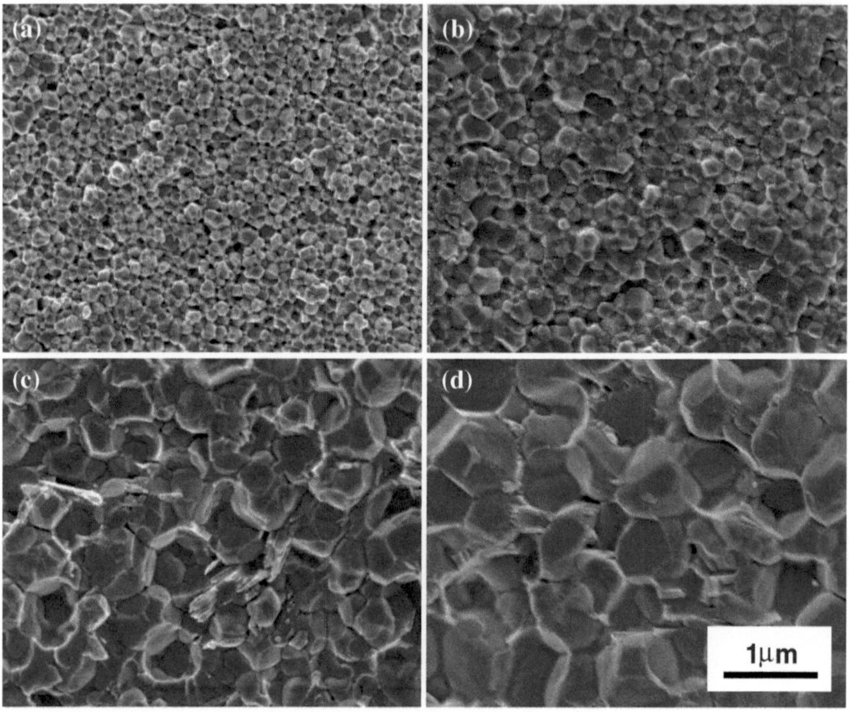

Fig. 3.13 Cross-sectional SEM images of the PLZT ceramics derived from the milled oxide mixture sintered for 4 h at different temperatures: **a** 900 °C, **b** 925 °C, **c** 950 °C and **d** 1000 °C. Reproduced with permission from [90]. Copyright © 2001, Elsevier

as-milled powder, showing that powder has a particle size at the scale of tens of nanometer. It is believed that the refined characteristic of the powders caused by the high-energy ball milling that makes it possible to obtain the fully dense PLZT ceramics at sintering temperature of as low as 1000 °C. It can be understood that the continuous milling would also prevent the growth of the particles.

Figure 3.13 shows cross-sectional SEM images of the samples sintered at different temperatures for 4 h. After sintering at 900 °C, the sample has an average grain size of only about 0.38 μm, even though the grains are well developed with clear grain boundaries. As the sintering temperature is increased to 925 °C, the average grain size slightly increases to 0.41 μm. The microstructures of the PLZT ceramics sintered at 900 and 925 °C are nearly identical. When the sintering temperature is increased to 950 °C, the average grain size of the sample is increased greatly to about 1.1 μm. Further increase in sintering temperature did not result in significant grain growth, with average grain sizes of 1.5 μm for 975 °C and 1.8 μm for 1000 °C. It means that 950 °C is a critical sintering temperature at which a sharp increment in grain size of the PLZT ceramics occurs. Figure 3.14 shows surface SEM images of the chemically etched PLZT ceramics, which were derived from the

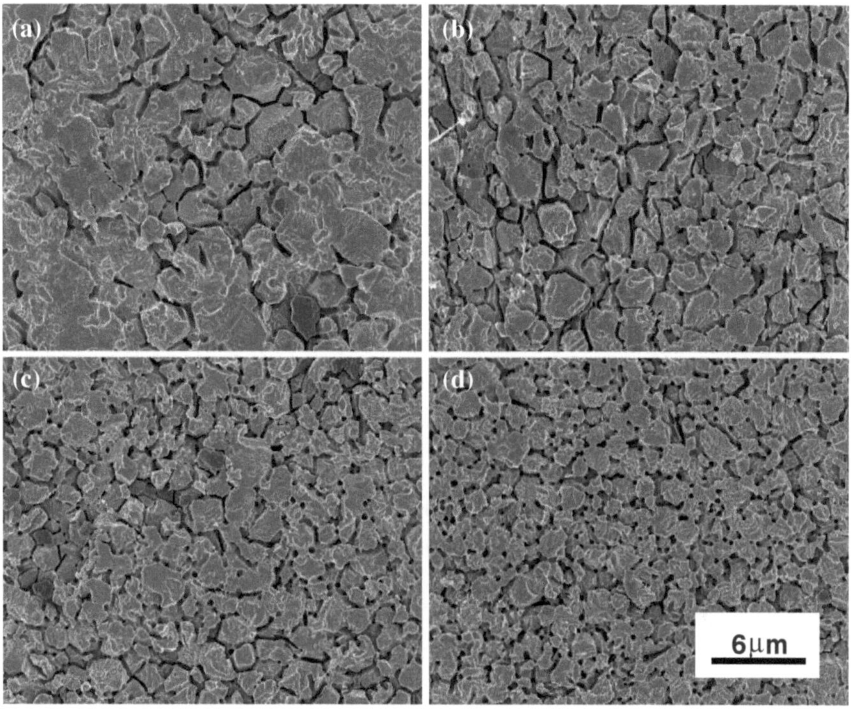

Fig. 3.14 Chemically etched surface SEM images of the 1125 °C-annealed PLZT ceramics that were previously sintered for 4 h at different temperatures: **a** 900 °C, **b** 925 °C, **c** 950 °C and **d** 1000 °C. Reproduced with permission from [90]. Copyright © 2001, Elsevier

sintered samples by repeatedly annealing at 1125 °C for 6 h for 4 times. Obviously, the grain size of the annealed samples decreases gradually with sintering temperature.

The variations of average grain size of the sintered and annealed PLZT ceramics, as a function of sintering temperature, are shown in Fig. 3.15. This observation can be explained, by considering the process as a two-step sintering. A two-step sintering process was used to control the grain growth of nanosized Y_2O_3, in which the Y_2O_3 was initially heated to 1310 °C without stop and then cooled to 1150 °C and kept at the temperature for 20 h [94]. Fully sintered Y_2O_3 ceramics without final-stage grain growth could be achieved. The suppression of the final-stage grain growth has been attributed to that the second-step sintering proceeded in a 'frozen' microstructure with slower kinetics. For the PLZT ceramics sintered at different temperatures, the 'frozen' degree in microstructure was different. The higher the sintering temperature at which the ceramics were obtained, the higher the 'frozen' degree in microstructure. Therefore, the PLZT ceramics sintered at low temperature with small grain size will lead to the annealed samples with large grain size. The annealed PLZT ceramics were all transparent, with optical properties similar to

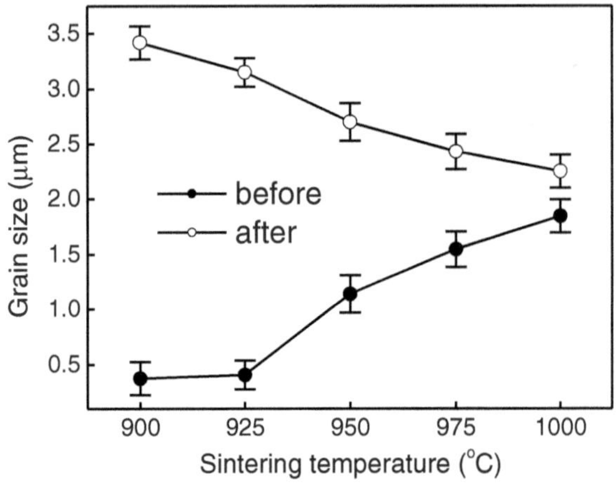

Fig. 3.15 Grain size variations as a function of sintering temperature of the PLZT ceramics before and after thermal annealing at 1125 °C. Reproduced with permission from [90]. Copyright © 2001, Elsevier

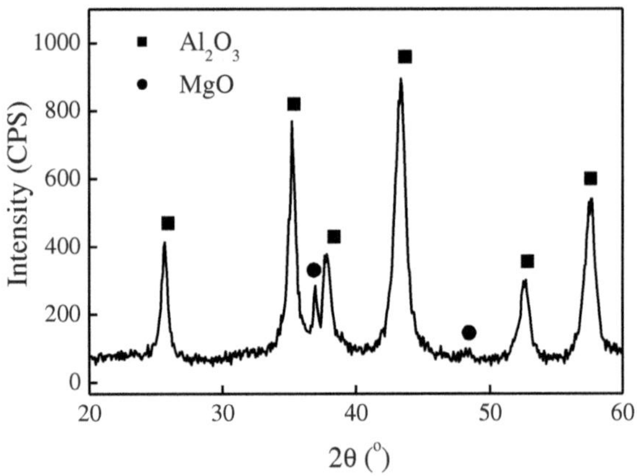

Fig. 3.16 XRD pattern of the mixture of MgO and Al_2O_3 (for $MgAl_2O_4$) milled for 12 h. Reproduced with permission from [37]. Copyright © 2002, Elsevier

those fabricated by using other methods reported in the open literature, demonstrating the efficiency and effectiveness of this mechanochemical synthesis technique.

Planetary high-energy ball milling has also been used to synthesize spinel and YAG powders [36, 37, 95]. Commercially available MgO (99.99+ % purity) and Al_2O_2 (99.9+ % purity) powders were mixed with the composition of $MgAl_2O_4$, which for 12 h, with milling parameters similar to those used for the above PLZT. Figure 3.16 shows XRD pattern of the mixture of MgO and Al_2O_3 that was milled for 12 h [37]. There is no other phase observed in the XRD pattern, which means that no reaction occurred during the milling process. However, the broadened and reduced diffraction peaks of Al_2O_3 and MgO imply that the grains of Al_2O_3 and MgO have been greatly refined as a result of the high-energy ball milling. This is evidenced by the SEM image of the milled mixture, as shown in Fig. 3.17. The milled mixture powder has grain sizes from 100 to 300 nm. Some small particles are within tens of nanometers. The refined structure of the milled mixture is directly

Fig. 3.17 SEM image of the mixture of MgO and Al_2O_3 (for $MgAl_2O_4$) milled for 12 h. Reproduced with permission from [37]. Copyright © 2002, Elsevier

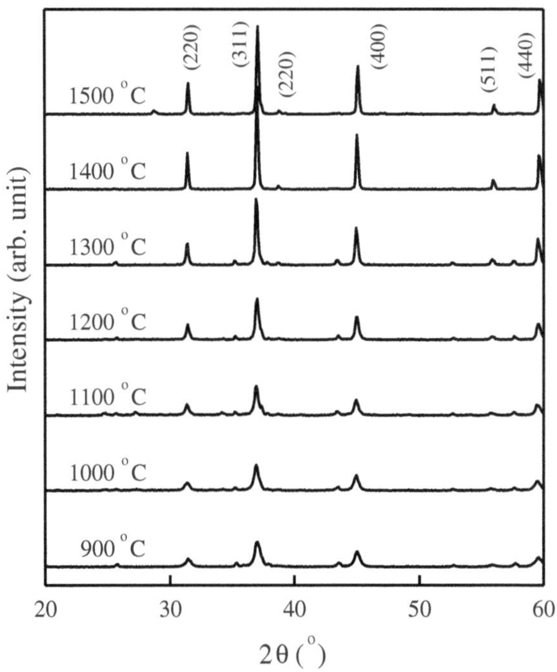

Fig. 3.18 XRD patterns of the mixture of MgO and Al_2O_3 (for $MgAl_2O_4$) milled for 12 h after calcination at various temperatures for 2 h. Reproduced with permission from [37]. Copyright © 2002, Elsevier

allows the low temperature formation of $MgAl_2O_4$ spinel phase from the mixture. In this case, the high-energy milling is not to facilitate the chemical reaction, but only to refine the powders so as to increase their reactivity and thus decrease the phase formation temperature.

Figure 3.18 shows XRD patterns of the milled mixture calcined at different temperatures for 2 h. The $MgAl_2O_4$ spinel phase was readily formed after calcining

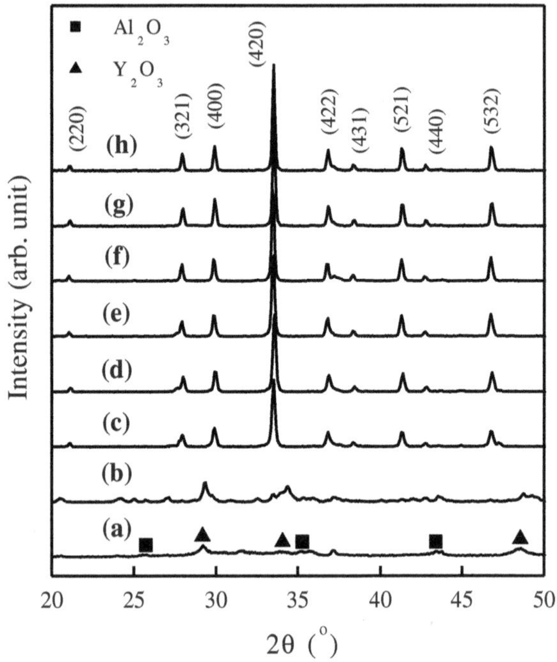

Fig. 3.19 XRD patterns of the mixture of Y_2O_3 and Al_2O_3 (for $Y_3Al_5O_{12}$) milled for 12 h after calcination at various temperatures for 2 h: **a** as-milled, **b** 900 °C, **c** 1000 °C, **d** 1100 °C, **e** 1200 °C, **f** 1300 °C, **g** 1400 °C and **h** 1500 °C. Reproduced with permission from [36]. Copyright © 2002, Elsevier

at 900 °C for 2 h, indicating the high reactivity of the MgO and Al_2O_3 mixture as a result of the high-energy ball milling. This temperature is significantly lower than required by the conventional solid-state reaction process where calcinations temperature is usually as high as 1300 °C. Therefore, this temperature (900 °C) for the phase formation of $MgAl_2O_4$ is comparable with those required by using most wet-chemical routes. However, the mechanical milling method is much simpler than the chemical ways, showing its potential application in the fabrication of transparent ceramics.

Similar results were observed for YAG [36]. Mixture of commercial Y_2O_3 and Al_2O_2 powders, with the nominal composition of $Y_3Al_5O_{12}$, was milled for 12 h. The milled mixture was then calcined for form YAG. Figure 3.19 shows XRD patterns of the milled powder which was calcined at different temperatures. YAG phase was formed after calcining at 1000 °C, which consisted of particles with an average particle size of 0.1–0.2 μm. Both the grain size and grain morphology of the YAG powders produced by using the high-energy ball milling are similar to those which were synthesized by using the chemical routes reported in the literature.

In another study, the refining effect of the high-energy milling has been confirmed by using TEM observation. Figure 3.20 shows TEM and high resolution (HR)TEM images of the Y_2O_3 and Al_2O_2 mixture ball milling for 20 h. Although agglomerated particles were formed due to the milling, they consisted of nanosized crystals, as shown in Fig. 3.20b. It is also found that there are defects, such as lattice

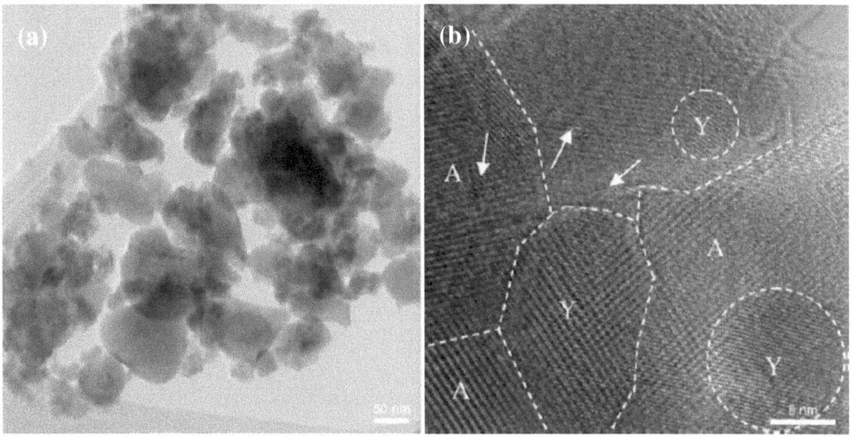

Fig. 3.20 TEM (**a**) and HRTEM (**b**) images of the mixture of Y_2O_3 and Al_2O_3 (for $Y_3Al_5O_{12}$) milled for 20 h (A for Al_2O_3 and Y for Y_2O_3). Reproduced with permission from [96]. Copyright © 2009, Elsevier

mismatch and nanocracks, at the boundaries of the small crystals. The presence of such defects, together with the nanosized crystals, has been responsible for the low temperature formation of YAG. It is expected that such powders should be suitable to produce transparent ceramics, which deserves further development.

Besides the examples discussed above, ball milling has also been used to modify or treat powders to have improved sintering behavior and thus optical properties, although the mechanisms behind are still not clearly identified [97–101].

A commercial jet mill was used to treat Yb^{3+} doped Lu_2O_3 powder that was synthesized by using co-precipitation method [97]. The effect of feed rate was specifically studied. It was found that jet milling is able to break large agglomerates without cross-contamination. Median agglomerates were refined from 8.74 to 1.06 μm, when the powder was jet milled at an optimized feed rate 0.75 lb h^{-1}. By using a sacrificial run, contamination could be effectively eliminated. The Yb: Lu_2O_3 ceramics made from the jet milled powders exhibited homogeneous microstructure and high transparency, with transmission very close to the theoretical limit. The jet mill was lined with tungsten carbide (WC).

A similar effect was observed when ball milling was used to treat commercial Lu_2O_3 powder, followed by sintering with spark plasma sintering method [98]. The application of ball milling could reduce the particle size of commercial Lu_2O_3 powder, so that the as-sintered samples derived from the milled powder had larger grain sizes than those from the unmilled powder. It was found that the former had much poorer optical transmittance, which was attributed to the assumption that ball milling introduced defects into the Lu_2O_3 powder, because no direct evidence provided in that study. However, after the two groups of samples were annealed at 1050 °C for 12 h, they became almost the same without obvious difference in appearance. Furthermore, the samples made with the milled powder had slightly higher optical transmittance in the visible range. This is because thermal annealing eliminated the defects in the samples from the milled powder and due to their large grain sizes they possessed better optical properties. Although this effect is not very significant, it could be of interest and importance when a slight increase in transparence may lead to great improvement in performance, especially for laser applications. The milling in this study was conducted with ZrO_2 media in ethanol.

The effect of ball milling has also been demonstrated in other transparent ceramics. It was shown that, when using Y_2O_3, α-Al_2O_3, and Nd_2O_3 to prepare Nd:YAG ceramics through solid-state reaction, ball milling time had a crucial effect on particle size, densification behavior, microstructure development of the powder, and optical properties of the final transparent ceramics [99]. It was found that 12 h is the critical milling time, below which porosity of the Nd:YAG ceramics sintered at given temperatures was decreased with increasing milling time. Above this milling time, further increase in milling time had no obvious effect. The main reason is the refinement of the coarse powders due to the ball milling. A planetary ball mill, with 10 mm diameter alumina (Al_2O_3) balls, was used to mill the powders in ethanol. The disc and bottle (vial or container) rotation speeds were 130 and 260 rpm, respectively. The balls-to-powder weight ratio was 3:1, while the solid loading of the ball milled slurry was about 1.8 g ml^{-1}.

A very interesting observation on the effect of ball milling on densification and optical properties of Ho:Y_2O_3 transparent ceramics prepared by using solid-state reaction method with commercial oxide powders [101]. Commercial Y_2O_3 (>99.99 %), Ho_2O_3 (>99.99 %), and ZrO_2 (>99 %) powders were mixed, according to the stoichiometric composition of $(Ho_{0.005}Y_{0.995})_2O_3$, with 3 wt% ZrO_2 as sintering aid. A planetary milling machine was used for the ball milling experiment, with milling of 15 h, speed of 140 rpm and agate balls (5 mm in diameter) in high purity ethanol. Powder and ball filling was fixed, with the weight ratio of powder: ball of 1:2. The solid content (C_S) of the slurry, or called slurry concentration, was adjusted between 33.2 and 17.4 vol%, i.e., C_S = 33.2, 23.7, 20.7, 19.5, 18.4 and 17.4 vol%, by controlling the amount of ethanol. The slurry was then dried at 80 °C for 24 h and sieved through a 140-mesh screen. After removing organic components by calcining at 800 °C for 3 h, the powders were uniaxially pressed into pellets at 15 MPa. The pellets were then further cold isostatically pressed (CIP) at 200 MPa. After CIP, the green bodies were sintered by using a vacuum furnace under vacuum ($P \leq 10^{-3}$ Pa) for various times at temperature range of 1400–1850 °C. It was found that the slurry concentration has an obvious effect on agglomeration behaviors of the powders, which 18.4 vol% is an optimized concentration.

Figure 3.21 shows relative densities of the Ho:Y_2O_3 ceramics as a function of sintering temperature, which were made with powders corresponding to different slurry concentrations. In the range of Cs = 33.2–18.4 vol%, the relative density of the sintered samples increased steadily with decreasing slurry concentration at every given sintering temperature. After sintering at 1850 °C for 2 h, the samples with 20.7–17.4 vol% slurry concentration were sintered to over 99 % relative density, while those with slurry concentration of 23.7 and 33.2 vol% had much lower bulk density, reaching only 98.1 and 97.3 %, respectively. This implies that

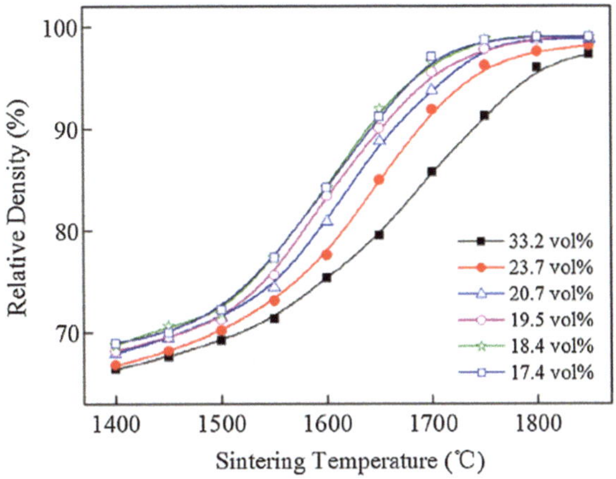

Fig. 3.21 Relative density versus sintering temperature for the ZrO_2-doped Ho:Y_2O_3 ceramics derived from the powders ball milled with different slurry concentrations. Reproduced with permission from [101]. Copyright © 2015, Elsevier

the agglomerated powders were difficult to be completely densified, as stated earlier. The solid concentration should be optimized in the range of 17.4–20.7 vol% in this case.

Figure 3.22 shows SEM images of the Ho:Y_2O_3 ceramics for different C_S values, which were sintered at 1800 °C for 2 h. Grain boundary pores can be observed in the 33.2 vol% ceramics, as shown in Fig. 3.22a. These pores could be attributed to the large pores present between the agglomerates. Pore grain boundary separation would have occurred during the sintering process, so that it was difficult to remove the pores that have been entrapped in the samples during the sintering. The presence of the residual pores embraced in the samples became the major factors that affected their optical transparency. With decreasing slurry concentration, grain growth was gradually promoted and pores were removed, as shown in Fig. 3.22b, c.

Figure 3.23 shows grain growth kinetics for the 33.2, 23.7, and 18.4 vol% samples sintered at 1800 °C for 0.25–30 h, according to grain growth kinetic rate equation. After trying n values of 2–4, it was found that $n = 3$ provided the best fitting for the data. It has been accepted that a grain growth exponent of 3 means that grain growth was controlled by either liquid phase mass transport or solute drag mechanisms in solid-state systems, if the mass migration was controlled boundary mechanism. SEM examination indicated that only a small number of isolated grain boundary pores were observed in the sample sintered at 1800 °C for 0.25 h, which suggests that the grain growth was most likely controlled by grain boundary migration instead of pore drag. Because there was no liquid phase in this case, the cubic grain growth kinetics should be due to solute drag. It also implies that agglomeration of the powders had no effect on the sintering mechanism, but greatly influenced the microstructural development.

In-line optical transmittance spectra of the ZrO_2-doped Ho:Y_2O_3 ceramics sintered at 1800 °C for 30 h, which were from the powders corresponding to different slurry concentrations, are shown in Fig. 3.24. All the absorption peaks are related to holmium ion (Ho^{3+}) absorption in the Y_2O_3 matrix. With decreasing slurry

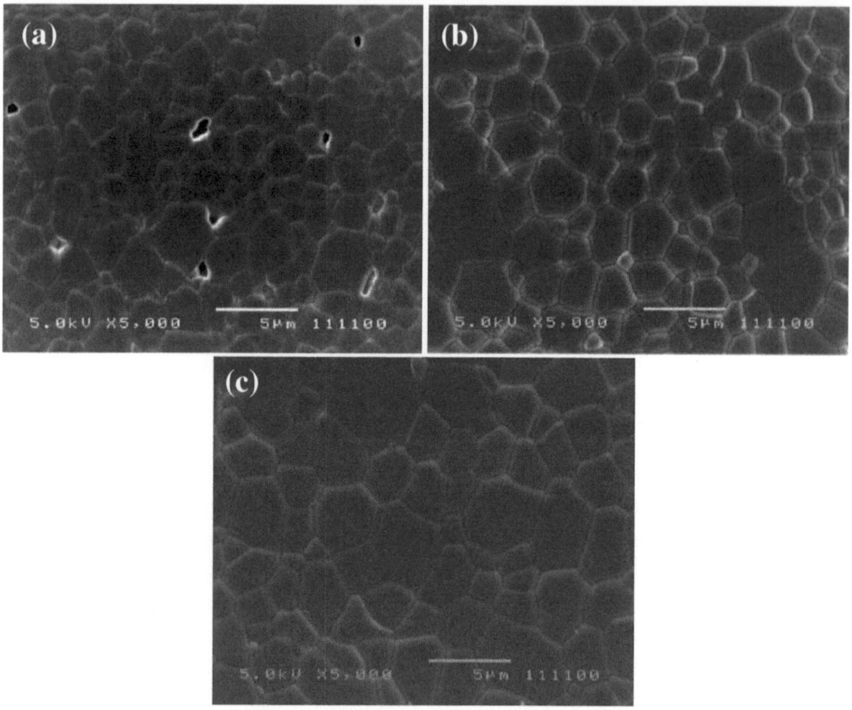

Fig. 3.22 Thermally etched surface SEM images of the ZrO_2-doped Ho:Y_2O_3 ceramics sintered at 1800 °C for 2 h in vacuum, corresponding to different slurry concentrations: **a** 33.2 vol%, **b** 23.7 vol% and **c** 18.4 vol%. Reproduced with permission from [101]. Copyright © 2015, Elsevier

concentration, transmittance steadily increased and reached the maximum in the sample with slurry concentration of 18.4 vol%. After that, it slightly dropped with further decrease in slurry concentration, implying again that the optimal slurry concentration was 18.4 vol%. It is understood that this optimization range should be different for different materials and different powder characteristics even for the same materials.

3.2.2 Wet-Chemical Routes

There are two main routes to produce ceramic powders from a solution: (i) evaporation of the liquid and (ii) precipitation by adding a chemical reagent that reacts with components in the solution. Other wet-chemical routes include: sol–gel, gel combustion, emulsion, and so on.

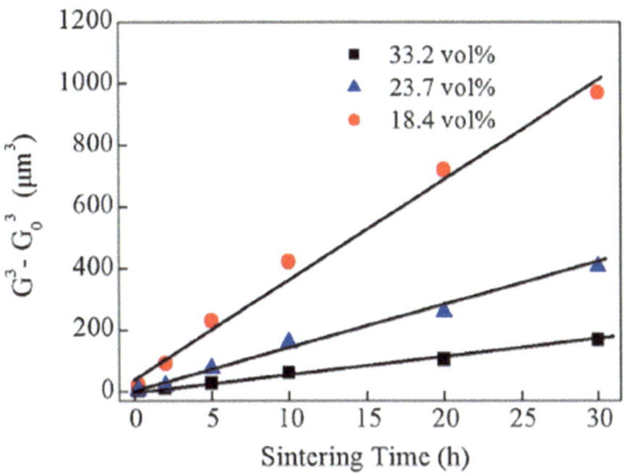

Fig. 3.23 G^3–G_0^3 vs sintering time for the ZrO_2-doped Ho:Y_2O_3 ceramics sintered at 1800 °C with different slurry concentrations. Reproduced with permission from [101]. Copyright © 2015, Elsevier

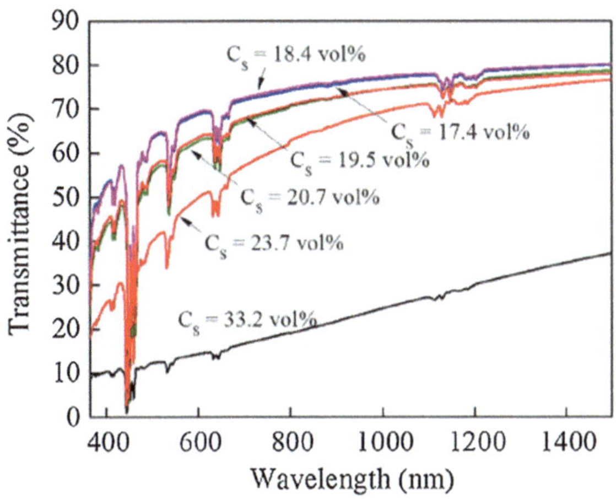

Fig. 3.24 In-line optical transmittance spectra of the ZrO_2-doped Ho:Y_2O_3 ceramics (2.14 mm thick) from the powders with different slurry concentrations. Reproduced with permission from [101]. Copyright © 2015, Elsevier

3.2.2.1 Chemical Precipitation/Co-precipitation

Mechanisms of Precipitation

The kinetics and mechanism of precipitation have been well studied and widely available in the open literature [8, 22, 23, 102–107]. Precipitation from solution involves two basic steps: (i) nucleation of fine particles and (ii) growth of the nuclei. The control of the powder characteristics is realized by controlling the reaction conditions for both nucleation and growth, as well as the extent of coupling between the two processes.

Nucleation Stage

The type of nucleation for precipitation from a solution belongs to homogeneous nucleation, because it occurs in an essentially homogeneous phase, with no foreign inclusions in the solution or on the walls of the reaction vessel. If these inclusions are present and assist the nucleation, the process is then called heterogeneous nucleation. Once heterogeneous nucleation is present, it is difficult to control the particle sizes. Therefore, heterogeneous nucleation should be avoided. However, heterogeneous nucleation can be applied to synthesize particles with desirable coatings.

The classical theories for vapor-to-liquid and vapor-to-solid transformations have been successfully used to describe the homogeneous nucleation of solid particles in solutions [108, 109]. For such a purpose, it is necessary to highlight the main features of the classical theories for vapor-to-liquid transformation. When there are random thermal fluctuations applied to a supersaturated vapor consisting of atoms or molecules, the density and free energy of the system will be fluctuated locally. Density fluctuations create clusters of atoms known as embryos, which can grow by absorbing atoms from the vapor phase. The embryos have a range of sizes in the vapor with vapor pressures that can be well described by the following Kelvin equation [110]:

$$\ln\left(\frac{p}{p_0}\right) = \frac{2\gamma v_l}{kTr}, \tag{3.19}$$

where p is the supersaturated vapor pressure, p_0 is the saturated vapor pressure, γ is the specific surface energy of the cluster, v_l is the volume of the molecule in a liquid drop that is formed by condensation of the vapor, k is the Boltzmann constant, T is the absolute temperature, and r is the radius of the embryo, which is assumed to be spherical. Due to the higher vapor pressures, small embryos evaporate back to the vapor phase. In this case, there is a critical radius r_c for the embryos. Those with radius smaller than this critical value cannot grow, while those with $r > r_c$ can grow into larger and larger ones. The formation of nuclei should overcome an energy barrier, which can be represented by the free energy change, due to the formation of a spherical nucleus with a radius r. The change in Gibbs free energy can be written as:

$$\Delta G_n = 4\pi r^2\gamma - \frac{4}{3}\pi r^3 \Delta G_v. \tag{3.20}$$

The first term on the right-hand side is the intrinsically positive contribution of the surface free energy, while the second term is due to the contribution by the bulk free energy change. With unit volume of the liquid, the free energy change G_v due to the transfer from vapor to liquid is given by [108]:

$$\Delta G_{\mathrm{n}} = \frac{kT}{v_{\mathrm{l}}} \ln\left(\frac{p}{p_0}\right), \tag{3.21}$$

where v_{l} is the volume per molecule in the liquid. Substituting for G_v in Eq. (3.20) yields:

$$\Delta G_{\mathrm{n}} = 4\pi r^2 \gamma - \frac{4}{3}\pi r^3 \frac{kT}{v_l} \ln\left(\frac{p}{p_0}\right). \tag{3.22}$$

Accordingly, if the supersaturation ratio $S = p/p_0 = 1$, the Gibbs free energy change ΔG_{n} increases monotonically as a parabola, due to the disappearance of the bulk term. For $S < 1$, the ΔG_{n} curve increases more rapidly, because a fractional S makes the second term on the right-hand side to be positive, enhancing the effect, owing to presence of the surface free energy barrier. For $S > 1$, the second term is negative, so that there is a maximum in ΔG_{n} at some critical radius r_{c}, as shown schematically in Fig. 3.25. The critical radius r_{c} is obtained by setting $\mathrm{d}(\Delta G_{\mathrm{n}})/\mathrm{d}r = 0$, which gives rise to:

$$r_{\mathrm{c}} = \frac{2\gamma v_{\mathrm{l}}}{kT \ln(p/p_0)}. \tag{3.23}$$

Putting Eq. (3.23) into Eq. (3.22), the height of the free energy activation barrier can be derived, which is:

$$\Delta G_{\mathrm{c}} = \frac{16\pi\gamma^3 v_l^2}{3[kT \ln(p/p_0)]^2} = \frac{4}{3}\pi r_c^2 \gamma. \tag{3.24}$$

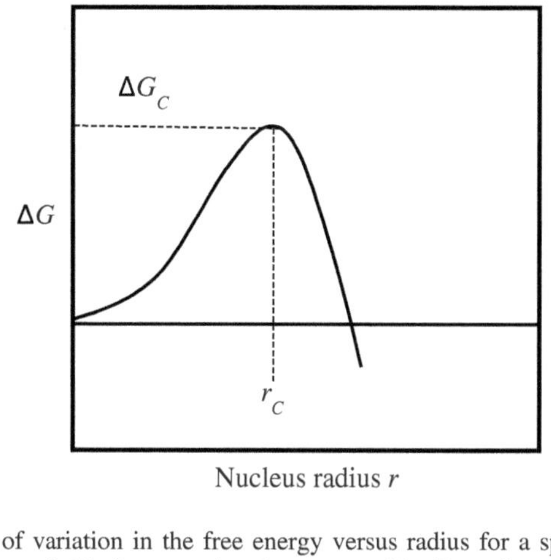

Fig. 3.25 Schematic of variation in the free energy versus radius for a spherical droplet. Some critical size must be exceeded before a nucleus becomes stable. Reproduced with permission from [111]. Copyright © 1950, American Chemical Society

Therefore, at this stage, sufficient increase in the supersaturation ratio $S = p/p_0$ results in an increase the atomic/molecular bombardment rate in the vapor, which leads to a reduction in ΔG_c and r_c, so that the probability of subcritical embryo growing to supercritical size in a short time approaches unity.

The nucleation rate I represents the rate of formation of critical nuclei, because only this kind of nuclei can grow into liquid droplets. According to the pseudo-thermodynamic treatment of vapor-to-liquid transformation, I is proportional to exp $(-\Delta G_c/kT)$, where k is the Boltzmann constant and ΔG_c is given by Eq. (3.24). The rate of the nuclei growth is also dependent on the frequency of atoms to join it, which can be written as $\nu\exp(\Delta G_m/kT)$, where ν is the characteristic frequency and G_m is the activation energy for atom migration. Putting $\nu = kT/h$, where h is Planck's constant, an approximate expression for the nucleation rate is given by [112]:

$$I \approx \frac{NkT}{h}\exp\left(\frac{-\Delta G_m}{kT}\right)\exp\left(\frac{-16\pi\gamma^3 v_l^2}{3kT[kT\ln(p/p_0)]^2}\right), \tag{3.25}$$

where N is the number of atoms per unit volume in the phase that is undergoing the transformation.

Homogeneous nucleation of particles from solution has been observed in various methods that are used to synthesize ceramic powders [113–116]. For example, in aqueous solutions, metal ions are usually hydrated [117, 118]. Embryos of hydrated metal ions are formed by progressive accumulation of ions to one another through a polymerization process. These polynuclear ions are the precursors for nucleation. When the concentration of the polynuclear ions is above some minimum supersaturation level, homogeneous nucleation starts to form solid nuclei. The nucleation rate of particles from solution can be expressed as [113]:

$$I \approx \frac{2N v_s (kT\gamma)^{1/2}}{h}\exp\left(\frac{-\Delta G_a}{kT}\right)\exp\left(\frac{-16\pi\gamma^3 v_l^2}{3k^3T^3[\ln(C_{ss}/C_s)]^2}\right), \tag{3.26}$$

where N is the number of ions per unit volume in the solution, v_s is the volume of a molecule in the solid phase, γ is the specific energy of the solid–liquid interface, ΔG_a is the activation energy for the transport of an ion to surface of the solid, C_{ss} is the supersaturated concentration, and C_s is the saturated concentration of the ions in the solution. The nucleation rate is strongly dependent on the supersaturation ratio C_{ss}/C_s.

Particle Growth Stage

Although nuclei are very small in size, they can grow at different rates and thus have different sizes, even during the very early stage of nucleation. Therefore, the starting point of growth of the nuclei into particles could take place at different times. Solute species, such as ions or molecules, are transported to the nuclei in a supersaturated solution, which accumulate at their surface and then are aligned with the nuclei to form particles. The rate-controlling step in the growth of the particles can be (i) diffusion toward the particle and (ii) addition of new material to the particle by surface reaction. The specific mechanisms and the coupling among them govern the final size and size distribution of the particles.

For diffusion-controlled growth, it is assumed that the nuclei are far away one another, so that each individual particle can grow at its own rate, the diffusion of solute species toward the particle can be described by Fick's first law. If the particles are assumed to be spherical with radius r, the flux J through any spherical shell of radius x is given by:

$$J = 4\pi x^2 D \frac{\mathrm{d}C}{\mathrm{d}x}, \tag{3.27}$$

where D is the diffusion coefficient for the solute through the solution and C is concentration of the solute. Assuming that the saturation concentration C_s is maintained at the particle surface and that the concentration of the solute far away from the particle is C, a concentration gradient is built up that approaches a stationary state in times of the order of r^2/D. In this stationary state, J is not dependent on x, and integration of Eq. (3.27) yields:

$$J = 4\pi r D(C_\infty - C_s). \tag{3.28}$$

The rate of increase in radius of the particle is then given by:

$$\frac{\mathrm{d}r}{\mathrm{d}t} = \frac{JV_s}{4\pi r^2} = \frac{DV_s(C_\infty - C_s)}{r}, \tag{3.29}$$

where V_s is the molar volume of the solid that is precipitating on the particle. Equation (3.28) can also be rewritten as:

$$\frac{\mathrm{d}(r^2)}{\mathrm{d}t} = 2DV_s(C_\infty - C_s). \tag{3.30}$$

This means that the square of the radius of all particles increases at the same constant rate, regardless the original size of the particles. Although the assumptions leading to Eq. (3.30) are oversimplified, it has been observed that if diffusion controlled growth is the governing mechanism, particles with different size have the same magnitude of $\mathrm{d}(r^2)/\mathrm{d}t$ [119].

If the absolute width of the particle size distribution is Δr for a mean radius r and Δr_0 for the mean radius r_0 of the system in the initial state, according to Eq. (3.30), there are:

$$\frac{\Delta r}{\Delta r_0} = \frac{r_0}{r}, \quad \frac{\Delta r}{r} = \left(\frac{r_0}{r}\right)^2 \frac{\Delta r_0}{r_0}. \tag{3.31}$$

According to Eq. (3.31), the absolute width of the size distribution becomes narrower in the ratio r_0/r and the relative width decreases even faster in the ratio $(r_0/r)^2$.

For surface-reaction-controlled growth, each new layer formed at surface of the particle has to be nucleated first by a process that is different from the homogeneous nucleation [120–124]. There are two types of growth mechanisms that have been observed: mononuclear growth and polynuclear growth. In the mononuclear growth mechanism, once a nucleation step is formed on the particle surface, there is sufficient time for the entire layer to form before a new step starts. In this case, the particles grow due to the increase in thickness of the layers on their surfaces. Due to the growth is realized layer by layer, the particle surface may have faceted appearance at the macroscopic scale. The particle growth can be described by the following equation:

$$\frac{dr}{dt} = K_1 r^2, \tag{3.32}$$

where K_1 is a constant. The relative width of the size distribution is given by:

$$\frac{\Delta r}{r} = \frac{r}{r_0} \frac{\Delta r_0}{r_0}, \tag{3.33}$$

which increases in the ratio r/r_0.

In the polynuclear growth mechanism, formation of nucleation steps on the particle surface is too fast, so that a new layer is created before the previous one has been completed. The growth rate is independent of the surface area of the particles, which is given by:

$$\frac{dr}{dt} = K_2, \tag{3.34}$$

where K_2 is a constant. In this case, the relative width of the distribution decreases according to the following equation:

$$\frac{\Delta r}{r} = \frac{r_0}{r} \frac{\Delta r_0}{r_0}. \tag{3.35}$$

Particle Size Distribution

The principles to obtain particles with a narrow size distribution by precipitation from solution have been well studied [20, 111, 125, 126]. The main features can be

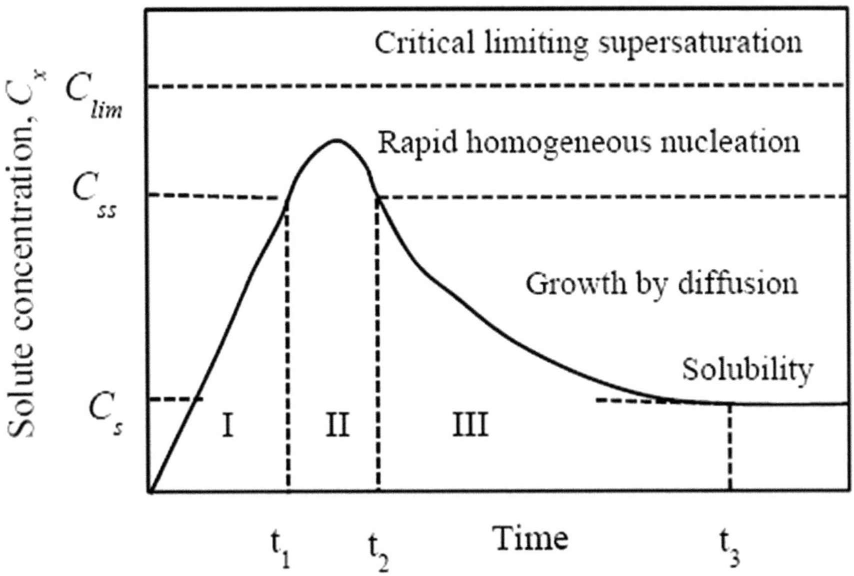

Fig. 3.26 Schematic representation of the solute concentration versus time in the nucleation and growth of particles from a solution. Reproduced with permission from [111]. Copyright © 1950, American Chemical Society

represented in the diagram shown in Fig. 3.26, which is often referred to as the LaMer diagram. With the progress of the reaction, the concentration of the solute to be precipitated, C_x, increases to or above the saturation value C_s. If the solution is assumed to be free of foreign inclusions and the container walls are clean and smooth, it is possible for C_x to largely exceed C_s, so as to form a supersaturated solution. Finally, a critical supersaturation concentration C_{ss} will be reached after some time t_1, where homogeneous nucleation and growth of solute particles will occur, leading to a decrease in C_x to a value below C_{ss} after a time t_2. Further growth of the particles continuously takes place due to diffusion of the solute through the liquid and then the precipitation onto surface of the particles. Eventually, particle growth stops after t_3 when $C_x = C_s$. To obtain particles with uniform size, the nucleation should be triggered in a short time interval, t_2–t_1. This could be achieved by using solutions with sufficiently low concentration of reactants. At the same time, the solute should be released slowly, to allow the diffusion mechanism to occur, so that solute will precipitate onto the particles without further nucleation.

Particle Aggregation

The particles synthesized by precipitation form solution could consist of aggregates of much finer primary particles [17–19, 127–132]. With DLVO theory for colloid stability, it is found that under identical surface charge densities, the barrier to aggregation increases with the size for two equal-sized particles, so that their rate of aggregation decreases exponentially. Also, fine particles aggregate more quickly with large particles than they do with themselves. According to the model, during a precipitation reaction, the first nuclei grow rapidly by aggregation to a size that is colloidally stable. These particles then sweep through the suspension, picking up freshly formed nuclei and smaller aggregates. The formation of particles with uniform size is thus possible through the size-dependent aggregation rates.

Ostwald Ripening

The growth of particles in liquids has also been observed, through a process called dissolution–precipitation, in which the smaller particles are dissolved and the solute precipitates on larger particles. This type of growth, also called coarsening, is known as Ostwald ripening [133–137]. The coarsening of precipitates in solid media can occur by a similar process. The theory of Ostwald ripening is also known as the Lifshitz–Slyozov–Wagner (LSW) theory. Matter transport from smaller particles to larger particles can be controlled by either (i) diffusion through the liquid or (ii) interface reaction. The average radius $\langle r \rangle$ of the particles assumed to be spherical is predicted to increase with time t, according to the following equation:

$$\langle r \rangle = \langle r_0 \rangle^m + Kt, \tag{3.36}$$

where $\langle r_0 \rangle$ is average radius of the particles at the initial state, K is a constant that obeys the Arrhenius relation and m is an exponent that is dependent on the mechanisms, with $m = 2$ for interface reaction control and $m = 3$ for diffusion control. With any initial particle size distribution, the final particle size distribution always approaches a self-similar distribution, because it is dependent only on $r/\langle r \rangle$ and independent on time. The maximum radius of the distribution is $2\langle r \rangle$ for the interface reaction mechanism and $(3/2)\langle r \rangle$ for the diffusion mechanism. Therefore, Ostwald ripening alone cannot be used to obtain monodispersed powders.

Chemical Precipitation

Various powders for transparent ceramics have been synthesized by using chemical precipitation methods [138–154]. Chemical precipitation starts when a metal salt solution is mixed with a precipitant solution. Metal salts that can be used to prepare solutions mainly include chlorides, nitrates sulfates and acetates, as well as alkoxides or other forms. The most widely used precipitants are ammonia water ($NH_3 \cdot H_2O$), ammonia carbonate ($(NH_4)_2CO_3$), ammonia bicarbonate (NH_4HCO_3), sodium carbonate (Na_2CO_3), potassium carbonate (K_2CO_3), sodium hydroxide ($NaOH$), potassium hydroxide (KOH), urea ($CO(NH_2)_2$), hydrazine (N_2H_4), ammonium oxalate ($(NH_4)_2C_2O_4$), ammonium citrate ($(NH_4)_2C_6H_6O_7$), and so on.

The mixing of the precipitated solutions and the precipitant solutions can be conducted in various ways. Therefore, chemical precipitation is a very feasible synthesis method, which can be carried out through various combinations of the ion salts and precipitants, as well as those processing parameters. However, only limited ways have been used in the open literature.

The products or precipitates are not oxides, instead, they are most likely hydroxides or carbonates, depending on the properties of the salts and the precipitants. The ionic reactions for precipitation of Mg^{2+} and Al^{3+} are as follows:

$$Mg^{2+} + 2OH^- \rightarrow Mg(OH)_2 \downarrow, \tag{3.37}$$

$$Al^{3+} + 3OH^- \rightarrow Al(OH)_3 \downarrow, \tag{3.38}$$

$$Mg^{2+} + CO_3^{2-} \rightarrow MgCO_3 \downarrow, \tag{3.39}$$

$$2Al^{3+} + 3CO_3^{2-} + 3H_2O \rightarrow 2Al(OH)_3 \downarrow + 3CO_2 \uparrow . \tag{3.40}$$

The precipitants are filtered, washed, and dried to make oxide powders. Oxide powders can be obtained by calcining the precipitants at a suitable temperature. Properties of the final oxide powders are determined by the properties of the precipitants, including particle size, size distribution, morphology, and so on. The properties of the precipitants can be controlled by various conditions, such as type of the metallic ions, concentration of the solutions, pH value, temperature, the way of mixing, and use of additives. Particle size of the derived powders sometimes is dependent on calcination temperature.

The precipitation of magnesium carbonate, from reaction of $MgCl_2$ with Na_2CO_3, with the presence of polyacrylamide (PAM), has been studied [155]. It was found that, various factors, such as such as PAM concentration, reaction temperature, and stirring speed, could affect the precipitation behavior of magnesium carbonate. The product weight increased slightly with increasing content of PAM, but decreased slightly with increasing temperature, while the stirring speed exhibited almost no effect. The precipitants obtained at 25 and 50 °C were nesquehonite, i.e., $MgCO_3 \cdot 3H_2O$. At 75 °C, the final product was an unidentified amorphous phase. Morphology of the magnesium carbonate was determined by the concentration of the polymer. Figure 3.27 shows SEM images of the precipitates, showing the effect of PAM content on the morphology of the precipitation products. With increasing PAM concentration, particle morphology of the precipitates was changed from plate block shape to fan-like shape. This effect of PAM was attributed to the adsorption of the polymer on surface of the precipitates during the precipitation. Obviously, $MgCO_3$ powders with such morphologies are not suitable to be used for fabrication of MgO transparent ceramics.

Another example is the synthesis of MgO nanoparticles from hydroxide precipitation in aqueous solutions [156]. The precipitation was conducted at a controlled temperature, while $MgCl_2$ solution at a concentration of 1 mol L^{-1} was mixed with NaOH solution with a concentration of 2 mol L^{-1}. The two solutions

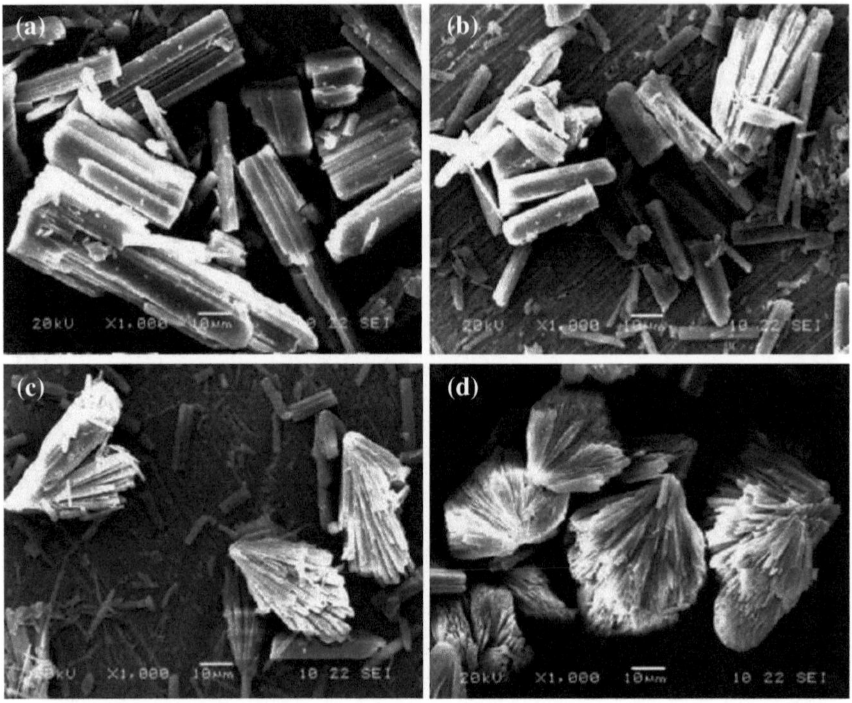

Fig. 3.27 SEM images of the powders made with different PAM concentrations (mg L^{-1}): **a** 0.2, **b** 0.4, **c** 0.6 and **d** 0.8. Reaction temperature was 25 °C, with stirring speed of 500 rpm. Reproduced with permission from [155]. Copyright © 2010, Elsevier

were mixed by simultaneously pouring into a 100 mL cylindrical container in a bath at 80 °C for 2 h. Vigorous stirring was applied during the mixing of the two solutions, while the stirring was continued during the aging of the precipitate. The suspension was aged at the reaction temperature for 2 h and then at room temperature for 1 day. The solid phase obtained was recovered by filtration, which was washed with deionized water and absolute alcohol and then air-dried at 60 °C for 4 h. The powders were finally subject to a hydrothermal treatment at 250 °C for 1 h, 370 °C for 2 h and 435 °C for 3 h, successively. Figure 3.28 shows SEM image of the MgO powder. Although the MgO powder was used for sorption of uranium, it is believed to be suitable as precursor powder transparent MgO ceramics.

There has been a report on the synthesis of nanocrystalline α-alumina powders, with a primary mean particle diameter of 10 nm, by using chemical precipitation from aluminum nitrate and ammonia solution [157]. The presence of ammonium nitrate, which was the by-product of the precipitation reaction in the $Al(OH)_3$ dry gel, was found to be able to decrease the formation temperatures of γ-, δ-, θ-, and α-Al_2O_3 during heating. With the presence of 5 wt% α-alumina seed crystals, with an average diameter of 100 nm, together with 44 % ammonium nitrate, the θ-Al_2O_3 → α-Al_2O_3 conversion temperature was decreased from 1200 to 900 °C.

Fig. 3.28 SEM image of the MgO powders. Reproduced with permission from [156]. Copyright © 2012, John Wiley & Sons

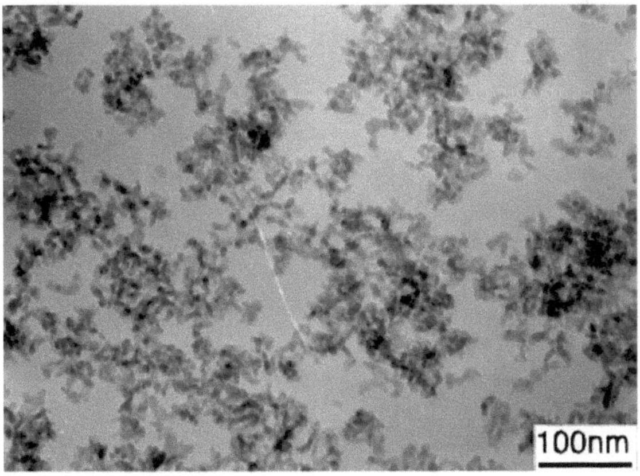

Fig. 3.29 TEM image of the α-Al_2O_3 powders before ball milling. Reproduced with permission from [157]. Copyright © 2000, Elsevier

The α-Al_2O_3 powder milled in anhydrous alcohol has an agglomeration strength of 76 MPa, which was soft agglomerate, whereas the sample milled in deionized water exhibited an agglomeration strength of 234 MPa, which belonged to hard agglomerate. Regardless to the state of agglomeration, the initial stage sintering of the two powders was controlled by grain boundary diffusion, with activation energies of 365 and 492 kJ mol^{-1}, respectively. The alumina ceramic produced by sintering the soft agglomerated powder at 1400 °C for 2 h showed a mean grain size of 0.93 μm, a mean flexural strength of 700 MPa, and a fracture toughness of 4.75 MPa $m^{1/2}$.

Figure 3.29 shows a TEM image of the as-synthesized α-Al_2O_3 powder, indicating that the diameter of the primary particles before milling is about 10 nm. Figure 3.30 shows SEM images of the α-Al_2O_3 powders milled in anhydrous

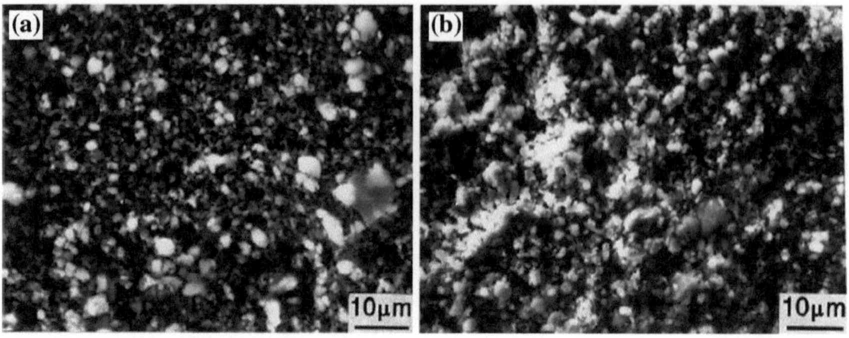

Fig. 3.30 SEM images of the α-Al_2O_3 powders milled in different solvents: **a** anhydrous alcohol and **b** deionized water. Reproduced with permission from [157]. Copyright © 2000, Elsevier

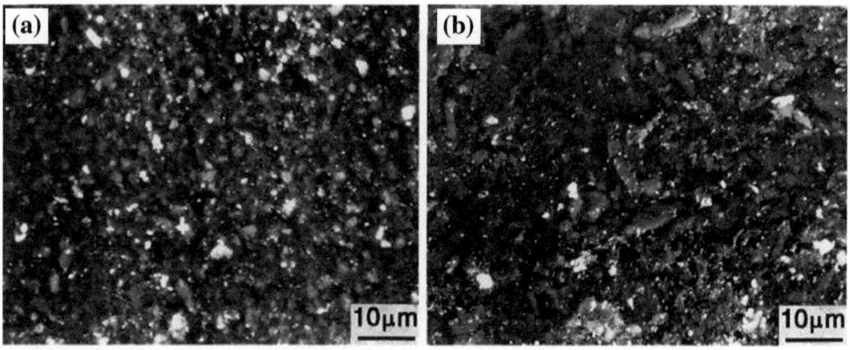

Fig. 3.31 SEM images of the green bodies from the α-Al_2O_3 powders with different agglomerations: **a** soft and **b** hard. Reproduced with permission from [157]. Copyright © 2000, Elsevier

alcohol and deionized water. After ball milling, both powders were agglomerated, with agglomerate diameters in the range from less than 0.1 μm to about 7 μm. For both powders, the green density versus compaction pressure curves consisted of two straight lines but with different slopes. By assuming that the change in slope is related to agglomeration strength, the agglomeration strengths of the two powders were estimated to be 76 and 234 MPa, for anhydrous alcohol and deionized water, respectively.

It has been accepted that the microstructural homogeneity of the green bodies is determined by the strength of agglomerates. Figure 3.31 shows SEM images of the green bodies from the two α-Al_2O_3 powders. Obviously, the sample from the powder milled in anhydrous alcohol more homogeneous than the one from the powder milled in deionized water. This is because the compaction pressure of ~200 MPa was higher than the agglomeration strength of the powder milled in anhydrous alcohol, so that the agglomerates could be broken. As stated above, this

Fig. 3.32 SEM images of the final ceramics sintered at 1400 °C for 2 h from the α-Al_2O_3 powders with different agglomerations: **a** soft and **b** hard. Reproduced with permission from [157]. Copyright © 2000, Elsevier

powder was soft agglomerated. In contrast, the powder milled in deionized water had hard agglomerates, which could not be crashed down. Relative densities of the green bodies for the soft and hard agglomerated powders were 36.62 and 28.94 %, respectively. The two green bodies showed different linear shrinkage behaviors, implying their different sintering behaviors due to the different agglomeration properties.

SEM images of the final ceramics derived from the two powders are shown in Fig. 3.32 [157]. The soft agglomerated powder led to a green body with homogeneous microstructure. The final ceramics reached a relative density of 99 %, with an average grain size of 0.93 μm. Due to the inhomogeneous green body, the sample with hard agglomeration experienced a differential sintering and a different internal stresses in the compact. Comparatively, the hard agglomerated powder could only be sintered to a relative density of 73.06 %. This study provided an important reference to the development of transparent ceramics.

Monodispersed porous α-Al_2O_3 nanoparticles with controllable sizes have been prepared by using a homogeneous precipitation method [158]. The precipitation solutions were prepared with $Al(NO_3)_3{\cdot}9H_2O$ and $Al_2(SO_4)_3{\cdot}18H_2O$ in distilled water. The Al^{3+} concentration in the solutions were 10 and 0.5 mmol L^{-1}. Precipitant was urea ($CO(NH_2)_2$), with a mole ratio of $CO(NH_2)_2/Al^{3+}$ to be 100. After vigorous stirring for 0.5 h, the reaction mixtures were heated to 95 °C at a constant heating rate of 2 °C min^{-1} and then kept at the temperature for 3 h for precipitation. The precipitates obtained were thoroughly washed and then dried at 60 °C. The dried precipitates were then calcined at different temperatures for 2 h at a heating rate of 5 °C min^{-1}. α-Al_2O_3 nanoparticles could be obtained after calcination at 1200 °C. Diameters of the porous α-Al_2O_3 nanoparticles obtained with an Al^{3+} concentration of 10 mmol L^{-1} were in the range of 200–500 nm, while diameters of those with an Al^{3+} concentration of 0.5 mmol L^{-1} were in the range of 40–70 nm. The porosity developed during the heating process was due to phase transformation and shrinkage of precursor particles. Figure 3.33 shows

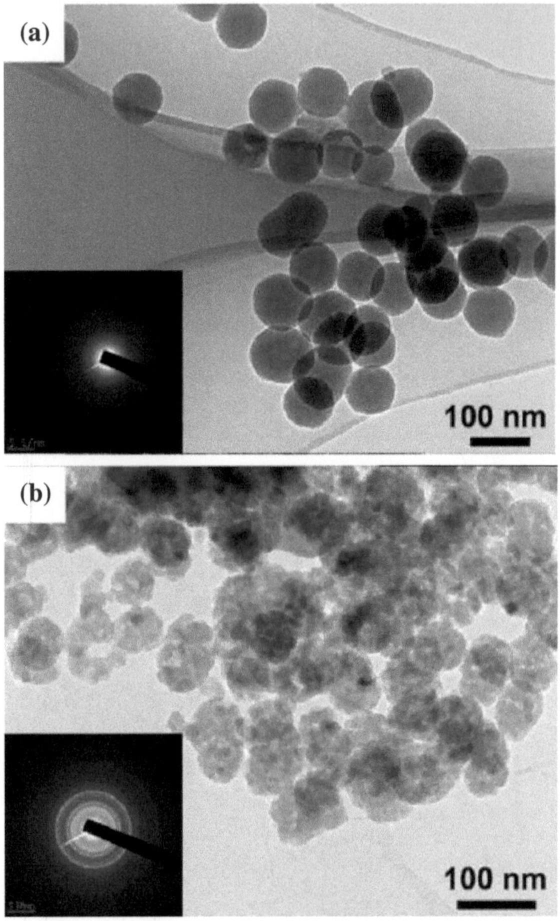

Fig. 3.33 TEM images of **a** amorphous $Al(OH)_3$ nanoparticles and **b** α-Al_2O_3 nanoparticles after a calcination at 1200 °C for 2 h. Reproduced with permission from [158]. Copyright © 2012, Elsevier

representative TEM images of the as-synthesized amorphous precipitate and nanoparticles after annealing at 1200 °C by using the precipitation method.

Ultrafine yttria particles with near-monodispersed size were synthesized solution by using a homogeneous precipitation method with urea as precipitant [159]. The effects of four experimental variables, including concentrations of yttrium ion and urea, reaction temperature, and solution pH value, on morphology of the particles and the reaction kinetics, have been studied in the system. It was found that the concentration of yttrium ion, which was varied in the range of 0.005–0.04 M, showed a significant effect on the average size of the particles, which monotonically increased from 65 nm to over 220 nm. Moreover, with increasing concentration of yttrium ion, the size distribution was gradually broadened and the particles started to form agglomeration at concentrations of 0.025 M and above, which was attributed to the reduction of zeta potential of the solutions. However, the yttrium ion concentration had no obvious effect on the rate of precipitation reaction. TEM

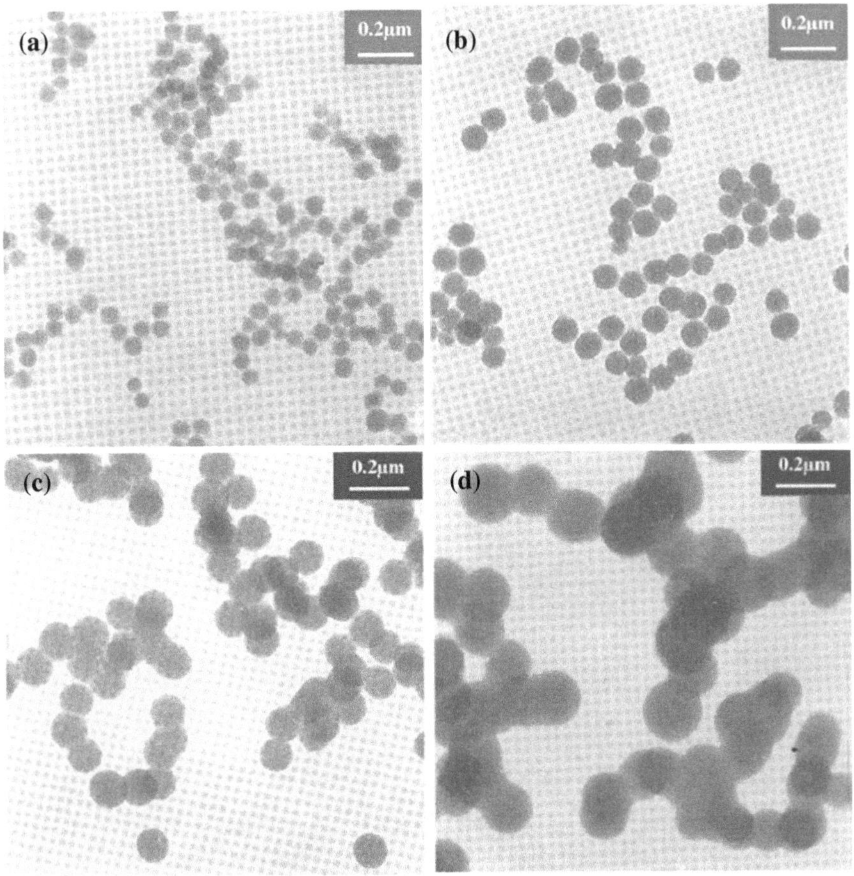

Fig. 3.34 TEM images of the precipitates from the solutions with different $Y(NO_3)_3$ concentrations: **a** 0.005 M, **b** 0.015 M, **c** 0.025 M and **d** 0.04 M. The urea concentration and pH were fixed at 1.0 and 6.0, while the reaction temperature was 95 ± 1 °C. Reproduced with permission from [159]. Copyright © 2004, Elsevier

images of representative precipitates from the solutions with different yttrium ion concentrations are shown in Fig. 3.34.

The effect of the urea concentration is somehow opposite to that of the yttrium ion concentration. As the urea concentration was increased from 0.04 to 4.0 M, the average particle size of the precipitates gradually decreased from 220 to 100 nm. However, too high urea concentration, e.g., 7.0 M, led to serious interparticle agglomeration. The rate of precipitation was increased with increasing urea concentration up to 3.0 M. Above 3.0 M, the effect of the urea concentration was weakened. Reaction temperature only affected the kinetics of precipitation. Figure 3.35 shows TEM images of representative precipitates derived from the solutions with different concentrations of urea [159].

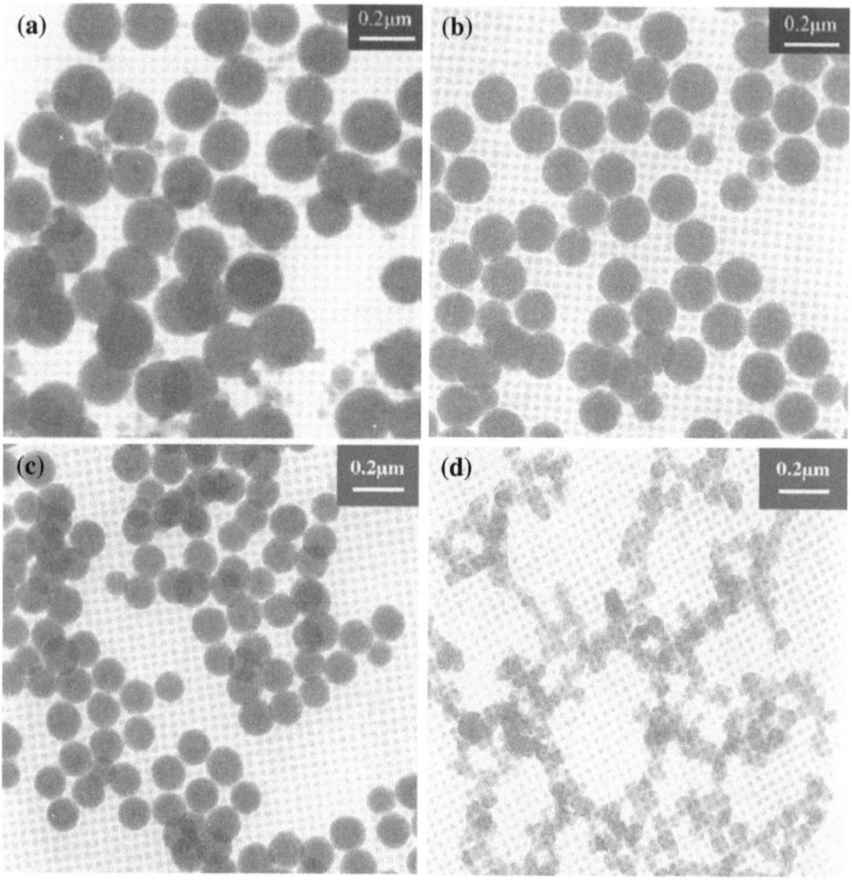

Fig. 3.35 TEM micrographs of yttrium compound prepared under different urea concentrations as **a** 0.04 M, **b** 0.5 M, **c** 2.0 M and **d** 7.0 M. The yttrium concentration and pH value were fixed at 0.02 and 6.0, while the reaction temperature was 95 ± 1 °C. Reproduced with permission from [159]. Copyright © 2004, Elsevier

Arrhenius-type reaction kinetics was used to estimate the activation energy of the precipitation reaction, resulting in a value of about 29 kcal mol^{-1}, which is comparable with the activation energies of urea decomposition, i.e., 28–32 kcal mol^{-1}. Therefore, the urea decomposition was a rate-determining step in the formation of the yttria particles for the homogeneous precipitation method. Particle morphology and the reaction kinetics were also influenced by pH value of the solutions. Low pH values, e.g., <2.0, particles showed severe agglomeration and the rate of precipitation reaction was slow. As the pH value was higher 3.0, near-monodisperse yttria particles could be achieved [159].

Complex oxides, such as YAG, spinel, mullite and ferroelectrics, contain more than one type of metallic element in their chemical formula [140, 160–175]. In this case, chemical precipitation is also called coprecipitation. Chemical coprecipitation is also applicable to simple oxides with various dopants [138, 139, 142, 144, 145, 176, 177]. The solutions for coprecipitation can be formed with mixed salts (including chlorides, nitrates and acetates), mixed alkoxides, or combinations of salts and alkoxides. The types of the anions can be different, but in most cases, they are same. The co-precipitated precursors have high level homogeneity and thus can be converted to complex oxide powders at temperatures that are much lower than that required by the conventional solid-state reaction method. One of the problems encountered in coprecipitation is the difference in hydrolysis rate of the different reactants in the solution, which leads to segregation of the precipitated materials. However, this problem can be readily addressed by controlling the conditions to achieve homogeneous precipitation.

Monodispersed yttrium aluminum garnet (YAG) powders have been synthesized by using a homogeneous precipitation method, from mixed solutions of yttrium nitrate, aluminum nitrate and ammonium aluminum sulfate, with urea as precipitant [15]. The focus of this study was on the effect of the molar ratio of aluminum precursors, i.e., aluminum nitrate to ammonium aluminum sulfate. It was found this molar ratio showed a significant effect on morphology and particle size of the precipitated powders. Spherical precipitate particles could be obtained when the molar ratio was 1:1. During the homogeneous precipitation process, aluminum ions were precipitated first in the form of monodispersed spherical particles. After that, yttrium ions were precipitated on the surface of the aluminum sphere particles. Monodispersed spherical YAG particles with a diameter of 500 nm were obtained by calcining the precipitated powder at 1100 °C for 5 h. The monodispersed spherical Nd:YAG powders showed high sinterability. Transparent YAG ceramics have been obtained after vacuum sintering at 1650 °C for 3.5 h. Figure 3.36 shows SEM images of the precipitates from the solutions with different molar ratios of aluminum nitrate to ammonium aluminum sulfate.

Hydrothermal Synthesis

Precipitation from solution under hydrothermal conditions has been widely used for the synthesis of fine crystalline particles of various oxides [178–180]. The process involves heating reactants, such as metal salts, oxide, hydroxide, or even metal powder, in the form of solution or suspension, at certain temperatures. Water is the most widely used solvent. In this case, the precipitation temperatures are set between the boiling and critical points of water, i.e., 100–374 °C, while the pressures are up to 22.1 MPa, which is the vapor pressure of water at the critical point. Due to the presence of high pressures, hardened steel autoclaves are usually used to carry out hydrothermal reactions. The autoclaves have inner surfaces of which are lined with a plastic, such as Teflon, to prevent corrosion of the vessels. Similar to chemical precipitation method, hydrothermal synthesis also offers almost unlimited flexibility in combination of types and concentrations of starting reactants, additives, pH levels, temperatures, time durations, and so on.

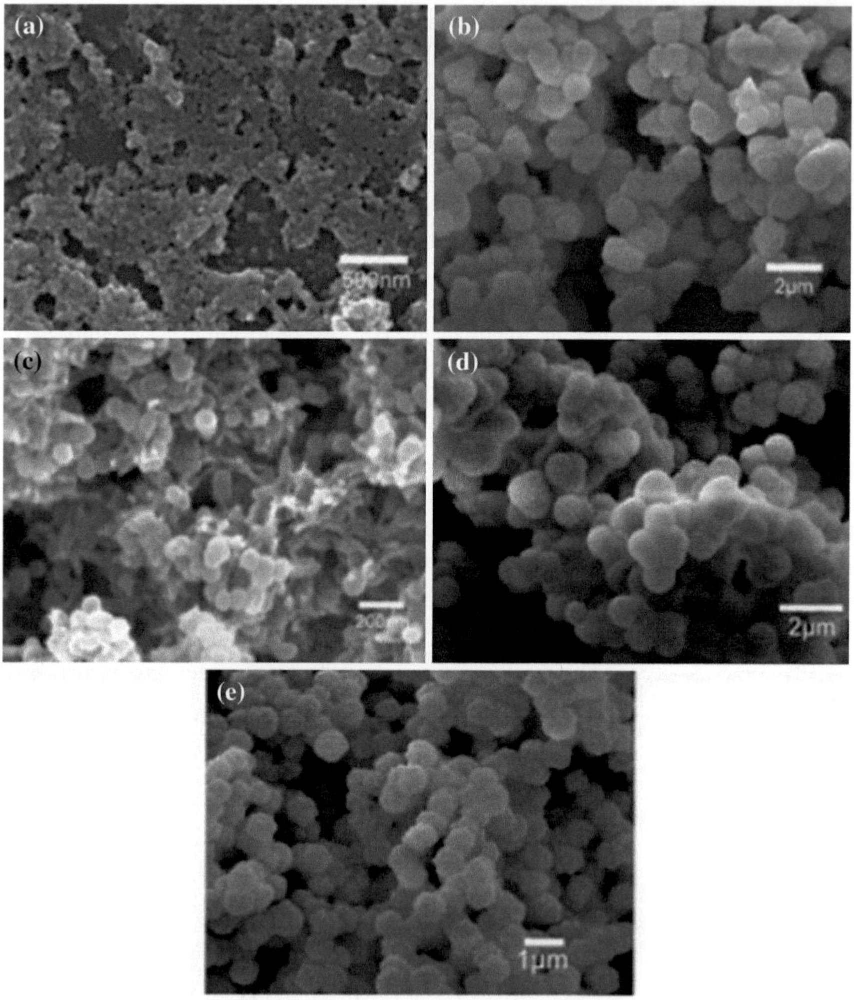

Fig. 3.36 SEM images of the precipitates from the solutions with different molar ratios of aluminum nitrate and ammonium aluminum sulfate: **a** 0.005:0, **b** 0:0.005, **c** 0.0033:0.0017, **d** 0.0017:0.0033 and **e** 0.0025:0.0025. Reproduced with permission from [15]. Copyright © 2012, John Wiley & Sons

Hydrothermal synthesis of $BaTiO_3$ powders has been extensively studied [181–186]. One example is to use the reaction between TiO_2 gels or fine anatase particles and $Ba(OH)_2$ in a strongly alkaline solution (pH 12–13) at 150–200 °C. The reaction mechanisms have been discussed in detail [187]. Depending on the reaction time and temperature, particles with average sizes in the range of 50–200 nm can be obtained. An alternative method is the crystallization of an amorphous gel of barium titanium acetate in a strongly alkaline solution of tetramethyl-ammonium hydroxide at 150 °C for 10–15 h. The process involves dissolution of the gel and precipitation of crystalline $BaTiO_3$ particles, coupled with Ostwald ripening of the particles, leading to $BaTiO_3$ with weak agglomeration and average particle sizes of 200–300 nm.

Powders synthesized by hydrothermal synthesis are crystallized phases, so that the calcination step is not required, which is different from the normal solution precipitation. The powders also have very fine particle/grain size (10–100 nm), narrow size distribution, single crystal particles, high purity, and good chemical homogeneity. However, the application of hydrothermal precipitation to synthesize transparent ceramic powders is still not very popular, which deserves further exploration. Several examples are presented as follows [188–192].

One example of hydrothermal synthesis is the preparation of Al_2O_3 powders [147, 188]. α-Al_2O_3 powders and α-Al_2O_3/γ-AlOOH mixtures with controlled contents of γ-AlOOH have been synthesized by using a hydrothermal synthesis [188]. This is a very comprehensive study. First, the hydrothermal reactions were conducted under alkaline or acidic conditions at 380–435 °C for 1–10 days, at pressures of 6.9–14.5 MPa, with concentrated precursors without stirring. Second, the precursors were formed by mixing various types of aluminum hydroxides with water, and optionally with α-Al_2O_3 seeds, hydrogen peroxide, sulfuric acid, dopants (i.e., $KMnO_4$), and/or other additives. Third, all the experiments were performed at industrial scale by using large production autoclaves.

Furthermore, the synthesized α-Al_2O_3 powders possessed up to 100 % phase purity, 99.98 % chemical purity, equiaxed morphology, low aggregation levels, narrow crystallite size distributions, with primary particle sizes ranging between 100 and 40 μm, as well as high reproducibility. It was clarified that, various parameters, including types of precursors, the use and properties of seeds, chemical additives, as well as reaction temperature and time duration of the hydrothermal synthesis, could be used to govern the properties of the final powders.

Different growth mechanisms for the growth of nanosized and rough powders were observed. The rough α-Al_2O_3 was developed by vapor phase transport and/or surface diffusion from the γ-AlOOH intermediate particles, while the submicrometer α-Al_2O_3 was formed by assembly of nanosized building blocks, resulting in crystallographically ideal mesocrystals. Figure 3.37 shows SEM images of representative samples derived from the hydrothermal synthesis under different conditions. It is believed that α-Al_2O_3 powders prepared through optimization of the hydrothermal synthesis should be suitable for fabrication of high quality transparent alumina ceramics.

Hydrothermal synthesis has also been used to prepare rare earth oxides, (Y_2O_3–Gd_2O_3):Eu^{3+} and Y_2O_3:(Yb^{3+}/Er^{3+}), which were used for phosphors applications [189]. The simple hydrothermal method can be used to control the 1D and 2D nanostructures of the oxides. The reaction was a conversion of nitrates mixture into carbonate hydrate phase, with the presence of ammonium hydrogen carbonate solution, at a reaction temperature of 200 °C for 3 h. Morphological architectures of the rare earth oxides obtained were controlled by the thermal treatment at

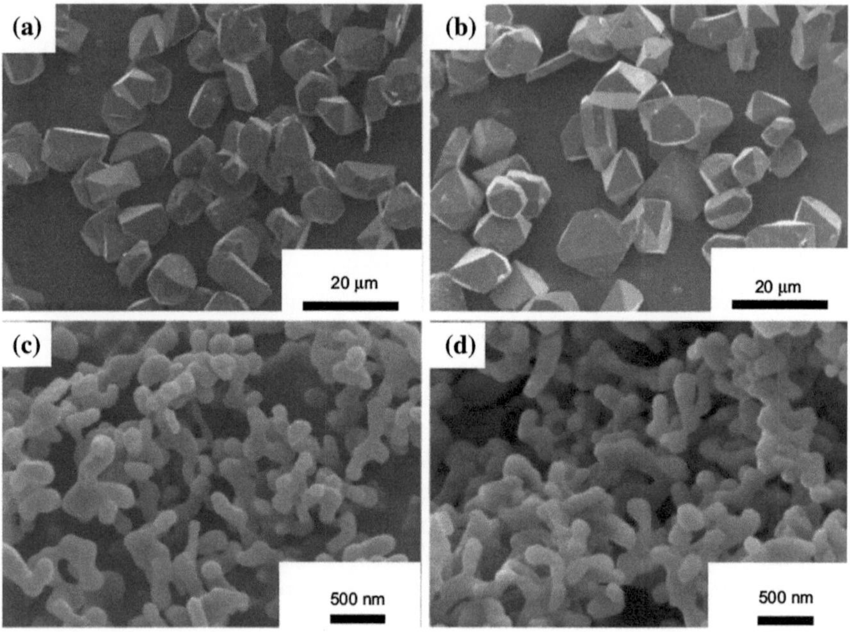

Fig. 3.37 SEM images of the hydrothermally synthesized phase-pure α-Al_2O_3 powders: **a** undoped 10 mm (Exp. No. 3), **b** Mn-doped 10 mm (Exp. No. 6), **c** undoped 250 nm (Exp. No. 9), **d** Mn-doped 250 nm (Exp. No. 11). All powders were in as-synthesized form. The Exp. Nos. can be found in the original reference. Reproduced with permission from [188]. Copyright © 2010, John Wiley & Sons

temperatures of 600–1100 °C and time durations of 3–12 h. The obtained powders exhibited high purity and homogeneous dopants distribution.

Another example is the synthesis of YAG:Ce (5 %) phosphors [190]. The hydrothermal reaction was carried out at 360 °C for various time durations, 6–26 h. Single-phase YAG:Ce (5 %) was obtained. The powder with reaction time of 22 h was treated by using spark plasma sintering (SPS) in vacuum and in air. No attempt was made to prepare transparent ceramics with the hydrothermally synthesized powders. However, these data could be used as a reference, if one wishes to employ hydrothermal synthesis to produce powders for transparent ceramics.

Core–Shell Structures

Coated particles, known as composite particles, are comprised of particles of one solid that are coated uniformly with a layer of another material, which have varying thickness, depending on composition requirement and other properties of the coated and coating phases [193–196]. Coated particles have several interesting advantages in terms of ceramic powder processing. For example, the presence of thin coatings

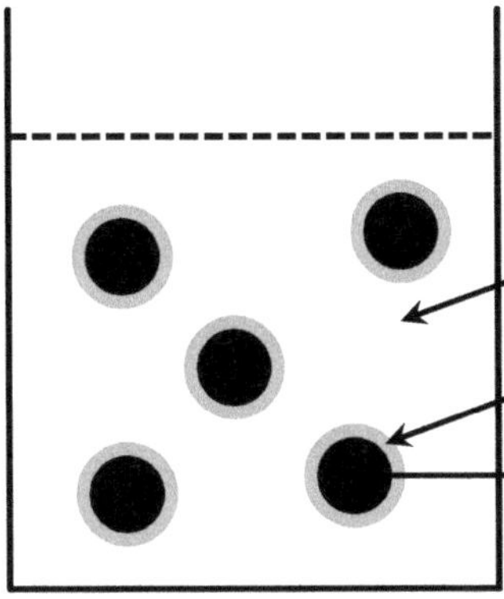

Fig. 3.38 Schematic of the preparation of coated particles by the precipitation from solution onto dispersed particles. Produced with permission from [5]. Copyright © 2003, CRC Press

is able to modify surface characteristics of colloidal dispersions, which allow for uniform incorporation of additives such as sintering aids and dopants. Particles coated with thick layers have significantly enhanced sintering behavior. This is particularly useful to fabrication of ceramic composites and synthesis of complex oxides.

There are several techniques that have been used to prepare coated particles, with a schematic shown in Fig. 3.38 [5]. Desirable coatings on particles can be achieved by controlling several parameters to facilitate proper interaction between the particulate suspension (A) and the material (B) that is to be precipitated on it from the solution [197].

Four types of A–B interactions could be observed in experiments. (i) B nuclei are formed homogeneously in the solution and then grow to form particles that have no interaction with A, which thus leads to particles of simple mixtures of A and B. (ii) Particles of B are homogeneously nucleated, which have weak interaction with particles of A and then deposit on A particles, but with rough and nonuniform morphology, especially when the B particles have large sizes. (iii) Homogeneously nucleated particles of B have weak interaction with particles of A at an early stage to form aggregates, on which the particles of B continuously grow, thus producing a particulate coating of B on A. The coating could be more homogeneous than that in case (ii), especially when the B particles are much smaller than the A particles. (iv) B particles are nucleated heterogeneously on surface of A particles, so that a uniform layer of B on A due to the growth of the B nuclei. This is the most desirable case, because smooth and uniform coatings can be formed.

There are several requirements that should be satisfied, in order to produce coated particles with uniform and smooth coating layers in the above case (iv). First, the nucleation and growth steps must be separated. Figure 3.39 shows a modified LaMer diagram for homogeneous precipitation of monodispersed particles [5]. The curve "a" in Fig. 3.39 is the case of a single burst of homogeneous nucleation followed by growth. When the A particles are present in the solution, heterogeneous nucleation of B can be triggered on their surfaces, when the solute concentration reaches C_h, which is the critical concentration for heterogeneous

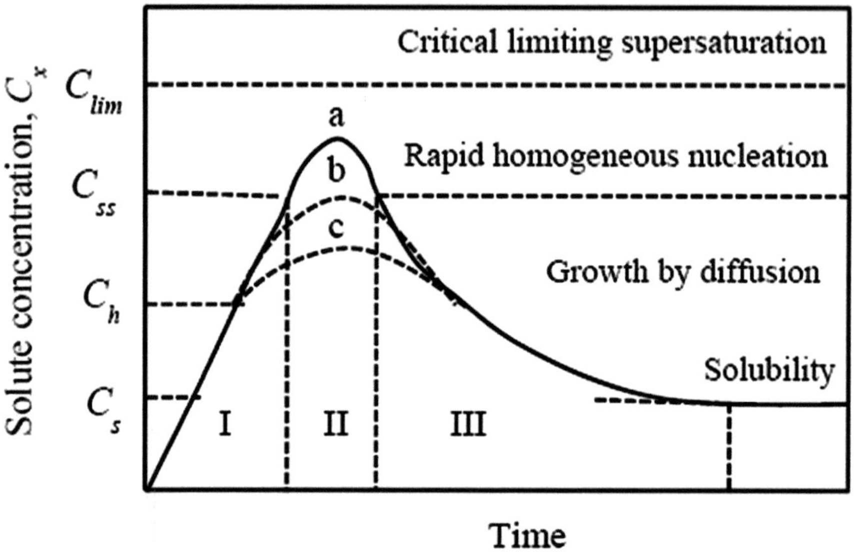

Fig. 3.39 Modified LaMer diagram for the formation of coated particles by precipitation from solution. Produced with permission from [5]. Copyright © 2003, CRC Press

nucleation. To produce coated particles with uniform coating layers, the presence of one rapid burst of heterogeneous nucleation, i.e., curve "c" in the figure, without reaching C_{ss} is a critical requirement.

Second, colloidal stability of the dispersion is important. During the nucleation and growth, the dispersion must be stable against flocculation and settling, in order to obtain well-dispersed and uniformly coated particles. The newly formed surface layers during this stage are likely to act as binder to form agglomerates, which once formed will have destroy the stability of the dispersion system.

Third, surface area of the core particles must be sufficiently large, so that the concentration of the solute will never reach C_{ss}. Otherwise, free particles of B will be precipitated, besides the coated particles, which is corresponding to curve "b" in Fig. 3.39. The surface area of the core particles has a direct influence on the rate of generation of the solute (r_g) by the reaction (nucleation) and the rate of removal of the solute (r_r) by precipitation/coating. The minimum surface area of the core particles that is available for deposition, A_{min}, is closely related to the maximum concentration of the solute, C_{max}. For a given r_g and assuming that the suspensions are sufficiently concentrated so that the interface reaction is rate-controlling step, A_{min} is defined as:

$$r_g = KA_{min}(C_{ss} - C_s), \quad (3.41)$$

where K is a constant and $C_{max} = C_{ss}$. The maximum surface area for deposition, A_{max}, means that the concentration of the solute should exceed C_h, otherwise only partial coating will be realized. As a result, the following equation must also be satisfied:

$$r_g = KA_{max}(C_h - C_s). \tag{3.42}$$

The maximum production rate and the latitude of the experimental conditions are then related to the maximum value of the ratio A_{max}/A_{min}. By equating Eqs. (3.41) and (3.42), there is:

$$\frac{A_{max}}{A_{min}} = 1 + \frac{C_{ss} - C_h}{C_h - C_s}. \tag{3.43}$$

In order to optimize the conditions, so that coating is formed in suspensions with high particle concentration, high production rates, and ease of processing, C_h should be as close as possible to C_s and C_{ss} and C_h should be separated as far as possible. This is because (i) C_h is close to C_s so that heterogeneous nucleation can start immediately after C_s is passed and (ii) C_{ss} is much greater than C_s so that homogeneous precipitation is far remote away from the onset of heterogeneous precipitation.

In practice, A_{min} is experimentally determined for a given r_g and for particles with a known size by decreasing the concentration of the particles in the suspension until precipitation occurs. For a low A_{min}, coating of the particles in a suspension should be highly possible, because $C_{ss} - C_s$ is relatively large, according to Eq. (3.41). If the conditions to prevent homogeneous precipitation cannot be achieves, it is possible to use catalyst to pretreat the surface of the particles. Various coated particles by using precipitation from solution have been reported in the literature [195, 198–205].

The degree of crystallinity of the deposited material can have a remarkable effect on morphology of the coating layers. In theory, the coatings can be amorphous, polycrystalline or even single crystal. Generally, smooth and uniform coatings are achieved more easily for amorphous coatings [194, 195], while polycrystalline deposits lead to rough layers [206]. However, even for amorphous layers, the morphology of the coatings can be controlled through adjusting experimental conditions. For example, when SiO_2 was coated on $Y(OH)CO_3$ particles at room temperature and at 80 °C, the coating was smooth and rough, respectively [207]. It means that precipitation temperature can be used as a factor that can affect morphology of the coatings.

Coated particle technique has been well employed to synthesize powders for transparent ceramics, especially for YAG. YAG powder retaining the morphology of Al_2O_3 powder has been synthesized by using a partial wet-chemical process, in order to form yttrium precipitate coated Al_2O_3 particles [208]. The formation of the so-called core–shell structure had two steps, including (i) direct precipitation of yttrium component at the surface of the Al_2O_3 particles and (ii) assembly of the yttrium precipitate. A spherical surface reaction process was illustrated. YAG phase can be obtained at a temperature that was about 300 °C lower than those required by the traditional solid-state reaction process.

Y-containing colloidal precursor particles were first synthesized. $Y(NO_3)_3{\cdot}6H_2O$ (99.99 %) and urea were dissolved in distilled water to make a transparent solution, with Y^{3+} concentration of 0.015 M and urea concentration of 0.5 M. α-Al_2O_3 (99.99 %) powder was then added into the mixed solution, which was dispersed by using ultrasonication. The mixed turbid liquid with Y/Al = 3/5 was homogenized by agitation, which was then heated to 90 °C and kept at that temperature for 3 h. After cooling down, the resulting precursor was collected with suction filtration and thorough washing. After rinsing with anhydrous ethanol, the powder was dried at 100 °C for 24 h and then calcined at various temperatures for 3 h to form YAG.

Morphology of the α-Al_2O_3 powder and the as-obtained precipitated precursor were observed by suing TEM, as shown in Fig. 3.40 [208]. The pure α-Al_2O_3

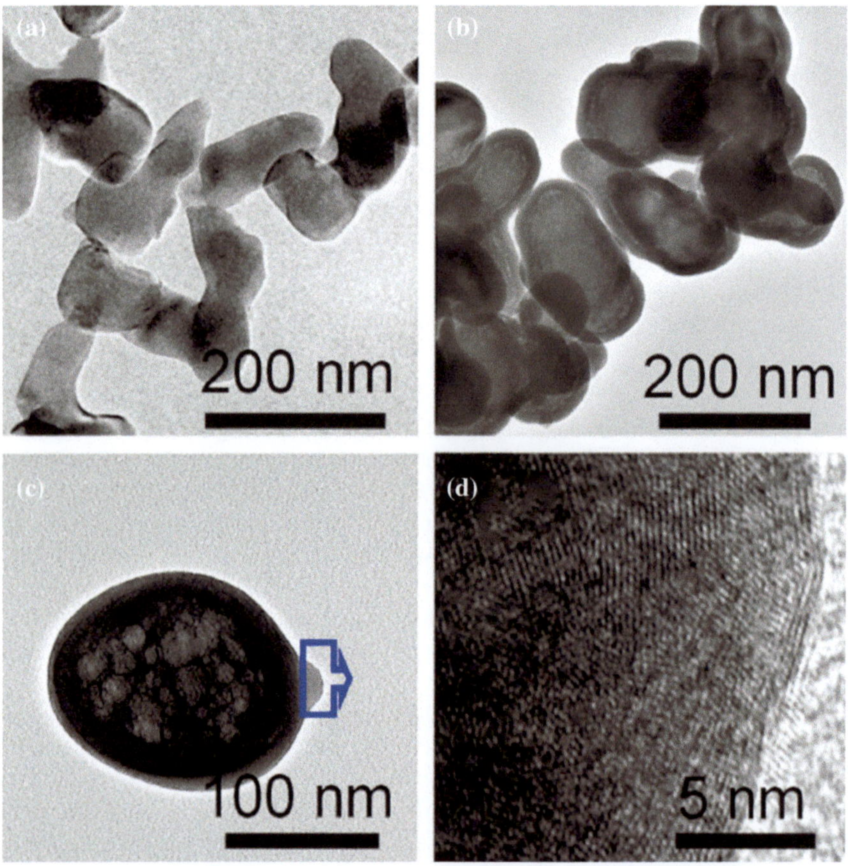

Fig. 3.40 TEM images of the precursors: **a** pure α-Al_2O_3, **b** Y-precipitate/Al_2O_3 with low electron exposure, **c** Y-precipitate/Al_2O_3 with high electron exposure and **d** the lattice image of the shell. Produced with permission from [208]. Copyright © 2013, Elsevier

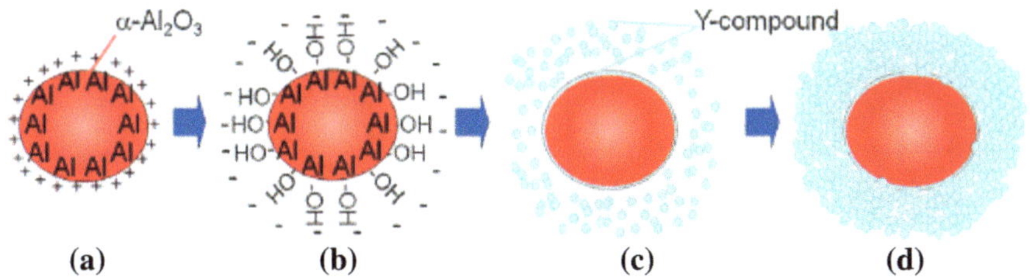

Fig. 3.41 Mechanism of the formation of the Y-precipitate/Al_2O_3 precursor in the urea solution. Produced with permission from [208]. Copyright © 2013, Elsevier

powder consisted of homogeneous particles with sizes of 100–200 nm. The powder was well crystallized with continuous and clear edges, as shown in Fig. 3.40a. The as-obtained precursor particles exhibited obvious interfaces between the core and the shell, as shown Fig. 3.40b. The Y-precipitate could be $Y(OH)CO_3$ or similar carbonate, which would decompose under the irradiation of the electrons. Figure 3.40d indicates that the compound of the shell had low crystallinity, consisting of tiny particles.

A potential mechanism for the formation process of the Y-precipitate/Al_2O_3 is shown in Fig. 3.41. Two stages were involved. At the initial stage, a thin layer of Y-compound was precipitated at the surface of the Al_2O_3 particles, so that the surface was modifies. At the second stage, explosive nucleation and assembly onto the Y-compound pre-coated Al_2O_3 particles occurred. In the presence of urea, there would be negative ions, such as OH^-, HCO_3^-, etc., which were adsorbed at the surface α-Al_2O_3, leading to negative zeta potential. The Y^{3+} ions were hydrated and adhered to the hydrolyzed Al_2O_3 surface to form yttrium hydroxide. Finally, a thin shell was formed.

Similarly, a mixed precipitation method was developed to synthesize powder for transparent YAG ceramics [172]. The YAG powders were obtained through calcining the precursors at 1050 °C for 2 h. Particle size of the powders was in the range of 200–500 nm. Yttrium nitrate hydrate ($Y(NO_3)_3{\cdot}H_2O$, purity >99.9 %), aluminum nitrate hydrate ($Al(NO_3)_3{\cdot}H_2O$, purity >99.9 %) and ammonium hydrogen carbonate (NH_4HCO_3, analytical reagent) were used as raw materials.

Flow chart of the mixed precipitation method is shown in Fig. 3.42. Briefly, alumina precursor powder was first synthesized by adding aluminum nitrate solution (0.15 M) dropwisely into the ammonium bicarbonate solution (1.5 M) under vigorous stirring. The as-prepared alumina precursor powder was then dispersed in yttrium nitrate solution to form uniform suspension. After that, ammonium bicarbonate solution (0.5 M) was added dropwisely into the suspension, so that yttrium precipitate was coated on the alumina precursor powder. The mixed precursor was aged, washed, dried, and then calcined to form yttrium aluminum garnet, with a phase formation temperature about 300 °C lower than that required by the traditional solid-state reaction process. Alternatively, it is also worth trying to coat yttria with alumina.

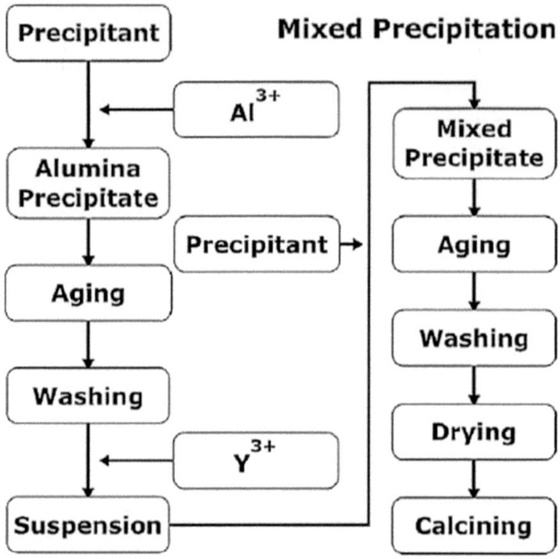

Fig. 3.42 Flow chart of the mixed precipitation process. Produced with permission from [172]. Copyright © 2012, Elsevier

3.2.2.2 Liquid Evaporation Methods

Evaporation of liquid is another method to make a solution to be supersaturated, so as to produce particles through the nucleation and growth. To synthesize powders with fine particles, nucleation must be fast and the growth is sufficiently slow. To achieve this, the supersaturation state must be reached in a very short time, so that a large number of nuclei are formed almost at the same time, without growing. Therefore, if a solution is broken into very tiny droplets, the surface area over which evaporation takes place can be increased tremendously. For solutions of two or more salts, the differences in concentration and solubility must be taken into account. This is because evaporation of such liquid solutions will cause different rates of precipitation for different salts, which will result in segregation of the final powders. The formation of very small droplets can be used to address this problem, because there is no mass transfer between individual droplets. Moreover, for a given size, the lower the concentration of the solutions, the smaller the particle size will be in the final powders. This means that the degree of segregation can be further decreased by using diluted solutions. Various techniques have been used to produce ceramic powders through the evaporation of liquid solutions.

Spray Drying

In spray drying, a solution is broken up into fine droplets by a fluid atomizer and sprayed into a drying chamber, as shown schematically in Fig. 3.43 [47]. Moisture is evaporated at the contacts between the spray and drying medium that is usually

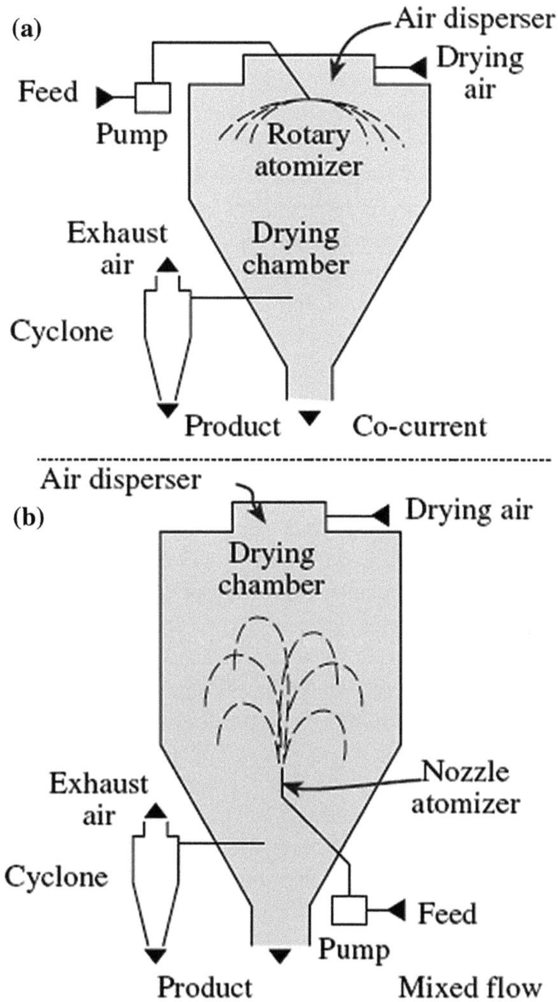

Fig. 3.43 Schematics of spray dryers: **a** centrifugal atomizer with concurrent air flow and **b** nozzle atomizer using mixed flow conditions. Reproduced with permission from [47]. Copyright © 2007, Springer

hot air. The products are dry particles of the metal salts, which are carried by the air stream to leave the chamber and are then collected by using bag collectors or cyclones. Various atomizers are available, which are usually categorized according to the manner in which energy is supplied to produce the droplets [209–211].

For example, in rotary atomization, also known as centrifugal atomization, the liquid is centrifugally accelerated to a high velocity through a spinning disk that is located at the top of the drying chamber, before it is discharged into the chamber. In pressure atomization, pressure nozzles atomize the solution by accelerating it through a large pressure difference and injecting the liquid into the chamber. Pneumatic atomization is to impact the solution by using a stream of high-speed gas from a nozzle. Ultrasonic atomization is to pass the solution over a piezoelectric vibrator that is vibrating at ultrasonic frequencies. Droplet sizes ranging from less than 10 μm to >100 μm can be produced by these atomizers.

Solutions for spray drying are usually aqueous solutions of metal salts, such as sulfates, chlorides and nitrides, due to their high water solubility. The evaporation rate of moisture from the droplet and the maximum temperature of generally >300 °C that the particles will experience in the drying chamber are determined by the temperature and flow pattern of the hot air, as well as the configuration of the chamber. Important solution parameters include the size of the droplets and the concentration and composition of the metal salts, which are dominant factors in controlling the size of the primary particles and the size and morphology of possible agglomerates. Morphology of the agglomerate is not very critical in spray drying of solutions, because the particle characteristics are mainly determined by the subsequent calcination and milling steps. With suitable conditions, the agglomerates are usually spherical, with primary particle sizes of <100 nm. The temperature in the drying chamber of spray is generally not sufficiently high to cause decomposition or solid-state chemical reaction, so spray-dried salt powders must be calcined to form designed compounds [211–215].

A simple strategy has been reported to prepare dense spherical Y_2O_3:Eu^{3+} phosphor particles with a narrow size distribution, by using a two-step spray drying, with a commercially available spray drying facility [216]. Hollow Y_2O_3:Eu^{3+} precursor particles were first prepared by spray drying an aqueous precursor solution containing citric acid. Nanosized particles were obtained from the precursor powders by using ball milling, which were then dispersed in water to form a colloidal suspension. The suspension was subjected to second-step spray drying, so that porous granules of dense spherical nanoparticles could be formed. Finally, highly crystalline Y_2O_3:Eu^{3+} powder was prepared by sintering the granules at temperatures of >1200 °C. The final Y_2O_3:Eu^{3+} particles were spherical and showed promising luminescence properties as a red phosphor. SEM images of representative samples are shown in Fig. 3.44.

Spray Pyrolysis

If a higher temperature and a reactive (e.g., oxidizing) atmosphere are used in the spray chamber, solutions of metal salts can be not only dried but also decomposed at the same time. This process is known as various names, such as spray pyrolysis, spray roasting, spray reaction and evaporative decomposition of solutions, among which spray pyrolysis is the most commonly used [217–223].

Figure 3.45 shows a schematic demonstrating the idealized stages involved in the formation of a dense particle from a droplet of solution [224, 225]. The droplet experiences evaporation, when the solute concentration in the outer layer increases to a value of above the supersaturation limit, thus resulting in the precipitation of fine particles. Precipitation is followed by a drying stage, in which the vapor phase diffuses through the pores in the precipitated layer. Decomposition of the

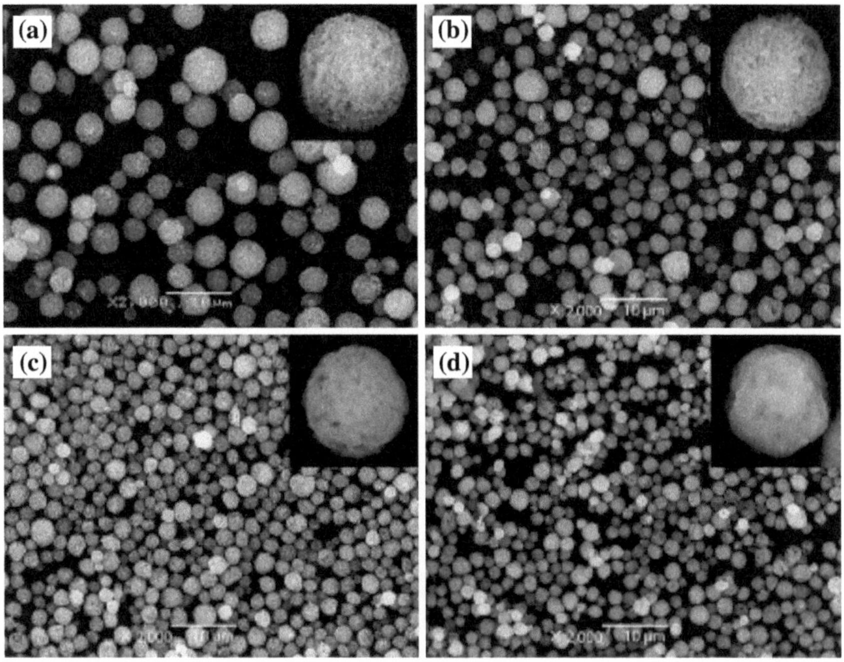

Fig. 3.44 SEM images of the **a** $Y_2O_3:Eu^{3+}$ powders prepared by second-step spray drying process, and post-treated $Y_2O_3:Eu^{3+}$ powders at the temperature of **b** 1200 °C, **c** 1300 °C, and **d** 1400 °C. Reproduced with permission from [216]. Copyright © 2014, The Royal Society of Chemistry

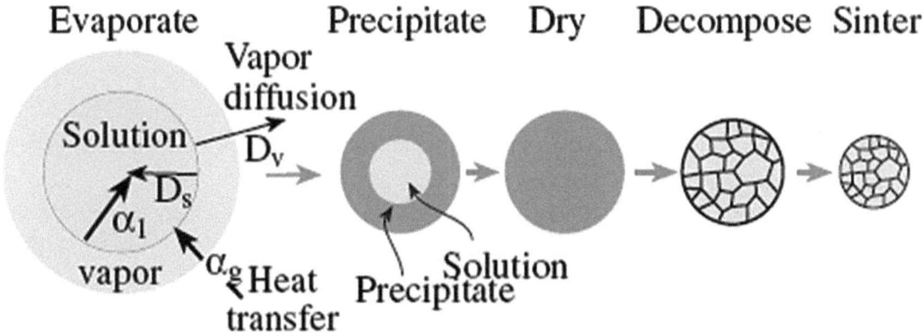

Fig. 3.45 Schematic of the stages in the spray pyrolysis process. Reproduced with permission from [224], Copyright © 1993, John Wiley & Sons. Reused with permission from [47]. Copyright © 2007, Springer

precipitated salts leads to a porous particle consisting of a certain number of very fine grains, which is finally heated to become a dense particle.

The spray pyrolysis process is able to produce particles with a variety of morphologies through controlling the experimental conditions. For the fabrication of transparent ceramics, dense particles are superior to those with highly porous or hollow shell-like morphologies, because a subsequent milling step is normally unnecessary. As a result, there will be less chance to introduce possible contamination.

The effects of the conditions on precipitation in the droplet and the solution chemistry on the morphology and microstructure of the final particles have been well investigated [224]. In order to obtain dense particles, the first step is to achieve homogeneous nucleation and growth in the droplet, which is known as volume precipitation. The prerequisite is a small droplet size and a slow drying process, so as to reduce the gradients in solute concentration and temperature. As mentioned earlier, a large difference between the supersaturation concentration C_{ss} and the saturation concentration C_s of the solute in solution means a high nucleation rate, as demonstrated in Fig. 3.39 and Eq. (3.25). A high C_s, i.e., a high solute solubility, and a positive temperature coefficient of solute solubility are also important, because there will be sufficient solute available to form filled agglomerates of the touching primary particles in this case. In addition, the precipitated particles should not be thermoplastic or melt during the decomposition stage, otherwise, they will be deformed and morphology will be negatively affected.

It is also found that multicomponent and composite particles with a variety of microstructural characteristics can be synthesized by controlling the spray conditions. The drying of a droplet containing fine precipitates is quite different from that of a pure liquid droplet. The fine precipitates act a resistance to the mass transport of the solvent vapor. As the temperature of the drying chamber is too high, boiling of the solution might occur, which leads to inflation or disintegration of the droplets. Moreover, high capillary stresses and fracture of the particle could be observed, if the fine pores inside the precipitates are combined and the drying rate of the droplet is too fast.

The dried salts must be completely decomposed before they are brought to the final sintering process. For small scale laboratory experiments, decomposition times are relatively short, nitrates and acetates are preferable to sulfates, due to their lower decomposition temperatures. In addition, acetates have a low solubility, while nitrates, acetates, and sulfates are easy to introduce impurities into the final powder. On the other hand, chlorides and oxychlorides are more likely used for industrial productions, owing to their high solubilities. However, the corrosive nature of the gases produced during the decomposition and the deleterious effect of residual chlorine on subsequent sintering should be considered carefully. It is better to sinter the particles in situ, so as to take full advantage of the spray pyrolysis process. If the process in conducted at a sufficiently high temperature, the fine pores between the primary particles and the short interparticle collision time in the process are beneficial to the formation of dense individual particles.

Nano-sized MgO particles with well-crystallized structures have been prepared by using spray pyrolysis method with solutions of citric acid [219]. The optimum concentration of citric acid was 0.4 M, in order to obtain micrometer-sized MgO particles. The function of citric acid in the spray solution was to alter the thermal properties of the solution and thus to control the morphologies of the particles. The micrometer-sized precursor particles with hollow porous structures which were obtained through the spray pyrolysis step could be broken into nano-sized MgO particles by using a post-treatment. Without the presence of critic acid, the average sizes of the MgO particles thermally treated at temperatures of 500 and 900 °C were 13 and 28 nm, respectively. If citric acid was added into the spray solutions, the average particle sized were decreased correspondingly to 9 and 22 nm.

A systematic study has been reported on the comparison between spray pyrolysis and solid-state reaction method to prepare YAG powders [222]. The influence of precursor characteristics and synthesis conditions on the formation of yttrium aluminum garnet, $Y_3Al_5O_{12}$(YAG), was investigated. Two types of sources were used in this study: (i) single-source precursors, i.e., cohydrolyzed yttrium and aluminum alkoxides and yttrium aluminum glycolates, and (ii) multiple-source precursors, i.e., mixtures of metal nitrates and mixtures of separately hydrolyzed yttrium and aluminum alkoxides. Phase-pure YAG was formed only in the solid-state thermal decomposition experiments. The lack of formation of YAG in all the spray pyrolysis experiments was ascribed to the short heating times and fast heating rates, which resulted in the formation of kinetic products. In the case of the metal nitrates, an additional factor that influenced product formation was the difference in thermal reactivity of the precursors. It was concluded that the formation of complex metal oxide materials by conventional or aerosol routes is not necessarily achieved by the use of a chemically homogeneous precursor, such as a single-source precursor. It also was necessary to ensure that the precursors and intermediates have similar thermal decomposition temperatures to avoid phase segregation in the initial stages of thermal decomposition.

A series of works have been reported on preparation of YAG based phosphor powders by using spray pyrolysis method, with spray solutions of different fluxes [226–229]. It has been demonstrated that spray pyrolysis method can be used to control the morphologies and sizes of the YAG based phosphor powders in a very feasible way. Highly spherical particles with well controlled size distribution can be readily synthesized by optimizing the processing parameters.

Another example is the synthesis of YAG:Ce phosphor powders, with a composition of $Y_{2.965}Al_5O_{12}:Ce_{0.035}$, by using spray pyrolysis, from spray solutions with BaF_2 flux [228]. The BaF_2 flux has been show feasible control in morphology and improvement in optical properties of the YAG:Ce phosphor powders. The optimal concentration was 9 wt% BaF_2 flux, which resulted in powders with regular morphology and without aggregation. The average size of the YAG:Ce phosphor powders was 1.73 μm.

A 1.7 MHz ultrasonic spray generator having six vibrators was used to generate sufficient droplets. The spray pyrolysis was carried out with flowing air as carrier gas at a flow rate of 40 L min^{-1}, at reaction temperature of 900 °C. The powders were stayed inside the reactor for 0.6 s. The overall solution concentration was 0.5 M, while concentrations of the BaF_2 flux were in the range of 3–15 wt% of YAG:Ce. All the as-prepared powders were post-treated at temperatures of 1300–1600 °C for 5 h in 10 % H_2/N_2 gas mixture. Figure 3.46 shows SEM images of representative samples. All the as-prepared powders showed a spherical shape and a dense structure, with an average particle size in the range of several micron meters, which was not significantly affected by the concentration of the BaF_2 flux. After thermally annealed at 1500 °C, the spherical shape of the powders was deformed. However, the powders derived from the solution without BaF_2 flux demonstrated higher agglomeration, while the particle size of the powders increased with increasing concentration of BaF_2. In terms of powder morphology, the optimal concentration of BaF_2 was 9 wt%.

Freeze-Drying

Similar to that in spray drying, in freeze-drying, a solution of metal salt is also broken up by an atomizer into fine droplets in the first step [230–232]. In contrast to spray dry or spray pyrolysis, in which high temperature is used, in freeze-drying, the droplets are frozen rapidly, by spraying them into a cold bath of immiscible liquid, such as hexane and dry ice or liquid nitrogen. The frozen droplets are then transferred into a cooled vacuum chamber, where the solvent is removed through vacuum evaporation. In some cases, the system needs to be heated slightly to aid the sublimation. Generally, spherical agglomerates of fine primary particles can be produced, with the agglomerate size to be the same as that of the frozen droplets. The size of the primary particles, in the range of 10–500 nm, is determined by the processing parameters, including the rate of freezing, the concentration of the metal salt in the solution, and the chemical composition of the salt. After the drying process, the powder of the salt is heated at high temperatures to obtain oxides.

Generally, the solubility of most salts decreases with decreasing temperature, so that the supersaturation state can be rapidly reached during the rapid cooling of the droplets in freeze-drying. In this case, particle nucleation is very fast while growth is very slow. As a result, freeze-drying is able to produce powder with very fine particle sizes. In this respect, freeze-drying is more efficient in producing fine particles than the other spray drying techniques, as demonstrated in various materials, including Al_2O_3 [233], Y_2O_3 [234], ZrO_2 [235–237], $MgAl_2O_4$ [238–241], YAG [7, 140, 242, 243] and mullite [244]. Selected examples are discussed in a more detailed way as follows.

High sinterability $MgAl_2O_4$ powder has been produced from alkoxide precursors by using a freeze-drying method, with clear alumina sol and magnesium methoxide as starting materials [241]. The tap density and sinterability of the spinel power are affected by the ball milling step. Highly dense transparent polycrystalline $MgAl_20_4$ ceramics were prepared from the powders by sintering and hot isostatic pressing (HIP).

The alumina sol was prepared from aluminum isopropoxide through hydrolysis and peptization, while magnesium methoxide was prepared through the reaction of magnesium pieces (99.99 %) in excessive methanol under reflux in an N_2 atmosphere for 24 h. The spinel sol was prepared by slowly mixing the alumina sol into a methanol solution of magnesium methoxide in stoichiometric composition so as to form spinel $MgAl_2O_4$. The spinel sol was then heated at 85 °C for 2 days to evaporate excessive water and organic solvents, which was then sprayed into a tray

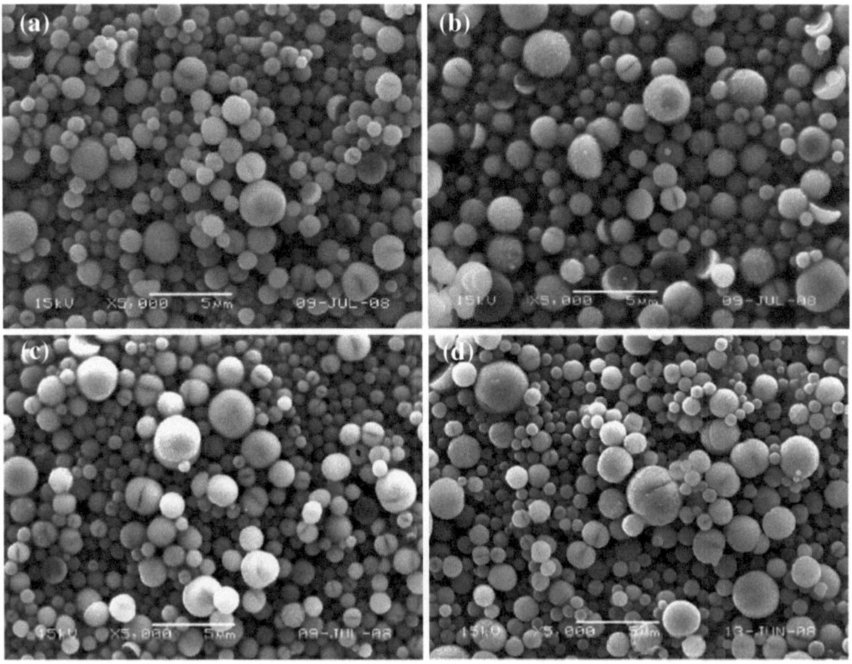

Fig. 3.46 SEM images of the precursor powders prepared by using the spray pyrolysis with and without the use of BaF_2 flux: **a** no flux, **b** BaF_2 = 3 wt%, **c** BaF_2 = 9 wt% and **d** BaF_2 = 15 wt%. Reproduced with permission from [228]. Copyright © 2009, Elsevier

containing liquid nitrogen through a two-fluid nozzle. After the freeze spraying, the frozen beads of spinel sol were loaded into a laboratory freeze-dryer with a tray dryer attachment, which was then gradually heated to 50 °C under 60 mTorr, i.e., ~8.0 Pa, until all the ice sublimed. The freeze-dried powder was calcined to for spinel phase. The spinel powders were then deagglomerated by wet- and dry milling steps with plastic bottles and yttria-stabilized-zirconia (YSZ) ceramic balls for 24 h.

Figure 3.47 shows XRD patterns of the as-freeze-dried powder and those calcined at different temperatures for 2 h are [241]. The as-freeze-dried powder consisted of a mixture of AlO(OH) and $MgAl_2(OH)_8$. Calcination at 900 °C resulted in the formation of spinel phase, while complete crystallization was observed after calcination at 1200 °C. The powder calcined at 1100 °C had an average particle size of about 50 nm.

Another example is the synthesis of Nd:Y_2O_3 nanosized powder by using freeze-drying, which was then mixed with commercial Al_2O_3 power to fabricate Nd:YAG transparent ceramics [7]. Yttrium nitrate hexahydrate, $Y(NO_3)_3{\cdot}6H_2O$, and neodymium nitrate hexahydrate, $Nd(NO_3)_3{\cdot}6H_2O$, both with purity of ≥99.99 %, were used as starting materials, while ammonium hydrogen carbonate, NH_4HCO_3, was used as precipitant. Aqueous solution with 0.25 M Y^{3+} and 2 at.% Nd^{3+} was

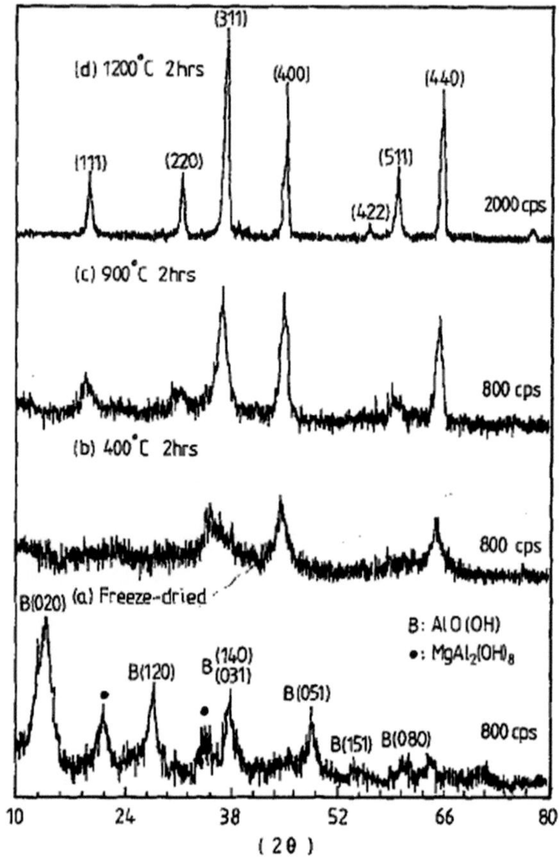

Fig. 3.47 XRD patterns of the as-prepared powder and the powders calcined at different temperatures. Reproduced with permission from [241]. Copyright © 1992, John Wiley & Sons

prepared by dissolving the metal nitrates in deionized water. After that, 2 M NH_4HCO_3 aqueous solution was added dropwisely into the metallic ion solution under mild agitation at room temperature. White precipitate was formed during the mixing. After aging for 2 days, the precipitate was washed thoroughly with deionized water, which was then frozen and dried through freeze-drying at −80 °C and 0.03 Torr. Nd:Y_2O_3 ultrafine powders were obtained by calcining the freeze-dried precipitates in air at temperatures of 500–1,100 °C for 4 h at a heating rate of 3 °C min^{-1}. Figure 3.48 shows SEM images of the powders after calcining at different temperatures. As the calcination temperature was increased from 900 to 1100 °C, the average particle size increased from ~25 to ~40 nm, without the presence of severe agglomeration. In contrast, the powders calcined by using normal furnace were heavily agglomerated. Highly transparent 2 at.% Nd:YAG ceramics were obtained from the freeze-drying derived Nd:Y_2O_3 powders and commercial Al_2O_3 powders by using vacuum sintering at 1750 °C for 5 h. Similar freeze-drying process was used to synthesis nanosized Nd:YAG powder to prepared transparent Nd:YAG ceramics [242, 243].

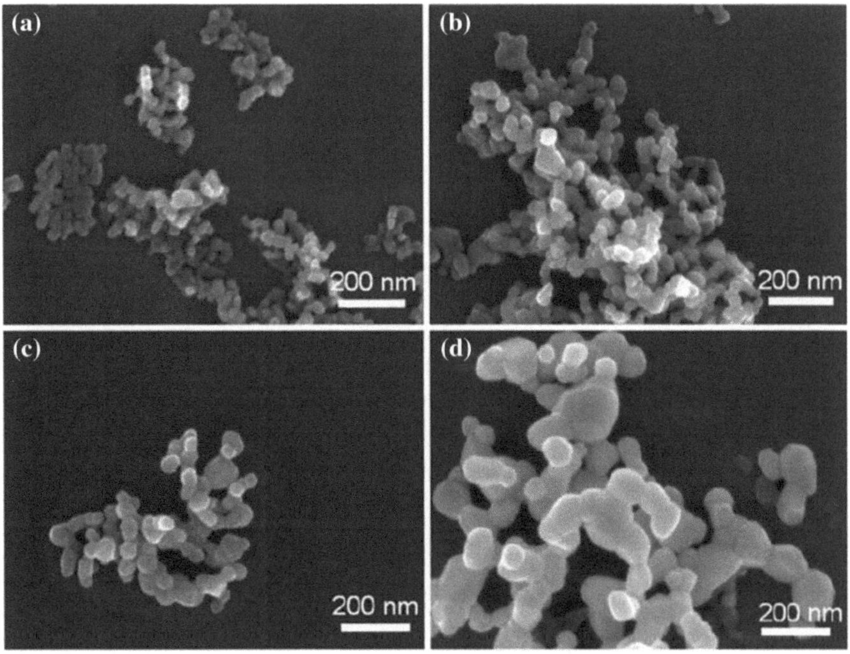

Fig. 3.48 SEM images of the freeze-dried power calcined at different temperatures: **a** 900 °C, **b** 1000°, **c** 1100 °C and **d** 1200 °C. Reproduced with permission from [7]. Copyright © 2011, Springer

Spray Drying of Suspensions

Spray drying is also applicable to suspensions of fine particles, which is usually knows as slurries, when they have relatively high concentrations. The key of this process is to limit the agglomeration of the dried powder to a scale that is equal to or less than the size of the droplet. Minimizing the degree of the agglomeration is required by the subsequent compaction and sintering steps [25, 245–250]. Spray drying of suspensions is used on a large scale industrially for granulating fine powders to control their flow and compaction characteristics during die pressing. Therefore, if large scale fabrication of transparent ceramics is considered, this technique will be very useful.

One earlier example is regard to Al_2O_3 transparent ceramics [251]. Both spray drying and freeze-drying processes were used and compared. The starting raw materials were reagent grade α- and γ- Al_2O_3 and magnesium sulfate as dopants. The impurities detected, such as copper, silicon, sodium and calcium, were less than 10 ppm. The α- Al_2O_3 had a surface area of 2 $m^2\ g^{-1}$ with an average grain size of 8 μm, while the γ- Al_2O_3 a possessed a surface area of 94 $m^2\ g^{-1}$ with an average grain size of 0.2 μm. The powders contained 0.25 wt% of MgO. The Al_2O_3 colloidal dispersions were made with Al_2O_3 contents of 10 wt%, required amount of magnesium sulfate for 0.25 wt% MgO. Stable colloidal dispersions could be obtained at pH = 2.5–3. Experimental results indicated that both types of powders, which were processed by both spray drying and freeze-drying, could be sintered to near theoretical density, with high optical transparency.

A study has been conducted to optimize the spray drying parameters in order to prepare transparent Yb:YAG ceramics [215]. Commercial oxides were used as the starting materials, with compositions of $Yb_{0.03}Y_{2.97}Al_5O_{12}$ and $Yb_{0.294}Y_{2.706}Al_5O_{12}$, corresponding to YAG doped with 1.0 and 9.8 at.% of Yb, respectively. The effects, of solid content in the spray solutions, drying medium flux, temperature and aspiration rate of the spray drying process, on optical properties of the final sintered ceramics, were systematically investigated. The samples were sintered by using vacuum sintering at temperatures in the range of 1650–1735 °C 16 h. Aspiration rate of 70 % and solid concentration of 16–24 wt% should be used in order to obtain well granulated powders without obstruction of the nozzles. Optimal processing parameters include inlet temperature of 70 °C and flux of 40 % of the drying medium, in terms of the optimization of optical transparency.

3.2.2.3 Gel Methods

Gel methods involve the formation of semirigid gels or highly viscous resins from liquid precursors, which appear as an intermediate step toward ceramic powders. This type of methods is particularly useful for the synthesis of complex oxides, due to the high level chemical homogeneity that can be achieved. Oxide powders are usually obtained by decomposing the gels or resins, followed by calcination and milling to control the particle characteristics. To form the gels or resins, the starting constituents are mixed at the atomic scale through polymerization process. If no constituent is volatilized during the decomposition and calcination steps, the cation composition of the powder will be identical to that in the original solution. Therefore, these methods are able to achieve good chemical homogeneity and controllable stoichiometry. Different gel processes are discussed as follows.

Sol–Gel Process

Sol–gel process involves the formation of polymeric gels by hydrolysis, condensation and gelation, which are then dried and ground into powders [252–255]. Dried gels with lower viscosity are easier to grind, so that the extent of contamination introduced during the milling is relatively low. If the liquid content is removed under supercritical conditions, there is no shrinkage for the dried gels, which have therefore low viscosity. Preferably, metal alkoxides are used for sol–gel process, because they have strong hydrolysis reactions. However, in some cases, metal salts also have hydrolysis behaviors.

Metal alkoxides have a general formula of $M(OR)_z$, where z is an integer equal to the valence of the metal M, and R is an alkyl chain. They can be considered as derivatives of either an alcohol, ROH, in which the hydrogen is replaced by the metal M, or a metal hydroxide, $M(OH)_z$, in which the hydrogen is replaced by an alkyl group. The reactions involve hydrolysis, which can be described by using the following equation:

$$M(OR)_z + xH_2O \rightarrow M(OR)_{z-x}(OH)_x + xROH, \quad (3.44)$$

which is followed by condensation and polymerization through dehydration reaction, as follows:

$$-M-OH + HO-M- \rightarrow -M-O-M- + H_2O. \quad (3.45)$$

Uniform particles have also been obtained through the hydrolysis of metal salt solutions [22, 23, 103, 127]. Comparatively, this method can produce powders with a wider range of chemical compositions. Besides oxides, hydrous oxides, sulfates, carbonates, phosphates, and sulfides can be prepared by using this method. However, more experimental parameters must be controlled in order to produce uniform particles, such as concentration of the metal salts, chemical composition of the starting materials, temperature, pH of the solution, and the concentrations of anions and cations that form intermediate complexes.

Metal ions are usually hydrated in aqueous solutions. The conditions for homogeneous precipitation of uniform particles can be realized by using a technique called forced hydrolysis. The key of this technique is to promote the deprotonation of hydrated cations by heating the solution at elevated temperatures (90–100 °C).

The soluble hydroxylated complexes produced by the hydrolysis reaction serve as the precursors of the nucleation. Nucleation and growth of uniform particles can be achieved by generating the precursors at proper rate, through the adjustment of temperature and pH value. A general approach is to age the solutions at elevated temperatures. It has been found that the process is very sensitive to a minor change in various conditions. Also, anions could play a decisive role in determining the outcome of the reaction. The conditions for precipitation of uniform particles should be specifically studied case by case. It is found that slow release of anions from organic molecules such as urea or formamide can be used provide the conditions for nucleation and growth of uniform particles in solution. Various oxides have been synthesized by using sol–gel process [253, 256–262].

A sol–gel process was developed to synthesize nanosized Y_2O_3 powders [256]. Various organic acids, including acetic acid, oxalic acid, malonic acid, tartaric acid and citric acid, were used as chelating agents, while yttrium nitrate hexahydrated was used as the source of yttrium, which was dissolved in water so that yttrium ion concentration was 0.5 mol L^{-1}. Organic acids were then slowly added in the form of aqueous solution with desired [chelating agent]/[Y] ratio. pH value was then adjusted to 2 with 30 % ammonia solution. The mixed solution was then heated at 115 °C for 12 h to form gels. For oxalic and tartaric acids, the gels should be filtrated and washed, because rigid gels could not be formed, due to the low viscosity of their sols. All the filtrated and chelated gels were calcined at two temperatures, i.e., 800 and 1100 °C, in a flowing oxygen atmosphere, at a heating rate of 200 °C h^{-1} for 4 h. Two [chelating agent]/[Y] ratios were studied, i.e., 2 and 20.

It was found the powders derived from acetic, oxalic, malonic and tartaric acids exhibited crystalline grains, while amorphous phase was obtained for citric acid. Y_2O_3 powders could be readily formed in the precursor gels at heating temperatures of >700 °C. The chelating agent had no effect on composition of the Y_2O_3 powders. Average grain sizes of the Y_2O_3 powders, sintered at 800 and 1100 °C, were about 30 and 60 nm, respectively. However, chelating agent has a significant effect on morphologies of the Y_2O_3 grains, which simply follow the morphologies of the precursor gels. This study provided a reference on selection of chelating agent when the morphology of the powder is important. As stated earlier, spherical shapes with narrow particle size distribution are desired.

Nano-sized cerium-doped lutetium aluminum garnet (LuAG:Ce) powders were prepared by using a sol–gel combustion process, which was started from a mixed solution of metal nitrates, with an organic glycine as fuel [263]. The obtained precursor was agglomerated and had a foamy-like morphology, consisting of porous crystallites with a uniform size of about 40 nm. High purity crystalline LuAG:Ce powder was achieved by calcining the combustion precursor at 1000 °C for 2 h. Using the prepared powder, LuAG:Ce transparent ceramics have been fabricated by using vacuum sintering at 1850 °C for 10 h, followed by post-annealing at 1450 °C for 20 h in air. The transparent ceramics exhibited an in-line light transmittance of about 50 % in the visible wavelength range and had a uniform microstructure with an average grain size of about 8 μm.

Crystalline LuAG:Ce powders doped with 1.0 mol% CeO_2 ($Lu_{2.97}Ce_{0.03}Al_5O_{12}$) were prepared by a sol–gel combustion process, using glycine as a fuel. The process involves the exothermic reaction between metal nitrates and organic fuel, e.g., glycine. The typical stoichiometric synthesis reaction is:

$$\begin{aligned} 6M(NO_3)_3 + 10NH_2CH_2COOH + 18O_2 &\rightarrow 3M_2O_3 \\ + 5N_2 + 18NO_2 + 20CO_2 + 25H_2O&, \end{aligned} \tag{3.46}$$

where M represents Lu, Ce, and Al. The characteristics of the prepared powders are greatly influenced by the reaction temperature, which can be controlled by adjusting the molar ratio of glycine to metal nitrates. In the current study, a stoichiometric molar ratio of nitrate to glycine, i.e. 1.67, was adopted to prepare LuAG:Ce powders.

Lutetium oxide (Lu_2O_3, 99.99 %), aluminum nitrate hydrate ($Al(NO_3)_3{\cdot}9H_2O$, analytical grade), cerium nitrate hydrate ($Ce(NO_3)_3{\cdot}6H_2O$, 99.99 %), nitric acid (HNO_3, excellent grade), glycine (NH_2CH_2COOH, analytical grade), and deionized water were used as starting materials in the present work. High purity Lu_2O_3 powders were first dissolved in nitric acid to form lutetium nitrate solution. After completely dissolving, according to the general formula composition

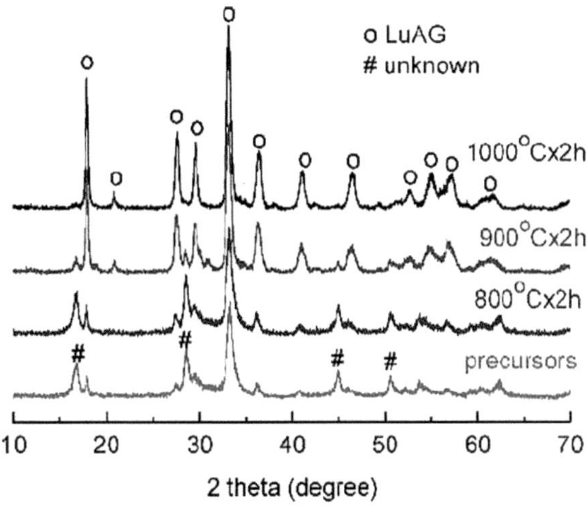

Fig. 3.49 XRD patterns of the as-prepared precursors and powders calcined at various temperatures. Reproduced with permission from [263]. Copyright © 2006, Materials Research Society

$Lu_{2.97}Ce_{0.03}Al_5O_{12}$, a stoichiometric cerium nitrate and aluminum nitrate were introduced to the resultant lutetium nitrate solution with appropriate dosage of glycine. The mixed solution was heated and continuously stirred using a magnetic agitator until a transparent sticky gel formed. Then the gel was rapidly heated, and an acute auto-combustion process took place at about 700 °C accompanied by the evolution of brown fumes. As a result, a brown and fluffy precursor was yielded. The precursor was then heat treated at 800–1000 °C for 2 h in a muffle furnace in air. The calcined powders were ball milled in a polyurethane container using 6 mm diameter ZrO_2 balls and ethanol as media for 2 h.

XRD patterns of the precursors prepared by the sol–gel combustion process and the powders calcined at various temperatures are shown in Fig. 3.49. It was unexpected that the precursors prepared by the present sol–gel combustion process using glycine as fuel were crystalline instead of amorphous, consisting of cubic LuAG and some unknown phase. Additionally, no distinct difference can be observed between the diffraction patterns of the resultant precursors and the powders calcined at 800 °C for 2 h. At 900 °C, the characteristic peaks of LuAG phase appeared with rather intense peaks and the unknown phase showed very weak peaks, indicating that mostly LuAG phase is produced at this temperature. At 1000 °C, continuing refinement of peak shapes and intensity was observed, indicating that crystalline LuAG grew with increasing calcination temperature.

Figure 3.50 shows typical EPMA and TEM micrographs of the as-prepared precursors and powders calcined at 1000 °C for 2 h, respectively. It can be seen that the precursors have a foamy-like and porous morphology (Fig. 3.50a), very similar to the one observed for the previously studied $(Gd,Y)_2O_3$ powders. The TEM micrograph of the precursors, shown in Fig. 3.50b, reveals the shape and size of primary particles. The as-prepared powders are agglomerated and consist of near-spherical crystallites of about 30-nm particle size estimated from TEM images. The TEM micrograph of the powders calcined at 1000 °C for 2 h, shown in Fig. 3.50c,

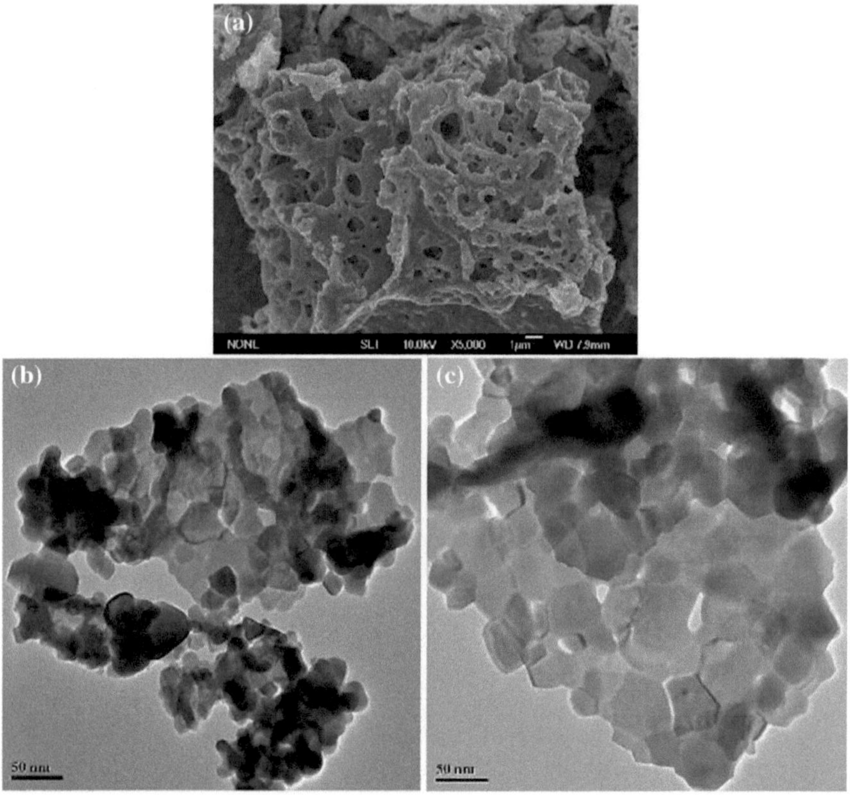

Fig. 3.50 EPMA and TEM micrographs of the precursors and powders calcined at 1000 °C for 2 h: **a** EPMA, precursors, **b** TEM, precursors and **c** TEM, powders calcined at 1000 °C for 2 h. Reproduced with permission from [263]. Copyright © 2006, Materials Research Society

reveals that the powders consist of pointed crystallites with fairly uniform size of about 40 nm.

Highly sinterable yttrium aluminum garnet (YAG) nanopowders have been synthesized by a sol–gel combustion method with various chelating agents and fuels, namely citric acid, tartaric acid, glycine, and ethylene diamine tetraacetic acid (EDTA) [260]. The preparation involved the thermal decomposition of a chelating agent (fuel)-nitrate gel and the formation of amorphous precursors. The as-synthesized precursors were studied by infrared spectroscopy (IR), thermogravimetric (TG) and differential scanning calorimetric (DSC) analyses. The nanopowders calcined at 1000 °C were characterized by X-ray powder diffraction (XRD) and transmission electron microscopy (TEM). It was found that the chelating agents and fuels used had a significant influence on the average grain size and agglomeration of YAG nanopowders. The rate of combustion reaction between chelating agent (fuel) and nitrate was responsible for the growth of the grains. Nanoparticles with the smallest size and high sinterability were obtained when using EDTA, which derived

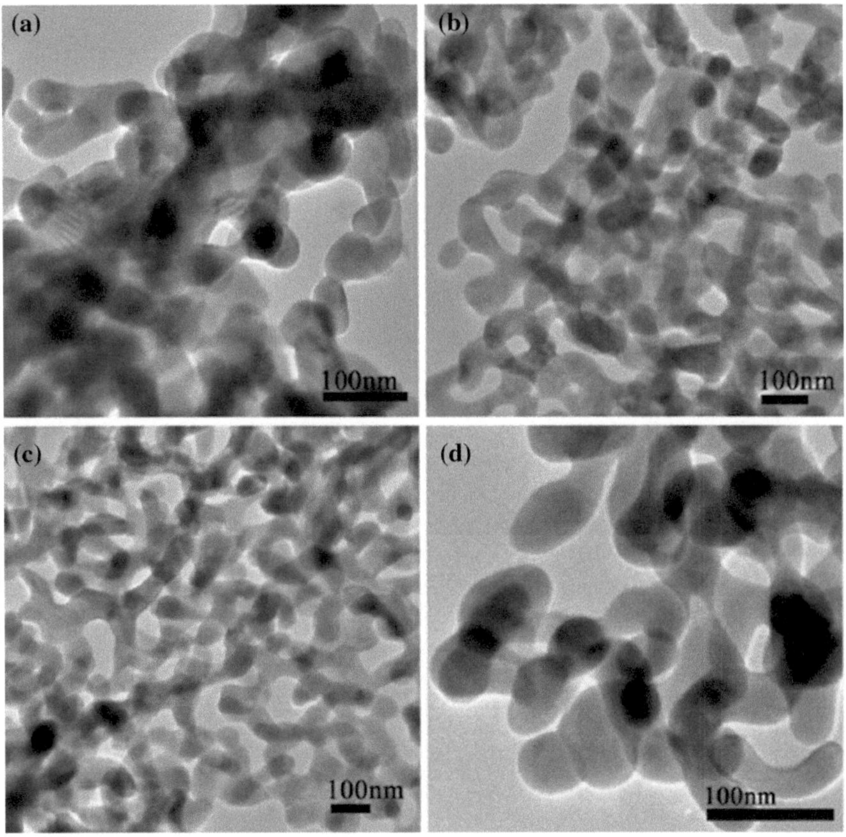

Fig. 3.51 TEM images of the YAG powders using chelating agents: **a** citric acid, **b** tartaric acid, **c** glycine and **d** EDTA. Reproduced with permission from [260]. Copyright © 2010, Elsevier

from the significant blocking of the diffusion path associated with the lowest combustion reaction rate.

Figure 3.51 shows TEM images of the YAG powders calcined at 1000 °C. Most particles had a semispherical shape with a relatively high porosity. Such porosity could be caused by the release of gas during the combustion process. By adjusting the content of the organic compounds, specific surface areas of the powders could be controlled. Average sizes of the nanopowders were 60, 80, 60 and 40 nm, corresponding to the chelating agents and fuels of citric acid, tartaric acid, glycine and EDTA, respectively. Citric acid and tartaric acid that showed the faster reaction led to larger average particle sizes. In contrast, smaller grains were observed when using EDTA. The faster the reaction, the fewer the residues would be in reaction process, which made the cations to diffuse more easily. Additionally, all the YAG nanopowders calcined at 1000 °C exhibited high crystallinity. Therefore, EDTA was most suitable to synthesize YAG nanopowders with smallest grain size, which is beneficial for the preparation of transparent YAG ceramics.

Mullite ($3Al_2O_3 \cdot 2SiO_2$) powder has been prepared by supercritical drying the gels derived from sol–gel process, which has high sinterability with fully dense mullite ceramics to be obtained at a temperature of below 1200 °C [264–268]. This temperature is significantly lower than that required by the conventional solid-state reaction method from the precursors of mixed oxide. The high sinterability has been attributed to the amorphous structure, high surface area, and molecular level homogeneity of the gel-derived powders.

The Pechini Method

In the Pechini method, metal ions from starting materials, such as carbonates, nitrates, and alkoxides, are complexed in an aqueous solution with α-carboxylic acids, such as citric acid. When heated with a polyhydroxy alcohol, such as ethylene glycol, polyesterification takes place. Once excess liquid is removed, a transparent resin is formed. The resin is then heated to decompose the organic constituents. After milling and calcination, oxide powders can be obtained accordingly. Various oxide nanopowders have been synthesized by using Pechini or modified Pechini methods [269–276], although less information is available on the fabrication of transparent ceramics with these powders.

Cubic nanocrystalline $Yb:Lu_2O_3$ powders were synthesized by modified Pechini sol–gel method [277]. Lu_2O_3 nanocrystals doped with 50 % Yb were prepared using the modified Pechini method. Figure 3.52 shows the flow chart for the synthesis process. Lu_2O_3 (99.9 %) and Yb_2O_3 (99.9 %) were dissolved in concentrate HNO_3 (~65 %) under stirring, which was heated to form nitrates. With the moles of the EDTA, [EDTA], the moles of the metal cations, [metal], and the moles of ethylene glycol, [EG], two groups of molar ratios were used: [EDTA]/[Metal] = C_M and [EDTA]/[EG] = C_E. C_M measured the degree of the chelation process of the metal ions with the EDTA. For the solutions with low values of C_M, no homogeneous chelation of the metals occurred, due to insufficient EDTA molecules in the solutions. However, too high C_M meant that more organics had to be removed. Therefore, the value of C_M should be optimized. Experimental results indicated that molar ratio $C_M = 1$ was suitable the preparation of the metals–EDTA complexes. During mixing and heating, the esterification agent was added, ethylene glycol, with a molar ratio C_E equal to 2. By this way, the solution was transformed into precursor resin. On the other hand, C_E accounted the degree of esterification between the EDTA and the ethylene glycol. Finally, the viscous gel obtained was calcined at temperatures of 300–1000 °C for 1–5 h to form the nanocrystals of $Yb:Lu_2O_3$. Yb:Lu_2O_3 powders with sizes in the range of 50–100 nm could be obtained accordingly.

A Pechini-type sol–gel process was used to synthesize YAG nanophosphors, YAG:Ce,Pr system, with concentration of Pr^{3+} from 0.125 to 2 mol% while maintaining the content of Ce^{3+} constant at 2 mol%, for white light-emitting diode (LED) applications [278]. Yttrium and aluminum nitrates and acids were mixed

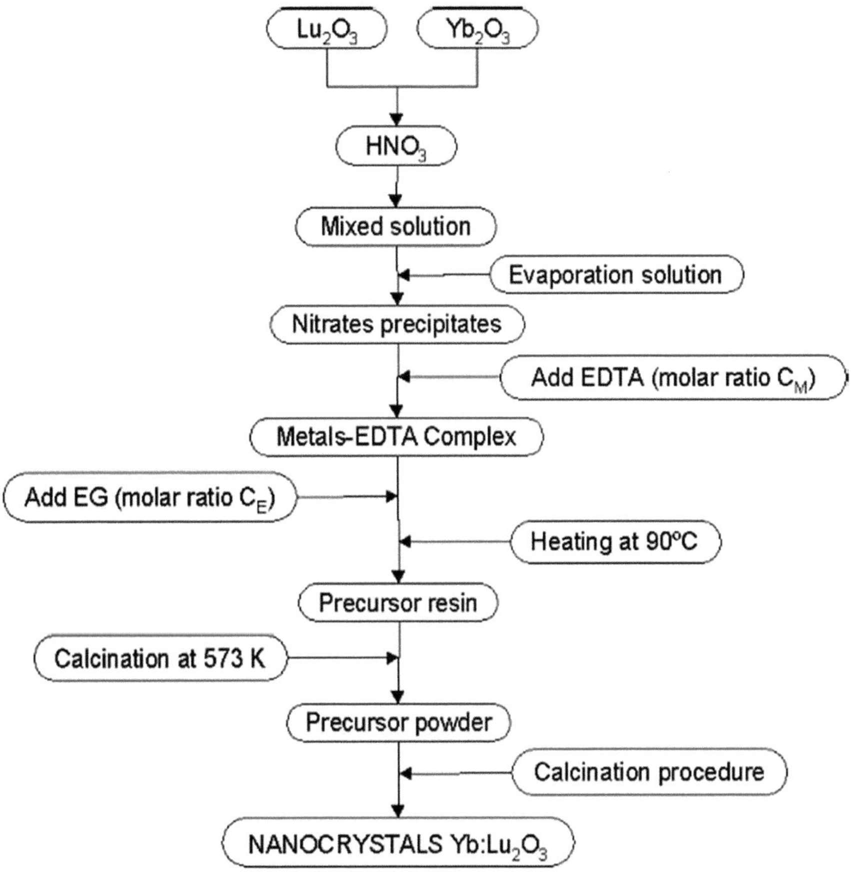

Fig. 3.52 Flow chart of synthesis of the $Yb:Lu_2O_3$ nanocrystals with a modified Pechini method. Reproduced with permission from [277]. Copyright © 2008, Elsevier

directly in a crucible at 100 °C. Solutions of lanthanides, with [Ce] = 24 mM and [Pr] = 6 mM, were prepared by dissolving the precursors in a mixture of HNO_3: H_2O = 1:1. After complete dissolution of the salts, EG was added and stirred until gels were formed. After that, the reaction mixtures were left on the hot plate overnight. Finally, the samples were calcined at 1000 °C for 4 h. The quantities of reagents were calculated according to the following reaction:

$$\begin{aligned} & xY(NO_3)_3 + 5Al(NO_3)_3 + yLn(NO_3)_3 + EG \\ & + CA \longrightarrow HNO_3 : H_2OY_xAl_5O_{12} : Ln_y \end{aligned} \tag{3.47}$$

The molar ratio among all salts, CA and EG was 1:3:4.5, while 20 mL HNO_3: H_2O solution was used for every 10 mmol of salts. The obtained nanopowders had well crystalline structures.

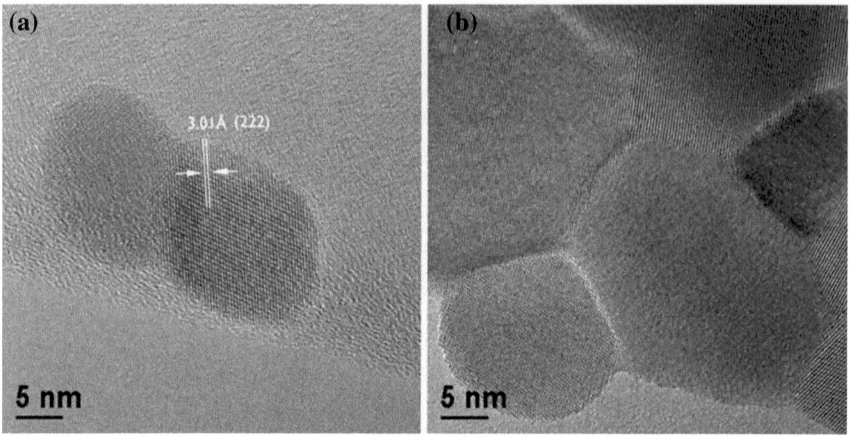

Fig. 3.53 HRTEM image showing single monocrystalline nanoparticles (sample Tb1Eu2, $Tb_{0.02}Eu_{0.04}Y_{1.94}O_3$) with crystal planes spacing of 3.01 Å, corresponding to the (222) crystal plane of Y_2O_3 (**a**), and TEM image of an agglomeration of the sample Y_2O_3 (**b**). Reproduced with permission from [279]. Copyright © 2013, Springer

Tb^{3+} and Eu^{3+} co-doped Y_2O_3 nanoparticles with a volume-weighted average size of about 30 nm have been synthesized by using a similar Pechini-type sol–gel process [279]. The nanopowders synthesized by using this method had an average particle size of about 35 nm. TEM images of representative samples are shown in Fig. 3.53.

A modified Pechini method was reported to synthesize nanosized $MgAl_2O_4$ spinel powders doped with Eu^{3+} ions [280]. XRD results indicated that the powders were single-phase spinel with high crystallinity. Concentration of the optically active Eu^{3+} ions has been set to 0.5–5 mol% in respect to the appropriate Mg^{2+} molar content. Starting materials used in the synthesis included high purity Eu $(NO_3)_3·6H_2O$, $AlCl_3·6H_2O$ and $Mg(NO_3)_2·6H_2O$. Stoichiometric amounts of the three salts were dissolved in deionized water to for homogeneous solution. Required quantity of citric acid and ethylene glycol were added into the solution, which was then heated to 80 °C to increase its viscosity. The obtained resin was calcined at temperatures of 700–1000 °C for 3 h. White powders of $MgAl_2O_4$ and doped with Eu^{3+} were thus obtained. The average spinel particle sizes were about 15 nm for sample calcined at 700 °C and 20 nm at 1000 °C, respectively.

Citrate Gel Method

The citrate gel method involves the dissolving of solution of various nitrates in citric acid [257, 262, 281–289]. Through controlling the pH values, no precipitate is in the solutions. After the solution is heated at a suitable temperature in air for a period of time, it becomes more and more viscous liquid containing polybasic chelates. Finally, amorphous solids are obtained, which are pyrolyzed in air to produce crystalline oxide powders.

A citrate gel combustion method has been reported to synthesize nanosized ytterbium doped yttria [290]. Nanocrystalline Yb^{3+}:Y_2O_3 powders single-phase cubic yttria crystal structure could be prepared by calcining the as-prepared precursors at 1100 °C for 3 h. Powders obtained were well dispersed with an average particle size of 60 nm. By using the obtained powders, nearly fully dense Yb^{3+}:Y_2O_3 ceramics, with promising optical properties, were produced by vacuum sintering at 1800 °C for 12 h.

Citrate gel combustion method is a modified sol–gel technique in which the gel is decomposed by a self-propagating high temperatures synthesis (SHS) process. It is a combination of sol–gel and SHS reaction processes. An aqueous nitrate solution of Y^{3+} and Yb^{3+} were prepared by dissolving Y_2O_3 (99.99 %) and Yb_2O_3 (99.99 %) in diluted nitric acid under stirring at 80 °C for 2 h. The metal nitrates were weighted according to the required proportion and mixed in a glass cylindrical beaker with doping of minor amount of ammonium sulfate ($(NH_4)_2SO_4$) as a dispersant. Double distilled water was used in the experiments for homogenous mixing of metal nitrates. Citric acid was added to the mixtures as a fuel and mixed thoroughly followed by a clear solution was obtained. It was calculated that 1.667 mol of citric acid was required to prepare 1 mol of yttria with complete combustion. The equivalence ratio, i.e., the ratio of the oxidizing valency to reducing valency (O/F) was maintained at unity (O/F = 1). The valency of nitrogen was not considered because of its conversion to molecular nitrogen (N_2) during combustion. The assumed complete combustion reactions can be written as:

$$\begin{aligned} &6RE(ON_3)_3 + 5CH_2COOHCOHCOOHCH_2COOH \\ &\rightarrow 3RE_2O_3 + 9N_2\uparrow + 30CO_2\uparrow + 20H_2O\uparrow \end{aligned} \tag{3.48}$$

The mixed solution in the glass beaker was kept in a water bath of 80 °C for 8 h until it transformed into a honey-like yellow transparent gel. The gel was dried at 120 °C for 24 h and became a deep yellow sticky gel. This gel was rapidly heated to 300 °C and at this stage an auto combustion process occurred accompanied with a brown fume, and finally yielded a fluffy precursor. The precursor was crushed and then heat treated at temperatures ranging from 600 to 1100 °C for 3 h in an oxygen atmosphere. For sintering, powders calcined at 1100 °C for 3 h were dry pressed into Ø15 mm pellets in a steel mold at 30 MPa and then cold isostatically pressed (CIP) at 200 MPa. The pellets were then sintered under vacuum at 1800 °C. The vacuum in the furnace was 10^{-3} Pa during the sintering period.

Figure 3.54 shows SEM morphologies of the Yb^{3+}:Y_2O_3 nanopowders calcined at 1100 °C for 3 h without (Fig. 3.54a) and with (Fig. 3.54b) ammonium sulfate doping. From Fig. 3.54b, it can be seen that particles with sulfate doping disperse uniformly, and have a relatively narrow size distribution. Most particles are spherical in shape with the average particle size of about 60 nm. The release of gas in the combustion process gave a significantly porous structure in the calcined

Fig. 3.54 SEM images of the Yb^{3+}:Y_2O_3 powders calcined at 1100 °C without sulfate ions doping (**a**) and doped with sulfate ions (**b**). Reproduced with permission from [290]. Copyright © 2011, Elsevier

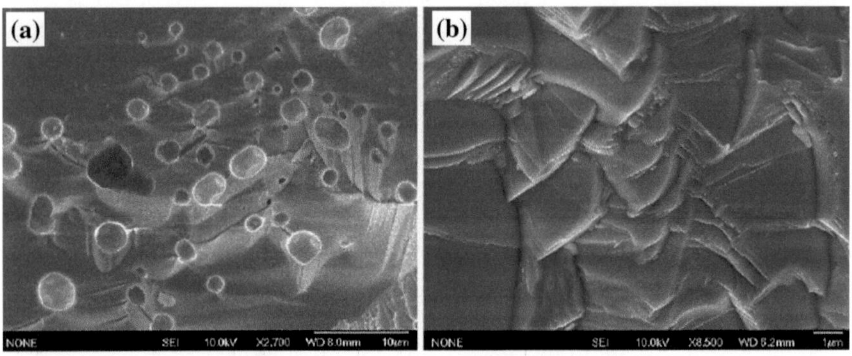

Fig. 3.55 Fracture surface SEM images of the Yb^{3+}:Y_2O_3 ceramics sintered at 1800 °C for: **a** 5 h and **b** 12 h. Reproduced with permission from [290]. Copyright © 2011, Elsevier

powders, and this is beneficial to the dispersion of powders. Compared with Fig. 3.54b, powder without sulfate doping has a lower dispersivity and a wide size distribution with a range from 50 to 90 nm, and is not so uniform with a mixture of spherical and rectangular in morphology.

Figure 3.55 is the SEM images of fracture surfaces of the sintered yttria ceramic samples by vacuum sintering process under 1800 °C for 5 h (Fig. 3.55a) and for 12 h (Fig. 3.55b). The presence of a great number of porosity at grain boundary junctions and as clusters in the grain interior can be seen in the sample sintered for 5 h. With the increase of sintering time, a significant grain growth and densification has occurred in the sample, it can be seen that nearly pore-free microstructure of the sample remained homogeneous without abnormal grain growth, and transgranular fracture indicates that no pore entrapment has occurred in the densified regions (Fig. 3.55b).

Glycine Nitrate Process

Glycine nitrate process is a combustion method for the preparation of ceramic powders [291–299]. A highly viscous mass is formed by evaporation of a solution of metal nitrates and glycine is ignited to produce powders. Glycine, an amino acid, is used to form complexes with the metal ions in the solution. It can increase the solubility of the nitrates and prevents the precipitation of the metal ions when water is evaporated. Because the thorough mixing of the starting constituents, the final powders have high chemical homogeneity. Glycine also serves fuel to enhance the ignition step, because it is oxidized by the nitrate ions. The reactions during the ignition are highly explosive, so that extreme care must be taken during this step. Large quantity production should be avoided at lab research scale. The final products are usually loose mass of extremely fine and crystalline powders. As compared with the Pechini method, milling and calcination of the products are not necessary. The ultrafine size and crystalline nature of the powders are attributed to the rapid reaction during the ignition at high temperatures.

Nanosized cubic yttria-stabilized zirconia (ZrO_2–8 mol% Y_2O_3) powder was synthesized by using a glycine–nitrate process, which was combined with high-energy ball milling [300]. Effect of calcination temperature on sintering activity of the powders was studied to identify the optimized calcination temperature, which was 900 °C, in terms of sinterability. Sintering of the nanopowder was carried out by using spark plasma sintering at 1200–1350 °C for 5 min. Transparent ceramics could be achieved at 1300 °C.

High purity $Zr(NO_3)_4{\cdot}5H_2O$ and $Y(NO_3)_3{\cdot}6H_2O$ were used as starting materials for synthesis. They were mixed according to the composition of 92 mol% ZrO_2 + 8 mol% Y_2O_3 and dissolved in deionized water. After that, glycine was added into the solution as a fuel, with a molar ratio of glycine to total metal ions (Zr + Y) of 2.5:1. Precursors were formed by adding ammonia water to the solution at a rate of 3 mL/min, until to pH = 8. Then, ethylene glycol (EG), with a molar ratio of glycine:EG to be 1:1, was added into the solution to trigger the esterification reaction. The resultant solution was heated at 75 °C with stirring to form gel precursors. After aging for 12 h, the gel was filtered and thoroughly washed with ethanol. Finally, the gel was dried at 150 °C and ignited to brown powders.

The brown powder was first ball milled with high purity zirconia balls for 12 h at a rotational speed of 200 rpm, in isopropyl alcohol with 3 wt% polyethylene glycol. The ball milled mixture was dispersed supersonically for 30 min and then dried by using a rotary evaporator. The dried powder was further calcined at different temperatures for 2 h. Finally, the calcined powders were ball milled for second time with zirconia balls for 12 h in isopropyl alcohol. The milled powders were dried and then sintered by using SPS, with graphite dies with an inner diameter of 15 mm at 50 MPa uniaxial pressure. The temperature was increased to 600 °C within 10 min, from where and onwards the pressure was increased to 100 MPa. The samples were sintered at 1200–1350 °C for 5 min at a heating rate of 100 °C min^{-1}.

Figure 3.56 shows SEM images of the powders in form of pellets, which were calcined 700 and 900 °C first and then annealed at 1100 °C for 1 h [300]. The

Fig. 3.56 SEM images of the powders calcined in air at **a** 700 °C and **b** 900 °C. Reproduced with permission from [300]. Copyright © 2012, Elsevier

sample from the 700 °C-calcined powder was still powder-like without obvious sign of sintering, while the one from the 900 °C-calcined powder was a porous ceramics, with obvious grain growth, as shown in Fig. 3.56b. It has been well known that low calcination temperature leads to porous powders, which cannot be compacted densely and uniformly. However, too high calcination temperature results in powder with reduced specific surface area and thus poor sinterability. Therefore, there is an optimized calcination temperature for different nanopowders, which was 900 °C for the 8YSZ nanopowder.

Figure 3.57 shows SEM images of the 8YSZ ceramics sintered at different temperatures [300]. The samples sintered at 1200 and 1250 °C had grain sizes still at the nanometric scale, with pores were mainly located at grain boundaries. At the early and intermediate stages of SPS process, creeping and sliding of grains are the dominating mechanism of densification, which is beneficial to break hard agglomerates and destroy large pores, without the occurrence of grain growth. Significant grain growth was observed as the sintering temperature was above 1250 °C. At the final stage of sintering at high temperatures, densification is dominated by grain boundary diffusion. It is well accepted that the grain growth of YSZ is closely related to the diffusion of oxygen vacancy, which is enhanced with increasing temperature. Therefore, at high temperatures, pores could be entrapped due to fast grain boundary migration, thus leading to pores that were changed to be from intergranular to intragranular, as shown in Fig. 3.57d.

3.2.2.4 Nonaqueous Liquid Reactions

Reactions in nonaqueous liquids are mainly used to synthesize nonoxide powders, such as Si_3N_4. One of the advantages of these methods is the high purity and fine particle size of the powder. There are at least four such reactions: (i) vapor phase reactions, (ii) gas–solid reactions, (iii) gas–liquid reactions and (iv) gas–gas reactions.

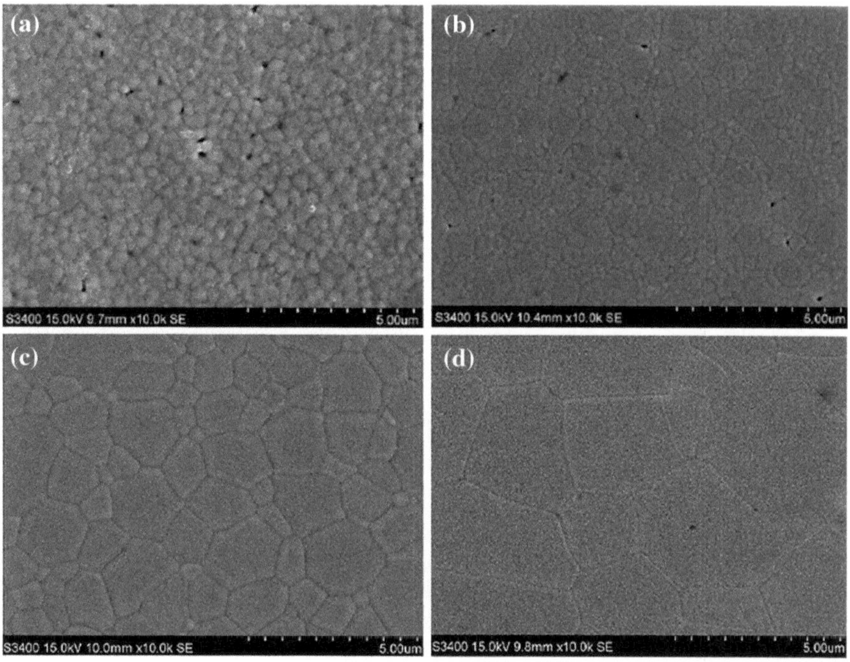

Fig. 3.57 SEM images of thermally etched surfaces of the samples sintered by using SPS at different temperatures: **a** 1200 °C, **b** 1250 °C, **c** 1300 °C and **d**1350 °C. Reproduced with permission from [300]. Copyright © 2012, Elsevier

Vapor phase reactions can be used synthesize oxide and nonoxide powders. The synthesis of Si_3N_4 and SiC powders will be used as examples to discuss these reactions [301–308]. Si_3N_4 has two hexagonal polymorphs, α-phase and β-phase, with the α-phase having slightly higher free energy at the formation temperature. Powders of α-Si_3N_4 have a more equiaxial particle shape and higher sinterability than β-Si_3N_4, because the particles of the latter grow into more elongated shape. Therefore, α-Si_3N_4 is more preferred.

One example of gas–solid reactions is nitridation. Direct nitridation of Si is a widely used method to synthesize Si_3N_4 powders, in which Si powder with particle sizes in the range of 5–20 μm is reacted with N_2 at temperatures of 1200–1400 °C for 10–30 h [309–314]. The Si_3N_4 powder is usually a mixture of α and β phases. The relative quantities of the two phases can be controlled by selecting the reaction temperature, the partial pressure of the N_2 gas in the nitriding atmosphere, and the purity of the Si powder. Si_3N_4 powder can also be produced through the carbothermic reduction of SiO_2 in a mixture of fine SiO_2 and C powders, which is followed by nitridation at 1200–1400 °C in N_2 [315]. Due to the widespread availability of pure fine SiO_2 and C, this method has become an attractive alternative to the direct nitridation of Si.

As a gas–liquid reaction, it has been showed that the reaction between liquid silicon tetrachloride ($SiCl_4$) and NH_3 gas in dry hexane at 0 °C can be used to prepare fine Si_3N_4 powder with very low levels of metallic impurities (<0.03 wt%) [316]. The powder synthesized through this reaction is amorphous, which can be crystallized to α-Si_3N_4 by heating at 1200–1400 °C for a prolonged time.

An example of gas–gas reactions is the formation of particles through the homogeneous nucleation and growth in the gas phase and is governed by the same equations for the nucleation of liquid droplets from a supersaturated vapor described above, i.e., Eqs. (3.18)–(3.24). Gas-phase reactions have been used to produce ceramic powders at both industrial scale and laboratory scale. Various heating techniques, including flame, furnace, plasma, and laser, are being used to the reactant gases for gas–gas reactions. Among these, flame synthesis, especially flame spray pyrolysis, has been emerged as a newly developed technique used to synthesize various oxide powders [317–326].

Figure 3.58 shows a schematic diagram of a representative flame assisted spray pyrolysis (FASP) reactor, which has been used to synthesize oxide nanoparticles [317]. Fuel, consisting of 1–12 mL min^{-1} of C_2H_2 of ⩾99.5 % or 12–30 mL min^{-1} of H_2 of ⩾99.999 %, was delivered through 12 holes (i.d. = 0.5 mm) equidistantly placed at the inner perimeter of a cylindrical torus ring (i.d. = 1.6 cm) coaxially to and at 1 cm above the nozzle issuing the precursor solution spray. A stainless steel tube (i.d. = 1.6 cm) was placed between ring and nozzle to shield the spray from air entrainment. Precursor solutions were fed through the central nozzle by a syringe pump at 2 mL min^{-1} and atomized by co-flowing 6 mL min^{-1} of oxygen (⩾99.5 %) through the surrounding annulus at 1.5–1.7 bar pressure drop. Additional sheath oxygen was supplied through 32 holes of 0.8 mm diameter each surrounding the nozzle at 1.5 cm radius. All gas flow rates were controlled by calibrated mass flow controllers. Using a vacuum pump, product particles were collected on water cooled glass microfiber filters (25.7 cm in diameter) placed at least 60 cm above the burner.

A flame spray pyrolysis (FSP) technique has been used to synthesize lutetium oxide (Lu_2O_3) nanoparticles from lutetium nitrate [323]. Optical quality transparent ceramics were prepared via hot pressing of the Lu_2O_3 nanoparticles formed using the FSP technique. A custom built flame spray pyrolysis system was used in this study. The system was composed of a two phase nozzle, consisting of a central capillary tube, surrounded by a ring of pilot flames, which were supplied with 1.5 L min^{-1} CH_4 and 3 L min^{-1} O_2. After ignition of the pilot flames, a co-flowing stream of O_2 and the liquid precursor are fed through the central capillary tube, where the oxygen was used as a dispersion gas to atomize the liquid precursor. All gases were supplied with flow rates regulated by using mass flow controllers.

The experiments were carried out by varying the O_2 dispersion gas flow rate from 2 to 5 L min^{-1}. The powders were collected on Whatman GF 6 glass fiber filter paper mounted in a water cooled stainless steel collection chimney equipped with a vacuum pump (Busch). The Lu_2O_3 samples were synthesized from a solution of $Lu(NO_3)_3{\cdot}6H_2O$ containing 100 g of the nitrate crystal dissolved in 500 ml DI water. The nitrate solution was then purified by crystallization and subsequent washes, followed by filtration through 0.8 μm filter media. Batches of 20 g of Lu

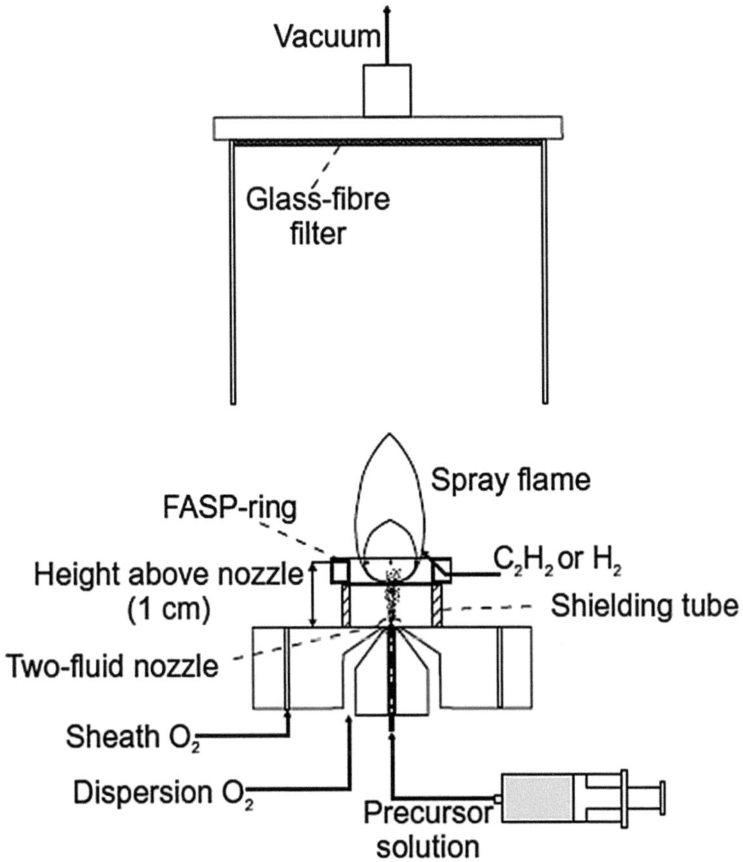

Fig. 3.58 Schematic of a flame assisted spray pyrolysis (FASP) reactor consisting of a two-fluid nozzle for atomizing the liquid precursor with oxygen and a cylindrical torus (FASP) ring with boreholes to inject and combust C_2H_2 with the precursor solution spray. The ring was located at 1 cm above the nozzle and the spray was shielded from air entrainment by a tube. Reproduced with permission from [317]. Copyright © 2013, Elsevier

$(NO_3)_3 \cdot 6H_2O$ were dissolved in 100 ml of ethanol to obtain the final FSP nanoparticle precursor solution.

Figure 3.59 is a transmission electron microscope image of a sample of cubic Lu_2O_3 nanoparticles, with the corresponding particle size distribution given in Fig. 3.59b. The mean particle diameter was found to be 15 nm. The sample was synthesized with an oxygen dispersion gas flow rate of 2 L min^{-1} and a precursor flow rate of 10 mL min^{-1}. It can be seen that the sample consists of agglomerates as well as primary particles that have been broken off during the sonication process. The BET surface area of the particles was found to be 28.9 m^2 g^{-1}, with an equivalent diameter of 22 nm. XRD indicated that the oxygen dispersion gas flow rate could be used control the phase content of the final Lu_2O_3 powders. For

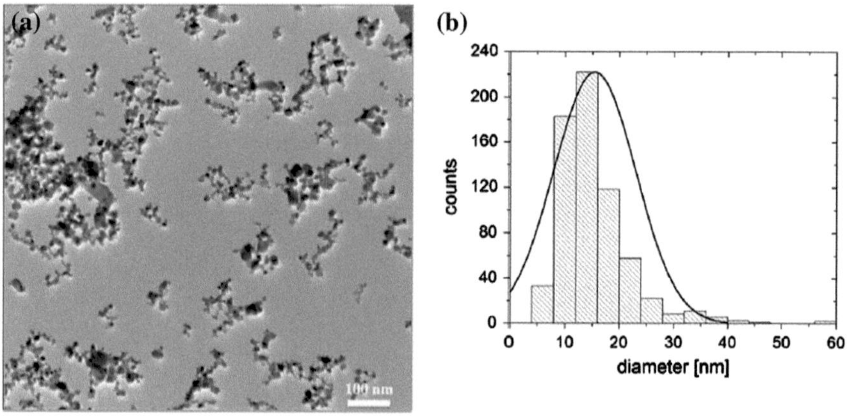

Fig. 3.59 **a** TEM image of the Lu_2O_3 nanoparticle and **b** its particle size distribution. Reproduced with permission from [323]. Copyright © 2012, Elsevier

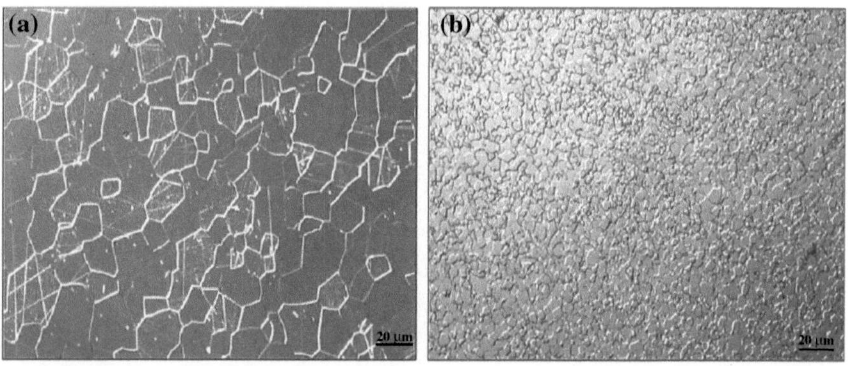

Fig. 3.60 SEM images of the cubic phase sample hot pressed at **a** 1625 °C and **b** the mixed phase hot pressed at 1450 °C. Reproduced with permission from [323]. Copyright © 2012, Elsevier

instance, the sample that was synthesized at 2 L min^{-1} was pure cubic Lu_2O_3, with space group of *Ia3*, whereas the one synthesized at 5 L min^{-1} contains a metastable monoclinic phase.

The powders were mixed with a sintering aid before sintering. Hot pressing densification was carried out in vacuum in a temperature range of 1450–1625 °C, at a pressure of ~8 kpsi, followed by hot isostatic pressing at 30 kpsi. The polished samples exhibited high transparency. Densification of ceramic powders is accomplished through atomic diffusion which is activated by thermal and shear energy in the case of hot pressing. Since hot pressing uses both thermal and mechanical energy sources, ceramics can be densified at lower temperatures which results in a finer grained microstructure.

Figure 3.60 shows SEM images of the samples made with the pure cubic phase powder that was hot pressed at 1625 °C and the mixed monoclinic phase pressed at

1450 °C [323]. The mixed phase sample had greatly reduced grain size, due to its lower sintering temperature. In other words, flame spray synthesis process offered an additional advantage to produce metastable phase, in this case (Lu_2O_3). The monoclinic Lu_2O_3 phase is a high free energy metastable phase, as compared with the cubic equilibrium phase. Because the high free energy, the presence of the monoclinic phase reduced the sintering temperature and pressure. The metastable phase was completely transferred to cubic phase during the hot pressing sintering.

3.3 Summary

A wide range of methods have been used to synthesize ceramic powders, with different compositions. Although most of these methods are applicable to transparent ceramic power synthesis, some of them have not been employed in the fabrication of transparent ceramics. Solid-state reaction is still the key method, while wet-chemical routes are increasingly used, whereas gas-phase reactions are only limited to nonoxide ceramics. For large scale applications, cost-effectiveness, environmental friendliness and less energy consumption, are all very important considerations, when selecting synthetic methods.

References

1. Biswas DR (1989) Development of novel ceramic processing. J Mater Sci 24:3791–3798
2. Kendall K (1989) Influence of powder structure on processing and properties of advanced ceramics. Powder Technol 58:151–161
3. Rice RW (1990) Ceramic processing—an overview. AIChE J 36:481–510
4. Sakai S, Ito S (1988) Problems on grinding process for ceramics. J Japan Soc Lubr Eng 33:509–513
5. Rahaman MN (2003) Ceramic processing and sintering, 2nd edn. CRC Press, New York
6. De Bonis A, Cultrone G, Grifa C, Langella A, Morra V (2014) Clays from the Bay of Naples (Italy): new insight on ancient and traditional ceramics. J Eur Ceram Soc 34:3229–3244
7. Gong H, Zhang J, Tang DY, Xie GQ, Huang H, Ma J (2011) Fabrication and laser performance of highly transparent Nd:YAG ceramics from well-dispersed Nd:Y_2O_3 nanopowders by freeze-drying. J Nanopart Res 13:3853–3860
8. Haruta M, Delmon B (1986) Preparation of homodisperse solids. J Chim Phys Phys Chim Biol 83:859–868
9. Hu S, Lu CH, Wang W, Ding MY, Ni YR, Xu ZZ (2013) Synthesis of monodisperse erbium aluminum garnet (EAG) nanoparticles via a microwave method. J Rare Earths 31:490–496
10. Huang YH, Jiang DL, Zhang JX, Lin QL, Huang ZR (2011) Synthesis of mono-dispersed spherical Nd:Y_2O_3 powder for transparent ceramics. Ceram Int 37:3523–3529
11. Li J, Li JP, Chen Q, Wu WJ, Xiao DQ, Zhu JG (2012) Effect of ammonium sulfate on the monodispersed $Y_3Al_5O_{12}$ nanopowders synthesized by co-precipitant method. Powder Technol 218:46–50
12. Li JS, Sun XD, Liu SH, Huo D, Li XD, Li JG et al (2013) Synthesis of dispersed Y_2O_3 nanopowder from yttrium stearate. In: Bao YW, Jiang DY, Gong JH (eds) Testing and evaluation of inorganic materials III. Trans Tech Publications Ltd, Stafa-Zurich, pp 3–7

13. Li XX, Wang WJ (2009) Preparation of uniformly dispersed YAG ultrafine powders by co-precipitation method with SDS treatment. Powder Technol 196:26–29
14. Xu GG, Zhang XD, He W, Liu H, Li H, Boughton RI (2006) Preparation of highly dispersed YAG nano-sized powder by co-precipitation method. Mater Lett 60:962–965
15. Xu XJ, Sun XD, Liu H, Li JG, Li XD, Huo D et al (2012) Synthesis of monodispersed spherical yttrium aluminum garnet (YAG) powders by a homogeneous precipitation method. J Am Ceram Soc 95:3821–3826
16. Dudnik EV, Zaitseva ZA, Shevchenko AV, Lopato LM (1995) Sintering of ultradisperse powders based on zirconium dioxide (Rev). Powder Metall Met Ceram 34:263–271
17. Aiken B, Hsu WP, Matijevic E (1988) Preparation and properties of monodispersed colloidal particles of lanthanide compounds, 3. Yttrium (III) and mixed yttrium (III)—cerium (III) systems. J Am Ceram Soc 71:845–853
18. Bogush GH, Tracy MA, Zukoski CF (1988) Preparation of monodisperse silica particles—control of size and mass fraction. J Non-Cryst Solids 104:95–106
19. Hsu WP, Ronnquist L, Matijevic E (1988) Preparation and properties of monodispersed colloidal particles of lanthanide compounds, 2. Cerium (IV). Langmuir 4:31–37
20. Lamer VK (1948) Monodisperse colloids and higher-order Tyndall spectra. J Phys Colloid Chem 52:65–76
21. Li J, Wu YJ, Tanaka H, Yamamoto T, Kuwabara M (2004) Preparation of a monodispersed suspension of barium titanate nanoparticles and electrophoretic deposition of thin films. J Am Ceram Soc 87:1578–1581
22. Matijevic E (1985) Production of monodispersed colloidal particles. Annu Rev Mater Sci 15:483–516
23. Matijevic E (1986) Monodispersed colloids—art and science. Langmuir 2:12–20
24. Pizette P, Martin CL, Delette G, Sornay P, Sans F (2010) Compaction of aggregated ceramic powders: from contact laws to fracture and yield surfaces. Powder Technol 198:240–250
25. Lee HM, Huang CY, Wang CJ (2009) Forming and sintering behaviors of commercial α-Al_2O_3 powders with different particle size distribution and agglomeration. J Mater Process Technol 209:714–722
26. Staiger M, Bowen P, Ketterer J, Bohonek J (2002) Particle size distribution measurement and assessment of agglomeration of commercial nanosized ceramic particles. J Dispersion Sci Technol 23:619–630
27. Bouvard D, Carry C, Chaix JM, Martin CM, Missiaen JM, Perier-Camby L et al (2001) Compression and sintering of powder mixtures: experiments and modelling. Adv Eng Mater 3:593–597
28. Olmos L, Martin CL, Bouvard D (2009) Sintering of mixtures of powders: experiments and modelling. Powder Technol 190:134–140
29. Johnson DW (1981) Non-conventional powder preparation techniques. Am Ceram Soc Bull 60:221–224
30. Hlavacek V, Puszynski JA (1996) Chemical engineering aspects of advanced ceramic materials. Ind Eng Chem Res 35:349–377
31. Kwade A (2001) Ultrafine comminution of ceramic raw materials—part 2: application. CFI-Ceram Forum Int 78:E39–E44
32. Kwade A (2001) Ultrafine comminution of ceramic raw materials part 1: fundamentals. CFI-Ceram Forum Int 78:E37–E45
33. Eskandari A, Aminzare M, Hesabi ZR, Aboutalebi SH, Sadrnezhaad SK (2012) Effect of high energy ball milling on compressibility and sintering behavior of alumina nanoparticles. Ceram Int 38:2627–2632
34. Yu SH (2001) Hydrothermal/solvothermal processing of advanced ceramic materials. J Ceram Soc Jpn 109:S65–S75
35. Roy R (1998) The sol-gel process in ceramic science: early history of discovery and subsequent development. In: Komarneni S, Sakka S, Phule PP, Laine RM (eds) Sol-Gel synthesis and processing. pp 3–36

36. Kong LB, Ma J, Huang H (2002) Low temperature formation of yttrium aluminum garnet from oxides via a high-energy ball milling process. Mater Lett 56:344–348
37. Kong LB, Ma J, Huang H (2002) $MgAl_2O_4$ spinel phase derived from oxide mixture activated by a high-energy ball milling process. Mater Lett 56:238–243
38. Dominey DA, Morley H, Young DA (1965) Kinetics of decomposition of nickel oxalate. Trans Faraday Soc 61:1246–1255
39. Tompkins FC, Young DA (1965) Decomposition of calcium azide. Trans Faraday Soc 61:1470–1480
40. L'Vov BV (2001) The physical approach to the interpretation of the kinetics and mechanisms of thermal decomposition of solids: the state of the art. Thermochim Acta 373:97–124
41. L'Vov BV (2002) Mechanism and kinetics of thermal decomposition of carbonates. Thermochim Acta 386:1–16
42. Salem IA, El-Maazawi M, Zaki AB (2000) Kinetics and mechanisms of decomposition reaction of hydrogen peroxide in presence of metal complexes. Int J Chem Kinet 32:643–666
43. Drozdetskaya GV, Gropyanov VM (1977) Calculation of kinetics of $CaCO_3$ decomposition under nonisothermal conditions. J Appl Chem USSR 50:480–483
44. Gallagher PK, Johnson DW (1976) Kinetics of thermal-decomposition of $CaCO_3$ in CO_2 and some observation on kinetic compensation effect. Thermochim Acta 14:255–261
45. Koga N, Tanaka H (1988) Kinetics of thermal-decomposition of $CaCO_3$ to MO, $SrCO_3$ to MO, $BaCO_3$ to MO. J Therm Anal 34:177–188
46. Kingery WD, Bowen HK, Uhlmann DR (1976) Introduction to ceramics, 2nd edn. Wiley, New York
47. Carter CB, Norton MG (2007) Ceramics materials: science and engineering. Springer, Berlin
48. Ewing J, Beruto D, Searcy AW (1979) Nature of CaO produced by calcite powder decomposition in vacuum and in CO_2. J Am Ceram Soc 62:580–584
49. Jander W (1927) Reactions in solid states at room temperature, I. Announcement the rate of reaction in endothermic conversions. Z Anorg Allg Chem 163:1–30
50. Jander W (1934) The reaction process in solid states. Angew Chem 47:0235–0238
51. Carter RE (1961) Kinetic model for solid-state reactions. J Chem Phys 34:2010–2015
52. Carter RE (1961) Mechanism of solid-state reaction between magnesium oxide and aluminum oxide and between magnesium oxide and ferric oxide. J Am Ceram Soc 44:116–120
53. Ikesue A, Furusato I, Kamata K (1995) Fabrication of polycrystalline transparent YAG ceramics by a solid-state reaction method. J Am Ceram Soc 78:225–228
54. Wang JQ, Yue YL, Tao WH, Yu QH, Tao ZD, Sun XD (2003) Preparation of transparent yttrium aluminum garnet ceramics by relatively low temperature solid-state reaction. Trans Nonferrous Metals Soc China 13:1096–1101
55. Wen L, Sun XD, Ma WM (2004) Fabrication of transparent Nd :YAG ceramics by a solid-state reaction method. J Inorg Mater 19:295–301
56. Li HL, Liu XJ, Huang LP (2005) Fabrication of transparent cerium-doped lutetium aluminum garnet (LuAG:Ce) ceramics by a solid-state reaction method. J Am Ceram Soc 88:3226–3228
57. Li HL, Liu XJ, Huang LP (2006) Fabrication of transparent Ce:LuAG ceramics by a solid-state reaction method. J Inorg Mater 21:1161–1166
58. Wu YS, Li J, Qiu FG, Pan YB, Liu Q, Guo JK (2006) Fabrication of transparent Yb, Cr:YAG ceramics by a solid-state reaction method. Ceram Int 32:785–788
59. Li J, Chen Q, Yang LL, Feng GY, Wu WJ, Zheng FS et al (2011) High transmittance of Nd-doped YAG transparent ceramics prepared by solid-state reaction method. Ferroelectrics 411:62–68
60. You YL, Qi LH, Miao HZ, Pan W (2013) Preparation of transparent YAG ceramics via a modified solid-state reaction method. Rare Metal Mater Eng 42:348–350
61. Li JL, Xu J, Shi Y, Qi HF, Xie JJ, Lei F (2014) Fabrication and microstructure of cerium doped lutetium aluminum garnet (Ce:LuAG) transparent ceramics by solid-state reaction method. Mater Res Bull 55:161–167

62. Ganesh I, Srinivas B, Johnson R, Saha BP, Mahajan YR (2004) Microwave assisted solid state reaction synthesis of $MgAl_2O_4$ spinel powders. J Eur Ceram Soc 24:201–207
63. Zhan W, Wang ZF, Zhang BB, Wang XT (2009) $MgAl_2O_4$ spinel synthesized by solid state reaction at low-heating temperature. Rare Metal Mater Eng 38:34–37
64. Zargar HR, Bayati MR, Rezaie HR, Golestani-Fard F, Molaei R, Zanganeh S et al (2010) Influence of nano boehmite on solid state reaction of alumina and magnesia. J Alloy Compd 507:443–447
65. Su XH, Li JG, Zhou ZJ (2012) Solid-state reaction preparation and sintering behavior of $MgAl_2O_4$ nanopowders. J Inorg Mater 27:991–996
66. Wang ZF, Liu H, Wang XT, Zhan W (2013) Sintering behavior of $MgAl_2O_4$ spinel powders prepared by solid state reaction at low temperature. Rare Metal Mater Eng 42:670–672
67. Wang NL, Zhang XY, Jiang HT, Dong TT, Yang D (2012) Fabrication of Er^{3+}/Yb^{3+} co-doped Y_2O_3 transparent ceramics by solid-state reaction method and its up-conversion luminescence. Mater Chem Phys 135:709–713
68. Yuan XY, Zhang F, Liu XJ, Zhang Z, Wang SW (2011) Fabrication of transparent AlON ceramics by solid-state reaction sintering. J Inorg Mater 26:499–502
69. Yuan XY, Liu XJ, Zhang F, Wang SW (2010) Synthesis of γ-AlON powders by a combinational method of carbothermal reduction and solid-state reaction. J Am Ceram Soc 93:22–24
70. Liu BL, Li J, Ivanov M, Liu WB, Liu J, Xie TF et al (2014) Solid-state reactive sintering of Nd:YAG transparent ceramics: the effect of Y_2O_3 powders pretreatment. Opt Mater 36:1591–1597
71. Kong LB, Zhang TS, Ma J, Boey F (2008) Progress in synthesis of ferroelectric ceramic materials via high-energy mechanochemical technique. Prog Mater Sci 53:207–322
72. Sepelak V, Duevel A, Wilkening M, Becker K-D, Heitjans P (2013) Mechanochemical reactions and syntheses of oxides. Chem Soc Rev 42:7507–7520
73. Fang F, Zhu M (2002) Mechanical alloying in immiscible alloy systems. Prog Nat Sci 12:170–174
74. Murty BS, Ranganathan S (1998) Novel materials synthesis by mechanical alloying/milling. Int Mater Rev 43:101–141
75. Ruiz-Navas EM, da Costa CE, Lopez FV, Castello JMT (2000) Mechanical alloying: a method to obtain metallic powders and composite materials. Rev De Metal 36:279–286
76. Suryanarayana C (2001) Mechanical alloying and milling. Prog Mater Sci 46:1–184
77. Cheng L, Liu P, Qu SX, Cheng L, Zhang HW (2015) Microwave dielectric properties of Mg_2TiO_4 ceramics synthesized via high energy ball milling method. J Alloy Compd 623:238–242
78. Bafrooei HB, Ebadzadeh T (2013) $MgAl_2O_4$ nanopowder synthesis by microwave assisted high energy ball-milling. Ceram Int 39:8933–8940
79. Cedeno-Mattei Y, Perales-Perez O, Uwakweh ONC (2013) Effect of high-energy ball milling time on structural and magnetic properties of nanocrystalline cobalt ferrite powders. J Magn Magn Mater 341:17–24
80. Zakeri M, Rahimipour MR, Abbasi BJ (2013) Synthesis of nanostructure tetragonal ZrO_2 by high energy ball milling. Mater Technol 28:181–186
81. Aguilar-Garib JA, Sanchez-de-Jesus F, Bolarin-Miro AM, Ham-Hernandez S (2011) Synthesis of $NiMn_2O_4$ assisted by high-energy ball milling of NiO-MnO powders. J Ceram Process Res 12:721–726
82. Sepelak V, Begin-Colin S, Le Caer G (2012) Transformations in oxides induced by high-energy ball-milling. Dalton Trans 41:11927–11948
83. Kong LB, Zhang TS, Ma J, Boey FYC (2007) Anisotropic grain growth in mullite powders as a result of high-energy ball milling. J Am Ceram Soc 90:4055–4058
84. Parashar SKS, Choudhary RNP, Murty BS (2005) Size effect of $Pb_{0.92}Nd_{0.08}(Zr_{0.53}Ti_{0.47})_{0.98}O_3$ nanoceramic synthesized by high-energy ball milling. J Appl Phys 98:104305

85. Kong LB, Ma J, Huang H, Zhang TS, Boey F (2003) Anisotropic mullitization in CuO-doped oxide mixture activated by high-energy ball milling. Mater Lett 57:3660–3666
86. Kong LB, Zhang TS, Ma J, Boey F (2003) Anisotropic grain growth of mullite in high-energy ball milled powders doped with transition metal oxides. J Eur Ceram Soc 23:2247–2256
87. Meitl MA, Dellinger TM, Braun PV (2003) Bismuth-ceramic nanocomposites with unusual thermal stability via high-energy ball milling. Adv Funct Mater 13:795–799
88. Kong LB, Ma J, Zhu W, Tan OK (2002) Transparent PLZT8/65/35 ceramics from constituent oxides mechanically modified by high-energy ball milling. J Mater Sci Lett 21:197–199
89. Kong LB, Ma J, Zhu W, Tan OK (2002) Translucent PMN and PMN-PT ceramics from high-energy ball milling derived powders. Mater Res Bull 37:23–32
90. Kong LB, Ma J, Zhu W, Tan OK (2001) Preparation and characterization of translucent PLZT8/65/35 ceramics from nano-sized powders produced by a high-energy ball-milling process. Mater Res Bull 36:1675–1685
91. Kong LB, Ma J, Huang HT, Zhu W, Tan OK (2001) Lead zirconate titanate ceramics derived from oxide mixture treated by a high-energy ball milling process. Mater Lett 50:129–133
92. Kong LB, Ma J, Zhang RF, Zhang TS (2002) Fabrication and characterization of lead lanthanum zirconate titanate (PLZT7/60/40) ceramics from oxides. J Alloy Compd 339:167–174
93. Kong LB, Ma J, Zhu W, Tan OK (2001) Preparation and characterization of PLZT ceramics using high-energy ball milling. J Alloy Compd 322:290–297
94. Chen IW, Wang XH (2000) Sintering dense nanocrystalline ceramics without final-stage grain growth. Nature 404:168–171
95. Zhang QW, Saito F (2003) Mechanochemical solid reaction of yttrium oxide alumina leading to the synthesis of yttrium aluminum garnet. Powder Technol 129:86–91
96. Huang H, Gong H, Tang D, Tan OK (2009) Synthesis and characterization of yttrium aluminum garnet by high-energy ball milling. Opt Mater 31:716–719
97. Kim W, Baker C, Villalobos G, Frantz J, Shaw B, Sadowski B et al (2014) Highly transparent ceramics obtained from jet milled sesquioxide powders synthesized by co-precipitation method. Optical Mater Express 4:2497–2503
98. An LQ, Ito A, Goto T (2011) Effects of ball milling and post-annealing on the transparency of spark plasma sintered Lu_2O_3. Ceram Int 37:2263–2267
99. Liu J, Lin L, Li J, Liu J, Yuan Y, Ivanov M et al (2014) Effects of ball milling time on microstructure evolution and optical transparency of Nd:YAG ceramics. Ceram Int 40:9841–9851
100. Chen XT, Lu TC, Wei N, Lu ZW, Zhang W, Ma BY et al (2014) Effect of ball-milling granulation with PVB adhesive on the sinterability of co-precipitated Yb:YAG nanopowders. J Alloy Compd 589:448–454
101. Wang J, Zhang J, Luo DW, Yang H, Tang DY, Kong LB (2015) Densification and microstructural evolution of yttria transparent ceramics: the effect of ball milling conditions. J Eur Ceram Soc 35:1011–1019
102. Wilhelmy RB, Patel RC, Matijevic E (1985) Thermodynamics and kinetics of aqueous ferric phosphate complex-formation. Inorg Chem 24:3290–3297
103. Matijevic E, Simpson CM, Amin N, Arajs S (1986) Preparation and magnetic properties of well-defined colloidal chromium ferrites. Colloids Surf 21:101–108
104. Haruta M, Lemaitre J, Delannay F, Delmon B (1984) Preparation and properties of colloidal spherical particles of molybdenum and cobal sulfides. J Colloid Interface Sci 101:59–71
105. Overbeek JTG (1977) Recent developments in understanding of colloid stability. J Colloid Interface Sci 58:408–422
106. Overbeek JTG (1982) Mondisperse colloidal systems, fascinating and useful. Adv Colloid Interface Sci 15:251–277
107. Overbeek JTG (1984) Interparticle forces in colloid science. Powder Technol 37:195–208

108. McDonald JE (1962) Homogeneous nucleation of vapor condensation, 1, thermodynamic aspects. Am J Phys 30:870–876
109. McDonald JE (1963) Homogeneous nucleation of vapor condensation, 2. Kinetic aspects. Am J Phys 31:31–41
110. Atkins P, de Paula J (2006) Physical chemistry, 8th edn. Oxford University Press, Oxford
111. Lamer VK, Dinegar RH (1950) Theory, production and mechanism of formation of monodispersed hydrosols. J Am Chem Soc 72:4847–4854
112. Christian JW (1975) The theory of transformations in metals and alloys: pergamon. Oxford
113. Wood GR, Walton AG (1970) Homogeneous nucleation kinetics of ice from water. J Appl Phys 41:3027–3036
114. Hodgson AW (1984) Homogeneous nucleation. Adv Colloid Interface Sci 21:303–327
115. Oxtoby DW (1992) Homogeneous nucleation—theory and experiment. J Phys-Condens Matter 4:7627–7650
116. Roldugin VI (2000) Homogeneous nucleation theory. Russ J Phys Chem 74:S522–S532
117. Holmes HF, Mesmer RE (1986) Thermodynamics of aqueous solutions of alkali-metal sulfates. J Solution Chem 15:495–518
118. Wesolowski D, Drummond SE, Mesmer RE, Ohmoto H (1984) Hydrolysis equilibria of tungsetn (VI) in aqueous sodium chloride solutions to 300 °C. Inorg Chem 23:1120–1132
119. Reiss H (1951) The growth of uniform colloidal dispersions. J Chem Phys 19:482–487
120. Nielsen AE (1957) Nucleation in barium sulfate precipitation. Acta Chem Scand 11:1512–1515
121. Nielsen AE (1958) The kinetics of crystal growth in barium sulfate precipitation. Acta Chem Scand 12:951–958
122. Nielsen AE (1959) The kinetics of crystal growth in barium sulfate precipitation, 2. Temperature dependence and mechanism. Acta Chem Scand 13:784–802
123. Nielsen AE (1959) The kinetics of crystal growth in barium sulfate precipitation, 3. Mixed surface reaction and diffusion controlled rate of growth. Acta Chem Scand 13:1680–1686
124. Nielsen AE, Bindra PS (1973) Effect of curvature on interfacial tension in liquid systems measured by homogeneous nucleation. Croat Chem Acta 45:31–52
125. Smellie RH, Lamer VK (1954) The electrokinetic properties of dilute monodisperse sufur hydrosols. J Phys Chem 58:583–591
126. Zaiser EM, Lamer VK (1948) The kinetics of the formation and growth of monodispersed sulfur hydrosols. J Colloid Sci 3:571–598
127. Matijevic E, Hsu WP (1987) Preparation and properties of monodispersed colloidal particles of lanthanide compounds, 1. Gadolinium, europium, terbium, samarium and cerium (III). J Colloid Interface Sci 118:506–523
128. Hsu WP, Matijevic E (1985) Optical properties of monodispersed hetatite hydrosols. Appl Opt 24:1623–1630
129. Bogush GH, Zukoski CF (1991) Uniform silica particle precipitation—an aggregative growth model. J Colloid Interface Sci 142:19–34
130. Sugimoto T (1987) Preparation of monodispersed colloidal particles. Adv Colloid Interface Sci 28:65–108
131. Park J, Privman V, Matijevic E (2001) Model of formation of monodispersed colloids. J Phys Chem B. 105:11630–11635
132. Matijevic E (1993) Preparation and properties of uniform size colloids. Chem Mater 5:412–426
133. Enomoto Y, Kawasaki K, Tokuyama M (1987) The time-dependent behavior of the Ostwald ripening for the finite volume fraction. Acta Metall 35:915–922
134. Marqusee JA, Ross J (1984) Theory of Ostwald ripening—competitive growth and its dependence on volume fraction. J Chem Phys 80:536–543
135. Jain SC, Hughes AE (1978) Ostwald ripening and its application to precipitates and colloids in ionic crystals and glasses. J Mater Sci 13:1611–1631
136. Kabalnov A (2001) Ostwald ripening and related phenomena. J Dispersion Sci Technol 22:1–12

137. Vengrenovitch RD (1982) On the Ostwald ripening theory. Acta Metall 30:1079–1086
138. Chen JY, Shi Y, Shi JL (2004) Synthesis of (Y, Gd)2O3: Eu nanopowder by a novel co-precipitation processing. J Mater Res 19:3586–3591
139. Chen JY, Shi Y, Shi JL (2004) Preparation and luminescence property of (Y, Gd)2O3: Eu nanopowder by complex precipitation. J Inorg Mater 19:1260–1266
140. Gong H, Tang DY, Huang H, Ma J (2009) Fabrication of yttrium aluminum garnet transparent ceramics from yttria nanopowders synthesized by carbonate precipitation. J Electroceram 23:89–93
141. Gong H, Tang DY, Huang H, Zhang TS, Ma J (2008) Effect of grain size on the sinterability of yttria nanopowders synthesized by carbonate-precipitation process. Mater Chem Phys 112:423–426
142. Huang Z, Guo W, Fei BJ, Li JT, Cao YG (2013) Influence of sulphate on synthesis of Nd: Y_2O_3 powders via coprecipitation route and fabrication of transparent ceramics. Mater Res Innovations 17:73–79
143. Huang ZG, Sun XD, Xiu ZM, Chen SW, Tsai CT (2004) Precipitation synthesis and sintering of yttria nanopowders. Mater Lett 58:2137–2142
144. Luo JM, Li YX, Deng LP (2008) Synthesis of Yb^{3+}:Y_2O_3 nanocrystalline by co-precipitation method and the properties of transparent ceramics. Chin J Inorg Chem 24:260–264
145. Wang NL, Zhang XY, Bai ZH, Liu QS, Lu LP, Mi XY et al (2010) Carbonate-precipitation synthesis of Yb^{3+}:Y_2O_3 nanopowders and its characteristics. Powder Technol 203:458–461
146. Wang PH, Wang NL, Zhang XY (2012) Carbonate co-precipitation synthesis of Lu_2O_3:Er^{3+} nano-powders and its property characterization. Chin J Inorg Chem 28:2335–2340
147. Ghanizadeh S, Bao X, Vaidhyanathan B, Binner J (2014) Synthesis of nano α-alumina powders using hydrothermal and precipitation routes: a comparative study. Ceram Int 40:1311–1319
148. Kong J, Chao B, Wang T, Yan Y (2012) Preparation of ultrafine spherical AlOOH and Al_2O_3 powders by aqueous precipitation method with mixed surfactants. Powder Technol 229:7–16
149. Hassanzadeh-Tabrizi SA, Taheri-Nassaj E (2009) Economical synthesis of Al_2O_3 nanopowder using a precipitation method. Mater Lett 63:2274–2276
150. Du X, Zhao S, Liu Y, Li J, Chen W, Cui Y (2014) Facile synthesis of monodisperse alpha-alumina nanoparticles via an isolation-medium-assisted calcination method. Appl Phys A-Mater Sci Process 116:1963–1969
151. Shen L, Hu C, Sakka Y, Huang Q (2012) Study of phase transformation behaviour of alumina through precipitation method. J Phys D-Appl Phys 45:215302
152. Zamorategui A, Sugita S, Zarraga R, Tanaka S, Uematsu K (2012) Evaluation of dispersability of gamma alumina prepared by homogeneous precipitation. J Ceram Soc Jpn 120:290–294
153. Wang W, Qiao X, Chen J (2008) The role of acetic acid in magnesium oxide preparation via chemical precipitation. J Am Ceram Soc 91:1697–1699
154. Yan C, Xue D, Zou L (2006) Fabrication of hexagonal MgO and its precursors by a homogeneous precipitation method. Mater Res Bull 41:2341–2348
155. Guo M, Li Q, Ye X, Wu Z (2010) Magnesium carbonate precipitation under the influence of polyacrylamide. Powder Technol 200:46–51
156. Camtakan Z, Erenturk S, Yusan S (2012) Magnesium oxide nanoparticles: Preparation, characterization, and uranium sorption properties. Environ Progr Sustain Energy 31:536–543
157. Li JG, Sun XD (2000) Synthesis and sintering behavior of a nanocrystalline alpha-alumina powder. Acta Mater 48:3103–3112
158. Su X, Chen S, Zhou Z (2012) Synthesis and characterization of monodisperse porous α-Al_2O_3 nanoparticles. Appl Surf Sci 258:5712–5715
159. Sohn S, Kwon Y, Kim Y, Kim D (2004) Synthesis and characterization of near-monodisperse yttria particles by homogeneous precipitation method. Powder Technol 142:136–153

160. Arabgari S, Malekfar R, Motamedi K (2011) Parameters effects on the surface morphology and structure of Nd:YAG nanopowders synthesized by co-precipitation method. J Nanopart Res 13:597–611
161. Guo W, Lu TC, Tong SH (2007) Effect of phase of YAG powder synthesized by co-precipitation on transparent ceramic sintering. In: Pan W, Gong JH (eds) High-performance ceramics IV, Pts 1-3. Trans Tech Publications Ltd, Stafa-Zurich, pp 2054–2057
162. Ji XB, Deng JG, Kang B, Huang H, Wang X, Jing W et al (2013) Thermal decomposition of $Y_3Al_5O_{12}$ precursor synthesized by urea homogeneous co-precipitation. Js Anal Appl Pyrol 104:361–365
163. Jiang FH, Zhao DG, Wang JQ (2013) Synthesis of YAG nano-powders and transparent ceramic by microwave homogeneous precipitation. Asian J Chem 25:5487–5489
164. Li HL, Liu XJ, Xie RJ, Zeng Y, Huang LP (2006) Fabrication of transparent cerium-doped lutetium aluminum garnet ceramics by co-precipitation routes. J Am Ceram Soc 89:2356–2358
165. Li HL, Liu XJ, Xie RJ, Zhou GH, Hirosaki N, Pu XP et al (2008) Cerium-doped lutetium aluminum garnet phosphors and optically transparent ceramics prepared from powder precursors by a urea homogeneous precipitation method. Jpn J Appl Phys 47:1657–1661
166. Li J, Chen F, Liu WB, Zhang WX, Wang L, Ba XW et al (2012) Co-precipitation synthesis route to yttrium aluminum garnet (YAG) transparent ceramics. J Eur Ceram Soc 32:2971–2979
167. Li XX, Zheng BY, Odoom-Wubah T, Huang JL (2013) Co-precipitation synthesis and two-step sintering of YAG powders for transparent ceramics. Ceram Int 39:7983–7988
168. Liu JH, Piao XQ, Cui HX, Guan XX, Li JL (2003) Preparation of Yb:YAG transparent laser ceramics with urea co-precipitation method. J Rare Earths 21:113–116
169. Marlot C, Barraud E, Le Gallet S, Eichhorn M, Bernard F (2012) Synthesis of YAG nanopowder by the co-precipitation method: Influence of pH and study of the reaction mechanisms. J Solid State Chem 191:114–120
170. Serantoni M, Costa AL, Zanelli C, Esposito L (2014) Crystallization behaviour of Yb-doped and undoped YAG nanoceramics synthesized by microwave-assisted urea precipitation. Ceram Int 40:11837–11844
171. You YL, Qi LH, Li XL, Pan W (2013) Preparation of YAG nano-powders via an ultrasonic spray co-precipitation method. Ceram Int 39:3987–3992
172. Wang L, Kou HM, Zeng YP, Li J, Pan YB, Guo JK (2012) Preparation of YAG powders and ceramics through mixed precipitation method. Ceram Int 38:4401–4405
173. Su CH, Zhang HS, Hui H, Qiong S, Hu HD, Zhang HB et al (2005) Preparation of neodymium-doped yttrium aluminum garnet transparent ceramics by homogeneous precipitation method. J Rare Earths 23:716–720
174. Li JG, Ikegami T, Lee JH, Mori T, Yajima Y (2001) A wet-chemical process yielding reactive magnesium aluminate spinel ($MgAl_2O_4$) powder. Ceram Int 27:481–489
175. Li JG, Ikegami T, Lee JH, Mori T, Yajima Y (2001) Synthesis of Mg-Al spinel powder via precipitation using ammonium bicarbonate as the precipitant. J Eur Ceram Soc 21:139–148
176. Lv GZ, Li XD, Di H, Sun XD, Chen SW, Tsai CT et al (2008) Fabrication of transparent YAG ceramics from co-precipitation synthesized nanopowders. In: Hu X, Fillery B, Qasim T, Duan K (eds) Structural integrity and failure. Trans Tech Publications Ltd, Stafa-Zurich, pp 271–276
177. Huang YH, Jiang DL, Zhang JX, Lin QL (2009) Precipitation synthesis and sintering of lanthanum doped yttria transparent ceramics. Opt Mater 31:1448–1453
178. Kriechbaum GW, Kleinschmit P (1989) Superfine oxide powders—flame hydrolysis and hydrothermal synthesis. Angewandte Chemie-Int Ed Engl 28:1416–1423
179. Sahraneshin A, Takami S, Minami K, Hojo D, Arita T, Adschiri T (2012) Synthesis and morphology control of surface functionalized nanoscale yttrium aluminum garnet particles via supercritical hydrothermal method. Prog Cryst Growth Charact Mater 58:43–50

180. Yang Q, Lu Z, Liu J, Lei X, Chang Z, Luo L et al (2013) Metal oxide and hydroxide nanoarrays: Hydrothermal synthesis and applications as supercapacitors and nanocatalysts. Prog Nat Sci-Mater Int 23:351–366
181. Ahn KH, Lee Y-H, Kim M, H-s Lee, Youn Y-S, Kim J et al (2013) Effects of surface area of titanium dioxide precursors on the hydrothermal synthesis of barium titanate by dissolution-precipitation. Ind Eng Chem Res 52:13370–13376
182. Avila HA, Ramajo LA, Reboredo MM, Castro MS, Parra R (2011) Hydrothermal synthesis of $BaTiO_3$ from different Ti-precursors and microstructural and electrical properties of sintered samples with submicrometric grain size. Ceram Int 37:2383–2390
183. Han J-M, Joung M-R, Kim J-S, Lee Y-S, Nahm S, Choi Y-K et al (2014) Hydrothermal synthesis of $BaTiO_3$ nanopowders using TiO_2 nanoparticles. J Am Ceram Soc 97:346–349
184. Kaur J, Kotnala RK, Verma KC (2012) Surfactant free hydrothermal synthesis, electrical, optical and ferroelectric properties of $BaTiO_3$ nanoparticles. J Optoelectron Adv Mater 14:219–223
185. Ozen M, Mertens M, Luyten J, Snijkers F, D'Hondt H, Cool P (2012) Hydrothermal synthesis of carbonate-free submicron-sized barium titanate from an amorphous precursor: synthesis and characterization. Ceram Int 38:619–625
186. Wang W, Cao L, Liu W, Su G, Zhang W (2013) Low-temperature synthesis of $BaTiO_3$ powders by the sol-gel-hydrothermal method. Ceram Int 39:7127–7134
187. Eckert JO, HungHouston CC, Gersten BL, Lencka MM, Riman RE (1996) Kinetics and mechanisms of hydrothermal synthesis of barium titanate. J Am Ceram Soc 79:2929–2939
188. Suchanek WL (2010) Hydrothermal synthesis of α-alumina (α-Al_2O_3) powders: study of the processing variables and growth mechanisms. J Am Ceram Soc 93:399–412
189. Mancic L, Lojpur V, Marinkovic BA, Dramicanin MD, Milosevic O (2013) Hydrothermal synthesis of nanostructured Y_2O_3 and $(Y_{0.75}Gd_{0.25})_2O_3$ based phosphors. Opt Mater 35:1817–1823
190. Huang BT, Ma YQ, Qian SB, Zou D, Zheng GH, Dai ZX (2014) Luminescent properties of low-temperature-hydrothermally-synthesized and post-treated YAG: Ce (5 %) phosphors. Opt Mater 36:1561–1565
191. Moore CA, McMillen CD, Kolis JW (2013) Hydrothermal growth of single crystals of $Lu_3Al_5O_{12}$ (LuAG) and its doped analogues. Cryst Growth Des 13:2298–2306
192. Qian S, Ma Y, Zan F, Zou D, Dai Z, Zheng G et al (2013) Fine YAG:Ce^{3+} nanoparticles synthesised by supercritical hydrothermal reaction. Micro Nano Lett 8:201–205
193. Garg A, Matijevic E (1988) Preparation and properties of uniformly coated inorganic colloidal particles, 3. Zirconium hydrous oxide on hematite. J Colloid Interface Sci 126:243–250
194. Garg A, Matijevic E (1988) Preparation and properties of uniformly coated inorganic colloidal particles, 2. Chromium hydrous oxide on hematite. Langmuir 4:38–44
195. Kratohvil S, Matijevic E (1987) Prepration and properties of coated, uniform, inorganic colloidal particles, 1. Aluminum (hydrous) oxide on hematite, chromia and titania. Adv Ceram Mater 2:798–803
196. Ocana M, Hsu WP, Matijevic E (1991) Preparation and properties of uniform-coated colloidal particles, 6. Titania Zinc-oxide. Langmuir 7:2911–2916
197. Garg AK, Dejonghe LC (1990) Microencapsulation of silicon-nitride particles with yttria and yttria-alumina precursors. J Mater Res 5:136–142
198. Okamura H, Barringer EA, Bowen HK (1986) Preparation and sintering of monosized Al_2O_3-TiO_2 composite powder. J Am Ceram Soc 69:C22–C24
199. Okamura H, Barringer EA, Bowen HK (1989) Preparation and sintering of narrow-sized Al_2O_3-TiO_2 composite powders. J Mater Sci 24:1867–1880
200. Garg AK, Dejonghe LC (1993) Metal-coated colloidal particles. J Mater Sci 28:3427–3432
201. Rahaman MN, Dejonghe LC (1991) Sintering of ceramic particlulate composites—effect of matrix density. J Am Ceram Soc 74:433–436
202. Aiken B, Hsu WP, Matijevic E (1990) Preparation and properties of uniform mixed and coated colloidal particles, 5. Zirconium compounds. J Mater Sci 25:1886–1894

203. Aiken B, Matijevic E (1988) Preparation and properties of uniform coated inorganic colloidal particles, 4. Yttrium basic carbonate and yttrium-oxide on hematite. J Colloid Interface Sci 126:645–649
204. Hu CL, Rahaman MN (1992) Factors controlling the sintering of ceramic particulate composites, 2. Coated inclusion particles. J Am Ceram Soc 75:2066–2070
205. Hu CL, Rahaman MN (1994) Dense Al_2O_3/ZrO_2 particulate composites by free sintering of coated powders. J Am Ceram Soc 77:815–819
206. Gherardi P, Matijevic E (1986) Interactions of precipitated hematite with preformed colloidal titania dispersions. J Colloid Interface Sci 109:57–68
207. Giesche H, Matijevic E (1994) Preparation, characterization and sinterability of well-defined silica/yttria powders. J Mater Res 9:436–450
208. Sang YH, Qin HM, Liu H, Zhao LL, Wang YN, Jiang HD et al (2013) Partial wet route for YAG powders synthesis leading to transparent ceramic: a core-shell solid-state reaction process. J Eur Ceram Soc 33:2617–2623
209. Mezhericher M, Levy A, Borde I (2015) Multi-scale multiphase modeling of transport phenomena in spray-drying processes. Drying Technol 33:2–23
210. Nandiyanto ABD, Okuyama K (2011) Progress in developing spray-drying methods for the production of controlled morphology particles: from the nanometer to submicrometer size ranges. Adv Powder Technol 22:1–19
211. Okuyama K, Abdullah M, Lenggoro IW, Iskandar F (2006) Preparation of functional nanostructured particles by spray drying. Adv Powder Technol 17:587–611
212. Hein JC, Rafflenbeul R, Beckmann M (1982) Progress in the technology of spray drying. Chem Ing Tech 54:787–792
213. Melnyk IV, Zub YL, Veron E, Massiot D, Cacciaguerra T, Alonso B (2008) Spray-dried mesoporous silica microspheres with adjustable textures and pore surfaces homogenously covered by accessible thiol functions. J Mater Chem 18:1368–1382
214. Pilarska A, Markiewicz E, Ciesielczyk F, Jesionowski T (2011) The influence of spray drying on the dispersive and physicochemical properties of magnesium oxide. Drying Technol 29:1210–1218
215. Serantoni M, Piancastelli A, Costa AL, Esposito L (2012) Improvements in the production of Yb:YAG transparent ceramic materials: Spray drying optimisation. Opt Mater 34:995–1001
216. Cho JS, Jung KY, Son MY, Kang YC (2014) Large-scale production of spherical $Y_2O_3:Eu^{3+}$ phosphor powders with narrow size distribution using a two-step spray drying method. Rsc Adv 4:62965–62970
217. Jung DS, Bin Park S, Kang YC (2010) Design of particles by spray pyrolysis and recent progress in its application. Korean J Chem Eng 27:1621–1645
218. Patil PS (1999) Versatility of chemical spray pyrolysis technique. Mater Chem Phys 59:185–198
219. Han JM, Jung DS, Lee SH, Kang YC (2008) Nano-sized MgO particles ranging from 13 to 28 nm synthesized by spray pyrolysis. J Ceram Process Res 9:140–145
220. Sohn JR, Kang YC, Park HD (2002) Morphological control of Y_2O_3:Eu phosphor particles by adding polymeric precursors in spray pyrolysis. Jpn J Appl Phys Part 1-Regul Papers Short Notes Rev Papers 41:3006–9
221. Marchal J, John T, Baranwal R, Hinklin T, Laine RM (2004) Yttrium aluminum garnet nanopowders produced by liquid-feed flame spray pyrolysis (LF-FSP) of metalloorganic precursors. Chem Mater 16:822–831
222. Nyman M, Caruso J, HampdenSmith MJ (1997) Comparison of solid-state and spray-pyrolysis synthesis of yttrium aluminate powders. J Am Ceram Soc 80:1231–1238
223. Mooney JB, Radding SB (1982) Spray pyrolysis processing. Annu Rev Mater Sci 12:81–101
224. Messing GL, Zhang SC, Jayanthi GV (1993) Ceramic powder synthesis by spray-pyrolysis. J Am Ceram Soc 76:2707–2726
225. Zhang SC, Messing GL, Borden M (1990) Synthesis of solid, spherical zirconia particles by spray pyrolysis. J Am Ceram Soc 73:61–67

226. Jung KY, Lee DY, Kang YC (2005) Morphology control and luminescent property of Y3Al5O12: Tb particles prepared by spray pyrolysis. Mater Res Bull 40:2212–2218
227. Jung KY, Lee HW (2007) Enhanced luminescent properties of $Y_3Al_5O_{12}$:Tb^{3+}, Ce^{3+} phosphor prepared by spray pyrolysis. J Lumin 126:469–474
228. Lee SH, Jung DS, Han JM, Koo HY, Kang YC (2009) Fine-sized $Y_3Al_5O_{12}$: Ce phosphor powders prepared by spray pyrolysis from the spray solution with barium fluoride flux. J Alloy Compd 477:776–779
229. Lee SH, Koo HY, Lee SM, Kang YC (2010) Characteristics of $Y_3Al_5O_{12}$: Ce phosphor powders prepared by spray pyrolysis from ethylenediaminetetraacetic acid solution. Ceram Int 36:611–615
230. Claussen IC, Ustad TS, Strommen I, Waide PM (2007) Atmospheric freeze drying—a review. Drying Technol 25:947–957
231. Geidobler R, Winter G (2013) Controlled ice nucleation in the field of freeze-drying: Fundamentals and technology review. Eur J Pharm Biopharm 85:214–222
232. Qian L, Zhang H (2011) Controlled freezing and freeze drying: a versatile route for porous and micro-/nano-structured materials. J Chem Technol Biotechnol 86:172–184
233. Tallon C, Moreno R, Nieto MI (2006) Synthesis of gamma-Al_2O_3 nanopowders by freeze-drying. Mater Res Bull 41:1520–1529
234. Huang Y-H, Jiang D-L, Zhang J-X, Lin Q-L (2008) Fabrication of transparent yttria ceramics through gel-freezing dry method. J Inorg Mater 23:1135–1140
235. Li RX, Yin S, Chen L, Yu KN, Li SH, Liang HZ et al (2003) Microstructure evolution of ZrO_2-(Al_2O_3, Fe_2O_3) materials synthesized by frozen-drying. J Inorg Mater 18:373–378
236. Tallon C, Moreno R (2009) Isabel Nieto M. Synthesis of ZrO_2 nanoparticles by freeze drying. Int J Appl Ceram Technol 6:324–334
237. Zhao Z, Chen D, Jiao X (2007) Zirconia aerogels with high surface area derived from sols prepared by electrolyzing zirconium oxychloride solution: Comparison of aerogels prepared by freeze-drying and supercritical CO_2(l) extraction. J Phys Chem C 111:18738–18743
238. Zhang X, Zhao HZ, Ma Q, Li XW, Wang HZ, Zhang WJ (2005) Preparation of spinel nano-powder with vacuum freeze drying. Rare Metal Mater Eng 34:78–81
239. McHale JM, Navrotsky A, Kirkpatrick RJ (1998) Nanocrystalline spinel from freeze-dried nitrates: Synthesis, energetics of produce formation, and cation distribution. Chem Mater 10:1083–1090
240. Tao CS, Pask JA (1973) Correlation of sintered spinel microstructure with characteristics of powder prepared by freeze-drying. Am Ceram Soc Bull 52:708
241. Wang CT, Lin LS, Yang SJ (1992) Preparation of $MgAl_2O_4$ spinel powders via freeze-drying of alkoxide precursors. J Am Ceram Soc 75:2240–2243
242. Rabinovitch Y, Bogicevic C, Karolak F, Tetard D, Dammak H (2008) Freeze-dried nanometric neodymium-doped YAG powders for transparent ceramics. J Mater Process Technol 199:314–320
243. Gong H, Tang DY, Huang H, Ma J (2009) Agglomeration control of Nd:YAG nanoparticles via freeze drying for transparent Nd:YAG ceramics. J Am Ceram Soc 92:812–817
244. Ding SQ, Zeng YP, Jiang DL (2007) Fabrication of mullite ceramics with ultrahigh porosity by gel freeze drying. J Am Ceram Soc 90:2276–2279
245. Fonseca AT, Vieira JM, Baptista JL (1986) Dependence of the densification on grain-growth and on agglomeration in sintering of dolomite. J De Physique 47:435–440
246. Kurajica S, Tkalcec E, Matijasic G, Curkovic L, Schauperl Z, Sipusic J et al (2011) Influence of agglomeration and contamination in the course of amorphous powder grinding on structure and microstructure of sintered mullite. Croat Chem Acta 84:63–71
247. Liu DM, Lin JT, Tuan WH (1999) Interdependence between green compact property and powder agglomeration and their relation to the sintering behaviour of zirconia powder. Ceram Int 25:551–559
248. Sacks MD, Pask JA (1982) Sintering of mullite-containing materials, 2. Effect of agglomeration. J Am Ceram Soc 65:70–77

249. Wang C, Chen S-H (2012) Factors influencing particle agglomeration during solid-state sintering. Acta Mech Sin 28:711–719
250. Zych L, Lach R, Wajler A (2014) The influence of the agglomeration state of nanometric $MgAl_2O_4$ powders on their consolidation and sintering. Ceram Int 40:9783–9790
251. Amato I, Baudrocco F, Martorana D (1976) Evaluation of freeze-drying and spray drying processes for preparing transparent alumina. Mater Sci Eng 26:73–78
252. Komarneni S (1996) Some significant advances in sol-gel processing of dense structural ceramics. J Sol-Gel Sci Technol 6:127–138
253. Li DY, Hui Y, Lian JS, Xie TT (2005) Progress in research on nanometer Al_2O_3 fabricated by sol-gel method. J Rare Earths 23:600–605
254. Shojaie-Bahaabad M, Taheri-Nassaj E, Naghizadeh R (2008) An alumina-YAG nanostructured fiber prepared from an aqueous sol-gel precursor: preparation, rheological behavior and spinnability. Ceram Int 34:1893–1902
255. Saha D, Mistry KK, Giri R, Guha A, Sensgupta K (2005) Dependence of moisture absorption property on sol-gel process of transparent nano-structured γ-Al_2O_3 ceramics. Sens Actuators B-Chem 109:363–366
256. Dupont A, Parent C, Le Garrec B, Heintz JM (2003) Size and morphology control of Y_2O_3 nanopowders via a sol-gel route. J Solid State Chem 171:152–160
257. Maia AS, Stefani R, Kodaira CA, Felinto MCFC, Teotonio EES, Brito HF (2008) Luminescent nanoparticles of $MgAl_2O_4$:Eu, Dy prepared by citrate sol-gel method. Opt Mater 31:440–444
258. Zhang HJ, Jia XL, Liu ZJ, Li ZZ (2004) The low temperature preparation of nanocrystalline $MgAl_2O_4$ spinel by citrate sol-gel process. Mater Lett 58:1625–1628
259. Gong H, Tang D-Y, Huang H, Han M-D, Sun T, Zhang J et al (2013) Crystallization kinetics and characterization of nanosized Nd:YAG by a modified sol-gel combustion process. J Cryst Growth 362:52–57
260. Guo K, Chen HH, Guo XX, Yang XX, Xu FF, Zhao JT (2010) Morphology investigation of yttrium aluminum garnet nano-powders prepared by a sol-gel combustion method. J Alloy Compd 500:34–38
261. Sun J, Zeng FM, Li JL, Cao YI, Wan YC, Liu JH (2005) Synthesis of Nd: GGG ultrafine polycrystalline powders by sol-gel method. J Rare Earths 23:188–190
262. Yu F, Yuan D, Cheng X, Duan X, Wang X, Kong L et al (2007) Preparation and characterization of yttrium gallium garnet nanoparticles by citrate sol-gel method at low temperature. Mater Lett 61:2322–2324
263. Liu XJ, Li HL, Xie RJ, Hirosaki N, Xu X, Huang LP (2006) Cerium-doped lutetium aluminum garnet optically transparent ceramics fabricated by a sol-gel combustion process. J Mater Res 21:1519–1525
264. Guo XZ, Li WY, Nakanishi K, Kanamori K, Zhu Y, Yang H (2013) Preparation of mullite monoliths with well-defined macropores and mesostructured skeletons via the sol-gel process accompanied by phase separation. J Eur Ceram Soc 33:1967–1974
265. Ivankovic H, Tkalcec E, Rein R, Schmidt H (2006) Microstructure and high temperature 4-point bending creep of sol-gel derived mullite ceramics. J Eur Ceram Soc 26:1637–1646
266. Jin XH, Gao L, Guo JK (2001) Preparation of mullite powders by sol-gel methods. J Inorg Mater 16:555–558
267. Leivo J, Linden M, Teixeira CV, Puputti J, Rosenholm J, Levanen E et al (2006) Sol-gel synthesis of a nanoparticulate aluminosilicate precursor for homogeneous mullite ceramics. J Mater Res 21:1279–1285
268. Moallemi-Nejad M, Heshmati-Manesh S (2008) Synthesis of nano-structured mullite by sol-gel method with intermediate mechanical activation. Int J Mod Phys B 22:2947–2954
269. Bhattacharyya D, Hazra C, Gangwar P, Sivakumar S, Mahalingam V (2012) Microstructural, optical and cytotoxicity studies of Eu^{3+}-doped Y_2O_3 nanoparticles prepared by simple control of the chain length of cross-linkers via pechini method. Sci Adv Mater 4:656–662
270. Hajizadeh-Oghaz M, Razavi RS, Estarki ML (2014) Large-scale synthesis of YSZ nanopowder by Pechini method. Bull Mater Sci 37:969–973

271. Kodaira CA, Stefani R, Maia AS, Felinto MCFC, Brito HF (2007) Optical investigation of $Y_2O_3:Sm^{3+}$ nanophosphor prepared by combustion and Pechini methods. J Lumin 127:616–622
272. Majid A, Tunney J, Argue S, Kingston D, Post M, Margeson J et al (2010) Characterization of CuO phase in SnO_2–CuO prepared by the modified Pechini method. J Sol-Gel Sci Technol 53:390–398
273. Saengkerdsub S, Im HJ, Willis C, Dai S (2004) Pechini-type in-situ polymerizable complex (IPC) method applied to the synthesis of Y2O3: Ln (Ln = Ce or Eu) nanocrystallites. J Mater Chem 14:1207–1211
274. Soisuwan P, Chambers DC, Trimm DL, Mekasuwandumrong O, Panpranot J, Praserthdam P (2005) Characteristics and catalytic properties of alumina-zirconia mixed oxides prepared by a modified Pechini method. Catal Lett 103:63–68
275. Zhang YW, Wang JJ, Yan ZG, Yan CH (2004) Effects of synthetic parameters in pechini method on the formation and properties of $(ZrO_2)_{0.92}(Sc_2O_3)_{0.08}$ solid solution. J Rare Earths 22:475–482
276. Zhang K, Hu W, Li J, Tang Y, Liu H (2009) Matrix induced synthesis of Y3Al5O12: Ce phosphor through the Pechini method. Int J Mater Res 100:238–242
277. Galceran M, Pujol MC, Aguilo M, Diaz F (2008) Synthesis and characterization of nanocrystalline $Yb:Lu_2O_3$ by modified Pechini method. Mater Sci Eng B-Solid State Mater Adv Technol 146:7–15
278. Marin R, Sponchia G, Riello P, Sulcis R, Enrichi F (2012) Photoluminescence properties of $YAG:Ce^{3+},Pr^{3+}$ phosphors synthesized via the Pechini method for white LEDs. J Nanoparticle Res 14:1–13
279. Back M, Boffelli M, Massari A, Marin R, Enrichi F, Riello P (2013) Energy transfer between Tb^{3+} and Eu^{3+} in co-doped Y_2O_3 nanocrystals prepared by Pechini method. J Nanoparticle Res 15:1–14
280. Wiglusz RJ, Grzyb T, Lis S, Strek W (2009) Preparation and spectroscopy characterization of $Eu:MgAl_2O_4$ nanopowder prepared by modified Pechini method. J Nanosci Nanotechnol 9:5803–5810
281. Courty P, Ajot H, Marcilly C, Delmon B (1973) Oxides, mixed or in slid-solution, highly dispersed obtained by thermal decomposition of amorphous precursors. Powder Technol 7:21–38
282. Marcilly C, Courty P, Delmon B (1970) Preparation of highly dispersed mixed oxides and oxide solid solutions by pyrolysis of amporphous orangic precursors. J Am Ceram Soc 53:56–57
283. Batoo KM (2011) Structural and electrical properties of Cu doped $NiFe_2O_4$ nanoparticles prepared through modified citrate gel method. J Phys Chem Solids 72:1400–1407
284. Xie G-Q, Lu J-Q, Zheng H-Y, Liu X-J, Luo M-F, Li X-N (2010) Effect of carbonization temperature on the textural properties of $Ce_{0.8}Zr_{0.2}O_2$ solid solution by an improved citrate sol-gel method. J Alloy Compd 493:169–174
285. Bao A, Tao CY, Yang H (2008) Luminescent properties of nanoparticles $LaSrAl_3O_7:RE^{3+}$ (RE = Eu, Tb) via the citrate sol-gel method. J Mater Sci-Mater Electron 19:476–481
286. Kim KM, Ryu JH, Mhin SW, Park GS, Shim KB (2008) Luminescence of nanocrystalline $Tb_3Al_5O_{12}:Ce^{3+}$ phosphors synthesized by nitrate-citrate gel combustion method. J Electrochem Soc 155:J293–J296
287. Soo MS, Joon YP, Byung JC (2003) Synthesis of yttrium aluminum garnet powder by citrate gel method. J Ceram Process Res 4:145–150
288. Li J, Pan Y, Qiu F, Wu Y, Guo J (2008) Nanostructured Nd:YAG powders via gel combustion: the influence of citrate-to-nitrate ratio. Ceram Int 34:141–149
289. Jia XL, Zhang HJ, Yan YJ, Liu ZJ (2004) Effect of the citrate sol-gel synthesis on the formation of $MgAl_2O_4$ ultrafine powder. Mater Sci Eng A-Struct Mater Prop Microstruct Process 379:112–118

290. Wang NL, Zhang XY, Bai ZH, Sun HY, Liu QS, Lu LP et al (2011) Synthesis of nanocrystalline ytterbium-doped yttria by citrate gel combustion method and fabrication of ceramic materials. Ceram Int 37:3133–3138
291. Birol H, Rambo CR, Guiotoku M, Hotza D (2013) Preparation of ceramic nanoparticles via cellulose-assisted glycine nitrate process: a review. Rsc Adv 3:2873–2884
292. Hajarpour S, Raouf AH, Gheisari K (2014) Structural evolution and magnetic properties of nanocrystalline magnesium- zinc soft ferrites synthesized by glycine-nitrate combustion process. J Magn Magn Mater 363:21–25
293. Valefi M, Falamaki C, Ebadzadeh T, Hashjin MS (2007) New insights of the glycine-nitrate process for the synthesis of nano-crystalline 8YSZ. J Am Ceram Soc 90:2008–2014
294. He TM, He Q, Wang N (2005) Synthesis of nano-sized YSZ powders from glycine-nitrate process and optimization of their properties. J Alloy Compd 396:309–315
295. Kim GD, Park JA, Lee HL, Lee DA, Moon JW, Kim JD (1999) Synthesis and sintering of $BaTiO_3$ powders by the glycine-nitrate process using metal carbonate and alkoxide. J Ceram Soc Jpn 107:691–696
296. Toniolo JC, Lima MD, Takimi AS, Bergmann CP (2005) Synthesis of alumina powders by the glycine-nitrate combustion process. Mater Res Bull 40:561–571
297. Wang Z, Yang B, Fu Z, Dong W, Yang Y, Liu W (2005) UV-blue photoluminescence from ZrO_2 nanopowders prepared via glycine nitrate process. Appl Phys a-Mater Sci Process 81:691–694
298. Komlev AA, Vilezhaninov EF (2013) Glycin-nitrate combustion synthesis of nanopowders based on nonstroichiometric magnesium–aluminum spinel. Russ J Appl Chem 86:1344–1350
299. Zhukov AV, Chizhevskaya SV, Klimenko OM, Merkushkin AO (2014) Glycine-nitrate synthesis of partially yttrium-stablized zirconium nanopowders for hard ceramics. Glass Ceram 70:400–403
300. Lei LW, Fu ZY, Wang H, Lee SW, Niihara K (2012) Transparent yttria stabilized zirconia from glycine-nitrate process by spark plasma sintering. Ceram Int 38:23–28
301. Durham BG, Murtha MJ, Burnet G (1988) Si_3N_4 by the carbothermal ammonolysis of silica. Adv Ceram Mater 3:45–48
302. Lin DC, Kimura S (1996) Kinetics of silicon monoxide ammonolysis for nanophase silicon nitride synthesis. J Am Ceram Soc 79:2947–2955
303. Ziegenbalg G, Breuel U, Ebrecht E, Holldorf H, Brink R (2001) Synthesis of alpha-silicon nitride powder by gas-phase ammonolysis of CH_3SiCl_3. J Eur Ceram Soc 21:947–958
304. Deng J, Su K, Wang X, Zeng Q, Cheng L, Xu Y et al (2009) Thermodynamics of the gas-phase reactions in chemical vapor deposition of silicon carbide with methyltrichlorosilane precursor. Theoret Chem Acc 122:1–22
305. Kim HI, Choi JM, Kim DJ, So MG (2002) Synthesis and crystallization of fine SiC-Si_3N_4 composite powders by a vapor phase reaction. J Ceram Process Res 3:82–85
306. Ning XG, Ye HQ, Liang Y, Zheng F (1992) Structural characterization of ultrafine SiC powder made from laser-heated vapor-phase reactions. J Mater Sci Lett 11:59–62
307. Moshtaghioun BM, Monshi A, Abbasi MH, Karimzadeh F (2011) A study on the effects of silica particle size and milling time on synthesis of silicon carbide nanoparticles by carbothermic reduction. Int J Refract Metal Hard Mater 29:645–650
308. Narisawa M, Okabe Y, Iguchi M, Okamura K (1998) Synthesis of ultrafine SiC powders from carbon-silica hybridized precursors with carbothermic reduction. J Sol-Gel Sci Technol 12:143–152
309. Bendovskii EB (2013) Silicon nitridation at different nitrogen pressures. Glass Ceram 69:324–329
310. Chung S-L, Chang C-W (2009) Carbothermal reduction and nitridation synthesis of silicon nitride by using solution combustion synthesized precursors. J Mater Sci 44:3784–3792
311. Hu HL, Zeng YP, Xia YF, Yao DX, Zuo KH (2014) High-strength porous Si_3N_4 ceramics prepared by freeze casting and silicon powder nitridation process. Mater Lett 133:285–288

312. Liu S, Ji H, Liang H, Ma Y (2008) Preparation of Si_3N_4 powders by the direct nitridation of silicon. Rare Metal Mater Eng 37:378–381
313. Yao DX, Zeng YP (2011) High flexural strength porous silicon nitride prepared via nitridation of silicon powder. J Inorg Mater 26:422–426
314. Yin S-w, Wang L, Tong L-g, Yang F-m, Li Y-h (2013) Kinetic study on the direct nitridation of silicon powders diluted with alpha-Si_3N_4 at normal pressure. Int J Miner Metallurgy Mater 20:493–498
315. Suri J, Shaw LL, Zawrah MF (2012) Tailoring the relative Si_3N_4 and SiC contents in Si_3N_4/SiC nanopowders through carbothermic reduction and nitridation of silica fume. Int J Appl Ceram Technol 9:291–303
316. Mazdiyas Ks, Cooke CM (1973) Synthesis, characterization and consolidation of Si_3N_4 obtained from ammonolysis of $SiCl_4$. J Am Ceram Soc 56:628–633
317. Rudin T, Wegner K, Pratsinis SE (2013) Towards carbon-free flame spray synthesis of homogeneous oxide nanoparticles from aqueous solutions. Adv Powder Technol 24:632–642
318. Yeh CL, Yeh SH, Ma HK (2004) Flame synthesis of titania particles from titanium tetraisopropoxide in premixed flames. Powder Technol 145:1–9
319. Yeh CL, Zhao E, Ma HK (2001) An experimental investigation of combustion synthesis of silicon dioxide (SiO_2) particles in premixed flames. Combust Sci Technol 173:25–46
320. Kathirvel P, Chandrasekaran J, Manoharan D, Kumar S (2014) Preparation and characterization of alpha alumina nanoparticles by in-flight oxidation of flame synthesis. J Alloy Compd 590:341–345
321. Tok AIY, Boey FYC, Zhao XL (2006) Novel synthesis of Al_2O_3 nano-particles by flame spray pyrolysis. J Mater Process Technol 178:270–273
322. Varatharajan K, Dash S, Arunkumar A, Nithya R, Tyagi AK, Raj B (2003) Synthesis of nanocrystalline α-Al_2O_3 by ultrasonic flame pyrolysis. Mater Res Bull 38:577–583
323. Baker C, Kim W, Sanghera J, Goswami R, Villalobos G, Sadowski B et al (2012) Flame spray synthesis of Lu_2O_3 nanoparticles. Mater Lett 66:132–134
324. Goldstein A, Goldenberg A, Yeshurun Y, Hefetz M (2008) Transparent $MgAl_2O_4$ spinel from a powder prepared by flame spray prolysis. J Am Ceram Soc 91:4141–4144
325. Laine RM, Marchal JC, Sun HP, Pan XQ (2006) Nano-α-Al_2O_3 by liquid-feed flame spray pyrolysis. Nat Mater 5:710–712
326. Teoh WY, Amal R, Maedler L (2010) Flame spray pyrolysis: an enabling technology for nanoparticles design and fabrication. Nanoscale. 2:1324–1347

Chapter 4
Powder Characterization and Compaction

4.1 Introduction

Once a ceramic powder is synthesized, there should be some parameters or data to demonstrate its properties or quality. The understanding of powder quality is important for two reasons: (i) quality control of the starting powders and (ii) microstructure monitoring or controlling of the final ceramics with desired properties and performances. A set of characterization experiments has to be conducted to characterize a ceramic powder after synthesis. A minor deviation in chemical composition and a tiny quantity of impurity will have significant influence on the microstructure and properties of transparent ceramics. Important characteristics of a ceramic powder can be categorized into four groups: (i) physical characteristics, (ii) chemical composition, (iii) phase composition, and (iv) surface characteristics. Because a large number of techniques have been developed for the characterization of solid materials, only those techniques that are most frequently and commonly used for ceramic powders will be discussed in detail, with principles of the characterization methods to be focused.

Since the microstructure of green body has an equally significant effect on the subsequent firing stage and microstructure of the final ceramics, a good control of the microstructure of the green body is also very important [1–4]. The homogeneous packing of powder is the desired goal of the consolidation step. Since the packing density controls the amount of shrinkage during firing, the achievement of high packing density is the first requirement. To understand the packing efficiency, it is necessary to understand the geometrical particle packing concept. The knowledge of colloids has been greatly helpful and useful in large-scale fabrication of ceramics, by using special forming techniques, such as tape casting, slipcasting, and gelcasting [5–7].

Mechanical consolidation with dies or molds is still the most widely used green body forming technology, especially in ceramic industries. Because of the particle–wall of die and particle–particle frictions, pressure cannot be transmitted uniformly

L.B. Kong et al., *Transparent Ceramics*, Topics in Mining, Metallurgy and Materials Engineering, DOI 10.1007/978-3-319-18956-7_4

to the powder compact. Therefore, the green body will have density inhomogeneity caused by the evenly applied stress. With the development in synthesis of nanosized powders, consolidation would encounter new challenges [8, 9]. Other forming techniques, such as extrusion and injection are rarely used for transparent ceramics [10–12]. The recently revived 3D printing or additive manufacturing, could find applications in the fabrication of transparent ceramics, where special shapes or architecture are required [13–17].

Additives play an important role in the production of ceramic green bodies. In most cases, the correct use of the additives can be vital to the success of the forming process. However, the additives must be removed from the green body before the sintering is started. For consolidation techniques, such as tape casting, slipcasting, and injection molding, where a large amount of additives have to be used, binder removal can be one of the limiting steps in the overall fabrication process.

4.2 Characterizations of Ceramic Powder

4.2.1 Physical Properties

Various terms have been used to characterize the physical properties of ceramic powders. In this book, the terminology proposed by Onoda and Hench [18] and later adopted by Rahaman [19] will be used with minor or without any modifications. A powder can be characterized as an assemblage of small units with distinct physical properties. The small units are usually known as particles, which can exhibit complicated structures.

4.2.1.1 Characteristics of Particles

Primary Particles

A primary particle is a discrete unit with relatively low porosity, which can be either a single crystal, a polycrystalline particle, or an amorphous/glass. They can be isolated if pores are present. A primary particle could be defined as the smallest unit of the powder with clearly defined surfaces. Therefore, it cannot be broken down into even smaller units by some physical agitations, such as ultrasonic agitation in a liquid. A polycrystalline primary particle consists of tiny crystals, which are also known as crystallites or grains.

Agglomerates

An agglomerate consists of a collection of primary particles, which are bound together by weak interactions, such as surface forces, liquid bridging, or solid bridging. Figure 4.1 shows a schematic diagram of an agglomerate containing dense polycrystalline primary particles [19]. Agglomerates usually exhibit porous microstructures, where the pores are interconnected in most cases. There are two types of agglomerates: soft agglomerates and hard agglomerates. Soft agglomerates are formed by weak surface forces and thus can be broken down into primary particles by ultrasonic agitation in a liquid or by dispersing with suitable disperse agents. In contrast, hard agglomerates are made of primary particles that are bonded by solid bridges similar to chemical bounding. Therefore, they cannot be broken down into primary particles by ultrasonic agitation. Hard agglomerates lead to the formation of microstructural defects when they are used to fabricate ceramics.

Particles

Particles are those small units that are movable as separate entities when the powder is dispersed by agitation and can be made of primary particles, agglomerates, or combinations of them.

Granules

The granules are large agglomerates, with sizes ranging from 100 μm to 1 mm. Granules are formed through the addition of polymer-based binders to a powder, followed by tumbling or spray drying. The purpose of having these nearly spherical large agglomerates is to improve the flowability of the powder during filling and compaction in die pressing.

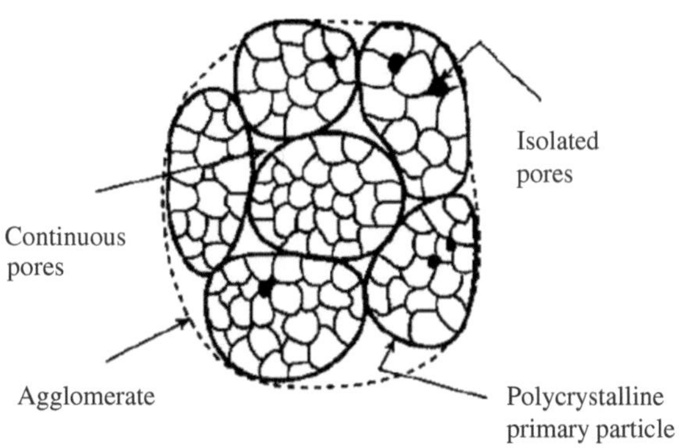

Fig. 4.1 Schematic diagram of an agglomerate containing dense polycrystalline primary particles. Reproduced with permission from [19]. Copyright © 2003, CRC Press

Flocs

Flocs are clusters of particles that are usually present in liquid suspensions. The clusters are formed when particles are held together through weak forces, such as electrostatic forces or bounding of organic polymers. As a result, these particles can be easily separated or redispersed by appropriately modifying the interfacial properties, so that the weak force interaction disappears. The presence of flocs will decrease the packing homogeneity of the consolidated body and thus is undesirable.

Colloids

A colloid means systems consisting of finely divided phases in a fluid state. A colloidal suspension, also known as sol in some cases, is made of fine particles that are uniformly dispersed in a liquid. These particles are called colloidal particles, which exhibit Brownian motion and almost have no sedimentation under normal gravity. The colloidal particles have sizes ranging from 1 nm to 1 μm.

Aggregates

An aggregate represents a coarse constituent in a mixture, which also contains a fine constituent acting as the bond. An example is the pebbles in concrete, in which the fine cement particles form the bond.

4.2.2 Particle Size and Size Distribution

Ceramic powders usually consist of particles with different sizes that are distributed over a certain range. Some powders have a very narrow size distribution, such as those prepared by chemical precipitation under well-controlled conditions, whereas others may exhibit a broad size distribution, such as those made through mechanical milling or solid-state reaction. Some particles have spherical, near spherical, or equiaxial shapes, while there are many cases where the powders are irregular in shape, including rod, wire, fiber, disk, plate, and so on. The importance of characterizing size and size distribution of ceramic particles is due to their effects in the consolidation and sintering behaviors of the powders.

4.2.2.1 Particle Size

Particle size of spherical particles is their diameter, while the size of particles with irregular shapes is more difficult to be determined. Therefore, it is important to have a definition of particle size. For particles with irregular shapes, their size can be

Table 4.1 Some definitions of particle size [19]

Symbol	Name	Definition
x_S	Surface diameter	Diameter of a sphere having the same surface area as the particle
x_V	Volume diameter	Diameter of a sphere having the same volume as the particle
x_{SV}	Surface volume diameter	Diameter of a sphere having the same surface area to volume ratio as the particle
x_{STK}	Stokes diameter (or equivalent spherical diameter)	Diameter of a sphere having the same sedimentation rate as the particle for laminar flow in a liquid
x_{PA}	Projected area diameter	Diameter of a circle having the same area as the projected area of the particle
x_C	Perimeter diameter	Diameter of a circle having the same Width of the minimum square aperture through which the particle will pass
x_A	Sieve diameter	Width of the minimum square aperture through which the particle will pass
x_F	Feret's diameter	Mean value of the distance between pairs of parallel tangents to the projected outline of the particle
x_M	Martin's diameter	Mean chord length of the projected outline of the particle

simply defined as the diameter of the sphere with same volume, although their volumes are difficult to measure in practice. Some definitions of particle size are given in Table 4.1 [19].

4.2.2.2 Average Particle Size

To describe the average particle size of a powder, it is assumed that the powder consists of N particles, having sizes x_1, x_2, x_3, …, x_N, respectively. A mean size, $\bar{x}$ can be calculated and the standard deviation in the mean, s can be obtained, by the following equations:

$$\bar{x} = \sum_{i=1}^{N} \frac{x_i}{N}, \tag{4.1}$$

$$s = \sum_{i=1}^{N} \left[\frac{(x_i - \bar{x})^2}{N} \right]^{1/2}. \tag{4.2}$$

The value of $\bar{x}$ is used as the average particle size of the powder, while s is a measure of the spread in the distribution of the particle size. In a random (Gaussian) distribution, about two thirds of the particles have sizes within the range of $\bar{x} \pm s$.

In most practical characterization techniques, a smaller number (n) of particle sizes are categorized, i.e., n is much smaller than N. With each size category, the number of particles will be counted, so that there are n_1 particles in a size category centered about x_1, n_2 of size x_2, …, and n_n of size x_n. Similarly, the mass or volume of the particles within each size category can also be used. Therefore, the number of particles within a size category will be considered, corresponding to a number-weighted average. The representation of the data in terms of mass or volume of the particles within a size category can be described in a similar way. The data can be used to determine the mean size and the standard deviation, given by the following equations:

$$\bar{x}_{\mathrm{N}} = \frac{\sum_{i=1}^{n} n_i x_i}{\sum_{i=1}^{n} n_i}, \tag{4.3}$$

$$S = \left[\frac{\sum_{i=1}^{n} n_i (x_i - \bar{x})^2}{\sum_{i=1}^{n} n_i}\right]^{1/2}. \tag{4.4}$$

Similarly, the volume-weighted average size is given by:

$$\bar{x}_{\mathrm{V}} = \frac{\sum_{i=1}^{n} v_i x_i}{\sum_{i=1}^{n} v_i} = \frac{\sum_{i=1}^{n} n_i x_i^4}{\sum_{i=1}^{n} n_i x_i^3}. \tag{4.5}$$

It is noticed that the mean particle size determined by using Eq. (4.3) is the arithmetic mean. Although it is not the only mean size that can be defined, it is the most significant if the powder has a normal particle size distribution. The geometric mean $\bar{x}_{\mathrm{g}}$ is the nth root of the product of the diameter of the particles with a number of n, which is given by:

$$\log \bar{x}_{\mathrm{g}} = \frac{\sum_{i=1}^{n} n_i \log x_i}{\sum_{i=1}^{n} n_i}. \tag{4.6}$$

If the distribution is lognormal, it has a particular value. There is also a harmonic mean, $\bar{x}_{\mathrm{h}}$, which is defined as the number of particles divided by the sum of the reciprocals of diameters of the individual particles and is given by:

$$\bar{x}_{\mathrm{h}} = \frac{\sum_{i=1}^{n} n_i}{\sum_{i=1}^{n} \frac{n_i}{x_i}}. \tag{4.7}$$

Because it is related to the specific surface area, this parameter will be important if the surface area of the powder is concerned.

The mean particle size determined by using Eq. (4.3) is also known as linear mean diameter, which is denoted as $\bar{x}_{\mathrm{NL}}$. It is used to distinguish from the surface mean diameter, $\bar{x}_{\mathrm{NS}}$, and the volume mean diameter $\bar{x}_{\mathrm{NV}}$. The surface and volume mean diameters are defined as:

$$\bar{x}_{\mathrm{NS}} = \frac{\left(\sum_{i=1}^{n} n_i x_i^2\right)^{1/2}}{\sum_{i=1}^{n} n_i}, \tag{4.8}$$

$$\bar{x}_{\mathrm{NV}} = \frac{\left(\sum_{i=1}^{n} n_i x_i^3\right)^{1/3}}{\sum_{i=1}^{n} n_i}. \tag{4.9}$$

4.2.2.3 Presentation of Particle Size

Histogram figure has been a simple and widely used way to present the particle size data. Rectangles are constructed over the size intervals, with their heights being proportional to the percent or fraction of the particles within the size intervals. In a given histogram, the rectangles are constructed in such a way that their areas are proportional to the number of the particles in the corresponding intervals. The total area under the histogram is equal to the number of the particles that have been counted. To compare the histograms for different powders, it is better to use 100 particles, so that the areas of the rectangles are equal to the percentages of the particles in the intervals.

More commonly, the data are plotted as cumulative size distributions, by summing the percent (or fraction) of particles finer than a given size, which is defined as the cumulative number percent finer (CNPF), or larger than a given size, which is defined as the cumulative number percent larger (CNPL).

In many cases, a more complete description of the particle size data should be provided, which is presented by using mathematical equations, with parameters that can be used to compare different powders in a more quantitative manner. If the number of particles is too large and the size intervals, Δx, are sufficiently narrow, the particle size distribution data can be fitted into smooth curves. Generally, the cumulative size distribution data is used as the starting point, rather than the histogram, while a smooth curve is fitted through the data. Therefore, it can be assumed that the cumulative size distribution is a function of x, which is $Q_{\mathrm{N}}(x)$. The fractional size distribution function, $q_{\mathrm{N}}(x)$, can then be obtained from $Q_{\mathrm{N}}(x)$ by using the following derivative:

$$q_{\mathrm{N}}(x) = \frac{\mathrm{d}}{\mathrm{d}x} Q_{\mathrm{N}}(x), \tag{4.10}$$

where $q_{\mathrm{N}}(x)\mathrm{d}x$ is the fraction of particles with sizes between x and $x + \mathrm{d}x$.

The function $q_{\mathrm{N}}(x)$ represents the measured size distribution, which is the description of the data in terms of a mathematical equation. In most cases, the measured size distribution function is fitted in terms of an expected size distribution, e.g., the normal distribution, given by:

$$q_{\mathrm{N}}(x) = \frac{1}{s\sqrt{2\pi}} \exp\left[-\frac{(x-\bar{x})^2}{2s^2}\right], \tag{4.11}$$

where $\bar{x}$ is the mean particle size and s is the standard deviation for the mean particle size. In practical fitting, the values of $\bar{x}$ and s need to be adjusted, so that the best fit to the data can be obtained. Because the data are assumed to be smoothly varying functions, the mean size and standard deviation can also be defined with equations for the normal distribution, which are expressed as:

$$\bar{x} = \frac{\int_{-\infty}^{\infty} x q_{\mathrm{N}}(x)\mathrm{d}x}{\int_{-\infty}^{\infty} q_{\mathrm{N}}(x)\mathrm{d}x}, \tag{4.12}$$

$$S = \left[\frac{\int_{-\infty}^{\infty} q_{\mathrm{N}}(x)(x-\bar{x})^2\mathrm{d}x}{\int_{-\infty}^{\infty} q_{\mathrm{N}}(x)\mathrm{d}x}\right]^{1/2}. \tag{4.13}$$

There is a serious problem with the normal distribution, i.e., it predicts a finite fraction of particles with sizes to be less than zero. Also, the largest particle size is not limited, i.e., it is infinite. As a result, the normal distribution usually cannot provide a good fit to real particle size data.

It has been found that the lognormal distribution is able to offer a good approximation to the size distribution of the powders prepared by using spray drying or mechanical milling, which is given by:

$$q_{\mathrm{N}}(\ln x) = \frac{1}{s\sqrt{2\pi}} \exp\left[-\frac{(\ln x-\bar{x})^2}{2s^2}\right], \tag{4.14}$$

where $\bar{x}$ is the mean of the natural logarithm of the particle sizes and s is the corresponding standard deviation. Although the lognormal distribution describes the particle sizes that are larger than zero, the largest size is unlimited.

It is further observed that there is an empirical function that can describe the particle size distribution of the powders prepared by using mechanical milling, which is the Rosin–Rammler equation. One of the modified forms of the equation is given by:

$$Q_{\mathrm{M}}(x) = \alpha b x^{\alpha-1} \exp(-b x^{\alpha}), \tag{4.15}$$

where $Q_{\mathrm{M}}(x)$ is the mass fraction of the particles with sizes between x and $x + \mathrm{d}x$, while α and b are empirical constants for the powders to be analyzed. The value of α is usually between 0.4 and 1.0. Another empirical distribution function for the size distribution of milled powders is the Gates–Gaudin–Schuhman equation, given by:

$$Q_M(x) = \alpha \frac{x^{\alpha-1}}{x_{max}^{\alpha}}, \tag{4.16}$$

where x_{max} is the maximum particle size and α is an empirical constant. Unlike the lognormal distribution, in both the Rosin–Rammler and Gates–Gaudin–Schuhman equations, the largest particles have a finite size.

4.2.3 Particle Shapes

Particle shapes of particles have significant influences on their properties and performances, such as flow properties, packing efficiency, and interaction with fluids in terms for example of viscosity of a suspension. The shapes of particles are usually qualitatively described as spherical, equiaxial, acicular, angular, fibrous, dendritic, flaky, and so on. If a particle does not have a simple geometry, like sphere, cube, or cylinder, it is difficult to give a quantitative characterization of particle shape. A shape factor is usually used to characterize the shape of a particle, which is a measure of the deviation from an ideal geometry, such as sphere or cube. For particles with elongated morphologies, aspect ratio is an effective parameter to represent the shape, which is defined as the ratio of the longest dimension to the shortest one. With sphere geometry as the reference, the shape factor ($1/\psi$) of a particle is defined as:

$$\frac{1}{\psi} = \frac{SA_{particle}}{SA_{sphere}}, \tag{4.17}$$

where $SA_{particle}$ is the surface area of the particle and SA_{sphere} is the surface area of a sphere with the same volume of the particle. Therefore, ψ is also known as sphericity. According to Eq. (4.17), the shape factor of a sphere is unity, whereas the shape factor for all other shapes is greater than unity. For example, the shape factor of a cube is given by $(6/\pi)^{1/3} \approx 1.24$. However, the shape factor or sphericity is not a unique parameter to reflect the real shape of particles, because particles with different shapes can share the same shape factor.

Shape coefficients are alternative parameters that have been used to characterize the shapes of particles. In shape coefficients, the measured particle size, x, is correlated to the measured particle surface area A or volume V, which are, respectively, given by:

$$A = \alpha_A x^2, \tag{4.18}$$

$$V = \alpha_V x^3, \tag{4.19}$$

where α_A called the area shape coefficient and α_V is the volume shape coefficient. According to these definitions, the area and volume shape coefficients for sphere are π and $\pi/6$, respectively, whereas the coefficients of cube are 6 and 1, respectively.

Therefore, it is clearly observed that a quantitative characterization of the particles with irregular shapes is a difficult and complicated task, with fairly low efficiency and effectiveness. Fortunately, ceramic processing prefers to use is spherical or equiaxial particles, because they provide high packing homogeneity in green body and high sinterability, especially for transparent ceramics that require almost 100 % densification. Nevertheless, this information is of usefulness as general knowledge for powder characterization.

4.2.4 Measurement of Particle Size and Size Distribution

Principles and practice of the measurements of particle size and size distribution can be readily found in the open literature [20–25]. The common methods that have been used to measure particle size and particle size distribution, as well as their size range capabilities, are listed in Table 4.2. The main features of these methods are described below.

4.2.4.1 Microscopy

Particle size measurement with microscopy is a direct observation method, where individual particles are observed directly, so that their shape and the degree of agglomeration can be identified at the same time. It provides straightforward information on a ceramic powder. There are three types of microscopies: optical microscopy, scanning electron microscopy (SEM), and transmission electron microscopy (TEM). Optical microscopes can be used for particle sizes larger than

Table 4.2 Methods for measurement of particle size

Method	Range of size (μm)
Microscopy	>1
Scanning electron microscopy	>0.1
Transmission electron microscopy	>0.001
X-ray line broadening	<0.1
Light scattering intensity	0.1–1000
Brownian motion	0.005–1
Sedimentation	0.1–100
Coulter counter	0.5–400
Sieving	20–10,000

1 μm, while TEM is able to identify particle of as small as 1 nm. To be accurate, a large number of particles, e.g., a few hundred or even thousand, should be measured. Particle size distribution is figured out based on the number of particles within an appropriate size range.

4.2.4.2 Sieving

Sieves are used to separate particles into fractions with different size ranges. The particles are classified according to their ability or inability to pass through an aperture with a controlled size. Sieves are constructed with wire mesh with openings between 20 μm and 10 mm, which are characterized by a mesh size and a corresponding aperture size. The wire mesh has square apertures, whose size is determined by the number of wires per linear dimension and the diameter of the wire. The mesh size is equal to the number of wires per inch linearly of the sieve screen, which is the same as the number of square apertures per inch. The relationships among mesh number M, aperture width a, wire diameter w, and the open area A can be described by the following equations:

$$M - \frac{1}{a+w}, \tag{4.20}$$

$$a = \frac{1}{M} - w, \tag{4.21}$$

$$A = \left(\frac{a}{a+w}\right)^2 = (\mathrm{Ma})^2. \tag{4.22}$$

Sieving can be conducted for dry powders or wet slurries/suspensions either manually or by machine. Sieving machines are designed to impart shaking, rotating, or vibrating motion to the powders on the screens. Sieves are stacked in such a way that the sieve with the coarsest mesh aperture is arranged at the top and that with the smallest mesh aperture is at the bottom. The powders to be sieved are placed on the top sieve. A sieve stack usually consists of four to six sieves that are arranged in a $\sqrt{2}$ progression of sizes. The fractions produced by sieving are weighed so that a particle distribution based on mass can be obtained. Sieving is mostly used in granulation of powders of transparent ceramics.

4.2.4.3 Sedimentation

When a spherical particle is falling in a viscous liquid at a sufficiently low velocity, velocity will reach constant, which is called the terminal velocity. In this case, the effective weight of the particle is balanced by the frictional force exerted on it by the liquid. The frictional force F on the particle is given by Stokes law:

$$F = 6\pi\eta a v, \tag{4.23}$$

where η is the viscosity of the liquid, a is the radius of the particle, and v is the terminal velocity. When the frictional force F is equal to the effective weight of the particle, there will be:

$$x = \left(\frac{18\eta v}{d_S - d_L}\frac{1}{g}\right)^{1/2}, \tag{4.24}$$

where x is the diameter of the sphere particle, g is the acceleration due to gravity, while d_S and d_L are the densities of the particle and the liquid, respectively. Equation (4.24) is usually known as the Stokes equation, which is different from the Stokes law given by Eq. (4.23). Therefore, measurement of the sedimentation rate can be used to determine the sphere diameter by using the Stokes equation. When particle shapes are not spheres, the obtained particle size is called the Stokes diameter, x_{STK}, or equivalent spherical diameter.

Measurement of particle size with Eq. (4.24) has a limited validity range. Stokes law is valid for laminar or streamline flow, i.e., there is no turbulent flow. There are also assumptions that the particles have no interparticle collisions or interactions. There is a critical velocity, v_c, at which transition from laminar to turbulent flow occurs. This critical velocity is given by:

$$v_c = \frac{N_R\eta}{d_L x}, \tag{4.25}$$

where N_R is a dimensionless number known as the Reynolds number. The transition from laminar to turbulent flow takes place when the critical value is about 0.2. As a result, following relations can be observed:

$$\frac{v d_L x}{\eta} < 0.2, \tag{4.26}$$

$$\frac{v d_L x}{\eta} > 0.2. \tag{4.27}$$

The maximum particle size for laminar flow can be found by substituting v from Eq. (4.24) into Eqs. (4.26) and (4.27).

When the particle size is too small to be measured by using sedimentation under gravity, its collisions with the molecules of the liquid will cause measurable displacement. This effect will increase the lower limit to about 1 μm, when the gravitational settling measurement is carried out using water as the liquid. If the suspension is centrifuged, the settling rate of the particles can be increased, so that the size range can be extended down to about 0.1 μm.

In this method, the powder can be introduced as a thin layer on top of a column of clear liquid, which is sometimes known as two-layer or line-start technique. It can also be uniformly dispersed in the liquid, called the homogeneous suspension technique. The particle size distribution is then measured according to the change in concentration or density of the suspension as a function of time, height along the suspension, or both of them. During the measurement, the suspension is contained in a glass cell, through which a light beam or an X-ray beam is projected at a given height, while the intensity of the transmitted beam is measured by a photocell or an X-ray detector located at the opposite side. According to Eq. (4.24), the higher the velocity of particles, the faster the particles will settle. The intensity I of the transmitted beam will increase according to the following equation:

$$I = I_0 \exp(-KACy), \tag{4.28}$$

where I_0 is the intensity of the incident beam, K is a constant called the extinction coefficient, A is the projected area per unit mass of particles, C is the concentration by mass of the particles, and y is the length of the light path through the suspension. The particle size distribution, such as the cumulative mass percent finer (CMPF) versus the Stokes diameter, is thus obtained.

4.2.4.4 Light Scattering

As a light beam is passed through a column of suspension of particles, its intensity will be decreased from the incident intensity I_0 to a lower value I [26–29]. The decrease in the intensity is mainly due to the scattering of the light by the particles, rather than absorption. The scattered light, with an intensity of I_s, can be observed from the side of the suspension in a darkened surrounding. Therefore, through the measurement of I_s, size of the particles can be characterized. The light beam is reflected when it encounters sufficiently large particles as compared with the wavelength of the incident light. However, if the particle sizes are comparable with the wavelength of the light, scattering occurs. The scattered light is usually toward the direction of the light beam, but deviated with small angles, as shown schematically in Fig. 4.2. Measuring I_s of this forward-scattered light as a function of the scattering angle is the principle of light scattering measurements.

Diffraction theory can be used to describe the dependence of I_s on the size of the particles. Considering the complexity of the general theory, it is more convenient to discuss three limiting cases, which are defined according to the ratio of the particle size x to the wavelength λ of the incident light, including (i) Rayleigh scattering theory for $x \ll \lambda$, (ii) Mie theory for $x \approx \lambda$, and (iii) Fraunhofer diffraction theory for $x \gg \lambda$.

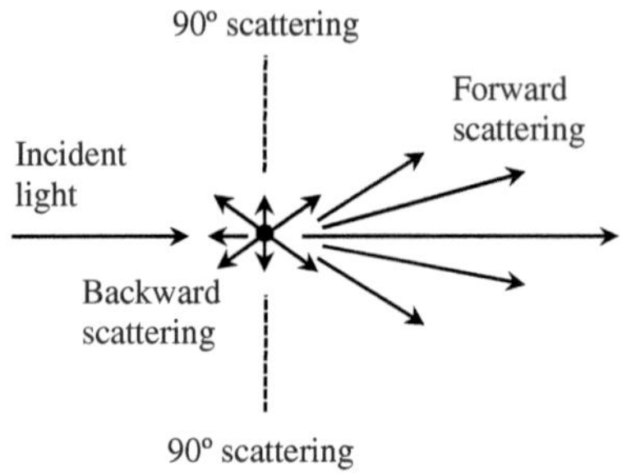

Fig. 4.2 Illustration of scattering of light by small particles

The Rayleigh theory is for scattering of small particles ($x \ll \lambda$), which indicates that I_s is proportional to square of the volume of the particle. The Fraunhofer diffraction theory is applicable to larger particles ($x \gg \lambda$), where the distribution of I_s is similar to that of diffraction from a single slit, with the size of the scattering particle comparable with the width of the slit. Therefore, in the Fraunhofer theory, I_s is proportional to square of the width of the slit, so that I_s is proportional to the projected area of the particles that scatter the light. According to diffraction theory, the scattering angle, θ, is inversely proportional to the particle size, which is given by:

$$\sin\theta \approx \frac{1.22\lambda}{x}. \tag{4.29}$$

Smaller particles scatter a small amount of light, while according to Eq. (4.29), they scatter the light through a larger angle.

Laser has been widely used for the measurement of particle size through light scattering [30]. Laser scattering measurements are very accurate and fast. The techniques based on the Fraunhofer diffraction theory can measure the size of particles in the range of 2–100 μm. The Mie theory can extend the measurable size range to 0.1–1000 μm, if special light collection systems are used.

Particles smaller than 1 μm tend to have Brownian motion when they are dispersed in liquids, because of their collisions with the molecules of the liquids. The speed of the motion increases with increasing temperature, due to the increase in momentum of the colliding molecules. It also increases with decreasing particle size, because of the increased probability of the collision on the particle with one side that is not matching that with the other size, as well as the enhanced response to the unbalanced collision of the particles with smaller sizes. Brownian motion also contributes to scattering of the incident light, thus leading to fluctuations in the average intensity. These fluctuations are usually recorded at right angles to the incident light by using light detectors. Particle size data can be determined based on the Stokes–Einstein equation:

$$a = \frac{kT}{3\pi\eta D}, \tag{4.30}$$

where a is the radius of the particle if it is assumed spherical, k is the Boltzmann constant, T is the absolute temperature, η is the viscosity of the liquid, and D is the diffusion coefficient. This technique is also known as dynamic light scattering or photo correlation spectroscopy, which can be used to measure particles with sizes in the range of 5 nm to 1 μm.

4.2.4.5 X-Ray Diffraction Broadening

Particle sizes of ceramic powders can also be measured by using the broadening of X-ray diffraction peaks, with particles sizes of less than 0.1 μm [31–33]. Generally, crystal size is used to describe the size obtained by the X-ray line broadening, because the size is the dimension of the crystals that are single crystals, poly-crystals, or their agglomerates. If the primary particles are single crystals, the crystal size determined by X-ray line broadening is comparable to the particle size measured by using other methods, such as electron microscopy. However, if the primary particles are polycrystalline or agglomerates, the obtained size is largely underestimated, which is much smaller than the values determined by using other methods.

The principle is based on the assumption that the width of the diffraction peak increases with decreasing crystal size. The approximate relation between the peak broadening and crystal size is given by the Scherrer equation:

$$x = \frac{C\lambda}{\beta\cos\theta}, \tag{4.31}$$

where x is the crystal size, C is a numerical constant ($\sim$0.9), λ is the wavelength of the X-rays, β is the effective broadening represented by the full width at half maximum (FWHM) in radians, and θ is the angle for the diffraction peak.

To get the effective broadening β, the measured broadening should be corrected by taking away the broadening due to (i) the instrument and (ii) residual stress in the materials. Broadening due to the instrument can be accurately determined by using a single crystal with a calibrated broadening. Residual stress is created when cooling or heating the sample, during which compressive or tensile stresses are produced due to the inhomogeneous or anisotropic thermal expansion of the materials. The peak broadening is dependent on the magnitude and sign of the stresses, increasing as tan θ. Because the broadening caused by the small crystal size given by Eq. (4.31) varies as a function of 1/(cos θ), the broadening due to the residual stress can be readily separated.

4.2.5 Surface Area

Surface area of powders is another important parameter that can be used to determine the average particle size, together with the consideration of shape of the particles and the presence or absence of pores. Surface area is usually measured basing on the gas adsorption phenomenon [34]. Adsorption means that gas molecules as adsorbate are condensed on the free surfaces of the solid powders as adsorbent, which is different from absorption, where the gas molecules penetrate into the mass of the absorbing solids. There are two types of adsorption: (i) physical adsorption and (ii) chemical adsorption.

In physical adsorption, the interaction between the gas molecules and the solid surface through physical forces, is a relatively weak force and similar to the van der Waals force for condensation of a vapor to a liquid. In contrast, chemical adsorption, also known as chemisorption, occurs when the adsorbed gases are strongly bonded on the surfaces of the particles through a strong chemical bonding.

Physical adsorption is the basis for the various techniques to measure surface area of ceramic powders. The surface area is determined in terms of the amount of the gas adsorbed by a given mass of solid powder at a given temperature, under different gas pressures p. In practice, gases with a fixed volume are used for the powder, so that the amount of gas adsorbed can be identified according to the decrease in pressure of the gas. The amount of gas adsorbed versus p, or p/p_0, when the gas is at pressures below its saturation vapor pressure p_0, can be plotted as a graph, which is known as the adsorption isotherm. Figure 4.3 shows the types of these isotherms, according to Brunauer, Emmett and Teller (BET) classification [35–38]. The Type VI isotherm is called stepped isotherm, which is relatively rarely observed, but has special theoretical interest. This isotherm offers the possibility to determine the monolayer capacity of a solid, which is defined as the amount of gas that is required to cover the surface of the unit mass of the solid with a monolayer, so as to calculate the specific surface area of the solid.

According to the kinetic model for adsorption, the surface of the solid is assumed to serve as effective adsorption sites [39, 40]. When the rate of condensation of the gas molecules on the unoccupied sites on the surface is equal to the rate of evaporation of the gas molecules from the occupied sites, a state of dynamic equilibrium is reached, which leads to the following equation:

$$\frac{V}{V_{\mathrm{m}}} = \frac{bp}{1 + bp}, \tag{4.32}$$

where V is the volume of the gas adsorbed per unit mass of the solid at a gas pressure p, V_{m} is the volume of the gas that is needed to cover the surface as a monolayer, and b is an empirical constant. Equation (4.32) is called the Langmuir adsorption equation for one monolayer coverage, which is represented by the Type I isotherms, as shown in Fig. 4.3.

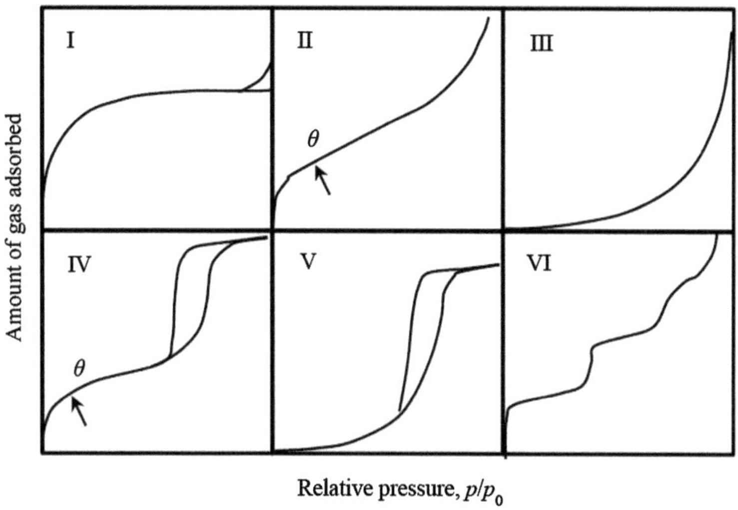

Fig. 4.3 Schematic adsorption isotherms, with Types I to V being grouped according to the original classification proposed by Brunauer et al., as well as the Type VI, known as the stepped isotherm. Reproduced with permission from [37, 38]. Copyright © 1940, American Chemical Socicty

For multilayer adsorption, i.e., the Type II isotherm in Fig. 4.3, Langmuir equation can be used after a modification with several assumptions [38]. The equation is usually known as the BET equation, which is given by:

$$\frac{p/p_0}{V(1-p/p_0)} = \frac{1}{V_{\mathrm{m}}c} + \frac{c-1}{V_{\mathrm{m}}c}\frac{p}{p_0}, \tag{4.33}$$

where c is a constant and the other terms are as defined previously. When experimental data are plotted with the term on the left-hand side of Eq. (4.33) versus p/p_0, a straight line with a slope s and an intercept i can be obtained as follows:

$$s = \frac{c-1}{V_{\mathrm{m}}c}, \tag{4.34}$$

$$i = \frac{1}{V_{\mathrm{m}}c}. \tag{4.35}$$

The monolayer volume V_{m} can therefore be derived as:

$$V_{\mathrm{m}} = \frac{1}{s+i}. \tag{4.36}$$

Therefore, specific surface area of the solid V_m can be calculated by using the following expression:

$$S_w = \frac{N_A \sigma V_m}{V_0}, \tag{4.37}$$

where S_w is the specific surface area ($m^2\ g^{-1}$), N_A is the Avogadro number, $6.023 \times 10^{23}\ mol^{-1}$, σ is the area of the adsorbed gas molecule, $16.2 \times 10^{-20}\ m^2$ for N_2, V_m is monolayer volume in $cm^3\ g^{-1}$ and V_0 is the volume of 1 mol of the gas at STP, 22,400 $cm^3\ mol^{-1}$.

For N_2 gas adsorption, with these values, Eq. (4.31) can give:

$$S_w = 4.35 V_m. \tag{4.38}$$

If the powder is assumed to be unagglomerated and the particles are dense spheres, the particle size can be estimated by the following equation:

$$x = \frac{6}{S_w d_s}, \tag{4.39}$$

where d_s is the density of the solid.

The BET equation is commonly acknowledged to be valid over a range of relative pressures (i.e., p/p_0 values) of 0.05–0.3. As the standard technique, nitrogen gas is used as the adsorbate, with the adsorption at the boiling point (77 K) of liquid nitrogen. The molecular area $\sigma = 16.2 \times 10^{-20}\ m^2$ (16.2 Å^2) has been well used for the measurement of surface area with the Type II isotherms. Argon gas, with $\sigma = 16.6 \times 10^{-20}\ m^2$, is also used as an alternative. It is chemically inert and is composed of spherically symmetrical monatomic molecules, with adsorption to be easily measured at the boiling point of liquid nitrogen.

4.2.6 Porosity of Particles

Individual primary particles could be dense, while agglomerates are most likely porous. Therefore, it is desirable to quantitatively characterize the porosity and pore size and distribution of the agglomerates. For accessible pores, i.e., those that are not completely isolated from the external surface, they can be characterized by using two methods: (i) gas adsorption, also known as capillary condensation and (ii) mercury intrusion porosimetry, also called mercury porosimetry for simplicity. The pore size can be diameter, radius, or width. Three types of pores have been classified according to their sizes: micropores (<2 nm), mesopores (2–50 nm), and macropores (>50 nm). Generally, gas condensation is applicable to the measurement in mesopores, whereas mercury porosimetry is more suitable to macropores.

4.2.6.1 Gas Adsorption

At lower gas pressures, e.g., $p/p_0 < 0.3$, the adsorbed gases cover the surface of the solid by forming multilayers, whereas at higher pressures, the gas could condense to liquid due to the capillary force of a porous solid, which can be used to measure the pore size and the size distribution. When the adsorbed gases are condensed, the isotherm will have a hysteresis loop for a cycle of an adsorption and a desorption, represented by the Type IV and V isotherms, as shown in Fig. 4.3.

If the capillaries are assumed to be cylindrical with a radius r, the relative gas pressure in the capillaries for condensation can be described by the Kelvin equation:

$$\ln\frac{p}{p_0} = -\frac{2\gamma_{LV}V_L\cos\theta}{RT}\frac{1}{r}, \tag{4.40}$$

where p is the gas pressure over a meniscus with radius r, p_0 is the saturation gas pressure of the liquid with a plane surface, γ_{LV} is the surface tension of the liquid–vapor interface, V_L is the molar volume of the liquid, θ is the contact angle between the liquid and the pore wall, R is the gas constant, and T is the absolute temperature. The Kelvin equation indicates that the vapor pressure p over a concave meniscus should be less than p_0. As a result, capillary condensation of a vapor to a liquid should occur within a pore at the pressure p that is determined by the radius r for the pore, which is lower than p_0.

When using the Kelvin equation to determine the pore size distribution, with the capillary condensation part of the Type IV isotherms, the adsorbing gas should be nitrogen, so that both surface area and pore size distribution can be obtained from one isotherm. If the volume of the gas adsorbed on external surface of the solid is smaller than that adsorbed inside the pores, the volume of the pores can be obtained, when the volume of the gas adsorbed V_g is converted to a liquid or condensed volume V_c. The relationship between V_g and V_c is given by:

$$V_c = \frac{M_w V_g}{\rho_L V_0}, \tag{4.41}$$

where M_w is the molecular weight of the adsorbate, ρ_L is the density of the liquefied gas at its saturated vapor pressure, and V_0 is the molar volume of the gas at STP, i.e., 22,400 cm^3. For nitrogen gas, $M_w = 28$ g and $\rho_L = 0.808$ g cm^{-3}, so that there is:

$$V_c = 1.547 \times 10^{-3}\ V_g. \tag{4.42}$$

For nitrogen adsorption at the boiling point of liquid nitrogen at 77 K, $\gamma_{LV} = 8.72 \times 10^{-3}$ N m^{-1}, $V_L = 34.68 \times 10^{-6}$ m^3 and θ is assumed to be zero, so that Eq. (4.40) becomes:

$$r\frac{-4.05\times 10^{-1}}{\log(p/p_0)}\,\text{nm.} \tag{4.43}$$

Because typical values of p/p_0 are in the range 0.5–0.95, the gas adsorption is corresponding to the pore radius with values in the range of 1–20 nm.

For any relative pressure p_i/p_0 on the isotherm, the cumulative volume of the pores with pore radius values up to r_i can be derived from V_{ci}, where V_{ci} and r_i are determined by using Eqs. (4.42) and (4.43), respectively. The pore size distribution curve, $v(r)$, as a function of r, can be obtained by differentiating the cumulative pore volume curve with respect to r, i.e., $v(r) = dV_c/dr$. Because this analysis does not take into account the capillary condensation, the obtained values of the pore size are only representative of the inner cores, because the pore walls are already covered with an adsorbed layer of gas.

4.2.6.2 Mercury Porosimetry

Because ceramic powders usually have macropores, mercury porosimetry is more suitable than gas adsorption. The principle of the technique is the phenomenon of capillary rise, as shown schematically in Fig. 4.4 [19]. When a liquid wets the walls of a narrow capillary, with contact angle, $\theta < 90°$, it will climb up the walls of the capillary. If the liquid does not wet the walls of a capillary, with contact angle, $\theta > 90°$, it will be depressed. When a nonwetting liquid is used, it is necessary to force the liquid to flow up the capillary to the level of the reservoir by applying a pressure. For a capillary with principal radii of curvature r_1 and r_2 in two orthogonal directions, the pressure can be obtained by using the Young and Laplace equation:

$$p = -\gamma_{LV}\left(\frac{1}{r_1}+\frac{1}{r_2}\right)\cos\theta, \tag{4.44}$$

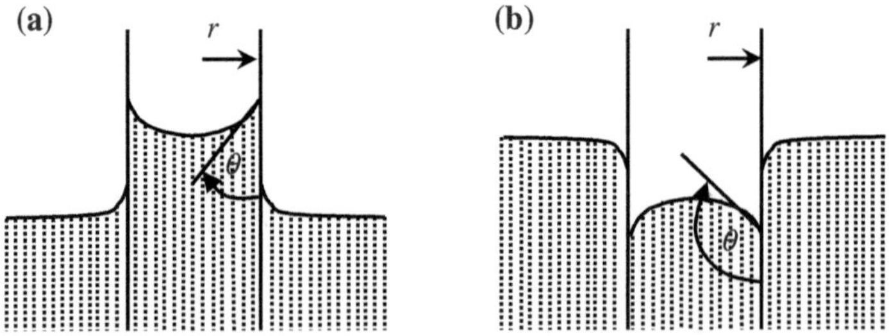

Fig. 4.4 Capillary rise phenomena of **a** a wetting liquid with contact angles $\theta < 90°$ and **b** a nonwetting liquid with contact angles $\theta > 90°$. Reproduced with permission from [19]. Copyright © 2003, CRC Press

where γ_{LV} is the surface tension of the liquid–vapor interface. For a cylindrical capillary with radius r, Eq. (4.44) becomes:

$$p = \frac{-2\gamma_{LV}}{r}\cos\theta. \tag{4.45}$$

The γ_{LV} and θ values of mercury exhibit slight variation, depending on its purity and the type of surface of the powders to be measured. Therefore, when pressure p has units of megapascals (MPa) and r in micrometers (μm), Eq. (4.45) becomes:

$$r = \frac{0.735}{p}. \tag{4.46}$$

When using mercury porosimetry, a sample holder is first partially filled with the powder to be tested, which is then evacuated and finally filled with mercury. The volume of mercury intruded into the powder sample, V_m, is measured as a function of the applied pressure p. If the pores are assumed to be cylindrical, the value V_{mi} at any applied pressure p_i leads to the cumulative volume of all pores that have a radius equal to or greater than r_i. This corresponds to that in gas adsorption where the cumulative pore volume is the volume of the pores with a radius to be less than or equal to r_i. Therefore, the cumulative pore volume decreases with increasing r, in mercury porosimetry, where it increases with increasing r, in gas adsorption. In both cases, the pore size distribution $v(r)$ is obtained by differentiating the cumulative pore volume curve with respect to r. The pore size distribution can also be obtained directly from the data of V_m versus p by using the following relation:

$$v(r) = \frac{dV_m}{dp}\frac{p}{r}. \tag{4.47}$$

The mercury porosimeters can be used to measure the pore sizes in the range of 5 nm to 200 μm.

4.2.6.3 Pycnometry

Pycnometry is used to determine the isolated porosity within particles. The apparent density of a particle, d_a, is defined as:

$$d_a = \frac{m_{solid}}{V_{solid} + V_{pore}}, \tag{4.48}$$

where m_{solid} is the mass of solid, while V_{solid} and V_{pore} are the volumes of the solid and isolated pores, respectively. The value of d_a can be measured experimentally. If d_t is equal to the theoretical density of the particles, i.e., the density of the particle solid without pores, the amount of the isolated porosity is equal to $1 - d_a/d_t$.

For coarser powders, i.e., particle size >10 μm, a liquid pycnometer is used. A calibrated bottle is weighed, with mass m_0, and the powder is added, with total mass m_1. A liquid with density, d_L, is then added, with total mass m_2. It the pycnometer bottle contains only the liquid, i.e., no powder, has a mass m_3, the apparent density of the particles d_a is given by:

$$d_a = \frac{m_1 - m_0}{(m_3 - m_0) - (m_2 - m_1)}(d_L - d_{air}) + d_{air}, \tag{4.49}$$

where d_{air} is the density of air. The particle surface should be well wetted by the liquid, in order to remove trapped air, by using suitable methods, such as boiling the liquid. Helium gas pycnometry is generally used for powders with sizes to be smaller than 10 μm. Because helium molecule has a small size, it is easy to penetrate into very small pores. The volume occupied by the solid is measured according to the volume of the gas displaced. The apparent density of the powder is then calculated by using the mass of the powder and the volume that has been measured.

4.2.7 Chemical Compositions

Chemical composition of key component determines the properties of the ceramics, while small quantities of other components will have significant effect on the processing and microstructural development of the materials. Low concentrations of dopants in the range of 0.1–10 at.% are usually used to enhance the sintering behavior and modify the microstructure. Trace impurity elements at concentrations less than a few hundred parts per million (ppm) are inevitably present even in the cleanest powders, which could also have significant effect, which has however often been overlooked. For lab-scale experiments of transparent ceramics, element analysis is generally not an indispensable step, because the main components and additives have all been predesigned.

Various techniques can be used for quantitative analysis of chemical composition, including (i) optical atomic spectroscopy (atomic absorption, atomic emission, and atomic fluorescence), (ii) X-ray fluorescence spectroscopy, (iii) mass spectrometry, (iv) electrochemistry, and (v) nuclear and radioisotope analysis [41]. Among these, optical atomic spectroscopy, involving atomic absorption (AA) or atomic emission (AE), has been the most widely used for chemical analysis of ceramic powders. It can be used to determine the contents of both major and minor elements, as well as trace elements, because of its high precision and low detection limits.

X-ray fluorescence spectroscopy (XRF) is another measurement technique for major and minor elemental concentrations, as well as trace elements. However, it has very low sensitivity to elements with atomic number $Z < 9$. Optical atomic spectroscopy is suitable to analyze solutions, which requires complete dissolution of the powder in a liquid, while XRF can be used to directly analyze solids and thus more suitable for ceramic powders that are generally not dissolvable to form solutions.

4.2.7.1 Atomic Absorption and Atomic Emission

The principle of optical atomic spectroscopy is the transitions of electrons between the outer energy levels in atoms. An atom with multiple electronic structure has many energy levels in the ground state, i.e., the state of lowest energy, as well as many unoccupied energy levels which the electrons can take when excited. When there are electrons that take such higher energy levels, the atom is in an excited state. Figure 4.5 shows the simplest case, with only two energy levels, a ground state with energy E_g and an excited state with energy E_e. The frequency of the radiation absorbed or emitted due to the transition is given by the Planck's law:

$$v = \frac{E_e - E_g}{h}, \tag{4.50}$$

where h is Planck's constant. Optical atomic spectroscopy uses the energy transitions with wavelength of the radiation in range of 10 nm to 400 μm.

At thermal equilibrium state at a given temperature T, the ratio of the number of atoms in the excited state, n_e and the number in the ground state, n_g, is given by the Boltzmann equation:

$$\frac{n_e}{n_g} = \exp \leq \left(-\frac{E_g - E_e}{kT} \right), \tag{4.51}$$

where k is the Boltzmann constant. The number of atoms in the ground state becomes to be less than the total number n, due to the excitation of the n_e atoms. However, if the temperature is not very high, n_g is always very close to n. Therefore, according to Eq. (4.51), the number of excited atoms n_e is proportional to n. Because the intensity of the absorbed or emitted radiation is proportional to n_e, it is thus proportional to the total number of atoms n. Each element has its own characteristic energy levels, so that the elemental composition can be

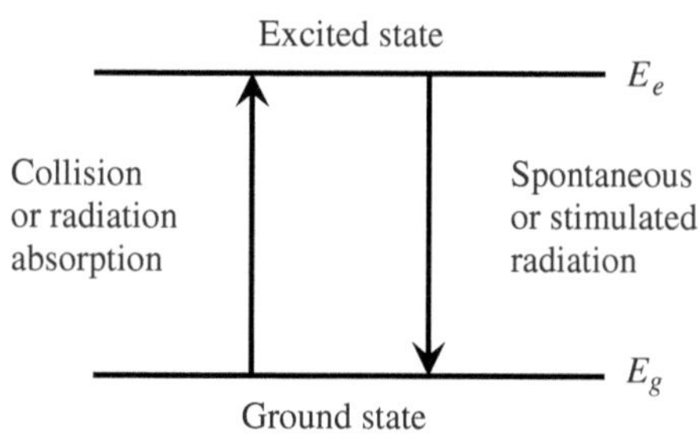

Fig. 4.5 Schematic of the transitions in optical atomic spectroscopy

determined by the wavelength or frequency of the absorbed or emitted radiation, while their concentrations can be derived from the intensity of the radiations. In practice, standard samples with known concentrations are usually used as reference for calibration curves, from which the concentration of the samples to be tested can be identified. The concentration of the sample to be tested, C_u, can also be determined through the following relation:

$$C_u = C_s \frac{I_u}{I_s}, \tag{4.52}$$

where C_s is the concentration of a standard, while I_u and I_s are the measured intensities of the unknown and standard samples, respectively.

As mentioned earlier, optical atomic spectroscopy is only able to analyze solution sample. As a result, ceramic powders to be tested should be made into solution. The solution is then broken into fine droplets and vaporized into individual atoms by heating, which is the step critical to the precision and accuracy of the analysis. Flame is generally used to vaporize the solution, which is therefore also known as flame atomic absorption spectrometry or flame AA.

4.2.7.2 X-Ray Fluorescence Spectroscopy

X-ray fluorescence spectroscopy (XRF) is also known as atomic XRF [42], which can be used for qualitative or quantitative analysis of elemental composition, through the measurement of the wavelength and intensities of the electron transitions from the outer to the inner energy levels of the atom. The energy associated with these transitions is in the rage of 0.6–60 keV, which is significantly higher than that of the transitions in optical atomic spectroscopy, which is at the order of eV. In XRF, energetic X-rays are usually used.

When a primary beam of monochromatic X-rays shines on a sample, the X-rays can be absorbed or scattered. Coherent scattering of X-rays by ordered scattering centers, such as the lattice in crystals, leads to diffraction, which is for determination of crystal structure, as discussed later. When atoms or ions absorb the X-rays, electrons could be ejected, which is known as the photoelectric effect, or excited to higher energy levels, as shown in Fig. 4.6a. The ejected electrons can be used to characterize the surface properties of particles, which is called X-ray photoelectron spectroscopy (XPS).

As the excited atoms transit back to their ground state, there could be two processes. The first process is due to a rearrangement of the electrons, leading to the ejection of other electrons from higher energy levels, which is known as Auger electrons, as shown in Fig. 4.6b. The second process involves the transitions of electrons from the outer levels to fill the vacant electron sites in the inner energy levels, leading to the emission of X-rays, as shown in Fig. 4.6c. The secondary beam of X-rays emitted from the powder sample is due to the interaction of the primary X-ray beam, which is the principle of XRF. It is the characteristics of the

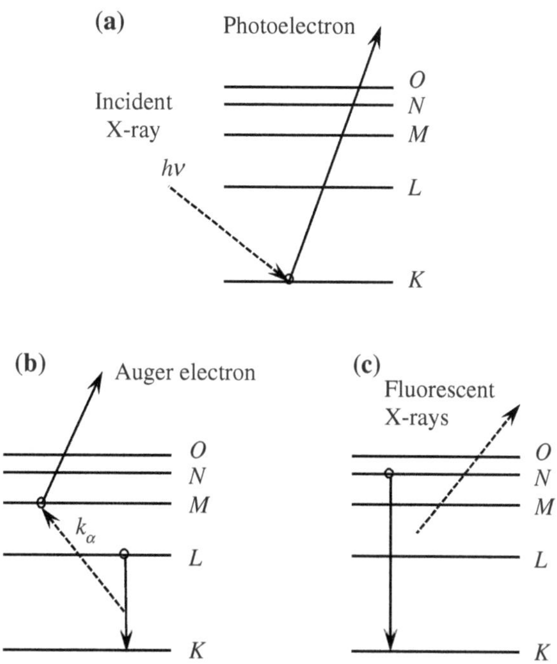

Fig. 4.6 **a** Interaction of X-rays with a solid leading to atomic excitation with the emission of a photoelectron, **b** followed by deexcitation with the emission of an Auger electron, or **c** by the emission of characteristic X-ray photons

elements and thus can be used for element identification. The relationship between the wavelength λ of the emitted X-rays and the atomic number Z of the element is given by:

$$\frac{1}{\lambda} = A(Z - \alpha)^2, \tag{4.53}$$

where A is a constant for each series of spectral lines and $\alpha \approx 1$.

4.2.8 Crystal Structure and Phase Composition

X-ray diffraction (XRD) has been the most widely used characterization technique to determine the structure and chemical composition of crystalline materials. It offers qualitative and quantitative characterization of crystalline compounds or phases in the form of solid materials. The requirements of X-ray diffraction are (i) the atomic spacing in the solid being comparable with the wavelength of the X-rays and (ii) the scattering centers being spatially distributed in an ordered way, e.g., the lattices in crystals. The diffraction of X-rays by crystals follows the Bragg's law, as shown in Fig. 4.7, with the condition for constructive interference being given by:

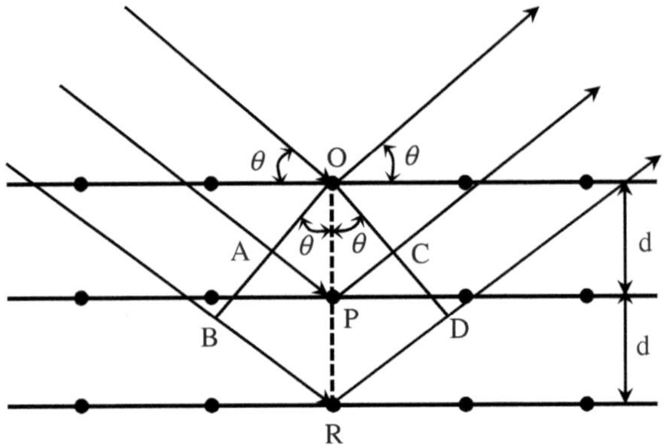

Fig. 4.7 Principle of X-ray diffraction by a crystal

$$2d \sin \theta = n\lambda, \tag{4.54}$$

where d is the spacing between the lattice planes of the crystal, θ is the angle of incidence and reflection, n is an integer, also known as the order of diffraction, and λ is the wavelength of the X-rays, e.g., $\lambda = 0.15406$ nm for Cu Kα radiation. When the reflection angle θ does not satisfy the relationship of Eq. (4.54), interference is destructive.

4.2.9 Surface Properties

Surface area of a powder increases geometrically with decreasing particle size, so that the volume fraction of the outermost layer of ions on the surface increase significantly, which has a significant effect on properties of the powder. With the development of nanotechnology, it is readily to synthesize powders with nanosized particles (1–100 nm). Therefore, characterization of surface properties becomes more and more important. Specifically for ceramics or transparent ceramics, the consolidation of fine ceramic powders with liquid suspensions to produce more uniform green bodies has been shown to play an important role in the fabrication ceramics, especially when special or complex structures are required. Because the quality of microstructure of the consolidated body is determined by the dispersion behavior of the powder and the interaction between the particles in the suspension, which is closely related to the surface properties of the particles, controlling the physical and chemical properties of particles is a critical to ceramics fabrication.

Furthermore, the surface properties of the powders have direct effect on the densification and microstructural evolution of green bodies during the firing process. This is because the surfaces of the particles will form interfaces and grain boundaries, which are actually the sources and sinks for the diffusion of various species within the materials. Therefore, the structure and composition of the interfaces and grain boundaries will determine the behaviors of densification and

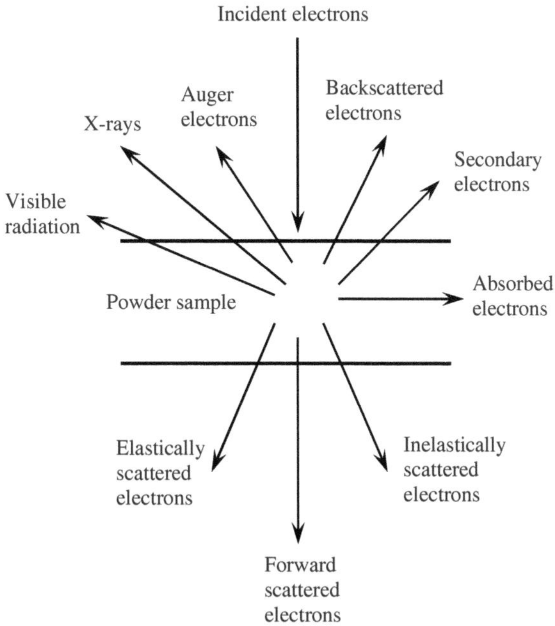

Fig. 4.8 Emissions due to the interaction of an electron beam with a solid powder sample

grain growth. The presence of impurities on the surfaces of the powders changes the composition of the grain boundaries and ultimately the microstructure of the final ceramic products.

Various techniques have been developed to characterize the surface properties of ceramic powders [43–45]. Generally, the principles of surface characterization techniques are to use the interactions of the samples with atomic particles, such as atoms, ions, neutrons, and electrons, or radiations, such as X-rays and ultraviolet rays. Various emissions are produced during the interactions, which are collected as signals to analyze the samples. Figure 4.8 shows the principal emissions caused by the interactions of an electron beam with solid particles.

Depending on the thickness of the solid and the energy of the electron beam, a certain fraction of electrons will be scattered in the forward direction, another fraction will be absorbed, and the rest fraction will be scattered in the backward direction. The forward-scattered electrons consist of electrons that have undergone elastic scattering, i.e., interactions with the atoms of the sample that result in a change in direction but virtually no loss in energy, and electrons that have undergone inelastic scattering, i.e., interactions that result in a change in direction and a reduction in the energy of the incident electrons. The elastically scattered electrons are much greater in number than the inelastically scattered electrons. They are used in the transmission electron microscope to produce diffraction effects in the determination of the structure of the sample. The backscattered electrons are highly energetic electrons that have been scattered in the backward direction. The majority of these will also have elastic collisions with the atoms of the sample.

When the incident electron beam shines on the sample, there are also secondary effects to be generated. For example, some electrons of the atoms could be knocked out of their orbits by the incident electrons. If the ejected electrons are near the surface of the sample within about 20 nm, they may have sufficiently high energy to escape from the sample, which are called secondary electrons. Scanning electron microscope (SEM) images are constructed with these secondary electrons. The transition can also produce radiations in the form of X-rays or visible light, which can be used to characterize the chemical composition of the powders. If the X-rays are absorbed by electrons in the outer orbitals of the atoms and they are very close to the surface of the sample, they may be able to escape from the sample. These escaped electrons are called Auger electrons which reveal the chemical information near the surface regions.

4.2.9.1 Structure of Surface

The surface of a particle is usually not smooth and homogeneous at microscopic or submicroscopic scale, with the presence of various types of irregularities, such as adatom, step adatom, terrace, terrace vacancy, kink, and so on. With agglomerates of finer primary particles, porosity and boundaries between the primary particles will be present on the surface. The atomic irregularities on a surface are distinguished by the number of nearest neighbors, which is called coordination number. The atoms in terraces are surrounded by the largest number of nearest neighbors. Atoms located in steps have fewer neighbors, while those in kink sites have even fewer. Kink, step, and terrace atoms have large equilibrium concentrations on real surfaces. Steps and kinks are known as line defects, while atomic vacancies and adatoms belong to point defects. These effects are heavily involved in atomic transport along the surface. The techniques for characterization of surface structure of solids include electron diffraction and scanning probe microscopy.

4.2.9.2 Chemistry of Surface

Electron, ion, and photon emissions from the outermost layers of the particle surface can be used to reveal the qualitative or quantitative information on chemical composition of the surface of ceramic powders. The most widely used techniques include (i) Auger electron spectroscopy (AES), (ii) X-ray photoelectron spectroscopy (XPS), which is also known as electron spectroscopy for chemical analysis (ESCA), and (iii) Secondary ion mass spectrometry (SIMS).

Figure 4.9a shows the principle of AES, which is based on a two-step process. As an incident electron beam shines a powder material, an electron could be emitted from the inner atomic orbital due to the collision, while the resulting vacant site can be filled by another electron from an outer orbital. The energy due to the transition could be released in the form of an X-ray photon, forming characteristic X-rays, which thus can be used to analyze compositions combined with electron microprobe techniques. The energy can be transferred to another electron in an outer orbital, which is ejected from the atom with a kinetic energy, E_K, given by:

$$E_K = E_1 - E_2 - E_3^*, \tag{4.55}$$

where E_1 and E_2 are the binding energies of the atom in the singly ionized state, while E_3^* is the binding energy for the doubly ionized state. The ejected electron, known as the Auger electron, escapes from the solid and loses its energy very soon due to the inelastic collision with the bound electrons. If the Auger electron is emitted sufficiently close to the surface, it is able to escape from the surface and thus can be detected by using an electron spectrometer. The number of such electrons, or the derivative of the counting rate, is generally plotted as a function of kinetic energy of the electron. Because each type of atom has its own characteristic electron energy levels, the elemental composition of the sample to be tested can be determined according to the peaks in the Auger spectrum, by comparing with standard Auger spectra for the respective element.

The principle of X-ray photoelectron spectroscopy (XPS) is a well-known photoelectric effect, in which the sample is irradiated by a source of low-energy X-rays that leads to the emission of electrons from the lower energy atomic orbitals, as shown schematically in Fig. 4.9b. The kinetic energy of the emitted photoelectrons, E_K, is given by:

$$E_K = h\nu - E_b - W, \tag{4.56}$$

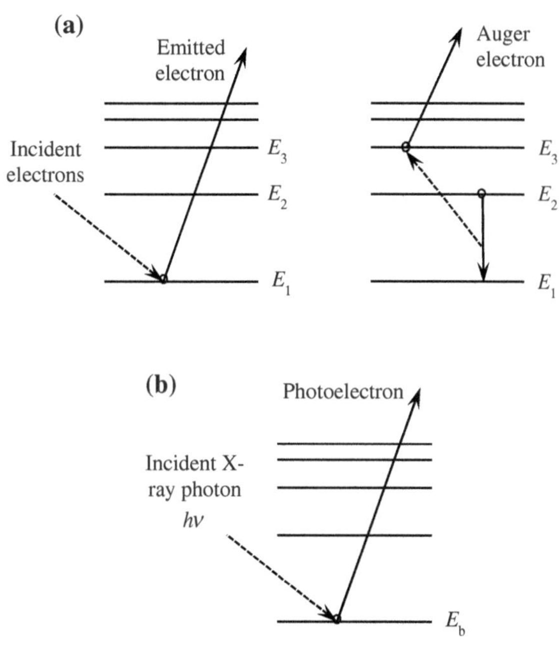

Fig. 4.9 Schematic of the electron emissions in **a** Auger electron spectroscopy and **b** X-ray photoelectron spectroscopy

where hv is the frequency of the incident X-ray photon, E_b is the binding energy of the photoelectron and W is the work function of the spectrometer, which is a factor used to correct for the electrostatic environment in which the electron is produced and measured. By measuring E_K in a spectrometer with a known value of W, the binding energy can be obtained with Eq. (4.56). The data are the number of emitted electrons plotted as a function of the binding energy. The binding energy of an electron is the characteristic property of the atom and orbital from which the electron is emitted. Besides the valence electrons that participate in chemical bonding, each atom (except hydrogen) has core electrons that are not involved in bonding, which are concerned in XPS. Every element in the periodic table has one or more energy levels, corresponding to different peaks in this range. By comparing the position of each peak in the spectrum with standard spectra, the presence of the elements can be determined.

For quantitative analysis, a principal peak is selected for each element, whose intensity is measured as the peak area excluding the background. The fractional atomic concentration of an element, A, is given by:

$$C_{\mathrm{A}} = \frac{I_{\mathrm{A}}/S_{\mathrm{A}}}{\sum_i I_i/S_i}, \tag{4.57}$$

where I_i is the measured peak intensity of the element i and S_i is the atomic sensitivity factor for that peak. The atomic sensitivity factors can be calculated theoretically or derived empirically, which are usually used as reference manuals provided by the manufacturer of the instrument. The accuracy of the quantitative analysis is less than ±10 %, similar to that for AES.

Secondary ion mass spectrometry (SIMS) is to measure the secondary ions, ionized clusters, atoms and atomic clusters, which are emitted from the surface of particles, when it is bombarded with a primary beam of ions, such as He^+, Ne^+, or Ar^+, with energies in the range of hundreds of eV to keV scale. The emitted ions and ionized clusters are analyzed directly by using a mass spectrometer. Therefore, chemical composition of the surface can be analyzed with the obtained accordingly. SIMS has two modes of analysis: (i) static and (ii) dynamic. Static SIMS uses an ion beam with low current density, so as to confine the analysis to the outermost layers. Dynamic SIMS uses beams of high current density, so that successive atomic layers can be eroded at a relatively high rate. Comparatively, the analytical conditions of dynamic SIMS are less suitable for surface analysis.

4.3 Compaction of Ceramic Powders

4.3.1 Packing of Particles

Packing of ceramic powders is the last step before sintering, which has significant effect on the efficiency and effectiveness of the sintering, thus attracting a lot interest [46–50]. There are two types of particle packing: (i) regular or ordered packing and (ii) random packing [51–55]. In ceramic processing, random packing is more general than regular packing [56–60]. Several parameters have been used to characterize the packing arrangement, two of which there are the most widely used. The first is called packing density, also known as the packing fraction or the fractional solids content, which is defined as:

$$\rho_{pd} = \frac{V_{solids}}{V_{green\ body}}, \tag{4.58}$$

where ρ_{pd} is packing density, V_{solids} and $V_{green\ body}$ are volumes of solids and total volume of the green body (solids + voids). The second parameter is coordination number, which is the number of particles in contact with any particle of interest.

4.3.1.1 Regular Packing of Monosized Spheres

The regular packing of monosized spheres is analogous to the packing of atoms in crystalline solids, with regularly repeated three-dimensional (3D) configurations, including simple cubic, body-centered cubic, face-centered cubic, and hexagonal close-packed structures, with packing densities 0.524, 0.680, 0.740, and 0.740, and coordination numbers 6, 8, 12, and 12, respectively.

To form three-dimensional packing configurations of particles, two steps are used: (i) packing spheres in two dimensions to form layered structures and (ii) stacking the layers to form three-dimensional structures. Figure 4.10 shows two types of layers, where the angles of intersection between the rows are 90° for a square layer and 60° for a simple rhombic or triangular layer, although other types of layers with angles of intersection between these two values are also possible [19]. Because there are three simple geometrical ways of stacking each type of the layer on top of one another, six packing arrangements will be available totally. Because two of the ways to stack the square layers are identical to two of the ways to stack

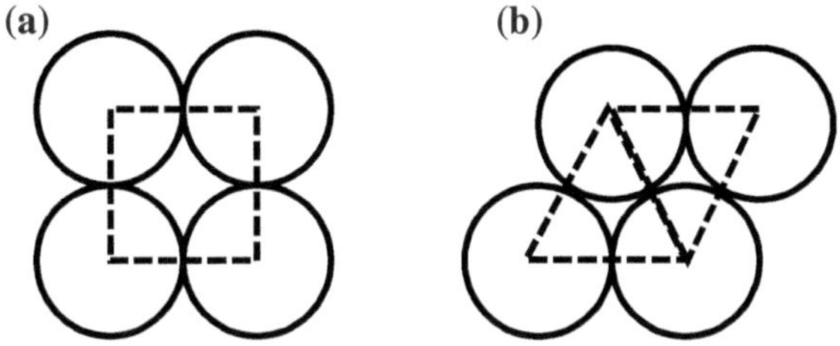

Fig. 4.10 Schematic of two types of layers for the regular packing of monosized spheres: **a** square and **b** rhombic or triangular. Reproduced with permission from [19]. Copyright © 2003, CRC Press

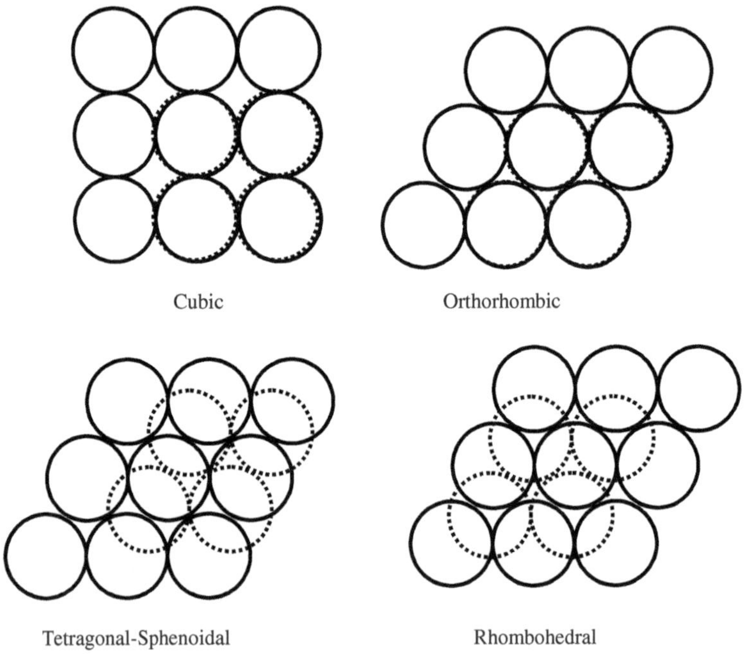

Fig. 4.11 The four packing arrangements produced by stacking square and triangular layers of monosized spheres. Reproduced with permission from [19]. Copyright © 2003, CRC Press

the simple rhombic layers, there are only four different regular packing arrangements, as shown schematically in Fig. 4.11 [19].

Generally, there are four typical packing arrangements, i.e., simple cubic, orthorhombic, tetragonal-sphenoidal, and rhombohedral, with packing densities 0.524, 0.605, 0.698, and 0.740, and coordination numbers 6, 8, 10, and 12, respectively. The rhombohedral packing configuration is the most stable, due to its highest packing density. However, in practice, these ideal packing configurations are hardly achieved over a large region, even using special consolidation methods, such as slow sedimentation. Those small regions with ordered packing are called domains, with domain boundaries in between domains [19]. Therefore, practical green bodies are expected to have random packed.

4.3.1.2 Random Packing

There are two states for random packing: (i) dense random packing and (ii) loose random packing. When the particles are poured into a container and then are settled through vibration, the resulting packing configuration will reach highest packing density, i.e., minimum porosity, which called dense random packing. In contrast, loose random packing is obtained if the particles are simply poured into the container, so that they are not rearranged and settled into a favorable position. The two packing densities are called poured density and tap density, respectively. Between these two end cases, there are many packing configurations [61].

Monosized Particles

According to experimental studies, by shaking hard spheres in a container, dense random packing of monosized spheres has an upper limit of the packing density, ranging from 0.635 to 0.640 [56], which is in a good agreement with computer simulation results of 0.637 [62]. In addition, the maximum packing density achieved by random packing of monosize spheres is independent of the size of the spheres. For loose random packing, theoretical simulations and experiment results of packing density are in the range of 0.57–0.61.

In practice, especially in industrial-scale production, it is very rarely to have spherical particles with monosized distribution. The surfaces of the particles are also not as smooth as assumed in modeling and simulation. Particles with rough surfaces and irregular shapes tend to form agglomeration because of the increased interparticle friction forces, thus leading to lower packing density, as compared with spheres. Although spherical particles are desirable to achieve high packing density, the highest density is obtained with cubic particles. Anisotropic particles can be packed to high packing density also, if the packing is highly ordered, while random packing leads to very low packing density.

Bimodal Mixtures of Spheres

It is obvious that the packing density of spheres with dense random packing can be increased by filling the interstitial voids with smaller spheres that are smaller than those of the original structure, with the packing density to be determined by (i) the ratio of the sphere diameters and (ii) the relative fraction of the mixed spheres. The packing density of the binary mixture can be maximized by filling the interstitial holes with a large number of very small spheres, which is about 0.868.

The packing of binary mixtures of spheres is also usually represented in terms of the apparent volume, i.e., total volume of the solid phase and porosity, occupied by unit volume of solid, which is given by [63]:

$$V_{\mathrm{a}} = \frac{1}{1-P}, \tag{4.59}$$

where P is the fractional volume of the voids, i.e., the porosity.

Binary Mixtures of Nonspherical Particles

The use of mixture of nonspherical particles with different sizes is also able to achieve high packing density, although it cannot reach that of spherical particles. The larger the surface roughness, the more irregular the shape and the larger the aspect ratio, and thus the lower the packing density will be. Similar to that for spherical particles, the packing density of mixed nonspherical particles also increases with increasing size ratio of the two types of particles, as well as the fraction of large to small particles. The composition that leads to the maximum packing density is dependent on the shape of the particles.

Ternary and Multiple Mixtures

The packing density of mixtures of spheres can be further increased by using multiple mixtures, such as ternary mixtures, quaternary mixtures, and so on. For instance, if all the interstitial holes in the binary mixture, with maximum packing density = 0.868, are filled with a large number of fine spheres in dense random packing, the maximum packing density could be as high as 0.952. If quaternary mixtures are used, the maximum packing density will be 0.983. However, such approaches are not practically meaningful. For example, for ternary mixtures, if the particle size ratio should be at least 7 to achieve optimum packing. If the finest particles are assumed to 1 μm in size, the medium and the large particles should be 7 and 49 μm, respectively. Such a large difference in particle size and the presence of the coarse particles make the powders not suitable for ceramic processing from the sintering and microstructure point of view.

Particles with Continuous Size Distributions

In almost all cases, ceramic powders consist of particles with a continuous size distribution. For mixture particles with discrete sizes, the packing density increases with increasing number of components the mixtures. Similarly, for powders with continuous particle size distributions, the wider the particle size distribution, the higher the packing density will be achieved. The packing density increases with increasing standard deviation of the distribution S increases, i.e., the spread of the distribution in the sizes, which can be expressed by the following equation [64–66]:

$$\rho_{\mathrm{pd}} = a - b/S, \tag{4.60}$$

where a and b are constants that are different for different powders. For example, spherical lead glass beads, sphericity $\psi = 1$ have $a = 0.96$, which is equal to predicted value by theory [66]. The packing density is also affected by the shape of the particles, which is usually relatively lower for more irregular particles.

It is also found that optimum packing is obtained when the particle size distribution can be described by a power law equation, which is known as the Andreasen equation [67]:

$$W = \left(\frac{D}{D_{\mathrm{L}}}\right)^n, \tag{4.61}$$

where W is the cumulative mass fraction of particles that are finer than a given size D, D_{L} is the largest particle size in the distribution, and n is an empirical constant used to fit the experimental particle size distribution. Experimental studies indicate that the highest packing density is observed when $n = 1/3 \sim 1/2$. In this equation, it has been assumed that there are all values of particle sizes of smaller than D_{L}, i.e., the particles could be infinitely small, which is not realistic in practice. To address issue, the equation is modified by introducing a finite minimum particle size D_{S} [51]:

$$W = \frac{D^n - D_{\mathrm{S}}^n}{D_{\mathrm{L}}^n - D_{\mathrm{S}}^n}, \tag{4.62}$$

where the exponent n is the same as that in Eq. (4.61).

It is possible to further increase the packing density by mixing two groups of powders, with continuous particle size distributions of different mean particle sizes. In addition, the particle size distribution of the powder smaller particles should be relatively wider. However, it is worth mentioning that the packing density itself is not a parameter for predicting the densification behavior and microstructural evolution during the sintering process. As a matter of fact, packing uniformity is more important for transparent ceramics. According to computer simulations, as the width of the particle size distribution is increased, the scale of density fluctuation is increased [68, 69]. The increased scale of density fluctuation will cause problems during the sintering stage. Therefore, in real applications, hard agglomerates and large particles should be removed before packing, by using various methods, among which sedimentation through colloidal technology is the most effective [70]. Figure 4.12 shows representative packing structures with potential defects in powder compacts [2].

4.3.2 Additives and Ceramic Forming

Various additives, generally with very low concentrations, have been used to control the characteristics of the feed materials, in order to achieve desired shapes and control the packing uniformity of the green body of ceramics. For some methods, such as tape casting and injection molding, the use of suitable additives is especially important. The additives can be either organic or inorganic materials.

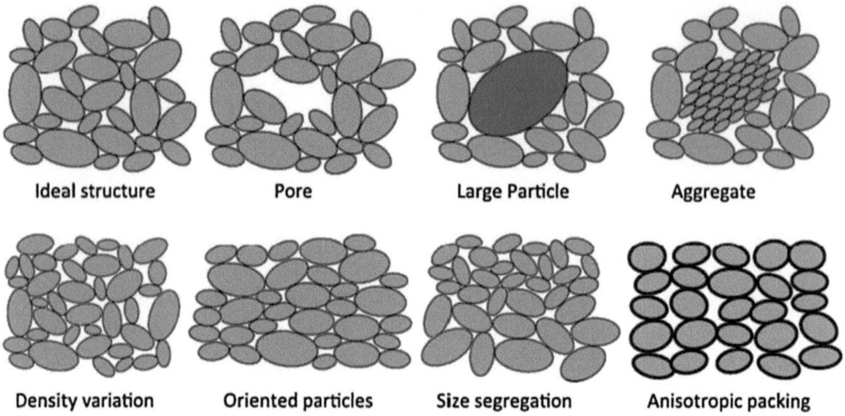

Fig. 4.12 Representative packing structures and defects in powder compacts. Reproduced with permission from [2]. Copyright © 2014, Elsevier

Organic additives are more widely used for such applications, because they can be removed more easily than inorganic ones.

According to their functions, the additives in ceramic process can be classified into four main groups: (i) solvents, (ii) dispersants, also known as deflocculants, (iii) binders, and (iv) plasticizers. Additives with other functions, such as lubricants and wetting agents, are also used in some occasions. Successful additives have often been found by trial-and-error approaches, although information is available on chemical principles and applications of additives [71, 72].

4.3.2.1 Solvents

Liquids are used in ceramic processing have two major functions: (i) providing fluidity for the powder to be processed during the forming and (ii) dissolving the additives to be introduced into the powder as solvents, so as to disperse the additives throughout the powder uniformly.

Water and various organic liquids can be used as solvents. Organic solvents often have higher vapor pressures, lower latent heat of vaporization, lower boiling points, and lower surface tensions than water, due to the strong hydrogen bonding in water molecules. Some organic solvents could also have lower values of viscosity than water. The selection of a liquid is dependent on application and based on several general considerations: (i) ability to dissolve other additives, (ii) rate of evaporation, (iii) ability to wet the ceramic powders, (iv) viscosity, (v) reactivity with respect to the powders, (vi) safety, and (vii) cost [73–75].

The capability of a liquid to wet a solid by is defined in terms of the contact angle, which is given by:

$$\cos\theta = \frac{\gamma_{SV} - \gamma_{SL}}{\gamma_{LV}}, \tag{4.63}$$

where γ_{SV}, γ_{SL}, and γ_{LV} are the surface tensions of the solid–vapor, solid–liquid, and liquid–vapor interfaces, respectively.

In practical applications, good wetting, i.e., low θ, is desirable, which is obtained by a low value of γ_{LV} if γ_{SL} does not change considerably. Organic solvents usually have lower values of γ_{LV}. If water has to be used, the surface tension can be reduced by adding surfactants, which are usually known as wetting agents. Due to their relatively high viscosity and the strong tendency to form hydrogen bonds with hydroxyl groups on the surfaces of oxide powders, the viscosity of the suspension increases rapidly with the concentration of particles, so that low particle concentration is necessary to maintain a high viscosity [76, 77]. However, water is more cost-effective and safe than organic solvents. Therefore, if allowed, it is suggested to use water.

4.3.2.2 Dispersants

Dispersants, also known as deflocculants, are used to stabilize slurries against flocculation, by increasing the repulsion between the particles dispersed [78–83]. The use of dispersants can significantly increase the particle concentration without increasing the viscosity of the slurries. The stabilization principles include electrostatic, steric, and electrosteric stabilizations. Dispersants have a wide range of chemical compositions [84]. According to their chemical structure, dispersants can be classified into three categories: (i) simple ions and molecules, (ii) short-chain polymers with a functional head or end group, also known as surfactants and, (iii) polymers with low to medium molecular weights.

Simple ions and molecule dispersants are most mainly used in aqueous suspensions. They are inorganic compounds, including salts, acids, and bases, which are also known as electrolytes. Selective adsorption of one type of ions onto the particle surface coupled with the formation of a diffuse layer of the counterions, i.e., ions with opposite charge, provides electrostatic stabilization, due to the repulsion between the double layers. The stability of the suspensions is influenced through control the repulsion force between the particles, by the valence and radius of the counterions. According to the Schulze-Hardy rule, the higher the valence of the counterions, the more effective they will be, while for ions with the same valence, the smaller the ions the more effective the dispersants are.

Surfactants are molecules with a short-chain organic tail, containing up to 50–100 carbon atoms, and a functional head group, either nonionic or ionic in nature. For nonionic surfactants, the head group could be polar, which, however, cannot be ionized to produce charged species, so that nonionic surfactants are usually effective in organic solvents. Their adsorption onto the surfaces of particles could be due to either stronger coordinate bonding or van der Waals attraction, with the former being much stronger than the latter. According to the Lewis acid-base concept, an atom in the surfactant functional with an unshared pair of electrons, such as N or O, might act as a Lewis base to a coordinate bond with an atom, like Al, with an incomplete shell of electrons, on the surface of the particles, acting as a Lewis acid. Stabilization is due to steric repulsion between the organic tails or micelles that are stretched out in the organic solvent. For example, menhaden fish oil is used to disperse oxides, such as Al_2O_3.

Ionic surfactants can be anionic, whose functional head group can be ionized to form a negatively charged species, or cationic, with a positively charged head group to be formed. They are effective for aqueous solvents. Usually, the dissociation of anionic surfactants leads to negatively charged oxygen species, which are adsorbed to positively charged surface of the particles due to the electrostatic attraction. The suspensions are therefore stabilized due to the electrostatic repulsion between the negative charges adsorbed on surface of the particles. Cationic surfactants are dissociated to form positively charged nitrogen species, whose mechanisms of adsorption and stabilization are similar to those of the anionic surfactants.

These polymer dispersants can also be either nonionic or ionic. Nonionic polymeric dispersants include poly(ethylene oxide) (PEO), poly(ethylene glycol) (PEG), poly(vinyl pyrrolidone) (PVP), poly(vinyl alcohol) (PVA), polystyrene (PS), and block copolymers of PEO/PS. Due to higher molecular weight, these polymers are more suitable as binders. Those with their chain segments containing OH groups or polar species are effective in water, while those without polar groups are suitable for organic solvents. In both cases, the adsorption of the polymers is due to weaker van der Waals force or stronger coordinate bonding. In aqueous suspensions, hydrogen bonding can cause very effective adsorption. Their stabilization mechanism is steric repulsion due their uncharged characteristics.

Ionic polymeric dispersants, also known as polyelectrolytes, have chain segments with groups that can be ionized. This polar characteristic makes ionic polymer dispersants more suitable for aqueous suspensions. The dissociation of the ionized groups can result in either negatively charged species, which are known as anionic polymers, or positively charged species, which are called cationic polymers. Sodium or ammonium salts of the polyacrylic acids have been successfully used for aqueous slurries of various oxide powders. However, for transparent ceramics, the use of the sodium salts is not recommended, because residual Na ions can create problems for microstructural control during sintering, even at very low concentrations. Poly(ethylene imine) (PEI) is an example of cationic polymers, which becomes positively charged in acidic environment, but remains a weak base under basic conditions. The mechanisms of adsorption are electrostatic attraction, while stabilization is due to the electrosteric effect.

Binders are used to strengthen the green body by forming bridges between the particles. Binders are usually long-chain polymers, with corresponding monomers, including vinyls, acrylics, and ethylene oxides (or glycols). The vinyls have a linear chain backbone in which the side groups are attached to every other *C* atom, whereas the acrylics have the same backbone structure but with one or two side groups attached to the *C* atom. The cellulose derivatives are naturally occurring binders, with molecules being made up of a ring-type monomer unit that has a modified α-glucose structure. The polymer can be modified by varying the side groups *R*.

The selection of a binder for a given forming process is based on several considerations, including (i) binder burnout characteristics, (ii) molecular weight, (iii) glass transition temperature, (iv) compatibility with the dispersant, (v) effect on the viscosity of the solvent, (vi) solubility in the solvent and (vii) cost. For industry applications, cost-effectiveness is the important priority. The binders should be completely burnout before sintering. This needs to be balanced with the high molecular weight for binder strength.

The other consideration of the use of binders is their effect on rheology of the suspensions. In the casting methods, such as tape casting, the addition of a binder should not lead to rapid increase in the viscosity of the suspension with increasing concentration, because the amount of powder that can be incorporated into the suspension will be limited due to the requirement of desirable viscosity. In contrast, a rapid increase in viscosity is desired in extrusion, so as to offer high strength of the green body. In this respect, binders are also grouped into low, medium, and high viscosity grades, which is mainly dependent on the size and structure of the polymer chains.

4.3.2.3 Plasticizers

Plasticizers are organic compounds with lower molecular weights than the binders. The primary function of the plasticizer is to soften the binder in the dry state, i.e., reduce the T_g of the binders, so as to increase the flexibility of the green bodies [85–87]. If the binders are used in solutions, the plasticizers should be soluble in the same liquid in which the binders are dissolved. At the same time, the binders and plasticizers are homogeneously mixed like a single substance in the dry state. The molecules of plasticizers are located between the polymer chains of the binders, so that the chain alignment is disrupted and the van der Waals bonding between adjacent chains is reduced. This will make the binders to be softened, which however also decreases the strength of the green bodies.

4.3.2.4 Other Additives

For a given forming process, the number of additives that are used should be minimized, so that the potential for undesirable interactions between the components will be decreased. However, in some cases, small amounts of other additives have to be used for special requirements. For example, to improve the wetting characteristics of the particles by the liquids, it is necessary to add some surfactants to reduce the surface tension of the liquids. To reduce the friction between the particles or between the particles and the walls of die, lubricants are commonly used in some processing, such as die compaction, extrusion and injection molding. In tape casting, homogenizers, such as cyclohexanone, are usually used to increase the mutual solubility of the components, thus improving the homogeneity of the powder mixtures.

4.3.3 Powder Compaction

4.3.3.1 Dry and Semidry Pressing

Pressings, including uniaxial pressing in a die and isostatic pressing, are the most widely used ways to compact dry powders, which usually contain water of <2 wt% [88–90]. Die compaction is used both in lab and industry, which provides the rapid formation of green bodies with relatively simple shapes and accurate dimensions. The packing density of the green bodies made in this way can be significantly varied, depending on the degree of agglomeration of the dry powders, coupled with the nonuniform transmission of the applied pressure during the compaction process, which limits the height-to-diameter ratio to 0.5–1.0 for simple shapes, like disks or pellets. Isostatic pressing offers better uniformity in packing density and can be used to form green bodies with complex shapes. In some cases, the green bodies have to be polished and machined, which require the green bodies to have sufficiently high mechanical strength.

Die Compaction

Powders or granular materials are simultaneously compacted and shaped at a uniaxial pressure in a rigid die, during the process of die compaction, which involves three steps: (i) filling of the die, (ii) compaction of the powder, and (iii) ejection of the green body. The compaction can be conducted mainly in three modes: (i) single action mode, (ii) double action mode, and (iii) floating die mode, which are named according to the relative motion of the die and the punches. Single action mode means that the top punch moves while the bottom punch and the die are fixed, whereas the double action mode means that both punches move while the die is fixed. As for the floating die mode, the top punch and the die move while the bottom punch is fixed. Among them, the double action mode is usually used in industry, due to its high effectiveness in terms of packing homogeneity.

As feeding materials, powders are usually mixed with a small quantity of binder, e.g., PVA solution with a concentration of <5 vol%. In industry, operations are controlled by machines in a continuous mode, which thus require high flow properties of the feeding materials, for efficient die filling, fast pressing rates, and reproducible green body properties. Due to their poor flowing property, fine powders cannot be compacted homogeneously. Therefore, they need to be granulated, by using spray drying of slurries. The slurries for spray drying are usually made with dispersants such as ammonium polyacrylate, binders such as poly(vinyl alcohol), plasticizer such as poly(ethylene glycol) and lubricants such as ammonium stearate.

The characteristics of the granules are determined by several factors, including the particle size and size distribution of the starting powders, the degree of flocculation of the slurries, the type of additives and the spray-drying parameters. Generally, commercial spray-drying equipment can produce granules with nearly spherical shapes and sizes in the range of 50–400 μm, with average sizes of 100–200 μm. The particle packing density of the granules is determined by the particle size and size distribution of the precursor powders, the particle concentration, and colloidal properties of the slurries, with values of 45–55 % [91, 92]. The packing homogeneity is dependent on the colloidal properties of the suspensions and the conditions of the spray-drying step. It is also related to the properties of the binders [93]. The strength of the granule is determined by the packing density and the nature of the binder. The higher the packing density or the harder the binder, the harder the granules will be. Granules should have smooth surfaces, so that the friction among them and between the granules and the die walls are minimized.

When the powders or granules have a wide size distribution, spherical shapes and smooth surface, they will have better flowing properties during the die filling [51, 94, 95]. The uniformity of the die filling has a direct effect on the packing homogeneity of the green bodies. In addition, the die filling uniformity is also related to the ratio of the granule size to the diameter of the dies. The narrower the dies the lower the overall packing density will be, because the packing density near the die walls is generally lower. According to simulations, when the diameter of the die is about 250 times the diameter of the granules, the effect of the die walls will be negligible [96–99].

The granules in the filled die are loosely packed, with large voids on the order of the granule size and small voids that are smaller than the granule size, as schematically shown in Fig. 4.13 [19]. The compaction involves two stages: (i) disappearance of the large voids due to the rearrangement of the granules and (ii) the filling of the small voids due to the deformation of the granules, depending on the amount and type of agglomerates in the starting powders. If the primary particles have monomodal pores, the compaction has only one stage, involving sliding and rearrangement, along with some fracture at higher pressures. The voids will be reduced due to the applied pressure. If the powder contains weak agglomerates with low-density, two stage compaction will be observed; rearrangement and sliding first to reduce the larger voids and by fracture of the agglomerates and further rearrangement and sliding to eliminate the smaller voids. If the particles experience elastic compression, the ejection of the compact from the die will be affected and crack-like defects in the green bodies could be produced.

The density of the compact as a function of the applied pressure is usually used to characterize the compaction behavior of a ceramic powder, which can be used for process optimization and quality control of the green bodies. If the density is plotted

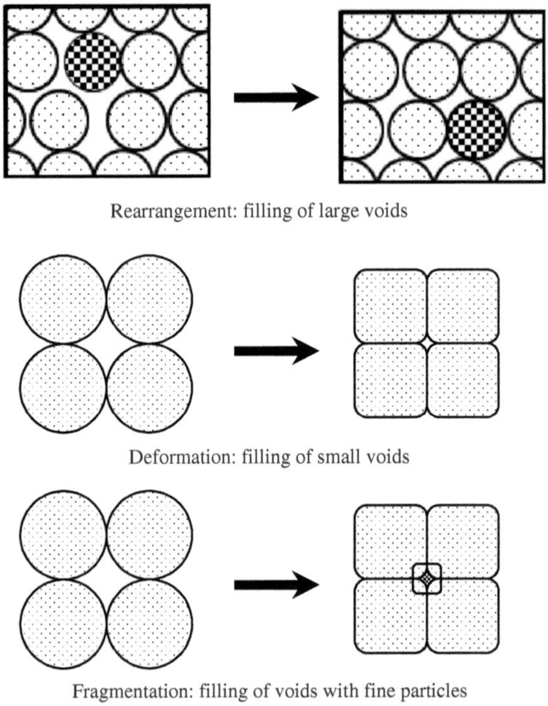

Fig. 4.13 Schematic diagram showing the stages of compaction of granules. Reproduced with permission from [19]. Copyright © 2003, CRC Press

versus the logarithm of the pressure, the compaction data of granules often demonstrate two straight lines, which are separated by a break point, corresponding to the low-pressure linear region due to the rearrangement of the particles and the high-pressure linear region due to the deformation of the particles, respectively. The break point is an indication of the hardness or the strength of the granules. Similarly, powders only consisting of primary particles have a single line, while powder with agglomerates will have two lines, with the break point corresponding to the strength of the agglomerates.

Until now, no sophisticated theory is available for compaction of ceramic powders due to its high complexity [100, 101]. Among various empirical equations, there is a simple and effective one, which is:

$$p = \alpha + \beta \ln\left(\frac{1}{1-\rho}\right), \tag{4.64}$$

where p is the applied pressure, ρ is the relative density, while α and β are constants that are dependent on starting density and nature of the materials.

Due to the friction between the powder and the wall of the die, the applied pressure cannot be transmitted to the powder uniformly, which causes pressure gradients and thus density gradients in the powder compact. When a uniaxial pressure p_z is applied to a powder compact, a radial stress p_r and a shear stress τ are generated at the die wall, both of which vary with the distance along the die, leading to nonuniform stress in the compact. The radial pressure coefficient p_r/p_z and the powder–wall friction coefficient τ/p_r can be measured to study the stresses related to die compaction [102–105]. There are several parameters that could be related, including mean powder particle size d_p, roughness of the wall of the die R_w, hardness of the particles of the powder H_p, hardness of the wall H_w, and properties of the lubricants used.

Densities of compacts of ceramic powders can be characterized with various techniques, such as microscopy, microhardness, X-ray tomography, X-ray radiography, ultrasonic testing, and nuclear magnetic resonance [106–114]. For example, X-ray tomography has been used to monitor the microstructural evolution of ceramic powder compacts, during the compaction process, by installing a compaction set-up was installed inside an X-ray tomography equipment [108].

The compaction process and the formation of heterogeneity in density are started from filling step and persisted during all the process. Density map showed the evolution of forehead densification that was observed near the advancing punch. The destruction mode of granules could be directly observed. At certain humidity level, the compaction could have a brittle or a plastic behavior. Rearrangement and/or destruction were found to start near the yield pressure (Py), which was in contrary to the general understanding that Py corresponds to the end of this microscopic process. It was also found that spherical cavities inside hollow granules were the most difficult to remove. Therefore, hollow granules should be avoided in order to improve the sintering behaviors of ceramic powders.

The presence of gradient density was identified from the distribution of the porosity in the height direction. There were two groups of defects or pores, i.e., >50 μm and in the range 20–50 μm. The >50 μm defects were detrimental, because they could not be removed or resorbed during the sintering process. Generally, the compaction of granules in dry atmosphere without binder led to green bodies with only a few small size defects. This could be the reason why industrial applications prefer to granules by using isostatic pressing without the use of binders. Plastification of granules due to the high level of humidity and the use of binder was detrimental to densification. Also, the reduction in the heterogeneity of density in the initial stage is not sufficient to decrease the detrimental impact of high humidity.

Density maps at pressures of 8 and 110 MPa are shown in Fig. 4.14 [108]. For both pressures, the most homogeneous density was obtained when the granules were made without the use of binder at low humidity, i.e., "without binder R.H. 5." For the granules with binders, density gradients appeared at low pressures, which were retained throughout the whole compaction process.

As shown in Fig. 4.15, at 8 MPa, denser regions (1) were observed near the corners of the top punch, i.e., upper part of the compacts. At high pressures, a region (3) with intermediate density appeared in the middle part of the compact, close to the bottom punch. Between these two regions, an arch-like region (2) with a lower density linked the lower corners. The arch shape of this region was attributed to the arching effects between particles, forming a connected network that was supported by the die wall. Humidification generated more severe gradient densities between the different regions. A pouring step in dry conditions, followed

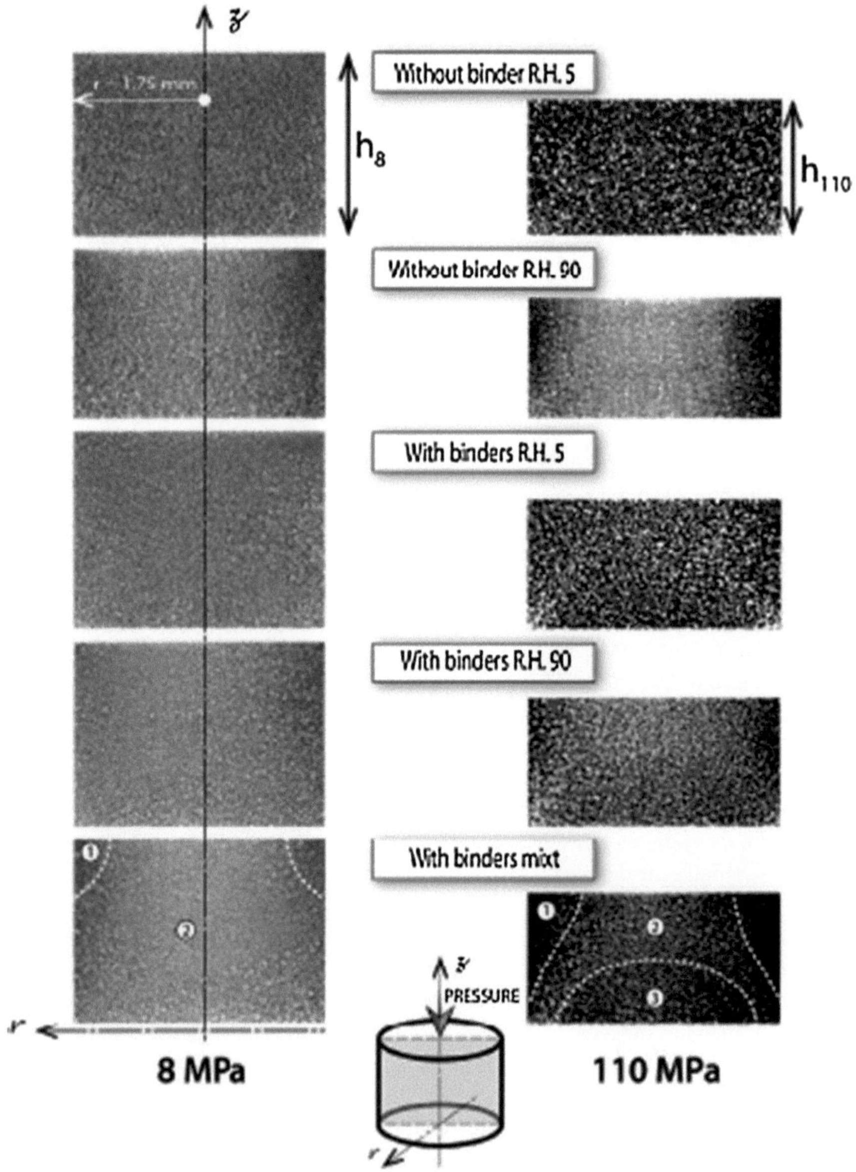

Fig. 4.14 Density maps on axial cross-sections of the pellets at 8 MPa (*left*) and 110 MPa (*right*), allowing the comparison of the density homogeneity throughout the pellets. Higher heterogeneities are always observed when the granules are conditioned in a high relative humidity (R.H. 90 %). Reproduced with permission from [108]. Copyright © 2013, Elsevier

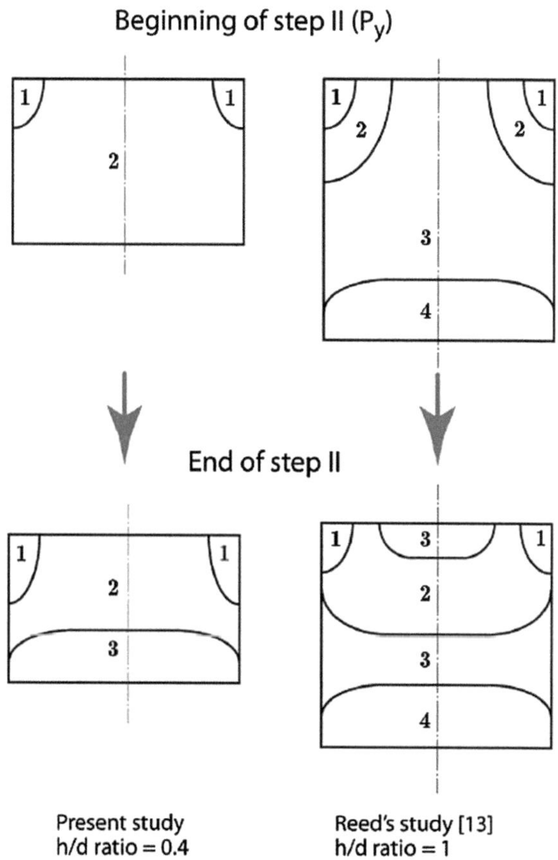

Fig. 4.15 Density evolutions of green body pressed uniaxialy. Reproduced with permission from [108]. Copyright © 2013, Elsevier

by humidification of granules before compaction, was unable to reduce the density heterogeneity, as compared with direct pouring wet granules. These results are useful because high homogeneity in density of green compacts is very important to achieve high sinterability.

Hardness of granules is the key factor to influence the compaction, which is determined by the packing behavior, size and size distribution of the particles, as well as the type of the binder used. Hard granules are easy to rearrange, while those with too high hardness are difficult to deform, thus leading to green bodies with large intergranular pores, which are difficult to be removed during the sintering. This will have a negative effect on density and microstructural development of the final sintered ceramics. On the other hand, soft granules are deformed easily, but exhibit insufficient rearrangement, so that large packing flaws could be present after the die filling, which lead to compacts with large density gradients. The presence of density gradients is detrimental to sintering. Therefore, the hardness of granules should be a trade-off value to ensure the occurrence of both rearrangement and deformation during the compaction.

The powder is subject to elastic compression during the compaction process, which leads to an expansion of the compact, as the applied pressure is released, due to the stored elastic energy. This expansion, also known as springback, strain recovery, or strain relaxation, is nearly instantaneous on release of the pressure. The quantity of the springback is determined by several factors, including (i) properties of the powder to be compacted, (ii) properties of the organic additives, (iii) magnitude of the applied pressure, (iv) rate of pressing, and (v) gas permeability of the powder compact. Usually, the more the organic additives used and the higher the applied pressure, the larger the springback will be. A small strain recovery is beneficial to the separation of the compacts from the punch, but if it is too large, defects or flaws could be produced in the compact. There is also friction between the compact and the die wall to resist the ejection of the compact, which can be minimized by using suitable lubricants. Desirable green bodies should be free of macroscopic defects and density gradients.

Isostatic Compaction

Isostatic pressing is a compaction technique through the application of a uniform hydrostatic pressure to the powder confined in a flexible rubber container. There are two modes of isostatic pressing: (i) wet-bag pressing and (ii) dry-bag pressing. In wet-bag isostatic pressing, the powder is filled in a flexible rubber mold, which is then submerged into a pressure vessel filled with oil as fluid, as shown

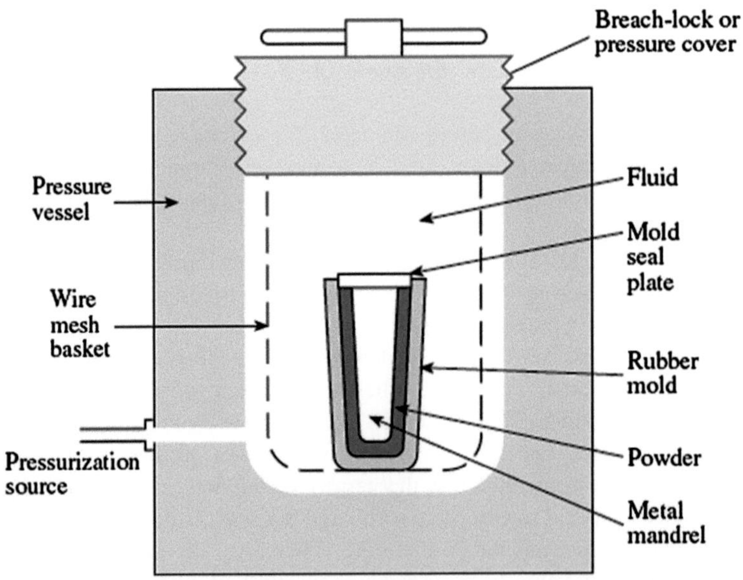

Fig. 4.16 Schematic of a wet-bag isostatic pressing system. Reproduced with permission from [115]. Copyright © 2007, Springer

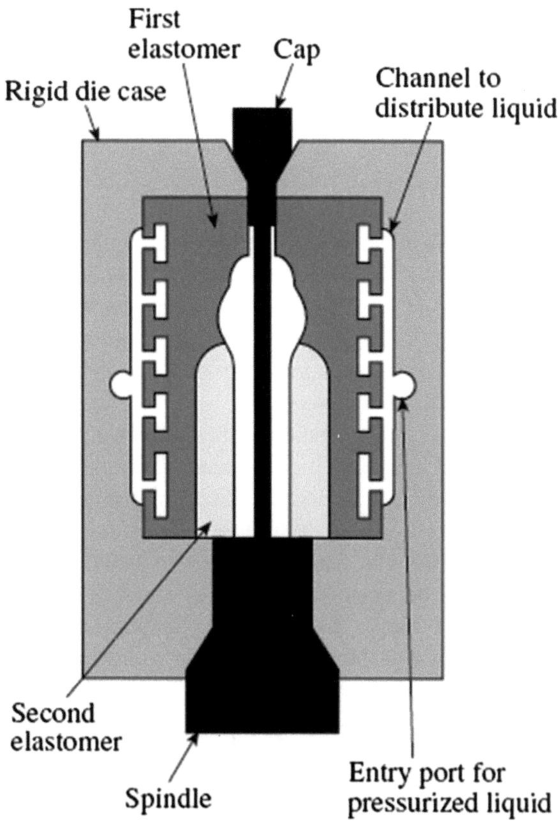

Fig. 4.17 Schematic of a die for dry-bag isostatic pressing of a spark plug insulator. Reproduced with permission from [115]. Copyright © 2007, Springer

schematically in Fig. 4.16 [115]. The green body is retrieved after the pressing. This mode is suitable for compacts with complex shapes and large sizes. Figure 4.17 shows a schematic of dry-bag isostatic pressing system, in which the mold is fixed in the pressure vessel [115]. The pressure is applied to the powder that is filled in the space between a relatively thick rubber mold and a rigid core. After release of the pressure, the powder compact can be collected from the mold.

In comparison with die compaction, isostatic pressing can provide compacts with much less defects or flaws. However, delamination and fracture caused by springback could be present if the pressure is released too fast. In addition, isostatic pressing has been rarely used to directly compact ceramic powders, due to its difficulty in handling powder samples. In most cases, it is used to further increase the density of green bodies either before or after the calcination step. Usually, the increase in density after isostatic compaction is less than 10 %, but that is sufficient to promote the densification of green bodies of most transparent ceramics.

4.3.3.2 Casting Methods

Casting methods include (i) slipcasting, (ii) pressure casting, (iii) tape casting, and (iv) gelcasting, depending on the colloidal systems in which removal of the liquid is used to consolidate particles that are suspended in a slurry. In slipcasting and pressure casting, consolidation of the particles is achieved during the suspension liquid to flow through over a porous medium with a pressure gradient. In tape casting, consolidation is caused by evaporation of the liquid. Gelcasting is a newly developed method, in which the particles in a slurry are immobilized by polymerization and cross-linking of a monomer solution to form a gel prior to evaporation of the liquid. Electrophoretic deposition (EPD) is a process in which ceramic bodies are deposited on an electrode by applying a direct current (DC) electric field to a colloidal suspension.

The green body, usually called the cast, should be homogeneous in particle packing and have a high green density. To meet these requirements, the slurry must have not only a high particle concentration but also appropriate rheological properties to ensure a smooth flowing and allow a sufficiently high casting rate. Therefore, several aspects should be taken in account, including (i) colloidal interactions between particles in a suspension, (ii) factors that control the rheological behaviors of a suspension, (iii) packing behaviors of the particles, and (iv) characteristics of the various organic additives.

Slipcasting

In the process of slipcasting, a slurry is poured into a microporous mold, in which a capillary suction pressure, with the order of 0.1–0.2 MPa, is created due to its porous nature [116]. This pressure draws the liquid from the slurry into the mold. A layer of solid is formed on the walls of the mold due to the consolidation, which is usually known as-cast or cake, as shown schematically in Fig. 4.18 [115]. After the cast reaches a sufficient thickness, the surplus slip is poured out, so that the mold and the cast can be dried. During the drying process, the cast shrinks away from the mold and can be easily removed. After being fully dried, the cast is heated to burnout the binders and additives and then sintered to obtain the final product.

The mechanism of slipcasting has been thoroughly studied [117–121]. The slipcasting process involves the flow of liquid through a porous medium, which is described by Darcy's law. In one dimension, Darcy's law is expressed as:

$$J = \frac{K}{\eta_L} \frac{dp}{dx}, \tag{4.65}$$

where J is the flux of liquid, K is the permeability of the porous medium, dp/dx is the pressure gradient in the liquid, and η_L is the viscosity of the liquid. In slipcasting, the pressure gradient that causes the flow is attributed to the capillary suction pressure of the porous mold. As the consolidation of the particles proceeds,

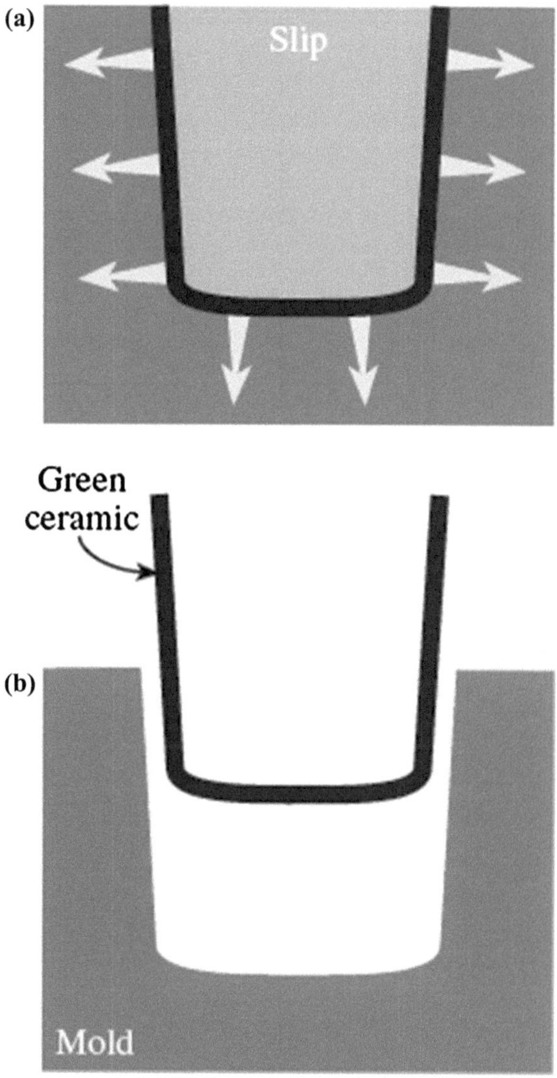

Fig. 4.18 Schematic diagram of the slipcasting system: and **a** initial system, **b** after the formation of a thin cast. Reproduced with permission from [115]. Copyright © 2007, Springer

the liquid, or known as filtrate, passes through two types of porous media: (i) the consolidated layer and (ii) the mold.

If the resistance of the mold to liquid flow is neglected and the liquid flow is considered to flow through the porous consolidated layer, the model of Adcock and McDowall can be used [117]. In this case, the thickness of the cast L_c increases with time t, following a parabolic relation, given by:

$$L_c^2 = \frac{2K_c pt}{\eta_L(V_c/V_s - 1)}, \tag{4.66}$$

where K_c is the permeability of the cast, p is the pressure difference across the cast that is generally assumed to be constant and equal to the suction pressure of the mold, V_c is the volume fraction of solids in the cast as it is assumed to be incompressible, and V_s is the volume fraction of solids in the slurry. It is found that the rate of consolidation decreases with time. Therefore, the thickness of the cast is limited.

However, in practice, both the consolidated layer and the mold have resistance to the flow of the slurry [119–122]. Therefore, the capillary suction pressure p is given by:

$$p = \Delta p_c + \Delta p_m, \tag{4.67}$$

where Δp_c and Δp_m are the hydraulic pressure drops in the cast and the mold, respectively, as shown in Fig. 4.19 [121]. If η_L, V_c, V_s, K_c, porosity P_m, and permeability K_m of the mold are constant with time, the pressure drops Δp_c and Δp_m are linear. The flux of the liquid must be the same through the cast and the mold, so that there is:

$$J = \frac{K_c}{\eta_L L_c}\Delta p_c = \frac{K_m}{\eta_L L_m}\Delta p_m, \tag{4.68}$$

where L_c is the thickness of the cast and L_m is the thickness of the mold saturated with liquid. Integration of Eq. (4.67) under appropriate boundary conditions also leads to a parabolic equation, which is given by [121]:

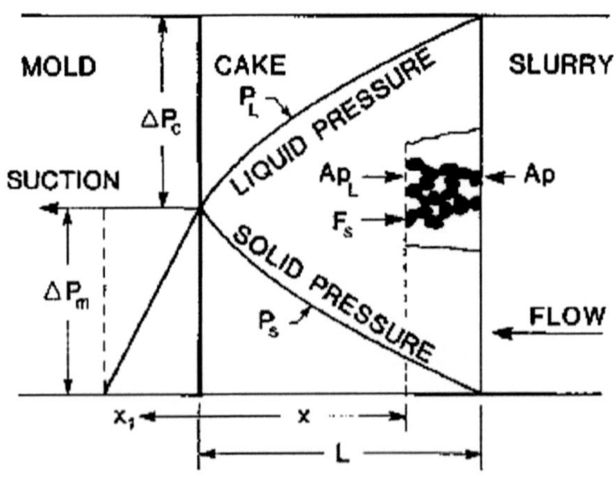

Fig. 4.19 Hydraulic pressure distribution across the cast and the mold in slipcasting. Reproduced with permission from [121]. Copyright © 1986, American Ceramic Society

$$L_c^2 = \frac{2Hpt}{\eta_L}, \quad (4.69)$$

where the function H is dependent on the properties of both the cast and the mold, which is given by:

$$H = \left[\left(\frac{V_c}{V_s} - 1\right)\left(\frac{1}{K_c} + \frac{V_c/V_s - 1}{P_m K_m}\right)\right]^{-1}. \quad (4.70)$$

When the resistance to liquid flow in the mold can be neglected, Eq. (4.66) is recovered from Eqs. (4.68) and (4.70). This situation applies when the term $(V_c/V_s - 1)/P_m K_m$ in Eq. (4.70) is very small as compared with the term $1/K_c$, i.e., there is:

$$\frac{K_c(V_c/V_s - 1)}{P_m K_m} \ll 1. \quad (4.71)$$

Equation (4.71) indicates that the hydraulic resistance of the mold can be neglected if the mold porosity P_m, the volume fraction of solids in the slurry V_s, and the ratio of the permeability of the cast to that of the mold K_c/K_m are all sufficiently high.

Equation (4.66) indicates that the rate of consolidation increases with increasing permeability of the cast K_c. It has been shown that the Carman–Kozeny equation is able to well describe the key parameters of slipcasting, which is given by:

$$K = \frac{P^3}{\alpha(1 - P)^2 S^2 \rho_s^2}, \quad (4.72)$$

where P is porosity, S is specific surface area of the solid phase, ρ_s is density of the solid phase, and α is a constant that is equal to 5 for many slurries and defines the shape and tortuosity of the channels. For a cast consisting of monodispersed spherical particles with a diameter of D, its permeability can be expressed as:

$$K_c = \frac{D^2(1 - V_c)^3}{180\, V_c^2}. \quad (4.73)$$

This equation means that, in order to increase K_c, it is necessary to increase D or decrease V_c. In terms of sintering of ceramics, it is undesirable to increase D, because an increase in D means a decrease in sintering rate. Therefore, decreasing V_c is more realistic to control the consolidation rate, which can be achieved by controlling the colloidal stability of the slurry, as discussed earlier.

Characteristics of the mold, such as porosity and pore size, have significant effects on the consolidation rate of the casting. According to Eqs. (4.66) and (4.69), the consolidation rate increases with increasing capillary suction pressure p of the mold. Therefore, if other effects are all kept constant, an increase in p would always lead to an increase in consolidation rate. Because the capillary suction pressure increases with decreasing pore radius of the mold, a decrease in the pore radius implies an increase in the casting rate. However, the permeability of the mold K_m will also decrease correspondingly. As a result, pore size of the mold should be optimized, so as to achieve the maximum rate of consolidation [121].

Colloidal stability of the slurry is the most important effect on microstructure of the cast. For example, flocculated slurries usually result in highly porous casts. Furthermore, the effective pressure in the cast p_s decreases rapidly and nonlinearly across the mold–cast interface, as shown in Fig. 4.19. In this case, due to the compressible nature of the highly porous cast, coupled with the varied p_s, the cast will suffer a rapid reduction in density, resulting in heterogeneous microstructure in the green body, which is undesirable for sintering [117]. A well-dispersed slip containing no agglomerates and stabilized by electrostatic or steric repulsion leads to the formation of a fairly dense cast with high microstructural uniformity.

It is also demonstrated in Eq. (4.66) that the consolidation rate increases with increasing concentration, V_s, of the solids in the slurry. For monodispersed particles, the effect of V_s on the resulting green body density of the cast is not significant. If the powder has a wide distribution of particle sizes, V_s has complicated effects on the green density. In this case, a low V_s could mean a low green density, if the sedimentation rate of the particles is pronounced as compared with the casting rate, which leads to segregation of the particles of different size. For suspensions with high V_s, the sedimentation becomes less pronounced, so that high green densities can be obtained, because the fine particles tend to fill the interstices formed by the large ones.

Other parameters such as size, size distribution, and shape of the particles can also influence the compressibility and thus the packing density of the cast. Figure 4.20 summarizes the effects of particle size, particle shape, and the degree of

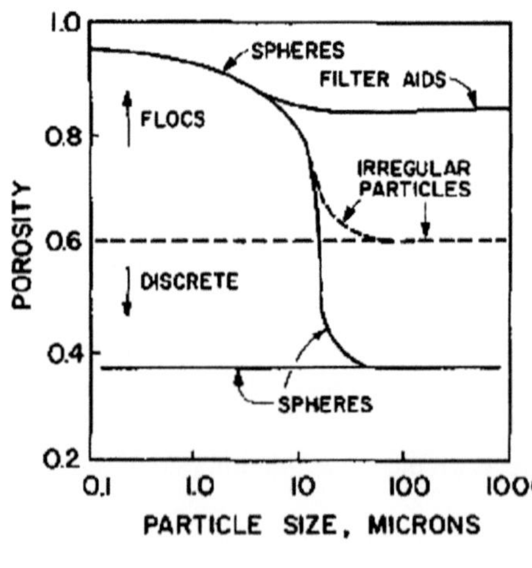

Fig. 4.20 Schematic diagram showing the effects of particle size, shape, and degree of flocculation on the porosity of the cast produced in slipcasting. Reproduced with permission from [121]. Copyright © 1986, American Ceramic Society

flocculation of the slurry on the packing density of the cast [121]. For relatively coarse particles, with sizes in the range of >10 μm, colloidal effects are insignificant and the degree of dispersion of the particles has no effect on the packing density.

Large spheres of the same size produce casts with a packing density of 0.60–0.65, close to that for dense random packing, while irregular particles produce casts with a lower packing density. As the particle size decreases below about 10 μm, colloidal effects control the packing density. At an extreme, well-dispersed slips produce casts with high packing density, whereas, at the other extreme, flocculated slips yield a low packing density. As stated above, particles with a distribution in size may produce casts with a packing density that is higher than that for monodisperse particles if segregation does not take place.

The rate of consolidation also increases with decreasing viscosity of the filtrate η_L, as shown by Eq. (4.66). A decrease in η_L is achieved by either increasing the temperature of the slurry or using a liquid with a lower viscosity. For aqueous slips, an increase in the temperature improves the stability of the slip, thus leading to a cast with a higher packing density. This decreases the consolidation rate, because the permeability of the cast K_c is decreased. However, the reduction in η_L with increasing temperature has a much greater effect, so that the overall result is an increase in the casting rate.

Several types of microstructural defects can be present in green articles formed by slipcasting, which could be formed during the casting and are usually dependent on the properties of the slurry. Large voids caused by air bubbles in the slurry are often observed, which can be avoided by improving the wetting characteristics of the particles by the liquid, proper deairing of the slurry and minimizing the turbulent flow of the slurry during the casting. Elongated or anisometric particles can be preferentially aligned along certain directions, usually parallel to the mold surface. Segregation, in which the larger particles settle faster than the smaller ones, can be alleviated by improving the colloidal stability of the slurry.

The defects formed during the slipcasting process could affect mechanical strength of the final ceramics, as shown in one example of slipcasted alumina [123]. A commercial low-soda alumina powder was used as the raw material. Figure 4.21 shows the procedure of sample preparation. The powders were dispersed with a commercial ammonium polycarboxylate dispersant in ion exchanged water at the solid loading of 50 vol%. The amount of dispersant added to Al_2O_3 powder was in the range of 0–2 mass%, with controlled pH values for the slurry of 0 mass%. The slurries were ball milled for 24 h with alumina media and then vacuum treated for 10 min for deairing. A plaster mold was used to form a compact with a dimension of 65 × 65 × 7 mm^3, by using slipcasting. An aluminum foil was also used to get a compact of 100 × 65 × 10 mm^3 through spontaneous drying. The compacts detached from the plaster and the foil were dried at 110 °C for 24 h and then calcined at 500 °C for 5 h to remove the dispersant. Finally, the compacts were sintered at 1600 °C for 2 h in air.

Figure 4.22 shows SEM images of the Al_2O_3 ceramics derived from the dispersed (0.2 mass% dispersant) and the flocculated (2 mass% dispersant) slurries [123]. Both samples seemly had a same microstructure, i.e., uniform and consisting

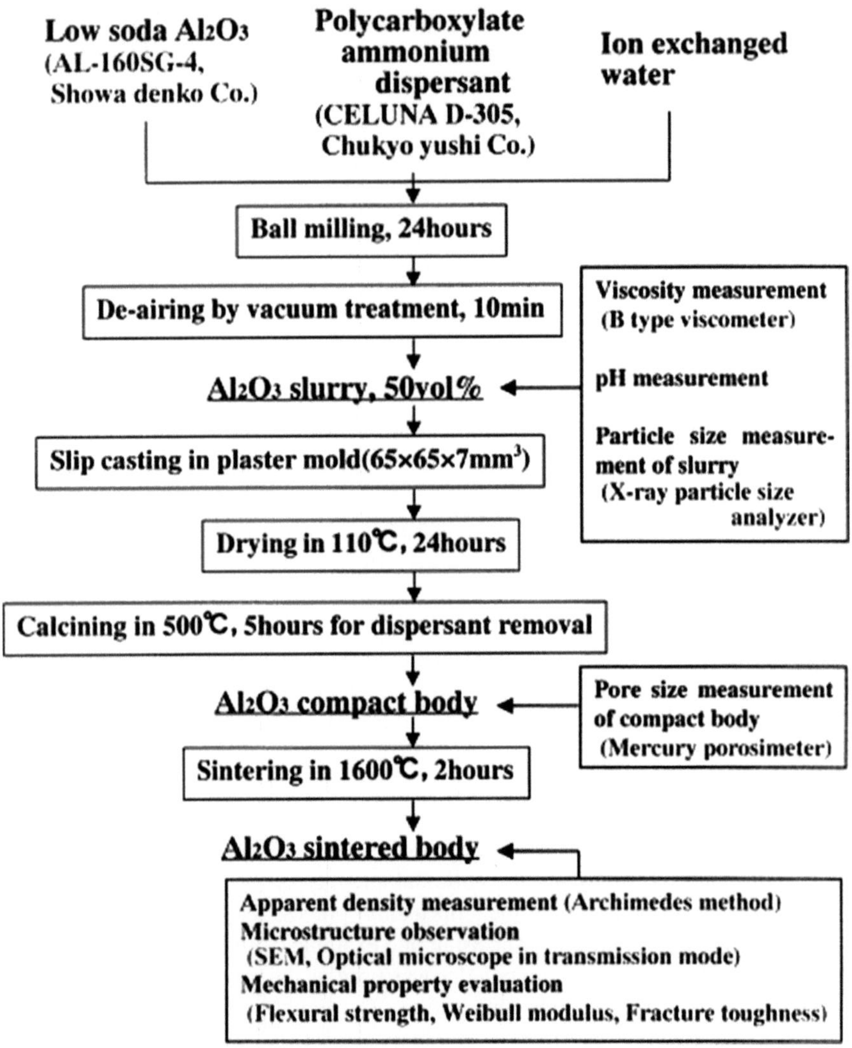

Fig. 4.21 Flow chart of the Al_2O_3 slipcasting process. Reproduced with permission from [123]. Copyright © 2000, Elsevier

of elongated alumina grains. Also, average grain sizes and relative densities of the two ceramics were 4.3 and 4.2 μm and 99.7 and 98.7 %, respectively, which were very close. There were small pores of about 1 μm both in grains and at grain boundaries. There was not abnormal grain growth.

Figure 4.23 shows transmission mode optical photomicrographs of the ceramics derived from the dispersed (0.2 mass% dispersant) and the flocculated (2 mass% dispersant) slurries. Both samples had some black features of elongated shapes, with uniform sizes of 10–15 μm long and a few micrometers wide. Additionally,

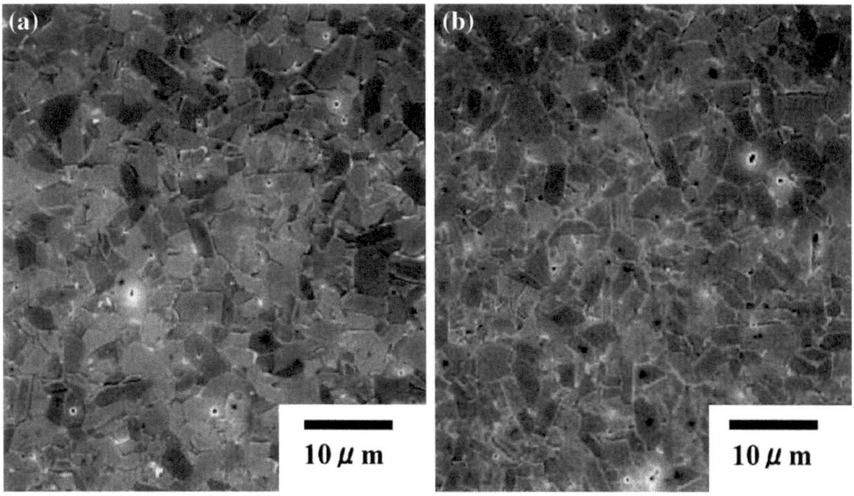

Fig. 4.22 SEM images of polished and thermally etched surfaces of the Al_2O_3 ceramics derived from the dispersed (0.2 mass% dispersant) and the flocculated (2 mass% dispersant) slurries. Reproduced with permission from [123]. Copyright © 2000, Elsevier

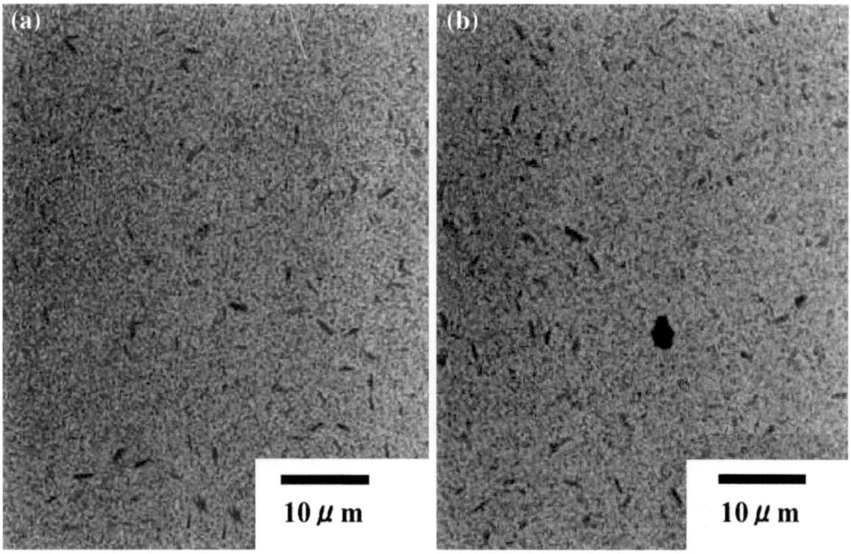

Fig. 4.23 Optical photomicrographs taken in the transmission mode of the Al_2O_3 ceramics derived from the dispersed (0.2 mass% dispersant) and the flocculated (2 mass% dispersant) slurries. Reproduced with permission from [123]. Copyright © 2000, Elsevier

besides the elongated pores, the sample derived from the flocculated slurry contained rounded features also, as shown in Fig. 4.23b. This kind of large feature was not found in the sample derived from the dispersed slurry.

The elongated pores were formed through the liquid flow during the casting process, because they were found in all slipcast samples but not observed in the spontaneously dried sample for which there was no occurrence of rigorous flow of water. The formation of these defects was insensitive to the slurry properties. The origin of spherical pores was likely attributed to the entrapped air bubbles during the deairing process. The removal of the air bubbles by deairing was easy for the dispersed slurry which had a low viscosity, but difficult for the flocculated slurry due to its high viscosity. The Weibull's plots for flexural strengths of the two ceramics were almost the same in the region of high strengths. However, because the sample derived from the flocculated slurry contained a higher concentration of the spherical pores, sometimes low strength could be encountered. Such lower strengths could be readily attributed to the presence of the detrimental defects, i.e., the spherical pores. Obviously, these features can be happening in any other materials. Slipcasting has been used to fabricate various transparent ceramics and related ceramic materials [124–136].

A commercial Y_2O_3 powder was used to fabricate highly transparent Y_2O_3 ceramics doped with ZrO_2 by using slipcasting combined with vacuum sintering [137]. The effects of the content of ZrO_2 on transparency, grain size, and lattice parameter of the Y_2O_3 ceramics were systematically investigated. The addition of ZrO_2 could increase the transparency and decrease the grain size of the final Y_2O_3 ceramics. The commercial Y_2O_3 powder (5 N purity) consisted of agglomerated platelet particles with an average particle size of 2.0 μm and a BET surface area of 3.44 $m^2\ g^{-1}$. The Y_2O_3 powder was mixed with various contents of ZrO_2 (3 N purity) as sintering aid and milled with ZrO_2 balls for 12 h in ethanol. The milled slurries were dried at 60 °C for 24 h. The as-obtained powders were calcined at 1000 °C for 2 h. The calcined powders exhibited an average particle size of 0.48 μm and a BET surface area was 5.40 $m^2\ g^{-1}$.

The calcined powders were milled with deionized water and dispersant to form suspensions with solid of 30 vol% Y_2O_3. The suspensions were slipcasted with a gypsum mold. After demolding and drying, the green bodies were heat treated at 900 °C for 2 h in order to burnout the organic additives. The green bodies with a relative density of about 45 % were sintered at 1730–1900 °C for 5–15 h in vacuum of 2×10^{-3} Pa in a furnace with tungsten heating element. The sintered samples were then annealed at 1500 °C for 10 h in air to obtained transparent Y_2O_3 ceramics. The Y_2O_3 ceramics doped with 5 mol% ZrO_2 have a grain size of about 5 μm and transmittance of 81.7 %.

Through coupling slipcasting of nanopowder suspensions with spark plasma sintering, highly transparent Nd:Lu_2O_3 ceramics without suing sintering aids were obtained [133]. The effect of shaping with slipcasting on residual microstructural defects after sintering was investigated. The presence of homogeneous and submicron pore size in the green compacts promoted the elaboration of highly transparent ceramics.

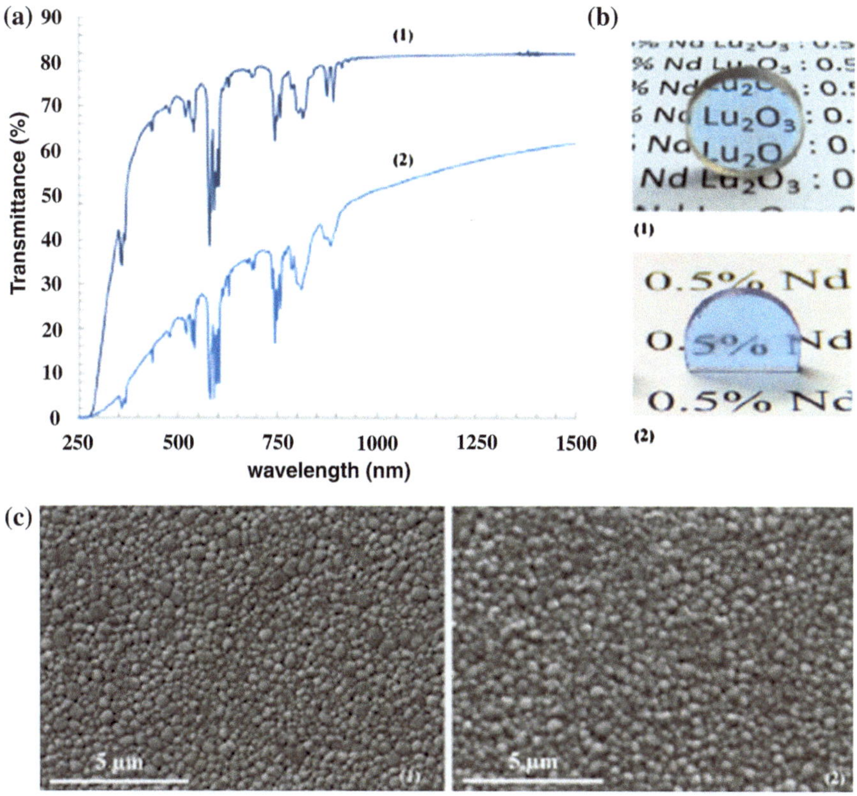

Fig. 4.24 Optical transmittance spectra (**a**), together with corresponding photographs (**b**) and SEM images (**c**) of the transparent Nd:Lu_2O_3 ceramics derived from slipcast (*1*) or unshaped (*2*) powders. Reproduced with permission from [133]. Copyright © 2014, Elsevier

Lu_2O_3 and Nd_2O_3 powders with purity of > 99.99 % were dissolved in nitric acid with an Nd/Lu atomic ratio of 0.5 at.%. The rare earth nitrate solution was mixed with NH_4HCO_3 solution to facilitate precipitation. The precipitate was dried and then calcined at 900 °C for 2 h to form phase pure crystalline Nd:Lu_2O_3 nanopowder with a particle size of 30 nm. The powder was then deagglomerated by using ball milling for 2 h in ethanol. The Nd:Lu_2O_3 nanopowder was compacted by using slipcasting, followed by sintering with SPS. SPS was carried out in a graphite die. The temperature was increased to 1400 °C for 15 min at a rate of 50 °C min^{-1}, while the pressure was progressively increased to 130 MPa simultaneously.

As shown in Fig. 4.24a, transmittance of the sample made with slipcast powder reached a value of 81.3 % at 1060 nm, which was much higher than that of the one derived from the unshaped powder. Figure 4.24b shows photographs of the corresponding Nd:Lu_2O_3 ceramics, where the slipcast sample was more transparent. SEM images of the two samples are shown in Fig. 3.25c. They possessed a similar

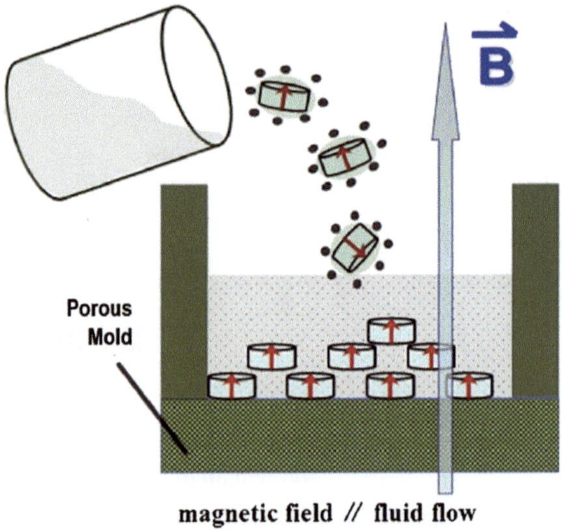

Fig. 4.25 Schematic diagram of slipcasting under magnetic field. Reproduced with permission from [141]. Copyright © 2012, Elsevier

microstructure, which was fully dense and almost pore-free, with a grain size of about 300 nm.

Transparent YAG ceramics have been fabricated by slipcasting an aqueous dispersed mixture of commercial Al_2O_3 and Y_2O_3 powders, with poly(acrylic acid) and citric acid [138]. Polyethylene glycol of 0.5 wt% (PEG 4000) and 0.5 wt% tetraethyl orthosilicate (TEOS) were used as binder and sintering aid, respectively. The dried samples were vacuum sintered at 1800 °C for 16 h. It was found that YAG ceramics cast from Newtonian suspensions were optically transparent, with optical transmittances of >80 % over 340–840 nm, which is comparable to the performance of YAG single crystals grown by using the Czochralski technique. The flocculated dispersions, with higher viscosity and non-Newtonian rheology, led to translucent samples with large pores and lower optical transmittances.

Slipcasting Al_2O_3 with magnetic field is a new development of this technology [139–141]. For instance, transparent Al_2O_3 ceramics with textured structures have been prepared by using slipcasting with the aid of magnetic fields, after sintering at 1850 °C for 5 h in vacuum [141]. In-line transmittance of the Al_2O_3 ceramic shaped under 12 T reaches as high as 70.3 % at 600 nm. High quality α-alumina powder with a purity of 99.99 %, a mean particle size about 400 nm and a BET specific surface area about 6 $m^2 g^{-1}$, was selected as raw material. 400 ppm MgO was used as sintering aid. Alumina suspension with about 25 vol% solid loading was prepared by mixing the raw powder with polyacrylate ammonium and deionized water, with ultrasonic dispersion for 30 min and then planetary ball milling for 10 h. Figure 4.25 shows the set-up for slipcasting at magnetic fields in the direction parallel to the casting direction, with strength of up to 12 T. After casting, the wet green bodies were dried at 120 °C for 24 h and calcined at 800 °C for 2 h in air.

Finally, the samples were pressurelessly sintered at 1850 °C for 5 h in vacuum at vacuum level of about 2×10^{-3} Pa.

XRD patterns of the ceramics from the green bodies that were slipcast at different magnetic fields are shown in Figs. 4.26 and 4.27, for top slices and side slices, respectively. The ceramics without magnetic fields exhibited a random microstructure. The application of magnetic fields led to grain orientation, forming textured microstructures. For top slices (Fig. 4.26), the diffraction intensity of peaks (006) related to *c*-planes (00l) was enhanced gradually with increasing strength of

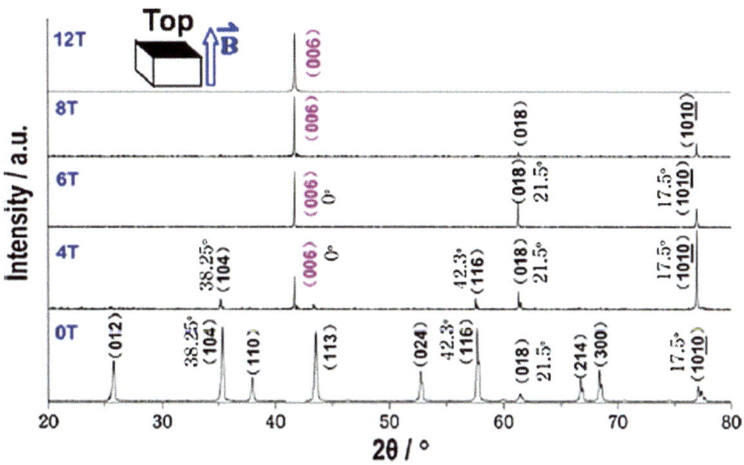

Fig. 4.26 XRD patterns of top slices of alumina ceramics shaped at magnetic fields of 0, 4, 6, 8, and 12 T. Reproduced with permission from [141]. Copyright © 2012, Elsevier

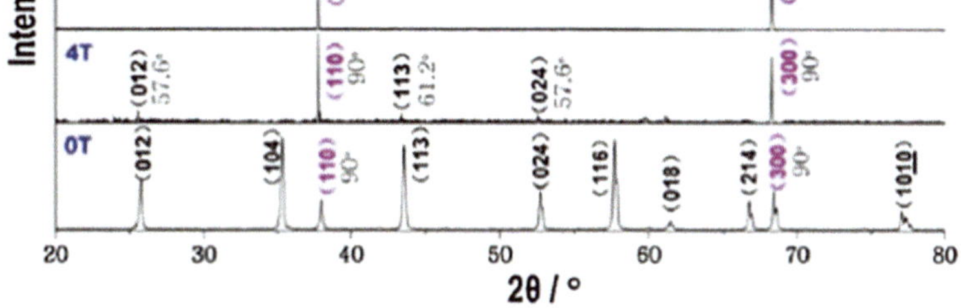

Fig. 4.27 XRD patterns of side slices of alumina ceramics shaped at magnetic fields of 0, 4, 6, 8, and 12 T. Reproduced with permission from [141]. Copyright © 2012, Elsevier

magnetic field. Comparatively, for side slices, the diffraction intensities of (hk0) planes perpendicular to *c*-planes, such as (110) and (300) planes, were enhanced with increasing magnetic field strength, as shown in Fig. 4.27.

The XRD characterization demonstrates that a complete crystalline texture with the *c*-axis parallel to the direction of magnetic field is obtained by slipcasting within a high assisted magnetic field and sintering. In other words, alumina ceramics fabricated via magnetic alignment method provides grains with *c*-axis parallel to each other and to the optical axis. Therefore, when light transmits through optical axis of grains, the physical environment remains the same; theoretically birefringence of the grain boundary can be eliminated. This is why in-line transmittance of the ceramics we prepared is greatly improved.

Figure 4.28 shows SEM images of polished and thermal etched surfaces of the transparent alumina ceramics, which were shaped at 12 T and sintered at 1850 °C for 5 h in vacuum. For the top slice, the grains appeared to be nearly isometric shape as shown in Fig. 4.28a, with an average grain size of ~40 μm. In contrast,

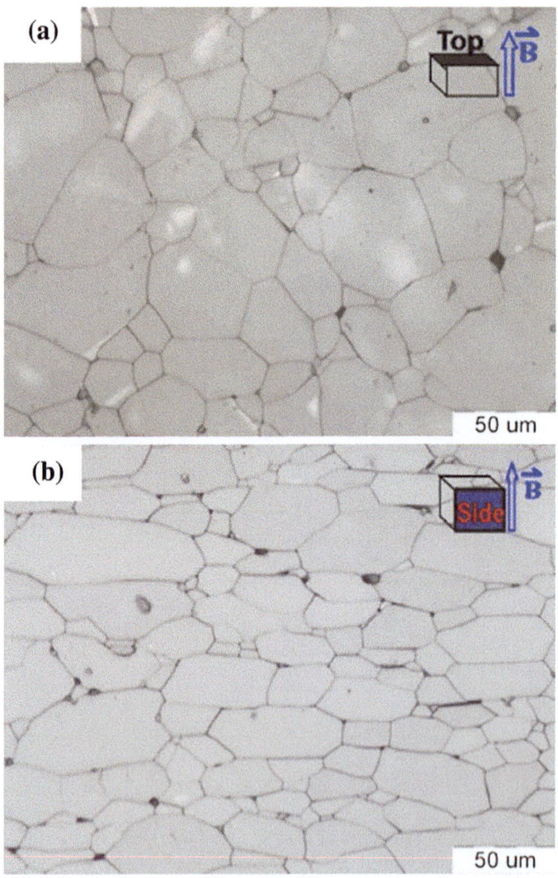

Fig. 4.28 Thermal-etched surface SEM images of the Al_2O_3 transparent ceramics shaped at 12 T and sintered at 1850 °C: **a** top slice and **b** side slice. Reproduced with permission from [141]. Copyright © 2012, Elsevier

grains of the side slice shown in Fig. 4.28b were elongated, due to the platelet grains that were aligned perpendicularly to the magnetic field. More importantly, the alumina ceramics with oriented grains exhibited much higher transmittance than that of the normal random polycrystalline sample. Therefore, the magnetic field assisted slipcasting is an effective way to prepare highly transparent alumina ceramics. This technology is especially useful, when anisotropic microstructure is required. In addition, textured ceramics have higher mechanical strength.

Pressure Casting

Equation (4.66) indicates that, for a given slurry, as the filtration pressure p is increased, the time taken to produce a given thickness of cast L is decreased. In other words, the casting can be speeded up by the application of an external pressure to the slurry [127, 142]. This is the principle of pressure casting, also referred to as pressure filtration or filter pressing [143]. The plaster of Paris molds used in slipcasting is too weak to withstand pressures of >0.5 MPa. Therefore, plastic or metal molds must be used in pressure casting. A schematic of main features of a laboratory-scale pressure casting device is shown in Fig. 4.29. Particles in the slurry form a consolidated layer, i.e., the cast, on the filter, as the liquid is

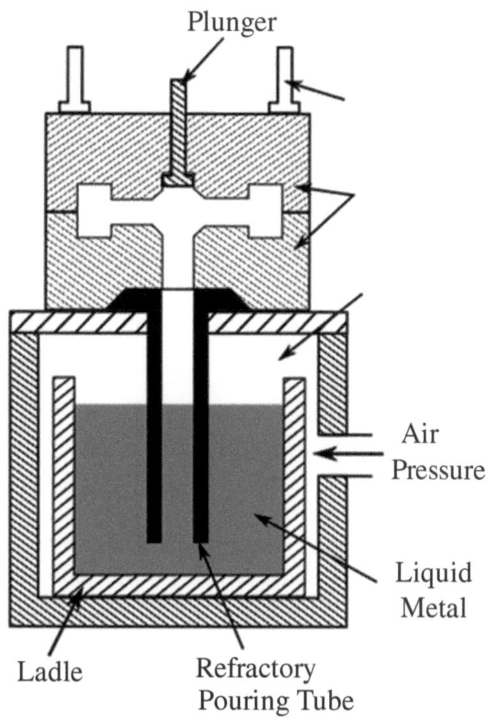

Fig. 4.29 A laboratory-scale pressure casting apparatus. Adopted from: http://commons.wikimedia.org/wiki/File:Low_pressure_permanent_mold_casting_schematic.svg

forced through the system. The cast provides a much greater resistance to the flow of the liquid when compared to the filter, so that the kinetics of pressure casting can be described by using Eq. (4.66).

Kinetics and mechanics of pressure casting have been systematically studied [118, 144, 145]. As observed in slipcasting, the colloidal stability of the slurry has the largest effect on the microstructure of the cast. It has been shown that the highest packing density is obtained with well-dispersed slurries, with which the packing density obtained is independent on the applied pressure, as the pressure is >0.5 MPa. The fact that high packing densities can be obtained with these dispersed slurries at such low applied pressures implies that rearrangement must be facilitated by the repulsive forces between the particles. In contrast, for flocculated slurries, the packing density is dependent on the applied pressure, with a nearly linear relationship between the relative density and the logarithm of the applied pressure, which is similar to the trend observed in the die compaction of dry powders.

Pressure-cast bodies of ceramics also have problems, when they are cast from either a dispersed or a flocculated slurry. As the last portion of the slurry is consolidated, the pressure gradient across the cast becomes zero, so that the total applied pressure is transferred to the cast. Therefore, upon the applied pressure is removed, the cast expands, i.e., undergoes strain recovery, due to the stored elastic energy. However, the nature of the strain recovery is different from that in the die compaction of dry powders, in a way that the strain recovery for the compacts of the pressure casting is time dependent. This time-dependent strain recovery is attributed to the fact that the fluid, either liquid or gas, must flow into the compact to allow the particle network to expand and relieve the stored strain. The magnitude of the recovered strain increases with increasing consolidation pressure nonlinearly, which can be described by the following Hertzian elastic stress–strain relation:

$$p = \beta \varepsilon^{3/2}, \tag{4.74}$$

where p is the stress, ε is the strain, and β is constant depending on system. Therefore, for each material, the processing should be specifically optimized. One of the important advantages of pressure casting is its higher productivity, due to the shortened consolidation time and the compacted facility [146–149]. However, the pressure casting molds are more expensive, as compared to those used in other casting methods.

A high-pressure colloidal slipcasting (HPCSC) method has been developed to compact nanopowder of Nd:YAG, with green density up to 60 % [142]. The Nd:YAG nanopowder was synthesized by using a chemical precipitation method, from nitrate solutions with ammonium bicarbonate as precipitant. Figure 4.30 shows schematic diagram of pressure slipcasting set-up. It consisted of a mold with a diameter of 27 mm, Al_2O_3 porous brick with a thickness of 3–5 mm, and a membrane with pore size and thickness of 470 nm and 100 μm. Compacts with thicknesses of 1–8 mm could be made by using the set-up. The loading pressure was 200 MPa. The regularity of particles packing for YAG nanopowders was significantly improved. Transparent Nd:YAG ceramics were obtained by sintering

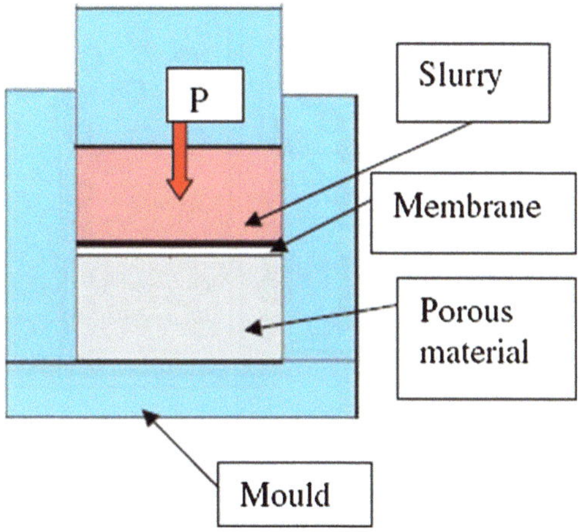

Fig. 4.30 Schematic of the high-pressure colloidal slipcasting set-up. Reproduced with permission from [142]. Copyright © 2009, Elsevier

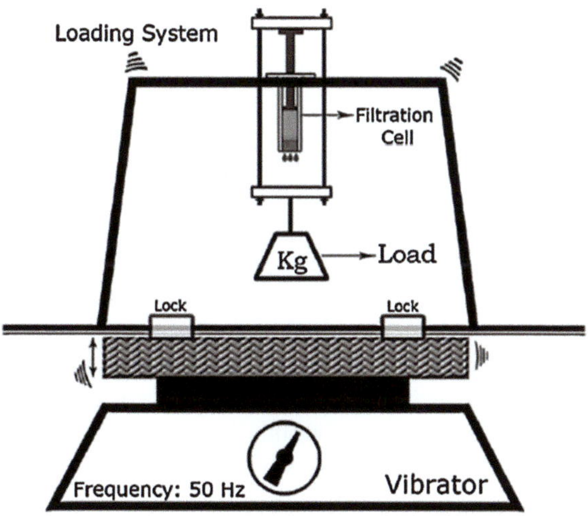

Fig. 4.31 Schematic of loading system with a vibration set-up. Reproduced with permission from [147]. Copyright © 2008, Elsevier

the compacts at 1700–1800 °C in vacuum. Laser effect was achieved with plate ceramics with longitudinal pumping by semiconductor laser.

A vibration method has been reported to improve particle packing in consolidation of submicrometer alumina powders when using pressure slipcasting [147]. In the experiment, a mechanical vibration was applied to the filtration cell, at a constant frequency of 50 Hz, with vibration amplitudes ranging from 0 (no vibration) to 2 mm. Figure 4.31 shows a schematic diagram of the slipcasting set-up, equipped with a vibrator. The influence of vibration on filtration characteristics, such as filtration rate, thickness, and green density of the consolidated samples, were studied.

It was found that the application of vibration could greatly increase the filtration rate, when the vibration amplitude was optimized. The optimized vibration

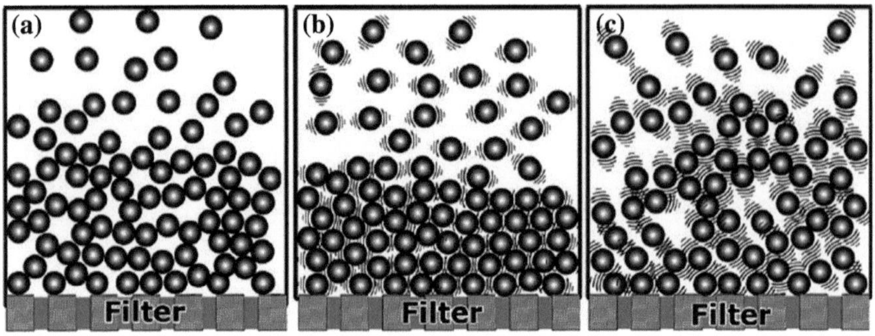

Fig. 4.32 Schematic of the reasoning behaviors for packing of the cake layer with different vibration intensities: **a** no vibration, **b** optimum vibration intensity, and **c** over vibration intensity. Reproduced with permission from [147]. Copyright © 2008, Elsevier

amplitude was increased as the quantity of the concentrated alumina slurries was increased. Figure 4.32 shows schematic diagram to explain the effect of the vibration [147]. The application of a vibration action could promote rearrangement of the particles and prevent the particle bridging during the formation of the cake, as shown in Fig. 4.32b. Static friction of the colloidal particles could be canceled out by the hydrodynamic vibration, so that the particles would be able to stay in more energy favorable positions, which led to denser compaction. However, if the vibration intensities were too high, the efficiency of the compaction was decreased, because the particles could not settle down easily. Over vibration led to bouncing of the particles out of their energy wells, so that large pores were entrapped due to the drift of other accumulating particles. In addition, the hydrodynamic drag force accompanied with higher flow rates on top surface of the growing cake would take the time that was required for particle rearrangement and densification. Therefore, there was an optimal vibration intensity at the applied frequency. If the vibration frequency is variable, there will be more options.

Tape Casting

The process principles and practical applications of tape casting have been well documented in the open literature [150–155]. Tape casting is also known as doctor blade process, in which the slurry is spread over a surface covered with a removable sheet of paper or plastic, by using a carefully controlled blade, called doctor blade. For large-scale production, the blade is kept stationary while the surface moves continuously, so as to produce long tapes, as shown in Fig. 4.33. In the laboratories, because the tapes are not necessarily too long, pulling the blade over a stationary surface is usually used. Dried tapes are obtained after all solvents are evaporated. The flexible green tapes are cut into desired lengths for subsequent processing. Thickness of the tapes can be readily controlled in a wide range of 10 μm to 1 mm.

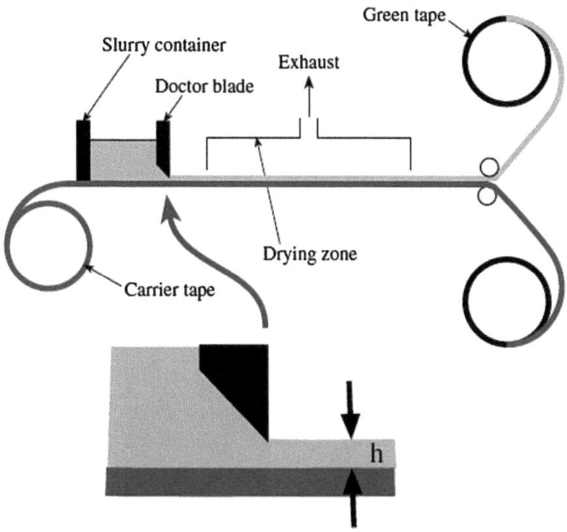

Fig. 4.33 Schematic diagram of the tape casting process. Reproduced with permission from [115]. Copyright © 2007, Springer Science + Business Media

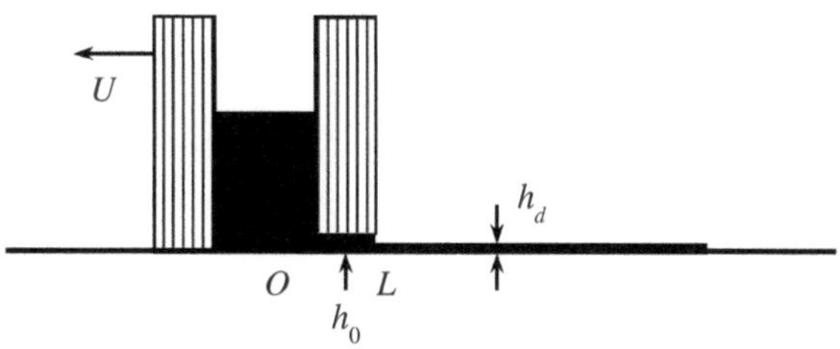

Fig. 4.34 Schematic of the key component in a tape casting machine

The quality of the tapes is determined by the quality of the slurry. The selection of solvent, dispersant, binder, plasticizer, and other additives is similar to that discussed earlier. Other factors include thickness of the tapes to be cast and the surface on which the cast is to be cast. Generally, highly volatile solvent systems, e.g., acetone or methyl ethyl ketone, are used for thin tapes, while slowly drying solvents, such as toluene, are necessary to cast thicker tapes, e.g., >0.25 mm. The most important organic additive is the dispersant, which determines the viscosity of the slurry and thus the particle concentration. Binder–plasticizer combination should also be carefully selected, due to their high contents in the tape casting slurries. In addition, it is important to ensure the strength and flexibility of the green tapes, while not neglecting the burnout of the additives.

As illustrated in Fig. 4.34, the key component of the tape casting facility is the assembly of the doctor blade. It consists of an adjustable doctor blade mounted in a frame with a reservoir to hold the slurry before it is cast under the blade to form the thin layer on the carrier surface. The flow behavior of the slurry during the tape casting has been analyzed theoretically to estimate the effect of various casting parameters on thickness of the tapes [152].

In a simple casting unit, a Newtonian viscous slurry and a laminar flow can be assumed, the thickness of the dry tape, h_d, shown in Fig. 4.34, is given by:

$$h_d = \frac{\alpha\beta}{2}\frac{\rho_w}{\rho_d}h_0\left(1 + \frac{h_0^2 \Delta p}{6\eta UL}\right), \tag{4.75}$$

where α (<1) and β (<1) are correction factors, ρ_w and ρ_d are the densities of the slurry and the dry tape, respectively, h_0 and L are the height and thickness of the doctor blade, respectively, Δp is the pressure difference that is determined by the height of the slurry in the reservoir, η is the viscosity of the slurry, and U is the velocity of the doctor blade relative to the casting surface. According to Eq. (4.75), if the second term in the brackets is $\ll$1, the thickness of the dried tape is proportional to the height h_0 of the doctor blade. For the height h_0 of <200 μm, this can be easily achieved. However, if the values of h_0 are too large, significant deviations from this relationship is observed, especially when the values of η, U, and L are relatively low.

Because the solvents are evaporated from the surface of the tapes and the tapes are tightly attached to the carrier surface, the shrinkage during the drying process is in the thickness direction of the tapes. Generally, if dried tapes contain 50 vol% ceramic powder, 30 vol% organic additives, and 20 vol% porosity, their thickness is about half the doctor blade height. Although the relatively density of the green tapes is relatively low, after the binder is burnout, they can be sintered to full density, as demonstrated by the examples of transparent ceramics later.

The main types of microstructural defects in green tapes are similar to those observed in the slipcast bodies. As the tapes are sufficiently thin, the rate of drying is sufficiently slow and the colloidal stability of the slurry is sufficiently high, segregation due to the settling of particles with different sizes is not a problem. Due to high contents, binder could be segregated to the surface of the tapes during drying which can lead to gradient in packing density. Therefore, it is possible to cause warping or cracking of the tapes during the sintering process. If there are elongated or anisometric particles in the slurry, preferential alignment of the particles along the direction of flow could be observed. This effect can be used to produce ceramics with textured microstructures [156–159]. Various transparent ceramics, including PLZT, alumina, multilayered YAG, composite YAG, and Al_2O_3 wafers, have been fabricated by using tape casting [160–166].

Ceramic wafers of Al_2O_3 were produced by using tape casting with aqueous slurries, followed by vacuum sintering [161]. The binder system used to form the tape casting slurry was a copolymer of isobutylene and maleic anhydride, which is environmentally friendly and marketed under the name ISOBAM. The rheological properties of the slurries were studied by varying solid loading and binder addition level. With optimal content of plasticizer, green tapes, with excellent plasticity and thicknesses of 240–740 μm, could be readily casted. The tapes displayed a post-sintering thickness of 150–660 μm. In-line transmittances of the unpolished Al_2O_3 transparent wafers, with a thickness of 660 μm, were 72 % at a wavelength of 5 μm and 26 % at a wavelength of 600 nm.

Commercial Al_2O_3 powders (D_{50} = 0.45 μm) were used as the starting materials. The commercial ISOBAM, wich a density of 1.3 g cm^{-3} and molecular weight of M.W. = 5,160,000–300,000, was used as both binder and dispersant. Tape casting slurries were prepared with the Al_2O_3 powder, ISOBAM and distilled water by using ball milling to achieve homogeneous dispersion of all the components. Slurries with different binder contents and solid loadings were studied. Viscosity measurements were performed by preshearing the slurry at 100 s^{-1} for 30 s and increasing the shear rate continuously from 0.001 to 200 s^{-1}. The experimental process was much simpler than many tape casting systems reported in the open literature. After degassing, the Al_2O_3 slurries were tape-casted, with a moving blade at a speed of 25 mm min^{-1} on a stationary polypropylene carrier film (Mylar), with blade gaps of 0.5 and 1.5 mm. After demolding and drying, the debinding and calcination of the green sheets were carried out at heating rates of 3–5 °C min^{-1} to 1000 °C for 3–5 h. The calcined samples were finally sintered in a vacuum furnace at 10^{-5}–10^{-6} MPa in the temperature range of 1750–1850 °C for 5 h at a heating rate of 10 °C/min^{-1}, so that transparent Al_2O_3 wafers would be obtained.

Figure 4.35 shows photographs of the green tapes casted with a solid loading of 68 wt% and binder content of 20 wt%. The green tapes were highly flexible and easy to handle, without any cracks upon freely bending. SEM observation indicated that Al_2O_3 particles are homogeneously distributed and encapsulated by the binders after tape casting, without any agglomerations, thus leading to improved density and strength of the green tapes.

Figure 4.36 shows photograph of a piece of transparent Al_2O_3 wafer, illustrating high-quality and smooth defect-free surfaces. SEM images of surface and fractured cross-section of the as-sintered transparent Al_2O_3 wafers are shown Fig. 4.37. The main grain sizes were in the range of 20–80 μm. No obvious differences were observed in the microstructure of the Al_2O_3 wafers in different directions of casting, i.e., transverse-to-casting and thickness directions. No pores were observed on the grains or at grain boundaries. In Fig. 4.37b, the transparent Al_2O_3 wafers were dominated by intergranular fracture with minor intracrystalline rupture. The thickness of the sample shown in the inset of Fig. 4.9b is about 660 μm.

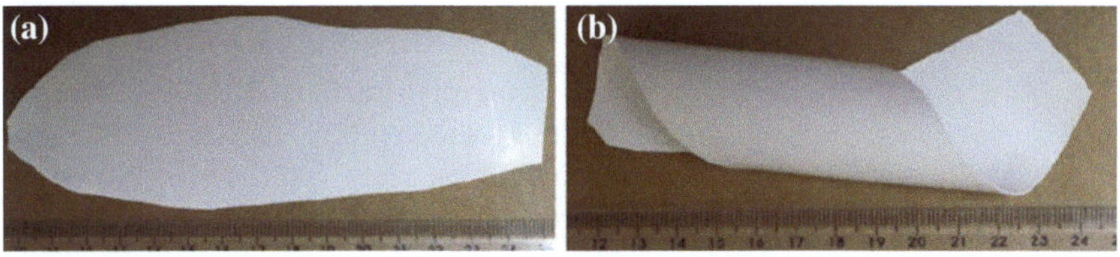

Fig. 4.35 Photographs of the Al_2O_3 green tapes after casting in air at room temperature: **a** flat green tape and **b** flexible green tape. Reproduced with permission from [161]. Copyright © 2014, Cambridge University Press

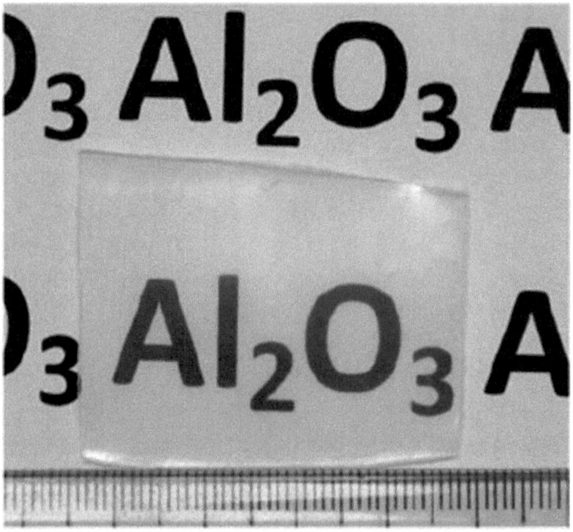

Fig. 4.36 Photograph of a piece of transparent Al_2O_3 wafer after sintering. Reproduced with permission from [161]. Copyright © 2014, Cambridge University Press

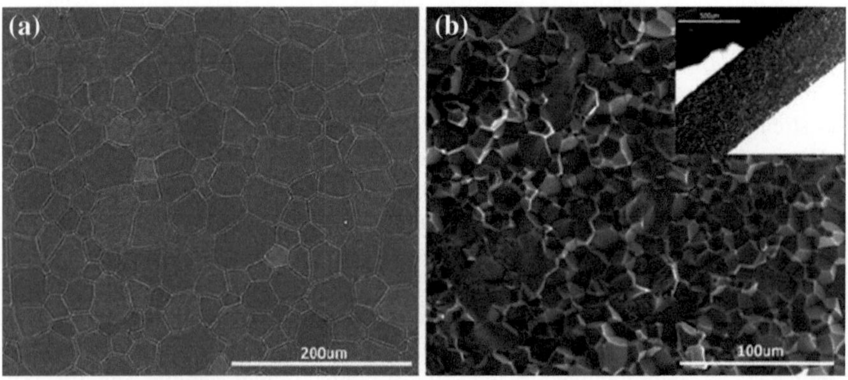

Fig. 4.37 Microstructures of transparent Al_2O_3 ceramics: **a** as-sintered surface and **b** fracture surface. The *inset* of **b** is a TEM image. Reproduced with permission from [161]. Copyright © 2014, Cambridge University Press

Another example is the fabrication of composite-structured $Y_3Al_5O_{12}$ transparent ceramics by using an aqueous tape casting method [162]. High-purity (99.99 %) Y_2O_3, Al_2O_3, tetraethyl orthosilicate (TEOS), MgO, polyacrylicacid (PAA, 25wt%), polyvinylalcohol (PVA-124), polyethyleneglycol (PEG-400, AR) were used as starting materials. Y_2O_3 and Al_2O_3 powders were weighed according to the stoichiometry of $Y_3Al_5O_{12}$. TEOS and MgO were used as sintering aids. The powder mixture was ball milled with the dispersant (PAA) and deionized water for

8 h or longer. Then, binder (PVA) and the plasticizer (PEG-400) were added. The new mixture was ball milled for another 24 h. After that, the slurry was sieved and deaired by using a vacuum facility. The gap height of the blade was controlled in the range of 300–500 μm, while the casting speed was 10 mm s^{-1}. The tapes were dried at room temperature for 36 h. Small pieces with circle shape were cut from the tapes and 30 layers were packed. High pressure was used to drive the air trapped in the tapes. The binder was burned out at 600 °C for 48 h. The green bodies were cold isostatic pressed at 200 MPa for 2 min to obtain homogeneous bodies with high densities. Finally, the green bodies were sintered by using solid-state reaction method (reactive sintering) in vacuum, at 1750 °C for 10 h, at pressure of below 1×10^{-3} Pa during the sintering process.

Representative as-cast tapes, single slices of the tape, stacked slices with 30 layers, and pressed slices are shown Fig. 4.38a [162]. The thickness of the tapes was about 100 μm. Figure 4.38b shows a green body and a sintered YAG transparent ceramics. After the binder removal and cold isostatic pressing procedures, the diameter of the disk was reduced from 20 mm to about 18 mm, which was further shrunk to about 15 mm after sintering. Composite-structured YAG ceramics with a thickness of 1.0 mm exhibited high optical quality.

Fig. 4.38 **a** Photographs of the tape, as-cut slice, stacked slices and as-pressed slices, **b** photographs of green body and sintered YAG composite structured ceramics, **c** fracture morphology of the green body, and **d** EPMA picture of the fracture surface of the YAG ceramics. Reproduced with permission from [162]. Copyright © 2013, Elsevier

Figure 4.38c shows fracture morphology of the green body, which had a very dense microstructure. No gap was observed across the fracture surface, implying the close compaction of the slices. Fracture morphology of polished surface of the YAG ceramics is shown in Fig. 4.38d. The sintered ceramics were nearly fully dense, without the presence of cracks on the surface, i.e., no delamination occurred during the binder removal and the sintering process. The average grain size of the YAG ceramics sintered at 1750 °C for 10 h was about 15 μm, while the in-line transmittance of the sample is 81.5 % at 1064 nm. Therefore, this is a flexible and feasible method to fabricate transparent ceramics with composite structures [163].

Gelcasting

Gelcasting is a recently introduced process that is based on ideas taken from the traditional ceramics industry and the polymer industry [167–173]. In summary, a slurry of ceramic particles dispersed in a monomer solution is poured into a mold, and the monomer is polymerized to immobilize the particles and to form a gel-like bonding phase (i.e., a binder). The system is removed from the mold while still wet, dried by evaporation of the liquid, heated to burnout the organic additives, and finally sintered.

In gelcasting, aqueous solvents are usually used, combined with processing methods, such as ball milling, to produce slurries with desired properties, including stability against flocculation, high particle concentration, and low viscosity. The key element of the process is the addition of an organic monomer to the solution, which is then polymerized in situ to form a strong and cross-linked gel. In addition to preventing segregation or settling of the particles, the gelling makes the body to be sufficiently strong to withstand the capillary stresses during drying, so that there is no limitation in thickness of the casts.

The gel formation has two stages: (i) initiation and (ii) polymerization. At the initiation stage, the viscosity does not change and no heat is generated. Because the initiator is added at room temperature, it takes a sufficiently long time of 30–120 min, so that deairing of the slurry and mold filling can be conducted without problems. The polymerization stage is generally carried out at slightly higher temperatures 40–80 °C to promote the gelation. Various materials can be used as mold for gelcasting, including aluminum (Al), glasses, polyvinylchloride (PVC), polystyrene (PS), and polyethylene (PE). Gelcasting has been used to fabricate various translucent and transparent ceramics [174–177].

A method of combining gelcasting in air and pressureless sintering in hydrogen was developed to prepare translucent alumina ceramics [174]. A stable suspension with 75 wt% solids was prepared by using ball milling. Trapped bubbles were removed completely before casting to obtain homogeneous green bodies. Final sintering was carried out in a hydrogen furnace at 1850 °C for 3 h. Homogeneous microstructures of both calcined and sintered bodies were achieved. The Al_2O_3 ceramics were translucent in the range of 190–1100 nm.

In this study, a commercial alumina powder, with 600 ppm MgO as dopant and a particle size of 0.45 μm (D_{50}), was used as starting material. Polyacrylic acid ammonium salt was used as dispersant with the amount of 1.0 mL for 100 g alumina powder. The suspension containing 75 wt% solids was prepared by mixing alumina powder and the aqueous premix of sorbitol polyglycidyl ether (SPGE, 15 wt% in water). 1 mL dipropylenetriamine per 3 g SPGE was added into the suspension, for cross-linking as a hardener. After thorough mixing, the suspension was immediately put into a vacuum chamber (0.05 atm) for degassing, in order to remove the trapped bubbles. After that, it was cast into molds. After consolidation and demolding, the wet green bodies were dried at room temperature for 24 h and then dried at 110 °C in an oven. The dried bodies were calcined at 1000 °C for 2 h, at a heating rate of 1 °C min^{-1}, which were then sintered at 1850 °C for 3 h in a hydrogen furnace.

Transparent Y_2O_3 ceramics have been fabricated by using aqueous gelcasting combined with vacuum sintering [175]. A nontoxic and water-soluble copolymer was used as both dispersant and gelling agent at room temperature in air. Surface treatment was conducted through a nucleophilic addition reaction to prevent hydrolysis of the Y_2O_3 powder. Optimized dispersant content and solid loading were obtained according to the rheological behavior of the Y_2O_3 slurries. Transparent Y_2O_3 ceramics were prepared after sintering at 1850 °C for 6 h in vacuum. The sample with a thickness of 1 mm showed an in-line transmittance of as high as 80.9 % in at 1100 nm. High-quality optical performance of the Y_2O_3 ceramics were attributed to the homogeneous microstructures of the calcined body and sintered product, as shown in Fig. 4.39.

A new gelcasting method has been reported to develop highly transparent yttrium aluminum garnet (YAG) ceramics [176]. The spontaneous gelcasting was carried out by using an aqueous mixture of commercial Al_2O_3 and Y_2O_3 powders and a water-soluble copolymer of isobutylene, as well as maleic anhydride, with a commercial name of Isobam, as both the dispersant and gelling agent. High-purity

Fig. 4.39 SEM images of fracture surfaces of the Y_2O_3 presintered body calcined at 1000 °C for 2 h (**a**) and the final ceramics sintered in vacuum at 1850 °C for 6 h (**b**), which were derived from the slurry of 81 wt% solids loading. Reproduced with permission from [175]. Copyright © 2014, Elsevier

powders of α-Al_2O_3 (99.99 %), Y_2O_3 (99.999 %), and Isobam 104# (a copolymer of isobutylene and maleic anhydride) were used as starting materials. The alumina and yttria powders were weighed according to the stoichiometric ratio of YAG ($Y_3Al_5O_{12}$), and then thoroughly mixed by using a ball milling in ethanol for 12 h, with 0.5 wt% tetraethyl orthosilicate (TEOS, 99.999 %) as sintering aid. The powder mixture was dried at 80 °C for 24 h and then sieved through a 200-mesh screen. The slurries, with the solid loadings of 68–74 wt% and the Isobam content of 0.3–1.0 wt%, were prepared by using ball milling for 0.5 h. The slurries were then degassed to remove air bubbles. After that, the slurries were cast into plastic molds for gelling. Once consolidation and demolding were finished, the green bodies were dried at room temperature for 24 h and then dried at 110 °C. Organics were burned out at 600 °C. The samples were finally sintered at 1780 °C for 6 h in a vacuum of around 10^{-3} Pa. The sintered products were then annealed at 1450 °C for 10 h in air.

The rheological behaviors of the slurries were affected by the Isobam concentration and solids loading. The optimized slurry had a solid loading of 68 wt% and Isobam content of 0.3 wt% (per powder). Figure 4.40 shows photographs and optical transmittance curves of the ceramic samples [176]. In-line transmittances of the YAG ceramic with a thickness of 7.5 mm were about 82.3 and 79.2 % at the wavelengths of 1000 and 400 nm, respectively. The transparent YAG ceramics exhibited a nearly pore-free microstructure, with an average grain size of about 15 μm, as shown in Fig. 4.41 [176]. The new gelling system should have special applications, where large and complexly shaped transparent ceramics are required.

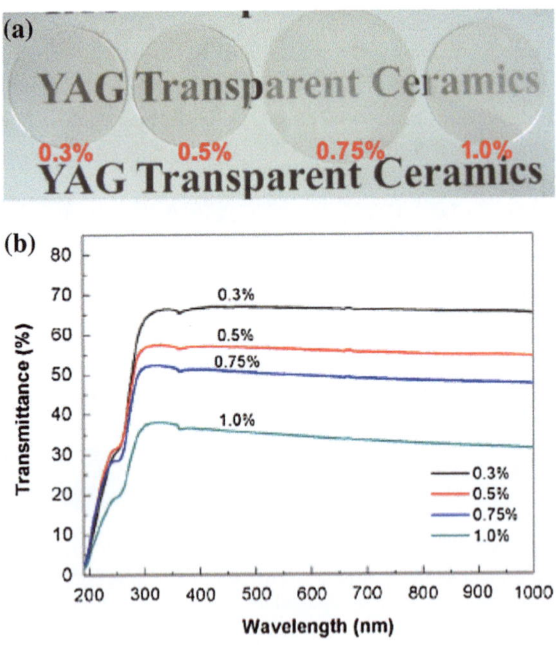

Fig. 4.40 Photographs (**a**) and in-line transmittances (**b**) of the mirror-polished YAG transparent ceramics (1 mm thick) with different Isobam contents (70 wt% solids loading). Reproduced with permission from [176]. Copyright © 2014, Elsevier

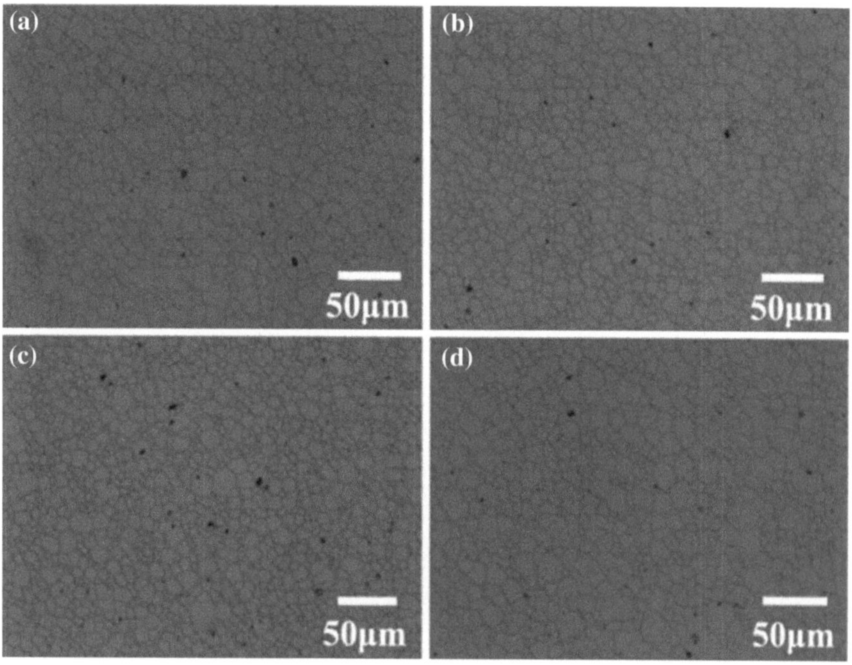

Fig. 4.41 SEM images of mirror-polished surface of the YAG ceramics with different contents of Isobam: **a** 0.3 wt%, **b** 0.5 wt%, **c** 0.75 wt%, and **d** 1.0 wt%. Reproduced with permission from [176]. Copyright © 2014, Elsevier

Electrophoretic Deposition

During the electrophoretic deposition (EPD), a DC electric field drives the charged particles in a colloidal suspension to migrate toward the oppositely charged electrode and deposit on it [178–180]. EPD involves a combination of electrophoresis and particle deposition on the electrode. Stable suspension is an important factor to ensure the deposit with high packing density. The presence of agglomerated particles in an unstable suspension leads to low-density deposits. The additional advantage of EPD is its controllability, because the rate and extent of deposition can be well controlled by adjusting electrical current and electric field. Mechanisms and kinetics of the deposition process of EPD have been well studied and documented in the open literature [181–183].

It has been found that, as a positively charged oxide particle with a double layer of counterions or lyosphere moves toward the cathode in an EPD cell, the electric field coupled with the motion of the charged particle through the liquid causes a distortion of the double-layer envelope, i.e., the envelope becomes thinner ahead of and thicker behind the particle [181]. Kinetics of EPD is important for controlling the thickness of the deposited layers. There are two modes to operate the system: (i) constant-voltage and (ii) constant-current. In constant-voltage EPD, the voltage between the electrodes is maintained constant. Because deposition requires a steeper electric field than electrophoresis, as the deposition thickness increases, the electric field decreases, since the electrical resistance is increased. Therefore, the particle motion rate, i.e., the rate of deposition is decreased. In constant-current EPD, the electrical current is kept constant by increasing the total potential difference between the electrodes, so the limited deposition problem in constant-voltage EPD can be addressed.

Assuming that the suspension is homogeneous and the decrease in concentration is only due to the EDP, the mass of particles m deposited on the electrode is equal to that consumed from the suspension, so that there is:

$$\frac{\mathrm{d}m}{\mathrm{d}t} = AvC, \tag{4.76}$$

where A is the area of the electrode, v is the velocity of the particle, C is the concentration of particles in the suspension, and t is the deposition time. For a concentrated suspension, the velocity of the particles is given by the Helmholtz–Smoluchowski equation:

$$v = \frac{\varepsilon\varepsilon_0\varsigma E}{\eta}, \tag{4.77}$$

where ε is the dielectric constant of the liquid, ε_0 is the permittivity of free space, ς is the zeta-potential of the particles, E is the applied electric intensity, and η is the viscosity of the liquid. If m_0 is the initial mass of the particles in the suspension, then there is:

$$m = m_0 - VC, \tag{4.78}$$

where V is the volume of the suspension. Combining Eqs. (4.76) and (4.77) subject to the boundary condition of Eq. (4.78), there is:

$$m = m_0(1 - e^{-\alpha t}), \tag{4.79}$$

$$\frac{\mathrm{d}m}{\mathrm{d}t} = m_0 \alpha e^{-\alpha t}, \tag{4.80}$$

where $\alpha = Av/V$. According to Eq. (4.80), the rate of deposition decreases exponentially with time and is controlled by the parameter α.

Alumina green bodies have been deposited from stabilized aqueous suspensions, with and without doping, by using an EPD method [184]. Commercial TM-DAR a-Al_2O_3 powder, with purity >99,99 %, D_{50} = 152 nm and a BET surface of 14,69 m^2 g^{-1}, was used as starting material. ZrO_2 nanopowder was monocline phase, with an average particle size D_{50} = 60 nm, while MgO nanopowder was cubic phase, with particle sizes of 25–35 nm and purity of >99.95 %. Ammonium polymethacrylate was used as dispersant, with the commercial name of Darvan C, as 25 mass% aqueous solution with pH values of 7.5–9. The content of the dispersant was 0.45 mg m^{-2} in terms of ammonium polymethacrylate. The effect of solid concentrations was studied.

To prepare the suspensions, dispersant and bidistilled water were mixed first. Then, the oxides used as dopants were added. Finally, alumina powder was gradually added and dispersed at a low temperature of 10 °C, with the aid of a dissolver (Dispermat N1/SIP). All the suspensions were ultrasonicated to destroy the agglomerates and then degassed at 200 mbar (20 kPa). Green compacts were formed by using membrane EPD. A cellulose membrane, which was permeable for ions but not for the Al_2O_3 particles, was used to divide the electrophoresis cell into two chambers, where the Al_2O_3 particles were deposited onto the membrane. The obtained green bodies were dried in air at room temperature and then were calcined at 800 °C in air to remove the organic additives. The calcined samples were finally sintered at temperatures of 1200–1300 °C for 2 h in air. Hot isostatic pressing (HIP) was used to densify the alumina samples at 160 MPa and 1200–1250 °C for 2 h in argon. The measured in-line transmission of the samples at 645 nm was more than 50 % and that is 58 % of the value of sapphire. 250 ppm was the optimal doping level for MgO and 450 ppm for ZrO_2, in terms of promoting densification of the green bodies, while however they had negative effect on optical transparency of the ceramics.

In a separate study, Al_2O_3, Y_2O_3, YAG, and LuAG transparent ceramics were obtained by using EDP [185]. First, colloidal dispersions of alumina, yttria, YAG, and LuAG were prepared [186]. Alumina and yttria nanoparticles were commercial powders, while YAG and LuAG nanopowders were synthesized by using glycothermal method, with aluminum triisopropylate and acetates of yttrium or lutetium. The powders were dispersed by using ball milling in the presence of stabilizing ligands. The suspensions were degassed by purging with nitrogen before use and used only once for EPD runs. Darvan C-N was used as stabilizer. For each system, the pH value was set as required. The set-up for EPD consisted of two vertically placed stainless-steel (316 L) electrodes, with the dimension of $35.0 \times 35.0 \times 1.0$ mm^3. The distance between cathode and anode was fixed at 40.0 mm. The electrodes were ultrasonically cleaned with deionized water and dried with compressed air before EPD. Care was taken to keep the electrode surface smooth and intact, without grooves and cracks, in order to avoid any texturing.

Figure 4.42 shows such a mold with dimension of $35.0 \times 35.0 \times 4.0$ mm^3. The mold had nine holes of 6.0 mm diameter, which allowed to have nine cylinders in a single EPD procedure, as shown in Fig. 4.43 [185]. The holes could be in other shapes, such as square, rectangle, or even circles. EPD power was supplied by using a computerized waveform generator controlling a power amplifier. The applied voltage was up to ±150 V, with deposition times of 3–15 min. Special dry process was used for different samples. The dried samples were then sintered at different conditions to obtain transparent ceramics.

Figure 4.44a shows surface SEM image of as-deposited Al_2O_3 green body, demonstrating the high packing density of the Al_2O_3 nanoparticles, with smooth surface and uniform cross-sectional microstructure. Well-developed interfaces

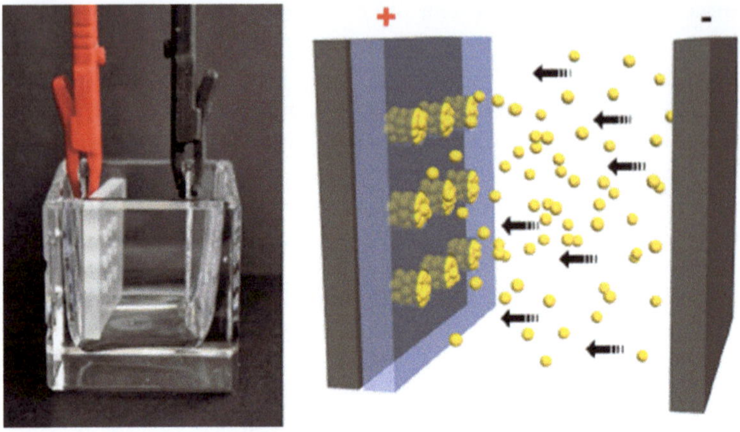

Fig. 4.42 EPD cell and schematic illustration of the EPD process with removable mold on electrode used to obtain cylinder samples. Reproduced with permission from [185]. Copyright © 2014, Springer

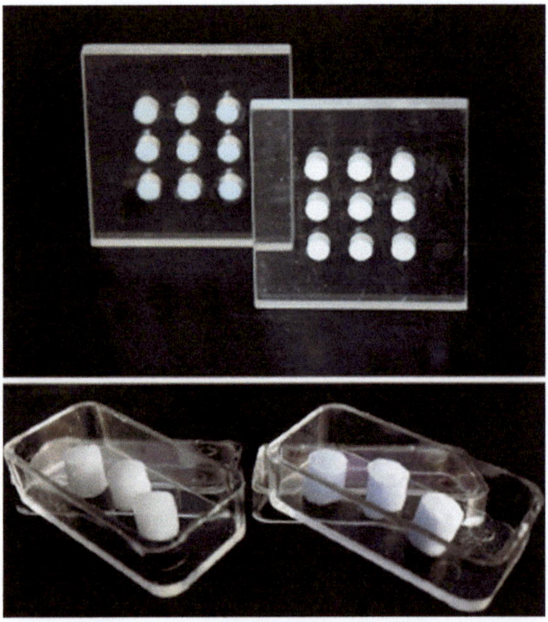

Fig. 4.43 Cylindrical green bodies in plastic mold after EPD (*top*) and as free-standing objects after drying (*bottom*): (*left*) yttria (*left*) and YAG (*right*). Reproduced with permission from [185]. Copyright © 2014, Springer

between the closely packed grains were formed after sintering, as shown in Fig. 4.44b. Because the primary powder had a very narrow particle size distribution, the inhomogeneous grain sizes were attributed to the residual inhomogeneities of the packed bodies caused during the EPD process. After hot isostatic pressing (HIP), there were some large pores observed in the sample, as shown in Fig. 4.44c.

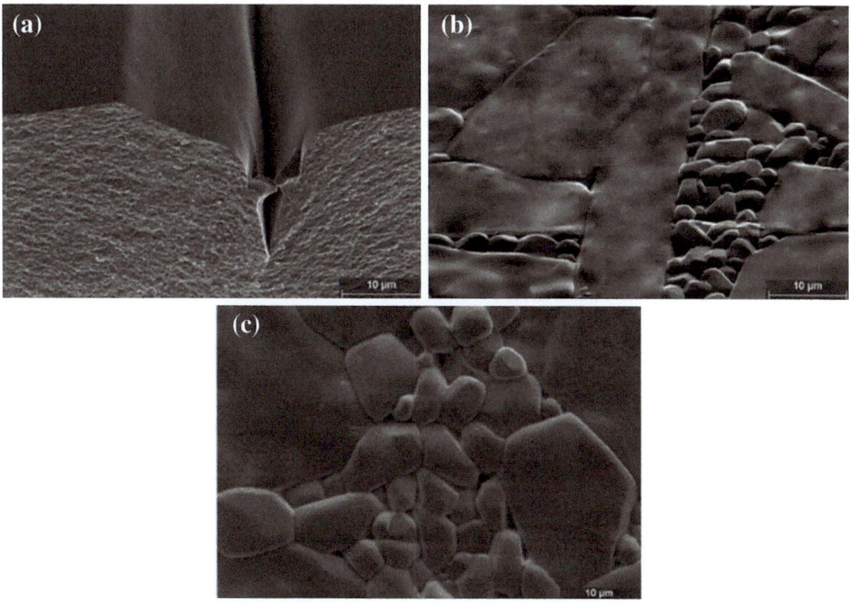

Fig. 4.44 SEM images (all scale bars 10 μm) of **a** EPD-formed alumina (from dispersion stabilized with polyacrylic acid at pH 5.4, fractured cross-section of green body with crack), **b** surface of EPD-deposited alumina sintered in air, and **c** surface of EPD-deposited alumina hot pressed on a piston–cylinder apparatus and sintered in air. Reproduced with permission from [185]. Copyright © 2014, Springer

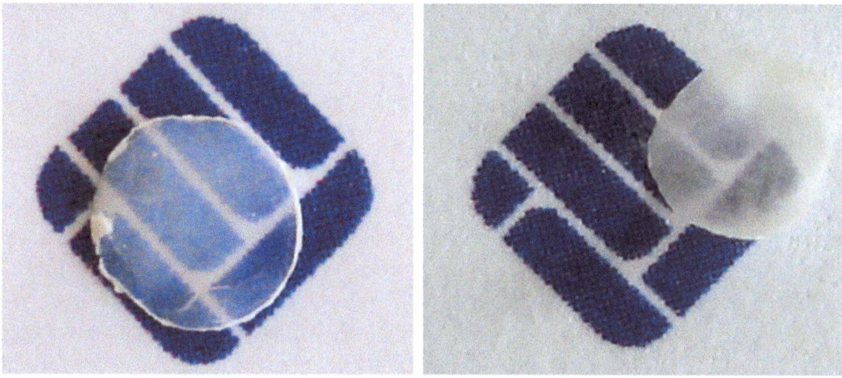

Fig. 4.45 Pictures of sintered alumina from **a** green body formed by EPD and **b** green body formed by EPD and additionally processed by hot isostatic pressing. Reproduced with permission from [185]. Copyright © 2014, Springer

Figure 4.45a shows a sintered sample derived from the green body deposited by using EPD, while Fig. 4.45b shows samples that were made from a green body further compacted by using hot isostatic pressing, but with lower translucency. This

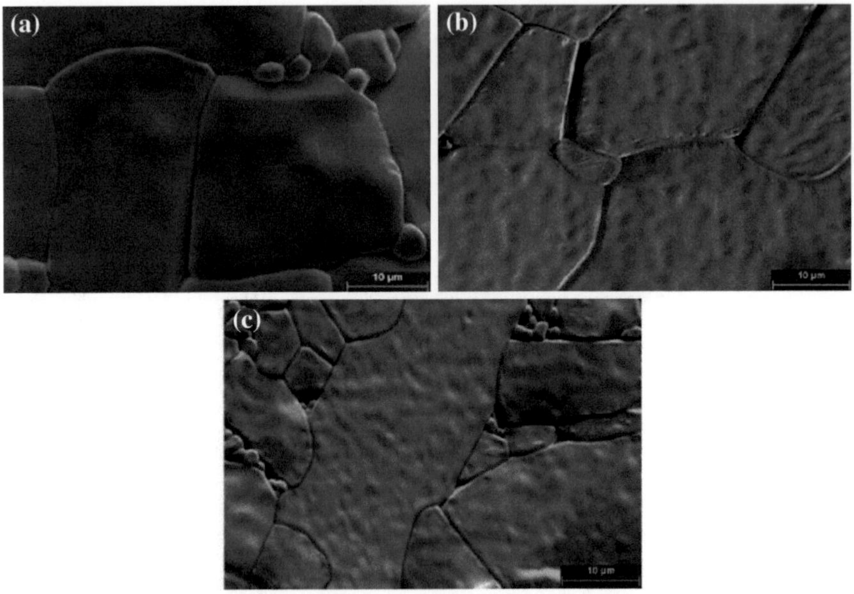

Fig. 4.46 SEM micrographs (all scale bars 10 lm) of surfaces of **a** EPD-deposited yttria sintered in air, **b** EPD-deposited YAG sintered in air, **c** EPD-deposited YAG additionally hot pressed and sintered in air. Reproduced with permission from [185]. Copyright © 2014, Springer

result means that densification of the EPD green bodies could not be enhanced by low-temperature hot isostatic pressing. In other words, the EPD green bodies could reach very high density, from the well-dispersed nanomaterials under proper deposition conditions, without any requirement of further processing.

SEM image of the yttria ceramics sintered in air is shown in Fig. 4.46a, while that of the YAG ceramics sintered in air is illustrated in Fig. 4.46b. Sintering led to microstructures with well-developed interfaces between adjacent grains, although there were some irregular particles and pores. The layered structures disappeared after sintering. Additional hot-pressing of the EPD-deposited YAG had no result in improvement of the microstructure. Instead, even more inhomogeneous grains sizes were observed, as shown in Fig. 4.46c, which is quite similar to the results of alumina deposits, as discussed above.

To deposit LuAG, ordinary DC-EPD at low cell voltage (<2.0 V) should be used [185]. Dense LuAG deposits were obtained only when CA or PAA was used as the stabilizer. LuAG particles were charged negatively, so that anodic deposition was observed. Due to low field, deposition for 15 min yielded a deposit with thicknesses of < 1.0 mm. Furthermore, above this time, the suspension became instable, so that further deposition could not be conducted. However, the thin layer LuAG had a very dense compaction, so that it was optically transparent, as shown in Fig. 4.47a. After sintering in air, the sample became less transparent, as shown in Fig. 4.47b, which was attributed to the formation of pores during the densification. This study

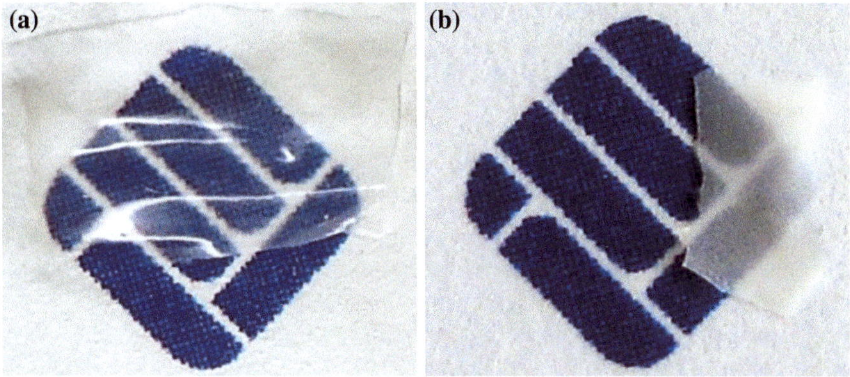

Fig. 4.47 Photographs of the EPD-deposited LuAG (dispersion stabilized with PAA): **a** green body and **b** sintered in air. Reproduced with permission from [185]. Copyright © 2014, Springer

has demonstrated that EPD is a feasible deposition technique for the fabrication of transparent ceramics, which thus deserves to be further explored in the future.

4.3.3.3 Plastic Forming Methods

Plastic deformation of a moldable powder–additive mixture is employed in several forming methods for ceramics [10, 187, 188]. Extrusion of a moist clay–water mixture is used extensively in the traditional ceramics sector for forming components with a regular cross-section (e.g., solid and hollow cylinders, tiles, and bricks). The method is also used to form some oxide ceramics for advanced applications, e.g., catalyst supports, capacitor tubes, and electrical insulators. Repeated coextrusion of a particle-filled thermoplastic polymer has been developed to produce textured microstructures or fine-scale structures. Plastic forming has two basic requirements. On the one hand, the mixture must flow plastically above a certain yield stress for the formation of the desired shape. On the other hand, the shaped article must be strong enough to resist deformation under the force of gravity or under stresses associated with handling. The selection of additives and the formulation of the mixture are critical steps in meeting these requirements.

Extrusion

Extrusion is a forming method, in which a powder mixture in the form of a stiff paste is compacted and shaped by forcing it through a nozzle in a piston extruder or a screw-fed extruder [189–193]. Piston extruder is simple in design, consisting of a barrel, a piston, and a die. In contrast, screw extruder is more complex, especially in the design of the extruder barrel and screw. The screw mixes the powder and other additives into a homogeneous mass and generates enough pressure to push the mixture against the resistance of the die. Shapes of the extruded bodies are determined by the head of the extruder screw and the die. The rotational flow of the mixture produced by the screw is changed by the extruder screw head into an axial flow for extrusion and uniform flow is generated in the die. When the body is released from the extruder, its cross-section is determined by the cross-section of the die.

The main approaches to obtain the required plastic properties for the feed materials include (i) controlling the properties of the powder–water system and (ii) proper use of binders [194, 195]. However, powders of advanced ceramics do not demonstrate the desirable plastic characteristics that are observed in clay–water systems, when they are mixed with water. Therefore, organic binders should be used to obtain the desired plastic characteristics. Either water or organic solvents can be used. In addition, the binder is generally selected from the medium to high viscosity grades, so as to ensure sufficient strength of the extruded body [196]. Aqueous systems are generally flocculated by using a small content of additives, e.g., $MgCl_2$, $AlCl_3$, or $MgSO_4$. Sometimes, lubricants are necessarily used to reduce die wall friction, e.g., stearates, silicones, or petroleum oils.

The flow pattern of the feed material through the extruder has effects on the quality of the extruded bodies. The flow pattern is related to the rheology of ceramic suspensions, which can be classified into four types: (i) ideal plastic, (ii) Bingham, (iii) shear thinning, and (iv) shear thickening. Bingham-type behavior is widely observed in ceramic suspensions and can be used for theoretical analysis [197]. It is characterized by the presence of differential flow with a central plug. The velocity is constant with radius at the central plug, while decreasing with radius from the central plug to the inner wall of the tube. In more extreme cases, the velocity of the material through the extruder is independent of the radius of the tube, which is called slippage flow or complete plug flow.

Extruded bodies can have various macroscopic defects, such as laminations, tearing, and segregation, as well as microscopic defects [198]. Laminations are cracks that usually formed, because of the incomplete re-knitting of the feed materials. Tearing contains mainly surface cracks that are formed as the material leaves the die, which is attributed either improper die design or low plasticity of the mixture. Segregation is due to the separation of the liquid and the solid phases in the mixture during the extrusion. Microscopic defects mainly include pores and contamination.

Coextrusion

Coextrusion is based on repeated coextrusion of a powder-filled thermoplastic polymer to form ceramics with textured microstructure or with fine-scale features [199, 200]. A mixture of ceramic particles, thermoplastic polymer, and processing aid, containing about 50 vol% ceramic powder is formed into a rod-like feed material, i.e., feed rod, with the required arrangement of the ceramic phase, through extrusion in a piston extruder. Extrusion of the feed rod through a die in a piston extruder produces an extrudate with a smaller cross-section, e.g., a diameter that is about 5 times smaller than the diameter of the feed rod, and a corresponding reduction in the scale of the structural features, which is usually conducted at temperatures in the range of 100–150 °C. The material produced after the first extrusion is cut into suitable lengths and reassembled to form a new feed rod. After repeated extrusion, an extrudate with the required small-scale structural features is produced. However, such a technique has not found application in transparent ceramics.

Injection Molding

The principles and practice of injection molding of metals and ceramics have been well documented in the open literature [201–208]. The production of a ceramic article by injection molding involves several steps: (i) selection of the powder and the binder, (ii) mixing the powder with the binder to form a homogeneous feed material in the form of granules, (iii) injection molding of the green body, (iv) removal of the binder or debinding at lower temperatures, and (v) sintering at higher temperatures to produce a dense ceramic article.

Various binder systems have been used in ceramic injection molding, which can be classified into five types: (i) thermoplastic compounds, (ii) thermosetting compounds, (iii) water-based systems, (iv) gelation systems, and (v) inorganics. Among these, thermoplastic compounds are the most widely used and understood, which are mainly commercial polymers. The ratio of powder to binder is a key parameter.

If f_m is the maximum packing density of the particles, which is defined as the volume fraction at which the particles touch one another, so that flow is not possible, then the actual particle volume fraction f used in injection molding is lower than f_m by 5–10 vol%. This means that in a well-dispersed suspension, the particles are separated from one another by a thin layer of polymer with a thickness of about 50 nm during the molding, so that the mixture is able to flow. Therefore, the volume fraction of particles f is determined by the particle size and distribution and the particle shape. In practice, the volume fraction of ceramic powders is determined from viscosity measurements by using a capillary rheometer. Data for the relative viscosity, i.e., the viscosity of the mixture divided by the viscosity of the unfilled polymer versus particle concentration can be well fitted by the following equation [209]:

$$\eta_r = \left(\frac{1 - 0.25f/f_m}{1 - f/f_m} \right)^2. \tag{4.81}$$

Plastic Forming of Ceramics

Various ceramics have been prepared by using extrusion and injection molding method [210–213]. For example, Al_2O_3 ceramics in the forms of bulk, sheet, and fiber have been fabricated through extrusion. Injection molding method has been used to develop translucent Al_2O_3 [214, 215]. Al_2O_3 [216] and Al_2O_3–YAG composite [217]. Representative examples are discussed in this part.

A new method has been reported to produce alumina ceramics by using extrusion method with hydraulic alumina (HA) as an inorganic binder [211]. The addition of a small quantity of HA offered great flowability and rigidity to the extruded alumina bodies. Alumina rods and tubes could be obtained at optimized processing conditions, including content of HA and water and curing time duration, by using a single-screw extruder, as shown in Fig. 4.48. Sintered bodies reached a relative density of 97 % after sintering at 1600 °C for 2 h. Photographs of representative rods and tubes are shown in Fig. 4.49. Dense alumina ceramics prepared with organic and inorganic binders had similar mechanical properties, i.e., Young's modulus of >380 GPa and H_v of >2000.

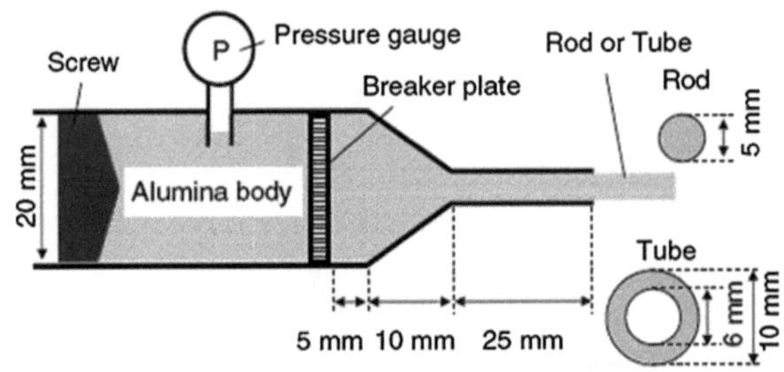

Fig. 4.48 Schematic of the single screw extruder. Reproduced with permission from [211]. Copyright © 2007, John Wiley & Sons

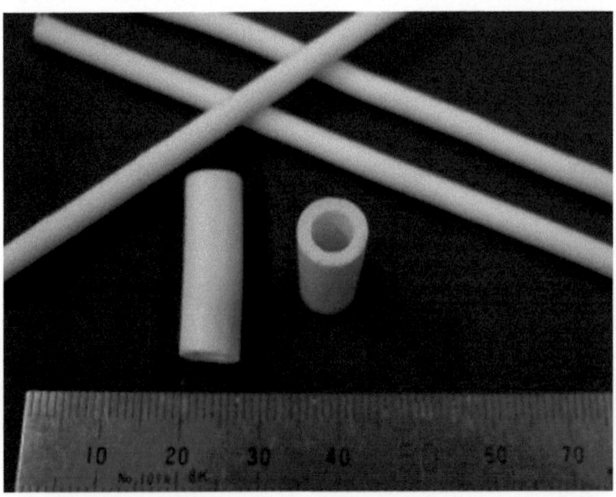

Fig. 4.49 Photographs of representative alumina rods and tubes after sintering at 1600 °C for 2 h. Reproduced with permission from [211]. Copyright © 2007, John Wiley & Sons

HA was mixture of γ-Al_2O_3 and χ-Al_2O_3, with a purity of 99.7 % and an average particle size of 6 μm, while α-alumina had a purity of 99.8 % purity, with a mean particle size of 0.55 μm and a BET surface area of 6 $m^2\ g^{-1}$. The HA powder was milled in methanol by using a planetary ball mill at 300 rpm for 2 h, in order to break down agglomerates and reduce particle size to D_{50} = 1.4 μm. The ball milled HA and the α-alumina powders were mixed by using a ball mill in a plastic container with alumina balls in methanol at 20 rpm for 24 h. The mixtures were dried at 60 °C in vacuum and then screened through a 150-mesh sieve. Distilled water was added to the powder mixture and mixed by using a mini shaker at 2500 rpm for 2 min. Then, the paste was kneaded manually with a wood stick. The paste was then put in a plastic bag and cured at room temperature for 2–4 days. Alumina body with methyl cellulose (AMC) was also prepared for comparison. Rods of 5 mm in diameter and tubes of 10 mm in outer diameter and 6 mm in inner diameter were fabricated at a extrusion speed of about 50 $mm{\cdot}s^{-1}$. The extruded green samples were dried at room temperature for 5 days in air. Dried samples were finally sintered at 1500–1600 °C for 2 h, at a heating rate of 50 °C h^{-1} to 600 °C and 300 °C h^{-1} thereafter.

Extrusion of suspensions from nanopowders could be a great challenge, because nanopowders would make it difficult to obtain high solid loadings due to the extremely high surface area, leading to interactions between particles and thus suspensions with too high viscosities for extrusion. There has been a report to address this challenge by using nanopowders to extrude YAG tubes [218].

Anhydrous ethanol and Nanotek γ-Al_2O_3, along with bicine, polyethylene glycol, and stearic acid, were used in this work. Al_2O_3 nanopowder was also derived from alumatrane that was synthesized by using a hydrolysis method [219]. Y_2O_3 nanopowder was made from yttrium propionate that was synthesized through the reaction of yttrium carbonate with propionic acid. Ethylene vinyl acetate hot-melt adhesive was a commercial product. To obtain Y_2O_3 and YAG nanopowders, yttrium propionate and the mixture with alumatrane were made into aerosols by using O_2 at 80 psi, with ceramic loading of <5 wt%, which were combusted by using a flame spray pyrolysis.

For reactive sintering, 60 g of the as-obtained Y_2O_3 nanopower was mixed with 750 mL ethanol, together with 2 wt% bicine as a dispersant. The suspension was ultrasonicated for 20 min by using an ultrasonic horn at a power of 100 W. After settling down for 24 h, the supernatant was decanted and dried. 60 g of NanoTek Al_2O_3 nanopowder was treated similarly. The dried nanopowders were mixed in the stoichiometric ratio of 3:5, together with 750 mL ethanol. After being ultrasonicated for 20 min at 100 W, the suspension was dried to form the powders for thermoplastic extrusion process. The YAG nanopowder was processed at the same conditions.

Powder mixtures for thermoplastic-ceramic extrusion consisted of 1 vol% polyethylene glycol (PEG) with a M_n = 3400, 1 vol% stearic acid, 3 vol% heavy mineral oil, 48–50 vol% nanopowders, and ethylene vinyl acetate-based hot-melt adhesive. The precursor mixtures were mixed by using a twin-screw shear mixer at a running speed of 60 rpm, while the mixing chamber was preheated to 120 °C.

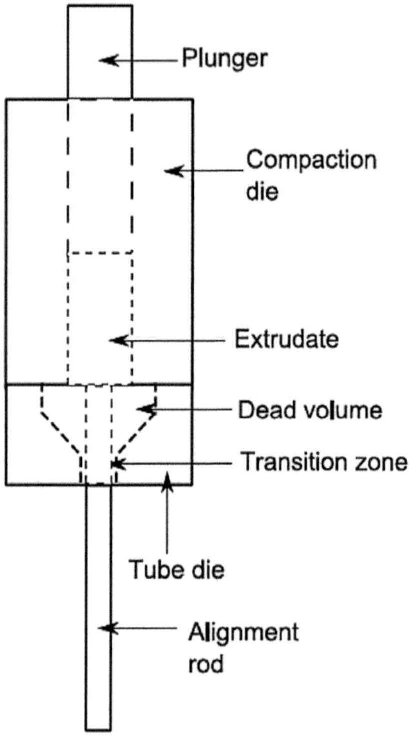

Fig. 4.50 Schematic of the extruder and the spider extrusion die. Reproduced with permission from [218]. Copyright © 2014, John Wiley & Sons

About 80 % of the hot-melt adhesive was added into the mixing chamber and allowed to melt. After that, the powder was slowly added. The rest 20 % hot-melt adhesive was used upon saturation of the powder, followed by the addition of all powder. Finally, the thermoplastic-ceramic melt was mixed for additional 15 min and then the mixture was allowed to cool down. The hardened melt was treated one more time to ensure homogeneity. Before extrusion, the thermoplastic–ceramic mixture was degassed at 110 °C for 2 h and at a vacuum level of about 10^{-3} Torr.

Extrusion was conducted by using a ram extrusion machine with a spider die of 31 mm in diameter. Figure 4.50 shows a schematic of the extruder and spider extrusion die. Before extrusion was started, the extrusion die was preheated to 140 °C. The granulated thermoplastic–ceramic mixture was then added into the die. The bottom of the die was blocked, while the ram was pressed to a pressure of 2.3 MPa. The die was cooled to 50 °C, while the stop was removed from the end of the die. A smaller heating mantle was then used to heat the spider region of the die. As the spider was heated to 120–130 °C, tubes with length of up to 0.7 m could be extruded out at a rate of about 2 mm min^{-1}, at die pressures of about 1 MPa.

The thermoplastic melts were loaded at 50 vol% powder, corresponding to a solid loading of 82 wt%. Powder loadings of <45 vol% were insufficient to maintain a tubular geometry through burnout, while higher ceramic loadings made it difficult to extrude defect-free bodies. The high content of binders required a three-step

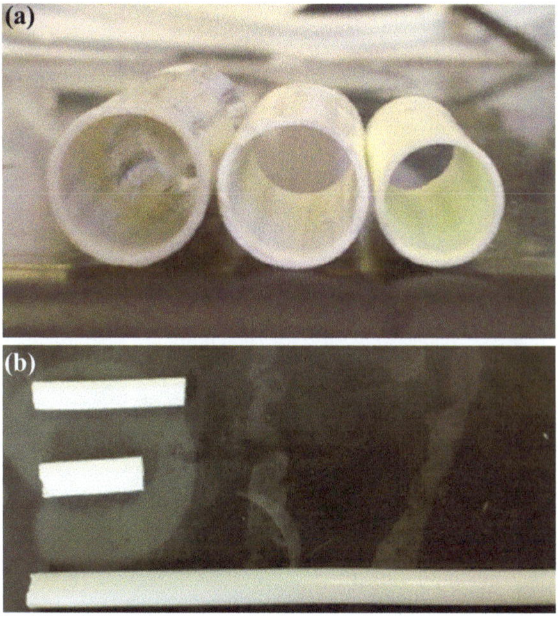

Fig. 4.51 **a** Optical photographs of the tubes from left to right: as-extruded, burnt-out, and 95 % densified. **b** Photographs of the tubes from top to bottom: 95 % densified, burned out, and as-extruded. Width of the figure is about 17 cm. Reproduced with permission from [218]. Copyright © 2014, John Wiley & Sons

burnout process, i.e., (i) in dry N_2 at 400 °C for 4 h and at heating rate of 3 °C min^{-1}, (ii) in dry air at 600 °C for 4 h and at heating rate of 3 °C min^{-1}, and (iii) in O_2 at 800 °C for 3 h and at a gas flow rate of 50 mL min^{-1}. Two sintering schedules were used to sinter the tubes, i.e., (i) 1500 °C for 0 h followed by 20–30 h at 1300 °C and (ii) 35 h at 1500 °C followed by 5 h at 1600 °C.

Figure 4.51a shows cross-sections of an as-extruded tube, a fully debindered tube, and a tube sintered to 95 % TD. Cross-sectional geometry was maintained from the as-extruded tubes through to the final sintered product. Figure 4.51b provides a view of longer extruded sections, along with a debindered tube and a sintered tube. Some deviations from linearity along the tube axis are found in the as-extruded tubes. However, the extrusion process did not use a guide rod, which is expected to improve overall straightness.

Hex-$Y_3Al_5O_{12}$ powder tubes sinter only to ~70 % TD using the same two-step scheme (Fig. 4.52a) and were only found to sinter on heating to 1500 °C for extended periods (Fig. 4.52b). After 35 h at 1500 °C in vacuum, sintering of hex-$Y_3Al_5O_{12}$ composition tubes stalled at ≈93 % TD. Further sintering at 1600 °C for 5 h in air brought the tubes to 98 % TD, with typical microstructures as shown in Fig. 4.6b. This microstructure is characterized by a distribution of grains much larger (3–5 μm) than that of the Y_2O_3–Al_2O_3 tube with two-step heat treatment to 1500/1300 °C (96 % TD). For optical applications, subsequent hot isotatic pressing is mandated by the need to eliminate the remaining closed porosity.

Liquid-feed flame spray pyrolysis provides highly dispersed and unaggregated nanopowders with average particle sizes of 20–70 nm, which are dependent on the processing conditions. Their chemical compositions can be controlled to ppm levels

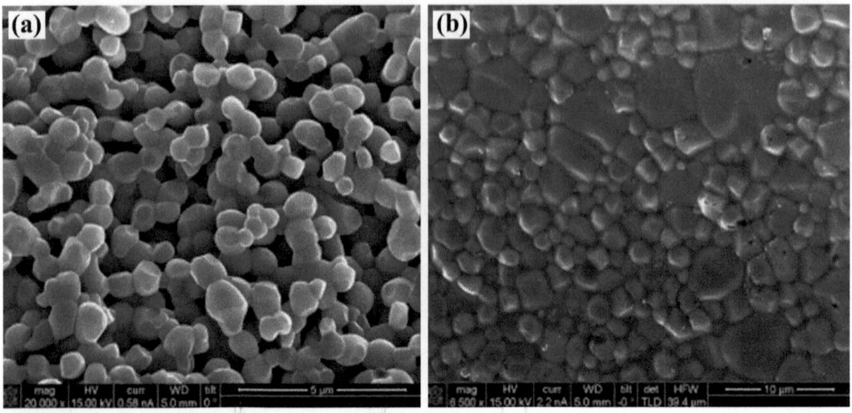

Fig. 4.52 SEM images of the samples with composition of $Y_3Al_5O_{12}$: **a** tube surface that was two-step heated to 1500/1300 °C (70 % TD) and **b** fracture surface after heating at 1600 °C for 5 h in air (98 % TD). Reproduced with permission from [218]. Copyright © 2014, John Wiley & Sons

via control of the initial precursor solution. In this paper, $Y_3Al_5O_{12}$ composition nanopowders are produced that are atomically mixed but offer a hexagonal crystal structure rather than a YAG structure. Y_2O_3 and δ-Al_2O_3 nanopowders are also produced and mixed to evaluate reactive sintering. It is shown that nanopowder/polymer mixtures permit the extrusion of tubes that retain their shape on debindering and sintering to 95 % theoretical density. More importantly, the sintering behavior of hex-$Y_3Al_5O_{12}$ is compared with that of tubes formed using 3:5 Y_2O_3:δ-Al_2O_3 mixtures to test the so-called bottom-up paradigm, which suggests that mixing on the finest length scales should provide optimal control of sintering rates, final densities, and grain sizes. Instead, it is found that reactive sintering is faster and offers better control of final grain sizes. Dense sintered tubes are translucent, and dimensional uniformity is maintained from extrusion through sintering.

Injection molding has been used in combination with pressureless sintering to fabricate translucent alumina ceramics [215]. Feedstocks for injection molding with excellent rheological properties were prepared by using ball milling. The powder was pretreated with stearic acid to have better dispersing behaviors. The ceramic powder was a commercial alumina with an average particle size of 0.35 μm and a BET specific area of 20 $m^2\ g^{-1}$. The powders were pretreated with 1.5 wt% stearic acid (SA) through ball milling with a planetary ball mill machine for 12 h. As sintering aids, 500 ppm MgO and 500 ppm Y_2O_3 were mixed with the powders, which were introduced in the form of nitrates. The thoroughly mixed powders were then dried at 40 °C in air for 12 h. After that, the powders were further mixed with binders with a twin-screw kneader at 150–170 °C for 1 h. The organic vehicles, i.e., binder, were polypropylene (PP, K8303), ethylene vinyl acetate copolymer (EVA, with VA content of 14 %), paraffin wax, SA and dibutyl phthalate, with designed compositions. Various solid loadings of 52–58 vol% were studied for optimization.

The samples were injected by using a commercial injection molding machine. The molds had geometries of 28.14 × 27.93 × 1.2 mm^3 for the slice and 4.47 × 4.64 × 3.10 mm^3 for the dental bracket. The metal halide lamp consisted of three parts, i.e., medium hollow spheroid with long axis of 11.45 mm, minor axis of 9.11 mm and both edges with hollow cylinder of 12.02 × Φ2.51 mm. The as-injected compacts were calcined at 450 °C for 2 h at a relatively slow heating rate in order to remove the organic binders. The calcined samples were finally sintered at 1830 °C for 100 min in a H_2 furnace. Optimized solid loading was identified according to the rheological data and the properties of the molded green bodies. The average grain size of the sintered body is 30–50 μm, with almost no pores and without abnormal grain growth. The translucent Al_2O_3 ceramics prepared in this way showed superior optical properties than those made by using other methods. Such a stable, efficient, and low-cost route should be extended to other transparent ceramics, especially when components with small sizes and precise dimensions are required.

4.3.4 Drying of Granular Ceramics

Moist granular ceramics produced by using the common forming methods, such as tape casting and extrusion, should be dried before the binder burnout and sintering. Drying is also a key step in solid free-form fabrication techniques that involve layer-by-layer deposition of colloidal suspensions.

4.3.4.1 Drying of Granular Layers

During the drying of liquid drops containing colloidal particles with dilute concentration of $\sim 10^{-4}$, migration of the particles was observed, which led to the formation of a ring [220–222]. The ring was formed due to the consolidation of the particles at the outer edge of the film, at the initial stage of drying. As the drying proceeded, liquid flows to the outer region to maintain a saturated state, which was driven by the capillary suction pressure of the ring. Particles in the liquid were deposited at the interface between the saturated ring and the supersaturated suspension. Therefore, the drying of the film is similar to consolidation in slipcasting.

4.3.4.2 Drying of Granular Solids

Ceramic green bodies made with cast and extrusion contain much less water than gels and thus experience much less shrinkage accordingly. In addition, granular ceramic articles have larger pores, thus having higher permeability. Therefore, drying of granular solids is not very difficult. Instead, the inhomogeneities in the green bodied could be a serious problem of granular solids. Physical process of drying and problems of cracking and warping are discussed in this part.

Regarding the evaporation rate, there is a constant rate period (CRP), which is followed by a falling rate period (FRP). Sometimes, two parts can be observed in the FRP, which are known as the first falling rate period (FRP1) and the second falling rate period (FRP2). Drying curves are usually presented as a function of the moisture content (MC) of the green bodies, in terms of percentage of the dry weight (w_{dry}) of the solid (dry basis) with respect to the wet weight (w_{wet}), given by:

$$\text{MC} = \frac{(w_{\text{wet}} - w_{\text{dry}}) \times 100\ \%}{w_{\text{dry}}} \tag{4.82}$$

The moisture content of the green bodies in casting or extrusion is usually in the range of 20–35 %.

During the CRP, the rate of evaporation is independent on the content of moisture. As evaporation starts, a dry surface region is formed, so that liquid tends to cover the dry region. As a result, tension is developed in the liquid, which could be balanced by the formation of compressive stresses in the solid phase. Because of the compressive stresses, the body tends to shrink, while the liquid meniscus remains at the surface. As drying proceeds, the particles form a denser packing, so that the body becomes stiffer. The liquid meniscus at the surface deepens and the tension in the liquid is increased. Eventually, the particles are surrounded by a thin layer of water and then the shrinkage stops. The moisture content at this point is also known as the leatherhard moisture content.

As the shrinkage stops, further evaporation pushes the meniscus into the body and the rate of evaporation starts to decrease, which corresponds to the so-called critical moisture content. It is found that the critical moisture content and the leatherhard moisture content are nearly the same. In the FRP1, the evaporation of liquid from the surface continues. There are pathways for the liquid to flow to the surface. Eventually, the liquid near the surface of the body is isolated into pockets. As the liquid stops flowing to the surface, it is removed by the diffusion of the vapor, so that the drying process enters the FRP2.

Another issue is cracking and warping during the drying process, which are caused by the differential strain due to (i) the pressure gradients in the liquid and (ii) the inhomogeneities in the body related to processing. A tension is developed in the liquid, as it tends to cover the dry region that is exposed due to the evaporation of the liquid. If the tension p in the liquid is uniform, the solid phase has no stress, whereas if p varies through the thickness, the body tends to shrink more at the locations where p is high. As a result, warping or cracking is caused due to the difference in strain.

If the evaporation rate is very high, the tension in the liquid can reach its maximum value. In this case, the total stress at the surface of the body is given by:

$$\sigma_{\text{x}} \approx \frac{2\gamma_{\text{LV}} \cos\theta}{a}, \tag{4.83}$$

where γ_{LV} is the surface tension of the liquid, θ is the contact angle and a is the pore radius. Ceramic green bodies containing a low level of binder are very weak, so that cracking is very likely to be caused by the stress, although it seems not too high.

During the CRP, the boundary condition at the surface of the body is given by:

$$\dot{V}_E = \frac{K}{\eta_L} \nabla p|_{surface}, \tag{4.84}$$

here $\dot{V}_E$ is the evaporation rate, K is the permeability of the body, η_L is the viscosity of the liquid, and ∇p is the pressure gradient in the liquid. According to Eq. (4.84), fast evaporation rate leads to high ∇p. To avoid cracking or warping, the green body should be dried as slowly as possible. At the same time, a tradeoff could be identified between safe drying and high production.

4.3.5 Binder Removal

The organic additives and other processing aids (referred to in this section simply as the binder) must be removed from the green body prior to sintering. This process of binder removal is commonly referred to as debinding. Ideally, we would like to remove the binder completely without disrupting the particle packing or producing any new microstructural defects in the green body. Residual contaminants (e.g., carbon and inorganic ions) and defects (e.g., cracks and large voids) generally have an adverse effect on the microstructural evolution during sintering and hence on the properties of the fabricated body. Debinding can be a critical step in ceramic processing, especially in the case of the forming methods where the binder content in the green body is relatively high. Debinding can be achieved by using three main methods, i.e., (i) extraction by capillary flow into a porous surrounding material, (ii) solvent extraction, and (iii) thermal decomposition. Among them, the thermal decomposition that is also known as thermal debinding is the most widely used method, due to its convenience and efficiency [223, 224].

Extraction by capillary flow is also known as wicking, in which the green body is heated in a packed powder bed or on a porous substrate that can absorb the molten binder. The net time t used to remove the binder through wicking is given by [225]:

$$t \approx \frac{5L^2 \eta V_s^2 D_w}{\gamma_{LV}(1 - V_s)^3 D(D - D_w)}, \tag{4.85}$$

where L is the thickness of the body, η and γ_{LV} are the viscosity and specific surface energy, respectively, of the molten binder, V_s is the particle packing density of the body, D is the particle size of the body, and D_w is the particle size of the powder bed. Therefore, small D_w and small L will ensure rapid removal of binders. Wicking is more useful to remove waxes than those polymers with high molecular weights, because the viscosity of the binder increases with increasing molecular weight. Furthermore, wicking is not able to remove all the binder. For example, as the molten binder enters the pendular state, i.e., it is trapped in isolated pockets, binder removal stops.

Solvent extraction involves immersing the component in a liquid in which at least one binder can be dissolved, leaving an open porous structure for subsequent binder burnout. Full debinding is desirable because the resulting powder compaction will have almost no strength. The time t for debinding is given by [225]:

$$t = \frac{L^2}{\alpha} \ln\left(\frac{V_B}{1 - V_s}\right) \exp\left(\frac{Q}{kT}\right), \tag{4.86}$$

where L is the thickness of the body, α is a factor that depends on the solubility of the binder in the solvent, V_B is the fraction of the binder to be removed, V_s is the particle packing density of the body, Q is the activation energy for the dissolution of the binder in the solvent, k is the Boltzmann constant, and T is the absolute temperature.

Thermal debinding is used to remove the binder as a vapor, by heating at ambient pressure in either an oxidizing or a nonoxidizing atmosphere or a partial vacuum. Both chemical and physical factors are involved in the process. The decomposition temperature and the products are determined by chemical compositions of the binder. Removal efficiency of the binder is determined by the physical properties, including heat transfer into the body and mass transport of the decomposition products out of the body. Binders used in ceramics are usually multiple component systems. These components have different physical and chemical properties. Moreover, their properties could be influenced by the ceramic powders. For most transparent ceramic processes, due to the small quantity of binders, binder removal usually carried out during the heating process of sintering. For the transparent ceramics processed by using tape casting or other casting methods, special binder removal steps are necessary.

4.3.6 Characterization of Green Bodies

As stated many times before, the microstructure of green body has a significant effect on the densification and microstructural evolution during the sintering. Therefore, characterization of the microstructure of green body could provide useful information that can be used to establish interrelationship between the characteristics of green bodies and the densification and microstructure of final ceramics. Generally, green body characteristics include relative density or porosity, pore size and size distribution, and microstructural homogeneity. Density cannot be used as the only parameter to predict the sintering characteristics of a ceramic powder compact.

The density of a green body can be directly measured from its mass and volume. Porosity and pore size distribution are normally measured by using mercury porosimetry or gas adsorption method. Microstructural homogeneity is generally observed by using SEM. There are also other techniques, which can provide additional information. For instance, impregnation of the green body with epoxy resin, followed by sectioning, polishing, and observation with SEM, could provide quasi-three-dimensional information of the green body microstructure. A liquid immersion technique can be used to characterize the internal structure of granules [93] and green bodies [226]. In this technique, the specimens are made transparent by immersion in a liquid with a matching refractive index and observed in an optical microscope under transmitted light. X-ray topography has been shown to be a powerful tool to observe microstructure of ceramic green bodies [227, 228]. However, all these special methods have rarely been reported for the characterization of transparent ceramic green bodies.

4.4 Summary

A wide range of techniques can be used to characterize ceramic powders. Proper characterization of the starting powders has critical importance in understanding their behavior during sintering process, so as to obtain transparent ceramics with desired performances. However, according to the open literature, only very limited number of techniques has been commonly used in real practices, probably because of the unavailability of some of the facilities in a general research laboratory. It is highly suggested to characterize ceramic powders by using as many techniques as possible, in order to accumulate database as complete as possible.

Powder consolidation plays an important role in determining sintering behaviors of the powders. Heterogeneities present in the green body cannot be easily removed during sintering, so that control of the green body microstructure will be very useful for microstructural control of the final ceramics. The selection of a forming method is dependent on the requirements in size and shape of the transparent ceramics. Sometimes, cost-effectiveness is an important consideration, especially for large-scale production. Particle packing and polymeric additives play an important role in powder consolidation. After forming, drying and debinding must be carefully controlled to avoid further formation of any microstructural defects.

References

1. Sigmund WM, Bell NS, Bergstrom L (2000) Novel powder-processing methods for advanced ceramics. J Am Ceram Soc 83:1557–1574
2. Uematsu K (2014) Processing defects in ceramic powders and powder compacts. Adv Powder Technol 25:154–162
3. Kendall K (1989) Influence of powder structure on processing and properties of advanced ceramics. Powder Technol 58:151–161

4. Uematsu K (2012) Process defects. In: Riedel R, Chen IW (eds) Ceramic science and technology. Wiley, New York
5. Lewis JA (2000) Colloidal processing of ceramics. J Am Ceram Soc 83:2341–2359
6. Tallon C, Franks GV (2011) Recent trends in shape forming from colloidal processing: a review. J Ceram Soc Jpn 119:147–160
7. Lange FF (1998) Colloidal processing of powder for reliable ceramics. Curr Opin Solid State Mater Sci 3:496–500
8. Ragulya AV (2008) Consolidation of ceramic nanopowders. Adv Appl Ceram 107:118–134
9. Mayo MJ (1996) Processing of nanocrystalline ceramics from ultrafine particles. Int Mater Rev 41:85–115
10. Blackburn S, Wilson DI (2008) Shaping ceramics by plastic processing. J Eur Ceram Soc 28:1341–1351
11. Evans JRG (2008) Seventy ways to make ceramics. J Eur Ceram Soc 28:1421–1432
12. Janssen R, Scheppokat S, Claussen N (2008) Tailor-made ceramic-based components—advantages by reactive processing and advanced shaping techniques. J Eur Ceram Soc 28:1369–1379
13. Deckers J, Vleugels J, Kruthl JP (2014) Additive manufacturing of ceramics: a review. J Ceramic Sci Technol 5:245–260
14. Travitzky N, Bonet A, Dermeik B, Fey T, Filbert-Demut I, Schlier L et al (2014) Additive manufacturing of ceramic-based materials. Adv Eng Mater 16:729–754
15. Heule M, Vuillemin S, Gauckler LJ (2003) Powder-based ceramic meso- and microscale fabrication processes. Adv Mater 15:1237–1245
16. Tay BY, Evans JRG, Edirisinghe MJ (2003) Solid freeform fabrication of ceramics. Int Mater Rev 48:341–370
17. Turner BN, Strong R, Gold SA (2014) A review of melt extrusion additive manufacturing processes: I. Process design and modeling. Rapid Prototyping J 20:192–204
18. Onoda GY Jr, Hench LL (1978) Ceramic processing before firing. Wiley, New York
19. Rahaman MN (2003) Ceramic processing and sintering, 2nd edn. CRC Press, New York
20. Black DL, McQuay MQ, Bonin MP (1996) Laser-based techniques for particle-size measurement: a review of sizing methods and their industrial applications. Prog Energy Combust Sci 22:267–306
21. Matyi RJ, Schwartz LH, Butt JB (1987) Particle-size, particle-size distribution, and related measurements of supported metal-catalysts. Catal Rev—Sci Eng 29:41–99
22. Miller BV, Lines RW (1988) Recent advances in particle-size measurements—a critical review. CRC Crit Rev Anal Chem 20:75–116
23. Allen T (1990) Particle size measurement, 4th edn. Chapman and Hall, London
24. Beddow JK (1980) Particulate science and technology. Chemical Publishing Co., New York
25. Staiger M, Bowen P, Ketterer J, Bohonek J (2002) Particle size distribution measurement and assessment of agglomeration of commercial nanosized ceramic particles. J Dispersion Sci Technol 23:619–630
26. Charalampopoulos TT (1992) Mophology and dynamics of agglomerated particulates in combustion systems using light-scattering techniques. Prog Energy Combust Sci 18:13–45
27. Xu RL (2015) Light scattering: a review of particle characterization applications. Particuology 18:11–21
28. Fan XF, Zheng WT, Singh DJ (2014) Light scattering and surface plasmons on small spherical particles. Light-Sci Appl 3
29. Brar SK, Verma M (2011) Measurement of nanoparticles by light-scattering techniques. TRAC-Trends Anal Chem 30:4–17
30. McGarvey M, McGregor D, McKay RB (1997) Particle size analysis by laser diffraction in organic pigment technology. Prog Org Coat 31:223–228
31. Cullity BD (1978) Elements of X-ray diffraction, 2nd edn. Addison-Wesley, Boston
32. Klug HP, Alexander LF (1974) X-ray diffraction procedures for polycrystalline and amorphous materials. Wiley, New York

33. Pecharsky V, Zavalij P (2009) Fundamentals of powder diffraction and structural characterization of materials, 2nd edn. Springer, Berlin
34. Gregg SJ, Sing KSW (1982) Adsorption, surface area and porosity, 2nd edn. Academic Press, New York
35. Brunauer S, Deming LS, Deming WE, Teller E (1940) On a theory of the van der Waals adsorption of gases. J Am Chem Soc 62:1723–1732
36. Brunauer S, Emmett PH (1937) The use of low temperature van der Waals adsorption isotherms in determining the surface areas of various adsorbents. J Am Chem Soc 59:2682–2689
37. Brunauer S, Emmett PH (1940) Chemisorptions of gases on iron synthetic ammonia catalysts. J Am Chem Soc 62:1732–1746
38. Brunauer S, Emmett PH, Teller E (1938) Adsorption of gases in multimolecular layers. J Am Chem Soc 60:309–319
39. Langmuir I (1916) The constitution and fundamental properties of solids and liquids, part I. Solids. J Am Chem Soc 38:2221–2295
40. Langmuir I (1917) The constitution and fundamental properties of solids and liquids, II. Liquids. J Am Chem Soc 39:1848–1906
41. Skoog DA, Holler FJ, Nieman TA (1998) Principles of instrumental analysis, 5th edn. Harcourt Brace, Philadelphia
42. Jenkins R (1999) X-ray fluoresence spectrometry, 2nd edn. Wiley, New York
43. Walls JM (1989) Methods of surface analysis. Cambridge University Press, Cambridge
44. Hudson JB (1992) Surface science: an introduction. Butterworth-Heinemann, Boston
45. Woodruff DP, Delchar TA (1994) Modern techniques of surface analysis, 2nd edn. Cambridge University Press, Cambridge
46. Galakhov AV (2014) Powder compact structure. Part 1. Particle packing inhomogeneity. Refract Ind Ceram 55:199–208
47. Galakhov AV (2014) Powder compact structure. Part 2. Methods for increasing particle packing uniformity. Refract Ind Ceram 55:209–219
48. Harthong B, Jerier JF, Richefeu V, Chareyre B, Doremus P, Imbault D et al (2012) Contact impingement in packings of elastic-plastic spheres, application to powder compaction. Int J Mech Sci 61:32–43
49. Zhong WZ, He KJ, Zhou ZY, Xia W, Li YY (2009) Physical model and simulation system of powder packing. Acta Physica Sinica 58:S21–S28
50. Zhou JH, Zhang YW, Chen JK (2009) Numerical simulation of random packing of spherical particles for powder-based additive manufacturing. J Manuf Sci Eng-Trans ASME 131
51. Dinger DR, Funk JE (1997) Particle-packing phenomena and their application in materials processing. MRS Bull 22:19–23
52. Cumberland DJ, Crawford RJ (1987) The packing of particles. Elsevier, New York
53. German RM (1989) Particle packing characteristics. Metal Powder Industries Federation, Princeton
54. Sethi G, Myers NS, German RM (2008) An overview of dynamic compaction in powder metallurgy. Int Mater Rev 53:219–234
55. Westman AER, Hugill HR (1930) The packing of particles. J Am Ceram Soc 13:767–779
56. Scott GD (1960) Packing of equal spheres. Nature 188:908–909
57. Scott GD, Charlesworth AM, Mak MK (1964) On random packing of spheres. J Chem Phys 40:611–612
58. Scott GD, Knight KR, Bernal JD, Mason J (1962) Radial distribution of random close packing of equal spheres. Nature 194:956–957
59. Scott GD, Mader DL (1964) Angular distribution of random close-packed equal spheres. Nature 201:382–383
60. Frost HJ (1982) Overview 17—cavities in dense random packings. Acta Metall 30:889–904
61. Khramtsov VD (2009) The packing density of the particles in powder mixtures of different dispersities. Russ J Non-Ferrous Metals 50:294–297

62. Frost HJ, Raj R (1982) Limiting densities for dense random packing of spheres. J Am Ceram Soc 65:C19–C21
63. McGeary RK (1961) Mechanical packing of spherical particles. J Am Ceram Soc 44:513–522
64. Wakeman RJ (1975) Packing densities of particles with log-normal size distributions. Powder Technol 11:297–299
65. Yu AB, Standish N (1993) A study of the packing of particles with a mixture size distribution. Powder Technol 76:113–124
66. Bierwage GP, Saunders TE (1974) Studies of effects of particle-size distribution on packing efficiency of particles. Powder Technol 10:111–119
67. Andreasen AHM, Andersen J (1930) The relationship between grain gradation and the clearance in products from loosening grains (with some experiments). Kolloid-Zeitschrift 50:217–228
68. Burk RC, Apte PS (1987) A packing scheme for real size distributions. Am Ceram Soc Bull 66:1389–1392
69. Burk RC, Zawidzki TW, Apte PS (1983) Particle-size distribution and its relation to sintering 1: a case study for UO_2 powders. J Am Ceram Soc 66:815–818
70. Roosen A, Bowen HK (1988) Influence of various consolidation techniques on the green microstructure and sintering behavior of alumina powders. J Am Ceram Soc 71:970–977
71. Moreno R (1992) The role of slip additives in tape casting technology, 2. Binders and plasticizers. Am Ceram Soc Bull 71:1647–1657
72. Moreno R (1992) The role of slip additives in tape casting technology, 1. Solvents and dispersants. Am Ceram Soc Bull 71:1521–1531
73. Cerrutti BM, de Souza CS, Castellan A, Ruggiero R, Frollini E (2012) Carboxymethyl lignin as stabilizing agent in aqueous ceramic suspensions. Ind Crops Prod 36:108–115
74. Leo S, Tallon C, Franks GV (2014) Aqueous and nonaqueous colloidal processing of difficult-to-densify ceramics: suspension rheology and particle packing. J Am Ceram Soc 97:3807–3817
75. Sikora M, Garcia VJ, Schilling CH, Tomasik P, Li CP (2004) Blends of maltodextrin and other polysaccharides as binders of aqueous alpha-alumina suspensions for ceramic processing. Starch-Starke 56:424–431
76. Cesarano J, Aksay IA (1988) Processing of highly concentrated aqueous alpha-alumina suspensions stabilized with poly-electrolytes. J Am Ceram Soc 71:1062–1067
77. Cesarano J, Aksay IA, Bleier A (1988) Stability of aqueous alpha-Al_2O_3 suspensions with poly(methacrylic acid) poly-electrolyte. J Am Ceram Soc 71:250–255
78. Dulina NA, Deineka TG, Yavetskiy RP, Sergienko ZP, Doroshenko AG, Mateychenko PV et al (2011) Comparison of dispersants performance on the suspension Lu_2O_3:Eu^{3+} stability and high-density compacts on their basis. Ceram Int 37:1645–1651
79. Khan AU, Mahmood N, Luckham PF (2012) Rheological characterization of alumina ceramic suspensions in presence of a dispersant and a binder. J Dispersion Sci Technol 33:1210–1217
80. Lv Y-H, Liu H, Sang Y-H, Liu S-J, Chen T, Qin H-M et al (2010) Electrokinetic properties of Nd:YAG nanopowder and a high concentration slurry with ammonium poly(acrylic acid) as dispersant. J Mater Sci 45:706–712
81. Mori T, Inamine I, Wada R, Hida T, Kiguchi T, Satone H et al (2009) Effects of particle concentration and additive amount of dispersant on adsorption behavior of dispersant to alumina particles. J Ceram Soc Jpn 117:917–921
82. Prabhakaran K, Sooraj R, Kumbhar CS, Melkeri A, Gokhale NM, Sharma SC (2010) Heterocoagulation moulding of alumina powder suspensions prepared using citrate dispersant. Ceram Int 36:1–8
83. Tanurdjaja S, Tallon C, Scales PJ, Franks GV (2011) Influence of dispersant size on rheology of non-aqueous ceramic particle suspensions. Adv Powder Technol 22:476–481
84. Bernhardt C (1988) Preparation of suspensions for particle-size analysis—methodical recommendations, liquids and dispersing agents. Adv Colloid Inter Sci 29:79–139

85. Balasubramanian S, Shanefield DJ, Niesz DE (2002) Effect of externally applied plasticizer on compaction behavior of spray-dried powders. J Am Ceram Soc 85:749–754
86. Ma LG, Huang Y, Yang JL, Le HR, Sun Y (2005) Effect of plasticizer on the cracking of ceramic green bodies in gelcasting. J Mater Sci 40:4947–4949
87. Sikora M, Schilling CH, Tomasik P, Li CP (2002) Dextrin plasticizers for aqueous colloidal processing of alumina. J Eur Ceram Soc 22:625–628
88. Ewsuk KG, Arguello JG, Bencoe DN, Ellerby DT, Glass SJ, Zeuch DH et al (2003) Characterizing powders for dry pressing, sintering. Am Ceram Soc Bull 82:41–47
89. Glass SJ, Ewsuk KG (1997) Ceramic powder compaction. MRS Bull 22:24–28
90. Sun L, Oguz B, Kwon P (2009) Powder mixing effect on the compaction capabilities of ceramic powders. Powder Technol 195:227–234
91. Takahashi H, Shinohara N, Okumiya M, Uematsu K, Junichiro T, Iwamoto Y et al (1995) Influence of slurry flocculation on the character and compaction of spray-dried silicon-nitride granules. J Am Ceram Soc 78:903–908
92. Takahashi H, Shinohara N, Uematsu K (1996) Influence of spray-dry slurry flocculation on the structure of sintered silicon nitride. J Ceram Soc Jpn 104:59–62
93. Uematsu K (1996) Immersion microscopy for detailed characterization of defects in ceramic powders and green bodies. Powder Technol 88:291–298
94. Funk JE, Dinger DR (1988) Slip control using particle-size analysis and specific surface-area. Am Ceram Soc Bull 67:890–894
95. Funk JE, Dinger DR (1994) Particle-size control for high-solid castable refractories. Am Ceram Soc Bull 73:66–69
96. Bortzmeyer D (1992) Modeling ceramic powder compaction. Powder Technol 70:131–139
97. Carlone P, Palazzo GS (2006) Computational modeling of the cold compaction of ceramic powders. Int Appl Mech 42:1195–1201
98. Kim HS, Oh ST, Lee JS (2002) Constitutive model for cold compaction of ceramic powder. J Am Ceram Soc 85:2137–2138
99. Schwartz EG (1965) Weinstei.As. Model for compaction of ceramic powders. J Am Ceram Soc 48:346–350
100. Aydin I, Briscoe BJ, Ozkan N (1997) Modeling of powder compaction: a review. MRS Bull 22:45–51
101. Aydin I, Briscoe BJ, Sanliturk KY (1996) The internal form of compacted ceramic components: a comparison of a finite element modelling with experiment. Powder Technol 89:239–254
102. Strijbos S (1977) Powder-wall friction—effects of orientation of wall grooves and wall lubricants. Powder Technol 18:209–214
103. Strijbos S, Vangroenou AB (1979) Recent progress in understanding die compaction of powders. J Am Ceram Soc 62:57–59
104. Vangroenou AB (1978) Pressing of ceramic powders—review of recent work. Powder Metall Int 10:206–211
105. Vangroenou AB (1981) Compaction of ceramic powders. Powder Technol 28:221–228
106. Lannutti JJ (1997) Characterization and control of compact microstructure. MRS Bull 22:38–44
107. Lannutti JJ, Deis TA, Kong CM, Phillips DH (1997) Density gradient evolution during dry pressing. Am Ceram Soc Bull 76:53–58
108. Cottrino S, Jorand Y, Maire E, Adrien J (2013) Characterization by X-ray tomography of granulated alumina powder during in situ die compaction. Mater Charact 81:111–123
109. Deville S, Maire E, Lasalle A, Bogner A, Gauthier C, Leloup J et al (2009) In situ X-ray radiography and tomography observations of the solidification of aqueous alumina particles suspensions. Part II: steady state. J Am Ceram Soc 92:2497–2503
110. Hondo T, Kato Z, Tanaka S (2014) Enhancing the contrast of low-density packing regions in images of ceramic powder compacts using a contrast agent for micro-X-ray computed tomography. J Ceram Soc Jpn 122:574–576

111. Moritz K, Moritz T (2010) ZrO_2 ceramics with aligned pore structure by EPD and their characterisation by X-ray computed tomography. J Eur Ceram Soc 30:1203–1209
112. Rousseau B, Gomart H, Zanghi D, Bernard D, Stampanoni M (2010) Synchrotron x-ray mu-tomography to model the thermal radiative properties of an opaque ceramic coating at T = 1000 K. J Mater Res 25:1890–1897
113. Xu F, Hu XF, Miao H, Zhao JH (2010) In situ investigation of ceramic sintering by synchrotron radiation X-ray computed tomography. Opt Lasers Eng 48:1082–1088
114. Maire E, Withers PJ (2014) Quantitative X-ray tomography. Int Mater Rev 59:1–43
115. Carter CB, Norton MG (2007) Ceramics materials: science and engineering. Springer, Berlin
116. Tsetsekou A, Agrafiotis C, Milias A (2001) Optimization of the rheological properties of alumina slurries for ceramic processing applications—Part I: slip casting. J Eur Ceram Soc 21:363–373
117. Adcock DS, McDowall IC (1957) The mechanism of filter pressing and slip casting. J Am Ceram Soc 40:355–362
118. Fennelly TJ, Reed JS (1972) Mechanics of pressure slip casting. J Am Ceram Soc 55:264–268
119. Kovac M, Havrda J, Oujiri F (1993) A model of slip casting. Ceram-Silik 37:67–72
120. Tiller FM, Hsyung NB (1991) Theory of filtration of ceramics, 2. Slip casting on radial surfaces. J Am Ceram Soc 74:210–218
121. Tiller FM, Tsai CD (1986) Theory of filtration of ceramics, 1. Slip casting. J Am Ceram Soc 69:882–887
122. Bergstrom L, Schilling CH, Aksay IA (1992) Consolidation behavior of flocculated allumina suspensions. J Am Ceram Soc 75:3305–3314
123. Takao Y, Hotta T, Nakahira K, Naito M, Shinohara N, Okumiya M et al (2000) Processing defects and their relevance to strength in alumina ceramics made by slip casting. J Eur Ceram Soc 20:389–395
124. Ganesh I (2011) Aqueous slip casting of $MgAl_2O_4$ spinel powder. Bull Mater Sci 34:327–335
125. Ganesh I, Sundararajan G, Ferreira JMF (2011) Aqueous slip casting and hydrolysis assisted solidification of $MgAl_2O_4$ spinel ceramics. Adv Appl Ceram 110:63–69
126. Sommer F, Kern F, El-Maghraby HF, Abou El-Ezz M, Awaad M, Gadow R et al (2012) Effect of preparation route on the properties of slip-casted Al_2O_3/YAG composites. Ceram Int 38:4819–4826
127. Kaminskii AA, Kravchenko VB, Kopylov YL, Bagayev SN, Shemet VV, Komarov AA et al (2007) Novel polycrystalline laser material: Nd^{3+}:$Y_3Al_5O_{12}$ ceramics fabricated by the high-pressure colloidal slip casting (HPCSC) method. Physica Status Solidi A—Appl Mater Sci 204:2411–2415
128. Li X, Li Q (2008) YAG ceramic processed by slip casting via aqueous slurries. Ceram Int 34:397–401
129. Mouzon J, Glowacki E, Oden M (2008) Comparison between slip casting and uniaxial pressing for the fabrication of translucent yttria ceramics. J Mater Sci 43:2849–2856
130. Olhero SM, Miranzo P, Ferreira JMF (2006) AlN ceramics processed by aqueous slip casting. J Mater Res 21:2460–2469
131. Promdej C, Areeraksakul S, Pavarajarn V, Wada S, Wasanapiarnpong T, Charinpanitkul T (2008) Preparation of translucent alumina ceramic specimen using slip casting method. J Ceram Soc Jpn 116:409–413
132. Ganesh I (2013) A review on magnesium aluminate ($MgAl_2O_4$) spinel: synthesis, processing and applications. Int Mater Rev 58:63–112
133. Boulesteix R, Epherre R, Noyau S, Vandenhende M, Maitre A, Salle C et al (2014) Highly transparent Nd:Lu_2O_3 ceramics obtained by coupling slip casting and spark plasma sintering. Scripta Mater 75:54–57
134. Ramavath P, Biswas P, Rajeswari K, Suresh MB, Johnson R, Padmanabham G et al (2014) Optical and mechanical properties of compaction and slip cast processed transparent polycrystalline spinel ceramics. Ceram Int 40:5575–5581

135. Zhou J, Pan YB, Li J, Zhang WX, Kou HM, Liu WB et al (2011) Fabrication of YAG transparent ceramics using slip casting with ethanol. J Inorg Mater 26:254–256
136. Ji XB, Deng JG, Kang B, Huang H, Wang X, Jing W et al (2013) Fabrication of transparent neodymium-doped yttrium aluminum garnet ceramics by high solid loading suspensions. Ceram Int 39:7921–7926
137. Jin LL, Zhou GH, Shimai SZ, Zhang JA, Wang SW (2010) ZrO_2-doped Y_2O_3 transparent ceramics via slip casting and vacuum sintering. J Eur Ceram Soc 30:2139–2143
138. Appiagyei KA, Messing GL, Dumm JQ (2008) Aqueous slip casting of transparent yttrium aluminum garnet (YAG) ceramics. Ceram Int 34:1309–1313
139. Mao XJ, Wang SW, Shimai S, Guo JK (2008) Transparent polycrystalline alumina ceramics with orientated optical axes. J Am Ceram Soc 91:3431–3433
140. Zhang L, Vleugels J, Van der Biest O (2010) Slip casting of alumina suspensions in a strong magnetic field. J Am Ceram Soc 93:3148–3152
141. Yi HL, Mao XJ, Zhou GH, Chen S, Zou XQ, Wang SW et al (2012) Crystal plane evolution of grain oriented alumina ceramics with high transparency. Ceram Int 38:5557–5561
142. Kopylov YL, Kravchenko VB, Bagayev SN, Shemet VV, Komarov AA, Karban OV et al (2009) Development of Nd^{3+}:$Y_3Al_5O_{12}$ laser ceramics by high-pressure colloidal slip casting (HPCSC) method. Opt Mater 31:707–710
143. Denker SP, Meyer J (1966) Pressure casting refractory ceramics. Rev Sci Instrum 37:679–680
144. Fennelly TJ, Reed JS (1972) Compression permeability of Al_2O_3 cakes formed by pressure slip casting. J Am Ceram Soc 55:381–383
145. Lange FF, Miller KT (1987) Pressure filtration—consolidation kinetics and mechanics. Am Ceram Soc Bull 66:1498–1504
146. Cho K-S, Lee SY (2013) Fabrication of large-size alumina by pressure-vacuum hybrid slip casting. J Korean Ceram Soc 50:396–401
147. Maleksaeedi S, Paydar MH, Saadat S, Ahmadi H (2008) In situ vibration enhanced pressure slip casting of submicrometer alumina powders. J Eur Ceram Soc 28:3059–3064
148. Betz T, Rieb W, Lehmann J, Ziegler G (1997) Important parameters in pressure casting of Al_2O_3. CFI-Ceramic Forum International 74:101–105
149. Schaffoener S, Aneziris CG (2012) Pressure slip casting of coarse grain oxide ceramics. Ceram Int 38:417–422
150. Hotza D, Greil P (1995) Aqueous tape casting of ceramic powders. Mater Sci Eng A—Struct Mater Prop Microstructure Process 202:206–217
151. Bohnleinmauss J, Sigmund W, Wegner G, Meyer WH, Hessel F, Seitz K et al (1992) The function of polymers in the tape casting of alumina. Adv Mater 4:73–81
152. Chou YT, Ko YT, Yan MF (1987) Fluid-flow model for ceramic tape casting. J Am Ceram Soc 70:C280–C282
153. Heinrich J (1986) Tape casting of oxidic and nonoxidic ceramic powders. Powder Metall Int 18:229–232
154. Huang XY, Liu CY, Gong HQ (1997) A viscoplastic flow modeling of ceramic tape casting. Mater Manuf Process 12:935–943
155. Tok AIY, Boey FYC, Lam YC (2000) Non-Newtonian fluid flow model for ceramic tape casting. Mater Sci Eng A—Struct Mater Prop Microstructure and Process 280:282–288
156. Chang YF, Poterala S, Yener D, Messing GL (2013) Fabrication of highly textured fine-grained α-alumina by templated grain growth of nanoscale precursors. J Am Ceram Soc 96:1390–1397
157. Hong SH, Messing GL (1999) Development of textured mullite by templated grain growth. J Am Ceram Soc 82:867–872
158. Ma S, Fuh JYH, Zhang YF, Lu L (2010) Synthesis of anisotropic lead titanate powders for templated grain growth of textured piezoelectric ceramics. Surf Rev Lett 17:159–164
159. Messing GL, Trolier-McKinstry S, Sabolsky EM, Duran C, Kwon S, Brahmaroutu B et al (2004) Templated grain growth of textured piezoelectric ceramics. Crit Rev Solid State Mater Sci 29:45–96

160. Zeng YP, Zimmermann A, Zhou LJ, Aldinger F (2004) Tape casting of PLZST tapes via aqueous slurries. J Eur Ceram Soc 24:253–258
161. Yang Y, Wu YQ (2014) Tape-casted transparent alumina ceramic wafers. J Mater Res 29:2312–2317
162. Ba XW, Li J, Zeng YP, Pan YB, Jiang BX, Liu WB et al (2013) Transparent $Y_3Al_5O_{12}$ ceramics produced by an aqueous tape casting method. Ceram Int 39:4639–4643
163. Tang F, Cao YG, Huang JQ, Guo W, Liu HG, Huang QF et al (2012) Multilayer YAG/RE: YAG/YAG laser ceramic prepared by tape casting and vacuum sintering method. J Eur Ceram Soc 32:3995–4002
164. Ba XW, Li J, Pan YB, Liu J, Guo JK (2013) Preparation of composite YAG transparent ceramics by aqueous tape casting route. Rare Metal Mater Eng 42:234–237
165. Tang F, Lin Y, Wang WC, Yuan XY, Chen J, Huang JQ et al (2014) High efficient Nd:YAG laser ceramics fabricated by dry pressing and tape casting. J Alloy Compd 617:845–849
166. Ba XW, Li J, Pan YB, Zeng YP, Kou HM, Liu WB et al (2013) Comparison of aqueous- and non-aqueous-based tape casting for preparing YAG transparent ceramics. J Alloy Compd 577:228–231
167. Janney MA, Omatete OO, Walls CA, Nunn SD, Ogle RJ, Westmoreland G (1998) Development of low-toxicity gelcasting systems. J Am Ceram Soc 81:581–591
168. Omatete OO, Janney MA, Nunn SD (1997) Gelcasting: from laboratory development toward industrial production. J Eur Ceram Soc 17:407–413
169. Omatete OO, Janney MA, Strehlow RA (1991) Gelcasting - A new ceramic forming process. Am Ceram Soc Bull 70:1641
170. Young AC, Omatete OO, Janney MA, Menchhofer PA (1991) Gelcasting of alumina. J Am Ceram Soc 74:612–618
171. Kokabi M, Pirooz A, Haghighi MN (1998) Gel-casting of engineering ceramics. Iran Polym J 7:169–175
172. Xiang JH, Huang Y, Xie ZP (2002) Study of gel-tape-casting process of ceramic materials. Mater Sci Eng A—Struct Mater Prop Microstructure Process 323:336–341
173. Xie ZP, Ma CL, Huang Y, Xiang JH (2002) Gel tape casting ceramic sheets. Am Ceram Soc Bull 81:33–37
174. Mao XJ, Shimai S, Dong MJ, Wang SW (2008) Gelcasting and pressureless sintering of translucent alumina ceramics. J Am Ceram Soc 91:1700–1702
175. Sun Y, Shimai S, Peng X, Zhou GH, Kamiya H, Wang SW (2014) Fabrication of transparent Y_2O_3 ceramics via aqueous gelcasting. Ceram Int 40:8841–8845
176. Qin XP, Zhou GH, Yang Y, Zhang J, Shu X, Shimai S et al (2014) Gelcasting of transparent YAG ceramics by a new gelling system. Ceram Int 40:12745–12750
177. Braun A, Wolff M, Falk G, Clasen R (2005) Transparent alumina ceramics with sub-microstructure by means of electrophoretic deposition. In: Brito ME, Filip P, Lewinsohn C, Sayir A, Opeka M, Mullins MW (eds) Developments in advanced ceramics and composites. American Ceramic Society, Westerville, pp 97–104
178. Boccaccini AR, Ferrari B (2014) Electrophoretic deposition of ceramic materials. Adv Appl Ceram 113:1–2
179. Hadraba H, Drdlik D, Chlup Z, Maca K, Dlouhy I, Cihlar J (2013) Layered ceramic composites via control of electrophoretic deposition kinetics. J Eur Ceram Soc 33:2305–2312
180. Olevsky EA, Wang X, Maximenko A, Meyers MA (2007) Fabrication of net-shape functionally graded composites by electrophoretic deposition and sintering: modeling and experimentation. J Am Ceram Soc 90:3047–3056
181. Sarkar P, Nicholson PS (1996) Electrophoretic deposition (EPD): mechanisms, kinetics, and application to ceramics. J Am Ceram Soc 79:1987–2002
182. Besra L, Liu M (2007) A review on fundamentals and applications of electrophoretic deposition (EPD). Prog Mater Sci 52:1–61
183. Corni I, Ryan MP, Boccaccini AR (2008) Electrophoretic deposition: from traditional ceramics to nanotechnology. J Eur Ceram Soc 28:1353–1367

184. Braun A, Falk G, Clasen R (2006) Transparent polycrystalline alumina ceramic with sub-micrometre microstructure by means of electrophoretic deposition. Materialwiss Werkstofftech 37:293–297
185. Bredol M, Micior J, Klemme S (2014) Electrophoretic deposition of alumina, yttria, yttrium aluminium garnet and lutetium aluminium garnet. J Mater Sci 49:6975–6985
186. Bredol M, Micior J (2013) Preparation and characterization of nanodispersions of yttria, yttrium aluminium garnet and lutetium aluminium garnet. J Colloid Inter Sci 402:27–33
187. Leo S, Tallon C, Stone N, Franks GV (2014) Near-net-shaping methods for ceramic elements of (body) armor systemsl. J Am Ceram Soc 97:3013–3033
188. Nieh TG, Wadsworth J (1998) Fine-structure superplasticity in materials. J Chin Inst Eng 21:659–689
189. Li MY, Tang L, Landers RG, Leu MC (2013) Extrusion process modeling for aqueous-based ceramic pastes-Part 2: experimental verification. J Manuf Sci Eng—Trans ASME 135:051009
190. Li MY, Tang L, Landers RG, Leu MC (2013) Extrusion process modeling for aqueous-based ceramic pastes-Part 1: constitutive model. J Manuf Sci Eng—Trans ASME 135:051008
191. Lu X, Lee Y, Yang S, Hao Y, Evans JRG, Parini CG (2010) Solvent-based paste extrusion solid freeforming. J Eur Ceram Soc 30:1–10
192. Mason MS, Huang T, Landers RG, Leu MC, Hilmas GE (2009) Aqueous-based extrusion of high solids loading ceramic pastes: process modeling and control. J Mater Process Technol 209:2946–2957
193. Scheying G, Wuhrl I, Eisele U, Riedel R (2004) Monoclinic zirconia bodies by thermoplastic ceramic extrusion. J Am Ceram Soc 87:358–364
194. Wight JF, Reed JS (2001) Polymer-plasticized ceramic extrusion, Part 2. Am Ceram Soc Bull 80:73–76
195. Wright JF, Reed JS (2001) Polymer-plasticized ceramic extrusion, part 1. Am Ceram Soc Bull 80:31–35
196. Schuetz JE (1986) Methylcellulose polymers as binders for extrusion of ceramics. Am Ceram Soc Bull 65:1556–1559
197. Reiner M (1926) Flow of elastic liquid by means of a capillary. Article on the theory of the viscosity measurements. Kolloid-Zeitschrift 39:80–87
198. Robinson GC (1952) Limitations imposed by raw mateials on firing schedules. J Am Ceram Soc 35:1–5
199. Van Hoy C, Barda A, Griffith M, Halloran JW (1998) Microfabrication of ceramics by co-extrusion. J Am Ceram Soc 81:152–158
200. Powell J, Assabumrungrat S, Blackburn S (2013) Design of ceramic paste formulations for co-extrusion. Powder Technol 245:21–27
201. Kryachek VM (2004) Injection moulding (review). Powder Metall Met Ceram 43:336–348
202. Medvedovski E, Peltsman M (2012) Low pressure injection moulding mass production technology of complex shape advanced ceramic components. Adv Appl Ceram 111:333–344
203. Rajabi J, Muhamad N, Sulong AB (2012) Effect of nano-sized powders on powder injection molding: a review. Microsyst Technol—Micro Nanosyst Inf Storage Process Syst 18:1941–1961
204. German RM (2013) Progress in titanium metal powder injection molding. Materials 6:3641–3662
205. German RM, Hens KF, Lin STP (1991) Key issues in powder injection-molding. Am Ceram Soc Bull 70:1294–1302
206. Hens KF, Lee DY, German RM (1991) Processing conditions and tooling for powder injection-molding. Int J Powder Metall 27:141
207. Zhang T, Evans JRG (1993) The solidification of large sections in ceramic injection molding, 2: modulated pressure molding. J Mater Res 8:345–351
208. Zhang T, Evans JRG (1993) The solidification of large sections in ceramic injection molding, 1: conventional molding. J Mater Res 8:187–194

209. Chong JS, Christia EB, Baer AD (1971) Rheology of concentrated suspensions. J Appl Polym Sci 15:2007–2021
210. Nagaoka T, Sato K, Hotta Y, Tsugoshi T, Watari K (2007) Extrusion of alumina ceramics with hydraulic alumina without organic additives. J Ceram Soc Jpn 115:191–194
211. Nagaoka T, Duran C, Isobe T, Hotta Y, Watari K (2007) Hydraulic alumina binder for extrusion of alumina ceramics. J Am Ceram Soc 90:3998–4001
212. Wegmann M, Gut B, Berroth K (1998) Extrusion of polycrystalline ceramic fibers. CFI-Ceramic Forum Int 75:35–37
213. Yu ZY, Huang Y, Wang CG, Ouyang S (2003) Freeform fabrication of Al_2O_3 ceramics by extrusion gelation. Rare Metal Mater Eng 32:167–170
214. Mannschatz A, Mueller-Koehn A, Moritz T, Klimke J, Krell A, Michaelis A (2013) Powder injection molding of submicron grained translucent alumina. CFI-Ceramic Forum Int 90: E39–E43
215. Liu W, Bo TZ, Xie ZP, Wu Y, Yang XF (2011) Fabrication of injection moulded translucent alumina ceramics via pressureless sintering. Adv Appl Ceram 110:251–254
216. Liu W, Xie Z (2014) Pressureless sintering behavior of injection molded alumina ceramics. Sci Sinter 46:3–13
217. el Ezz MA, El-Maghraby HF, Kern F, Sommer F, Awaad M, Gadow R et al (2013) Ceramic injection moulding of alumina-10 vol.-%YAG composites. Adv Appl Ceram 112:125–130
218. Taylor NJ, Laine RM (2014) Extrusion of YAG tubes shows that bottom-up processing is not always optimal. Adv Funct Mater 24:1125–1132
219. Baranwal R, Villar MP, Garcia R, Laine RM (2001) Flame spray pyrolysis of precursors as a route to nano-mullite powder: powder characterization and sintering behavior. J Am Ceram Soc 84:951–961
220. Deegan RD, Bakajin O, Dupont TF, Huber G, Nagel SR, Witten TA (1997) Capillary flow as the cause of ring stains from dried liquid drops. Nature 389:827–829
221. Deegan RD, Bakajin O, Dupont TF, Huber G, Nagel SR, Witten TA (2000) Contact line deposits in an evaporating drop. Phys Rev E 62:756–765
222. Guo JJ, Lewis JA (1999) Aggregation effects on the compressive flow properties and drying behavior of colloidal silica suspensions. J Am Ceram Soc 82:2345–2358
223. Lograsso BK, Bose A, Carpenter B, Chung C, Hens K, Lee D et al (1988) Baseline system for powder injection-molding: iron-polyethylene wax. J Metals 40:10–11
224. Lograsso BK, Bose A, Carpenter BJ, Chung CI, Hens KF, Lee D et al (1989) Injection molding of carbonyl iron with polyethylene wax. Int J Powder Metall 25:337–348
225. German RM, Bose A (1997) Injection molding of metals and ceramics. Metal Powder Industries Federation, Princeton
226. Uematsu K, Miyashita M, Kim JY, Kato Z, Uchida N (1991) Effect of forming pressure on the internal structure of alumina green bodies examined with immersion liquid technique. J Am Ceram Soc 74:2170–2174
227. Yang S, Fang W, Chi Y, Khan DF, Zhang R, Qu X (2014) Bulk observation of aluminum green compacts by way of X-ray tomography. Nucl Instr Methods Phys Res Sect B—Beam Inter Mater Atoms 319:146–153
228. Tanner BK (1977) Crystal assessment by X-ray topography using synchrotron radiation. Prog Cryst Growth Charact Mater 1:23–56

Chapter 5
Sintering and Densification (I)—Conventional Sintering Technologies

5.1 Introduction

It is well known that, to produce ceramics, green bodies must be sintered at a certain high temperature for a given time duration to develop required microstructure and thus desired properties. In particular, transparent ceramics must be fully dense to achieve maximum optical transmittance. Sintering process is governed by a number of parameters, which can be used to build up interrelationships among processing, microstructure, properties, and performance. Sintering behavior and microstructure development have been extensively studied. Qualitative understandings include driving forces of sintering, the mechanisms of densification, controlling factors, such as particle size of precursor powders, sintering temperature, time duration and applied pressure, electrical current, and so on.

Thermodynamics indicates that, during sintering process, there is always a decrease in the free energy of the system. Driving forces of sintering are mainly derived from the curvature of free surfaces of the precursor powders. External pressures if applied could also be important driving force. Kinetics of matter transport is concerned, because it determines the sintering time. Matter transport takes place predominantly through the diffusion of atoms, ions or other charged species. The paths of diffusion, corresponding to the mechanisms of diffusion, determine the mechanisms of the sintering. The rate of diffusion is controlled the type and concentration of defects. In this respect, defect chemistry, including defect structures and concentrations of the defect, is an important subject related to sintering. It is also necessary to predict the defect chemistry for different sintering conditions, such as temperature, gaseous atmosphere and dopants. Fundamentals of sintering of ceramics have been described systematically in the literature [1–18].

L.B. Kong et al., *Transparent Ceramics*, Topics in Mining, Metallurgy and Materials Engineering, DOI 10.1007/978-3-319-18956-7_5

5.2 Fundamental Aspects of Sintering

5.2.1 Driving Forces of Sintering

Sintering process is an irreversible process, during which the free energy of the system is decreased. The sources responsible for such a decrease are called the driving forces of sintering. There are three potential driving forces, (i) the curvature of particle surfaces, (ii) externally applied pressure, and (iii) chemical reaction.

5.2.1.1 Surface Curvature

Surface curvature is always present as the driving force of sintering, while external stress and chemical reaction may be absent. Therefore, surface curvature is the primary driving force of sintering. For one mole of ceramic powder, comprising of spherical particles with a radius a, the number of particles is given by:

$$N = \frac{3M}{4\pi a^3 \rho} = \frac{3V_m}{4\pi a^3}, \tag{5.1}$$

where ρ is density of the particles (materials), which are assumed to be pore-free, M is molecular weight and V_m is molar volume. The surface area of the powder is given by:

$$S_A = 4\pi a^2 N = \frac{3V_m}{a}. \tag{5.2}$$

If γ_{SV} is used to stand for the specific surface energy, i.e., the surface energy per unit area, of the particles, the surface free energy of the powder can be calculated by:

$$E_S = \frac{3\gamma_{SV} V_m}{a}. \tag{5.3}$$

Therefore, E_S is the quantity of the decrease in surface free energy of the powder, if a fully dense body is formed from the one mole of particles, i.e., it is the driving force of sintering.

5.2.1.2 External Pressure

External pressure can be applied to provide the driving force for sintering in addition to surface curvature, which could be comparatively smaller in this case. The pressure should be applied during the key state of sintering. There are two ways to apply pressure: hot pressing and hot isostatic pressing. By applying an external pressure, work is done on compact of the powder. For 1 mol of particles, the work done on the compact can be estimated by:

$$W = p_{\mathrm{a}} V_{\mathrm{m}}, \tag{5.4}$$

where p_{a} is the applied pressure and V_{m} is molar volume of the materials. W represents the driving force for densification due to the application of an external pressure.

5.2.1.3 Chemical Reaction

Chemical reactions can provide driving force for sintering only when they have contribution to the densification process. The change in free energy of a chemical reaction is given by

$$\Delta G^0 = -RT \ln K_{\mathrm{eq}}, \tag{5.5}$$

where R is the gas constant (8.3 J mol^{-1}), T is the absolute temperature, and K_{eq} is the equilibrium constant of the reaction. The value is much higher than the driving force of external pressure.

5.2.2 *Defects in Crystalline Solids*

The driving forces for sintering are attributed to the transport of matter, through the diffusion of atoms, ions or molecules in crystalline solids, due to the presence of various defects related to their structural imperfections. Properties and concentrations of defects control the rates of matter transport and thus determine the rates of processes, such as sintering, grain growth, and creep. The presence of defects in crystalline solids has structural reasons, i.e., the atoms or ions are not arranged in their lattice sites, and chemical reasons, i.e., the deviation from the stoichiometric composition. Structural defects in crystalline solids can be classified into three groups: (i) point defects, (ii) line defects, and (iii) planar defects. Point defects are associated with one lattice site and its close surroundings, such as missing atoms called vacancies, interstitial atoms at the interstices between atoms, and substitutional atoms on the sites that would normally be occupied by other types of atoms, as shown schematically in Fig. 5.1 for an elemental solid, e.g., pure metal [1]. The point defects in pure crystals, i.e., vacancies and interstitials, are called intrinsic or native defects. Because ceramics can be treated as ionic solids, point defects have more contributions than others to the defect chemistry.

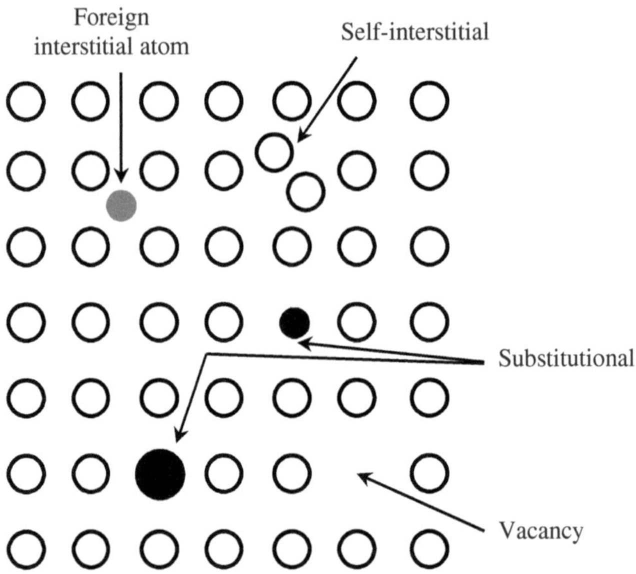

Fig. 5.1 Schematic showing various types of point defects in an elemental solid. Reproduced with permission from [1]. Copyright © 2003, CRC Press

5.2.2.1 Point Defects and Defect Chemistry

Defects in ceramics can be charged, which are different from those in metals. For a simple pure ionic oxide, with a stoichiometric formula of MO, consisting of a metal (M) with valence of +2 and an oxygen (O) with valence of −2, the types of point defects could be vacancies and interstitials of both the M and O, which can be either charged or neutral. Besides the single defects, it is also possible for the defects to associate with one another to form defect clusters. Electronic defects or valence defects, consisting of quasi-free electrons or holes, are also observed in crystalline solids. If there are impurities, e.g., solute atoms Mf, substitutional or interstitial defects of Mf could be formed, which can also be either charged or neutral.

As multicomponent compounds, the compositions of ceramics can be nonstoichiometric, if they are annealed in certain gaseous atmospheres, such as oxygen with controlled partial pressures, which is driven by the tendency to approach an equilibrium state with the surrounding environment. The consequence of the equilibration is the deviation of the composition from stoichiometry and the variation in concentration of the defects. For instance, if MO is annealed at an environment of low oxygen partial pressure, either oxygen-deficient oxide, i.e., MO_{1-x}, or metal excess oxide, i.e., $M_{1+y}O$, could be formed. However, if it is annealed at higher oxygen partial pressures, metal deficient oxide, i.e., $M_{1-y}O$, could be obtained, with metal vacancies to be the predominant defect. However, In this case, it is quite difficult to form oxygen interstitial, due to the large size of the oxygen ions.

Charged defects in solids have interactions with one another, which are similar to the interactions among ions, sometimes involving electrons, in solutions. In the solid-state cases, the crystals are considered to be a neutral medium, which can dissolve the charged defects. Defect chemistry is to study the properties and behaviors of point defects in solid materials. As stated earlier, according to their originality, there are two types of defects: (i) intrinsic and (ii) extrinsic defects. Intrinsic defects are thermally generated in pure compounds, while extrinsic defects are produced by introduction impurities or treatment in gaseous atmospheres.

5.2.2.2 Kroger-Vink Notation

The description of point defects in ionic solids is well represented by a standard notation, known as the Kroger-Vink notation. In this notation system, a defect, which is defined with respect to the perfect lattice, is described by three parts: the main symbol, a subscript, and a superscript. For instance, in the notation, $M_{\mathrm{L}}^{\mathrm{C}}$, the main symbol M represents the particular atom, or for a vacancy, it is V. The subscript L means the lattice site at which the defect is located. The superscript C describes the effective charge or relative charge of the defect, which is equal to the difference in valence between the species on the L site and the valence of the atom that occupies the L site in the perfect lattice before the formation of the defect. The effective charge could be: (i) positive, C = $\cdot$, (ii) negative, C = $'$ or (iii) neutral, C = $\times$. For electronic defects, a quasi-free electron is represented as e', while a missing electron or hole is represented as $h^{\cdot}$. Possible defects in Al_2O_3 are listed in Table 5.1, by using the Kroger-Vink notation. Square brackets are used to represent the concentration of defects, such as $[\mathrm{V}_{\mathrm{O}}^{\cdot\cdot}]$ and $[\mathrm{Ti}_{\mathrm{Al}}^{\cdot}]$. The concentrations of electrons and holes, $[e']$ and $[h^{\cdot}]$, are usually simplified as n and p, respectively.

5.2.2.3 Defect Reactions

Similarly to chemical reactions, the formation of defects or defect reactions can be constructed according to the following three conservation rules, i.e., (i) mass conservation or mass balance, (ii) electroneutrality or charge balance, and (iii) site ratio conservation or site balance.

Table 5.1 Kroger-Vink notation of potential defects in Al_2O_3

Notation	Defect
$\mathrm{Al}_{\mathrm{i}}^{\cdot\cdot\cdot}$	Al^{3+} ion at interstitial site
$\mathrm{V}_{\mathrm{Al}}'''$	Al^{3+} vacancy
$\mathrm{V}_{\mathrm{O}}^{\cdot\cdot}$	O^{2-} vacancy
$\mathrm{Mg}_{\mathrm{Al}}'$	Mg^{2+} replacing Al^{3+} at lattice site
$\mathrm{Ti}_{\mathrm{Al}}^{\cdot}$	Ti^{4+} replacing Al^{3+} at lattice site
e'	Electron
$h^{\cdot}$	Hole

A mass balance should be maintained so that mass is neither created nor destroyed in the defect reaction. Vacancies have zero mass, while the effect of electronic defects on the mass balance is neglected. The crystal must be electrically neutral in the defect reaction, which means that, the sum of the positive effective charges should be equal to the sum of the negative effective charges, for the whole reaction. The ratio of the number of regular cation sites to the number of regular anion sites in the crystal must be constant in the defect reaction. For example, in the compound MO_2, the ratio of the regular M and O sites must always have the ratio of 1:2. Sites may be created or destroyed in the defect reaction, but site ratio in the regular lattice cannot be changed.

For instance, if MgO is used to dope Al_2O_3, because the ionic radii of Mg^{2+} and Al^{3+} with coordination number of six are very close, the Mg ions can enter the lattice of Al_2O_3 to form solid solution as substitutional defects. Al_2O_3 has the corundum structure, in which one-third of the octahedral sites formed by the close-packed O ions are vacant, so that it is also highly possible for the Mg ions to sit on the interstitial sites. The defects with lower energy are more favorable. In Al_2O_3, the cation sites and anion sites have a number ratio of 2:3. If substitutional defects are formed, every two Mg atoms on cation sites will replace two Al sites and two O sites are involved. In this case, the third O site should be a vacancy for site conservation. Therefore, on the basis of mass and site balance, the defect reaction is given by:

$$2\mathrm{MgO} \xrightarrow{\mathrm{Al_2O_3}} 2\mathrm{Mg_{Al}} + 2\mathrm{O_O} + \mathrm{V_O}. \tag{5.6}$$

When the defects are fully ionized, with conservation of electroneutrality, there is:

$$2\mathrm{MgO} \xrightarrow{\mathrm{Al_2O_3}} 2\mathrm{Mg'_{Al}} + 2\mathrm{O_O^{\times}} + \mathrm{V_O^{\cdot\cdot}}. \tag{5.7}$$

Therefore, all mass, charge, and site ratio are balanced.

If Al interstitials are formed, instead of O vacancies, the defect reaction is given by:

$$3\mathrm{MgO} \xrightarrow{\mathrm{Al_2O_3}} 3\mathrm{Mg'_{Al}} + 3\mathrm{O_O^{\times}} + \mathrm{Al_i^{\cdot\cdot\cdot}}. \tag{5.8}$$

In this equation, the mass and site ratio are not balanced, so that it is invalid. If the Mg ions take interstitial sites, there is the following defect reaction:

$$3\mathrm{MgO} \xrightarrow{\mathrm{Al_2O_3}} 3\mathrm{Mg_i^{\cdot\cdot}} + 3\mathrm{O_O^{\times}} + 2\mathrm{V_{Al}'''}. \tag{5.9}$$

Furthermore, if Mg is self-compensated to form both substitutional and interstitial defects, the reaction is given by:

$$3\text{MgO} \xrightarrow{\text{Al}_2\text{O}_3} 2\text{Mg}'_{\text{Al}} + \text{Mg}_{\text{i}}^{\cdot\cdot} + 3\text{O}_{\text{O}}^{\times}. \tag{5.10}$$

Although the formation of interstitials does not create new crystal sites, they must be considered for mass and charge conservations.

5.2.2.4 Concentration of Defects

The concentration of defects can be derived from statistical thermodynamics point of view, but it is more convenient treat the formation of defects as a chemical reaction, so that equilibrium constant of mass action can be applied. For a general reaction, in which the reactants *A* and *B* lead to products *C* and *D*, the equation is given by:

$$aA + bB \leftrightarrow cC + dD, \tag{5.11}$$

where *a*, *b* are the numbers of moles of the reactants while *c* and *d* are those of the products. At a given temperature, when the reaction reaches an equilibrium state, if it is assumed that the activities of the reactants and products are equal to their concentrations, according to the law of mass action, there is an equation that governs the interrelationship among the reactants and products, as following:

$$K = \frac{[C]^c[D]^d}{[A]^a[B]^b}, \tag{5.12}$$

where the square brackets are used to represent the concentrations and *K* is called the equilibrium constant. The constant *K* is given by the Arrhenius equation as follows:

$$K = \exp\left(\frac{-\Delta G}{RT}\right), \tag{5.13}$$

where ΔG is the Gibbs free energy change of the reaction, *R* is the gas constant, and *T* is absolute temperature. In defect reactions, concentrations of the defects are usually expressed as fractions of sites, when using Eqs. (5.12) and (5.13). If the concentrations of the defects are sufficiently low, concentrations of the ions at their regular lattice sites are taken as unity. When a reaction involves gas phases, their partial pressures will be used as the concentrations of the gases.

5.2.2.5 Intrinsic Defects

There are two typical intrinsic defects in ionic crystals: (i) Schottky defect and (ii) Frenkel defect.

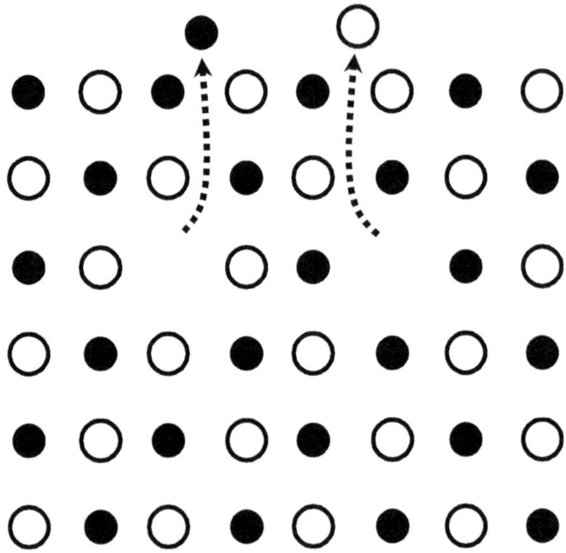

Fig. 5.2 Schematic diagram illustrating the formation of a Schottky defect. Reproduced with permission from [1]. Copyright © 2003, CRC Press

When a cation and an anion are taken away from their regular lattice sites to an external surface, an extra perfect crystal is formed and a pair of vacancies is left behind, which are called Schottky defect, as shown schematically in Fig. 5.2 [1]. To meet the requirement of electroneutrality balance of the crystal, vacancies must be formed in the stoichiometric ratio.

For simple oxide MO, if the defects are assumed to be fully ionized, the defect formation reaction can be written as:

$$\mathrm{M_M^\times + O_O^\times \leftrightarrow V_M'' + V_O^{\cdot\cdot} + M_M^\times + O_O^\times}. \tag{5.14}$$

Here, $\mathrm{M_M^\times}$ and $\mathrm{O_O^\times}$ on both sides of the equation can be canceled out, so that the net reaction can be simplified as:

$$0 \leftrightarrow \mathrm{V_M'' + V_O^{\cdot\cdot}}, \tag{5.15}$$

where 0 means that the defects are created from a perfect lattice, which also can be written as null of nil. When the reaction reaches equilibrium, according to the law of mass action, there is:

$$K_\mathrm{S} = [\mathrm{V_M''}][\mathrm{V_O^{\cdot\cdot}}] = \exp\left(\frac{-\Delta G_\mathrm{S}}{RT}\right), \tag{5.16}$$

where K_S is the equilibrium constant and $\Delta \mathrm{G_S}$ is the Gibbs free energy change of the formation of Schottky defects. For electroneutrality balance, there is $[\mathrm{V_M''}] = [\mathrm{V_O^{\cdot\cdot}}]$, so that the concentrations of the defects are given by:

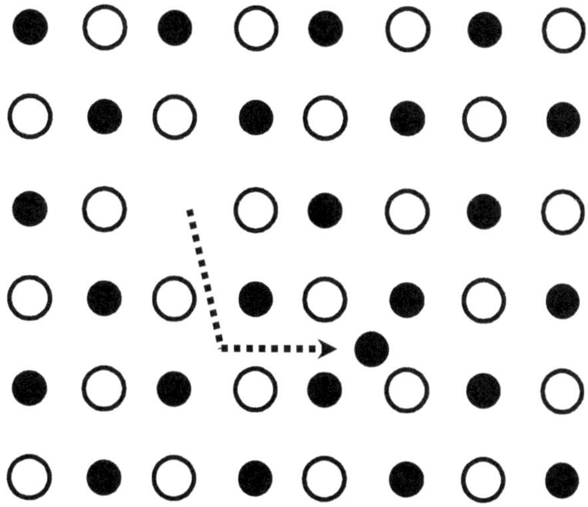

Fig. 5.3 Schematic diagram showing the formation of a Frenkel defect. Reproduced with permission from [1]. Copyright © 2003, CRC Press

$$[\mathrm{V}_{\mathrm{M}}''] = [\mathrm{V}_{\mathrm{O}}^{\cdot\cdot}] = \exp\left(\frac{-\Delta G_{\mathrm{S}}}{2RT}\right). \tag{5.17}$$

When an ion leaves its regular lattice site and stays at an interstitial site, leaving behind a vacant site, Frenkel defect is formed, as shown schematically in Fig. 5.3 [1]. For a simple oxide MO, if the defects are assumed to be fully ionized, the formation of a Frenkel defect at the cation site can be expressed as:

$$\mathrm{M}_{\mathrm{M}}^{\times} \leftrightarrow \mathrm{M}_{\mathrm{i}}^{\cdot\cdot} + \mathrm{V}_{\mathrm{M}}''. \tag{5.18}$$

The equilibrium constant of the reaction is given by:

$$K_{\mathrm{F}} = [\mathrm{M}_{\mathrm{i}}^{\cdot\cdot}][\mathrm{V}_{\mathrm{M}}''] = \exp\left(\frac{-\Delta G_{\mathrm{F}}}{RT}\right). \tag{5.19}$$

Eletroneutrality balance requires that the concentration of metallic interstitial should be equal to that of the metallic vacancy, i.e., $[\mathrm{M}_{\mathrm{i}}^{\cdot\cdot}] = [\mathrm{V}_{\mathrm{M}}'']$. In this case, the concentrations of the defects can be calculated by the following equations:

$$[\mathrm{M}_{\mathrm{i}}^{\cdot\cdot}] = [\mathrm{V}_{\mathrm{M}}''] = \exp\left(\frac{-\Delta G_{\mathrm{F}}}{2RT}\right). \tag{5.20}$$

Similar equations can be obtained as the Frenkel defects are formed at anion sites. However, anion Frenkel defects are rarely encountered, because anions have relatively large sizes, i.e., there are no interstitial sites for them to stay. Also, the formation of cation and anion Frenkel defects is not controlled by the requirement of electroneutrality balance, so that the concentration of cation interstitial may not be necessarily equal to that of anion interstitial. In other words, cation and anion can form interstitials independently.

5.2.2.6 Extrinsic Defects

As stated earlier, extrinsic defects are formed due to external influences, such as gaseous atmosphere environment which is usually responsible for nonstoichiometry in composition and impurities which can be intentionally incorporated such as dopants or unintentionally introduced such as impurities.

Nonstoichiometry

Equilibration of ionic solids with an ambient gaseous environment, e.g., metal oxide versus oxygen, can have a significant effect on the structure and behavior of defects. Oxides with elements of fixed valency, such as MgO, Al_2O_3, and ZrO_2, are not easily to deviate from the stoichiometry, while those with elements having variable valence, such as NiO, CoO, FeO, SnO_2, and TiO_2, could be deviated from the stoichiometry.

At a given temperature and composition, oxides are always in equilibrium with a specific oxygen partial pressure or tend to give out or take up oxygen until new equilibrium is reached. When oxides release oxygen, reduction reaction occurs, oxygen vacancies could be created in the lattice, in which with the electrons are liberated within the solid. The overall reaction can be written as:

$$\mathrm{O_O^\times} \leftrightarrow \frac{1}{2}\mathrm{O_2}(g) + \mathrm{V_O^{\cdot\cdot}} + 2e'. \tag{5.21}$$

The creation of the oxygen vacancies will change the number ratio to cation to anion, i.e., nonstoichiometry is present. The equilibrium constant for the reaction is given by:

$$K_\mathrm{R} = n^2[\mathrm{V_O^{\cdot\cdot}}]p_{\mathrm{O_2}}^{1/2} = K_\mathrm{R}^0 \exp\left(\frac{-\Delta G_\mathrm{R}}{RT}\right), \tag{5.22}$$

where the concentration of the oxygen gas is the partial pressure, K_R^0 is the equilibrium constant, and ΔG_R is the free energy change of the reduction. For electroneutrality balance, there is $n = 2[\mathrm{V_O^{\cdot\cdot}}]$, so that the concentration of oxygen vacancy can be derived from Eq. (5.22), given by:

$$[\mathrm{V_O^{\cdot\cdot}}] = \left(\frac{K_\mathrm{R}}{4}\right)^{1/3} p_{\mathrm{O_2}}^{-1/6}. \tag{5.23}$$

When the oxides take up oxygen, oxidation reaction takes place, which can be treated as the consumption of oxygen vacancies, with the charge being balanced by combining with an electron, thus leading to a missing electron or hole in the valence band, with defect reaction being expressed as:

$$\frac{1}{2}\mathrm{O}_2(g) + \mathrm{V}_\mathrm{O}^{\cdot\cdot} \leftrightarrow \mathrm{O}_\mathrm{O}^{\times} + 2h^{\cdot}. \tag{5.24}$$

The equilibrium constant is given by:

$$K_\mathrm{O} = \frac{p^2}{[\mathrm{V}_\mathrm{O}^{\cdot\cdot}]p_{\mathrm{O}_2}^{1/2}} = K_\mathrm{O}^0 \exp\left(\frac{-\Delta G_\mathrm{O}}{RT}\right). \tag{5.25}$$

Because oxidation and reduction are reversed reactions each other, which are actually the same process from the thermodynamic point of view, their reaction equations are dependent mutually. For example, if the reduction reaction of Eq. (5.21) is combined with the intrinsic electronic defect equilibrium:

$$0 = e' + h^{\cdot}, \tag{5.26}$$

which leads to the oxidation reaction of Eq. (5.24). Therefore, the oxidation and reduction reactions can be expressed in a number of different ways, the use of them is determined mainly by the consideration of convenience. For instance, oxidation can also be treated as the creation of oxygen lattice and cation vacancies. When their charges are combined with electrons, holes are formed in the valence band. For simple oxide MO, the reaction can be written as:

$$\frac{1}{2}\mathrm{O}_2(g) \leftrightarrow \mathrm{O}_\mathrm{O}^{\times} + \mathrm{V}_\mathrm{M}'' + 2h^{\cdot}. \tag{5.27}$$

Equation (5.27) can be obtained by adding the Schottky reaction to Eq. (5.24).

Effects of Dopants

Dopants, also known as solutes, with low concentrations, can significantly influence the sintering of ceramics. Dopants sometimes are necessary to create functionalities of ceramics. When the cation valence is different from that of the host cation, the dopant is called aliovalent dopant, whereas if the cation has the same valence as that of the host, it is called an isovalent dopant. For aliovalent dopants, when the valence of the solute cation is higher than that of the host cation, the dopant is known as a donor, otherwise, it is called a acceptor. Therefore, if Al_2O_3 is host, TiO_2 and MgO are donor and acceptor, respectively.

When Al_2O_3 is incorporated into MgO, due to the similarity in ionic radii, it is assumed that the Al will substitute for Mg, with charge balanced achieved by the creation of a vacant Mg site, through the following equation:

$$Al_2O_3 \xrightarrow{MgO} 2Al^{\cdot}_{Mg} + V''_{Mg} + 3O^{\times}_{O}. \tag{5.28}$$

If the intrinsic defects in MgO are assumed to be Schottky defects, according to Eq. (5.16), there is:

$$[V^{\cdot\cdot}_{O}][V''_{Mg}] = K_S. \tag{5.29}$$

The condition for charge balance is given by:

$$[Al^{\cdot}_{Mg}] + 2[V^{\cdot\cdot}_{O}] = 2[V''_{Mg}]. \tag{5.30}$$

At very low concentration of Al_2O_3, $[Al^{\cdot}_{Mg}] \ll [V''_{Mg}]$, so that there is:

$$[V^{\cdot\cdot}_{O}] = [V''_{Mg}] = K_S^{1/2}, \tag{5.31}$$

which means that the concentration of the intrinsic defects are independent on the concentration of Al_2O_3. In other words, the intrinsic defects are dominant.

When the concentration of Al_2O_3 in solid solution is increased, the extrinsic defects begin to be dominant. The charge balance is given by:

$$[Al^{\cdot}_{Mg}] = 2[V''_{Mg}]. \tag{5.32}$$

If the Al_2O_3 is completely incorporated, the concentration of Al in solid solution is equal to the total atomic concentration of Al, so that there is:

$$[Al^{\cdot}_{Mg}] = [Al]. \tag{5.33}$$

Because Eq. (5.29) is applicable for the cation and anion vacancies, combination of Eq. (5.29) with Eqs. (5.32) and (5.33) yields:

$$[V^{\cdot\cdot}_{O}] = \frac{2K_S}{[Al]}. \tag{5.34}$$

Furthermore, according to Eqs. (5.32) and (5.33), there also is following equation:

$$[V''_{Mg}] = \frac{1}{2}[Al]. \tag{5.35}$$

Brouwer Diagram

The variation in concentration of defects as a function of temperature, oxygen partial pressure or dopant concentration can be described semi-quantitatively in terms of a double logarithmic plot, which is known as a Brouwer diagram. Figure 5.4 shows a Brouwer diagram for the effects of Al_2O_3 dopant on the defect chemistry of MgO, as discussed above [1]. Similar plots can be constructed for the variations in concentration of defect with temperature or oxygen partial pressure. The Brouwer diagram shows the net effect, when different defect reactions are considered at the same time. Although there are unlimited number of defect reactions that can be derived for a given system, the number of defects that must be used to describe a given property that is related to defects, such as electrical conductivity or sintering, is actually limited. Also, it is necessary to identify the majority defects that determine the properties of the materials, as well as minority defects that could be relevant to a specific property of interest. The main defects that are usually considered in writing defect reactions include: (i) major intrinsic ionic defects (Schottky or Frenkel), (ii) intrinsic electronic defects, (iii) oxidation or reduction, and (iv) dopants. In addition, an equation for the electroneutrality condition is always necessary.

5.2.2.7 Defect Chemistry and Sintering

With the above discussion, it is summarized that the concentration of the point defects and thus the rate of matter transport through in crystal solids can be controlled by controlling three variables: (i) temperature, (ii) oxygen partial pressure, i.e., the gaseous atmosphere, and (iii) concentration of dopants. In the system of Al_2O_3-doped MgO discussed above, if the sintering rate of MgO is controlled by the diffusion of oxygen vacancies, the addition of Al_2O_3 will decrease the sintering rate, according to Eq. (5.34). However, if the diffusion of Mg vacancy is the rate-

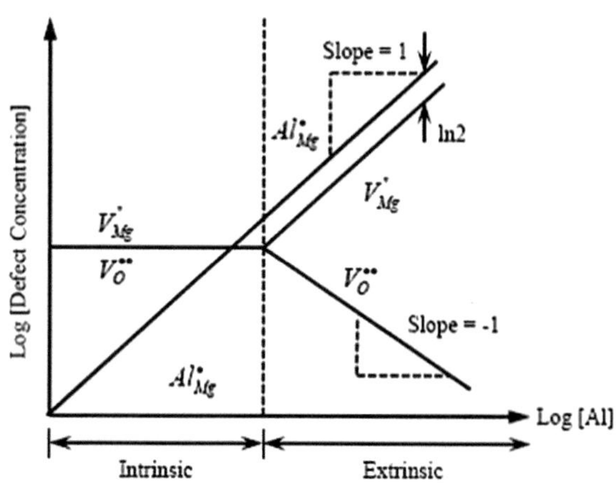

Fig. 5.4 Brouwer diagram for MgO doped with Al_2O_3. The defect concentrations are shown as a function of the concentration of Al. Reproduced with permission from [1]. Copyright © 2003, CRC Press

controlling mechanism, the presence of Al_2O_3 will increase the sintering rate, as demonstrated by Eq. (5.35). The factors that affect the sintering rate could be much more complicated in real systems. Also, the mechanisms of rate-controlling of sintering in most cases have not been clarified until now. Furthermore, there could be more than one mechanism during the sintering of a given material system and the rates of transport related to the different mechanisms can be changed with various conditions. There are also other factors, such as particle packing homogeneity, particle size, and size distributions, which could have significant effects on the rates of sintering.

5.2.3 Diffusion in Crystalline Solids

5.2.3.1 Fick's Laws of Diffusion

In the theory of solid-state diffusion, it is usually assumed that the movement of a diffusing species is driven by the gradients in concentration, irrespective with the atomic nature of crystal structure and atomic defects. The concentration can vary as a function of distance and time. When the concentration is changed with time, the diffusion process is described mathematically by the Fick's first law, i.e., the flux of the diffusing species J, which is defined as the number of items crossing per unit area normal to the direction of flux per second, is proportional to the concentration gradient $\mathrm{d}C/\mathrm{d}x$, which takes place in the direction of decreasing concentration. The one-dimensional Fick's first law is given by:

$$J_x = -D\frac{\mathrm{d}C}{\mathrm{d}x}. \tag{5.36}$$

The proportionality D is a constant, which is known as the diffusion coefficient or diffusivity, with a unit of $\mathrm{m^2\ s^{-1}}$ (SI) or $\mathrm{cm^2\ s^{-1}}$. The diffusion coefficient is a property of materials, which is the most useful parameter to characterize the rate of diffusive mass transport, showing a strong dependence on temperature. Although it is also a function of composition, if the diffusing species are significantly diluted, it can be assumed to be independent on the composition.

In practical experiments, it is difficult to maintain a concentration to be independent of time. Therefore, it is more often to characterize the change in concentration as a function of time t, thus leading to Fick's second law. For one dimension, the Fick's second law is given by:

$$\frac{\mathrm{d}C}{\mathrm{d}t} = D\frac{\mathrm{d}^2C}{\mathrm{d}x^2}. \tag{5.37}$$

The Fick's second law can be derived from the first law, together with the principle of matter conservation. For one-dimension, a region between the two

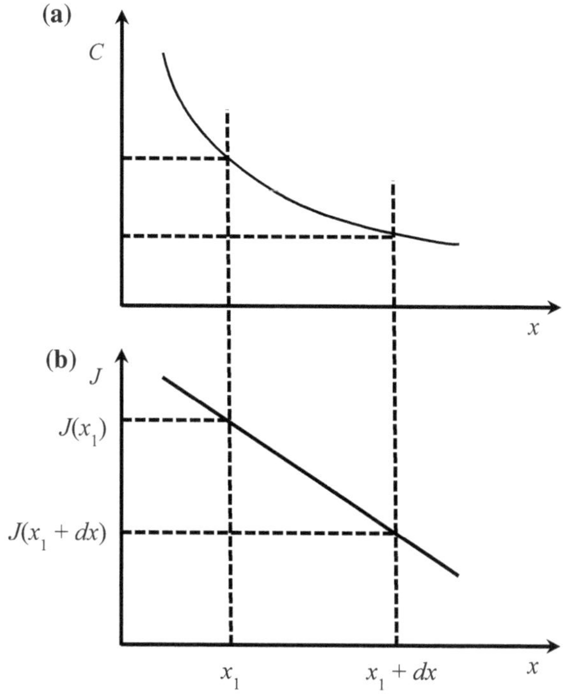

Fig. 5.5 Concentration C versus distance x (**a**) and the resulting flux J versus distance (**b**). Reproduced with permission from [1]. Copyright © 2003, CRC Press

planes $[x_1, (x_1 + \mathrm{d}x)]$ can be considered, as shown in Fig. 5.5 [1]. Figure 5.5a shows the solute concentration C as a function of distance x. Because $\mathrm{d}C/\mathrm{d}x$ at x_1 is larger than $\mathrm{d}C/\mathrm{d}x$ at $(x_1 + \mathrm{d}x)$, $J(x_1)$ will be larger than $J(x_1 + \mathrm{d}x)$, as shown schematically in Fig. 5.5b. Also, because $J(x_1) > J(x_1 + \mathrm{d}x)$, if matter conservation is applied, the solute concentration in the region between x_1 and $x_1 + \mathrm{d}x$ should increase. Considering a volume element with unit area normal to the x axis and $\mathrm{d}x$ in thickness, the rate of change of concentration is given by:

$$\left(\frac{\mathrm{d}C}{\mathrm{d}t}\right)_{x_1}\mathrm{d}x = J(x_1) - J(x_1 + \mathrm{d}x). \tag{5.38}$$

If $\mathrm{d}x$ is small, $J(x_1 + \mathrm{d}x)$ can be related to $J(x_1)$ through the following equation:

$$J(x_1 + \mathrm{d}x) = J(x_1) + \left(\frac{\mathrm{d}J}{\mathrm{d}x}\right)_{x_1}\mathrm{d}x. \tag{5.39}$$

By substituting Eq. (5.39) into Eq. (5.38) and using Eq. (5.36) for J, there is:

$$\frac{\mathrm{d}C}{\mathrm{d}t} = -\frac{\mathrm{d}J}{\mathrm{d}x} = -\frac{\mathrm{d}}{\mathrm{d}x}\left(-D\frac{\mathrm{d}C}{\mathrm{d}x}\right) = D\frac{\mathrm{d}^2C}{\mathrm{d}x^2}. \tag{5.40}$$

Equation (5.37) can be solved with certain boundary conditions that are determined by experiments.

For instance, a common technique to measure diffusion coefficient D is to deposit a very thin layer of a radioactive isotope (or mass isotope) on a flat surface of a thick sample, which is then annealed at a given temperature for a given time duration. By measuring the concentration of the diffusing species as a function of distance, the diffusion coefficient can be determined. In this case, the experimental system is a semi-infinite solid. If the initial thickness of the radioisotope layer is sufficiently small as compared with the diffusing distance of the radioisotope, the solution of Eq. (5.37) is given by:

$$C = \frac{C_0}{(\pi D^* t)^{1/2}} \exp\left(\frac{-x^2}{4D^* t}\right), \tag{5.41}$$

where C is the concentration at a distance x from the surface, t is the annealing time, and C_0 is the initial concentration in moles per unit area of the radioisotope, i.e., at $t = 0$. D^* is the diffusion coefficient of the radioisotope, which is known as the tracer diffusion coefficient.

5.2.3.2 Atomistic Diffusion Processes

Figure 5.6 schematically shows the diffusion of an atom at atomic level, the atom undergoes periodic jumping from one lattice site to another site via an intermediate stage of higher energy, which separates the sites that the atom sits before and after the diffusion process [1]. The energy barrier that must be overcome by the atom at the intermediate state before the jumping takes place is called activation energy. This periodic jumping experienced by the atoms is similar to a Brownian motion, which occurs in a random way over the lattice sites. Therefore, it is also called random diffusion, which can be described similarly to a random-walk problem, in order to determine the relationship between the macroscopic diffusion coefficients and the jump frequencies and jump distances of the atoms.

There are two adjacent planes, A and B, apart with a distance λ, in a crystalline solid that has a concentration gradient along the x axis, as shown schematically in Fig. 5.7 [1]. There are n_A diffusing atoms per unit area in plane A and n_B in plane B. Only he jumps to the left and right, i.e., those causing a change in position along the x axis, are considered. In this case, the probability P that an atom will have sufficiently high energy to overcome the energy barrier q is given by:

$$P = \exp\left(-\frac{q}{kT}\right), \tag{5.42}$$

where k is the Boltzmann constant and T is the absolute temperature. The number of atoms per unit area of plane A that have sufficient energy to overcome the energy barrier from A to B at any instant is $n_A P = n_A \exp(-q/kT)$. If the atoms vibrate about

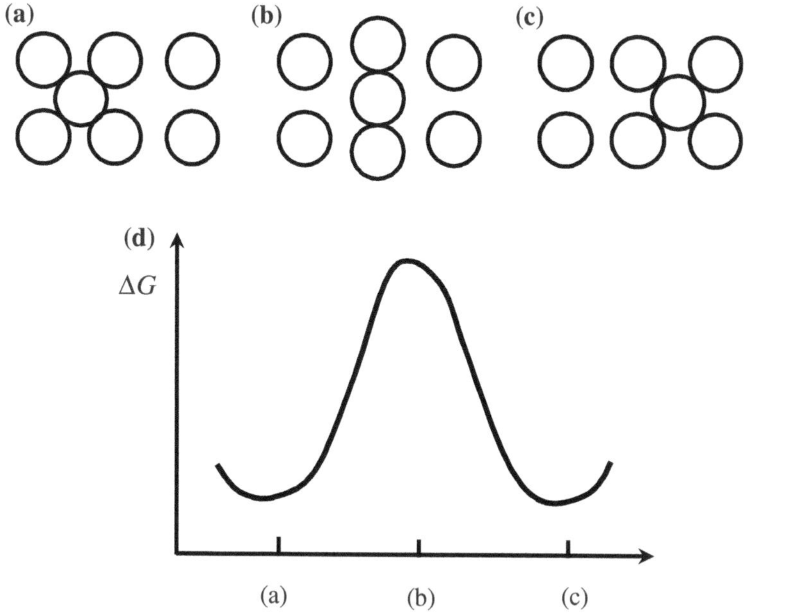

Fig. 5.6 Schematic diagram showing the sequence of configurations when an atom jumps from one lattice site to another (**a**–**c**) and the corresponding change in the free energy of the lattice (**d**). Reproduced with permission from [1]. Copyright © 2003, CRC Press

Fig. 5.7 Planes of atoms with a gradient in their concentration. Reproduced with permission from [1]. Copyright © 2003, CRC Press

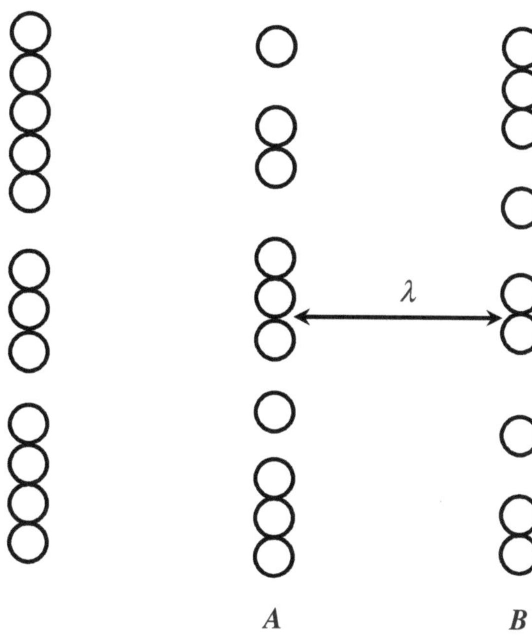

their mean positions at a frequency v, and assuming that the vibration frequency is the same in all six orthogonal directions, the flux of the atoms, i.e., the number jumping across unit area per second, from A to B is $(v/6)n_A\exp(-q/kT)$. Similarly, the flux of atoms from B to A is $(v/6)n_B\exp(-q/kT)$. Therefore, the net flux of atoms from A to B is therefore given by:

$$J = \frac{v}{6}(n_A - n_B)\exp\left(\frac{-q}{kT}\right). \tag{5.43}$$

If λ is the distance between the two adjacent planes, $n_A - n_B$ can be related to their concentrations, which is the number of the atoms per unit volume, with $n_A/\lambda = C_A$ and $n_B/\lambda = C_B$. In this case, the flux is given by:

$$J = \frac{v\lambda}{6}(C_A - C_B)\exp\left(\frac{-q}{kT}\right). \tag{5.44}$$

If it is assumed that the change in C is sufficiently slow, there will be $C_A - C_B = \lambda(\mathrm{d}C/\mathrm{d}x)$, so that Eq. (5.44) becomes:

$$J = \frac{v\lambda^2}{6}\exp\left(\frac{-q}{kT}\right)\frac{\mathrm{d}C}{\mathrm{d}x}. \tag{5.45}$$

This equation is identical to the Fick's first law, with the diffusion coefficient D to be given by:

$$D = \frac{v\lambda^2}{6}\exp\left(\frac{-q}{kT}\right). \tag{5.46}$$

Because q is a too small quantity for most diffusing atoms, larger quantities $Q = N_A q$ and $R = N_A k$ are usually used, where Q is the activation energy per mole, N_A is the Avogadro number, and R is the gas constant. Additionally, the term $v\lambda^2/6$ is usually represented by D_0, so that there is:

$$D = D_0\left(\frac{-Q}{RT}\right). \tag{5.47}$$

In this equation, D has an exponential dependence on temperature, which also has activation energy in the units of J mol^{-1}. Data for the diffusion coefficients of various ceramics can be found in open literatures and handbooks.

Note that the activation energy q is equal to $\Delta h - T\Delta s$, where Δh is the enthalpy and Δs is the entropy for atomic diffusion, Eq. (5.46) then becomes:

$$D = \frac{v\lambda^2}{6}\exp\left(\frac{\Delta s}{k}\right)\exp\left(\frac{-\Delta h}{kT}\right), \tag{5.48}$$

which is compared with Eq. (5.47), leading to the following expression:

$$D = \frac{v\lambda^2}{6} \exp\left(\frac{\Delta s}{k}\right). \tag{5.49}$$

Generally, the vibration frequency of atoms about their mean positions, which is usually known as the Debye frequency, is about 10^{13}–10^{14} s^{-1} and $\lambda \approx 0.2$ nm, so $v\lambda^2/6 \approx 10^{-7}$ $m^2\ s^{-1}$. The values of D_0 for diffusion by the interstitial and vacancy mechanisms are in the range of 10^{-7}–10^{-3} $m^2\ s^{-1}$ [1]. In addition, it has been reported that, for a given group of materials, the values of D_0 and Q/RT_m, with T_m being the melting temperature, are almost the same [19].

5.2.3.3 Diffusion Mechanisms

The types of defects determine the path of matter transport, while the diffusion along the major paths is responsible for the major mechanisms of matter transport. There are three main diffusion types: (i) lattice diffusion (also called volume or bulk diffusion), (ii) grain boundary diffusion, and (iii) surface diffusion.

Lattice Diffusion

Lattice diffusion is related to the movement of point defects over the lattice of crystal solids. Lattice diffusion has vacancy mechanism or interstitial mechanism, depending on the type of defects, vacancy or interstitial, which are most important lattice diffusion mechanisms.

Figure 5.8 shows an example of diffusion by vacancy mechanism, where an atom at a normal lattice site diffuses by exchanging its position with a vacant site [1]. The movement direction of the atom is opposite to that of the vacancy, so that the diffusion of the atom can be tracked, so is the diffusion of the vacancy. Although the diffusion coefficients of the atoms and the vacancies are closely related, they are not necessarily equal to each other. This is because an atom can only jump if a vacancy is located at a lattice site adjacent to it, whereas a vacancy can jump to any of the nearest neighbor sites. As a result, the number of atomic jumps is proportional to the fraction of the sites occupied by vacancies, C_v. The atomic diffusion coefficient D_a and the vacancy diffusion coefficient D_v are related by the following equation:

$$D_a = C_v D_v. \tag{5.50}$$

As discussed above, the concentration of vacancy is determined by temperature, atmosphere and dopants. It is worth mentioning that the flux of vacancies must be compensated by an equal and opposite flux of atoms for vacancy diffusion. Otherwise, the vacancies will be accumulates and form pores in the crystal solids.

In fact, pores can be formed during the interdiffusion of two atoms that have significantly different diffusion coefficients.

Sufficiently small dopants or regular atoms can take interstitial sites of the lattice. When they move to neighboring interstitial sites, interstitial diffusion takes place, as shown schematically in Fig. 5.8b, which can be described by the following equation:

$$D_a = C_i D_i, \tag{5.51}$$

where D_i is now the interstitial diffusion coefficient and C_i is the concentration of the interstitial atoms. This is similar to that of Eq. (5.50).

When the lattice is heavily distorted, interstitial diffusion becomes difficult, so that another diffusion mechanism is present, which is known as interstitialcy mechanism. Interstitialcy diffusion is facilitated by exchanging position between an atom at the regular lattice site with a neighboring interstitial atom, as shown schematically in Fig. 5.8. They can be different types of atoms.

There is another type of lattice diffusion mechanism, called direct exchange or ring mechanism. As shown schematically in Fig. 5.8d, during this diffusion, atoms exchange their positions through rotation in a circle way. There are no any defects involved in this diffusion. This is a multiatomic action. However, this mechanism is

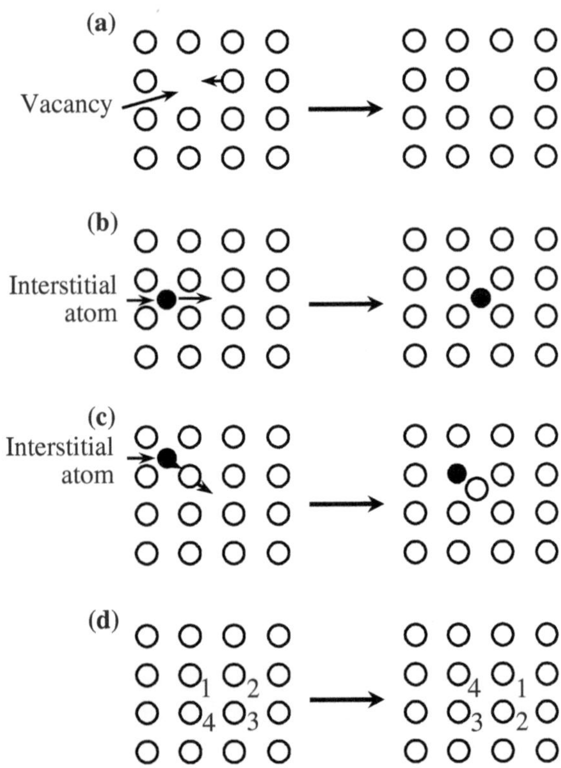

Fig. 5.8 Lattice diffusion by **a** vacancy mechanism, **b** interstitial mechanism, **c** interstitialcy mechanism, and **d** ring mechanism. Reproduced with permission from [1]. Copyright © 2003, CRC Press

rarely observed in ionic crystalline solids, due to the tremendous momentary distortion coupled with the large energy changes arising from the electrostatic repulsion.

Grain Boundary Diffusion

The presence of grain boundaries is one of the most distinct characteristics of polycrystalline ceramics. Grain boundaries are regions that separate the crystals or grains from one another, with lattice mismatching and disordering. Due to the highly defective nature of the grain boundaries, they are expected to have faster diffusion than lattice diffusion in the adjacent grains. Grain boundaries usually have widths of 0.5–1 nm. If the width of grain boundaries is assumed to be a constant, the volume fraction of the grain boundaries will increase with decreasing grain size. Therefore, the rate of grain boundary diffusion has been related to grain size.

Surface Diffusion

As discussed before, the free surface of a crystalline solid is not a perfectly flat plane, which could contain vacancies, terraces, kinks, edges, and adatoms. The migration of vacancies and the movement of adatoms facilitate the mechanisms of surface diffusion. The diffusion process is usually confined to a thin layer near the surface with a thickness of 0.5–1 nm.

Comparison of Diffusion Coefficients

The atoms on the surface are less tightly bound, because they have fewer neighbors than those within the bulk of the lattice. As a result, the activation energy for surface diffusion is less than that for lattice diffusion, which has been confirmed by the available experimental data. Due to its lower activation energy, surface diffusion becomes more and more important as the temperature is decreased, which has a significant effect on sintering. It has often been demonstrated that the diffusion coefficients of lattice diffusion D_l, grain boundary diffusion D_{gb}, and surface diffusion D_s increase in the order of $D_l < D_{gb} < D_s$, while the corresponding activation energies decrease in the order of $Q_l > Q_{gb} > Q_s$. However, these relations are only used as a general trend, which may not be correct for all cases or any case.

5.2.3.4 Types of Diffusion Coefficients

In the literature, there are diffusion coefficients that have been used to describe the diffusion characteristics of a particular species, such as atom, interstitial, or vacancy, a particular diffusion path, such as lattice, grain boundary, or surface diffusion, or a particular process, such as chemical diffusion or ambipolar diffusion.

There are various diffusion coefficients: (i) self-diffusion coefficient, (ii) tracer diffusion coefficient, (iii) lattice diffusion coefficient, (iv) grain boundary diffusion coefficient, (v) surface diffusion coefficient, (vi) defect diffusion coefficients, and (vii) chemical, effective or interdiffusion coefficient.

Self-diffusion coefficient D_{self} is the diffusion coefficient of the host atoms or ions in random diffusion, which is one of the important diffusion coefficients related to sintering. For lattice diffusion by the vacancy or interstitial mechanisms, D_{self} is given by Eq. (5.50) or (5.51).

Tracer diffusion coefficient D_{self} is usually difficult to directly measure. As stated earlier, it is more convenient to measure the diffusion coefficient of a radioactive or mass isotope. This measured coefficient is called the tracer diffusion D^*, which is close to but not the same as D_{self}, because the motion of the tracer atoms is not in a completely random manner. Successive jumps are mutually correlated, so that every jump is dependent on the previous one. Therefore, D_{self} and D^* can be related by the following equation:

$$D^* = fD_{self}, \tag{5.52}$$

where f is a correlation factor that depends on the crystal structure and the diffusion mechanism. The values for f are in the range 0.6–1.

Lattice diffusion coefficient D_l represents any diffusion process over the lattice of crystal solids. Grain boundary diffusion coefficient, D_{gb} or D_b, describes the diffusions in the grain boundaries. Surface diffusion coefficient D_s is to characterize the diffusions over a free surface. Defect diffusion coefficient is used to describe the diffusion of a particular type of defect, such as D_v for diffusion coefficient of a vacancy and D_i for diffusion coefficient of an interstitial. The diffusion coefficient of a defect is given by Eq. (5.46) or (5.48), which is independent of its concentration. Chemical diffusion coefficient is also known as effective or interdiffusion coefficient. This coefficient $\tilde{D}$ is for the diffusion due to the presence of a gradient in chemical composition or chemical potential, simply because ions will migrate to response any chemical gradient. Interdiffusion takes place when two components of a solid solution migrate in opposite directions down their coupled chemical potential gradients.

5.2.4 Chemical Potential

As discussed previously, matter transport is due to the flux of atoms or vacancies driven by gradients in the concentration, which can be described by using the Fick's first law. This special case of mass transport is not applicable to those with other types of driving forces, such as gradients in pressure, electric potential, and so on. To address this issue, it is necessary to use chemical potential, instead of concentration gradients, as driving force of the diffusions. Definition and description of chemical potential can be found in various textbooks.

5.2.4.1 Chemical Potential

For a phase with a given amount of mass and composition, at variable temperature T and pressure p, as an infinitesimal reversible process occurs, the change in the Gibbs free energy is given by:

$$\mathrm{d}G = \left(\frac{\partial G}{\partial T}\right)_p \mathrm{d}T + \left(\frac{\partial G}{\partial p}\right)_T \mathrm{d}p = -S\mathrm{d}T + V\mathrm{d}p, \tag{5.53}$$

where S is the entropy and V the volume of the system. For a phase of variable compositions, consisting of m chemical constituents, in which there are n_1 mol of the substance A_1, n_2 mol of A_2, …, n_m moles of A_m, the change in the Gibbs free energy is now given by:

$$\begin{aligned}\mathrm{d}G = & \left(\frac{\partial G}{\partial T}\right)_{p,n_1,n_2,\ldots,n_\mathrm{m}} \mathrm{d}T + \left(\frac{\partial G}{\partial p}\right)_{T,n_1,n_2,\ldots,n_\mathrm{m}} \mathrm{d}p \\ & + \left(\frac{\partial G}{\partial n_1}\right)_{p,T,n_2,n_3,\ldots,n_\mathrm{m}} \mathrm{d}n_1 + \cdots + \left(\frac{\partial G}{\partial n_\mathrm{m}}\right)_{p,T,n_2,n_3,\ldots,n_{m-1}} \mathrm{d}n_\mathrm{m}.\end{aligned} \tag{5.54}$$

Because the first two terms on the right-hand side of the equation are at constant mass and composition, Eq. (5.53) can be used. In this case, when a small amount of one constituent, e.g., $\mathrm{d}n_\mathrm{k}$ moles of the kth constituent, is introduced into the phase, with T, p, and the other n's remaining constant, the effect on the Gibbs free energy can be expressed as:

$$\mu_\mathrm{k} = \left(\frac{\partial G}{\partial n_\mathrm{k}}\right)_{p,T,n_1,n_2,\ldots,n_{k-1}n_{k+1},\ldots,n_\mathrm{m}}, \tag{5.55}$$

where μ_k is called the chemical potential of the kth constituent. Equation (5.54) can now be rewritten as:

$$\mathrm{d}G = -S\mathrm{d}T + V\mathrm{d}p + \sum_\mathrm{i}^{m} \mu_\mathrm{i} \mathrm{d}n_\mathrm{i}. \tag{5.56}$$

If the number of moles of a phase is increased, while T, p, and the composition are kept unchanged, Eq. (5.56) becomes:

$$\mathrm{d}G_\mathrm{T,p} = \sum_\mathrm{i} \mu_\mathrm{i} \mathrm{d}n_\mathrm{i}. \tag{5.57}$$

Because the μ_i is dependent only on T, p, and composition and they must be kept to be constant, Eq. (5.57) can be integrated to yield:

$$G = \sum_i \mu_i n_i. \tag{5.58}$$

For a pure substance, Eq. (5.58) can be reduced to:

$$G = \mu n, \tag{5.59}$$

which means that the chemical potential is the Gibbs free energy per mole at the given temperature and pressure.

5.2.4.2 Chemical Potential of Gas Mixtures

For 1 mol ideal gas at constant temperature T, there is:

$$\left(\frac{\partial T}{\partial p}\right)_T = V = \frac{RT}{p}, \tag{5.60}$$

where R is the gas constant and p is the pressure.

Integration of Eq. (5.60) yields:

$$G(T,p) = G_0(T) + RT \ln p, \tag{5.61}$$

where G_0 is the Gibbs free energy at standard pressure at the temperature, which is used as a reference to calculate Gibbs free energy change. The general expression for chemical potential now is given by:

$$\mu(T,p) = \mu_0(T) + RT \ln p. \tag{5.62}$$

For a mixture of ideal gases at constant temperature T and a constant total pressure, following Eq. (5.62), we can write for each component:

$$\mu_i(T,p) = \mu_{0,i}(T) + RT \ln p_i, \tag{5.63}$$

where p_i is the partial pressure of the ith component. To be more useful, the Gibbs free energy or the chemical potential of a particular component is usually related to its concentration, which is defined as:

$$C_i = \frac{n_i}{\sum n_i} = \frac{p_i}{p}, \tag{5.64}$$

where n_i is the number of moles of each component in the mixture and p is the total gas pressure of the system. Equation (5.63) can now be rewritten as:

$$\mu_i(T,p,C_i) = \mu_{0,i}(T,p) + RT \ln C_i, \quad (5.65)$$

Real gases have deviation from the behavior of ideal gas. If the deviation is sufficiently small, Eqs. (5.63)–(5.65) can be used for satisfactory estimations.

5.2.4.3 Chemical Potential of Solids and Liquids

The chemical potential of solid and liquid solutions can be defined as an expression similar to Eq. (5.63), in which the p_i is replaced by a new term a_i, called the activity. In this case, the chemical potential is given by:

$$\mu_i = \mu_{0,i} + RT \ln a_i. \quad (5.66)$$

The activity of pure liquids and solids, under some specified standard conditions of temperature and pressure, is considered to be unity. The activity of a component in mixture systems is expressed as:

$$a_i = \alpha_i C_i, \quad (5.67)$$

where α_i is called the activity coefficient and C_i is the concentration of the component, which is usually described as mole fraction. The chemical potential of the ith species in liquid or solid solutions is therefore given by:

$$\mu_i = \mu_{0,i} + RT \ln(\alpha_i C_i). \quad (5.68)$$

For ideal solutions, $\alpha_i = 1$.

5.2.4.4 Chemical Potential of Atoms and Vacancies in Crystals

It is assumed that vacancies are the only defects in a crystal of a pure element. If the numbers of the atoms and vacancies are N_a and n_v, the total number of lattice sites in the crystal is $N = N_a + n_v$. As a result, the Gibbs free energy of the crystal can be expressed as:

$$G = U + n_v g + pV - TS, \quad (5.69)$$

where U is the internal energy of the crystal and g is the energy required form the vacancy. The configurational entropy of the crystal can be obtained by using the Boltzmann relation, which in this case can be written as:

$$S = k \ln\left[\frac{(N_a + n_v)!}{N_a! n_v!}\right], \quad (5.70)$$

where k is the Boltzmann constant. The chemical potential of the atoms is defined as:

$$\mu_{\mathrm{a}} = \left(\frac{\partial G}{\partial N_{\mathrm{a}}}\right)_{T,p,n_{\mathrm{v}}}. \tag{5.71}$$

The factorials in Eq. (5.70) can be simplified by using the Stirling approximation, which is then applied to Eq. (5.71), leading to the following equation:

$$\mu_{\mathrm{a}} = \mu_{0,\mathrm{a}} + p\Omega_{\mathrm{a}} + k\ln\left(\frac{N_{\mathrm{a}}}{N_{\mathrm{a}} + n_{\mathrm{v}}}\right), \tag{5.72}$$

where $\mu_{0,\mathrm{a}}$ is a reference value or standard value and Ω_{a} is the volume of an atom, which is defined as:

$$\Omega_{\mathrm{a}} = \left(\frac{\partial V}{\partial N_{\mathrm{a}}}\right)_{T,p,n_{\mathrm{v}}}. \tag{5.73}$$

If C_{a} is the fraction of lattice sites occupied by the atoms in the crystal, then Eq. (5.71) can be written as:

$$\mu_{\mathrm{a}} = \mu_{0,\mathrm{a}} + p\Omega_{\mathrm{a}} + k\ln C_{\mathrm{a}}. \tag{5.74}$$

This equation means that the chemical potential of atoms in a crystal is dependent on the pressure and the atomic concentration. It is also noted that if the concentration of the vacancy is sufficiently low, the last term on the right-hand side of Eqs. (5.72) and (5.74) can be neglected.

The chemical potential of the vacancies is defined as:

$$\mu_{\mathrm{v}} = \left(\frac{\partial G}{\partial n_{\mathrm{v}}}\right)_{T,p,N_{\mathrm{a}}}. \tag{5.75}$$

Similar to the procedure used for μ_{a}, the chemical potential of the vacancies is given by:

$$\mu_{\mathrm{v}} = \mu_{0,\mathrm{v}} + p\Omega_{\mathrm{v}} + k\ln C_{\mathrm{v}}, \tag{5.76}$$

where C_{v} is the concentration of the vacancy, which is the fraction of lattice sites occupied by the vacancies. Although the volume of a vacancy Ω_{v}, can be different from the volume of an atom Ω_{a}, the rigid lattice approximation can be used to assume that $\Omega_{\mathrm{a}} = \Omega_{\mathrm{v}} = \Omega$, which makes it more convenient for further analysis.

5.2.4.5 Curved Surface

Chemical potentials of the atoms and vacancies beneath curved surfaces are different when the curvatures of the surfaces are different. This difference in chemical potential provides a driving force for the diffusional flux of atoms, in order to reduce the free energy of the system. The understanding of the dependence of chemical potential of atoms on curvature is especially important to the understanding of sintering.

For a pure element solid with adjoining convex and concave surfaces, it is assumed that vacancies are the only type of point defects in the solid. The surface area of the convex surface will be decreased, if the volume of the region beneath it is reduced, which can be achieved by reducing the concentration of the vacancies. Because a decrease in surface area means a decrease in contribution of the surface to the total free energy of the system, the concentration of the vacancy under convex surface is lower than that beneath a flat surface. Similarly, the concentration of the vacancy beneath the concave surface is higher than that under a flat surface. Therefore, the difference in the concentration of the vacancy results in a diffusional flux of vacancies from the concave region to the convex region, while a diffusional flux of atoms from the convex region to the concave region will be observed accordingly.

On a smoothly curved surface, when an infinitesimal hump is creased on the surface by taking atoms from beneath the surface and adding them to the surface, the change in surface free energy is approximately given by:

$$\delta\left(\int \gamma_{SV} \mathrm{d}S\right) = \int \delta(\gamma_{SV}) \mathrm{d}S_0 + \int \gamma_{SV} \delta(\mathrm{d}S), \tag{5.77}$$

where δ means a small change in a quantity, γ_{SV} is the specific surface energy, and dS is the change in surface area. It is assumed that the surface is uniform and isotropic, so that the first term on the right-hand side of Eq. (5.77) is zero, while the term δ(dS) is given by:

$$\delta(\mathrm{d}S) = (\sec\psi - 1)\mathrm{d}S_0 + \delta z\left(\frac{1}{R_1} + \frac{1}{R_2}\right)\mathrm{d}S_0, \tag{5.78}$$

where R_1 and R_2 are the two principal radii of curvature at dS_0. If Δy is small, R_1 and R_2 can be considered to be constant over the hump. The first term on the right-hand side of Eq. (5.78) can be neglected. When it is put into Eq. (5.77), the following equation can be obtained:

$$\delta\left(\int \gamma_{SV} \mathrm{d}S\right) = \gamma_{SV}\left(\frac{1}{R_1} + \frac{1}{R_2}\right)\delta_v, \tag{5.79}$$

where δ_v is the volume of the hump.

The change in the volume term of the free energy due to the creation of the small hump is given by:

$$\delta G_{\mathrm{v}} = -p\delta_{\mathrm{v}} + \mu_{\mathrm{v}}\frac{\delta_{\mathrm{v}}}{\Omega}, \tag{5.80}$$

where p is the mean hydrostatic pressure in the crystal just beneath the surface, μ_{v} is the chemical potential of a vacancy, and $\delta_{\mathrm{v}}/\Omega$ represents the number of vacancies created by the transfer of atoms to the hump. In equilibrium, the sum of the energy changes defined by Eqs. (5.79) and (5.80) should be zero, so that the following equation is derived:

$$\mu_{\mathrm{v}} = p\Omega + \gamma_{\mathrm{SV}}\left(\frac{1}{R_1} + \frac{1}{R_2}\right)\Omega. \tag{5.81}$$

If the curvature K of the surface is defined as:

$$K = \frac{1}{R_1} + \frac{1}{R_2}, \tag{5.82}$$

where $K > 0$ for a convex surface. Putting K into Eq. (5.81) yields:

$$\mu_{\mathrm{v}} = (p + \gamma_{\mathrm{SV}}K)\Omega. \tag{5.83}$$

Generally, chemical potential is measured relative to a reference value $\mu_{0,\mathrm{v}}$, and μ_{v} contains a term of vacancy concentration. The chemical potential of vacancy μ_{v}, with curvature, pressure and concentration, is given by:

$$\mu_{\mathrm{v}} = \mu_{0,\mathrm{v}} + (p + \gamma_{\mathrm{SV}}K)\Omega + kT\ln C_{\mathrm{v}}. \tag{5.84}$$

Similarly, the chemical potential of the atoms can be expresses by:

$$\mu_{\mathrm{a}} = \mu_{0,\mathrm{a}} + (p + \gamma_{\mathrm{SV}}K)\Omega + kT\ln C_{\mathrm{a}}. \tag{5.85}$$

Because C_{a} is usually is very small, the last term can be taken as zero.

In Eqs. (5.84) and (5.85), because C_{a} and C_{v} are generally very low, the last terms can be neglected. Therefore, μ_{a} and μ_{v} are determined essentially by the hydrostatic pressure in the solid and the curvature of the surface. Since the curvature term $\gamma_{\mathrm{SV}}K$ has the same units as pressure or stress, the curvature, and applied pressure effects can be treated with same formulation for the analysis of sintering.

5.2.5 Diffusional Flux Equations

For theoretical analysis of sintering, it is necessary to establish the equations for diffusional mass transport. These equations can be solved when subject to appropriate boundary conditions. There are two equivalent formulations for sintering: diffusion of atoms or diffusion of vacancies.

In an elemental solid, if the influence of the flux of the neutral atoms can be neglected, the flux of the atoms in one dimension can be expressed as:

$$J_x = -L_{ii}\frac{d\mu}{dx}, \tag{5.86}$$

where the coefficients L_{ii} are called transport coefficients and μ is the chemical potential. By comparing this equation with the Fick's first law, i.e., Equation (5.36), there is:

$$L_{ii} = D\frac{dC}{d\mu}. \tag{5.87}$$

According to the relation between chemical potential and concentration given in Eq. (5.68), there is:

$$\mu = \mu_0 + kT\ln(\alpha C), \tag{5.88}$$

where α is the activity coefficient. For this equation, $dC/d\mu$ can be derived, which is put into Eq. (5.87), there is:

$$L_{ii} = \frac{D_i C}{kT}, \tag{5.89}$$

where D_i is given by:

$$D_i = D\left(1 + \frac{d\ln\alpha}{d\ln C}\right)^{-1}. \tag{5.90}$$

For an ideal system, α is independent of concentration, i.e., ln α does not vary with ln C, so that the second term in the brackets in Eq. (5.90) is zero. By putting Eq. (5.89) into Eq. (5.86), the atomic flux equation can be derived as follows:

$$J_x = -\frac{D_i C}{kT}\frac{d\mu}{dx}, \tag{5.91}$$

where D_i is given by Eq. (5.90).

5.2.5.1 Atomic Flux

In a pure elemental solid, if the point defects are only vacancies, the total number of lattice sites will not be changed, when the atoms or vacancies diffuse from one region to another. Within a given region, the changes in number of atoms and the number of vacancies are equal and in opposite direction. The diffusional flux is dependent on the gradients, $\mu_a - \mu_v$, which follows Eq. (5.91), given by:

$$J_a = -\frac{D_a C_a}{\Omega kT}\frac{d(\mu_a - \mu_v)}{dx}, \tag{5.92}$$

where D_a is the atomic self-diffusion coefficient, k is the Boltzmann constant and C_a is the fraction of lattice sites occupied by the atoms. With appropriate boundary conditions, the diffusion response can be derived from Eq. (5.92).

5.2.5.2 Vacancy Flux

Because the atomic flux in sintering is equal and opposite to the vacancy flux, there is:

$$J_a = -J_v, \tag{5.93}$$

where J_v is given by:

$$J_v = -\frac{D_v C_v}{\Omega kT}\frac{d\mu_v}{dx} = -\frac{D_v}{\Omega}\frac{dC_v}{dx}, \tag{5.94}$$

where D_v is the vacancy diffusion coefficient. To determine J_v, it is necessary to have an expression of C_v. If C_v is the equilibrium concentration, there is:

$$\left(\frac{\partial G}{\partial n_v}\right)_{T,p,N_a} = 0, \tag{5.95}$$

so that $\mu_v = 0$. It is assumed that no externally applied pressure is applied, by putting $\mu_v = 0$ into Eq. (5.84), there is:

$$C_v = C_{0,v}\exp\left(-\frac{\gamma_{SV}K\Omega}{kT}\right), \tag{5.96}$$

where $C_{0,v}$ is a reference value, which the vacancy concentration under a flat surface. If $\gamma_{SV}K\Omega \ll kT$, Eq. (5.82) becomes:

$$C_v = C_{0,v}\left(1 - \frac{\gamma_{SV}K\Omega}{kT}\right). \tag{5.97}$$

Under the equilibrium condition, there is:

$$J_a = \frac{D_v}{\Omega}\frac{dC_v}{dx}, \tag{5.98}$$

where C_v is given by Eq. (5.97), which is a fraction of the lattice sites that are occupied by the vacancies.

5.2.6 Vapor Pressure of Curved Surfaces

During sintering process, there is an important matter transport through evaporation and condensation. The rate of the transport is proportional to the equilibrium vapor pressure over the surface, which has been related to the value of $C_a - \mu_v$ beneath the surface. If a number dN_a of atoms is taken away from the vapor and added to the surface, with a corresponding decrease in the number of vacancies beneath the surface. The free energy change due to this virtual operation must be zero, so that there is:

$$\mu_{vap} = \mu_a - \mu_v, \tag{5.99}$$

where μ_{vap} is the chemical potential of the atoms in the vapor phase. The vapor pressure is proportional to $\exp(\mu_{vap}/kT)$, so that there is:

$$p_{vap} = p_0 \exp\left(\frac{\mu_a - \mu_v - \mu_0}{kT}\right), \tag{5.100}$$

where p_0 is a reference value of the vapor pressure, corresponding to the standard value of chemical potential, μ_0, which is generally the value over a flat surface. According to Eqs. (5.84), (5.85) and (5.96), there is:

$$\mu_a - \mu_v = \mu_0 + \gamma_{SV} K\Omega. \tag{5.101}$$

When Eq. (5.101) is put into Eq. (5.100), there is:

$$p_{vap} = p_0 \exp\left(\frac{\gamma_{SV} K\Omega}{kT}\right), \tag{5.102}$$

This is called the Kelvin equation. If $\gamma_{SV} K\Omega \ll kT$, it becomes:

$$p_{vap} = p_0 \left(1 + \frac{\gamma_{SV} K\Omega}{kT}\right). \tag{5.103}$$

This equation means that, the vapor pressure of a given system under isothermal conditions increases with increasing curvature of the surface.

5.2.7 Diffusion in Ionic Crystals—Ambipolar Diffusion

In above discussion, all the diffusing species have been assumed to be uncharged atoms or vacancies, so that electrostatic effects on the diffusion are not involved, while in practical polycrystalline ceramics, the motion of charged species, such as ions, has significant contribution to matter transport. It is also expected that different ions will have different diffusion rates. When a matter transport occurs from a given source to a given sink, the stoichiometry and electroneutrality of the solid must be preserved in the different regions of the solid, as shown in Fig. 5.9 [1]. Other effects may also have influences on the diffusion. For instance, if the system is subject to an external electric field, the ions will migrate in response to the electric field, besides the diffusion along the concentration gradient. In fact, even no external field is applied, external fields could be formed by the ions themselves, which in turn can influence their motion.

A diffusing species can have a charge z_i, e.g., $z_i = +2$ for a doubly charged magnesium ion, $z_i = +3$ for a triply charged aluminum ion, and $z_i = -2$ for a doubly charged oxygen ion, and so on. In a region with an electric potential of ϕ, the chemical potential of an ion is increased by an amount of $z_i e\phi$, where e is the magnitude of the electron charge. Without the presence of an externally pressure, according to Eq. (5.74), the chemical potential of an ion is given by:

$$\mu_i = \mu_{0,i} + kT \ln C_i + z_i e\phi, \tag{5.104}$$

where C_i is the fraction of sites occupied by the ions in the crystal. When this equation is differentiated with respect to x and is put into Eq. (5.91), the flux of the ions is given by:

$$J_{ix} = -D_i \frac{dC_i}{dx} - B_i C_i z_i eE, \tag{5.105}$$

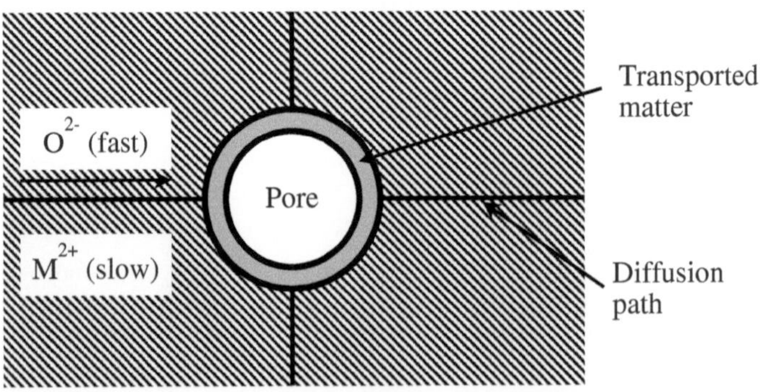

Fig. 5.9 Schematic diagram illustrating that the diffusion of ions in an ionic solid must be coupled to preserve the stoichiometry and electroneutrality of the solid. Reproduced with permission from [1]. Copyright © 2003, CRC Press

where D_i is the diffusion coefficient of the ions, $E = \mathrm{d}\phi/\mathrm{d}x$ is the electric field strength, and $B_i = D_i/kT$ is the ionic mobility. The first term on the right-hand side of Eq. (5.105) is the diffusion term given by the Fick's first law, while the second term is due to the ion migration caused by the electric field.

Equation (5.105) can be applied to a system, which has two different types of diffusing ions. One type has a positive charge z_+ and the other type has a negative charge z_-, corresponding to metal ions and oxygen ions in a metal oxide respectively. If no net current flows through the system, the electrical current density should be zero, so that there is:

$$z_+J_+ = -z_-J_-. \tag{5.106}$$

By considering J_+ and J_-, according to Eq. (5.105), there is:

$$-\left(z_+D_+\frac{\mathrm{d}C_+}{\mathrm{d}x} + B_+C_+z_+^2eE\right) = z_-D_-\frac{\mathrm{d}C_-}{\mathrm{d}x} + B_-C_-z_-^2eE. \tag{5.107}$$

Rearranging Eq. (5.107) leads to an equation for E, given by:

$$E = -\frac{1}{e(B_+C_+z_+^2 + B_-C_-z_-^2)}\left(z_+D_+\frac{\mathrm{d}C_+}{\mathrm{d}x} + z_-D_-\frac{\mathrm{d}C_-}{\mathrm{d}x}\right). \tag{5.108}$$

According to electroneutrality, there is:

$$C_+z_+ = -C_-z_-. \tag{5.109}$$

By differentiating Eq. (5.109) and multiplying by D_+, there will be:

$$z_+D_+\frac{\mathrm{d}C_+}{\mathrm{d}x} = -z_-D_+\frac{\mathrm{d}C_-}{\mathrm{d}x}. \tag{5.110}$$

When Eq. (5.110) is put into Eq. (5.108), there is:

$$E = -\frac{1}{e(-B_+z_+C_-z_- + B_-z_-C_-z_-)}(D_- - D_+)\frac{\mathrm{d}C_-}{\mathrm{d}x}. \tag{5.111}$$

When Eq. (5.111) for E is put into Eq. (5.105), the flux of the negative ions is given by:

$$J_- = -\frac{B_-z_-D_+ - B_+z_+D_-}{B_-z_- - B_+z_+}\frac{\mathrm{d}C_-}{\mathrm{d}x}. \tag{5.112}$$

By replacing the ionic mobility B_i with D_i/kT, Eq. (5.112) becomes:

$$J_- = -\frac{D_+ D_- (z_+ - z_-)}{D_+ z_+ - D_- z_-} \frac{dC_-}{dx}. \tag{5.113}$$

When Eqs. (5.106) and (5.109) are put into Eq. (5.113), there is:

$$J_+ = -\frac{D_+ D_- (z_+ - z_-)}{D_+ z_+ - D_- z_-} \frac{dC_+}{dx}. \tag{5.113}$$

By analogy with the Fick's first law, an effective diffusion coefficient can be defined, which is given by:

$$\tilde{D} = \frac{D_+ D_- (z_+ - z_-)}{D_+ z_+ - D_- z_-}. \tag{5.115}$$

If all ions are fully ionized in Al_2O_3, the diffusing species are Al^{3+} and O^{2-}, so that the effective diffusion coefficient is given by:

$$\tilde{D} = \frac{5 D_{Al^{3+}} D_{O^{2-}}}{3 D_{Al^{3+}} + 2 D_{O^{2-}}}. \tag{5.116}$$

Now, if $D_{O^{2-}} \gg D_{Al^{3+}}$, there is:

$$\tilde{D} = \frac{5}{2} D_{Al^{3+}}, \tag{5.117}$$

while if $D_{Al^{3+}} \gg D_{O^{2-}}$, there is:

$$\tilde{D} = \frac{5}{3} D_{O^{2-}}. \tag{5.118}$$

Equations (5.117) and (5.118) imply that the rate of matter transport is determined by the ion that diffuses more slowly, while the motion of the slower ion is accelerated by faster diffusing ions. Therefore, the concentration gradient of the faster diffusing ion is decreased faster than that of the slower diffusing ion. Generally, only a small amount of such diffusion can build up a large potential gradient. The potential gradient has the same sign as the concentration gradient, so that it decreases the transport rate of the faster ions and increases the transport rate of the slower ions. The potential occurs is set up at the point, where the two fluxes are related by Eq. (5.106). The coupled diffusion of charged species is also called ambipolar diffusion. It has played important roles in determining the mass transport process during the sintering, as well as other mass transport processes, such as creep behaviors of materials.

On the one hand, the slower or slowest diffusing species determine the rate of matter transport. On the other hand, if each ion has multiple diffusion paths, e.g., lattice diffusion and grain boundary diffusion, matter transport will take place through the fastest path. Therefore, the rate-controlling mechanism is determined by the slowest diffusing species along the fastest path. Other factors include the differences in the effective area and path length for mass transport, which should be considered when evaluating the rate-controlling mechanism of a diffusion process.

If the lattice and grain boundary coefficients are additive, for a pure oxide M_xO_y, the effective or ambipolar diffusion coefficient is given by [20–23]:

$$\tilde{D} = \frac{(x+y)(D_l^M + \pi\delta_{gb}D_{gb}^M/G)(D_l^O + \pi\delta_{gb}D_{gb}^O/G)}{y(D_l^M + \pi\delta_{gb}D_{gb}^M/G) + x(D_l^O + \pi\delta_{gb}D_{gb}^O/G)}, \tag{5.119}$$

where δ_{gb} is the width of the grain boundary and G is the grain size. Both the effective lattice and grain boundary diffusion coefficients for one of the species are included. Equation (5.119) indicates that the slowest effective diffusion coefficient is the rate-controlling factor, whereas within each term, the faster one of the two paths dominates, i.e., the rate-controlling mechanism is determined by the slowest diffusing species coupled with its fastest path.

5.3 Solid-State and Viscous Sintering

5.3.1 Brief Description

Sintering processes can be classified into four groups: (i) solid-state sintering, (ii) liquid-phase sintering, (iii) viscous sintering, and (iv) vitrification. Sintering phenomena in polycrystalline materials are much more complex than those in viscous sintering of amorphous materials, because of the possibility of matter transport paths and the presence of grain boundaries. Matter transport in solid-state sintering can occur through at least six different paths, which determine the mechanisms of sintering. In practice, more than one mechanism could be present at any stage of sintering. Therefore, analysis of sintering rates and the determination of the sintering mechanisms becomes more difficult. The most important occurrences related to the grain boundaries are grain growth and pore growth during sintering, which is as coarsening. The coarsening process provides an alternative route to facilitate the reduction in free energy of the powder system. As a consequence, it reduces the driving force for densification. Therefore, there is a competition between sintering (densification) and coarsening.

5.3.2 Sintering Mechanisms

Diffusional transport of matter along specific paths facilitates the sintering of polycrystalline ceramics, through corresponding mechanisms. Matter is transported from the regions of higher chemical potential to the regions of lower chemical potential, with the former and latter known as the source and the sink of matter. There are at least six different mechanisms that govern the sintering of polycrystalline ceramics. Figure 5.10 shows potential mechanisms for the sintering of three spherical particles [24]. The consequence of the sintering is the bonding and growth of the necks between adjacent particles, leading to strengthening of the powder compact.

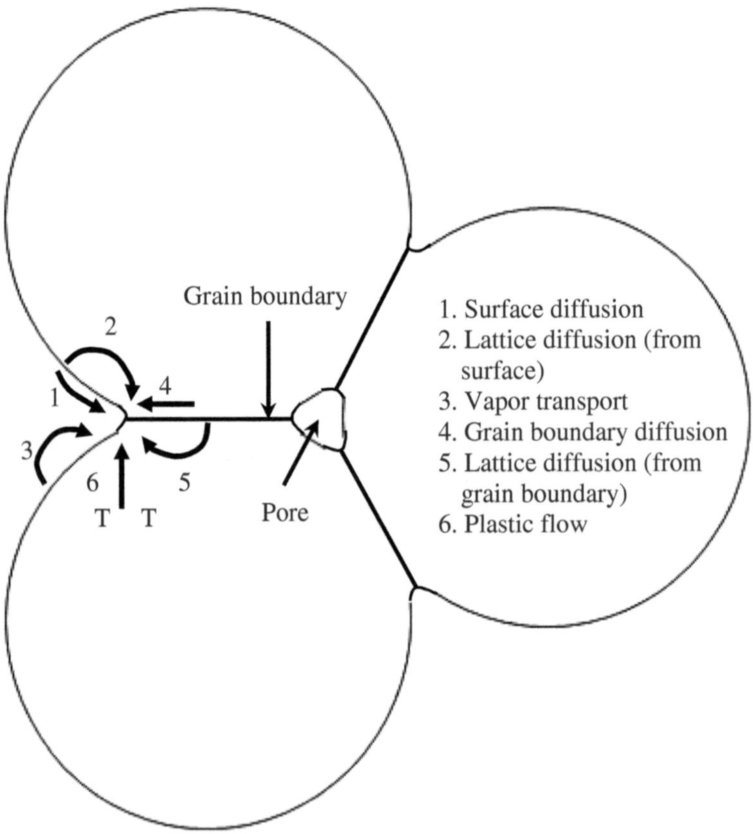

Fig. 5.10 Six distinct mechanisms can contribute to the sintering of a consolidated mass of crystalline particles: *1* surface diffusion (*SD*), *2* lattice diffusion from the surface, *3* vapor transport, *4* grain boundary diffusion, *5* lattice diffusion from the grain boundary, and *6* plastic flow. Only mechanisms *1* to *3* lead to densification, but all cause the necks to grow and so influence the rate of densification. Reproduced with permission from [24]. Copyright © 2007, Springer

All the mechanisms can be classified into two groups: (i) densifying mechanisms and (ii) nondensifying mechanisms, depending on whether they can cause shrinkage or densification of the compact. Nondensifying mechanisms include surface diffusion, lattice diffusion from the particle surfaces to the neck, and vapor transport, i.e., mechanisms 1, 2, and 3, which lead to neck growth without densification. In contrast, grain boundary diffusion and lattice diffusion from the grain boundary to the pore, i.e., mechanisms 4 and 5, are densifying mechanisms, which are the most important contributions to densification of polycrystalline ceramics. Diffusion from the grain boundary to the pore is benefit to both neck growth and densification. Similarly, plastic flow by dislocation motion, i.e., mechanism 6, also leads to neck growth and densification, but it is more pronounced in the sintering of metal powders. The nondensifying mechanisms reduce the curvature of the neck surface, i.e., the driving force for sintering, so as to reduce the rate of the densifying mechanisms. Therefore, they must be taken into account when discussing the sintering of ceramics. The diffusion of the different ionic species discussed above also has effect on the rate of densification.

Because amorphous materials, such as glasses, have no grain boundaries, their neck growth and densification are caused by viscous flows and the deformation of the particles. In practice, the paths of matter flows are not clearly defined. The geometrical changes caused by the viscous flow could be complex, in which the equations for matter transport can only be established with significantly simplified assumptions. The sintering mechanisms of polycrystalline and amorphous solids are summarized in Table 5.2.

5.3.3 Grain Boundary Effects

Polycrystalline materials have grain boundaries, while amorphous materials haven't, which has been one of the most distinct difference form each other. The grain boundaries determine the equilibrium shapes of the pores and the grains in polycrystalline ceramics. Figure 5.11 shows a hypothetical pore that is enclosed by three grains [1]. The forces must balance at the junctions where the surfaces of the pores meet the grain boundary. These forces are usually represented by the tension at the interfaces, i.e., the tension in the solid–vapor interface and the tension in the

Table 5.2 Sintering mechanisms of polycrystalline and amorphous solid materials

Mechanism	Source of matter	Sink of matter	Densifying	Nondensifying
Surface diffusion	Surface	Neck		×
Lattice diffusion	Surface	Neck		×
Vapor transport	Surface	Neck		×
Grain boundary diffusion	Grain boundary	Neck	×	
Lattice diffusion	Grain boundary	Neck	×	
Plastic/viscous flow	Dislocations	Neck	×	

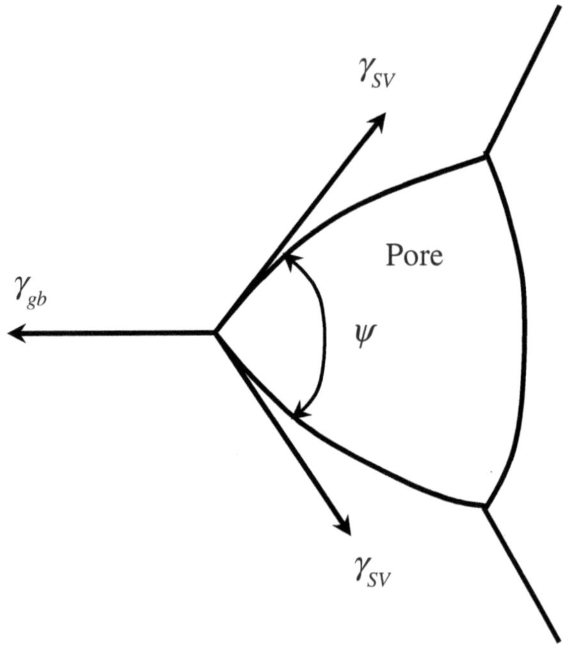

Fig. 5.11 The equilibrium shapes of the pores in polycrystalline solids are governed by the balance between the surface and interfacial forces at the point where the grain boundary intersects the pore. γ_{SV} is the surface tension, γ_{gb} is the grain boundary tension, and ψ is the dihedral angle. Reproduced with permission from [1]. Copyright © 2003, CRC Press

grain boundary. Similar to the surface tension of liquids, a tension is present due to the increase in energy caused by the increase in the area of the interfaces. At the junction, the tension at the solid–vapor interface is tangential to the interfaces, while that in the grain boundary is along the plane of the boundary. The balance of the forces leads to the following expression:

$$\gamma_{gb} = 2\gamma_{SV} \cos\left(\frac{\psi}{2}\right), \tag{5.120}$$

where ψ is the dihedral angle. While some sintering models consider such dihedral angles [3], it is more common to assume that the pores are sphere holes, i.e., $\psi = 180°$ or $\gamma_{gb} = 0$.

During the sintering of polycrystalline materials, the decrease in energy due to the elimination of free surface area is partly compensated by the energy due to the creation of new grain boundary area, as shown in Fig. 5.10. Therefore, the driving force of sintering is lower than that calculated by using Eq. (5.3), where the grain boundaries are neglected. If ΔA_{SV} and ΔA_{gb} are used to represent the changes in the free surface area and grain boundary area, respectively, the change in energy of the system is given by:

$$\Delta E = \gamma_{SV}\Delta A_{SV} + \gamma_{gb}\Delta A_{gb}, \tag{5.121}$$

where γ_{SV} is the specific surface energy and γ_{gb} is the specific grain boundary energy. The system will evolve toward an equilibrium state governed by the condition of $|\gamma_{gb}\Delta A_{gb}| = |\gamma_{SV}\Delta A_{SV}|$. If there is $|\gamma_{gb}\Delta A_{gb}| > |\gamma_{SV}\Delta A_{SV}|$, the change in energy ΔE in Eq. (5.120) is positive. As a result, the solid–vapor surface area tends to increase, i.e., the pores tend to grow.

The presence of the grain boundaries provides an additional opportunity for the system to decrease energy, which is the decrease in the total grain boundary energy. Grain growth is usually accompanied by pore growth, with the overall process being termed as coarsening, which is present concurrently with sintering. Due to the extreme complexity, there is still no a theory of sintering that is able to analyze three-dimensional behaviors of particles and interaction among the particles, as well the concurrent densification and coarsening related to the various transport mechanisms. In most cases, densification and grain growth are analyzed separately.

5.3.4 Theory of Sintering

Various approaches have been used in the theoretical analysis of sintering. Analytical models are among the earlier attempts to develop quantitative modelings of sintering [21]. The analytical models have provided the basis for the current understanding of sintering, although simple geometry and single mechanism are often assumed. The scaling laws provide a reliable guidance for understanding the dependence of sintering mechanisms on particle size, i.e., scale [25]. Numerical simulations are effective methods, because they can be used to analyze more realistic and complicated geometries and the occurrence of multiple mechanisms. However, numerical methods have high requirement in calculations. The topological models can be used to understand the evolution of microstructure, due to their capability to predict the sintering kinetics. The statistical models and the phenomenological equations have received much less attention [20]. Sintering maps are attempted to demonstrate the changes in the sintering behavior and mechanisms under different conditions, such as temperature and particle size, which have similar limitations to the analytical models.

5.3.5 Scaling Laws

The scaling laws deal with the effect of change in scale on microstructural development during sintering [25]. During the sintering of a powder compact, the particle size is the most fundamental scaling parameter. The scaling laws do not assume a specific geometrical model, but with assumptions that (i) the particle size of the powder system remains the same and (ii) the geometrical changes remain similar. Two systems are geometrically similar, so that the linear dimension of all of the features, such as grains and pores, of system 1, $D_{sys(1)}$ is equal to a numerical factor times the linear dimension of the corresponding features in system 2, $D_{sys(2)}$, i.e., there is:

$$D_{sys(1)} = \lambda D_{sys(2)}, \tag{5.122}$$

where λ is the numerical factor. Therefore, a simple magnification can be used to relate one system to the other, as shown schematically in Fig. 5.12 [1].

5.3.5.1 Derivation

Figure 5.13 shows a simple system with two spheres in contact, to demonstrate the derivation of the scaling laws [1]. If it takes a time Δt_1 to produce a certain microstructural change, e.g., the growth of a neck to a certain radius X_1, in system 1, it is necessary to derive the time Δt_2 taken by system 2 to produce a geometrically similar change. For geometrically similar changes, the initial radius of the particle and the neck radius of the two systems are related by:

$$R_2 = \lambda R_1, \quad X_2 = \lambda X_1. \tag{5.123}$$

The time taken to produce a certain change by diffusional flow of matter can be expressed as:

$$\Delta t = \frac{V}{JA\Omega}, \tag{5.124}$$

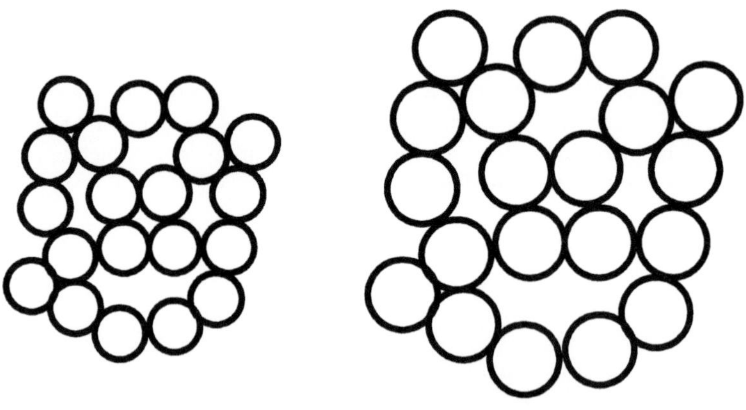

Fig. 5.12 An example of two geometrically similar systems consisting of a random arrangement of *circles*. The systems differ only in scale and involve a simple magnification of one relative to the other. Reproduced with permission from [1]. Copyright © 2003, CRC Press

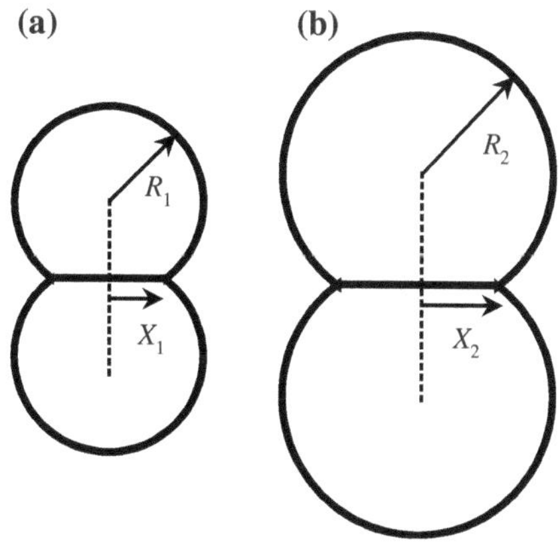

Fig. 5.13 Geometrically similar models consisting of two spheres in contact. The linear dimensions of system 2 are a factor λ times those of system 1. **a** System 1, **b** system 2. Reproduced with permission from [1]. Copyright © 2003, CRC Press

where V is the volume of the matter transported, J is the flux, A is the cross-sectional area over which the matter is transported, and Ω is the atomic volume. Accordingly, there is:

$$\frac{\Delta t_2}{\Delta t_1} = \frac{V_2 J_1 A_1}{V_1 J_2 A_2}. \tag{5.125}$$

Equation (5.125) can be used to consider matter transport by volume diffusion.

The volume of the matter transported is proportional to R^3, where R is the radius of the sphere shown in Fig. 5.13 [1]. As a result, V_2 is proportional to $(\lambda R)^3$, or $V_2 = \lambda^3 V_1$. For lattice diffusion, the area over which the matter diffuses is proportional to R^2. Therefore, A_2 is proportional to $(\lambda R)^2$, or $A_2 = \lambda^2 A_1$. The flux J is proportional to $\nabla\mu$, which is the gradient of chemical potential. For a curved surface with a radius of curvature r, μ varies as a function of $1/r$. Therefore, J varies with $\nabla(1/r)$ or $1/r^2$. Because J_2 is proportional to $1/(\lambda r)^2$, there is $J_2 = J_1/\lambda^2$. Therefore, the parameters for lattice diffusion are:

$$V_2 = \lambda^3 V_2, \quad A_2 = \lambda^2 A_1, \quad J_2 = J_1/\lambda^2, \tag{5.126}$$

which are put into Eq. (5.125), yielding:

$$\frac{\Delta t_2}{\Delta t_1} = \lambda^3 = \left(\frac{R_2}{R_1}\right)^3. \tag{5.127}$$

Therefore, according to Eq. (5.127), the time taken to produce geometrically similar changes by a lattice diffusion mechanism increases as the cube of the particle size. The scaling laws for the other mass transport mechanisms can be derived in a similar manner [25]. The laws can also be expressed in the following general form:

$$\frac{\Delta t_2}{\Delta t_1} = \lambda^m = \left(\frac{R_2}{R_1}\right)^m, \tag{5.128}$$

where m is an exponent that depends on the mechanisms of sintering. The values of m for different sintering mechanisms are listed in Table 5.3.

5.3.5.2 Application and Limitation

The scaling laws can be used to determine the dependence of the relative rates of sintering by the different mechanisms on the particle size of the powder, which useful in controlling the microstructures during the sintering of ceramics. As mentioned earlier, some mechanisms lead to densification while others do not. Therefore, in order to achieve high density, it is necessary to increase the rates of the densifying mechanisms and decrease those of the nondensifying mechanisms.

To determine the relative rates of the different mechanisms, it is more useful to express Eq. (5.128) in terms of rate. For a given change, the rate is inversely proportional to the time, so that Eq. (5.128) can be written as:

$$\frac{(\text{Rate})_2}{(\text{Rate})_1} = \lambda^{-m}. \tag{5.129}$$

For a given powder system, if it is assumed that grain boundary diffusion and vapor transport (evaporation/condensation) are the dominant mass transport mechanisms, the rates of sintering by these two mechanisms vary with the scale of the system with the following relations:

$$(\text{Rate})_{\text{gb}} \sim \lambda^{-4}, \tag{5.130}$$

and

$$(\text{Rate})_{\text{ec}} \sim \lambda^{-2}. \tag{5.131}$$

Table 5.3 Exponents for Herring's scaling laws described by Eq. (5.128)

Sintering mechanism	Exponent (m)
Surface diffusion	4
Lattice diffusion	3
Vapor transport	2
Grain boundary diffusion	4
Plastic flow	1
Viscous flow	1

The variation of the rates of sintering with λ for the two mechanisms is shown in Fig. 5.14 [1]. The crossover point of the two lines is arbitrary, but it does not affect the validity of the results. For small λ, i.e., as the particle size is small, the rate of sintering by grain boundary diffusion is higher than that by vapor transport. However, the rate of sintering by vapor transport dominates for large λ, i.e., for large particle sizes. According to the scaling laws, small particle size is beneficial to densification, when grain boundary diffusion and vapor transport are the dominant mechanisms. If surface diffusion and lattice diffusion are the dominant mechanisms, similar treatment indicates that surface diffusion is enhanced as the particle size is decreased. This principle is applicable to other combinations, such as lattice diffusion versus grain boundary diffusion and surface diffusion versus grain boundary diffusion.

In the derivation of the scaling laws, on general approach and simple physical principles are employed. Because the geometric details of the powder system are not involved in the derivation, the laws are applicable to particles of any shape and to all stages of the sintering process. However, the scaling laws also have limitations. During the derivation, the particle sizes of each powder system are assumed to be unchanged during the sintering, while the microstructural changes are geometrically similar in the two systems. The second assumption is a key limitation of the scaling laws, because it is difficult to be achieved in real powder systems. For the scaling laws to be valid, the two systems should be identical in chemical composition so that the mass transport coefficients are the same.

It is noted that the exponent m in Eq. (5.128) or Eq. (5.129) is related to the mechanism of sintering, the measurement of m could provide information on the mechanism of sintering. In practice, determination of the mechanism is a challenging, due to several reasons. One of the major problems is the presence of multiple mechanisms. Therefore, the exponent that is measured experimentally could be an overall result of several mechanisms. Moreover, the sintering mechanism can vary with the size of the particles.

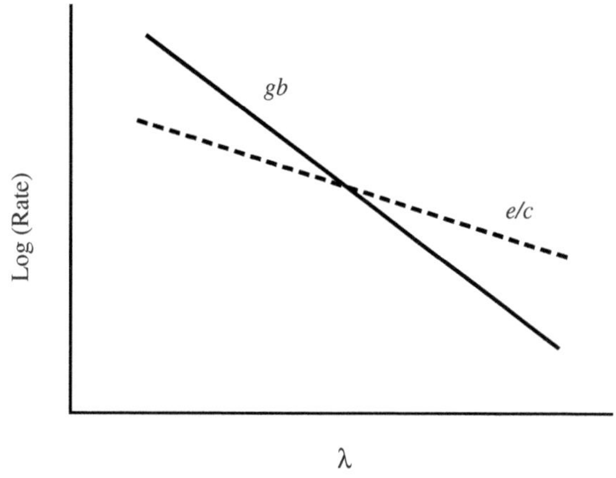

Fig. 5.14 Schematic diagram of the relative rates of sintering by grain boundary diffusion and by evaporation/condensation as a function of the scale (i.e., particle size) of the system. Reproduced with permission from [1]. Copyright © 2003, CRC Press

5.3.6 Analytical Methods

In the analytical models, the mass transport equations of different mechanisms are solved analytically to derive the equations of the sintering kinetics, in which a relatively simple and idealized geometry for the powder system is assumed. However, the microstructure of a real powder compact changes continuously, especially during the sintering. Therefore, it is difficult to find a single geometrical model that can describe the entire process and ensure the simplicity of analysis. To address this issue, the sintering process has to be conceptually divided into different independent stages. At each stage, an idealized geometry that has a rough similarity with the microstructure of the powder system can be assumed.

5.3.6.1 Stages of Sintering

Sintering process can be divided into three sequential stages: (i) initial stage, (ii) intermediate stage and (iii) final stage. In some analyses, an extra stage, zero stage, is included, which describes the instantaneous contact of the particles, when they are first brought together due to the elastic deformation in response to surface energy reduction at the interfaces [26]. A stage represents an interval of time or density, over which the microstructure is assumed to be reasonably well defined. For polycrystalline materials, Fig. 5.15 shows the idealized geometrical structures to represent the three stages [27, 28].

Initial Stage

The initial stage is characterized by a rapid interparticle neck growth through diffusion, vapor transport, plastic flow or viscous flow. The large differences in initial surface curvature among different particles are eliminated at this stage. Also, shrinkage or densification could be observed if the neck growth is due to the densifying mechanisms. For a powder with spherical particles, the initial stage is represented as the transition from Fig. 5.15a, b [27]. The initial stage is not finished until the radius of the neck between the particles reaches 0.4–0.5 of the particle radius. For a powder compact with an initial density of 0.5–0.6 of the theoretical density, this stage will contribute a linear shrinkage of 3–5 %, i.e., the density is increased to about 0.65 of the theoretical, if the densifying mechanisms are dominant.

Intermediate Stage

The intermediate stage starts when the pores reach their equilibrium shapes, according to the surface and interfacial tensions. The pores are still continuous. The

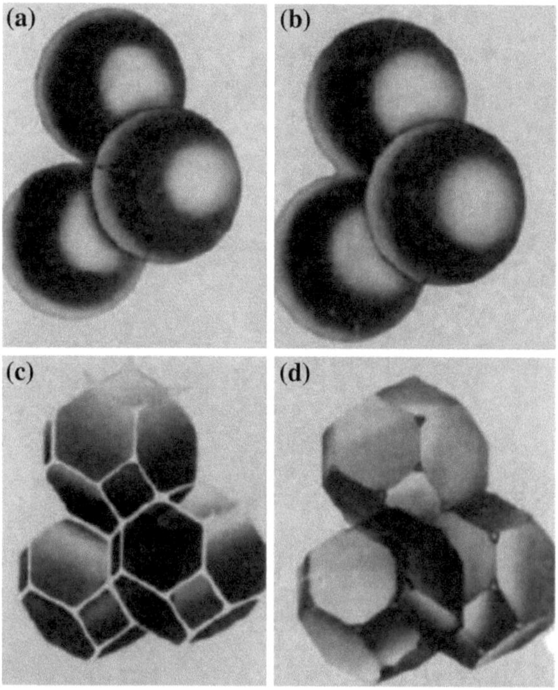

Fig. 5.15 Idealized models for the three stages of sintering. **a** Initial stage: Model structure represented by spheres in tangential contact. **b** Near the end of the initial stage: Spheres have begun to coalesce. The neck growth illustrated is for center-to-center shrinkage of 4 %. **c** Intermediate stage: Dark grains have adopted the shape of a tetrakaidecahedron, enclosing white pore channels at the grain edges. **d** Final stage: Pores are tetrahedral inclusions at the corners where four tetrakaidecahedra meet. Reproduced with permission from [27]. Copyright © 1961, American Institute of Physics

structure is usually idealized, by assuming that the pores are arranged as a spaghetti-like array and located along the grain edges, as shown in Fig. 5.15c [27]. Densification is assumed to take place only due to the shrinkage of the pores. Finally, the pores become unstable and pinch off and only those isolated pores are retained in the sintered body. At this point, the final stage is about to begin. The intermediate stage is the main contribution to the sintering process, which is ended at the relative density of 0.9.

Final Stage

The microstructural development in the final stage can occur in several ways. In a simplest case, the final stage begins when the pores pinch off and there are only isolated pores located at corners of the grains, as shown schematically in Fig. 5.15d.

Table 5.4 Parameters related to the three sintering stages of polycrystalline solids

Stage	Microstructure	Relative density	Model
Initial	Rapid interparticle neck growth	~0.65	Two monosized spheres in contact
Intermediate	Equilibrium pore shape with continuous porosity	0.65–0.9	Tetrakaidecahedron with cylindrical pores of the same radius along the edges
Final	Equilibrium pore shape with isolated porosity	>0.9	Tetrakaidecahedron with spherical monosized pores at the corners of grains

In this case, the pores are assumed to have shrunk continuously and most likely disappear altogether. Main parameters of the three sintering stages are summarized in Table 5.4.

5.3.6.2 Analytical Models

In analytical models, it is assumed that the particles in the initial powder compact are spherical with the same size and uniform packing, which is called the geometrical model [6]. With appropriate boundary conditions, the remainder of the powder system is considered as a continuum, having the same macroscopic properties, such as shrinkage and densification rate, as the isolated unit. The equations of the sintering kinetics can be derived from the established mass transport equations, which are solved under appropriate boundary conditions.

5.3.6.3 Initial Stage Models

Geometrical Parameters

The model of two equal-sized spheres, with radius of a, in contact is used for the initial stage, which is known as the two-sphere model. Two slightly different geometries are used represent the nondensifying mechanisms and the densifying mechanisms, respectively. The two-sphere model for the densifying mechanisms shows shrinkage, i.e., interpenetration of the two spheres, and thus neck growth. The neck is assumed to be circular with a radius X and a surface having a circular cross section with a radius r, which means that that the grain boundary energy is assumed to be zero. The main geometrical parameters of the model are the principal radii of curvature of the neck surface r and X, the area of the neck surface A, and the volume of material transported into the neck V, with $r \approx X^2/4a$, $A \approx \pi^2X^3/2a$, and $V \approx \pi X^4/8a$. The parameters of the nondensifying model are only slightly different from those of densifying the model, with $r \approx X^2/2a$, $A \approx \pi^2X^3/a$, and $V \approx \pi X^4/2a$.

Kinetic Equations

Matter transport through diffusion can be described in terms of the flux of atoms or the counterflow of vacancies. The approach based on the counterflow of vacancies driven by a vacancy concentration gradient has been used predominantly in the early development of sintering theory, which is discussed in this subsection first. More general approach based on the flux of atoms driven by a chemical potential gradient will be presented later. The following discussion is started with the mechanism of grain boundary diffusion.

According to Eq. (5.98), the flux of atoms into the neck can be determined, and thus the volume of matter transported into the neck per unit time is given by

$$\frac{\mathrm{d}V}{\mathrm{d}t} = J_{\mathrm{a}} A_{\mathrm{gb}} \Omega, \tag{5.132}$$

where A_{gb} is the cross-sectional area over which the diffusion takes place. Grain boundary diffusion is assumed to occur over a constant thickness δ_{gb}, so that $A_{\mathrm{gb}} = 2\pi X \delta_{\mathrm{gb}}$, where X is the radius of the neck. By combining Eqs. (5.98) and (5.132) and substituting A_{gb} with $2\pi X \delta_{\mathrm{gb}}$, there is:

$$\frac{\mathrm{d}V}{\mathrm{d}t} = D_{\mathrm{v}} 2\pi X \delta_{\mathrm{gb}} \frac{\mathrm{d}C_{\mathrm{v}}}{\mathrm{d}x}. \tag{5.133}$$

Because the neck radius increases radially in the direction orthogonal to a line joining the centers of the spheres, a one-dimensional solution is adequate. It is assumed that the concentration of vacancy between the neck surface and the center of the neck is constant, so that $\mathrm{d}C_{\mathrm{v}}/\mathrm{d}x = \Delta C_{\mathrm{v}}/X$, where ΔC_{v} is the difference in vacancy concentration between the neck surface and the center of the neck. If the concentration of vacancy at the center of the neck is assumed to be equal to that under a flat stress-free surface $C_{\mathrm{v},0}$, according to Eq. (5.97), there is

$$\Delta C_{\mathrm{v}} = C_{\mathrm{v}} - C_{\mathrm{v},0} = \frac{C_{\mathrm{v},0} \gamma_{\mathrm{SV}} \Omega}{kT} \left(\frac{1}{r_1} + \frac{1}{r_2} \right), \tag{5.134}$$

where r_1 and r_2 are the two principal radii of curvature of the neck surface. Because there is $r_1 = r$ and $r_2 = -X$ and if it is also assumed that $X \gg r$, Eq. (5.133) becomes

$$\frac{\mathrm{d}V}{\mathrm{d}t} = \frac{2\pi D_{\mathrm{v}} C_{\mathrm{v},0} \delta_{\mathrm{gb}} \gamma_{\mathrm{SV}} \Omega}{kTr}, \tag{5.135}$$

By using the relations of V and r, and letting the grain boundary diffusion coefficient D_{gb} to be equal to $D_{\mathrm{v}} C_{\mathrm{v},0}$, there is

$$\frac{\pi X^3}{2a}\frac{\mathrm{d}X}{\mathrm{d}t} = \frac{2\pi D_{\mathrm{gb}}\delta_{\mathrm{gb}}\gamma_{\mathrm{SV}}\Omega}{kT}\frac{4a}{X^2}. \tag{5.136}$$

Rearranging the equation yields

$$X^5\mathrm{d}X = \frac{16D_{\mathrm{gb}}\delta_{\mathrm{gb}}\gamma_{\mathrm{SV}}\Omega a^2}{kT}\mathrm{d}t. \tag{5.137}$$

By integrating and applying the boundary conditions $X = 0$ at $t = 0$, Eq. (5.137) becomes

$$X^6 = \frac{96D_{\mathrm{gb}}\delta_{\mathrm{gb}}\gamma_{\mathrm{SV}}\Omega a^2}{kT}t. \tag{5.138}$$

Equation (5.138) can also be rewritten as

$$\frac{X}{a} = \left(\frac{96D_{\mathrm{gb}}\delta_{\mathrm{gb}}\gamma_{\mathrm{SV}}\Omega}{kTa^4}\right)^{1/6}t^{1/6}. \tag{5.139}$$

Equations (5.138) and (5.139) predict that the ratio of the neck radius to the sphere radius increases as a function of $t^{1/6}$.

For this densifying mechanism, the linear shrinkage, defined as the change in length ΔL divided by the original length L_0, can also be found, which is given by

$$\frac{\Delta L}{L_0} = -\frac{h}{a} = -\frac{r}{a} = -\frac{X^2}{4a^2}, \tag{5.140}$$

where h is half the interpenetration distance between the two spheres. With Eq. (5.139), the following equation can be obtained:

$$\frac{\Delta L}{L_0} = -\left(\frac{3D_{\mathrm{gb}}\delta_{\mathrm{gb}}\gamma_{\mathrm{SV}}\Omega}{2kTa^4}\right)^{1/3}t^{1/3}, \tag{5.141}$$

which means that the shrinkage increase as a function of $t^{1/3}$.

The mechanism of viscous flow is discussed as follows. For this mechanism, matter transport is governed by the concept of Frenkel's energy balance, i.e., the rate of energy dissipation by the viscous flow is equal to the rate of the energy gained by the reduction in surface area, i.e., $(\mathrm{Rate})_{\mathrm{dis}} = (\mathrm{Rate})_{\mathrm{gain}}$.

With the parameters shown in Fig. 5.16 and assuming that the radius of the sphere remains roughly constant during the viscous flow, the decrease in the surface area of the two spheres is given by

$$S_0 - S = 8\pi a^2 - 4\pi a^2(1 + \cos\theta), \tag{5.142}$$

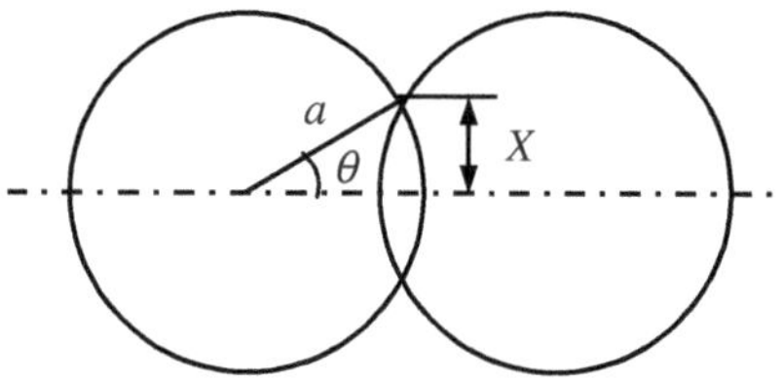

Fig. 5.16 Geometrical parameters of the two-sphere model used in the derivation of the initial stage equation for viscous sintering by Frenkel. Reproduced with permission from [1]. Copyright © 2003, CRC Press

which means that the material removed from the plane of contact is uniformly distributed over the surface of the sphere, instead of accumulating at the neck. For small values of θ, i.e., for small neck radius, $\cos\theta \approx 1 - \theta^2/2$, so that Eq. (5.142) becomes

$$S_0 - S = 2\pi a^2\theta^2. \tag{5.143}$$

The rate of change in energy due to the reduction in surface area can be expressed as

$$\dot{E}_s = -\gamma_{SV}\frac{dS}{dt} = 4\pi a^2\gamma_{SV}\frac{d}{dt}\left(\frac{\theta^2}{2}\right), \tag{5.144}$$

where γ_{SV} is the specific surface energy of the solid–vapor interface. According to Frenkel, the rate of energy dissipation by the viscous flow between the two spheres is given by

$$\dot{E}_v = \frac{16}{3}\pi a^3\eta\dot{u}^2, \tag{5.145}$$

where η is the viscosity of the glass state and $\dot{u}$ is the velocity of motion for the viscous flow which is given by

$$\dot{u} = \frac{1}{a}\frac{d}{dt}\left(\frac{a\theta^2}{2}\right) = \frac{d}{dt}\left(\frac{\theta^2}{2}\right). \tag{5.146}$$

This equation is derived with the assumption that the flow is uniform along the axis joining the centers of the spheres, rather than concentrated near the neck. By putting Eq. (5.146) into Eq. (5.145) for $\dot{u}$, with $\dot{E}_s = \dot{E}_v$, there is:

$$\frac{16}{3}\pi a^3\eta\dot{u}\frac{d}{dt}\left(\frac{\theta^2}{2}\right) = 4\pi a^2\gamma_{SV}\frac{d}{dt}\left(\frac{\theta^2}{2}\right). \tag{5.147}$$

Accordingly, the following relationship is obtained:

$$\dot{u} = \frac{3}{4}\frac{\gamma_{SV}}{\eta a}. \tag{5.148}$$

By putting Eq. (5.148) into Eq. (5.147) for $\dot{u}$ and integrating with the boundary conditions of $\theta = 0$ at $t = 0$, there is

$$\theta^2 = \frac{3}{2}\left(\frac{\gamma_{SV}}{\eta a}\right)t. \tag{5.149}$$

Because $\theta = X/a$, where X is the neck radius, from Eq. (5.149), there is

$$\frac{X}{a} = \left(\frac{3\gamma_{SV}}{2\eta a}\right)^{1/2} t^{1/2}. \tag{5.150}$$

The equation for the shrinkage by this densifying mechanism can be also readily derived.

Equations for initial stage of sintering have been intensively studied in the open literature [20, 21, 29–32]. With the equations for neck growth and the densifying mechanisms, shrinkage can be expressed in the following general forms:

$$\left(\frac{X}{a}\right)^m = \frac{H}{a^n}t, \tag{5.151}$$

$$\left(\frac{\Delta L}{L_0}\right)^{m/2} = -\frac{H}{2^m a^n}t, \tag{5.152}$$

where m and n are numerical exponents that are dependent on the mechanisms of the sintering and H is a function that describes the geometrical and material parameters of the powder system. With the assumptions made in each model, the values for m, n, and the numerical constant in H have been derived for various models [33, 34].

According to the equations of neck growth, it is observed that the plot of log (X/a) versus log(t) is a straight line, with the slope to be $1/m$. Therefore, so by fitting experimental data with the theoretical models, the values of m will be obtained. Similarly, shrinkage can be analyzed, if it takes place during the sintering process. Usually, simple systems, such as two spheres, one sphere on a plate, or two wires for neck growth and compacts of spherical particles for shrinkage, are used to collect experimental data for validation of the theoretical models. Because m is dependent on the mechanism of sintering, it is easily concluded that the value of m can be used to identify the mechanism of sintering. However, due to the assumption of single dominant mass transport mechanism in the models, this is prediction is not valid for practical material systems.

Also, the descriptions derived with the two-sphere geometry model cannot be simply applied to real powder compacts, unless the particles are spheres with the same size that are packed uniformly. The most effective approach is the assumption of a linear array of spheres for the powders with size distributions [35]. For the dihedral angle model, only when the dihedral angle is larger than 150°, the assumption that that the grain boundary energy is zero can be used as a good approximation [36]. Other assumptions, such as the redistribution of matter transport to the neck, have also been confirmed to be too simple for real sintering process [37].

5.3.6.4 Intermediate Stage Models

Geometrical Model

For the geometrical model of the intermediate stage, the powder is assumed to consist of a space-filling array of equal-sized tetrakaidecahedra as the particles [27]. The pores are cylindrical, with axis of the cylinder coinciding with the edge of the tetrakaidecahedra, as shown in Fig. 5.15c. The tetrakaidecahedron with cylindrical pores along its edges is taken as the unit cell of the structure. Figure 5.17 shows a tetrakaidecahedron, which is derived from an octahedron by trisecting each edge and joining the points to remove the six edges. The resulting structure has 36 edges, 24 corners, and 14 faces of 8 hexagonals and 6 squares. The volume of the tetrakaidecahedron is given by

$$V_t = 8\sqrt{2}l_p^3, \tag{5.153}$$

where l_p is the edge length of the tetrakaidecahedron. Because each pore is shared by three tetrakaidecahedras, if r is the radius of the pore, the total volume of the pores per unit cell is given by

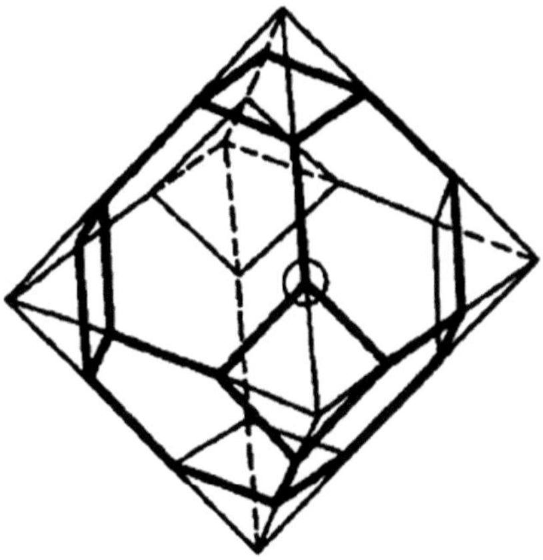

Fig. 5.17 A tetrakaidecahedron from a truncated octahedron shown by the *dark lines*. Reproduced with permission from [27]. Copyright © 1961, American Institute of Physics

$$V_p = \frac{1}{3}(36\pi r^2 l_p). \tag{5.154}$$

The porosity of the unit cell, V_p/V_t, is therefore given by

$$P_c = \frac{3\pi}{2\sqrt{2}}\left(\frac{r}{l_p}\right)^2. \tag{5.155}$$

Because it has been assumed in the model that the pore geometry is uniform, the nondensifying mechanisms are not present. This is attributed to the fact that the chemical potential is the same everywhere on the surface of the pores. Therefore, only two densifying mechanisms should be considered: lattice diffusion and grain boundary diffusion.

Sintering Equations of Lattice Diffusion

The cylindrical pores along the edges enclose each face of the tetrakaidecahedron, as shown in Fig. 5.18a. Because the vacancy flux from the pores terminates on the faces of the boundary, as shown in Fig. 5.18b, it can be assumed the diffusion is radial from a circular vacancy source while the shape effects on the corner of the tetrakaidecahedron is neglected [27]. To remain the boundary to be flat, the vacancy flux per unit area of the boundary should be the same over the whole boundary. The diffusion flux field can be treated as that of the temperature distribution in a surface-cooled and electrically heated cylindrical conductor. The flux per unit length of the cylinder can be expressed as

$$\frac{J}{l} = 4\pi D_v \Delta C, \tag{5.156}$$

where D_v is the vacancy diffusion coefficient and ΔC is the difference in concentration of vacancy between the pore or source and the boundary or sink. There are also several other assumptions. First, the flux equation with the dependence on the radius of the pore is not affected by the convergence of the flux to the boundary. Second, the width of the flux field, which is equivalent to l in Eq. (5.156), should be to the diameter of the pore. Third, the flux is increased by a factor of 2, due to the freedom of the vacancy diffusion flux that diverges initially, so as to provide additional available area.

With these assumptions, Eq. (5.156) becomes

$$J = 2(4\pi D_v \Delta C)2r. \tag{5.157}$$

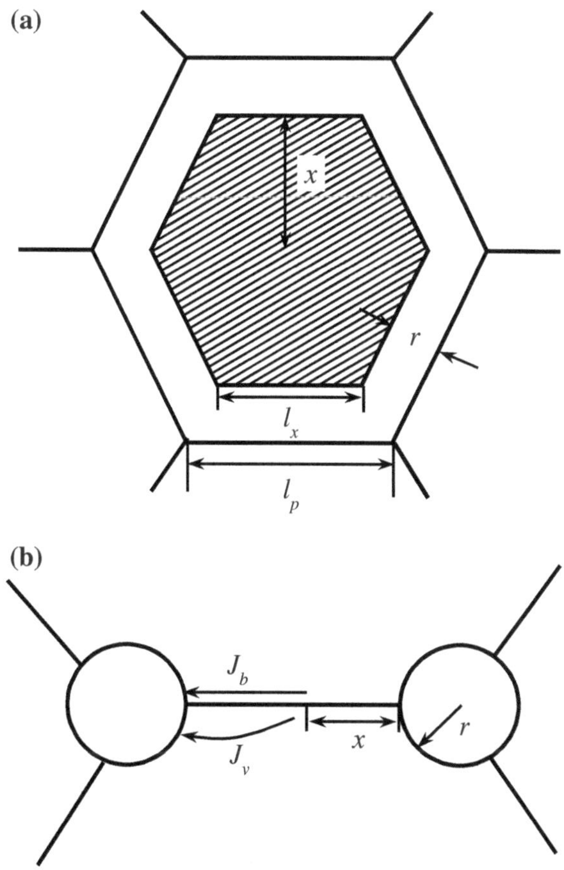

Fig. 5.18 **a** Schematic of the hexagonal neck for intermediate stage sintering equations of polycrystalline ceramics. **b** The section shows a cut through the neck with the atomic flux paths for grain boundary and lattice diffusion. Reproduced with permission from [1]. Copyright © 2003, CRC Press

Because there are 14 faces in a tetrakaidecahedron and each face is shared by two grains, the volume flux per unit cell will be given by

$$\frac{dV}{dt} = \frac{14}{2}J = 112\pi D_v \Delta C. \tag{5.158}$$

Because of the cylindrical pore in the intermediate stage, the two principal radii of curvature are r and ∞, so that ΔC is given by

$$\Delta C = \frac{C_{v,0}\gamma_{SV}\Omega}{kTr}. \tag{5.159}$$

By putting Eq. (5.159) into Eq. (5.158), $D_l = D_v C_{v,0}$, where D_l is the lattice diffusion coefficient, there is

$$\mathrm{d}V = \frac{112\pi D_{\mathrm{l}}\gamma_{\mathrm{SV}}\Omega}{kT}\mathrm{d}t. \tag{5.160}$$

The integral of $\mathrm{d}V$ is equal to the porosity given by Eq. (5.154), which is

$$\int \mathrm{d}V = 12\pi r^2 l_{\mathrm{p}}\Big|_{r_0}^{r}. \tag{5.161}$$

By comparing with Eq. (5.160), there is

$$r^2\Big|_{r}^{0} \approx -10\frac{D_{\mathrm{l}}\gamma_{\mathrm{SV}}\Omega}{l_{\mathrm{p}}kT}t\Big|_{\mathrm{t}}^{t_{\mathrm{f}}}, \tag{5.162}$$

where t_{f} is the time when the pore disappears. By dividing both sides of the equation by l_{p}^2 and evaluating the integrand, the following expression can be obtained:

$$P_{\mathrm{c}} \approx \frac{r^2}{l_{\mathrm{p}}^2} \approx \frac{10D_{\mathrm{l}}\gamma_{\mathrm{SV}}\Omega}{l_{\mathrm{p}}^3kT}(t_{\mathrm{f}} - t). \tag{5.163}$$

The model valid until the pores are pinched off and then isolated.

The sintering equations are commonly expressed with densification rates. Porosity P and relative density ρ are related by $P = 1 - \rho$, therefore, by differentiating Eq. (5.163) with respect to time, there is

$$\frac{\mathrm{d}}{\mathrm{d}t}(P_{\mathrm{c}}) = -\frac{\mathrm{d}\rho}{\mathrm{d}t} \approx -\frac{10D_{\mathrm{l}}\gamma_{\mathrm{SV}}\Omega}{l_{\mathrm{p}}^3kT}. \tag{5.164}$$

If l_{p} is assumed to be equal to the grain size G and the densification rate is expressed in the form of volumetric strain rate, Eq. (5.164) becomes

$$\frac{1}{\rho}\frac{\mathrm{d}\rho}{\mathrm{d}t} \approx \frac{10D_{\mathrm{l}}\gamma_{\mathrm{SV}}\Omega}{\rho G^3kT}. \tag{5.165}$$

This equation indicates that the densification rate at a given density is dependent inversely on the cube of the grain size, which is in agreement with the prediction by the Herring's scaling law.

Sintering Equations of Grain Boundary Diffusion

The geometrical model for lattice diffusion can be modified to derive the flux equations for grain boundary diffusion, which is

$$P_c \approx \frac{r^2}{l_p^2} \approx \left(\frac{2D_{gb}\delta_{gb}\gamma_{SV}\Omega}{l_p^4 kT}\right)^{2/3} t^{2/3}. \tag{5.166}$$

Similarly, Eq. (5.166) can be expressed as

$$\frac{1}{\rho}\frac{d\rho}{dt} \approx \frac{4}{3}\frac{D_{gb}\delta_{gb}\gamma_{SV}\Omega}{\rho(1-\rho)^{1/2}G^4 kT}. \tag{5.167}$$

At a given density, the densification rate due to the grain boundary diffusion is inversely proportional to the fourth power of the grain size, which is also coincident with predictions given by the scaling law.

5.3.6.5 Final Stage Models

For the final stage sintering models, the powder system is treated as an array of equal-sized tetrakaidecahedra with spherical pores with the same size located at the corners, as shown in Fig. 5.15d. The tetrakaidecahedron has 24 pores, with one at each corner, Because each pore is shared by four tetrakaidecahedra, the pore volume for every one tetrakaidecahedron is $V_p = (24/4)(4/3)\pi r^3$, with r being the radius of the pore. According to Eq. (5.153), the porosity per tetrakaidecahedron is given by

$$P_s = \frac{8\pi r^3}{8\sqrt{2}l_p^3} = \frac{\pi}{\sqrt{2}}\left(\frac{r}{l_p}\right)^3. \tag{5.168}$$

To be more convenient, the unit cell of the idealized structure can be treated as a thick-walled spherical shell of solid material, with a single pore of radius r to be located at the center, as shown in Fig. 5.19 [1]. The outer radius b of the spherical shell is defined in such a way that the average density of the unit cell is equal to the density of the powder system, which is

$$\rho = 1 - \left(\frac{r}{b}\right)^3. \tag{5.169}$$

The volume of the solid phase in the unit cell is $(4/3)\pi(b^3 - r^3)$. Because a unit cell contains only one pore, the number of pores per unit volume of the solid phase is given by

$$N = \frac{3}{4\pi}\frac{1-\rho}{\rho r^3}. \tag{5.170}$$

Similar assumptions are made for the final sintering stage, so that nondensifying mechanisms can be neglected, which is similar to that in the intermediate stage model [27, 38]. The procedure to derive the sintering equations for the intermediate

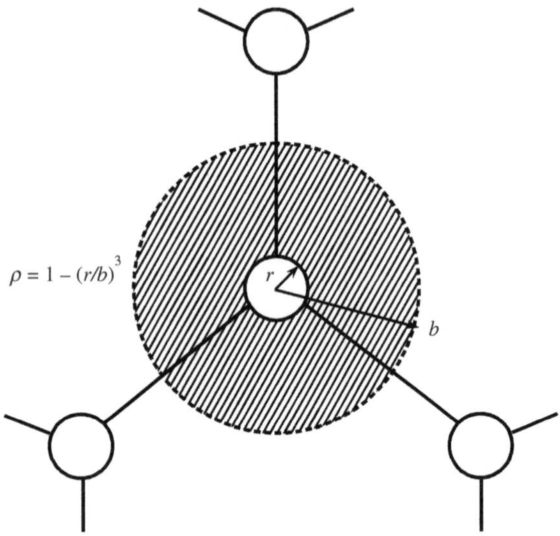

Fig. 5.19 A porous solid during the final stage of sintering can be modeled by constructing *a* spherical shell centered on a single pore. The outer radius *b* is chosen such that the density of the shell matches that of the porous solid. Reproduced with permission from [1]. Copyright © 2003, CRC Press

sintering stage can be used for the final stage, but with the atomic flux equation being assumed to be due to the diffusion between concentric spherical shells. The equation in terms of porosity is given by

$$P_s = \frac{6\pi}{\sqrt{2}} \frac{D_l \gamma_{SV} \Omega}{l_p^3 kT} (t_f - t), \tag{5.171}$$

where P_s is the porosity at a time t, D_l is the lattice diffusion coefficient, γ_{SV} is the specific energy of the solid–vapor interface, Ω is the atomic volume, l_p is the edge length of the tetrakaidecahedron that is assumed to be equal to the grain size, k is the Boltzmann constant, T is the absolute temperature, and t_f is the time when all the pores disappear. Equation (5.171) is almost the same as Eq. (5.163), with only a very small difference in the value of the numerical constant. As for the final stage sintering equation due to the grain boundary diffusion, the equation should be similar to Eq. (5.166).

5.3.7 Numerical Simulation of Sintering

Because analytical models are based on intensive assumptions, numerical simulations have been developed to provide a more comprehensive description of sintering. For example, in analytical models, it is assumed that the cross section of the neck surface is a circle to be tangential to both the grain boundary and the spherical surface of the particle, which usually known as circle approximation. In practice, the surface curvature is always changed at the point of tangent between the neck surface and the spherical surface of the particle, so that there should be an abrupt change in chemical potential. Also, analytical models cannot deal with the sintering governed by two or more mechanisms.

5.3.7.1 Sintering

Numerical simulations of sintering with diffusion mechanisms, such as surface diffusion and rain boundary diffusion, have been conducted and reported in the literature [39–43]. In such simulations, the three-dimensional real powder systems are simplified to two-dimensional geometrical models.

Figure 5.20 shows a model consisting of a row of cylinders with same radius, whose neck contours are shown in Fig. 5.21 [40]. Two situations are simulated for matter transport by (i) surface diffusion only and (ii) both surface diffusion and grain boundary diffusion, as shown in Fig. 5.21a, b, respectively. It is found that numerical simulation predicts undercutting and a continuous change in curvature of the neck surface, which is different from that given by the analytical models with the circle approximation. The region of the neck surface influenced by the matter transport also extends far beyond that predicted by the circle approximation. However, if both the surface diffusion and grain boundary diffusion are considered, the extension is less pronounced.

There have also been reports on numerical simulation of the initial stage of sintering, involving grain boundary, lattice, and surface diffusion at the same time [36]. However, this simulation also uses the circle approximation for the neck surface. Furthermore, the chemical potential gradients for the grain boundary diffusion and the lattice diffusion are assumed to be the same. The validity of numerical simulation has been compared with other methods [44, 45].

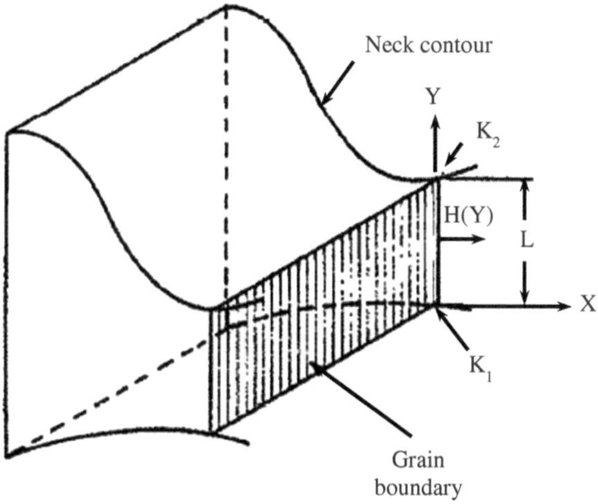

Fig. 5.20 Geometry of the neck formed between two cylindrical particles during sintering. Reproduced with permission from [40]. Copyright © 1979, Elsevier

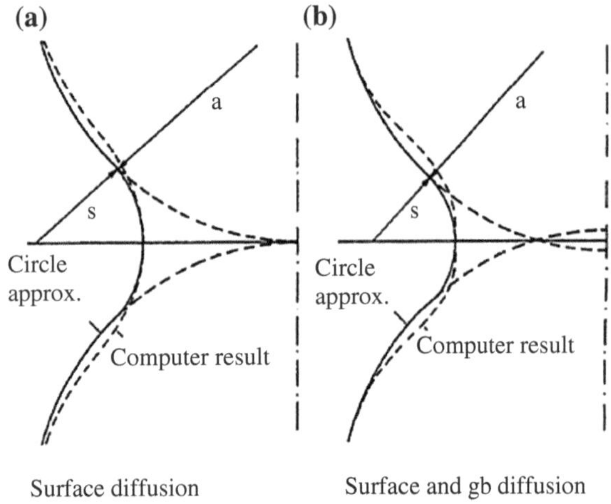

Fig. 5.21 Contours of necks between cylinders for sintering by **a** surface diffusion and **b** simultaneous surface and grain boundary diffusion. From Ref. [27]. Reproduced with permission from [40]. Copyright © 1979, Elsevier

Numerical simulation with finite element modeling has been used to study viscous sintering [42, 43, 46–50]. It is predicted that the flow is axially downward and radially outward near the neck, with most of the energy dissipated near the neck, which is in a qualitative agreement with the Frenkel model [46, 48]. Another example is the simulation of the sintering of two rigid particles that are coated with an amorphous layer, as shown in Fig. 5.22 [51]. Similar finite element modeling is used in this study. The simulation results indicate that full density can be achieved at a rate comparable to a packing of particles without a rigid core, if the coatings are sufficiently thick.

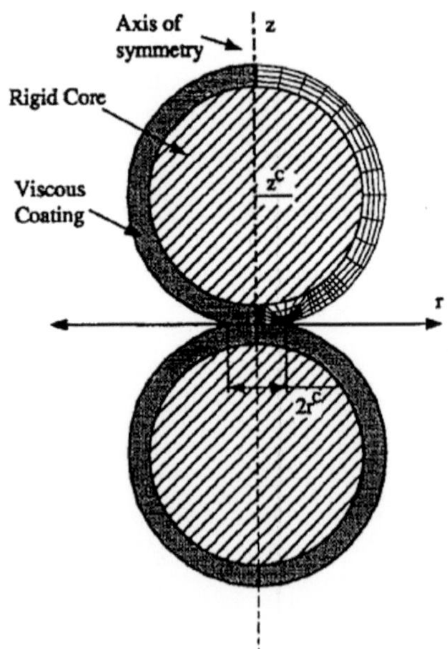

Fig. 5.22 Finite element simulation of viscous sintering of two rigid particles coated with an amorphous layer. Reproduced with permission from [51]. Copyright © 1994, John Wiley & Sons

5.3.8 Phenomenological Sintering Equations

Phenomenological approach is the development of empirical equations that can be used to fit the experimental data of sintering, in terms of density or shrinkage, as a function of time. Phenomenological sintering equations are useful to some numerical models, when equations for densification of a powder system are necessary. There is a simple expression that can be used to fit sintering and hot pressing data, which is given by

$$\rho = \rho_0 + K \ln \frac{t}{t_0}, \tag{5.172}$$

where ρ_0 is the density at an initial time t_0, ρ is the density at time t, and K is a temperature-dependent parameter, which is known as semi-logarithmic law. Using rate equation of Coble's intermediate or final stage model for sintering through lattice diffusion, e.g., Eq. (5.165), there is [27]

$$\frac{\mathrm{d}\rho}{\mathrm{d}t} = \frac{AD_1\gamma_{SV}\Omega}{G^3kT}, \tag{5.173}$$

where A is a constant that is dependent on the stage of sintering. If it is assumed that the grains grow following a cubic law, i.e.,

$$G^3 = G_0^3 + \alpha t \approx \alpha t, \tag{5.174}$$

and that the initial grain size is very small, i.e., $G^3 \gg G_0^3$, Eq. (5.173) can be simplified as

$$\frac{\mathrm{d}\rho}{\mathrm{d}t} = \frac{K}{t}, \tag{5.175}$$

where $K = AD_1\gamma_{SV}/(\alpha kT)$. Integration of Eq. (5.175) leads to Eq. (5.172), which is expected to be valid for both the intermediate and final stages of sintering, because Eq. (5.174) has the same form for both sintering stages.

In most cases, if grain growth is significantly prohibited, shrinkage data can be fitted by the following equation:

$$\frac{\Delta L}{L_0} = Kt^{1/\beta}, \tag{5.176}$$

where K is a temperature-dependent parameter and β is an integer. This equation has the same form as the initial stage shrinkage equations derived from the analytical models.

There are also various empirical sintering equations that can be found in the open literature. One of them is as follows:

$$\frac{V_0 - V_t}{V_0 - V_\mathrm{f}} = Kt^n, \tag{5.177}$$

where V_0 is the initial volume of the powder compact, V_t is the volume after sintering for time t, V_f is the volume of the fully dense solid, and K is a temperature-dependent parameter. Depending on materials, n has values between 0.5 and 1.0. Another equation is as follows:

$$\frac{V_t^\mathrm{P}}{V_0^\mathrm{P}} = (1 + C_1 mt)^{-1/m}, \tag{5.178}$$

where V_0^P is the initial pore volume of the compact, V_t^P is the pore volume after sintering for time t, while C_1 and m are constants. Although these empirical equations are successful in fitting various experimental sintering data, it is difficult for the fitted parameters to have physical meanings [52]. Therefore, it is necessary to be careful when using these empirical equations, before any meaningful understanding is available.

5.3.9 Sintering Diagrams

As discussed earlier, more than one mechanism commonly operates simultaneously during the sintering of polycrystalline systems. Besides the numerical simulations that can provide theoretical frameworks for the analysis of sintering with multiple mechanisms, sintering diagrams are even more practical approaches [33, 34]. Sintering diagrams demonstrate the dominant mechanism of sintering and the net rate of neck growth, for a given temperature and neck size, or density, to show the densification.

To construct a sintering diagram, it is necessary to have neck growth equations for specified geometrical models, usually geometrical models discussed above, and data of the material properties, such as diffusion coefficients, surface energy, and atomic volume, which included in the equations [33]. Numerical methods are used to construct the sintering diagrams, with the assumption that the total neck growth rate is the sum of all the neck growth rates for the individual mechanisms. In this respect, sintering diagrams are very close to the analytical models discussed earlier. In practice, a slight variation in the characteristics of the powders can cause significant changes in the material parameters, e.g., diffusion coefficients, so that each system should have its specific sintering diagrams.

5.3.10 Sintering at Pressures

The application of an external stress or pressure to the powder system during heating has been widely used to promote the sintering of ceramics. Various techniques have been developed for such a purpose, which can be classified into three types: (i) hot pressing, (ii) hot isostatic pressing, and (iii) hot forging. There have been models for hot pressing by considering the diffusional mass transport under the driving forces of surface curvature and applied stress [53].

5.3.10.1 Hot Pressing Models

Analytical sintering models are based on the idealized models for the three stages of sintering shown in Fig. 8.8. Because the vacancy concentration under the neck surface is not affected by the external stress, Eq. (5.134) is still valid, i.e., there is

$$\Delta C_v = C_v - C_{v,0} = \frac{C_{v,0}\gamma_{SV}\Omega}{kT}\left(\frac{1}{r_1}+\frac{1}{r_2}\right) = \frac{C_{v,0}\gamma_{SV}\Omega K}{kT}, \tag{5.179}$$

where K is the curvature of the pore surface. For the initial stage of sintering, $K = 1/r = 4a/X^2$, while for the intermediate and final stages, $K = 1/r$ and $2/r$, respectively, where r is the radius of the pores, a is the radius of the particles, and X is the radius of the neck. The stress p_a applied to the powder compact produces a stress p_e on the grain boundaries. Due to the presence of the pores, p_e is higher than p_a, which can be expressed as

$$p_e = \phi p_a, \tag{5.180}$$

where ϕ is a factor, which is also known as the stress intensification factor. The compressive stress on the grain boundary means that the concentration of vacancy is less than that of a flat stress-free boundary, so that there is

$$\Delta C_{v,b} = -\frac{C_{v,0}p_e\Omega}{kT} = -\frac{C_{v,0}\phi p_a\Omega}{kT}. \tag{5.181}$$

In the two-sphere model at the initial stage of sintering, ϕ is usually assumed to be equal to the area of the sphere projected onto the punch of the hot pressing die divided by the cross-sectional area of the neck, i.e., $\phi = 4a^2/\pi X^2$, while in those at the intermediate and final stages, it is supposed that $\phi = 1/\rho$, where ρ is the relative density of the body to be sintered.

Using the parameters for K and ϕ, the variations of $\Delta C_{v,p}$ and $\Delta C_{v,b}$ can be obtained [53]. For hot pressing, the difference in the concentration of vacancy between the neck surface and the grain boundary is given by $\Delta C = \Delta C_{v,p} - \Delta C_{v,b}$, therefore, for the initial stage, there is

$$\Delta C = \frac{C_0 \Omega 4a}{kTX^2}\left(\gamma_{SV} + \frac{p_a a}{\pi}\right). \tag{5.182}$$

This equation indicates that, at the initial stage, ΔC for hot pressing is similar to that for normal sintering, but $\gamma_{SV} + p_a a/\pi$ is used instead of γ_{SV}. Because p_a and a are constant, hot pressing equations can be derived from the sintering equations by simply changing γ_{SV} into $\gamma_{SV} + p_a a/\pi$.

Creep equations can be appropriately modified as models of intermediate and final stages of sintering. For simplicity, the matter transport during creep of a dense solid is considered first. For a pure single crystal solid with cubic structure, which is a rod with a cross section of length L. Normal stresses p_a are applied to the rod on the sides, as shown in Fig. 5.23a. It is assumed that self-diffusion within the crystal

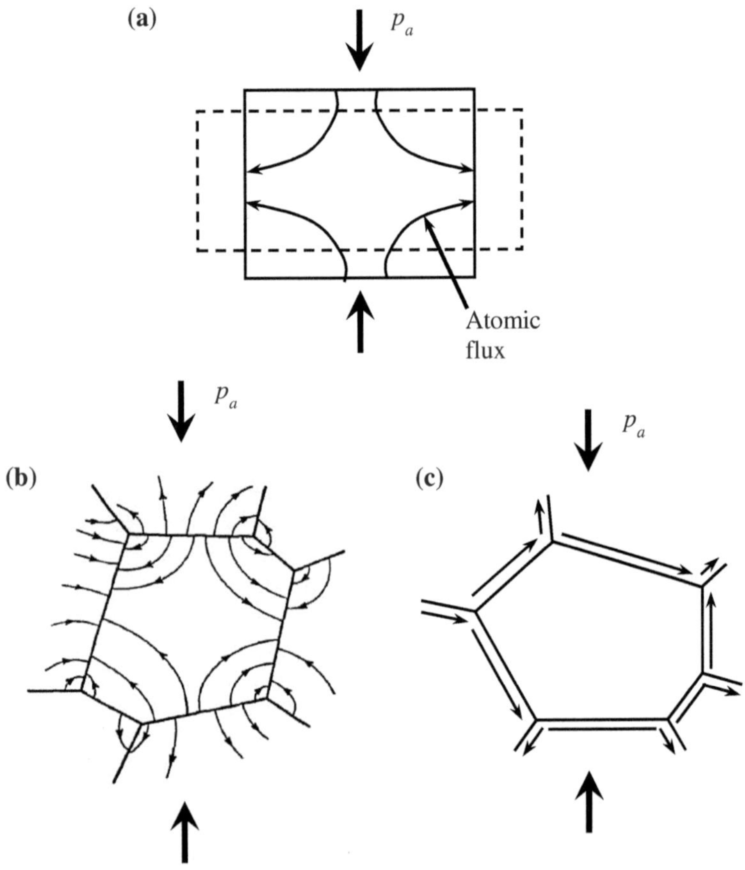

Fig. 5.23 **a** A single crystal subjected to a uniaxial stress, showing the direction of atomic flux. A representative grain in polycrystalline solids, showing the expected atomic flux by lattice diffusion (**b**) and grain boundary diffusion (**c**). **a**, **c** Reproduced from [1]. Copyright © 2003, CRC Press. **b** Reproduced with permission from [54]. Copyright © 1950, American Institute of Physics

solid causes it to deform, i.e., creep, so as to relieve the stresses [54]. During the creep, the atoms diffuse from interfaces subjected to a compressive stress, where they have a higher chemical potential, toward those regions subjected to a tensile stress, where they have lower chemical potential. This concept of creep can be extended to polycrystalline ceramics, i.e., the self-diffusion within the individual grains causes atoms to diffuse from the grain boundaries under compression toward the boundaries under tension, as shown in Fig. 5.23b. Creep by lattice diffusion is usually known as Nabarro-Herring creep, with the creep rate being given by

$$\dot{\varepsilon}_c = \frac{40}{3}\frac{D_l \Omega p_a}{G^2 kT}, \tag{5.183}$$

where D_l is the lattice diffusion coefficient, Ω is the atomic volume, p_a is the applied stress, G is the grain size, k is the Boltzmann constant, and T is the absolute temperature. The creep rate is a linear strain rate, defined as $(1/L)\mathrm{d}L/\mathrm{d}t$, where L is the length of the solid and t is time.

As shown in Fig. 5.23c, creep in ceramics can also be caused by the diffusion along the grain boundaries, with the creep equation being given by [55, 56]:

$$\dot{\varepsilon}_c = \frac{47.5 D_{gb} \delta_{gb} \Omega p_a}{G^3 kT}, \tag{5.184}$$

where D_{gb} is the grain boundary diffusion coefficient and δ_{gb} is the thickness of the grain boundaries. In Eqs. (5.183) and (5.184), the creep rate is linearly dependent on p_a, which has different dependences on the grain size and different numerical constants.

If the applied stress is sufficiently high, dislocations could be produced in some ceramics, which can facilitate matter transport as a mechanism of sintering, with creep rate given by

$$\dot{\varepsilon}_c = \frac{AD\mu b}{kT}\left(\frac{p_a}{\mu}\right)^n, \tag{5.185}$$

where A is a numerical constant, D is diffusion coefficient, μ is the shear modulus, and b is the Burgers vector, while the exponent n is dependent on the mechanism of the dislocation motion, with values in the range of 3–10.

Hot pressing equations can be derived from the creep equations with certain modifications. In practices, sintering data are usually present as density versus time to obtain densification rate. Because sintering compacts are porous before they are fully densified, the creep equations developed for dense solids should be modified to be suitable for hot pressing equations. First, the creep rate that is a linear strain rate should be related to the densification rate which is a volumetric strain rate. Second, the porosity should be compensated. In hot pressing, the mass M of the powder and the cross-sectional area A of the die, are constants, while the density

D increases and the sample thickness L decreases, with the relation of D and L being given by

$$\frac{M}{A} = LD = L_0 D_0 = L_f D_f, \tag{5.186}$$

where the subscripts 0 and f stand for the initial and final states. Differentiating Eq. (5.186) with respect to time gives

$$L\frac{dD}{dt} + D\frac{dL}{dt} = 0. \tag{5.187}$$

Rearranging Eq. (5.187), there is:

$$-\frac{1}{L}\frac{dL}{dt} = \frac{1}{D}\frac{dD}{dt} = \frac{1}{\rho}\frac{d\rho}{dt}, \tag{5.188}$$

where ρ is relative density. Equation (5.188) indicates that the densification rate is related to the linear strain rate of the body during the hot pressing. The linear strain rate can be measured directly by monitoring movement of the punch during the hot pressing.

In Coble's initial stage hot pressing model, the effective stress applied to the grain boundaries, p_e, is related to the externally applied stress, p_a, which is by Eq. (5.180). In that model, the total driving force (*DF*) is a linear combination of two effects: (i) external applied stress and (ii) surface curvature, which determines the densification rate, and it is

$$DF = p_e + \gamma_{SV}K = p_a\phi + \gamma_{SV}K, \tag{5.189}$$

where K is the curvature of the pore, which is equal to $1/r$ for intermediate stage model and $2/r$ for final stage model of the sintering. When this *DF* is used to replace the applied stress p_a in the creep equations for dense solids, hot pressing equations for sintering can be derived.

The modified creep equations should be further elaborated for real scenario of hot pressing sintering of ceramics, due to the differences between the two processes, such as those in the atomic flux field and the path length for diffusion. For example, in the creep models, the atomic flux is terminated at the grain boundaries under tension, while in hot pressing it is terminated at the surfaces of pores. During the creep of dense solids, because the grain size is not changed, the areas of grain boundaries are kept unchanged. However, during the hot pressing sintering, but both the areas of the grain boundaries and the path length for diffusion are increased with time.

5.3.10.2 Hot Pressing Mechanisms

There is no significant difference in mechanism between the normal sintering and hot pressing sintering. However, due to the application of an external stress, the densifying mechanisms are largely enhanced, while the nondensifying mechanisms are almost not affected. Therefore, it is not necessary to consider all the nondensifying mechanisms. This difference in the effect of the applied stress can be used to identify the mechanisms of densification, on the one hand, while on the other hand, new densification mechanisms could be brought out.

Figure 5.24 shows a schematic cross-sectional view of a representative system, consisting of three grains with an ideal hexagonal shape [1]. This diagram can be used to describe the overall shape change of powder compacts during hot pressing. Within the hot pressing die with fixed diameter, the powder compact would contract only in the direction of the external pressure, thus leading to flattening of the grains, as shown in Fig. 5.24, which is accompanied by sliding of grains and grain boundaries. Therefore, the major mechanisms that govern the sintering during hot pressing include rearrangement, lattice diffusion, grain boundary diffusion, viscous flow, plastic deformation through dislocation motion and grain boundary sliding.

If the applied stress is much higher than the driving force due to the curvature, the densification rate during hot pressing can be expressed as

$$\frac{1}{\rho}\frac{\mathrm{d}\rho}{\mathrm{d}t} = \frac{HD\phi^n}{G^m kT}p_\mathrm{a}^n, \tag{5.190}$$

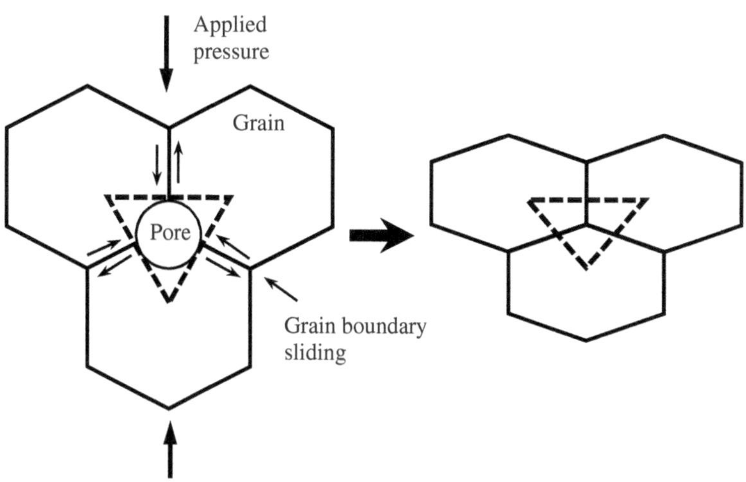

Fig. 5.24 Sketch illustrating the change in shape of the grain that occurs during hot pressing. The grains are flattened in the direction of the applied pressure. When matter transport occurs by diffusion, grain boundary sliding is necessary to accommodate the change in grain shape. Reproduced with permission from [1]. Copyright © 2003, CRC Pressing

where H is a numerical constant, D is the diffusion coefficient of the rate-controlling species, ϕ is the stress intensification factor, G is grain size, k is the Boltzmann constant, T is absolute temperature, while the exponents m and n are dependent on the mechanism of the densification. For example, lattice diffusion mechanism has $m = 2$ and $n = 1$, while for grain boundary diffusion, there are $m = 3$ and $n = 1$.

Equation (5.190) indicates that the plot of densification rate versus p_a can be used to derive the exponent n, so that the mechanism of densification can be identified. The applied pressures for commonly used hot pressing are 10–50 MPa, with which $n \approx 1$ is usually observed for many ceramics, corresponding to the diffusion mechanism for densification [57, 58]. High values of n are occasionally observed, indicating the possibility of dislocation mechanisms [59, 60]. The dominant densification mechanism can be varied with the applied pressure, temperature, and grain size. Similar to the sintering maps discussed above, there are also hot pressing maps to demonstrate the dominant mechanisms versus processing conditions.

5.3.11 Stress Intensification Factor

As mentioned before, there is an important factor, ϕ, in the hot pressing models, which is known as stress intensification factor or stress multiplication factor. It is used to relate the mean stress applied to the grain boundary, p_e, to the externally applied stress, p_a. The significance of ϕ is such that while p_a is the stress that is measured, p_e is its counterpart that influences the rate of matter transport. The factor ϕ is geometrically dependent on the porosity and the shape of the pores of the ceramic compacts.

When a hydrostatic pressure p_a is applied to the external surface of a powder compact, it can be represented by using a model, as shown schematically in Fig. 5.25a [1]. In this case, the applied pressure exerts a load on the surface of the solid, which is $F_a = A_T p_a$, where A_T is the total external cross-sectional area of the solid, including areas pores. Due to presence of porosity at the grain boundaries, the actual grain boundary area A_e is smaller than the total external area. If a force balance is maintained across any plane of the solid, the following expression is valid:

$$p_a A_T = p_e A_e. \tag{5.191}$$

Therefore, stress intensification factor is given by

$$\phi = \frac{p_e}{p_a} = \frac{A_T}{A_e}. \tag{5.192}$$

If the pores are assumed to be spherical and randomly distributed in a porous solid, the factor ϕ can be obtained through the following steps. It is assumed that there is

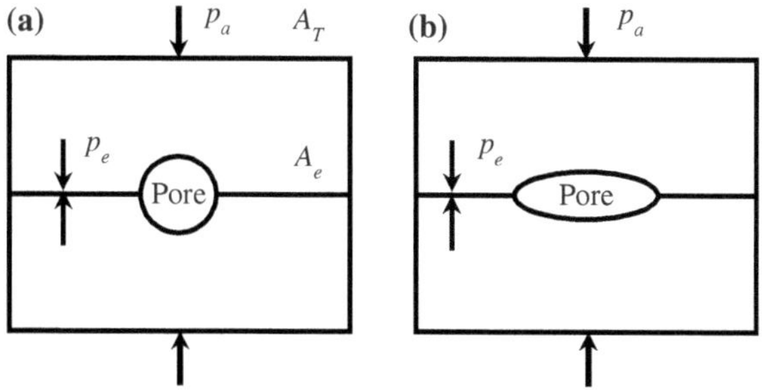

Fig. 5.25 **a** The effective stress applied to the grain boundary p_e is higher than the externally applied stress p_a, due to the presence of the pores at the grain boundaries. The stress intensification factor ϕ is defined as the ratio of the total external area to the actual grain boundary area. **b** ϕ is dependent on the porosity and the pore shape, with stronger dependence on porosity for pores with smaller dihedral angles, e.g., elliptical pores. Reproduced with permission from [1]. Copyright © 2003, CRC Press

an arbitrary plane through the solid, with the area fraction of porosity in the plane being equal to the volume fraction of porosity in the solid. If A_T is unity, there is $A_e = 1 - P$, where P is the porosity of the solid. Because $1 - P$ is also equal to the relative density, there is the following relation:

$$\phi = \frac{1}{\rho}. \tag{5.193}$$

This equation is valid for polycrystalline ceramics with equilibrium shapes of the isolated pores to be nearly spherical, i.e., dihedral angles are larger than 150°. Once the pores become nonspherical, the expression of ϕ could have very complicated form. If the shapes of the pores are dramatically changed, as shown in Fig. 5.25b [1], although the volume is the same, the value of ϕ could be changed, because it is now dependent on not only the porosity but also the shape of the pores. When the pore shape is not spherical, the dihedral angles are reduced, so that the actual area of the grain boundary is decreased, therefore, ϕ will be decreased.

An analysis of simulation results of a continuous network of pores with equilibrium shapes indicates that ϕ can be approximated using a simple expression, which is [59, 61, 62]:

$$\phi = \exp(\alpha P), \tag{5.194}$$

where α is a factor that is dependent on the dihedral angle and P is the porosity. The factor α can be determined with simulation results, while the equation has been verified by some experimental results. However, further studies are necessary to increase the significance of ϕ [63–66].

5.3.12 Sintering Stress

The densification rate for the diffusion-controlled mechanisms may be expressed as a general form, which is given by

$$\frac{1}{\rho}\frac{\mathrm{d}\rho}{\mathrm{d}t} = -\frac{3}{L}\frac{\mathrm{d}L}{\mathrm{d}t} = \frac{3}{\eta_\rho}(p_\mathrm{a}\phi + \sigma), \tag{5.195}$$

where $(1/L)\mathrm{d}L/\mathrm{d}t$ is the linear strain rate of the sintering compact, η_ρ has the dimensions of a viscosity that can be called densification viscosity, while σ is the effective stress applied to the atoms under the pore surface, which is given by the Young and Laplace equation:

$$\sigma = \gamma_\mathrm{SV}\left(\frac{1}{r_1} + \frac{1}{r_2}\right), \tag{5.196}$$

where r_1 and r_2 are the two principal radii of curvature of the pore surface. The quantity σ is the thermodynamic driving force for sintering. It has the units of a pressure or stress, which is usually known as sintering pressure or sintering potential. The equation of σ is actually more complex for polycrystalline ceramics, because the pores are in contact with the grain boundaries. It has two contributions, one from the pores and the other from the grain boundaries [67]. For an idealized final stage microstructure, where the pores and grains are assumed to be spherical in shape, σ is given by [68]

$$\sigma = \frac{2\gamma_\mathrm{gb}}{G} + \frac{2\gamma_\mathrm{SV}}{r}, \tag{5.197}$$

where γ_gb is the specific energy of the grain boundary, G is the grain size, and r is the radius of the pore. The driving force for sintering is also influenced by the dihedral angle and the mass transport mechanism, which has been calculated for a simple geometry, consisting of a line of spherical particles [69].

Another form of Eq. (5.195) is

$$\frac{1}{\rho}\frac{\mathrm{d}\rho}{\mathrm{d}t} = \frac{3\phi}{\eta_\rho}(p_\mathrm{a} + \Sigma), \tag{5.198}$$

where $\Sigma = \sigma/\phi$ has the units of stress, which known as sintering stress. Because Σ has a linear relationship with the externally applied stress p_a, it can be used to represent the equivalent externally applied stress, with an effect on sintering as same as those of the curved surfaces of pores and grain boundaries.

5.3.13 Alternative Sintering Equations

Besides those discussed above, the sintering equations can be derived in an alternative way by solving the differential equations of the atomic flux that is subject to appropriate boundary conditions [70].

5.3.13.1 Grain Boundary Diffusion

A simple geometrical model consisting of spherical particles with a diameter a, arranged as a simple cubic pattern, as shown in Fig. 5.26a, is used to derive the sintering equations for the mechanism of grain boundary diffusion [70]. It is

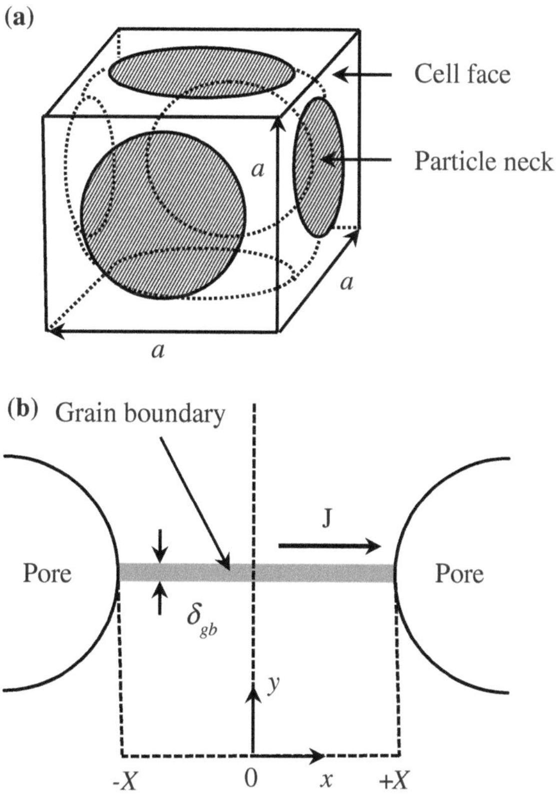

Fig. 5.26 **a** Schematic of the cell surrounding a spherical particle showing the neck or grain boundary section with an adjoining particle. Reproduced with permission from [70]. Copyright © 1991, John Wiley & Sons. **b** Geometrical parameters for matter transport through grain boundary diffusion. Reproduced with permission from [1]. Copyright © 2003, CRC Press

assumed that the grain boundary is flat, with a constant width δ_{gb}, as shown in Fig. 5.26b [1]. From Eq. (5.92), the flux of atoms as a function of distance along the neck can be expressed as

$$j(x) = -\frac{D_{gb}}{\Omega kT}\nabla\mu, \tag{5.199}$$

where D_{gb} is the grain boundary diffusion coefficient, Ω is the atomic volume, k is the Boltzmann constant, T is the absolute temperature, and μ is the chemical potential of the atoms. The total number of atoms crossing the neck at a radius x per unit time is given by

$$J(x) = 2\pi x\delta_{gb}j(x) = -\frac{2\pi x D_{gb}\delta_{gb}}{\Omega kT}\nabla\mu. \tag{5.200}$$

Because the displacement of the boundary must be independent of x, the rate to approach of the centers of the particle, dy/dt, is related to $J(x)$ through the following expression:

$$J(x) = \frac{\pi x^2}{\Omega}\frac{dy}{dt}. \tag{5.201}$$

From Eqs. (5.200) and (5.201), there is

$$\frac{d\mu}{dx} = -2Ax, \tag{5.202}$$

where

$$A = -\frac{kT}{4D_{gb}\delta_{gb}}\frac{dy}{dt}. \tag{5.203}$$

Integrating Eq. (5.202) yields:

$$\mu(x) = -Ax^2 + B, \tag{5.204}$$

where B is a constant. The chemical potential μ is related to the normal stress applied to the boundary σ by $\mu = \sigma\Omega$, so that Eq. (5.204) becomes

$$\sigma(x) = \frac{-Ax^2 + B}{\Omega}, \tag{5.205}$$

where the constants A and B can be derived from the corresponding boundary conditions. For the first boundary condition, the stresses must be balanced at $x = X$ and $x = -X$. According to the definition of sintering stress, the effects of the pores and grain boundaries are replaced by using an equivalent external stress Σ. Therefore, the stress applied to the surface of the pore should be equal to zero. In this case, the second boundary condition is $\sigma = 0$ at $x = \pm X$. Substituting either of them into Eq. (5.205) yields

$$B = Ax^2. \tag{5.206}$$

The second boundary condition also means that the average stress applied to the grain boundary is equal to $\phi\Sigma$, where ϕ is the stress intensification factor. This condition can be expressed as

$$\frac{1}{\pi X^2}\int_0^{2\pi}\int_0^{x} \sigma(x)\mathrm{d}x\mathrm{d}\theta = \phi\Sigma. \tag{5.207}$$

Substituting $\sigma(x)$ and integrating the equation yield

$$-\frac{AX^2}{2} + B = \Omega\phi\Sigma. \tag{5.208}$$

Together with Eq. (5.206), there are

$$A = \frac{\Omega\phi\Sigma}{X^2}, \quad B = 2\Omega\phi\Sigma. \tag{5.209}$$

From Eq. (5.200), the total flux at the surface of the neck between two particles is thus given by

$$J(X) = \frac{8\pi D_{\mathrm{gb}}\delta_{\mathrm{gb}}\phi\Sigma}{kT}. \tag{5.210}$$

When $J(X)$ is related to the shrinkage of the system, the total volume transported out of one neck in a time Δt is given by

$$\Delta V = -J(X)\Omega\Delta t = \pi X^2\Delta a, \tag{5.211}$$

where Δa is the corresponding change in the center-to-center distance between the particles. The total volumetric shrinkage in all three orthogonal directions is given by

$$\frac{\Delta V}{V} = \frac{3\Delta a}{a}, \tag{5.212}$$

and because $V = a^3$, the rate of change of the cell volume is given by

$$\frac{\mathrm{d}V}{\mathrm{d}t} = \frac{\Delta V}{\Delta t} = 3a^2\frac{\Delta a}{\Delta t}. \tag{5.213}$$

From Eq. (5.211), $\Delta a/\Delta t$ is obtained, which is put into Eq. (5.213), there is

$$\frac{\mathrm{d}V}{\mathrm{d}t} = -3a^2 \frac{J(X)\Omega}{\pi X^2}. \tag{5.214}$$

By definition, $\phi = a^2/(\pi X^2)$, so that the instantaneous volumetric strain rate is given by

$$\frac{1}{V}\frac{\mathrm{d}V}{\mathrm{d}t} = -\frac{3J(X)\Omega\phi}{a^3}. \tag{5.215}$$

The linear densification strain rate $\dot{\varepsilon}_\rho$ is defined as $(-1/3\ V)\ \mathrm{d}V/\mathrm{d}t$, which is used to substitute $J(X)$ in Eq. (5.210), there is

$$\dot{\varepsilon}_\rho = \frac{8\pi D_{\mathrm{gb}}\delta_{\mathrm{gb}}\Omega}{a^3 kT}\phi^2\Sigma. \tag{5.216}$$

5.3.13.2 Lattice Diffusion

The sintering equation for lattice diffusion can be derived in a similar way. The linear densification strain rate can also be derived from Eq. (5.216) by replacing $\pi D_{\mathrm{gb}}\delta_{\mathrm{gb}}$ with $2XD_{\mathrm{l}}$, where D_{l} is the lattice diffusion coefficient [71]. Because there is $\phi = a^2/(\pi X^2)$, the linear densification strain rate is given by

$$\dot{\varepsilon}_\rho = \frac{16 D_{\mathrm{l}}\Omega}{\pi^{1/2} a^2 kT}\phi^{3/2}\Sigma. \tag{5.217}$$

5.3.13.3 General Isothermal Sintering Equation

For matter transport by diffusion, a general equation for the linear densification strain rate can be expressed as

$$\dot{\varepsilon}_\rho = \frac{H_1 D\Omega\phi^{(m+1)/2}}{G^m kT}(\Sigma + p_{\mathrm{h}}), \tag{5.218}$$

where H_1 is a numerical constant that is dependent on the geometry of the model, p_{h} is the hydrostatic component of an externally applied stress, G is the particle or grain size, k is the Boltzmann constant, T is the absolute temperature, and Ω is the atomic volume. At the same time, $D = D_{\mathrm{gb}}\delta_{\mathrm{gb}}$ and $m = 3$ for grain boundary diffusion, while $D = D_{\mathrm{l}}$ and $m = 2$ for lattice diffusion.

5.4 Liquid-Phase Sintering

5.4.1 Brief Introduction

The sintering processes discussed above are all solid-state sintering, in which the materials remain in solid state throughout the sintering process. However, if there is a liquid phase present during the sintering, the sintering process can be significantly enhanced, which is known as liquid-phase sintering [72]. Usually, liquid-phase sintering is used to enhance densification rates, achieve accelerated grain growth, or produce specific grain boundary properties. The distributions of the liquid phase and the resulting solidified phases produced on cooling after densification are critical to achieving the required properties of the sintered ceramics. In most cases, the amount of liquid formed during sintering should be small, which can make it quite difficult to precisely control the composition of the liquid.

Liquid-phase sintering is more effective for ceramics that have a high degree of covalent bonding, and thus are difficult to densify through solid-state sintering. A disadvantage of liquid-phase sintering is the formation of glassy intergranular phase due to the solidification of the liquid phase, which could degrade the properties of the final ceramics. Because optical properties are especially sensitive as compared with other properties, it should be very careful when using liquid-phase sintering to process transparent ceramics.

A related process is called activated sintering, in which the additives segregate at grain boundaries to enhance the mass transport rates and thus accelerate densification. If sufficient amount of liquid is present with contents in the range of 25–30 vol.%, rearrangement of the solid phase coupled with liquid flow can lead to a fully dense material. Such large volume fractions of liquid are rarely used in transparent ceramics. In this case, the sintering process is known as vitrification.

5.4.2 Characteristics of Liquid-Phase Sintering

5.4.2.1 Densification Enhancement

The enhancement in densification during the liquid-phase sintering can be attributed to the (i) enhanced rearrangement of particles due to surface wetting effect and (ii) enhanced matter transport through the liquid due to its low viscosity. The microstructural characteristics of liquid-phase sintering can be compared with those of solid-state sintering by using the idealized two-sphere model, as shown in Fig. 5.27 [1]. If the liquid phase can wet the particles and has sufficient quantity to cover the particle surfaces, all the particles are separated one another by a liquid layer, which can greatly decrease the friction between every two adjacent particles. As a result, the particles would have higher mobility, thus making their rearrangement more easily. In contrast, solid-state sintering has no such effect. For grain boundary

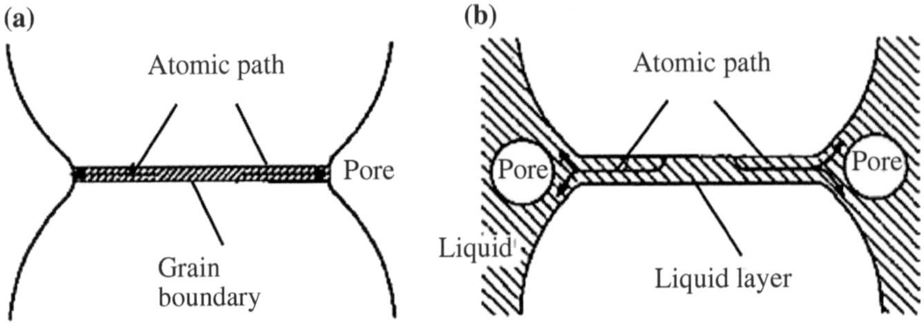

Fig. 5.27 Schematic diagrams of the idealized two-sphere model showing microstructural characteristics of **a** solid-state sintering and **b** liquid-phase sintering. Reproduced with permission from [1]. Copyright © 2003, CRC Press

diffusion, the rate of diffusion for solid-state sintering is controlled by the product of the grain boundary diffusion coefficient D_{gb} and the grain boundary thickness δ_{gb}, whereas that for liquid-phase sintering is determined by the product of the diffusion coefficient D_L of the species in the liquid and the thickness of the liquid layer δ_L. Because $\delta_L > \delta_{gb}$ and $D_L > D_{gb}$, $\delta_L D_L \gg \delta_{gb} D_{gb}$, i.e., the liquid can promote matter transport.

5.4.2.2 Driving Force for Densification

Because the liquid wets and spreads over the solid surfaces, pores will be formed in the liquid. The reduction of the liquid–vapor interfacial area provides the driving force for shrinkage or densification of the compact. If the pore in the liquid is assumed to be spherical with radius of r, the pressure difference across the curved surface is given by the Young and Laplace equation:

$$\Delta p = -\frac{2\gamma_{LV}}{r}, \tag{5.218}$$

where γ_{LV} is the specific surface energy of the liquid–vapor interface. The pressure applied to the liquid phase is lower than that applied to the pore, which leads to a compressive capillary stress to the particles. Therefore, the effect of the compressive stress due to the liquid is equivalent to the application of an external hydrostatic pressure to the system, as the driving force for sintering.

5.4.2.3 Formation of Liquid Phase

The formation of liquid phase during the liquid-phase sintering is schematically shown in Fig. 5.28 [73]. It is started from a mixture of two powders, i.e., the major component and an additive. The additive is molten to liquid phase or reacts with the

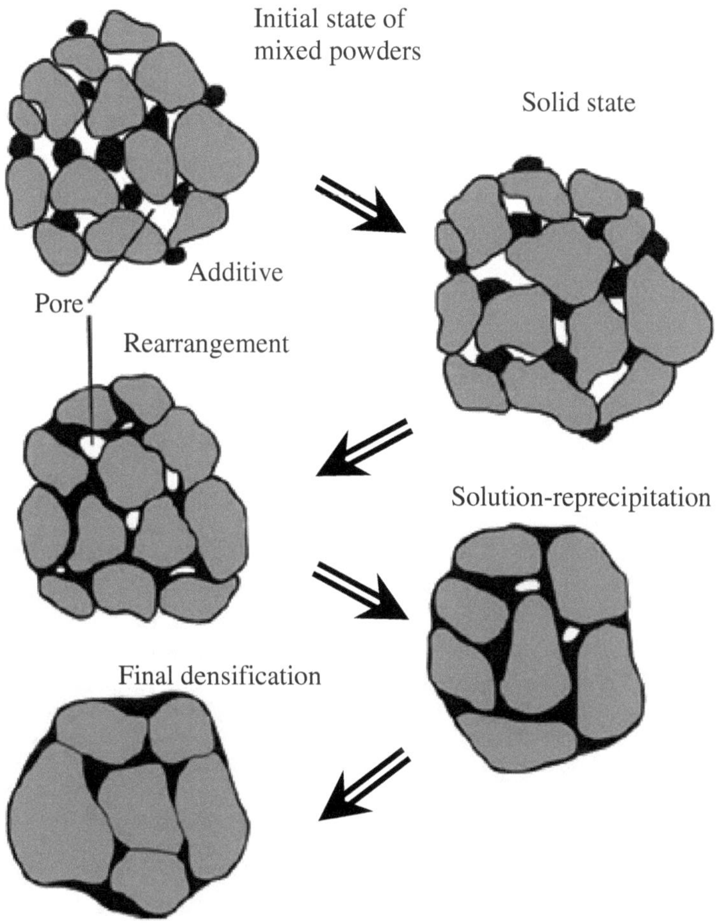

Fig. 5.28 Schematic of the microstructure variation in a liquid-phase sintering, starting with mixed powders, and pores between the particles. During heating the particles are sintered, but when a melt forms and spreads, the solid grains rearrange. Subsequent densification is accompanied by coarsening. In many cases, there is pore annihilation as diffusion in the liquid to accelerate the change in grain shape, so that pores are removed. Reproduced with permission from [73]. Copyright © 2009, Springer

major component to form a eutectic liquid. If it is preferred to form eutectic liquids, the additive and the firing conditions should be properly selected. Due to the small quantity of the liquid phase, the structure of the compact can be well remained, because the particles are held together by the capillary stress created by the liquid. However, the system has lower effective viscosity than its counterparts without a liquid phase. If the liquid is present throughout the sintering process, it is called persistent liquid-phase sintering, where the liquid precipitates at grain boundaries as glass phase after the sintering. Otherwise, it is called transient liquid-phase sintering, in which the liquid might disappear due to (i) incorporation into the solid phase to form solid solutions, (ii) formation of crystal phase through crystallization, or (iii) evaporation.

5.4.2.4 Microstructures

The ceramics fabricated using liquid-phase sintering contains two phases, i.e., (i) the major phase of crystalline grains and (ii) the grain boundary phase resulting from the liquid. In most cases, the grain boundary phase is amorphous. The quantity of the liquid phase determines the characteristics of the grain boundaries and thus the properties of the grains, such as grain shape, size, and morphology.

5.4.3 *Stages of Liquid-Phase Sintering*

Because chemical reactions between the particles and the liquid phase can be neglected, the sintering rate is mainly controlled by the interfacial energies. In this case, liquid-phase sintering generally has three stages, as shown in Fig. 5.28 [73]. The first stage is the rearrangement of the particles caused by the capillary stress gradients due to the redistribution of the liquid phase. At the second stage, the compact experiences densification, through the solution-re/precipitation process. Finally, the compact undergoes grain coarsening and further densification before the whole sintering is finished.

5.4.4 *Thermodynamic and Kinetic Factors*

Several kinetic and thermodynamic factors, as well as processing parameters, determine the microstructures of the ceramics by using liquid-phase sintering.

5.4.4.1 Wetting and Spreading of the Liquid

The capability of the liquid to wet and spread over the surface of the solid particle determines the effectiveness of liquid-phase sintering. The wetting capability of a liquid is dependent on its surface tension, i.e., lower the surface tension, the high the wetting capability the liquid has, which is usually characterized by a parameter, known as contact angle.

The contact angle is determined by the various interfacial energies of the solid–liquid–vapor systems, which is demonstrated using a droplet of liquid on the surface of a flat solid, as shown schematically in Fig. 5.29. When the specific energies

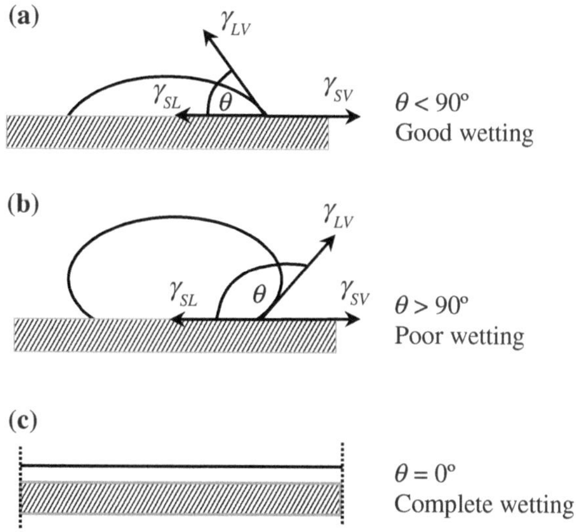

Fig. 5.29 Wetting behavior between a liquid and a solid showing **a** good wetting, **b** poor wetting, and **c** complete wetting for a liquid with a contact angle of θ

of the liquid–vapor, solid–vapor, and solid–liquid interfaces are represented with γ_{LV}, γ_{SV}, and γ_{SL}, respectively, at equilibrium state, there is

$$\gamma_{SV} = \gamma_{SL} + \gamma_{LV} \cos\theta. \tag{5.219}$$

Therefore, the contact angle is given by

$$\cos\theta = \frac{\gamma_{SV} - \gamma_{SL}}{\gamma_{LV}}. \tag{5.220}$$

5.4.4.2 Dihedral Angle

Figure 5.30 shows a two-dimensional diagram for a liquid in contact with the corner of two grains, where the grain boundaries intersect the surface of the liquid [73]. The angle between the solid–liquid interfacial tensions is defined as dihedral angle. Similar to Eq. (5.220), the dihedral angle can be obtained by the following expression:

$$\cos\frac{\psi}{2} = \frac{\gamma_{SS}}{2\gamma_{SL}}. \tag{5.221}$$

The solid–solid interfacial tension γ_{SS} is the same as the interfacial tension in the grain boundary γ_{gb} in the solid-state sintering.

The penetration of the liquid into the grain boundary is described by the dihedral angle ψ, which is determined by the ratio γ_{SS}/γ_{SL}. With $\gamma_{SS}/\gamma_{SL} < 2$, the dihedral angle has values of 0–180° and the liquid cannot penetrate the grain boundary, as shown in Fig. 5.31 [74]. In this case, solid-state processes are actually dominant.

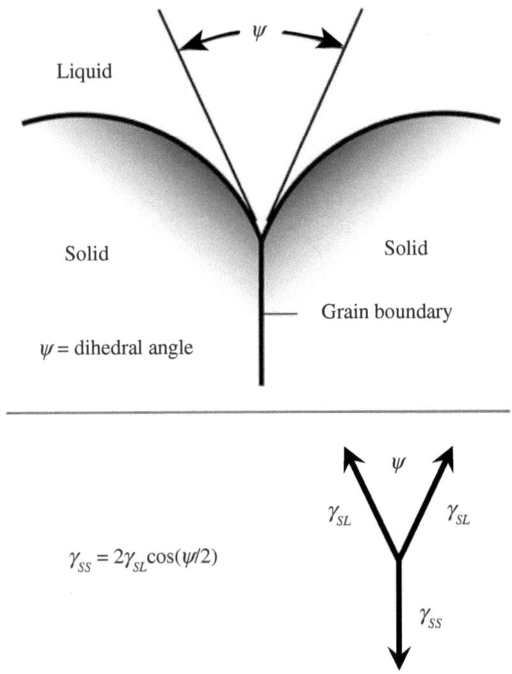

Fig. 5.30 Dihedral angle ψ for a liquid at a grain boundary. Reproduced with permission from [73]. Copyright © 2009, Springer

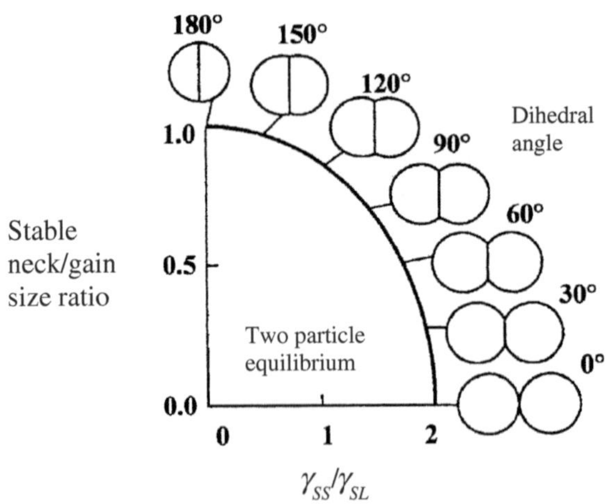

Fig. 5.31 Sketches of two particle contact geometries for various dihedral angles (and interfacial energy ratios). During liquid-phase sintering, the stable neck size to grain size ratio is determined by the equilibrium dihedral angle. Reproduced with permission from [74]. Copyright © 2001, Springer

Therefore, $\gamma_{SS}/\gamma_{SL} = 2$ is the critical value for complete penetration of the liquid into the grain boundary. If $\gamma_{SS}/\gamma_{SL} > 2$, Eq. (5.221) is not valid, indicating an entire penetration of the grain boundary. In other words, $\gamma_{SS}/\gamma_{SL} > 2$ corresponds to the fact that the total specific energy of the two solid–liquid interfaces is lower than that of the solid–solid interface. As a result, the overall energy of the system is decreased due to the penetration of the grain boundary by the liquid. The consequence of the complete penetration is the reduction in mechanical strength of the system.

Shape of the liquid and the grains are closely related to the dihedral angle. It has been accepted that the interfacial tensions at the three-grain junctions in a granular structure are in a balanced state, which determines the equilibrium distribution of second phases [75]. If the system is assumed to have no porosity, the shape of the liquid phase at equilibrium can be derived [61, 62, 76, 77]. Figure 5.32 shows two-dimensional shapes of a small quantity of liquid phase within the three-grain junctions with different dihedral angles [73]. With $\psi = 0$, the liquid entirely penetrates the grain boundary, so that there is no solid–solid contact. With increasing ψ, the penetration depth of the liquid phase along the grains decreases, while the

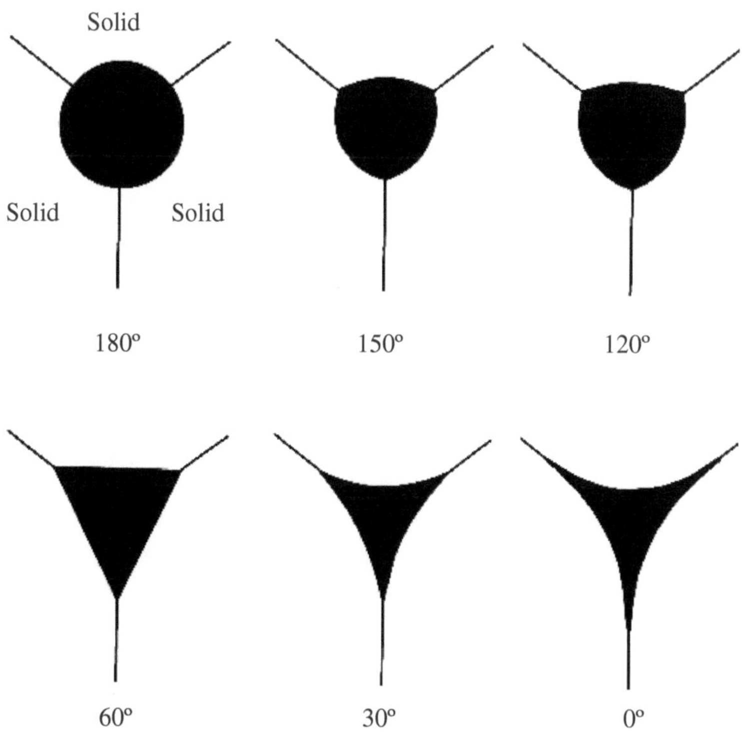

Fig. 5.32 At low liquid contents, the liquid phase forms pockets at the triple points where three-grain boundaries meet. The shape of the liquid pocket is dependent on the dihedral angle. Reproduced with permission from [73]. Copyright © 2009, Springer

amount of solid–solid contact, i.e., grain boundary area, increases accordingly. At $\psi = 60°$, the liquid is still able to be present along the three-grain edges to a very low degree. Once ψ is larger than 60°, the liquid phase is entirely isolated at the junction region of the grains.

5.4.4.3 Effect of Solubility

In a solid–liquid two phase system, the solid can be dissolved in the liquid and the liquid can also be dissolved in the solid. A high solubility of the solid in the liquid will facilitate potentially a high mass transport rate through the liquid layer, which is the basic requirement of liquid-phase sintering. However, the dissolving of the liquid in the solid is totally prevented, otherwise the compact would be severely swelling due to the formation of transient liquid phases. The effect of solubility on densification and swelling of the compact is demonstrated schematically in Fig. 5.33 [73].

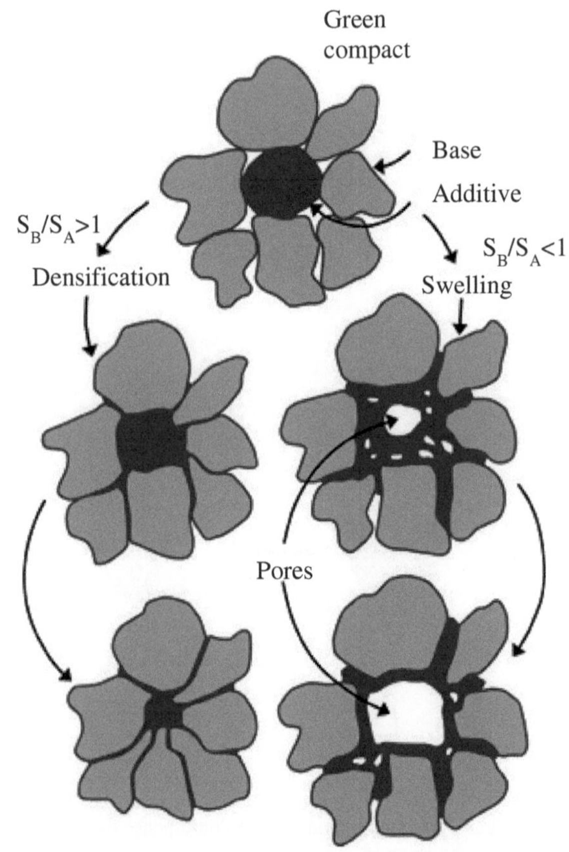

Fig. 5.33 Schematic diagram comparing the effects of solubility on densification or swelling during liquid-phase sintering, with S_A and S_B being solubility of the base and additive, respectively. Reproduced with permission from [73]. Copyright © 2009, Springer

The solubility is also affected by particle size of the compact. If the solubility is assumed to be equivalent to concentration, the relationship between the solute concentration and radius of the particle is given by

$$\ln\left(\frac{S}{S_0}\right) = \frac{2\gamma_{SL}\Omega}{kTa}, \tag{5.222}$$

where S is the solubility of the particle with a radius a in the liquid, S_0 is the equilibrium solubility of the solid in the liquid at planar interface, γ_{SL} is the specific energy of the solid–liquid interface, Ω is the atomic volume, k is the Boltzmann constant, and T is the absolute temperature. Equation (5.222) clearly indicates that the solubility increases with decreasing radius of the particle, due to the formation of a mass reservoir. This is the reason why matter transport always takes place from small particles to large particles. Such a process is known as Ostwald ripening. Additionally, rough regions of the particles have smaller radius of curvature and thus tend to be dissolved, whereas irregularities or defects between adjacent particles, such as pits and crevices, have negative radii of curvature, which have significantly low solubility, thus leading to precipitate in those regions.

5.4.4.4 Capillary Forces

When a liquid completely wets a solid, a compressive stress created due to the pressure deficit in the liquid that is defined by Eq. (5.218) will be applied to the particle. It has been shown that several factors, including contact angle, volume of the liquid, separation of the particles, and the particle size, have influences on the magnitude and nature of the compressive stress. The effects of these factors on the capillary force applied by the liquid can be evaluated by the idealized two-sphere model. The two-sphere particles in the model have same radius a, which are separated by a distance h by a liquid with a contact angle, as shown in schematically in Fig. 5.34 [73]. The shape of the liquid meniscus is called a nodoid. The analytical solution for the shape is in terms of elliptical integrals. Because the calculations of the meniscus shape and the capillary force are very complicated, it is assumed to have a circular shape, with the liquid meniscus being part of a circle. In this case, the pressure difference across the liquid–vapor meniscus is given by

$$\Delta p = \gamma_{LV}\left(\frac{1}{Y} - \frac{1}{r}\right), \tag{5.223}$$

where Y and r are the principal radii of curvature of the meniscus.

The force applied to the two spheres is the resultant contribution of two factors: (i) the pressure difference Δp across the liquid–vapor meniscus and (ii) the surface

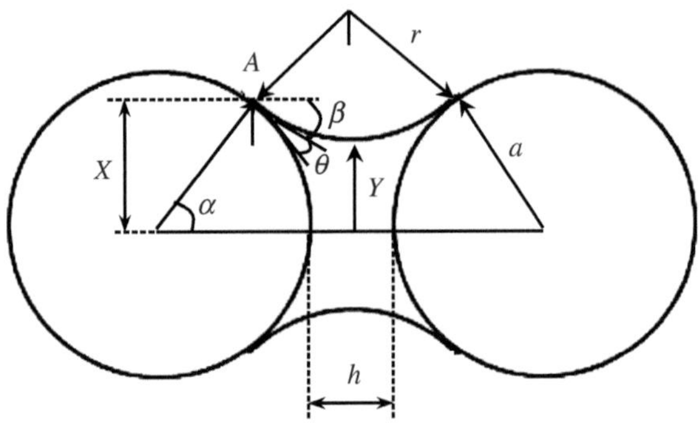

Fig. 5.34 Geometrical parameters for an idealized model of two spheres separated by a liquid bridge. Reproduced with permission from [73]. Copyright © 2009, Springer

tension of the liquid [61, 78, 79]. Therefore, at point A in Fig. 5.34, the force equation can be expressed as:

$$F = -\pi X^2 \Delta p + 2\pi X \gamma_{LV} \cos \beta, \tag{5.224}$$

where F is positive when the force is compressive.

Putting Δp Eq. (5.224) and letting $X = a\sin\,\alpha$, there is

$$F = -\pi a^2 \gamma_{LV} \left(\frac{1}{Y} - \frac{1}{r} \right) \sin^2 \alpha + 2\pi a \gamma_{LV} \sin \alpha \cos \beta. \tag{5.225}$$

The distance between the two spheres is given by

$$h = 2[r \sin \beta - a(1 - \cos \alpha)], \tag{5.226}$$

while the angles are related by the following expression:

$$\alpha + \beta + \theta = \pi/2. \tag{5.227}$$

By putting Eq. (5.227) into Eq. (5.226) and rearranging, there is

$$r = \frac{h + 2a(1 - \cos \alpha)}{2 \cos(\theta + \alpha)}. \tag{5.228}$$

The positive radius of curvature of the meniscus is given by

$$Y = a \sin \alpha - r[1 - \sin(\theta + \alpha)]. \tag{5.229}$$

Putting Eq. (5.228) into Eq. (5.229) yields

$$Y = a\sin\alpha - \frac{h + 2a(1 - \cos\alpha)}{2\cos(\theta + \alpha)}[1 - \sin(\theta + \alpha)]. \quad (5.230)$$

The volume of the liquid bridge is thus given by

$$V = 2\pi(r^3 + r^2Y)\{\cos(\theta + \alpha) - [\pi/2 - (\theta + \alpha)]\} + \pi Y^2 r\cos(\theta + \alpha). \quad (5.231)$$

Equation (5.224) reveals that F is a function of the interparticle distance h for a given volume fraction of the liquid, which can be used to identify the inter-relationship between the contact angle and the efficiency of the particle rearrangement processes. It is found that a small contact angle is always desirable to achieve high degree of densification.

5.4.5 Grain Boundary Films

As discussed above, the presence of the liquid layer applies a compressive capillary force to the particles. The liquid layer evolves during the sintering process. Once the densification through the solution-precipitation mechanism starts, the thickness of the liquid layer gradually decreases with time. When the liquid capillary becomes too narrow for the liquid to flow, the solution-precipitation nearly stops. In this case, the rate of reduction in thickness of the liquid layer due to viscous flow will compete with the rate of densification. Therefore, because the flow of the liquid though the capillary is sufficiently slow, there is a thin layer of the liquid to be remained after densification, which has a thickness of 0.5–2 nm for most ceramics.

The theory for two flat plates separated by a liquid layer has been used to calculate the rate of reduction in thickness of the liquid layer that separates two-sphere particles during the liquid-phase sintering [80]. The rate of approach of two plates separated by a Newtonian viscous liquid is expressed as [81]

$$\frac{dh}{dt} = -\frac{2\pi h^5}{3\eta A^2 h_0^2}F, \quad (5.232)$$

where h is the thickness of the liquid layer at time t, h_0 is the initial thickness of the liquid layer, F is the compressive force applied to the plates, η is the viscosity of the liquid, and A is the contact area between the liquid and the plate. The application of Eq. (5.232) to the liquid-phase sintering is shown as the model in Fig. 5.35, in which two spheres are held together through a liquid bridge due to the capillary force [1]. In practice, it is more convenient to use the time that is taken by the liquid to be depleted. Letting $y = h/h_0$ and integrating Eq. (5.232), the time for y to be zero is given by

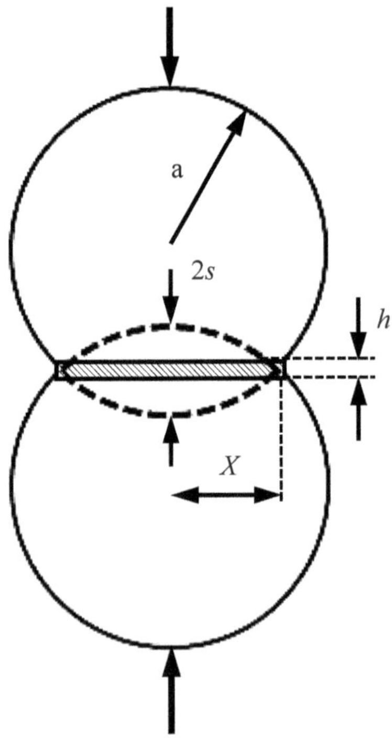

Fig. 5.35 Parameters of an idealized two-sphere model separated by a liquid layer of thickness h used to analyze the change in the thickness of the layer during sintering. Reproduced with permission from [1]. Copyright © 2003, CRC Press

$$t_{\mathrm{f}} = \frac{3\eta A^2}{8\pi h_0^2 F}\left(\frac{1}{y^4} - 1\right). \tag{5.233}$$

This equation indicates that the time for $y = 0$ is $t_{\mathrm{f}} = \infty$. In other words, there will always be liquid to be remained between the particles. However, when the liquid layer is too thin, e.g., several nm, other effects, such as structural and chemical forces and charge interactions, will become dominant instead of viscous flow. The equilibrium thickness of the liquid layer between adjacent grains can be explained in terms of the balance between two forces, i.e., (i) the attractive van der Waals forces of the grains and (ii) the short-range repulsive forces due to the resistance to deformation of the liquid phase [82].

5.4.6 Mechanisms of Liquid-Phase Sintering

Similarly, the mechanisms of liquid-phase sintering can be described by dividing the process into three stages. The first stage involves the redistribution of the liquid phase and rearrangement of the particles, the second stage is densification process through solution reprecipitation, while the third/last stage is known as Oswald ripening.

5.4.6.1 Stage 1

The capillary pressures developed during the liquid-phase sintering of fine particles forces the particles to rearrange as the viscosity of the system is still sufficiently low. Due to the presence of gradients in capillary pressure, the liquid phase will flow from regions with larger pores to regions with smaller pores, which is called liquid-phase redistribution, which has been confirmed experimentally in various systems [83]. Generally, the liquid phase fills from small pores to large pores.

A two-dimensional model of circular particles has been developed to describe the liquid redistribution behaviors, which assumes that the chemical potential of the liquid in all the pores in a given particle array is the same at equilibrium [84]. When an atom is located under the surface of a liquid–vapor meniscus, with an average radius of curvature r, its chemical potential is given by

$$\mu = \mu_0 + \frac{\gamma_{LV}\Omega}{r}, \tag{5.234}$$

where μ_0 is the chemical potential of the atom under a flat surface, γ_{LV} is the liquid–vapor surface energy, and Ω is the atomic volume. Therefore, the assumption of same chemical potential implies that the radius of the liquid menisci should be the same.

Figure 5.36 shows the two-dimensional model of liquid-phase redistribution, which is a regular array of circles with threefold coordinated pores and without shrinkage occurs [84]. The liquid phase can be redistributed in two ways. At very low volume fractions, the liquid phase is usually distributed evenly in the necks formed by the particles, as shown in Fig. 5.36a. There is a critical volume fraction, below which the distribution of the liquid phase is not affected the by the volume fraction and above which some of the pores are entirely filled with the liquid, as shown in Fig. 5.36b. This distribution is attributed to the requirement of minimizing the surface area. Therefore, the amount of the liquid phase located at the necks is not increased with increasing volume fraction of the liquid phase, because the increased amount of the liquid phase is all used to fill the pores. The range of the liquid volume for even distribution of the liquid in the necks decreases with increasing contact angle. Once some of the pores are filled, it is highly possible to form inhomogeneous liquid-phase distributions, such as that shown in Fig. 5.36c, which cannot be back to the homogeneous distribution.

Figure 5.37 shows the model with pores of both threefold and sixfold coordinations, which is used to mimic the inhomogeneously packed powder systems that are encountered in practical applications [84]. The results of free energy

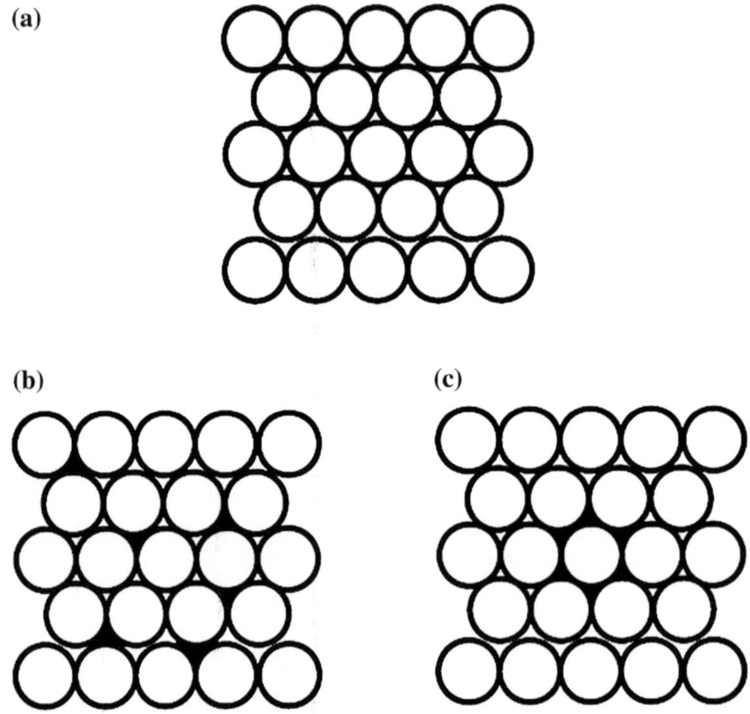

Fig. 5.36 Possible equilibrium configurations that can be adopted by a liquid in a close-packed two-dimensional array of particles with same size. **a** Isolated necks filled with liquid, **b** fraction of pores completely filled with liquid with homogeneous distribution, and **c** with inhomogeneous distribution. Reproduced with permission from [84]. Copyright © 1986, John Wiley & Sons

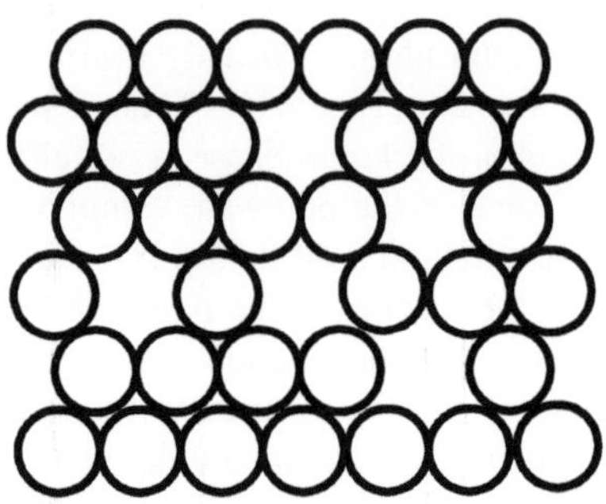

Fig. 5.37 Example of a two-dimensional arrangement of pores that contains threefold and sixfold coordinated pores. Reproduced with permission from [84]. Copyright © 1986, John Wiley & Sons

calculations, e.g., the fraction of the particles forming the sixfold coordinated pores as a function of the volume fraction of the liquid phase, can be used to describe the liquid-phase distribution behaviors.

It is found that the pores in the simple model are sequentially filled. In a real powder compact, the pores have a wide size distribution. In this case, the filling of the pores is started from those with smallest coordination number, because they have highest surface to volume ratios, so that largest solid–vapor interfacial areas are eliminated for a given volume of liquid. However, in practice, not all small pores can be filled, because the liquid might not be able to access to some of them. In addition, if particles to create the liquid phase are too large, huge voids will be left when the particles are molten. Therefore, in this respect, it is desirable to have powders with uniform particle size and homogeneous mixing of basic components and additives.

Once the liquid phase is formed, particles will start to rearrange almost immediately, which is finished in a few minutes. This rearrangement is responsible for the initial densification and affects the initial microstructure of the compact, which could have significant influence on further densification and microstructure development. Theoretical analysis of the rearrangement, especially for compacts with randomly packed particles, remains a challenge until now [85–88]. A simple kinetic relationship to describe the shrinkage as a function of time has been developed, by assuming that the surface tension forces to drive the densification are balanced by the viscous forces that resist the rearrangement, which is given by [89, 90]

$$\frac{\Delta L}{L_0} \sim t^{1+y}, \tag{5.235}$$

where ΔL is the change in length, L_0 is the original length, and y is a positive constant that is less than one.

The rearrangement process usually experiences two stages: primary rearrangement and secondary rearrangement. Primary rearrangement is a rapid process nearly immediately after the formation of the liquid phase, due to the surface tension forces of the liquid bridge applied to the particles. At this stage, if $\gamma_{SS}/\gamma_{SL} > 2$, the liquid can penetrate the grain boundaries, leading to fragmentation of the particles. The fragmented particles will further experience a rearrangement process, which is known as the secondary rearrangement. The secondary rearrangement is slower than the primary one, because it is dependent on the dissolving rate of the grain boundaries.

Full densification can be achieved through the rearrangement, if the content of the liquid phase is sufficiently high. The relative quantities of the solid and the liquid required for full densification is determined by the rearranged density of the particles. For a powder compact with a relative density of 60 %, i.e., 40 % porosity, if the particles rearrange to 64 % relative density through the rearrangement, full densification is achieved, when the liquid volume fraction is 36 %. Otherwise, further processes, e.g., solution-precipitation, are necessary to obtain full densification. In practice, rearrangement to full densification is quite rare, because the liquid-phase content is usually low, e.g., <5 vol.%. Moreover, the particles generally have irregular shapes.

5.4.6.2 Stage 2

With the gradual diminishment of the rearrangement process, solution precipitation starts to be dominant process, leading to further densification and coarsening. The densification and coarsening take place concurrently, which could be accompanied by accommodation in grain shape if the liquid phase has a small volume fraction. At the same time, coarsening and grain-shape accommodation are contributed by the coalescence of small grains in contact with large grains, besides the solution-precipitation. There are two models for densification: (i) contact flattening and (ii) Ostwald ripening.

Due to presence of the compressive capillary force, the solubility at the contact points between the particles is higher than that at other surfaces of the particles [89, 90]. Because of this difference in solubility, matter transport occurs from the contact points to other places, so as to form a flat contact region, as shown in Fig. 5.38 [1]. With increasing radius of the contact region, the stress along the interface decreases, so that the densification slows down. The rate of the matter transport is controlled by the slower one of the two mechanisms: (i) diffusion through the liquid layer and (ii) solution-precipitation via interface reaction.

The rate control by diffusion through the liquid layer can be described by using the two spherical particle model. The two particles have the same radius a, if each sphere is dissolved away along the center-to-center line by a distance h to form a circular contact region with radius X, these parameters can be related through the following expression:

$$h \approx \frac{X^2}{2a}. \tag{5.236}$$

The volume of material removed from each sphere is given by $V \approx X^2h/2$, which is combined with Eq. (5.236), there is

$$V \approx \pi a h^2. \tag{5.237}$$

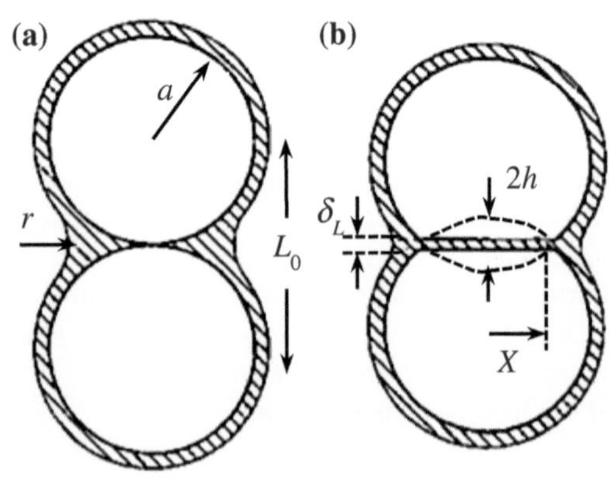

Fig. 5.38 Idealized two-sphere model for densification by contact flattening. Reproduced with permission from [1]. Copyright © 2003, CRC Press

Similar to the intermediate stage of solid-state sintering, the flux from the boundary per unit thickness is given by

$$J = 4\pi D_{\mathrm{L}} \Delta C, \tag{5.238}$$

where D_{L} is the diffusion coefficient for the solute atom in the liquid and ΔC is the difference in solute concentration between the contact region, C and at flat stress-free surface, C_0. If the thickness of the liquid layer is δ_{L}, the rate of consumption of the solid is given by

$$\frac{\mathrm{d}V}{\mathrm{d}t} = \delta_{\mathrm{L}} J = 4\pi D_{\mathrm{L}} \delta_{\mathrm{L}} \Delta C. \tag{5.239}$$

If ΔC is small, there is

$$\frac{\Delta C}{C_0} = \frac{p\Omega}{kT}, \tag{5.240}$$

where p is the local stress applied to an atom, Ω is volume of the atom, k is the Boltzmann constant, and T is absolute temperature. The capillary pressure Δp due to a spherical pore in the liquid is given by Eq. (5.218), which is equivalent to an externally applied hydrostatic pressure so that the resulting interparticle force is equal to that produced by the liquid layer. Because the area of the contact region is smaller than that of the rest of the particles, the local pressure is magnified at the contact area. If a simple force balance is assumed, the local pressure p at the contact region is given by

$$pX^2 = k_1 \Delta p a^2, \tag{5.241}$$

where k_1 is a geometrical constant. When Eq. (5.241) is combined with Eqs. (5.218) and (5.236), there is

$$p = k_1 \frac{\gamma_{\mathrm{LV}} a}{rh}. \tag{5.242}$$

It is assumed that the radius of the pore is proportional to the radius of the sphere, i.e., $r = k_2 a$, where k_2 is assumed to remain constant during sintering, Eq. (5.242) becomes

$$p = \frac{k_1}{k_2} \frac{\gamma_{\mathrm{LV}}}{h}. \tag{5.243}$$

According to Eqs. (5.238), (5.240) and (5.243), there is

$$\frac{dV}{dt} = \frac{4\pi k_1 D_L \delta_L C_0 \Omega \gamma_{LV}}{k_2 hkT}. \tag{5.244}$$

Because dV/dt is also equal to $(2\pi ah)\ dh/dt$ according to Eqs. (5.237) and (5.244) can also be rewritten as

$$h^2 dh = \frac{2k_1 D_L \delta_L C_0 \Omega \gamma_{LV}}{k_2 akT} dt. \tag{5.245}$$

With the boundary condition of $h = 0$ at $t = 0$, integration of Eq. (5.245) yields

$$h = \left(\frac{6k_1 D_L \delta_L C_0 \Omega \gamma_{LV}}{k_2 akT}\right)^{1/3} t^{1/3}. \tag{5.246}$$

Because there is $h/a = \Delta L/L_0 = -(1/3)\Delta V/V_0$ for small $\Delta L/L_0$, where $\Delta L/L_0$ and $\Delta V/V_0$ are the linear shrinkage and the volumetric shrinkage of the powder compact, respectively, the following expression can be obtained

$$-\frac{\Delta L}{L_0} = -\frac{1}{3}\frac{\Delta V}{V_0} = \left(\frac{6k_1 D_L \delta_L C_0 \Omega \gamma_{LV}}{k_2 a^4 kT}\right)^{1/3} t^{1/3}. \tag{5.247}$$

This equation indicates that, if the diffusion through the liquid layer is the rate-controlling step, the shrinkage increases as functions of time as $t^{1/3}$, starting particle size as $a^{-4/3}$ and thickness of the liquid layer as $\delta_L^{1/3}$.

If the interface reaction is the rate-controlling mechanism, in which the solid phase is dissolved into the liquid phase through reaction, the rate of mass transport is proportional to the contact area and the increased activity of the solid phase at the contact region due to the capillary pressure, which is given by

$$\frac{dV}{dt} = k_3 \pi X^2 (a - a_0) = 2\pi k_3 ha(C - C_0), \tag{5.248}$$

where k_3 is the rate constant of the phase boundary reaction, while a and a_0 are activities of the solid phase at the contact region and at flat surface, which are equal to their corresponding concentrations. In a similar way for the diffusion control, the shrinkage of the interface reaction control is given by

$$-\frac{\Delta L}{L_0} = -\frac{1}{3}\frac{\Delta V}{V_0} = \left(\frac{2k_1 k_3 C_0 \Omega \gamma_{LV}}{k_2 a^2 kT}\right)^{1/2} t^{1/2}. \tag{5.249}$$

This equation means that the shrinkage is proportional to $t^{1/2}$ and $1/a$.

The second mechanism of densification, i.e., densification accompanied by Ostwald ripening, can be described with an idealized model as shown in Fig. 5.39

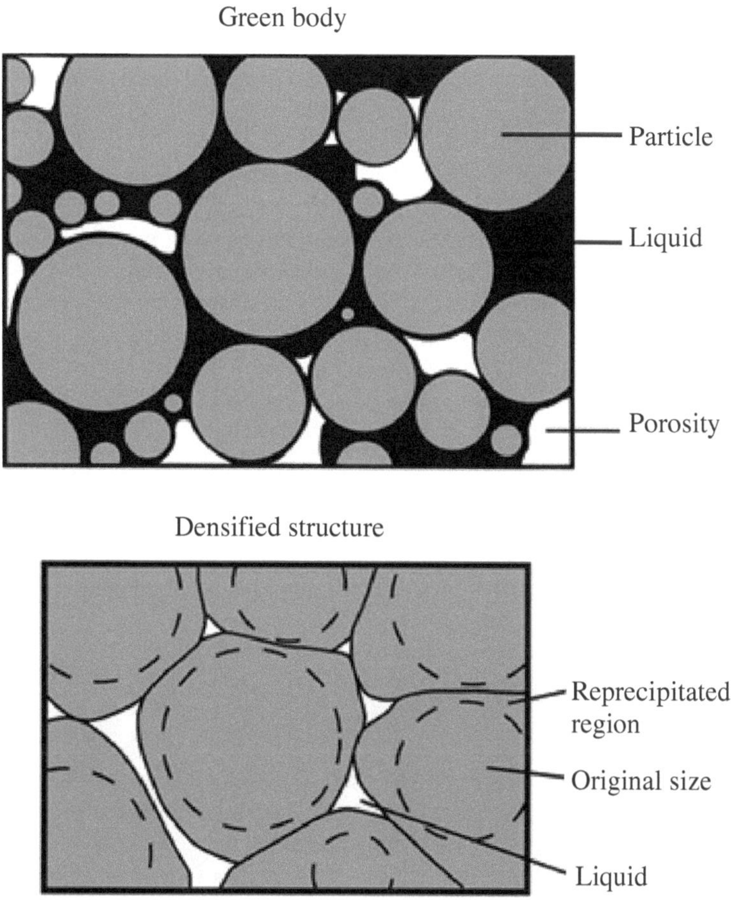

Fig. 5.39 Conceptual outline of the changes associated with solution-reprecipitation densification where both grain growth (Ostwald ripening) and grain-shape accommodation act to release liquid to fill residual pores. Reproduced with permission from [73]. Copyright © 2009, Springer

[73]. The smaller grains are dissolved into the liquid phase and then precipitate onto the larger grains from the contact points, thus facilitating shrinkage or densification. By assuming the Ostwald ripening to be diffusion controlled, the shrinkage rate can be expressed by the following expression [61]:

$$-\frac{\Delta L}{L_0} = -\frac{1}{3}\frac{\Delta V}{V_0} = \left(\frac{48 D_{\rm L} C_0 \Omega \gamma_{\rm LV}}{a^3 kT}\right)^{1/3} t^{1/3}. \tag{5.250}$$

The content of the liquid phase plays a key role in determining whether there is grain-shape accommodation or not in a liquid-phase sintering. In a given system, there is a critical content of the liquid phase, below which grain-shape accommodation occurs, while above the grain-shape accommodation is absent. When the content of the liquid phase is not sufficiently high to fill all the pores between the particles, if the particles have rounded or spheroidal shape, grain-shape accommodation takes place, which produces particles with polyhedral shapes with flat contact surfaces, so that the particles will have a more efficient packing. In this case, the liquid phase will be released from the well-packed regions to flow into the pores.

From energy point of view, grain-shape accommodation only occurs when the energy of the system is decreased because of it. For a given volume, the compact made of particles with polyhedral shapes has a larger surface area than that consisting of spherical particles. Therefore, only when the decreased interfacial energy caused by the filling of the pores is more than the increased interfacial energy due to the formation of the polyhedral particles, the grain-shape accommodation can take place. If the amount of the liquid phase is sufficiently high, the capillary pressure is too low to drive contact flattening. As a result, shape accommodation is much less pronounced, so that the particles remain to be spheroidal.

Coalescence-induced coarsening takes place when grains are pulled together due to the capillary force produced by the liquid phase [69, 91]. Coalescence usually experiences three steps: (i) contact formation between the grains, (ii) neck growth, and (iii) migration of the grain boundary, as shown schematically in Fig. 5.40 [73].

As shown in Fig. 5.41 [73], this curvature r depends on the dihedral angle ψ, and grain sizes G_1 and G_2 ($G_1 > G_2$) as

$$r = \cos\left(\frac{\psi}{2}\right)\left(\frac{G_1 G_2}{G_1 - G_2}\right) \tag{5.251}$$

Different mechanisms have been observed in coalescence, including solid-state grain boundary migration, liquid-film migration, and solution-precipitation through the liquid, as shown in Fig. 5.42 [73]. For small dihedral angle, liquid only partially penetrates the grain boundary. In this case, the grain boundary energy is increased during the movement of the boundary, so that coalescence is impeded. However, the small grain is absorbed by large grain through solution precipitation, which overcomes the energy barrier. For large dihedral angles, the liquid cannot penetrate the grain boundary, so that coalescence will become dominant. The larger the difference in size between the particles, the stronger the coalescence will be. Therefore, coalescence is most pronounced during the earlier period of the solution-precipitation process.

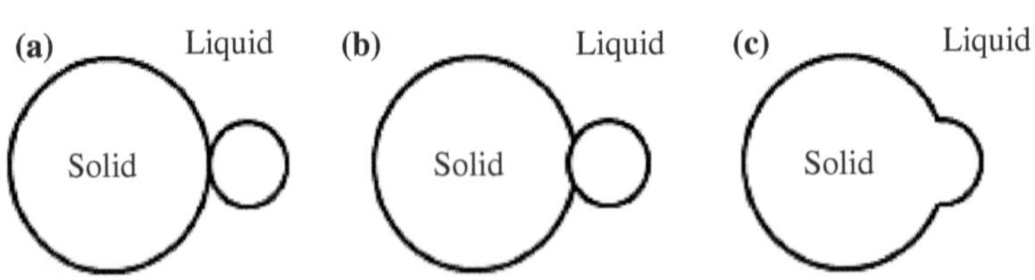

Fig. 5.40 Schematic diagram showing the grain growth by coalescence of small and large grains. Reproduced with permission from [73]. Copyright © 2009, Springer

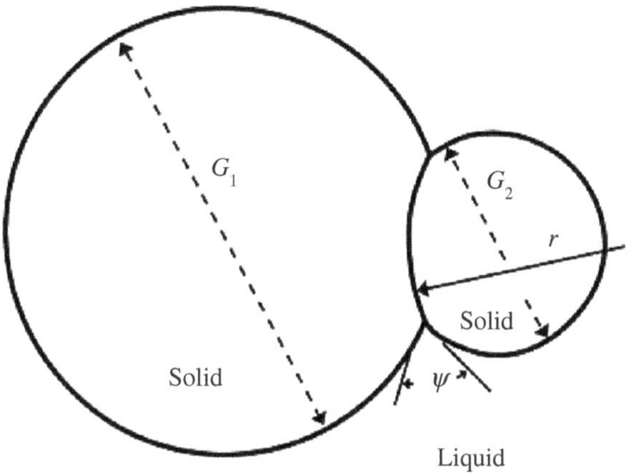

Fig. 5.41 The radius of curvature r of the grain boundary between contacting grains depends on the dihedral angle ψ and the grain size ratio (G_1/G_2). A large ratio induces a high curvature that aids rapid grain coalescence during liquid-phase sintering. Reproduced with permission from [73]. Copyright © 2009, Springer

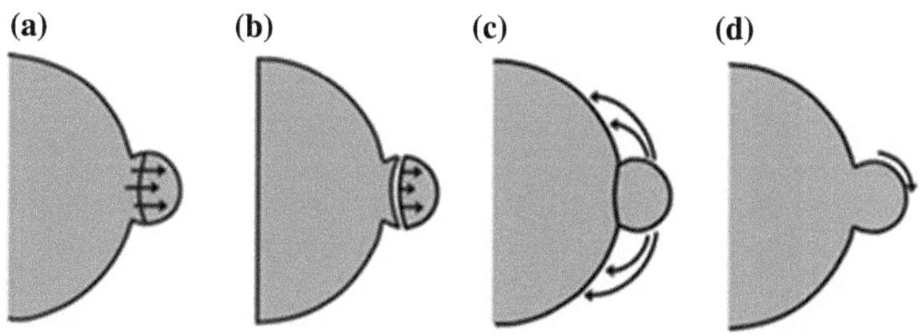

Fig. 5.42 Coalescence between two contacting grains by possible mechanisms: **a** solid-state grain boundary motion of curved grain boundaries, **b** liquid-film migration with diffusion across the film from the small to large grain, **c** solution reprecipitation from the small grain to the large grain through the surrounding liquid, and **d** small grain rotation to a lattice coincidence orientation where there is no grain boundary. Reproduced with permission from [73]. Copyright © 2009, Springer

Contact flattening is the first mechanism, which can be sketched in Fig. 5.43a [73]. A compressive force at the grain contacts from the wetting liquid pulls the grains together. This capillary stress causes preferential dissolution of solid at the contact point with reprecipitation at regions away from the contact. Densification results from the grain center-to-center motion [92]. The key step is solid diffusion in the liquid to areas away from the contact zone. For small grains, the contact zone stresses are quite large, so contact flattening tends to dominate LPS [93]. However, contact flattening does not explain grain growth and the decrease in the number of grains. When grain growth is inhibited there is less grain-shape accommodation [94–96].

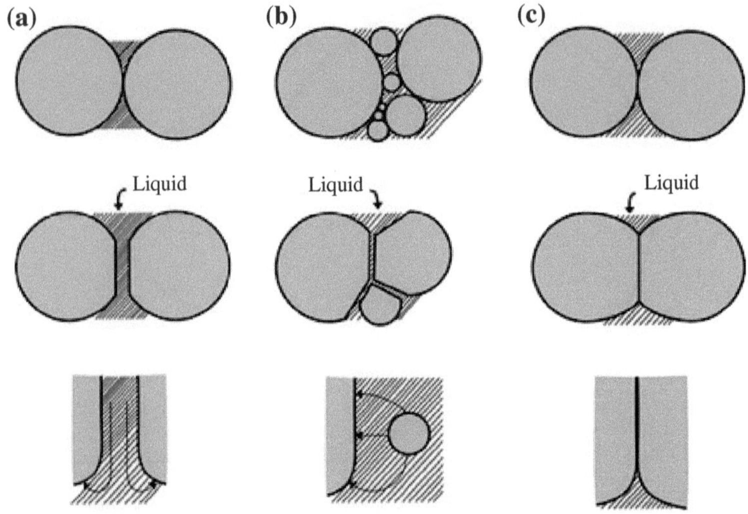

Fig. 5.43 The three mechanisms of grain-shape accommodation and neck growth during solution reprecipitation controlled densification of liquid-phase sintering: **a** contact flattening, **b** dissolution of small grains, and **c** solid-state bonding. Reproduced with permission from [73]. Copyright © 2009, Springer

The second densification mechanism involves dissolution of small grains with reprecipitation on large grains. Small grains disappear while large grains grow and undergo shape accommodation. Diffusion in the liquid is the controlling transport mechanism, as sketched in Fig. 5.43b [73]. This mechanism does not involve shrinkage, so it is not an explanation for densification, except that grain-shape accommodation enables better packing of the solid.

The third mechanism involves growth of the intergrain contact by diffusion along the liquid-wetted grain boundary [97–99], which is shown representatively in Fig. 5.43c [73]. The contact zone enlarges to change the grain shape with simultaneous shrinkage of the grains. This does not involve grain coarsening, but it does require a cooperative redistribution process of the mass deposited where the grain boundary intersects the liquid [37].

These three mechanisms are different in the source of the solid and in the detailed transport path, but they together can be used to explain grain-shape accommodation, grain growth, and densification. Grain growth takes place along with densification. Grain size and density tend to follow a common trajectory for most liquid-phase sintering systems, showing more rapid grain growth when pores are eliminated. Although neck growth is initially active, it is not sufficient to explain all microstructural developments. In addition, contact flattening and small grain dissolution are coupled to explain the microstructure and density progression that are typically observed during liquid-phase sintering.

5.4.6.3 Stage 3

At stage 3, densification slows down while coarsening becomes the dominant process. For $\gamma_{SS}/\gamma_{SL} > 2$, the grains are completely separated by the liquid layer. If $\psi > 0$, a rigid skeleton is gradually formed due to the solid–solid contacts. As a result, the isolated pores present in the liquid phase cannot be eliminated. The formation of the solid–solid contacts triggers solid-state sintering and coarsening. However, solution precipitation is still dominant over solid-state transport processes, because the matter transport through the liquid is much faster. Two events are dominant at this stage: (i) densification through pore filling and (ii) microstructural coarsening.

At low concentrations of the liquid phase, the elimination of isolated pores is continued at a reduced rate, due to the solution precipitation and grain-shape accommodation. If the volume fraction of the liquid is sufficiently high, the filling of the isolated pores could be discontinuous [95, 100]. In this case, grain growth, rather than grain-shape accommodation, is responsible for the filling of the pores.

As shown in Fig. 5.44, a large pore remains unfilled because of the preferential wetting of the necks between the particles [101]. As the grains continue to grow, the liquid phase reaches a favorable condition for filling the pore as determined by the curvature of the liquid–vapor meniscus [61, 76, 101]. From Fig. 5.45, the radius of curvature of the meniscus is given by [101]

$$r_{\mathrm{m}} = a\frac{1-\cos\alpha}{\cos\alpha}, \tag{5.252}$$

where a is the radius of the grain that is assumed to be spherical, so that r_{m} increases with increasing grain size. For zero contact angle, the critical point for pore filling to occur is that r_{m} is equal to the pore radius r, because after this point r_{m} cannot increase, so that the capillary pressure surrounding the pore will be reduced. Liquid is drawn from the numerous menisci at the surface of the sample, with very slight change in radii because only small volume of liquid is required to

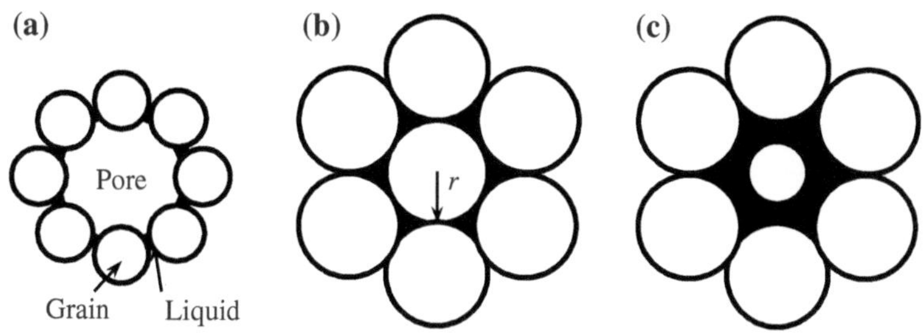

Fig. 5.44 Pore filling during the grain growth. A large pore is stable until grain growth increases the liquid meniscus radius sufficiently for capillary refilling of the pore. Reproduced with permission from [101]. Copyright © 1984, Springer

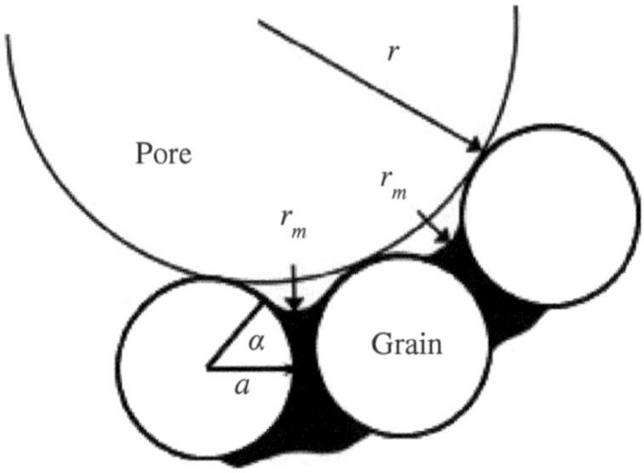

Fig. 5.45 Calculation model for pore refilling based on spherical grains surrounding the pore. Pore refilling depends on the liquid meniscus radius exceeding the pore radius. Reproduced with permission from [101]. Copyright © 1984, Springer

fill the pore. If the contact angle is larger than zero, r_{m} must lager than r to facilitate the pore filling.

The most important theory to describe the Ostwald ripening mechanism for microstructural coarsening in the liquid-phase sintering is the Lifshitz-Slyozov-Wagner theory, usually known as the LSW theory. According to the LSW theory, the increase in the average grain size G with time t is given by

$$G^m = G_0^m + Kt, \tag{5.253}$$

where G_0 is the initial average grain size, K is a temperature-dependent constant, and the exponent m is dependent on the grain growth mechanism: m = 3 for diffusion through the liquid layer and m = 2 for interface reaction. For many ceramics, the exponent m is close to 3, implying that grain growth of ceramics is most likely governed by diffusion coarsening mechanism.

According to Arrhenius equation, the solubility of the solid in the liquid and the diffusion of the dissolved species through the liquid layer would increase with increasing temperature. Therefore, the rate of coarsening usually increases with increasing temperature. An increase in the dihedral angle leads to a decrease in the contact area between the solid and the liquid, so that the grain boundary area is increased accordingly. Obviously, matter transport through liquid is faster than through solid-state diffusion. Therefore, a reduction in the solid–liquid contact area means a reduced solution-precipitation process, thus leading to a reduction in the rate of grain growth.

Given that the content of the liquid phase is sufficiently high to isolate the grains/ particles, decrease in volume fraction of the liquid phase means a decrease in thickness of the liquid-phase layer and thus shorter diffusion distance. Therefore, it is expected that the rate of matter transport would increase with decreasing content of the liquid phase, so that the rate of grain growth is increased correspondingly

[102]. Theories have been proposed to explain the dependence of grain growth on volume fraction of liquid phase, one of them is as follows [103, 104]:

$$K = K_{\mathrm{i}} + \frac{K_{\mathrm{L}}}{V_{\mathrm{L}}^{2/3}}, \tag{5.254}$$

where K_{I} and K_{L} are the rate constant at infinite dilution, K_{L} is a microstructure-dependent parameter, and V_{L} is the volume fraction of the liquid phase.

5.4.7 Hot Pressing with Liquid Phase

Liquid-phase sintering has also been applied to ceramics with hot pressing. Because chemical potential of the atoms under the contact surfaces increases with increasing stress, matter transport from the contact regions to the pores is enhanced, so that the densification rate is increased [105].

The densification kinetics, with a solution-precipitation mechanism in which diffusion through the liquid layer is rate-controlling, can be described similarly by using the Coble's equation for grain boundary diffusion-controlled solid-state sintering, as discussed earlier. In this case, it is only necessary to replace the grain boundary thickness δ_{gb}, the grain boundary diffusion coefficient D_{gb} and the solid–vapor interfacial energy δ_{SV} with the thickness of the liquid layer δ_{L}, the diffusion coefficient for the solute in the liquid D_{L} and the liquid-vapor interfacial energy δ_{LV}, respectively. As the applied pressure p_{a} is much higher than the capillary pressure due to the liquid meniscus, the densification rate is given by

$$\frac{1}{\rho}\frac{\mathrm{d}\rho}{\mathrm{d}t} = \frac{AD_{\mathrm{L}}\delta_{\mathrm{L}}\Omega}{G^3kT}p_{\mathrm{a}}\phi, \tag{5.255}$$

where A is a geometrical constant, which is equal to 47.5 for the intermediate stage of sintering, Ω is the atomic volume of the rate-controlling species, G is the grain size, k is the Boltzmann constant, T is the absolute temperature, and ϕ is the stress intensification factor.

5.4.8 Phase Diagrams in Liquid-Phase Sintering

Phase diagrams play an important role in selecting the powder compositions and sintering parameters for liquid-phase sintering. Because diagrams predict phases that are at equilibrium state, the reaction kinetics during liquid-phase sintering are often too fast for equilibrium to be achieved, so the phase diagrams should therefore serve only as a guide.

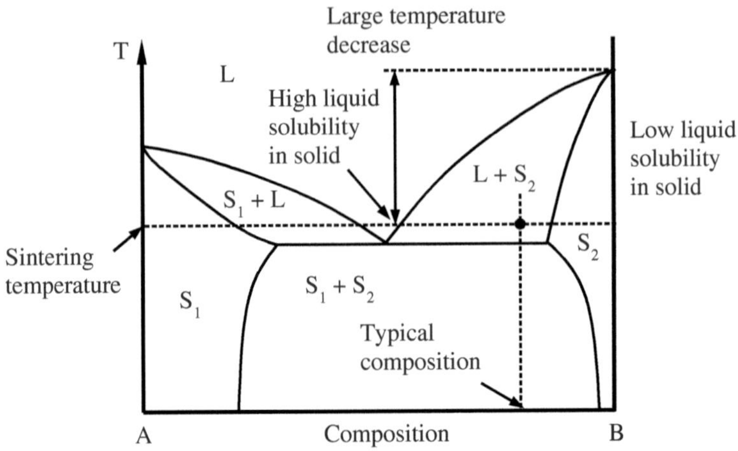

Fig. 5.46 Model binary phase diagram showing the composition and sintering temperature associated with liquid-phase sintering in the $L + S_2$ phase field. The favorable characteristics for liquid-phase sintering include a suppression of the melting temperature, high solid solubility in the liquid, and low liquid solubility in the solid. Reproduced with permission from [73]. Copyright © 2009, Springer

Figure 5.46 shows an idealized binary phase diagram, consisting of a major component, i.e., *B* and a liquid-producing additive, i.e., *A*, in which desirable composition and temperature characteristics for liquid-phase sintering have been indicated [73]. Importantly, there should be a large difference in melting temperature between the eutectic and the major component *B*. The system should have compositions away from the eutectic point, so that the volume of the liquid phase is increased gradually with temperature, so as to avoid the formation of all the liquid in the short time span. Usually, the sintering temperature is slightly above the eutectic temperature, which the composition of the system is in the $(L + S_2)$ region, e.g., the dot shown in the figure.

5.4.9 Activated Sintering and Vitrification

When powder compacts containing small amounts of a eutectic-forming additive are sintered, enhanced densification rates, as compared with that of the pure powder, could be observed well before the eutectic temperature is reached. This effect is generally known as activated sintering, which has been observed in various materials with a wide range of additives [106–110]. Currently, there is no clear difference in principles involved in activated sintering and liquid-phase sintering from the activated systems with a small content of additives, e.g., <1 wt%, at a sintering temperature below the eutectic temperature [111]. Although the mechanism of activated sintering has not been clearly identified, if the additive segregates at the grain boundaries, enhanced subeutectic densification rates could be observed, grain boundary transport rates are enhanced by the additive. The additive should also be able to form low-melting phases or eutectic with the major component and the major component should have a high solubility in the additive. When the grain boundaries are rich in the eutectic-forming additive, their transport rates will be enhanced, because of their relative melting points. For example, when ZnO was added with Bi_2O_3, Bi-enriched intergranular amorphous films with a thickness of ~ 1 nm thick were formed. Because these amorphous films accelerated mass transport, activated sintering was triggered [109].

If the densification of liquid-phase sintering is achieved due to the viscous flow of a liquid that is able to fill up the pore spaces between the solid grains, it is called vitrification [112–114]. The driving force for vitrification is the reduction of solid–vapor interfacial energy, because the flow of the liquid covers the surfaces of the solid. Traditional clay-based ceramics are usually densified through vitrification. However, it is very unlikely to be observed in the processing of transparent ceramics, because the content of liquid phase must be controlled to a limited level.

5.5 Concluding Remarks

Matter transport during the sintering of polycrystalline ceramics takes place through a thermally activated process, known as diffusion, with various paths in the solid, corresponding to the different mechanisms of diffusion, i.e., (i) lattice, (ii) grain boundary, and (iii) surface diffusion. The rate of atomic or ionic diffusion is dependent on temperature and the concentration of defects in the solid. The concentration of defects can be controlled by doping or introducing impurities. Mass transport during the sintering involves the flux of atoms and ions or the counterflow of vacancies. The flux of the diffusing species is attributed to the gradients in concentration in particular or in the chemical potential in general. The rate of sintering and other mass transport processes are controlled by the slowest diffusing species along its fastest path.

The sintering phenomena of polycrystalline ceramics are very complex, with mass transport having at least six paths that determine the mechanisms of sintering. There are also densifying mechanisms and nondensifying mechanisms. Analytical models have been developed to describe the dependence of the sintering rate on various variables, including particle size, temperature, and external pressure. However, due to the drastic simplification used in their development, the models can provide only a qualitative description of sintering. Numerical simulations are more effective in analyzing the complexities of sintering.

The presence of liquid phase could enhance densification through grain rearrangement and faster matter transport. Liquid-phase sintering involves three overlapping stages, including (i) rearrangement of grains/particles, (ii) solution precipitation, and (iii) Ostwald ripening. The efficiency of liquid-phase sintering is dependent on several parameters, e.g., low contact angle, low dihedral angle, high solid solubility of the solid in the liquid, homogeneous packing of the particulate solid, uniform distribution of the liquid-producing additive, and fine particle size. Activated sintering and vitrification are unlikely encountered in transparent ceramics due to the limited quantity of additives.

References

1. Rahaman MN (2003) Ceramic processing and sintering, 2nd edn. CRC Press, New York
2. Kemethmueller S, Hagymasi M, Stiegelschmitt A, Roosen A (2007) Viscous flow as the driving force for the densification of low-temperature co-fired ceramics. J Am Ceram Soc 90:64–70
3. Pino-Munoz D, Bruchon J, Drapier S, Valdivieso F (2014) Sintering at particle scale: an Eulerian computing framework to deal with strong topological and material discontinuities. Arch Comput Methods Eng 21:141–187
4. Chaim R, Levin M, Shlayer A, Estournes C (2008) Sintering and densification of nanocrystalline ceramic oxide powders: a review. Adv Appl Ceram 107:159–169
5. Fang ZZ, Wang H (2008) Densification and grain growth during sintering of nanosized particles. Int Mater Rev 53:326–352
6. German RM (2002) Computer modeling of sintering processes. Int J Powder Metall 38:48–66
7. Green DJ, Guillon O, Roedel J (2008) Constrained sintering: a delicate balance of scales. J Eur Ceram Soc 28:1451–1466
8. Hareesh US, Johnson R (2007) Rate controlled sintering: a unique concept for microstructural control. Trans Indian Ceram Soc 66:157–166
9. Lu K (2008) Sintering of nanoceramics. Int Mater Rev 53:21–38
10. Pan JZ (2003) Modelling sintering at different length scales. Int Mater Rev 48:69–85
11. Wakai F (2006) Modeling and simulation of elementary processes in ideal sintering. J Am Ceram Soc 89:1471–1484
12. Bordia RK, Scherer GW (1988) On constrained sintering, 1. Constitutive model for a sintering body. Acta Metall 36:2393–2397
13. Bordia RK, Scherer GW (1988) On contrained sintering, 2. Comparison of constitutive models. Acta Metall 36:2399–2409
14. Bordia RK, Scherer GW (1988) On contrained sintering, 3. Rigid inclusions. Acta Metall 36:2411–2416
15. Dudnik EV, Zaitseva ZA, Shevchenko AV, Lopato LM (1995) Sintering of ultradisperse powders based on zirconium dioxide (review). Powder Metall Met Ceram 34:263–271
16. Haviar M (1985) The mechanisms involved in solid-phase sintering. Silikaty 29:363–377
17. Kuang X, Carotenuto G, Nicolais L (1997) A review of ceramic sintering and suggestions on reducing sintering temperatures. Adv Perform Mater 4:257–274
18. Olevsky EA (1998) Theory of sintering: from discrete to continuum. Mater Sci Eng R-Rep 23:41–100
19. Brown AM, Ashby MF (1980) Correlations for diffusion constants. Acta Metall 28:1085–1101
20. Kuczynski GC (1949) Self-diffusion in sintering of metallic particles. Trans Am Inst Min Metall Eng 185:169–178
21. Shaler AJ, Udin H, Kuczynski GC, Bever M (1949) Self-diffusion in sintering metallic particles—discussion. Trans Am Inst Min Metall Eng 185:896–897
22. Gordon RS (1973) Mass-transport in diffusional creep of ionic solids. J Am Ceram Soc 56:147–152

23. Hg W (1973) Gordon RS. Effect of oxygen partial-pressure on creep of polycrystalline Al_2O_3 doped with Cr, Fe or Ti. J Am Ceram Soc 56:140–147
24. Carter CB, Norton MG (2007) Ceramics materials: science and engineering. Springer, Berlin
25. Herring C (1950) Effect of change of scale on sintering phenomena. J Appl Phys 21:301–303
26. Johnson KL, Kendall K, Roberts AD (1971) Surface energy and contact of elastic solids. Proc Ro Soc London Ser A-Math Phys Sci 324:301–313
27. Coble RL (1961) Sintering crystalline solids. 1. Intermediate and final state diffusion model. J Appl Phys 32:787–793
28. Coble RL (1961) Sintering crystalline solids. 2. Experimental test of diffusion models in powder compacts. J Appl Phys 32:793–799
29. Johnson DL, Cutler IB (1963) Diffusion sintering. 1. Initial stae sintering models and their application to shrinkage of powder compacts. J Am Ceram Soc 46:541–545
30. Johnson DL, Cutler IB (1963) Diffusion sintering. 2. Initial sintering kinetics of alumina. J Am Ceram Soc 46:545–550
31. Coble RL (1958) Initial sintering of alumina and hematite. J Am Ceram Soc 41:55–62
32. Kuczynski GC (1949) Study of the sintering of glass. J Appl Phys 20:1160–1163
33. Ashby MF (1974) First report on sintering diagrams. Acta Metall 22:275–289
34. Swinkels FB, Ashby MF (1981) Overview 11—A 2nd report on sintering diagrams. Acta Metall 29:259–281
35. Coble RL (1973) Effects of particle-size distribution in initial-stage sintering. J Am Ceram Soc 56:461–466
36. Johnson DL (1969) New method of obtaining volume grain-boundary and surface diffusion coefficients from sintering data. J Appl Phys 40:192–200
37. Swinkels FB, Ashby MF (1980) Role of surface redistribution in sintering by grain-boundary transport. Powder Metall 23:1–7
38. Coleman SC, Beere WB (1975) Sintering of open and closed porosity in UO_2. Phil Mag 31:1403–1413
39. Nichols FA, Mullins WW (1965) Morphological changes of a surface of revolution due to capillarity-induced surface diffusion. J Appl Phys 36:1826–1835
40. Bross P, Exner HE (1979) Computer-simulation of sintering processes. Acta Metall 27:1013–1020
41. Exner HE, Bross P (1979) Material transport rate and stress-distribution during grain-boundary diffusion driven by surface-tension. Acta Metall 27:1007–1012
42. Ross JW, Miller WA, Weatherly GC (1981) Dynamic computer-simulation of viscous-flow sintering kinetics. J Appl Phys 52:3884–3888
43. Ross JW, Miller WA, Weatherly GC (1982) Computer-simulation of sintering in powder compacts. Acta Metall 30:203–212
44. Svoboda J, Riedel H (1995) Quasi-equilimbrium sintering for coupled grain-boundary and surface-diffusion. Acta Metall Mater 43:499–506
45. Svoboda J, Riedel H (1995) New solution describing the formation of interparticle necks in solid-state sintering. Acta Metall Mater 43:1–10
46. Jagota A, Dawson PR (1988) Micromechanical modeling of powder compacts 1. Unit problmes for sintering and traction induced deformation. Acta Metall 36:2551–2561
47. Jagota A, Dawson PR (1988) Micromechanical modeling of powder compacts 2. Truss formulation of discrete packings. Acta Metall 36:2563–2573
48. Jagota A, Dawson PR (1990) Simulation of the viscous sintering of two particles. J Am Ceram Soc 73:173–177
49. Djohari H, Martinez-Herrera JI, Derby JJ (2009) Transport mechanisms and densification during sintering: I. Viscous flow versus vacancy diffusion. Chem Eng Sci 64:3799–3809
50. Martinezherrera JI, Derby JJ (1995) Viscous sintering of shperical-particles via finite-element analysis. J Am Ceram Soc 78:645–649
51. Jagota A (1994) Simulation of the viscous sintering of coated particles. J Am Ceram Soc 77:2237–2239

52. Pejovnik S, Smolej V, Susnik D, Kolar D (1979) Statistical-analysis of the validity of sintering equations. Powder Metall Int 11:22–23
53. Coble RL (1970) Diffusion models for hot pressing with surface energy and pressure effects as driving forces. J Appl Phys 41:4798–4807
54. Herring C (1950) Diffusional viscosity of a polycrystalline solid. J Appl Phys 21:437–445
55. Coble RL (1963) A model for boundary diffusion controlled creep in polycrystalline materials. J Appl Phys 34:1679–1682
56. Paladino AE, Coble RL (1963) Effect of grain boundaries on diffusion-controlled processes in aluminum oxide. J Am Ceram Soc 46:133–136
57. Harmer MP, Brook RJ (1980) The effect of MgO additions on the kinetics of hot-pressing in Al_2O_3. J Mater Sci 15:3017–3024
58. Beere W (1975) Diffusional flow and hot-pressing—study on MgO. J Mater Sci 10:1434–1440
59. Vieira JM, Brook RJ (1984) Kinetics of hot-pressing—the semilogarithmic law. J Am Ceram Soc 67:245–249
60. Vieira JM, Brook RJ (1984) Hot-pressing high-purity magnesium-oxide. J Am Ceram Soc 67:450–454
61. Beere W (1975) Unifying theory of stability of penetrating liquid-phase and sintering pores. Acta Metall 23:131–138
62. Beere W (1975) Second stage sintering kinetics of powder compacts. Acta Metall 23:139–145
63. Helle AS, Easterling KE, Ashby MF (1985) Hot-isostatic pressing diagrams—new development. Acta Metall 33:2163–2174
64. Oyane M, Shima S, Tabata T (1978) Consideration of basid equations and their application in forming of metal powders and porous metals. J Mech Working Technol 1:325–341
65. Shima S, Oyane M (1976) Pasticity theory for porous metals. Int J Mech Sci 18:285–291
66. Dutton RE, Shamasundar S, Semiatin SL (1995) Modeling the hot consolication of ceramic and metal powders. Metall Mater Trans A-Phys Metall Mater Sci 26:2041–2051
67. Kingery WD, Niki E, Narasimhan MD (1961) Sintering of oxide and carbide-meal compositions in presence of a liquid phase. J Am Ceram Soc 44:29–35
68. Huppmann WJ, Riegger H (1977) Liquid-phase sintering of model system W-Ni. Int J Powder Metall 13:243–247
69. Kaysser WA, Takajo S, Petzow G (1984) Particle growth by coalescence during liquid-phase sintering of Fe-Cu. Acta Metall 32:115–122
70. Chu MY, Rahaman MN, Dejonghe LC, Brook RJ (1991) Effect of heating rate on sintering and coarsening. J Am Ceram Soc 74:1217–1225
71. Raj R, Ashby MF (1975) Intergranular fracture at elevated-temperature. Acta Metall 23:653–666
72. German RM (1990) Supersolidus liquid-phase sintering, 1. Process review. Int J Powder Metall 26:23–34
73. German RM, Suri P, Park SJ (2009) Review: liquid phase sintering. J Mater Sci 44:1–39
74. Liu JX, German RM (2001) Microstructure effect on dihedral angle in liquid-phase sintering. Metall Mater Trans A-Phys Metall Mater Sci 32:165–169
75. Smith CS (1948) Grains, phases, and interfaces—an interpretation of microstructure. Trans Am Inst Min Metall Eng 175:15–51
76. Park HH, Kwon OJ, Yoon DN (1986) The critical grain-size for liquid flow into pores during liquid-phase sintering. Metall Trans A-Phys Metall Mater Sci 17:1915–1919
77. Park HH, Yoon DN (1985) Effect of dihedral angle on the morphology of grains in a matrix phase. Metall Trans A-Phys Metall Mater Sci 16:923–928
78. Hwang KS, German RM, Lenel FV (1987) Capilllary forces between spheres during agglomeration and liquid-phase sintering. Metall Trans A-Phys Metall Mater Sci 18:11–17
79. Zovas PE, German RM, Hwang KS, Li CJ (1983) Activated and liquid-phase sintering—process and problems. J Metals 35:28–33

80. Lange FF (1982) Liquid-phase sintering—are liquids squeezed out from between compressed particles. J Am Ceram Soc 65:C23–C24
81. Eley DD (1961) Adhesion. Oxford University Press, Oxford
82. Clarke DR (1987) On the equilibrium thickness of intergranular glass phases in ceramic materials. J Am Ceram Soc 70:15–22
83. Kwon OJ, Yoon DN (1981) Closure of isolated pores in liquid-phase sintering of W-Ni. Int J Powder Metall 17:127–134
84. Shaw TM (1986) Liquid redistribution during liquid-phase sintering. J Am Ceram Soc 69:27–34
85. Huppmann WJ, Riegger H (1975) Modeling of rearrangement processes in liquid-phase sintering. Acta Metall 23:965–971
86. Huppmann WJ, Riegger H, Kaysser WA, Smolej V, Pejovnik S (1979) Elementary mechanisms of liquid-phase sintering. 1. Rearrangement. Zeitschrift Fur Metallkunde 70:707–713
87. Huppmann WJ (1979) Elementary mechanisms of liquid-phase sintering 2. Solution-reprecipitation. Zeitschrift Fur Metallkunde 70:792–797
88. Lee SM, Chaix JM, Martin CL, Allibert CH, Kang SJL (1999) Computer simulation of particle rearrangement in the presence of liquid. Metals Mater Korea 5:197–203
89. Kingery WD (1959) Densification during sintering in the presence of a liquid phase 1. Theory. J Appl Phys 30:301–306
90. Kingery WD, Narasimhan MD (1959) Densification during sintering in the presence of a liquid phase 2. Experimental. J Appl Phys 30:307–310
91. Takajo S, Kaysser WA, Petzow G (1984) Analysis of particle growth by coalescence during liquid-phase sintering. Acta Metall 32:107–113
92. Marion JE, Hsueh CH, Evans AG (1987) Liquid-phase sintering of ceramics. J Am Ceram Soc 70:708–713
93. Eremenko VN, Naidich YV, Lavrinenko IA (1985) Liquid phase sintering. Consultants Bureau, New York
94. Yoon DN, Huppmann WJ (1979) Grain-growth and densification during liquid-phase sintering of W-Ni. Acta Metall 27:693–698
95. Yoon DN, Huppmann WJ (1979) Chemically driven growth of tungsten grains during sintering in liquid nickel. Acta Metall 27:973–977
96. Kaysser WA, Zivkovic M, Petzow G (1985) Shape accomodation during grain-growth in the presence of a liquid-phase. J Mater Sci 20:578–584
97. Gessinge GH, Fischmei HF (1972) Modified model for sintering of tungsten with nickel additions. J Less-Common Metals 27:129–141
98. Gessinge GH, Fischmei HF, Lukas HL (1973) Model for second-stage liquid-phase sintering with a partially wetting liquid. Acta Metall 21:715–724
99. Gessinge GH, Fischmei HF, Lukas HL (1973) Influence of a partially wetting second-phase on sintering of solid particles. Powder Metall 16:119–127
100. Kang SJL, Kim KH, Yoon DN (1991) Densification and shrinkage during liquid-phase sintering. J Am Ceram Soc 74:425–427
101. Park HH, Cho SJ, Yoon DN (1984) Pore filling process in liquid-phase sintering. Metall Trans A-Phys Metall Mater Sci 15:1075–1080
102. Kang TK, Yoon DN (1978) Coarsening of tungsten grain in liquid nickel-tungsten matrix. Metall Trans A-Phys Metall Mater Sci 9:433–438
103. German RM (1995) Microstructure of the gravitationally settled region in a liquid-phase sintered dilute tungsten heavy alloy. Metall Mater Trans A-Phys Metall Mater Sci 26: 279–288
104. Liu YX, Heaney DF, German RM (1995) Gravity-induced solid grain packing during liquid-phase sintering. Acta Metall Mater 43:1587–1592
105. Bowen LJ, Weston RJ, Carruthers TG, Brook RJ (1978) Hot-pressing and alpha-beta phase-transformation in silicon nitride. J Mater Sci 13:341–350

106. Hu SC, Dejonghe LC (1981) Pre-eutectic densification in MgF_2-CaF_2. Am Ceram Soc Bull 60:385
107. Hu SC, De Jonghe LC (1983) Pre-eutectic densification in MgF_2-CaF_2. Ceram Int 9:123–126
108. Wu SJ, Dejonghe LC, Rahaman MN (1985) Subeutectic densification and second-phase formation in Al_2O_3-CaO. J Am Ceram Soc 68:385–388
109. Luo J, Wang HF, Chiang YM (1999) Origin of solid-state activated sintering in Bi_2O_3-doped ZnO. J Am Ceram Soc 82:916–920
110. German RM, Munir ZA (1976) Enhanced low-temperature sintering of tungsten. Metall Trans A-Phys Metall Mater Sci 7:1873–1877
111. Zovas PE, German RM, Hwang KS, Li CJ (1983) Activated and liquid-phase sintering—progress and problems. J Metals 35:28–33
112. Shakhparonov MI, Durov VA (1979) Theory of collective reaction in liquid-phase 5. Collective reaction and vitrification. Zh Fiz Khim 53:2451–2455
113. Ruiz-Valdes JJ, Gorokhovsky AV, Escalante-Garcia JI (2005) Vitrification in the BaO-B_2O_3-Al_2O_3-TiO_2 system containing small admixtures of PbO. J Non-Cryst Solids 351:2036–2041
114. Yang HT, Yang GT, Yuan RZ (1998) Vitrification and devitrification of MgO during sintering of Si_3N_4-MgO-CeO_2 ceramics. Mater Chem Phys 57:178–181

Chapter 6
Sintering and Densification (II)—New Sintering Technologies

6.1 Introduction

Besides the conventional sintering techniques discussed in the previous chapter, various new sintering methods, such as microwave (MW) sintering, spark plasma sintering (SPS), and flash sintering, have been developed for ceramic processing. These methods have also been used to fabricate transparent ceramics. In this chapter, two of them, i.e., SPS and MW, will be presented in detail. It is necessary to indicate that SPS has a number of other names, such as electrical current activated/assisted sintering (ECAS) and field-assisted sintering (FAST). When these names are used in this chapter, they are meant to be the same, unless otherwise mentioned [1]. Additionally, because there are no theoretical studies on sintering phenomena specifically for transparent ceramics by using these new sintering technologies, simulations or modelings are all discussed for all types of materials in general.

6.2 Electric Current Activated/Assisted Sintering (ECAS)

6.2.1 *Brief Description*

Comprehensive description and latest development of the SPS (or ECAS) have been presented in several recent review papers [1–4]. During experiments, loose powders or cold-formed compacts to be consolidated are loaded into a container, which is heated to a targeted temperature and then held at the temperature for a given period of time, while a pressure is applied and maintained at the same time. Thermal energy is supplied by applying an electrical current that flows through the powders and/or their container, thus facilitating the consequent Joule heating effect.

L.B. Kong et al., *Transparent Ceramics*, Topics in Mining,
Metallurgy and Materials Engineering, DOI 10.1007/978-3-319-18956-7_6

SPS is a newly developed method for obtaining fully dense and fine-grained transparent ceramics at low temperatures within short time durations [5].

As compared with the conventional sintering methods, this technology possesses various technological and economical advantages, such as high heating rate, low sintering temperature, and short sintering time duration. Moreover, it can be used for consolidation of powders that are difficult to be sintered by the conventional sintering methods. In addition, it is not necessary to use sintering aids. In some case, the step of cold compaction can be skipped if necessary. Also, it is less sensitivity to the properties of the initial powders.

Due to the relatively low sintering temperature and short time durations offered by this technique, it is possible to fabricate ceramics with ultrafine or even nano-sized grains, because grain growth can be avoided or suppressed. Also, the rapid processing makes it possible to consolidate metastable powders without variation in physical and chemical properties. Furthermore, due to the fast consolidation rate, the processing can be conducted in air atmosphere, without the need to control environments. Additionally, it is also possible to prevent undesirable phase transformations or reactions in the initial powders of the processed materials, owing to the relatively short processing time. More importantly, short processing time means short production cycle and thus high productivity.

6.2.2 Working Principles

Figure 6.1 shows a schematic diagram of a representative ECAS experimental setup [1]. An electrical current, together with a mechanical pressure, is applied during the consolidation of the sample. The applied electrical current and mechanical pressure either can be constant throughout the sintering process or can be varied at selected stages of the densification, thus offering a high flexibility in varying sintering conditions or processing parameters.

The powders to be consolidated can be either electrically conducting or insulating. They are placed in a container, consisting of die, tube, and so on, which are then heated by applying an electrical current. If the powers are conductive, either conductive or insulative container can be used. Otherwise, the container must be conductive, so that the electrical circuit is guaranteed to be closure. Similarly, all electrodes, blocks, spacers, and plungers used in the circuit should be made of electrically conducting materials, such as copper (Cu), graphite, and stainless steel. Conducting powders are heated due to the Joule heating effect and the heat transferred from the container and electrodes, whereas nonconductive powders are heated only due the heat from the container and electrodes, without the presence of Joule heating. Conductive containers can be graphite, conductive ceramics, or steel. In ECAS processes reported in the open literature, graphite containers are the most widely used. The use of graphite limits the mechanical pressure levels that can be applied, which generally is not higher than 100 MPa. However, the presence of graphite makes the sintering environment to be reductive.

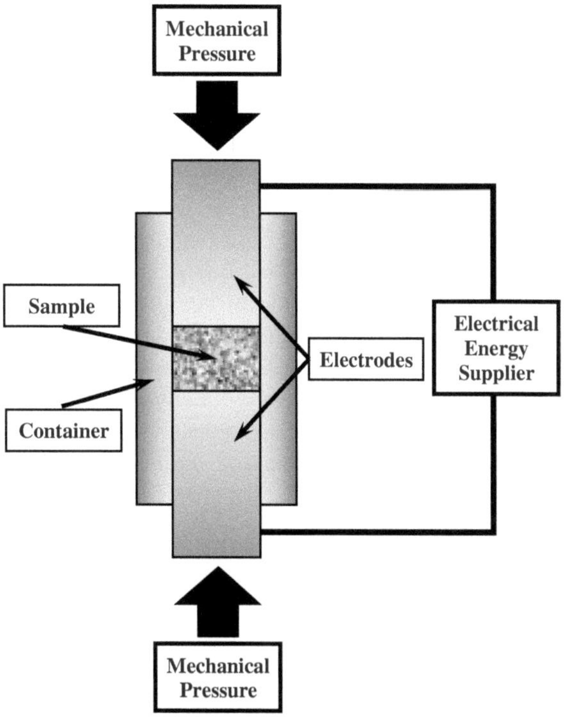

Fig. 6.1 Schematic diagram of a representative electric current assisted sintering (*ECAS*) experimental setup

In terms of the application of a mechanical pressure, ECAS process shares a similarity with hot pressing (HP). However, the application of the electrical current makes ECAS process to be much rapid and efficient. The heating rate during the ECAS process is dependent on the electric power supplied, as well as the geometry of the container/sample ensemble and the thermal and electrical properties of all the materials. The heating rates can be as high as 1000 °C min^{-1}. As a consequence, the processing time ranges from fraction of seconds to minutes depending on the characteristics of the material, size of the sample, configuration of the setup, and the capacity of the whole equipment.

In comparison, in the conventional HP process, the sample container is heated due to the radiation from the enclosing furnace through external heating elements and convection of inert gases if used. In other words, the sample is heated through the heat transfer occurring by conduction from the external surface of the container to the powders. Therefore, the heating rate is much slower, and thus, the process can last hours or even longer. In addition, its total efficiency is much lower than that of ECAS process, because a lot of heat is wasted when the whole volume of space is heated while the sample compact indirectly absorbs heat from the hot environment. Due to the high thermal efficiency, ECAS processes can consolidate samples within a very short time duration, especially when electrically insulative containers are used.

However, ECAS process currently has its own problems, such as inhomogenous temperature distribution when the electrical conductance of the powders is not adequate. Actually, the electrical current and the consequent temperature distribution within the samples are very sensitive to the homogeneity of density distribution of the final products. Furthermore, significant density spatial variation, especially at the beginning of the current flow, may result in high local overheating or even melting of the samples. This is the reason why almost all the samples studied until now have either cylindrical or rectangular shape. This problem should be addressed so that the ECAS process will find more practical applications.

Depending on the characteristics of the power supplies, the electrical current delivered the samples during ECAS processes can be different in intensity and waveform. However, regardless of the waveform of the applied current during the ECAS processes, it is necessary to emphasize that the Joule heating, which in turn determines the level of temperature experienced by the samples, is related to the root-mean-squared instantaneous current intensity, which is defined by the following equation:

$$I_{\mathrm{RMS}} = \sqrt{\frac{1}{\tau}\int_{t}^{t+\tau} I^2(t)\mathrm{d}t}, \tag{6.1}$$

where I is the instantaneous current and t the sampling time.

The mechanical load during ECAS processes is typically applied in the uniaxial direction in most of the currently available facilities. However, specific experimental setups have been designed to apply isostatic or quasi-isostatic [6] pressures to the samples. Ultrahigh isostatic pressure of up to 8 GPa has been reported [7, 8]. According to the characteristics of the applied currents, ECAS processes can be classified in several ways, with two essential types, i.e., resistance sintering (RS) and electric discharge sintering (EDS) [7, 8].

RS is usually observed when the electric power supply has a low voltage, with the order of few tens of volts, but a large current, with the order of thousands of amps, with current waveforms that can be direct current (DC), alternate current (AC), rectified current (RC), pulsed current, and so on. For the occurrence of EDS, the power supply is electrical energy that is stored in a capacitor bank, which is released to the powder compacts as a pulse. Due to this pulsed release mode, EDS process offers higher voltages and higher currents to the samples, as compared with RS. Additionally, the change of current versus time, dI/dt, during the EDS process may trigger electromagnetic phenomena in the compaction. Another significant difference between EDS and RS processes is in characteristic processing time. Generally, the processing time of EDS is in the range 10^{-5}–10^{-2} s, whereas that of RS is of the order of 1–1000 s. In practice, RS is far more frequently observed than EDS. The two types of process are described as follows.

6.2.2.1 Electric Discharge Sintering (EDS)

Electric discharge sintering (EDS) is also known as electric discharge compaction (EDC) or environmental electrodischarge sintering (EEDS), during which electrical energy is discharged from a capacitor bank so that it passes through a column of powders contained in an electrically nonconductive tube container [9]. The high transient current flowing through the column causes heating and sintering of individual particles of the powdery sample. At the same time, this current generates an intense magnetic field in the azimuthal direction, which makes the body of the powder to shrink radially. Therefore, after the discharge, the sample is compacted and the sample can be easily taken out after the sintering. Because of this, the containers can be used for serval times, which is one of the most important advantages of EDS [10].

The first electrical discharge unit is a magnetic forming machine. The capacitor bank generally consists of several capacitors with a total capacitance up to 25 mF, while a charging voltage up to 30 kV can be achieved [11]. The capacitor bank can be charged by using a variable transformer, together with a rectification and smoothing unit. The column of the metal powder acts as a short-circuit resistance across the capacitor bank. The applied discharge current density and intensity can be as high as 2800 MA m^{-2} and 90 kA, respectively. During experiments, the current decreases monotonically with increasing length of the sample, whereas it increases with increasing diameter of the sample.

The typical waveform of the current that flows through the powder column can be in critical, overdamped, and underdamped states. Sometimes, only two stages are observed [12]. During the first stage, the powders are compacted due to the application of the high-voltage pulse. At the second stage, sintering of the sample takes place, at a current density with the order of 10^2–10^3 A cm^{-2}. The reason for the presence of the multistage (two-stage) sintering is electrical resistance of the samples could decrease by nearly six orders of magnitude during the EDS processes. Such process can be conducted with or without the application of mechanical pressure. If mechanical pressures are applied, they can be either static or dynamic. Static mechanical pressures of up to 710 MPa have been reported. When dynamic axial pressure and electrical discharge are applied at the same time, the time duration between the onset of the discharge and the onset of the maximum axial force can be optimized to maximize the densification of the samples.

It has been found that, in some cases, in order to achieve high density, e.g., >95 % of theoretical density, the green density should be higher than 80 %, thus requiring cold pressing step [13]. Also, a minimum level of pressure is necessary for effective discharge to occur. The application of the pressure is actually to compact the powdery sample to a desired density. This is because when discharge is passed through a loose powdery compact, densification does not take place, but instead, intense sparking occurs between adjacent particles, which could lead to rupture of the container. EDS has been extensively used to process metallic materials, especially ferrous materials [13]. Various parameters, including electrical circuit parameters, properties of the powders and dimensions, and geometries of the samples, have influence on the effectiveness of compaction at a given energy input. Generally, if no mechanical pressure is applied, it is difficult to achieve sufficiently high density of the final products. In this case, a swaging and a sintering step are necessary to obtain dense samples.

Porous products processed by using EDS on the other hand have also various potential applications, such as debris filtration, fluid flow control, capacitors, catalyst supporter, and pressure surge protection units. It has been observed that tensile strengths of the EDS-processed porous structures are much greater than those of the samples prepared by using isostatic press of the same density. In fact, isostatically formed samples cannot be used to measure their mechanical strengths. The high strengths of EDS-processed samples are attributed to the formation of interparticle welds due to the heating effect caused by the electrical discharge that passes through the metallic powders, so that strong metal–metal bonds between the base metallic particles are developed. If metallic powders have relatively large particles, e.g., up to 150 mm, the very short sintering time of EDS cannot trigger sufficient diffusion to accomplish homogenization or reactions, even at a temperature that is slightly below the point at which a liquid phase is formed. In other words, it is necessary to provide a liquid phase in order to achieve a considerable degree of alloying or reaction between the metal powders in the mixture. In some cases, continuous axial wires or fibrous structures that are aligned in the direction of current flow can be obtained, due to the interparticle welding.

To form bars through EDS process, the container is first filled with the powders to be consolidated, which are then plugged between two copper electrodes that are pushed into contact with the powder column. To maintain an effective discharge, the powders must be confined in electrically insulating containers. Therefore, Pyrex glass tubes are usually used as the containers, because they can be used repeatedly without pitting in the inner surface and cracks or deformation. Electric energy is charged into capacitors, which is then discharged directly to the loose powder column through a high-voltage switch. The energy injected into the powder column due to the discharge is determined by the current. The density, distribution, and period of the current can be controlled by the capacitance, resistance, and inductance of the circuit.

Usually, the change in inductance during the process is very insignificant, but the variation in resistance of the circuit, due to the heating and progressive welding of particles, can be very pronounced. Also, the higher the current that flows in the circuit, the larger the magnetic body forces will be, which is beneficial to increase green density and strength of the samples. In this respect, the density of the instantaneous current passing through the specimen is the most significant parameter that determines the effectiveness of the compaction process. Therefore, it is expected that the degree of compaction is increased with increasing current density. However, it is necessary to mention that welding and compaction occur only within a certain range of discharge energy, which is depending on the dimension of the powder column and the type of the materials processed.

There is also a minimum voltage that is required to maintain the sample to be sufficiently strong. If the energy discharge is too low, the induced magnetic field will not be strong enough to reduce the diameter of the powder body. As a result, it will be difficult to remove the sample after the processing. This can be understood in terms of electrical resistance of the compact. If the resistance is too high, the current will be too low to facilitate the sintering. In contrast, if the resistance is too low, sintering will not be effective, because the current through the cross section of the compact is not uniform. In this case, a very low current will first pass through the area of least resistance in the compact, which will be overheated to be broken down and more current will then flow through.

EDS is also known as high-energy high-rate (HEHR) consolidation, which converts stored rotational kinetic energy into electrical energy by using the Faraday effect, with a homopolar generator (HPG) [14]. The HPG is a low-voltage (5–25 V) high-current device that is operated in a pulsed mode. In this case, pulse-resistive heating is produced at the interparticle interfaces, so that bonding takes place to achieve the consolidation. The applied current pulse can be as high as 250 kA with a few hundreds of milliseconds, with current densities ranging over 100–500 MA m^{-2}. The process happens within 3 s, with most of the energy to be delivered in the first 0.5 s. At the onset of the pulse, sufficient pressure is applied and maintained for 3–5 min. The specific energy input varies in the range of 0.4–14.25 kJ g^{-1}.

6.2.2.2 Resistance Sintering (RS)

RS process has been extensively studied in terms of different electrical current waveforms that have been applied, and various names and acronyms have been employed to designate the RS process. It is worth mentioning that although the word "plasma" has often been used to designate one of the RS processes, the samples are actually not in an external plasma environment, as in the real plasma sintering or microwave sintering. In addition, sometimes, different names and acronyms have been used to describe the processes, in which, however, the same electric current waveform is applied or the same facility is used. Also, in some cases, RS processes with different electric current waveforms or different equipment have identical names. Therefore, it is suggested to present all parameters or conditions clearly when describing whatever RS process, in order to avoid any unnecessary confusion. It is well known that the application of an electric current is the key feature of RS processes, which can be used for classification. Figure 6.2 shows the types of current waveforms that have been most widely used in the open literature [1].

The simplest current waveform is the constant DC, as shown in Fig. 6.2a, which is characterized by one parameter, i.e., the current intensity, I. The AC current waveform shown in Fig. 6.2b is determined by a maximum current intensity, I_{max}, and its frequency, ω. The third waveform in Fig. 6.2c is pulsed DC, which has four independent parameters, including maximum current, I_{max}, pulse duration τ_{pulse}, and the on- and off-time, τ_{ON} and τ_{OFF}. In the RS process of Fig. 6.2d, the electrical current is applied at two stages. A pulsed electrical current is imposed in the first

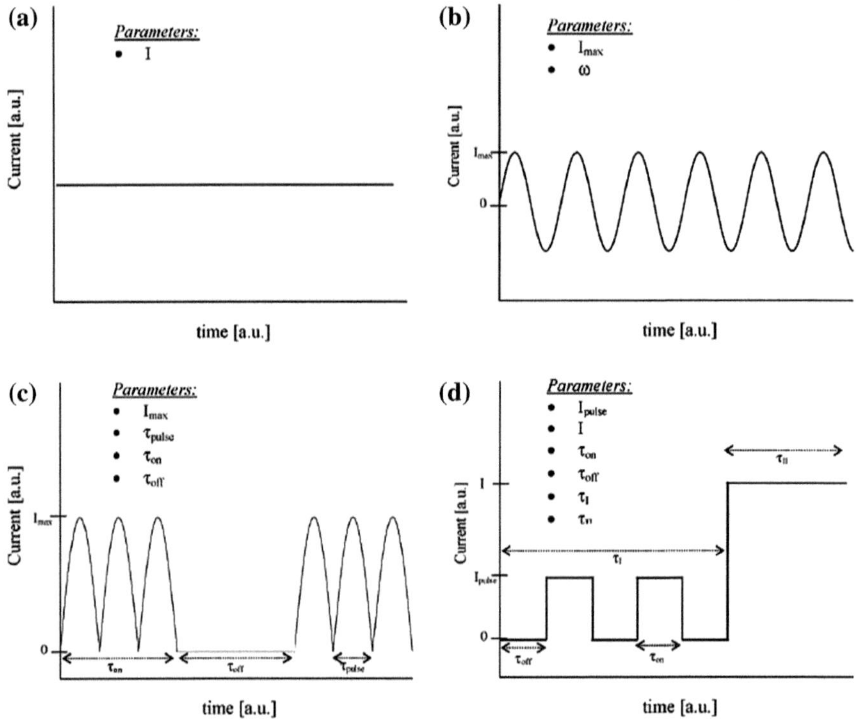

Fig. 6.2 Typical electric current waveforms applied in the RS processes: **a** constant DC, **b** AC, **c** pulsed DC, and **d** pulsed DC + DC. Reproduced with permission from [1]. Copyright © 2009, Elsevier

stage, while a constant DC is followed in the second stage. The first stage is defined with parameters of maximum current, I_{max}, the on-time, τ_{ON}, and the off-time, τ_{OFF}, while the second stage has only one parameter, i.e., current intensity, *I*. In addition, the relative durations, τ_{I} and τ_{II}, of the two stages must be clearly indicated. The squared pulses shown in Fig. 6.2d have been used in experiments, as illustrated in Fig. 6.3 [15]. Therefore, a precise description of the level of the current applied, together with a complete set of all operating parameters, is important to characterize a RS process [1].

In practice, there are more possibilities to synthesize current waveforms, through the combination of the basic ones shown in Fig. 6.2. Also, the waveforms or pulses can be further modified or changed by having different shapes. In this respect, the room for operation for RS process is unlimited. Different types of commercially available equipment, which are capable of applying different types of current waveforms, have been studied and can be found in the open literature [16, 17].

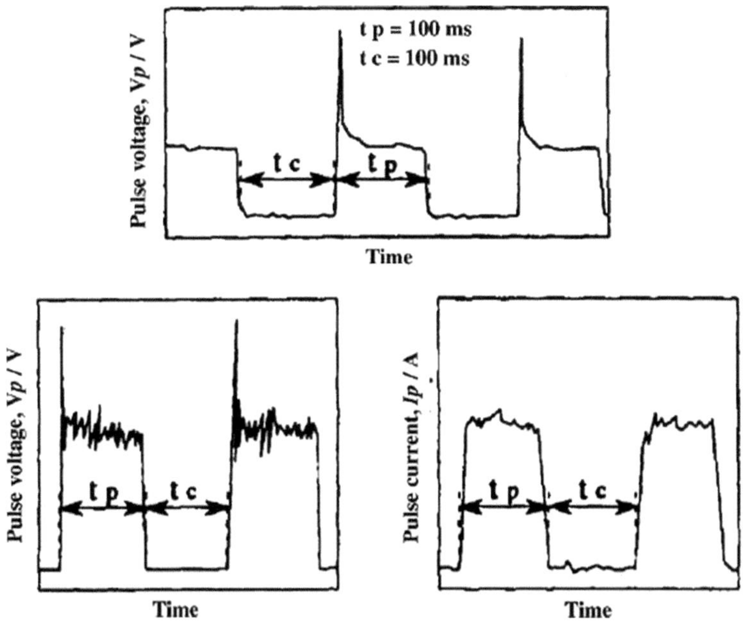

Fig. 6.3 Experimental measurements of the electrical currents with the type of the waveform shown in Fig. 6.2d. Reproduced with permission from [15]. Copyright © 1996, Elsevier

6.2.3 Brief History of ECAS Processes

The first example could be a US patent in early 1920s could be the first example of ECAS, regarding the production of dense articles starting from oxide powders to be sintered through the application of an electrical current [1]. The idea of simultaneously applying a uniaxial mechanical pressure and an electrical current to sinter metal powders reported in later 1920s. The mechanical pressure is applied through the electrodes to the confined WC/Co powder mixture and a direct alternate electrical current is simultaneously supplied through the electrodes connected to an external electric circuit, thus heating the powdery material to a temperature that is sufficiently high to sinter the sample thoroughly. The processing is finished within minutes. This method was used to consolidate elemental powder mixture of W, C and Co. There are the earliest examples of reactive sintering assisted by electrical current. Complete densification of the WC/Co cemented carbide in about one second was reported in early 1930s. The reason why such a shorter processing time is possible is because the apparatus is capable of discharging a pulse of direct current to the samples followed by imposition of an alternate current. The electric discharge was realized by using a condenser connected to a 2.5 kV source of direct current, which is in parallel with the sample to be sintered. The development of the technique was continuously progressed.

Various applications of ECAS process were patented later on. Progress in commercialization of ECAS process was achieved in the 1960s. The proposed ECAS method combines pressing and sintering of powdery metal parts in one shot due to the impulsive force of an unspecified "spark energy," and therefore, it is called spark sintering. This process had been claimed to be capable of sintering materials that cannot be processed by using the conventional sintering methods, thus having versatility and almost unlimited capabilities.

In 1966, a patent of ECAS process was issued, which is able to apply simultaneously an electrical current and a mechanical pressure of lower than 10 MPa. The electrical current is applied through the superimposition of a periodic current on a direct current. The light contact between the particles due to the relatively low pressure, along with the imposition of the periodic/alternate current, leads to the formation of spark discharges within the initial voids within the sample to be sintered. These electric sparks have a power of hundreds to even thousands of joules, which allow the particles to bond one another with a pressure that is even higher than the applied mechanical pressure. In earlier ECAS techniques, high pressure is required to reduce the contact resistance of the powdery samples. As the contact resistance is substantially equal to the internal resistance of the particles, a high current could be passed through the sample to develop the necessary bonding heat. In contrast, in the new design, the contact resistance must be greater than the internal resistance, preferably several times as great, so that at least during the initial stages of the sintering, most of the applied energy is in the form of spark discharge, instead of being dissipated in resistive heating of the particles. Since the sintering takes place immediately upon the spark discharges, densification can be achieved within seconds, as compared with earlier methods requiring tens of minutes and even hours to complete the sintering.

The process, named as spark sintering instead of electric discharge sintering, was presented as the latest method for producing powder metal parts, due to its capability of combining pressing and sintering in a single operation. In addition, it was already advanced to industrial scale since a semi-automatic system was developed. It has been reported that spark sintering could be a very effective process for consolidating alloy powders that have high creep resistance at elevated temperature because the processing parameters can be readily adjusted to produce a temperature that lies within the very narrow range between the onset of plasticity and melting. Considerable progress had also been achieved in forming and simultaneously bonding powders into solids by this technique. Complex configurations with minimized machining can be produced [18].

Although spark sintering has been realized as a practical and cost-effective process, a fundamental understanding on the real role of the electric current/voltage is lacking. There are also some problems related to the ECAS processes [19]. For example, the application of this process may be limited by the typical use of graphite dies, which are characterized by low wear resistance with a consequent increase in the processing costs. Therefore, the use of wear resistance alumina dies has been proposed to replace the graphite ones. Also, the use of short-time high-current pulses cannot produce dense parts of conducting powders in nonconducting dies, due to the inhomogeneous current distribution within the powder compact. To address this problem, a strategy has been developed so that alternating and direct current can be applied simultaneously. It is believed that continuous development of ECAS process will be advanced in the future.

6.2.4 Modeling and Simulation

Because RS represents the dominant majority, modeling considerations only related to this process will be presented and discussed in this subsection. RS process can be taken as a thermo-electromechanical system, so that the corresponding modeling is meant to deal with the complex coupling of thermal, electrical and mechanical phenomena involved in the powders inside the experimental setup, in which there are several elements characterized by different geometries and material properties. Generally, properties of the materials exhibit a nonlinear dependence on temperature and pressure. Therefore, modeling of such a system is particularly challenging, because nonlinearities are accompanied by model uncertainties. At the same time, the difficulties to obtain reliable experimental data may cause masking effects that can be macroscopically misinterpretable.

Usually, the description of the mechanical behavior of RS process has been completely neglected. Thus, the analysis is limited to solid (fully dense) or presintered compact specimen instead of sintering powders. In this respect, the results obtained are generally not applicable to RS-processed powder compacts, because the mechanics of sintering and densification is strongly coupled to the corresponding thermal and electrical phenomena. Furthermore, even the analysis is limited to a thermoelectric perspective suitable for compact samples, too many assumptions are necessary when developing the models, thus making it difficult to provide quantitative results. In almost all cases, simulation results more or less have deviations from actual operating conditions experimentally used, which thus are used as a qualitative reference of the behavior of the real system.

In practice, modeled results are not used to directly compare with experimental data. Even though comparison is necessary, it is only limited to temperature profiles. In contrast, electrical variables, such as measured voltage and current, are not used for comparisons. The purpose of modeling is to provide a quantitative track for these variables. If temperature profiles can be well modeled, it means that the modeling is reliable. This is because temperature distribution is the main factor having effect on heterogeneity of the sintered product, so that all other variables and their effects can be neglected.

Heat transfer and generation, e.g., Joule effect, in an electrode, such as punch–specimen–electrode (punch) system without a die, have been well modeled [20–23]. In this case, 1D axial thermal balances, i.e., for steel (50S2G or 20KhN3A) electrode/(punch) and for electrically conductive powder (VK16 or VK8VK hardalloys) samples, coupled through continuity boundary conditions at the interface contacts, as shown in Fig. 6.4, are considered [20]:

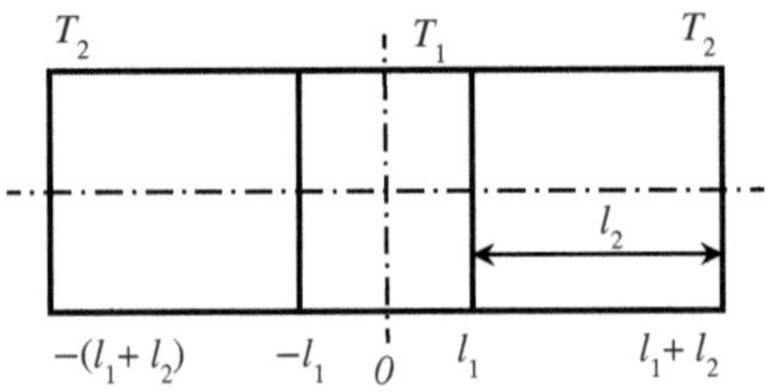

Fig. 6.4 Integration domain and boundary conditions of the model. Reproduced with permission from [20]. Copyright @ 1989, Springer

$$\rho_{\mathrm{i}} C_{\mathrm{p,i}} \frac{\partial T}{\partial t} = k_i \frac{\partial^2 T}{\partial z^2} + \rho_{\mathrm{el,i}} j^2, \quad \mathrm{i} = \begin{cases} \text{specimen powder}, \\ \text{electrode steel}. \end{cases} \tag{6.2}$$

The heat in the model is generated uniformly in such a way that it is proportional to the corresponding electrical resistivity, $\rho_{\mathrm{el,i}}$, and the square of same current density, j, that flows in series through the system of electrode (punch)–specimen–electrode (punch). During the modeling, axial symmetry is assumed, so that it is necessary to consider only half of the real domain of integration. A constant temperature that is equal to the initial value, T_0, can be assigned, because the extreme ends of the electrode (punch) are cooled, due to the contact with a high thermal capacity of the rest of the RS system. In order to obtain analytical solutions of the model, it is assumed that the current density and all thermophysical parameters, including density, ρ_{i}, thermal capacity, $C_{\mathrm{p,i}}$, thermal conductivity, k_{i}, and electric resistivity $\rho_{\mathrm{el,i}}$, are constant.

The effect of actual changes in thermophysical parameters with temperature has been evaluated when solving the model by starting with different initial temperatures, with which a representative mean value is derived and used. The sintering behavior of the powder sample can be modeled by using effective thermophysical parameters that are corrected with the corresponding values of its fully dense solid counterpart, for a given porosity (θ), which are expressed by the following equations:

$$\rho_{\mathrm{powder}} = \rho_{\mathrm{solid}}(1 - \theta), \tag{6.3}$$

$$C_{\mathrm{p,powder}} = C_{\mathrm{p,solid}}(1 - \theta), \tag{6.4}$$

$$k_{\mathrm{powder}} = k_{\mathrm{solid}}(1 - \frac{3}{2}\theta), \tag{6.5}$$

$$\rho_{\mathrm{el,powder}} = \rho_{\mathrm{el,solid}}(1 - \frac{3}{2}\theta)^{-1}. \tag{6.6}$$

Modeling results and experimental data have been compared, in terms of temperature temporal profile, for the system of VK8VK powder sample and 20KhN3A steel electrode, at two constant current densities of 429 and 784 A cm^{-2} [24]. The modeling results are derived with the thermophysical parameters evaluated at 20 and 500 °C, respectively. The temperature profile along the axis of the system inside the electrically conductive powdery specimen predicted by the model is relatively flat, whereas the temperature gradient through the electrode (punch) according to the simulation is higher than experimental observation, which can be attributed to the effect of cooled ends and a lower thermal conductivity of the materials.

A new model has been developed, dealing with the current density and the consequent Joule heat distribution between the specimen and the die [25]. Thermal balances, as given in Eq. (6.2), where Joule heat is expressed in terms of voltage gradient, are coupled to the current density balances, i.e., Kirchhoff law with distributed parameters in a 2D cylindrical coordinate system:

$$\frac{1}{r}\frac{\partial[(1/\rho_{\mathrm{el,i}})(r\partial\varphi/\partial r)]T}{\partial r}+\frac{\partial[(1/\rho_{\mathrm{el,i}})(\partial\varphi/\partial z)]T}{\partial z}=0,\quad \mathrm{i}=\begin{cases}\text{specimen,}\\ \text{graphite die,}\end{cases}\tag{6.7}$$

where φ is the electric potential, i.e., current density vector is proportional to voltage gradient according to Ohm's law, $\bar{j}=-(1/\rho_{\mathrm{e.1}})\overline{\nabla}\varphi$. Radiation heat losses from outer surface of the die are included in the boundary conditions, while the effective heat conduction dissipation in axial direction from the plunger to the electrodes is not modeled. Because the PDS apparatus chamber is operated in vacuum, the convection heat losses from outer surface of the die can be neglected. It has been assumed both temperature and voltage are continuous, i.e., the voltage and current density between the contacting sides are the same. Thermophysical parameters of solid graphite or copper powder specimen are used in the model, which is solved by using a numerical technique based on the method of fundamental solutions, so that the voltage and temperature spatial gradients and their discontinuity at the junctions of different materials can be captured easily.

Modeling results have been available about the axial profiles of the graphite solid and copper powder sample, in terms of voltage and current density under a uniform temperature distribution at 303 K, with a voltage drop of 1 V. For isotropic thermophysical properties, the copper powder sample is assumed to have a constant porosity. It can be concluded that the electrical current flows through the sample instead of the surrounding graphitic die when copper powders are used instead of solid graphite, because copper has higher electrical conductivity than graphite. Therefore, at same voltage drop, current density of the copper sample is higher than that of graphite one. As a result, axial voltage drop is concentrated through graphite plungers when a copper powder sample is inserted into the graphitic die, whereas an almost uniform distribution over the whole system is observed when the sample is graphitic. It is also found that heat generation is concentrated on the plungers, due to their smallest cross-sectional area, as demonstrated by theoretical potential, Joule heat, and temperature distribution contour for graphite solid and copper powder when a steady state is arrived at voltage drops of 0.8 and 0.6 V to the system. Therefore, temperature is almost uniform in the sample.

A different result has been reported in simulating TiB_2/BN composite solid that is inserted in a graphitic die [26, 27]. The difference in the simulation results between this study and the previous one could be attributed to the difference in properties between the two materials. Different from that of copper powder, electric resistivity of TiB_2/BN is almost equal to that of graphite, but with a lower thermal conductivity. The system simulated in this case has relatively larger dimensions. Thermal balances for the different materials are taken into account only in 2D cylindrical coordinate system. Thermophysical parameters are assumed to be constant in order to obtain steady-state analytical solution through an original derivation which is based on radial and axial symmetry.

The solution of the original 2D problem is represented by the product of the solutions of the two 1D problems, as shown in Fig. 6.5 [28]. The radial profile is derived by coupling thermal conduction and Joule heat generation inside the two infinitely long coaxial cylinders with radiuses equal to those of the real system, whereas axial profile is obtained by using sandwiched slabs of infinite surface area with heights equal to those of real system. Continuous boundary conditions at contact surfaces are assumed, with radiation heat losses from lateral outer surface of the external cylinder to mimic lateral outer surface of the die. The temperature of the exposed surfaces of slabs is assumed to be constant, i.e., ambient temperature, which simulates axial cooling from plungers to rams.

Steady-state temperature difference from the center of the sample along radial direction is predicted by using the model, while experimental temporal profiles of temperature measured by using two thermocouples positioned at center of the sample, which are radially apart, at plunger/die contact interfaces [28]. In this study, conditions for the experimental data and the parameters used for the simulations, such as electric current densities and corresponding heat generation, are not mentioned. Thermophysical data of the materials and height of sample are also not available. Nevertheless, the model predicts large differences in radial temperature,

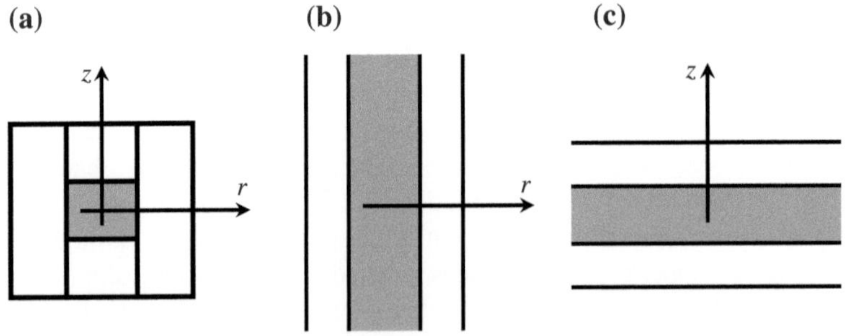

Fig. 6.5 Integration domain and boundary conditions of the model for TiB_2/BN composite solid in a graphite die: **a** 2D configuration, **b** r-direction, and **c** z-direction. Reproduced with permission from [28]. Copyright @ 2002, Elsevier

350 °C inside the sample and 100 °C inside the die, i.e., the TiB_2/BN sample is hotter than the graphite. The temperature difference inside the sample measured experimentally is 450 °C. Therefore, the modeling results are in a qualitative agreement with the experimental data. Steady state is not reached during the experiments.

Similar simulation results have been reported for insulating powder compaction by using SPS, for example, Al_2O_3 powder [27]. Temperature distribution is predicted by considering 1D radial thermal balance with conduction and Joule heat generation for a sample of 2 cm diameter with finite element method (FEM). The necessary boundary condition of thermal balance inside the sample is determined by taking surrounding graphitic die into account. The temperature at the sample/die interface is evaluated according to the Joule heat developed by the current flowing inside the die as a function of its electrical conductivity and cross-sectional area, which are compared with those of the sample. The sample and the die are considered as resistors connected in parallel. In this study, thermophysical parameters and system geometry are not described and boundary conditions are not clearly defined. Based on the predicted temperature distribution, grain growth of the powder is mathematically described by using Monte Carlo simulations through moving grain boundaries and peak points of a fine cell structure. Radial temperature profiles inside the Al_2O_3 sample at two steady states have been simulated, with boundary temperature at the sample/die contact be 1450 and 1650 °C, respectively. The temperature at center of the sample is always lower than that at its outer surface, with a maximum difference of 77 °C. This is because most of the electrical current flows through the graphite die, due its lower resistivity as compared with that of the sample. In this case, Al_2O_3 powder is heated by thermal conduction from the die, which is reflected by the distribution of grain growth.

A modeling study to simulate direct synthesis of $MoSi_2$–SiC composites from element mixture of Mo, Si, and C by using field activation has been reported [29]. The experimental process of spark plasma synthesis (SPS) is simulated, with simultaneous synthesis and densification of the materials, including also nanophases. A parallel combination is used to determine the distribution of flowing current density, between the electric resistances of the inner sample and the surrounding die. The reaction is highly exothermic. A schematic of the modeled sample is shown in Fig. 6.6, in which the cylindrical coordinate thermal balances for the sintering/synthesizing powder sample and graphitic die are coupled with the corresponding current density balances for properly evaluating Joule heat distribution [29]. The diameters of the sample are 2.0–8.0 cm, while the thickness of the sidewall of the graphite die is about 1.0 cm.

The reaction times as function of voltage for four sample sizes of 1–4 cm have been studied [29]. The formation of SiC is always slower than that of $MoSi_2$. This difference becomes less pronounced at high voltages or for large samples. There is a threshold voltage, below which ignition cannot be triggered within 200 s. This voltage is also different between $MoSi_2$ and SiC within one sample. Smaller samples correspond to faster reactions. As the voltage is increased, the 50 % reaction time decreases. The reaction time decreases with decreasing sample size.

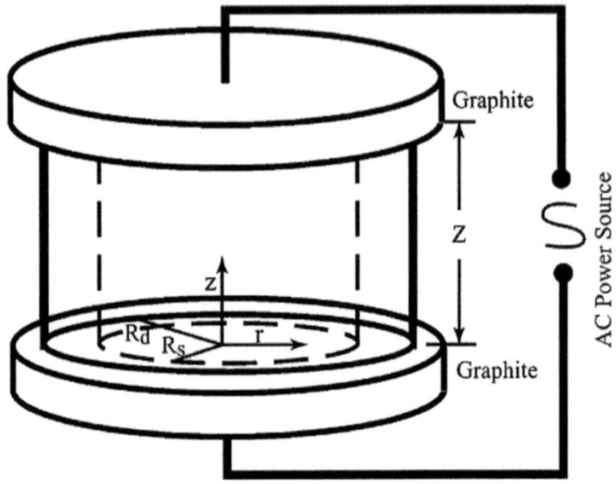

Fig. 6.6 Schematic of the cylindrical sample and die. *Top* and *bottom* graphite plates are modeled as boundary conditions for the sample and die. Reproduced with permission from [29]. Copyright © 2002, Elsevier

Above certain values of voltage, the effect of voltage becomes less pronounced and the difference in reaction time between the samples of different sizes diminishes.

The completion of the reaction is examined at the same point in space, i.e., halfway between the center and the surface of the sample and halfway between the top and the bottom [29]. The time, at which the completion of the reaction is measured, is either 200 s or the time at which less than 5 % of the original amount of is silicon remained at the outer edge of the sample. It is clearly demonstrated that the completion of $MoSi_2$ is always earlier than that of SiC. At higher voltages, for SiC, the larger the samples, the higher the degree of completion of reaction will be.

For the sample with 80 % $MoSi_2$, the 50 % reaction times for both $MoSi_2$ and SiC are much faster with the dies with higher conductivity [29]. According to total current in a cross section of the die and the sample, as a function of time, the conversion data of $MoSi_2$ and SiC are at the usual point, i.e., halfway between the center and the surface and halfway up. In the beginning, the current is carried nearly completely by the sample. The amount of current carried by the sample increases as the amount of $MoSi_2$ is increased, reaching a maximum at a point where the reaction of $MoSi_2$ is almost completed.

At that point, the die starts to carry the current, which increases gradually, and at the same time, the reaction to SiC also increases. Similar results are obtained for a die with lower conductivity, in which the current is carried by the sample throughout the whole reaction. For the 20 % $MoSi_2$ samples, the 50 % reaction times are also much faster with the dies of higher conductivities. According to plots of current distribution of the samples with this composition, it is observed that the difference between these two plots is pronounced. With a factor of 5.0, the sample carries almost no current throughout the reaction, while for the factor of 0.5, the sample carries the current transiently, with a peak at about 30 s, following which the current is carried primarily by the die.

However, in this study, plungers are not simulated. The thermal balance inside the powder compact is characterized by an additional source term, which is related to the enthalpy released during the highly exothermic reaction between starting powders. The rate of the corresponding heat generation is taken into account by coupling material balances of the reacting components with pseudo-homogeneous second-order kinetics and the kinetic constants of Arrhenius-type relation. Thermophysical parameters, such as specific heat, thermal, and electric conductivities, as a function of temperature, are determined by using the mixing rules, with the thermophysical properties of fully dense reactants, according to the composition that is varied as the combustion reaction is progressed. However, porosity variations during the sintering under mechanical loads are not evaluated.

Radiation and convective losses from all exposed surfaces of the dies are considered when defining the boundary conditions for thermal balances. Different effective heat transfer coefficients and emissivities are used to express heat losses from top to bottom or from lateral surfaces of the dies. This choice is made in order to simulate the presence of contact resistances at top and bottom surfaces of the dies. Boundary condition at contact surfaces between the sample and die is not clearly defined, instead they are assumed to be continuous. In boundary conditions for current density balances, electrical current is confined into the conducting materials and a voltage drop between the top and bottom equipotential surfaces of the die is applied. By the way, comparison with experimental data is not available and the model clearly predicts volume or wavelike combustion reaction, depending on sample size, initial powder composition, and electric conductivity of graphite die. The simulation results are obtained for the application of 15–40 V voltages along graphitic dies, which are relatively high values in SPS.

All the relevant qualitative results reached by modeling activities described so far, including hot spots in plungers, electrical current, and related Joule effect distribution field as a function of electric resistances of sample and die in parallel combination, have been confirmed [30]. Spark sintering behavior of presintered Ti powders (electrical conductive material) and alumina powders (electrical insulator material) are compared with model results obtained by considering the SPS system, as schematically shown in Fig. 6.7 [31]. Compacts of Ti or Al_2O_3 powders, with average particle sizes of 35 and 0.4 μm, are tested. The upper punch, die, and presintered compacts are drilled with holes of 3 mm diameter, and temperatures are measured by using thermocouples at fixed positions, as shown in Fig. 6.8, for 3.6 ks in the continuous current discharging conditions, at pressure of 37.5 MPa and current densities of 290 and 700 A for the punch–die–Ti and the punch–die–Al_2O_3 compact systems, respectively.

This is a simulation for the first time in which pulses of electrical current, with rectangular shape and duration of 100 ms, and mechanical load applied during the SPS experiments are mentioned and considered [32]. Besides the usual thermal balance, the model also considers heat conduction and Joule effect, coupled to current density conservation equation for the presintered powder compact samples, graphitic die, and plungers. The change in relative density of the presintered powder sample is taken into account when evaluating the effective thermophysical properties throughout the processing. The corresponding thermophysical properties of the full-dense materials, together with empirical functions, i.e., third-order

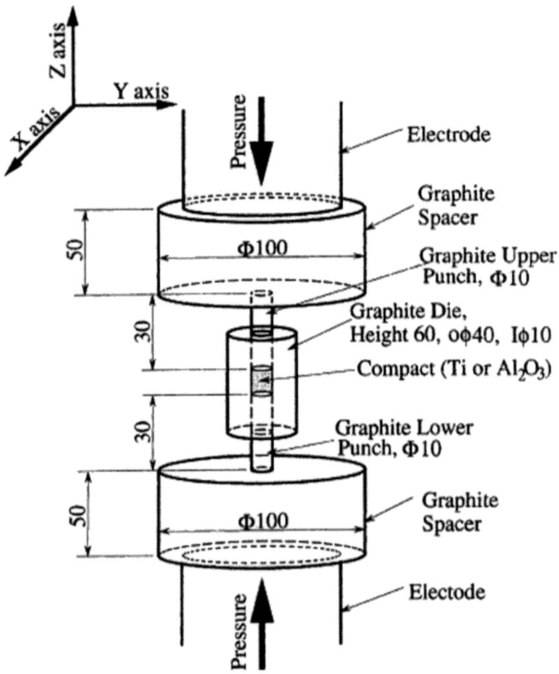

Fig. 6.7 Schematic diagram of the punch–die–compact system for modeling, with all units in mm. Reproduced with permission from [31]. Copyright © 2003, Elsevier

polynomials of relative density (D), have been used for the simulation. The density is treated as a function of temperature (T) and applied nominal pressure (P) according to the following expression [1]:

$$D(T, P) = D_m + aP + b(T - T_0)c, \tag{6.8}$$

where D_m is the initial relative density of the presintered powder sample and T_0 is the initial temperature of the system, whereas a, b, and c are the parameters that can be adjusted accordingly. Their values are fitted through the comparison of Eq. (6.8) and the relative density values determined experimentally. The density is estimated by measuring longitudinal displacements between the two electrodes, with the assumption that temperature and pressure distributions inside the sample are uniform.

Thermal and electric contact resistances are taken into account as a function of local temperature and are referred to a specific contact surface (horizontal and vertical) between the different elements of equipment, i.e., die/compact, punch/compact, punch/die, and graphite-spacer/punch, as shown in Fig. 6.7 [31]. As a result, the obtained contact conductances and thermal conductivities are in the range of 5×10^3–7×10^6 S m^{-2} and 3×10^{-2}–2×10^5 W m^{-2} K^{-1}, respectively.

The comparisons of the experimental data and modeling results, for the two samples, are shown in Figs. 5.9 and 5.10, respectively [31]. In these figures, temperature contour plot for Ti and Al_2O_3 compacts at steady state at a pressure of

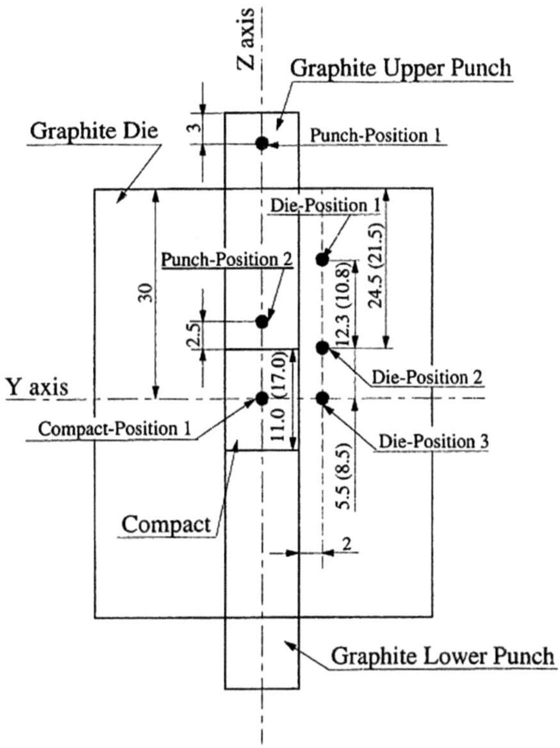

Fig. 6.8 Positions of measuring temperature for the punch–die–titanium and punch–die–alumina compacts in the system, with the values within the parenthesis for those in the punch–die–alumina compact and unit in mm. Reproduced with permission from [31]. Copyright © 2003, Elsevier

37.5 MPa and constant currents of 290 and 700 A are compared with the temperatures experimentally measured by using thermocouples at six different positions, with two in plungers, three in die, and one at center of the sample. According to Fig. 6.9, the matching for the Ti sample, with maximum deviation of 90 °C that corresponds to an error of about 10 %, is better than that for the Al_2O_3 sample. For the Al_2O_3 sample that is an insulator, current flows through the die. As a result, temperature is lower inside the sample than inside the die, with a difference of about 10 °C. In other words, the Al_2O_3 sample is heated due to the thermal conduction from the surrounding die. In contrast, Ti is a conductor, so that an opposite direction of heat flux is observed, whose temperature is higher than that of die, with a difference of about 30 °C. It means that the temperature gradient for conductive sample is larger for insulative one. In addition, show hot spots in plungers as a result of their smallest cross-sectional area in both cases (Fig. 6.10).

A punch–die–two step system has been developed to further increase the homogeneity of the sintered product with complex shapes [33]. The strategy proposed is the reduction in heterogeneity of temperature field distribution inside the samples, as shown in Fig. 6.11. The material is pure Ti powder with average particle of 35 μm. The compact has a length of 10 cm, with two steps of 15 and 30 mm in diameter, at a relative density of 100 %. Ti powder is filled into the

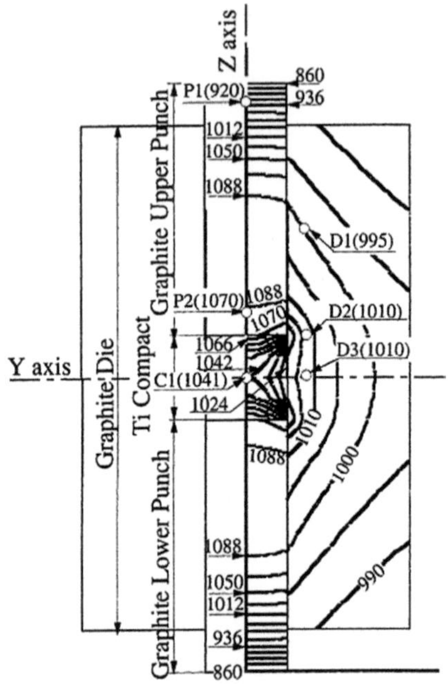

Fig. 6.9 Comparison between experimental data and model results in terms of temperature distribution (Kelvin scale) of Ti compact at t = 3600 s subjected to 37.5 MPa and 390 A. The measurements by six thermocouples at positions P_1, P_2, D_1, D_2, D_3, and C_1 are indicated in *brackets*. Reproduced with permission from [31]. Copyright © 2003, Elsevier

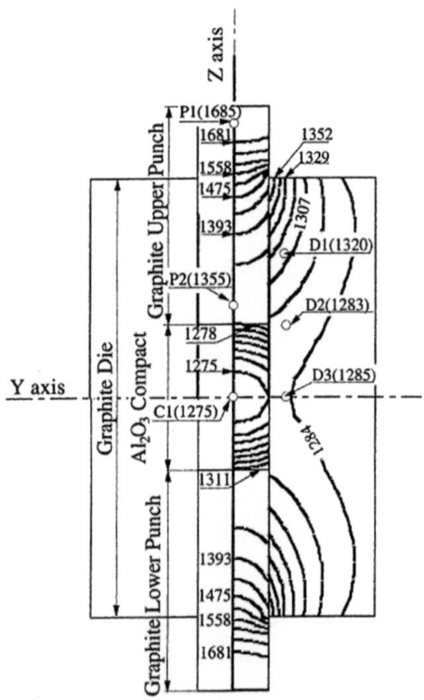

Fig. 6.10 Comparison between experimental data and model results in terms of temperature distribution (Kelvin scale) of Al_2O_3 compact at t = 3600 s subjected to 37.5 MPa and 700 A. The measurements by six thermocouples at positions P_1, P_2, D_1, D_2, D_3, and C_1 are indicated in *brackets*. Reproduced with permission from [31]. Copyright © 2003, Elsevier

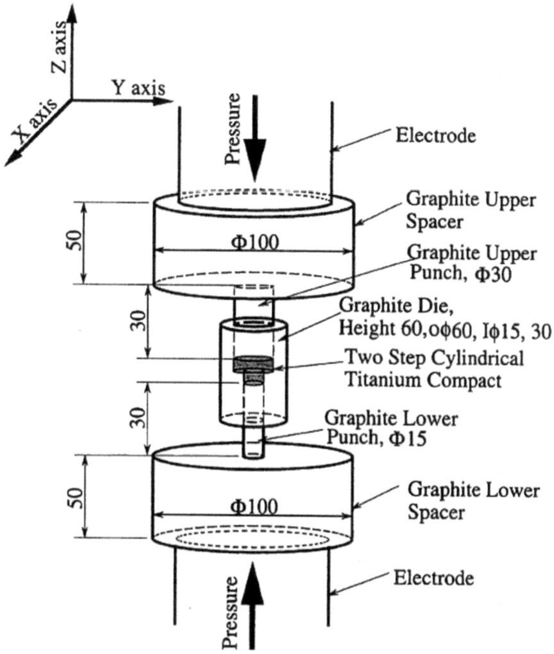

Fig. 6.11 Schematic of the punch-die-two step cylindrical titanium compact system, with units in mm. Reproduced with permission from [33]. Copyright © 2003, Elsevier

graphite die, which is plugged at both ends with graphite punches. Spark sintering is conducted with a combination of electric pulse discharging and continuous electric current discharging, at upper and lower punch pressures of 10 and 38 MPa, respectively. The current density on top surface of the upper graphite punch is 0.14 A mm^{-2}, while that on bottom surface of the lower punch is 0.56 A mm^{-2}. The temperatures are measured with thermocouples at eight positions, including two in plungers, three in die, and three in sample axis, as shown in Fig. 6.12, for 3.6 ks under continuous current discharging conditions of 900 A and constant upper and lower punch pressure of 10 and 38 MPa, respectively [33]. The current densities, on top and bottom surfaces of the upper and lower punches, are 1.3 A mm^{-2} and 5.1 A mm^{-2}, respectively. Three types of relative density are used to characterize the compact, including (i) mean density throughout the compact according to Archimedes' principle, (ii) density measured by using an image analyzer on polished section, and (iii) apparent density according to the longitudinal displacement measured directly between the electrodes during the continuous current discharge.

The first step is called presintering stage, which is carried out for 15 min with pulsed current, i.e., rectangular pulses of 100 A and duration 100 ms, 1 ON and 1 OFF operations, and constant mechanical load equal to 7 kN, which is about 10 MPa on top plunger and 38 MPa on bottom plungers, respectively. The second step then conducted for 1 h, with constant continuous current of 900 A, while mechanical load is kept unchanged. The whole process is modeled and compared with experimental results. In Figs. 5.13 and 5.14, the experimental data and

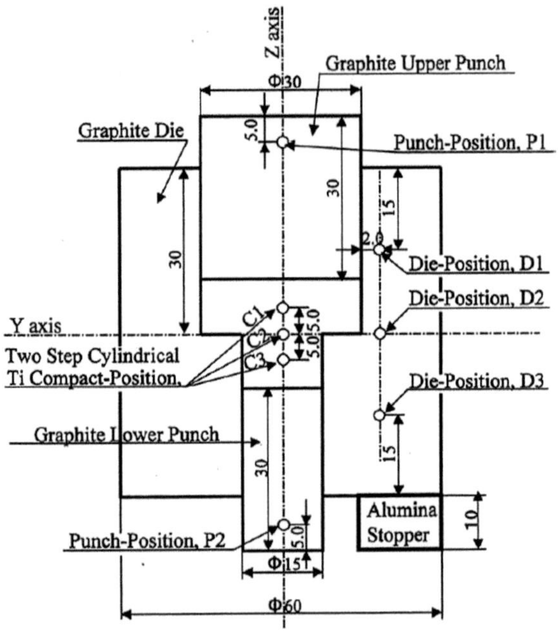

Fig. 6.12 Temperature measurement locations, with units in mm. Reproduced with permission from [33]. Copyright © 2003, Elsevier

modeling results in terms of temperature contour plot at 3 and 10 min, correspondingly, are compared [33]. The temperatures experimentally measured with thermocouples at the eight positions are indicated in brackets. A very agreement can be observed between experimental and simulation results, with maximum difference to be about 3 %. It is found that temperature reaches maximum value inside the lower plunger, i.e., at the smallest cross-sectional area, and the Ti sample is always hotter than the surrounding die in the radial direction (Figs. 6.13 and 6.14).

The presence of thermal conduction heating from the surrounding die to an insulator-like material sample has been further confirmed when analyzing SPS behavior of BN powder [34]. Densification behavior of the powder is well modeled, with the assumption that the SPS process consists of three stages. At the first stage, no obvious shrinkage can be observed although temperature increases continuously. When the process enters the second stage, there is a sharp increase in density due to the abrupt shrinkage, while the temperature is kept almost constant. At final stage, both temperature and density of sample are a constant. Therefore, only the first stage is modeled. The usual 2D cylindrical coordinate thermal and current density balances for the graphite die, plungers, and the powder compact are solved by FEM method. Boundary conditions for thermal balances, i.e., radiation losses from all exposed surfaces, assigned temperature at the plungers' ends mimicking a perfect cooling system in the SPS system and continuity condition at every contact surface are specified. Natural convection is neglected, because the processing is performed in vacuum conditions.

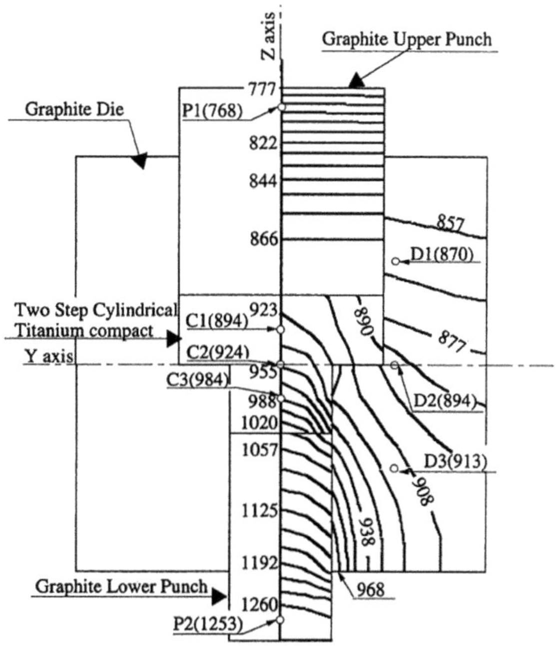

Fig. 6.13 Temperatures in the *y-z* plane at locations, P1, P2, D1, D2, D3, C1, C2 and C3 at the holding time of 0.18 ks, and calculated isothermal contours under constant current discharge conditions, with nits in K. The values within parentheses were the temperatures measured by the thermocouples. Reproduced with permission from [33]. Copyright © 2003, Elsevier

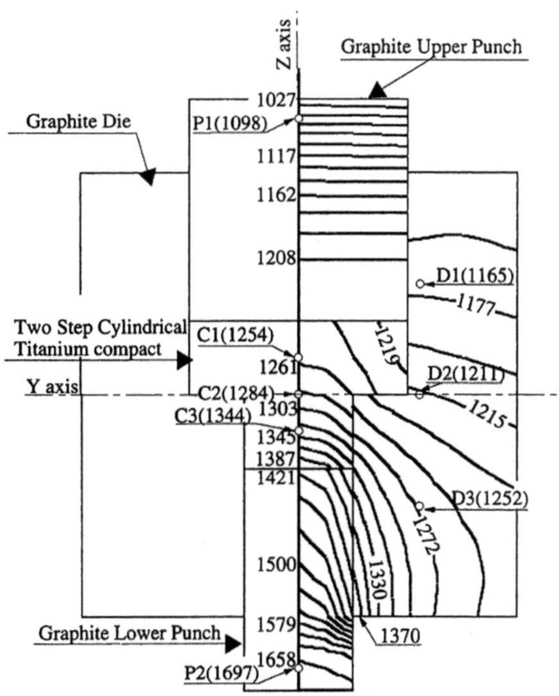

Fig. 6.14 Temperatures in the *y-z* plane at positions, P1, P2, D1, D2, D3, C1, C2 and C3 at the holding time of 0.6 ks, and calculated isothermal contours under constant current discharge conditions, with units in K. The values within parentheses were the temperatures measured by the thermocouples. Reproduced with permission from [33]. Copyright © 2003, Elsevier

Model results are obtained at 3 V voltage drop between the plungers' ends, as temperature distribution at four different times, for a quarter of domain that is due to the cylindrical symmetry of the system. The intensive heat generation in the plungers regulates the heat flux direction inside the sample and die, especially in the beginning of the heating process. Because the BN powder has higher resistivity than the graphite, the temperature of the sample is lower than that of the die at earlier stages. This is because electrical current preferentially flows through the less resistive die. As a result, thus Joule heat effect is more pronounced in the die than inside powder sample. The radial gradient of temperature is reversed near the end of the process, due to the radiation losses from outer surfaces of the die, which increases with increasing temperature. In this case, temperature radial profile inside the powder sample possesses a wavy shape.

Generally, temperature difference in radial direction inside the sample is lower than inside die, which could be about 10 and 40 °C, respectively. The modeling results have been qualitatively validated by the experimentally measured temperatures at different positions inside the sample. However, the experiments are carried out with operative conditions that are different from those ones used for the simulations. In experiments, the temperature is controlled by adjusting electric power input, whereas a constant voltage drop is used for the modeling. Nevertheless, it is suggested that temperature gradient inside the sample should be as low as possible, so that the sample will have a relative density distribution and minimized heterogeneity. This can be realized by lowering the heating rate [34].

The model has also been used to simulate the densification behavior of a conductor-like material, which is TiB_2/BN powder [35]. In this study, electrical current flows through the sample, so that the sample experiences a temperature higher than that of the die. In the axial direction, the temperatures experienced by the plungers continuously increase to the highest values in the whole system, due to the intensive Joule heat generation. Therefore, the sample is heated axially through the thermal conduction from the plungers, while heat is also losed to the die in the radial direction. A high radial difference in temperature inside the sample has been predicted, which is about 150 °C. However, the voltage drop between the rams that is applied as an input to the model is not mentioned and thus has not been compared with experimental data. Similarly, the transient spatial temperature differences inside the sample can be decreased, by lowering the heating rates, which is beneficial to increasing homogeneity of the final sintered products.

Electroconsolidation® is a process that can be used to densify preformed materials with complex-shaped to near-net-shaped by using electrically conductive particulate solids as a pressure transmitting medium [36]. The densification of preformed materials is realized by passing an electrical current through a bed of particulate graphite that surrounds the preform, while transmitting pressure is applied simultaneously by using a double-acting hydraulic press. The surrounding graphite powder, heated up by Joule heating effect, facilitates the preform to densify, due to the increase in temperature through thermal conduction and the pressure applied.

A 2D cylindrical coordinate model, which is developed with respect to the system geometry, consisting of three modules, is used for simulation. The first module describes the density and electrical resistivity of the graphite particulate medium, as a function of the magnitude of mechanical pressure applied. It consists of two empirical algebraic equations, which are determined by fitting the experimental data. A die with sufficiently large diameter is used to obtain the data, so that the effect of powder–die wall friction can be minimized. With this module, together with the temperature-dependent electric resistivity of fully dense graphite, the value of the graphitic powder, as a function of pressure and temperature, is obtained. With the results obtained in the first module, the resistive heating, voltage, and current distribution field inside the whole system can be derived in the second module. After that, the distribution field of resistive heating can be provided to the third module.

The innovative contribution of the model is the consideration of friction between the powder and the internal die, when determining the axial pressure distribution inside the particulate graphite medium [1, 36]. As a result, model results can be used to predict the temporal profiles, which are in a good agreement with the measured values by using thermocouples at two different positions inside the particulate graphite medium.

The relevance of contact resistances in SPS process has been simulated and confirmed, with the simulation shown in Fig. 6.15 as an example [4]. The system simulated is a Model 1050-Sumitomo SPS, where a solid graphitic cylinder is inserted into the die. The 2D cylindrical coordinate system of coupled thermal and electrical problems is numerically solved by using Abaqus (FEM). The heat losses due to radiation from all exposed surfaces, except those on the ends of the rams, have been considered, where a constant temperature of 25 °C is used for the simulation. Thermophysical parameters of all materials are available in that study. A proportional feedback controller based on the outer surface temperature of the die is modeled, in order to determine the voltage drop applied at two ends of the rams. This controller is used to imitate the actual proportional integral derivative (PID), which is observed in real SPS facilities. It is used to apply electric power input to the system when experiments are conducted in terms of temperature controlling.

Boundary conditions of both the horizontal and vertical contacting surfaces, as shown in Fig. 6.15, have been mathematically described with the assumption that all contact conductances are kept to be constant [4]. It is well known that the temperature and the electrical potentials are continuous functions across a perfect interface. Practically, it is difficult to have perfect contact when two real surfaces are considered, due to the geometrical irregularities and surface deposits. Therefore, in a real system, both temperature and electric potentials are discontinuous instead. In other words, a finite temperature ($\Delta\theta$) or voltage difference ($\Delta\varphi$) should be considered at the interface, as shown in Fig. 6.16, where the thermal flux and electric current density across it can be expressed as:

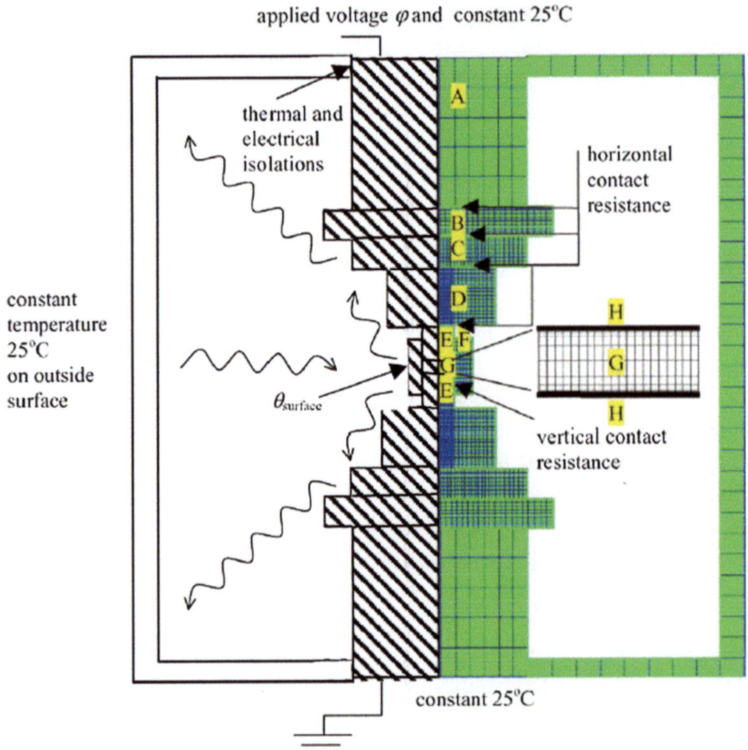

Fig. 6.15 Geometry of the SPS machine, the corresponding finite element mesh, (*right*) and schematic boundary conditions (*left*) for the simulations. *Curved arrows* represent potential radiations. Contact resistances are also marked. $\theta_{\mathrm{surface}}$ is the controlling temperature, and φ is the controlled potential applied to the system. Actual calculations are performed in one half of the geometry shown due to axial symmetry. The components referred by the letters *A–G* can be found in the original publication. Reproduced with permission from [4] by Zavaliangos et al. Copyright © 2004, Elsevier

$$\dot{q}_{\mathrm{c}} = h_{\mathrm{g}}(\theta_1 - \theta_2), \tag{6.9}$$

$$j = \sigma_{\mathrm{g}}(\varphi_1 - \varphi_2), \tag{6.10}$$

where h_{g} and σ_{g} are the thermal and electrical conductances of a gap, respectively. Besides the localized drop of temperature, the heat flux is also discontinuous, because Joule heat is generated due to the electrical contact resistance at the interface, given by:

$$\dot{q}_{\mathrm{ec}} = j(\varphi_1 - \varphi_2) = \sigma_{\mathrm{g}}(\varphi_1 - \varphi_2)^2. \tag{6.11}$$

Therefore, the heat flux at the contacting surfaces of the parts (1) and (2), as shown in Fig. 3.22, can be expressed as:

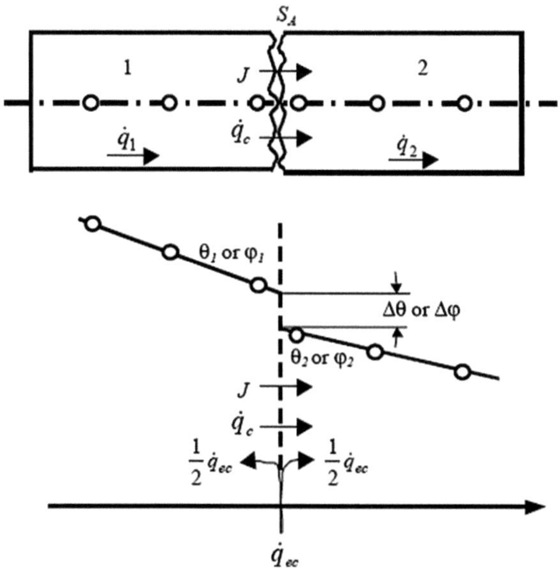

Fig. 6.16 Drops of temperature or voltage at an imperfect interface with apparent contact area (*SA*), due to the thermal contact resistance or electrical contact resistance as heat ($\dot{q}_c$), or electric current flux (*j*), which flows across the interface, respectively. Joule heat ($\dot{q}_{ec}$) is also generated and evenly distributed into two parts. $\dot{q}_1$ and $\dot{q}_2$ are heat flux in parts 1 and 2, respectively. Reproduced with permission from [4] by Zavaliangos et al. Copyright © 2004, Elsevier

$$\dot{q}_1 = \dot{q}_c - \frac{1}{2}\dot{q}_{ec} = g_h(\theta_1 - \theta_2) - \frac{1}{2}\sigma_g(\varphi_1 - \varphi_2)^2, \tag{6.12}$$

$$\dot{q}_1 = \dot{q}_c - \frac{1}{2}\dot{q}_{ec} = g_h(\theta_1 - \theta_2) + \frac{1}{2}\sigma_g(\varphi_1 - \varphi_2)^2, \tag{6.13}$$

where the Joule heat due to the electric contact resistance is assumed to be evenly divided between the two contacting surfaces. The contact conductances (h_g and σ_g) can be fitted by comparing the model results and the experimental data, which are obtained in the system shown in Fig. 6.21, at a pressure of 15 MPa and a heating rate of 15 °C min^{-1}. Electrical conductances are 1.25×10^7 and 7.5×10^6 S m^{-2}, while thermal conductances are 2.4×10^3 and 1.32×10^3 W m^{-2}, for horizontal and vertical interfaces, respectively. Therefore, the vertical contact resistances are larger than horizontal ones, which have been attributed to the fact that the mechanical load is applied in the horizontal direction, thus leading to relatively "perfecter" contacts.

One of the significances of this model is its ability to differentiate between the horizontal and vertical contact resistances, which are not considered in the previous simulations. Figure 6.17 shows experimental data and simulated results for the differences in temperature at the center of the sample and surface of the die [4]. The model results, with and without considering the thermal contact resistances, have also been compared. Without considering the thermal contact resistance, the

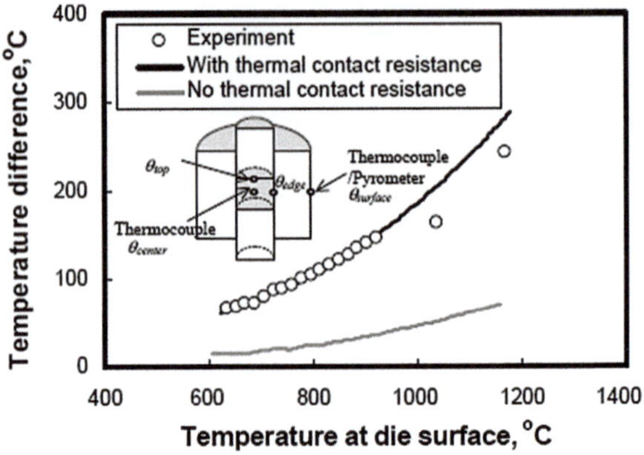

Fig. 6.17 Comparison between experimental data and model results in terms of temperature difference at the center of the sample and surface of the die. Reproduced with permission from [4] by Zavaliangos et al. Copyright © 2004, Elsevier

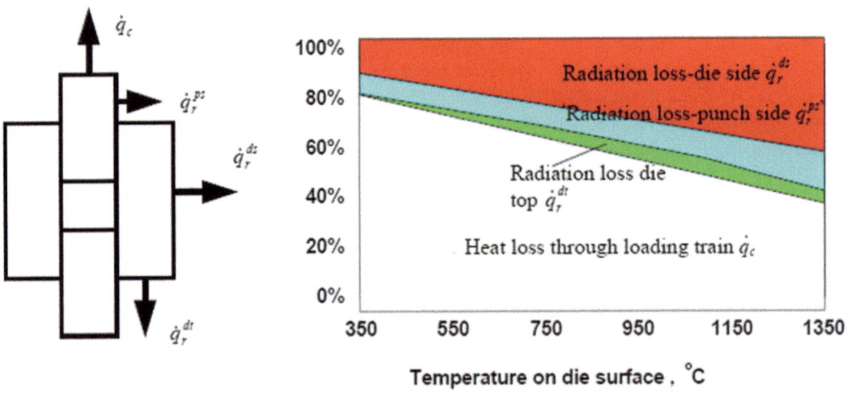

Fig. 6.18 Relative contributions of various mechanisms to the heat losses as a function of the controlling temperature. Reproduced with permission from [4] by Zavaliangos et al. Copyright © 2004, Elsevier

simulated results are much lower than the experimental observations. Experimental results indicate that the temperature difference can be more than 200 °C, with the center of the sample being much hotter than outer die, even though the conducting materials used for the two SPS elements are the same.

Fig. 6.18 shows relative contributions through all the possible heat loss mechanisms, as a function of temperature at surface of the die [4]. The temperature evolution in the plungers/sample/die assembly is a reflection of the synergistic effects of Joule heat generation and heat transfer. The heat is lost from the assembly to the loading train and eventually the water-cooled electrodes through thermal conduction and to the wall of the SPS chamber through radiation. It has been

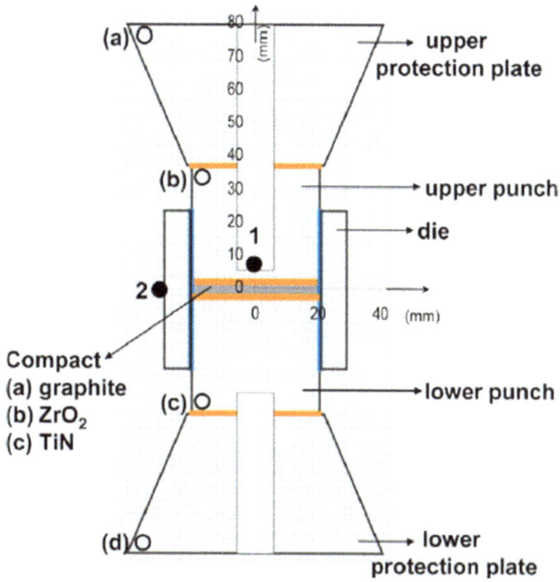

Fig. 6.19 Relative contributions of various mechanisms to the heat losses as a function of the controlling temperature. Reproduced with permission from [20] by Vanmeensel et al. Copyright © 2005, Elsevier

demonstrated that, at low temperatures, most of the heat is lost due to the thermal conduction ($\dot{q}_c$) to the loading train, whereas the loss due to the thermal radiation ($\dot{q}_r^{ps} + \dot{q}_r^{dt} + \dot{q}_r^{ds}$) becomes more and more significant, at higher temperatures. Among all mechanisms, the loss due to thermal conduction to the loading train always counts for a significant portion, which is in a range from 40 to 80 %. These conclusions have been supported by the simulation results on the cooling water system, with the assumption that the temperature at ends of the rams is the ambient temperature and constant. Therefore, the heat loss by conduction to the loading train should be included and considered, whenever heat loss is simulated.

Contact resistances have also been well studied in a simulation of the system, as shown in Fig. 6.19 (Model HP D 25/1-FCT, Germany), where a solid cylinder specimen is inserted into the die [20]. The system has two boreholes, where the upper one is used to measure the temperature near the sample with a pyrometer and the lower one is used to fix and position the whole assembly and maintain symmetry. In the balance equations, radiation heat losses are considered from the lateral surfaces. At the same time, an effective convective heat transfer to simulate the axial cooling directly from the ends of the spacers is also included. Experiments are carried out by controlling the temperatures measured at bottom of the upper borehole as close as possible to the center of the sample. The model is solved by using Abaqus (FEM), assuming a total current flowing through the assembly. In order to properly evaluate the Joule heating effect when a pulsed DC is applied, the root-mean-squared (RMS) values of the current and voltage are given by:

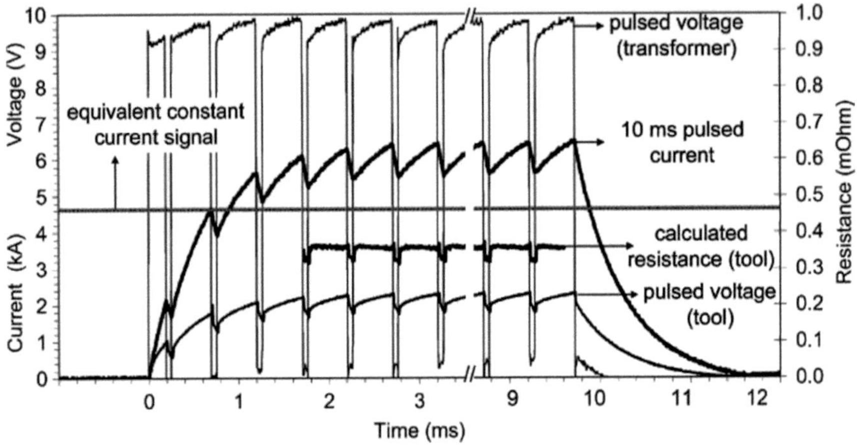

Fig. 6.20 Typical "on"-part of a 10/5 ms on/off pulsed current signal provided by the FCT system. The pulsed voltages over the transformer and over the tool are shown. The calculated resistance over the tool is obtained by dividing the pulsed voltage over the pulsed current. The constant DC current signal used during the FEM calculations generating the same power as the pulsed DC signal is also indicated. Reproduced with permission from [20] by Vanmeensel et al. Copyright © 2005, Elsevier

$$I_{\mathrm{RMS}} = \sqrt{2/3} I_{\mathrm{peak}}, \tag{6.14}$$

$$V_{\mathrm{RMS}} = \sqrt{2/3} V_{\mathrm{peak}}, \tag{6.15}$$

where I_{peak} and V_{peak} are the peak values of the corresponding pulsed signals experimentally measured as independent variables which are demonstrated in Fig. 6.20, where a pulse sequence of 10 ms ON and 5 ms OFF is used. It means that 10 ms is the duration of direct pulses, which is followed by a period of 5 ms without pulses [20].

Thermal and electric resistances at contacts between the adjacent elements of the die–plunger–sample setup have been evaluated experimentally by comparing the thermal and electrical responses of three different graphite configurations, with increasing complexity, so as to allow for demonstrating the influence of horizontal vertical resistances of the contacts. The dummy geometries used are schematically shown in Fig. 6.21, where graphite sheets with anisotropic thermophysical properties are inserted at the horizontal and vertical contacts [20]. In the case of the full dummy geometry (F), there is no contact, and the electrical resistivity of the solid graphite ($\rho_{\mathrm{el,solid\ graphite}}$) has been experimentally verified, when comparing the modeled results with the experimental data, while fitting the effective heat transfer coefficient at ends of the spacers.

The electrical resistances of the horizontal contacts between the plungers and the spacers as a function of temperature can be evaluated, by subtracting the experimentally measured electrical resistances of the *F* geometry from those of the SPP

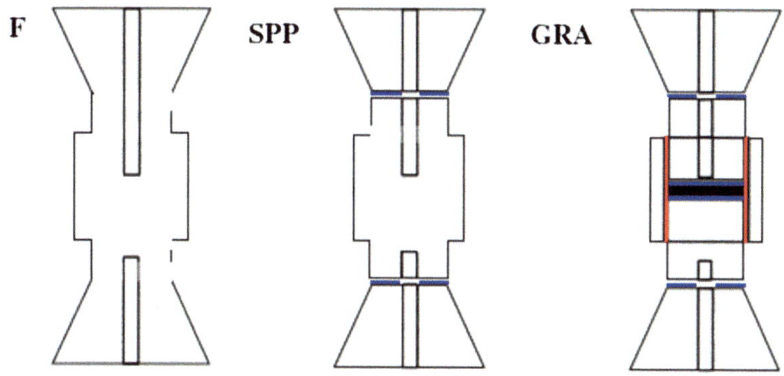

Fig. 6.21 Dummy geometries used to establish the influence of horizontal and vertical contact resistances. *Thick horizontal lines* indicate the presence of horizontal graphite papers, while *thick vertical lines* indicate the presence of the vertical graphite papers. Reproduced with permission from [20] by Vanmeensel et al. Copyright © 2005, Elsevier

geometry (Fig. 6.21) at various steady states. These contact resistances are used to determine the electrical resistivity of the graphite paper ($\rho_{\text{el,graphite paper}}^{\text{horizo tan}}$) sandwiched at horizontal contacts between the spacers and the plungers, which are modeled by using the thin-film approximation. The corresponding thermal conductivity of the sandwiched graphite paper ($\kappa_{\text{el,graphite paper}}^{\text{horizo tan}}$) is empirically given by:

$$\kappa_{\text{graphite paper}}^{\text{horizontal}}(T) = \mu\kappa_{\text{solid graphite}}(T)\frac{\rho_{\text{el,solid graphite}}}{\rho_{\text{el,graphite paper}}^{\text{horizontal}}(T)}, \tag{6.16}$$

where the value of the proportional constant μ is 2.85, which is derived from the fitting results through the comparison of modeled results and the experimental data of the configuration with the SPP dummy geometry. This empirical expression means that the ratio of the thermal and electrical conductivities of the graphite paper sandwiched at the horizontal contacts is proportional to that of solid graphite at any temperature.

Thermal and electrical resistivities of the graphite paper at the invertical contacts between the plungers and the die are given by:

$$\rho_{\text{el,graphitepaper}}^{\text{vertical}}(T) = \alpha_{\text{el}}\rho_{\text{el,graphitepaper}}^{\text{horizontal}}(T), \tag{6.17}$$

$$\kappa_{\text{el,graphite paper}}^{\text{vertical}}(T) = \alpha_{\text{therm}}\frac{1}{\kappa_{\text{graphite paper}}^{\text{horizontal}}(T)}, \tag{6.18}$$

where the values of the proportional constants, α_{el} and α_{therm}, are equal to 7, which is obtained through the fitting procedure, in which the modeled results are compared with the experimental data of the configuration with the GRA dummy geometry (Fig. 6.21) [20]. Therefore, the vertical, thermal, and electrical contact resistances are higher than horizontal ones. However, in this paper, the temperature dependence of contact resistances is determined. In this study, the effect of the applied mechanical load is not simulated, although preset mechanical load temporal changes are used during the experiments, which can be a subject of further study.

The effect of contact resistances can be readily observed, when comparing the radial temperature differences between the center of the sample and external surface of the die as a function of time, for both simulated results and the experimentally measured data. The model results are in a good agreement with the experimental data. The effect and contribution of vertical contact resistances has been clearly observed. The temperature difference increases from 100 to 200 °C, when the configuration is changed from the F and SPP geometries to the GRA one. The higher temperature difference of the GRA geometry is attributed to the combined effect of the vertical thermal contact resistances and vertical electrical resistances. In this case, the electrical current can be confined to flow into the plungers and the sample in series combination. Therefore, Joule heating effect is enhanced by reducing the cross-sectional area [20].

After evaluation with the solid graphitic samples, the model is applied to a fully dense conducting TiN and an insulating ZrO_2. Without experimental data, the modeling results indicate that the radial temperature difference inside the conducting sample (80 °C) is larger than that (25 °C) inside the insulating one. This has been ascribed to the difference in distribution of the current density and thus the Joule heating, due to the parallel combination of the sample and the surrounding die. Also, the conductive heat transfer rates between the two elements (the sample and the die) and the heat loss due to the radiation from outer surface of the die have played an important role. The overall effect is also closely related to the relative sizes of the different elements. Therefore, the measured temperature at the outer surface can be used to represent the sintering temperature of insulating samples, at least for the specific system geometry used in the study. When the samples are conductor-like materials, the real sintering temperature of the samples is significantly different from the temperature measured at outer surface of the die. In this case, if the surface of the die is covered with an insulation layer of porous graphite, the temperature measurement would be more reliable. This is because thus, the use of the porous insulation graphite layer can minimize the heat losses due to the radiation.

The model has also been applied to the FAST of composite powder, ZrO_2–TiN, with compositions from 65/35 to 10/90 vol% [37]. Mixture rules are adopted for the evaluation of thermophysical properties of the sintering powder. According to the electrical (σ) and thermal (κ) conductivities of fully dense ZrO_2 and TiN as a function of temperature, the Polder–Van–Santen (PVS) mixture rule is used to obtain the thermophysical parameters of fully dense ZrO_2–TiN composite, given by:

$$\kappa_{\text{mix}} = \frac{1}{4}\left\{2\kappa_{\text{m}} - \kappa_{\text{p}} + 3V_{\text{p}}(\kappa_{\text{p}} - \kappa_{\text{m}}) + \sqrt{\left[2\kappa_{\text{m}} - \kappa_{\text{p}} + 3V_{\text{p}}(\kappa_{\text{p}} - \kappa_{\text{m}})\right]^2 + 8\kappa_{\text{m}}\kappa_{\text{p}}}\right\}, \tag{6.19}$$

$$\sigma_{\text{mix}} = \frac{1}{4}\left\{2\sigma_{\text{m}} - \sigma_{\text{p}} + 3V_{\text{p}}(\sigma_{\text{p}} - \sigma_{\text{m}}) + \sqrt{\left[\sigma_{\text{p}} - 2\sigma_{\text{m}} + 3V_{\text{p}}(\sigma_{\text{p}} - \sigma_{\text{m}})\right]^2 + 8\sigma_{\text{m}}\sigma_{\text{p}}}\right\}. \tag{6.20}$$

The samples are treated as mixtures consisting a continuous matrix phase (m) and a secondary phase as solute or filler (p). The solute phase has particles with spherical shape which are homogeneously dispersed in the matrix. V_{p} is volumetric concentration of the solute phase in the mixture, which is either ZrO_2 or TiN, depending on the composition of ZrO_2–TiN mixtures. Thermophysical parameters of the fully dense ZrO_2–TiN composites with different compositions, as a function of temperature, have been studied [37].

The PVS mixture rule is also applied to study the effect of residual porosity on sintering behaviors of the ZrO_2–TiN composite powders, by incorporating porosity as a third phase with zero conductivity. With this assumption, Eqs. (6.19) and (6.20) are still applicable, by using a reduced volume concentration of the secondary phase, V_{p}^*, due to the presence of the pores. The reduced volume concentration can be related to relative density (D) of the sintering powder compacts, given by $V_{\text{p}}^* = D - V_{\text{m}}^*$, where V_{m}^* is the volume fraction of the matrix phase in partially sintered compacts.

Heat capacity of the sintering powder is determined by directly multiplying the temperature and the composition-dependent heat capacity of the fully dense material with the relative density. The relative density of the sintered powder compact (D) varies with time during the FAST process depending on composition of the samples. It can be measure by using the Archimede's method. Figure 6.30 shows experimental results of representative samples [37]. The corresponding experimental conditions, such as heating rate and pressure cycle, are demonstrated in the inset of Fig. 3.30. It has been assumed that the distribution of temperature inside the sample is uniform.

Thermal and electrical conductivities of the ZrO_2–TiN powder composites with different compositions can be well described as a function of temperature or relative density of the powder compact [37]. It is found that the samples with TiN contents of <70 vol% all experience a sharp increase in electrical conductivity, which is similar to percolation. In other words, during the FAST process, the samples change from an insulator to a conductor, due to the increase in relative density during the sintering. In this case, electrical current and Joule heat generation take place inside the sample rather than in the surrounding die.

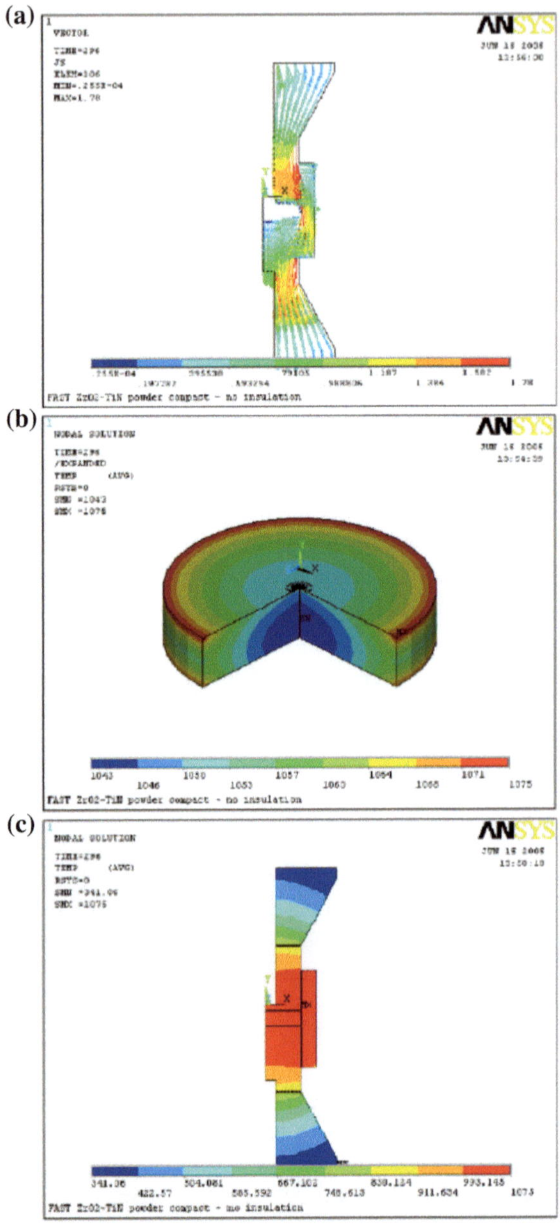

Fig. 6.22 Current density distribution (**a**), temperature distribution inside the sample (**b**), and inside the system (**c**), for the SPS of the ZrO_2–TiN composites (60/40), at a controlling temperature of 1050 °C at a pressure of 28 MPa. Reproduced with permission from [38]. Copyright © 2007, Elsevier

It is found that percolation occurs for the sample at temperatures between 1200 and 1300 °C. Figures 5.22 and 5.23 show current density distribution, temperature distribution inside the sample and that inside the system of the sample processed under two sets of processing conditions [38]. The model predicts that the current density is concentrated inside the die when the sample approaches a temperature of

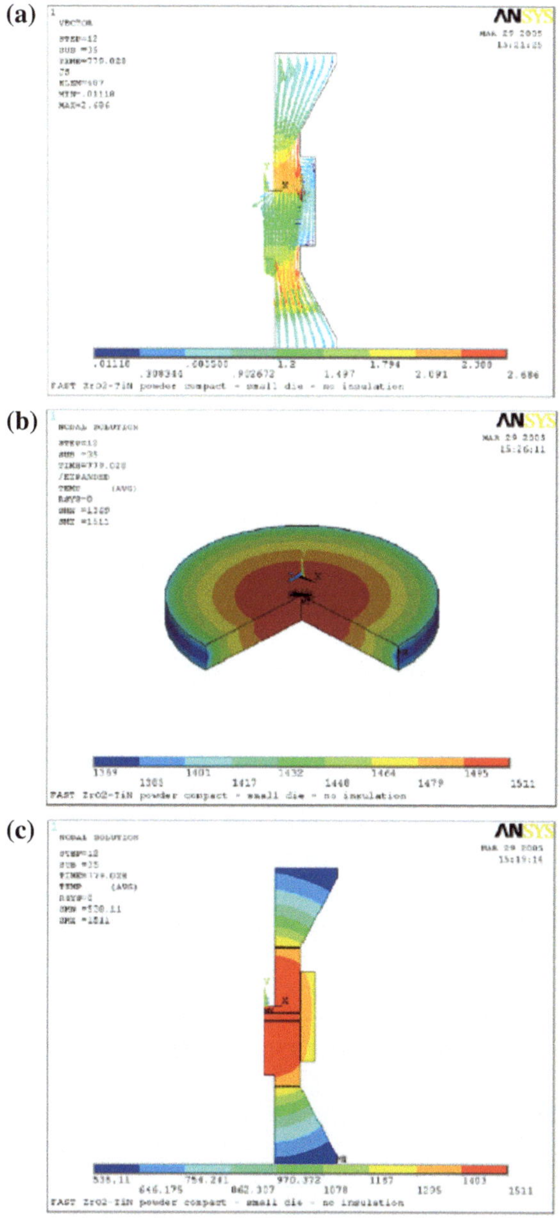

Fig. 6.23 Current density distribution (**a**), temperature distribution inside the sample (**b**), and inside the system (**c**), for the SPS of the ZrO_2–TiN (60/40) composite, at a controlling temperature of 1500 °C at a pressure of 56 MPa. Reproduced with permission from [38]. Copyright © 2007, Elsevier

1050 °C, as shown in Fig. 6.22a. When the sample reaches a higher temperature of 1500 °C, the current flows through the sample (Fig. 6.23a). By inspecting Figs. 5.23 b, c and 5.33b, c, it is readily observed, at the earlier stage of sintering, that ZrO_2–TiN powder composite behaves as an electrical insulator, and the temperature at center of the sample is lower than that of the surrounding die. In this case, the sample

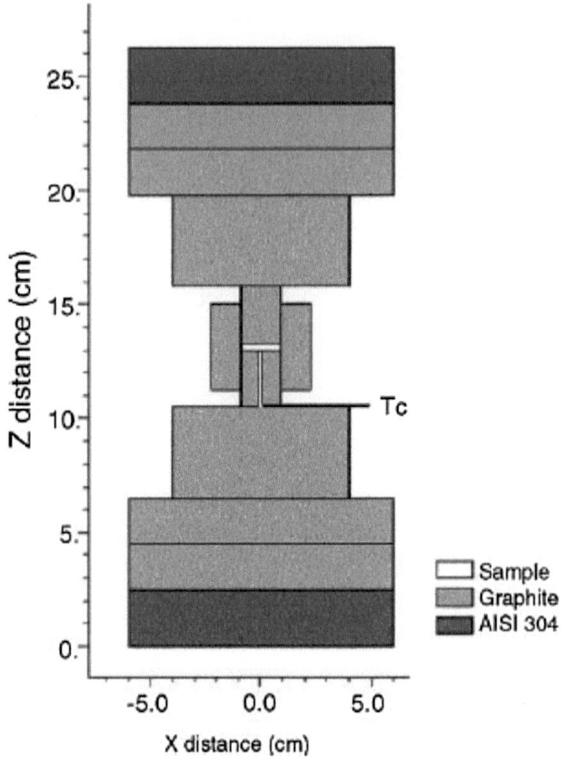

Fig. 6.24 Schematic of the portion of the spark plasma sintering apparatus used for modeling. Reproduced with permission from [23]. Copyright © 2005, Elsevier

is heated due to thermal conduction from the die. As the sintering is proceeded, the sample center becomes hottest, with a temperature much higher than that side of the die. This is because, at this stage of sintering, the sample experiences a transition from insulator to conductor, so that the current flows through it and thus the Joule heating becomes to be dominant. At the same time, the heat loss due to radiation from the exposed surface of the die is more and more pronounced. As a result, high temperature differences in radial direction are observed inside both the sample and the die, which are about 150 °C and about 400 °C, respectively.

Densification behaviors of electrical conducting Cu and insulator Al_2O_3 with SPS have been compared and modeled, with the portion of the system schematically shown Fig. 6.24 [39]. In this modeling study, stainless steel electrodes (rams) are included in the model. The boundary conditions of axial cooling and the consequent temperature distribution change dramatically. The mathematical equations used in the modeling, i.e., the enthalpy and current density conservation equations in cylindrical coordinates applied to the elements, are solved in full 3D version through CFD-Ace (FEM). Temperature dependences of thermophysical properties for all materials have been considered.

During the simulation, radiation heat losses from the exposed surfaces are considered, with the assumption that the ambient temperature, the voltage drop, or total flowing current at end surfaces of rams are constant, while contact resistances at the interfaces between different adjacent elements are neglected. The simulation results on the distribution of flowing current density are similar to those that can be found in the open literature for RS. The behavior of the Sumitomo SPS DC pulse generator is analyzed and monitored by using an oscilloscope [23]. The complex temporal profiles of the voltage between the ends of the rams are measured under the 8/2 ON/OFF operation. The current pulse pattern consists of eight direct voltage pulses of fixed time duration of 3 ms (ON), followed by two pulses of time duration of 3 ms with zero voltage (OFF). This operation leads to a low-frequency signal, which is less than 100 Hz, as described by the Fourier analysis of the pulse pattern.

In this case, the relevant skin effect would occur at higher frequencies due to the electrical conductivities and dimensions of the different elements used in the SPS system, because skin depth is given by $\delta = 1/\sqrt{f\mu\sigma}$, where f is frequency, μ is magnetic permeability, and σ is electrical conductivity. Therefore, under the operation with the 8/2 current pulse pattern, the electrical current would not be confined within a thin layer of the external surfaces due to the electromagnetic induction, but instead, it is distributed along the cross-sectional areas, according to the relative electrical conductances in series or parallel combination. This allows the model to consider just ohmic behavior of the materials, so that all the assumptions for the modeling are justified and that the current density simulated with the conservation equations would have well represented the real scenario of the experiments.

Figure 6.25 shows current distribution profiles of the Al_2O_3 and Cu samples at room temperature, which are calculated with a constant voltage of 4 V that is applied across the entire assembly [23]. The samples have a thickness of 3 mm. Contact resistances are assumed to be zero during the simulation. A distinct difference in the current distribution can be observed, especially the currents flowing through the samples. Because Al_2O_3 is an insulator, no Joule heating is expected at this temperature. In other words, the initial heating of the sample is not due to the current. In contrast, Cu is a conductor, so that Joule heating starts immediately. In both cases, the highest current densities are observed along the exposed portions of the plungers [23].

The simulated results on dependences of the fraction of the total current that flows through the sample, as a function of electrical conductivity at room temperature, over a range of eighteen orders of magnitude, have been available [39]. The lower end corresponds to the electrical conductivity of Al_2O_3, while the upper end the conductivity of Cu. Two samples have thicknesses of 3 and 10 mm, while the rest parameters of the system geometry are kept constant. It is found that no current flows through the samples whose conductivity is less than about 10^3 S m^{-1}. For 10 mm thick sample, almost all current is carried by the conducting sample (such as Cu). This can be simply attributed to the relative resistances of the two current paths, either through the sample (for conducting samples) or the surrounding die (for insulating samples). Because Cu has higher conductivity than

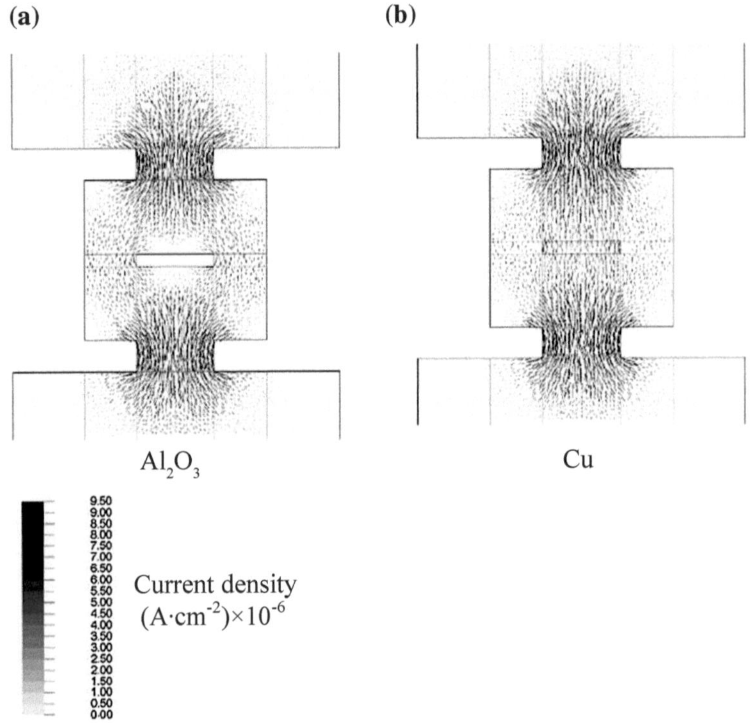

Fig. 6.25 Schematic of the portion of the spark plasma sintering apparatus used for modeling. Reproduced with permission from [23]. Copyright © 2005, Elsevier

graphite of the die, when the thickness of the sample is increased, the increase in resistance of the die is more significant than that of the sample (Cu), and the portion of current flows through the sample is increased with increasing thickness.

Electrical resistances of samples with a thickness of 3 mm measured by using a four-point method at room temperature across the die with the plungers in place for Al_2O_3 and Cu have been examined, as a function of applied mechanical pressure. There is a critical point of pressure that is about 50 MPa, below which the resistance of the case with Cu in side the tie is lower, whereas above which the resistances are the same for the two types of samples. In addition, the measured resistances decrease with increasing pressure for both Al_2O_3 and Cu samples. Theoretically, the overall resistance for the assembly of die and plungers, evaluated based on the room temperature resistivity of graphite, without including the contact resistance, is about 1.5×10^{-3} Ω. As the applied pressure is increased, the resistance approaches this theoretical value for both samples. This observation implies that contact resistance has a stronger effect than the resistivity of the samples and the contact resistances can be safely excluded, if all experiments are carried out at sufficiently high pressure levels, e.g., ≥100 MPa.

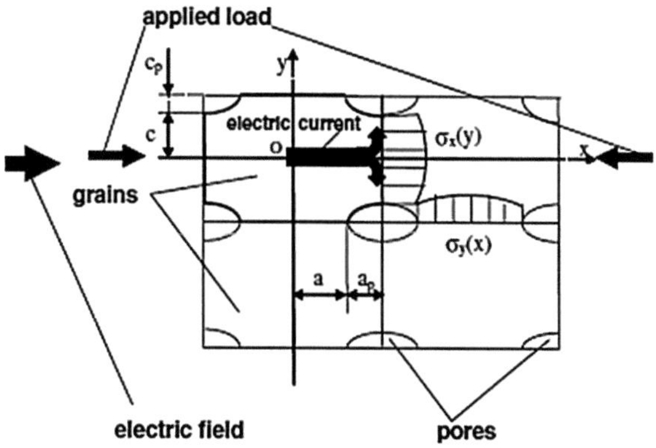

Fig. 6.26 Representative unit cell including rectangular grains and ellipsoidal pores in quadra-junctions. The pore–grain structure is subjected to a simultaneous action of external load, surface tension, and electromigration. The electric field is macroscopically unidirectional and can branch locally. Reproduced with permission from [40]. Copyright © 2006, Elsevier

Experimental data and modeling results of the Al_2O_3 sample have been compared, in terms of the temporal profiles of the temperatures at the center of the sample and surface of the die, when the system is subjected to a 400 s constant current step at 1000 A [23]. It is found that modeling results underestimate the real experimental results by about 200 °C. The modeled temperature at the center of the sample is lower than that at the surface of the die, which is opposite to the experimental observation. Therefore, further study is necessary to improve the occurence and capability of the modeling.

A constitutive model has been developed to model SPS behaviors of compacts of conducting powdery materials, such as Al powder [40]. The model is able to take care of the effect related to the flowing of electrical current, which is different from Joule heating. In this model, a constitutive law is proposed for sintering and densification of the poser, which is stress–strain mechanics. Grain-boundary diffusion, power-law creep driven by externally applied loading, sintering stress due to surface tension, and steady-state electromigration are all taken into account when developing the model. Schematic diagram of the system for modeling is shown in Fig. 6.26 [40]. The sample is a simply packed powder compact, with rectangular grains of semi-axes a and c and elliptical pores located at the grain quadra-junctions. Here, the electromigration is to represent the contribution of electric field to the diffusive mass transfer during the sintering process of the powder compacts.

According to Nernst–Einstein equation, matter flux $\bar{J}$ caused due to the grain-boundary diffusion is driven by stress and electromigration, which can be described as:

$$\bar{J} = C_{\mathrm{E}}\bar{E} + C_{\sigma}\bar{\nabla}\sigma, \tag{6.21}$$

where $\bar{E}$ is electric field and $\bar{\nabla}\sigma$ is gradient of stresses normal to grain boundary, whereas C_{E} and C_{σ} are temperature-dependent parameters related to mass diffusivity of the grain boundary.

The electric field is used an input of the model, which is evaluated as $E = V/l$, where V is the voltage drop along a characteristic length of the grain, l. $\bar{\nabla}\sigma$ can be obtained by solving the equilibrium mechanical problem, which allows to simulate the stress field distribution due to the surface tension (free-sintering) and the sintering under external loading to the powder compact. After that, linking the strain rate $\dot{\varepsilon}$ to the matter flux $\bar{J}$ in orthogonal direction leads to the shrinkage kinetics due to the diffusion of grain boundary, $\dot{\varepsilon}_{\mathrm{gb}}$, which is given by:

$$\dot{\varepsilon}_{\mathrm{gb}} = \dot{\varepsilon}_{\mathrm{gb}}^{\mathrm{E}} + \dot{\varepsilon}_{\mathrm{gb}}^{\mathrm{st}} + \dot{\varepsilon}_{\mathrm{gb}}^{\mathrm{load}}, \tag{6.22}$$

which includes the three contributions of electromigration, surface tension, and external load. The total shrinkage rate $\dot{\varepsilon}$ is the summation of the shrinkage rate due to the grain-boundary diffusion and the power-law creep mechanism, thus given by:

$$\dot{\varepsilon} = \dot{\varepsilon}_{\mathrm{gb}} + \dot{\varepsilon}_{\mathrm{creep}}. \tag{6.23}$$

The shrinkage rate due to power-law creep, $\dot{\varepsilon}_{\mathrm{creep}}$, can be evaluated based on the continuum theory of sintering, which deals with the dependences of normalized shear and bulk viscosity modules on porosity, θ, along with the effective sintering.

It has been found that the contribution of external loading to grain-boundary diffusion shrinkage kinetics, $\dot{\varepsilon}_{\mathrm{gb}}^{\mathrm{load}}$, can be neglected, so that it only has influence on the power-law creep phenomenon. A densification map for Al powder has been established, in which electromigration, $\dot{\varepsilon}_{\mathrm{gb}}^{\mathrm{E}}$, has a significant contribution to the total shrinkage, when the modeled powders have relatively low porosity and small grain size [40]. For micron-sized powders, power-law creep is the dominant mechanism for the shrinkage kinetics.

In a 2D axial coordinate system, x and y, porosity rate is related to strain rate, which can be expressed as:

$$\dot{\theta} = (1 - \theta)(\dot{\varepsilon}_x + \dot{\varepsilon}_y). \tag{6.24}$$

It is assumed that, during the SPS process, powder compact is pressed in a rigid die, i.e., $\dot{\varepsilon}_y = 0$, the densification kinetics can be given by:

$$\dot{\theta} = (1 - \theta)(\dot{\varepsilon}_{\mathrm{gb},x}^{\mathrm{E}} + \dot{\varepsilon}_{\mathrm{gb},x}^{\mathrm{st}} + \dot{\varepsilon}_{\mathrm{gb},x}^{\mathrm{load}} + \dot{\varepsilon}_{\mathrm{creep},x}). \tag{6.25}$$

With this equation, the temporal profile of porosity during the SPS of Al powder can be derived numerically, with the available experimentally measured temporal profile of temperature, together with the applied voltage gradient and constant pressure. It is found that the model satisfactorily predicts the shrinkage kinetics of the powder compacts.

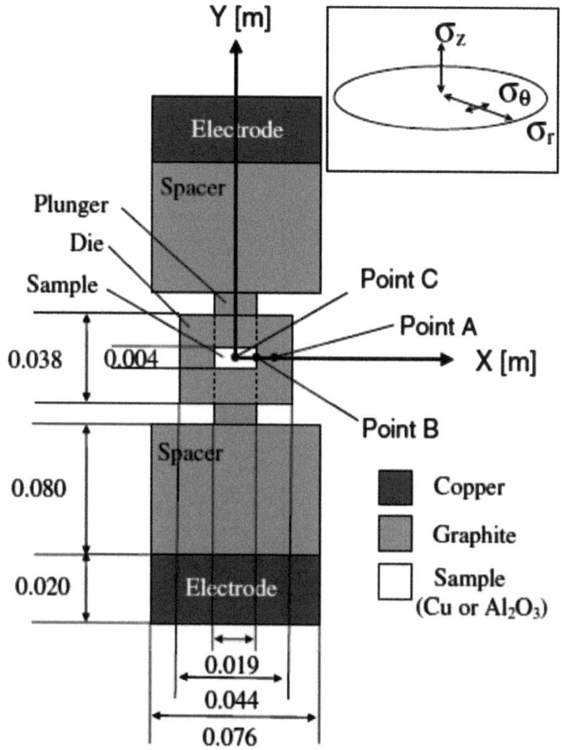

Fig. 6.27 Schematic of the current-activated sintering apparatus, showing dimensions and coordinate system, with the origin to be at the center of the sample. The *inset* in the *top right-hand corner* shows the relationship of the three-dimensional stresses in the radial, angular, and vertical directions. Reproduced with permission from [41]. Copyright © 2007, Elsevier

ECAS behaviors of electrical conducting Cu and electrical insulating Al_2O_3 by using the system shown in Fig. 6.27 have been modeled [41]. In this system, copper electrodes with a height of 2 cm are considered. In this ECAS study, a DC generator is used to provide the electrical current, while a mechanical loading is applied on the top ram, whereas the bottom ram is blocked by its end at a fixed vertical position. The model is for the first time to couple the usual enthalpy and current density conservation equations to the stress–strain mechanical equilibrium equations. The 2D cylindrical coordinates, with FEM minimization of the total mechanical potential energy related to elastic and thermal expansion, are solved to determine the stress distribution field inside the system, due to the mechanical and thermal loads. For boundary conditions of the thermal and electrical equations, radiation heat losses from all the exposed surfaces have been considered, while the ambient temperature and the voltage drop or total current flowing the cross section of the rams are assumed to be constant, whereas all contact resistances at the interfaces between different elements of the system are neglected. The later assumption could be valid, because during the experiments, a constant pressure of 140 MPa is applied, as discussed previously. This model is similar to that in the previous studies. A pretty good agreement can be observed between the experimental and the modeled results during the heating and cooling process [41].

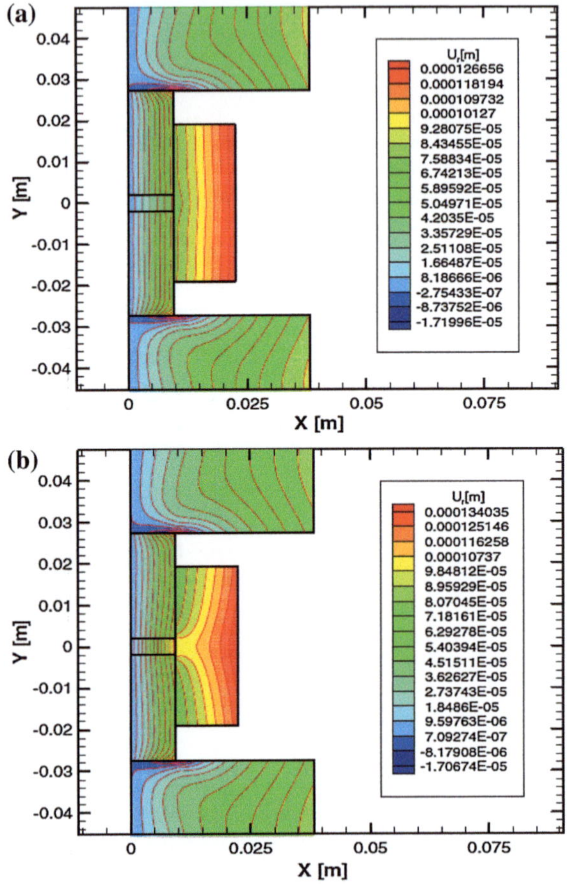

Fig. 6.28 Steady-state contour plots at 1000 A current input: **a** radial displacements of sample–die area of the Al_2O_3 sample (≈705 °C at center of sample) and **b** radial displacements of sample–die area of the Cu sample (≈692 °C at center of sample). Reproduced with permission from [41]. Copyright © 2007, Elsevier

However, for the conducting Cu sample, simulation results indicate that the temperature is usually lower than that of Al_2O_3 case, whereas the temperature difference inside the sample at steady state is higher for Cu than for Al_2O_3. Figure 6.28 shows the radial mechanical displacement distribution at steady state, as the system is applied with a current of 1000 A and a pressure of 140 MPa for the case of alumina and copper sample. The corresponding radial and axial components of the stress distribution field, due to thermal and Poisson-type expansion, are illustrated in Fig. 6.29, for the Al_2O_3 (a and c) and Cu (b and d) samples [41].

In the plunger–die–sample region, the radial displacement increases relatively uniformly with increasing radius, and thus, the highest displacement is observed near the surface of the die. These radial displacements are due to the synergistic effect of the thermal expansion and a Poisson-type expansion due to the pressure applied in the vertical direction. In contrast, the radial distribution, in the case of the Cu sample (Fig. 6.29b), does not increase uniformly with the radial position. Instead, the portion where the copper is in contact with the die wall has larger

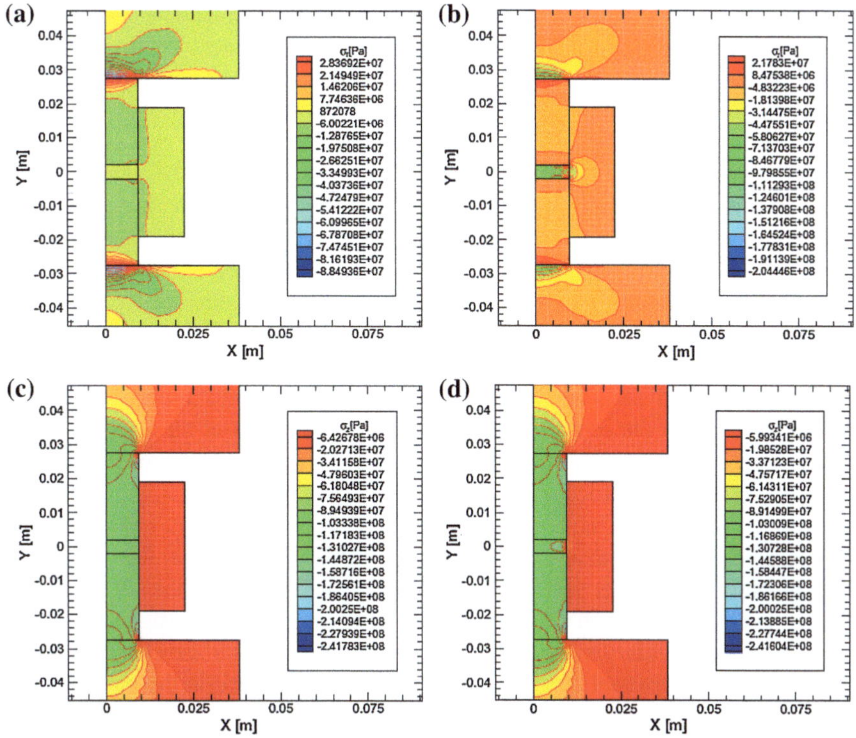

Fig. 6.29 Steady-state contour plots at 1000 A current input: **a** radial stress of sample–die area of the Al_2O_3 sample (≈705 °C at center of sample), **b** radial stress of sample–die area of the Cu sample (≈692 °C at center of sample), **c** vertical stress of sample–die area of the Al_2O_3 sample (≈705 °C at center of sample), and **d** vertical stress of sample–die area of the Cu sample (≈692 °C at center of sample). Reproduced with permission from [41]. Copyright @ 2007, Elsevier

displacement than the portions where the plunger is in contact with the die wall, i.e., graphite on graphite. The top portions have higher displacement due to the uniform pressure applied to the top electrode and the boundary condition that the vertical displacements of the lower end of the apparatus are set to be zero [41].

Under steady-state conditions, the differences in temperature inside the sample in radial direction are modest, with a maximum difference of 2 °C, corresponding to a variation of 1 %, whereas those in stress are relatively higher, with a maximum difference of 50 MPa, corresponding to a variation of 33 %. Therefore, the heterogeneity of the final sintered product is more due to the uneven distribution of stress than the distribution of temperature. As a result, the nominal sintering pressure, i.e., that obtained by the mechanical load applied over the cross-sectional area of plungers, and the sintering temperature recorded by using the pyrometer on the outer surface of the die may not be proper indications of the operative conditions of the experiments.

It is obvious that quantitative comparisons with experimental data are important to demonstrate the reliability modeled results. The agreement between the real experimental parameters and conditions, such geometry of system and properties of materials involved, and corresponding representations used for the simulation should be maintained. An actual modeling could be a complicated process, because various adjustable parameters for the mathematical description of the control system have to be used during the modeling. Therefore, it is more convenient to provide quantitative comparisons with experimental data in temperature than in voltage/current and mechanical displacement. It is therefore suggested to model the ECAS processes by separately analyzing a gradually increasing complexity according to the system behavior, so that the modeling could be conducted in the framework of a step by step procedure. Various physicochemical phenomena can be gradually included, along with their unknown modeling parameters. As a result, it is possible to independently fit the complete set of unknown parameters of the comprehensive ECAS model, thus avoiding the masking effect given by the various phenomena involved in the whole process [1].

These considerations have been accomplished in an example to model the SPS behavior of graphitic elements that are inserted between the two stainless steel rams used in the model 515S system (Sumitomo), which is schematically shown in Fig. 6.30 [42]. The geometry of the graphitic elements has been designed in such a way that it not necessary to consider the effect of vertical interfaces between them. As a result, the horizontal contact resistances can be excluded in the mathematical models. A 2D model in cylindrical coordinates based on the usual enthalpy conservation equation that takes into account the Joule heat generation is developed, which is coupled to density current balances expressed in terms of the RMS portion of the electric potential and the mechanic equilibrium equations due to elastic behavior and thermal expansion of the materials. Thermophysical properties of

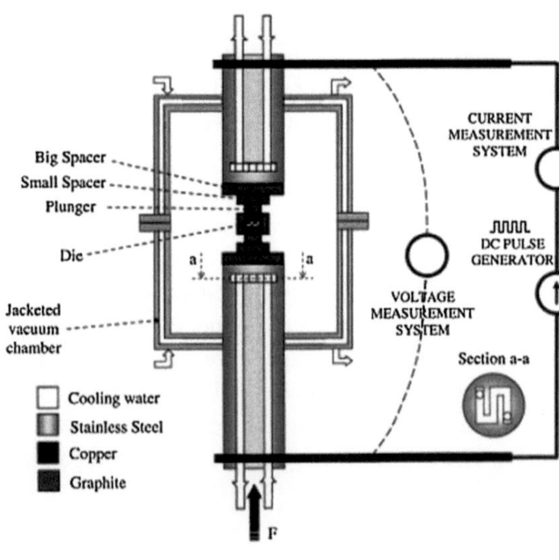

Fig. 6.30 Schematic of the SPS experimental setup (not to scale). Reproduced with permission from [42]. Copyright © 2007, John Wiley & Sons

graphite and stainless steel are used to model different graphitic elements inserted between the two rams. Monolithic graphitic elements are used to mimic the plungers positioned in place within the die.

In the real geometry of the system shown in Fig. 6.30, stainless steel rams and the corresponding water cooling system with different lengths have been modeled [42]. Because of this, the consideration of only one quarter of the integration domain is not applicable, because axial direction is not symmetric. To solve the model with FEM method, boundary conditions related to the thermal problem, including the radiation losses from all exposed surfaces, a constant ambient temperature at ends of the rams, and the forced convective losses from the rams along the contact interface with the cooling water, are adopted. To maintain the radial direction to be symmetric for simplicity of simulation, an hypothetical geometry of water cooling circuit according to that in the real system is calculated, with the assumption that the heat flow rate loss and the Joule heat generation inside the rams are equal, whereas the real heat transfer coefficient is determined by using the typical correlations of forced convections in closed ducts.

Only the horizontal thermal contact resistances between the stainless steel rams and the graphitic spacers are considered. Experimental results indicate that the horizontal thermal contact resistances between the graphitic elements can be neglected. Therefore, the continuity condition in terms of temperature and conductive heat flux is valid.

Regarding boundary conditions for the electrical problem, since current-controlled experimental runs are performed, the RMS portion of voltage drop between the two ends of the rams is determined by assigning the RMS value of the total electrical current flowing through the system, i.e., an integral equation of the gradient of RMS portion of electric potential is applied. Similarly, horizontal electrical contact resistances are considered to be present only between the stainless steel rams and the graphitic spacers, because also horizontal electrical contact resistances between graphitic elements are experimentally found to be negligible. Accordingly, the continuity condition for the resistive portion of the RMS electric potential and current density is applied. In particular, by simultaneously fitting experimental data in terms of temporal profiles of RMS portion of electric potential and temperature, thermal and electrical contact conductances are determined as a function of temperature and applied mechanical pressure through the following relationships [42]:

$$C_{\mathrm{T}}(T,P) = \alpha_{\mathrm{T}} \kappa_{\mathrm{Harm}} \left(\frac{P}{H_{\mathrm{Harm}}}\right)^{\beta_{\mathrm{T}}}, \tag{6.26}$$

$$C_{\mathrm{E}}(T,P) = \alpha_{\mathrm{E}} \sigma_{\mathrm{el,Harm}} \left(\frac{P}{H_{\mathrm{Harm}}}\right)^{\beta_{\mathrm{E}}}, \tag{6.27}$$

where k_{Harm} and $\sigma_{\mathrm{el,Harm}}$ are the parameters that have included temperature dependence of the contact conductances, which are expressed by using the harmonic mean of the individual thermal and electrical conductivities of graphite (G) and stainless steel (SS), expressed as:

$$\kappa_{\text{Harm}} = \frac{2\kappa_{\text{G}}(T)\kappa_{\text{SS}}(T)}{\kappa_{\text{G}}(T) + \kappa_{\text{SS}}(T)}, \tag{6.28}$$

$$\sigma_{\text{el,Harm}} = \frac{2\sigma_{\text{el,G}}(T)\sigma_{\text{el,SS}}(T)}{\sigma_{\text{el,G}}(T) + \sigma_{\text{el,SS}}(T)}. \tag{6.29}$$

The use of the harmonic mean is the requirement of the series combination between the two interfacing materials, in order to characterize the properties of the real contacts. The parameter H_{Harm} is the harmonic mean between the hardness of graphite and stainless steel, given by:

$$H_{\text{Harm}} = \frac{2H_{\text{G}}H_{\text{SS}}}{H_{\text{G}} + H_{\text{SS}}}. \tag{6.30}$$

α_{T}, β_{T}, α_{E}, and β_{E} are adjustable parameters, which are dependent on surface roughness and plastic behavior at contact interface between the graphite and stainless steel elements. They can be derived from fitting procedures. The corresponding electrical and thermal conductances in the range of temperature and mechanical load in that study are 0.25–1.7 × 10^6 Ω^{-1} m^{-2} and 0.007–2 × 10^3 W m^{-2} K^{-1}, respectively, which are comparable with the literature values [43]. The experimental data used for the fitting procedure are obtained from the system, with only one graphitic element of 8 cm in diameter between the two rams, which is subject to a RMS current of 1200 A with varying mechanical loadings of 0.6–10 MPa [42].

Figure 6.30 also shows that a Hall effect transducer and a voltage isolation amplifier are used in the system, which provides independent measurement of the pulsed current and voltage drop between the two rams, respectively. The RMS values of the electrical variables are important for the correct evaluation of the Joule heat distribution, which is demonstrated by the comprehensive analysis of the system behavior from the electromagnetic perspective. With these instantaneous measurement results, the corresponding average values of various parameters can be calculated as [42]:

$$\bar{I} = \frac{1}{\tau}\int_{t}^{t+\tau} I(t)\text{d}t, \tag{6.31}$$

$$\bar{V} = \frac{1}{\tau}\int_{t}^{t+\tau} V(t)\text{d}t, \tag{6.32}$$

$$I_{\mathrm{RMS}} = \sqrt{\frac{1}{\tau}\int_{t}^{t+\tau} I^2(t)\mathrm{d}t}, \tag{6.33}$$

$$V_{\mathrm{RMS}} = \sqrt{\frac{1}{\tau}\int_{t}^{t+\tau} V^2(t)\mathrm{d}t}, \tag{6.34}$$

A magnetic self-inductive response due to the pulsed electrical loading could be observed, depending on current level, pulse cycle, and system geometry. As stated earlier, a low-frequency (<100 MHz) pulse is not able to induce skin effect on electrical current path through the system [23]. The 515S SPS machine by Sumitomo has a built-in measurement system, providing the average values of the current flowing through the system and the average values of voltage drops between the rams. Depending on the pulse cycle adopted and the level of current flowing through the system, the RMS values of the current can be more than two times higher than the corresponding average values. According to circuit theory, this behavior is mainly caused by nonideal inductor, which is a series combination of resistance (R)–inductance (L), whose relationship is given by [42]:

$$V = V_{\mathrm{R}} + V_{\mathrm{L}} = \mathrm{R}I + \mathrm{L}\frac{\mathrm{d}I}{\mathrm{d}t}, \tag{6.35}$$

where V and I are the instantaneous voltage and electric current, respectively.

This observation can be readily attributed to the fact that SPS systems involve time varying electromagnetic fields. However, in this case, the assumption of current density conservation is not valid. As a result, the Maxwell's equations should be developed in such a way that the magnetic induction and the corresponding Joule heat distribution are included. Moreover, the whole model should be solved in the timescale of the generated electric pulses (fractions of millisecond), while temperature and displacement change in a much larger timescale of at least serval seconds. Obviously, such a mathematical description could have a very high degree of stiffness and its solution could be very time-consuming. To address this problem, a macroscopic nonideal inductor has been used, through a distributed parameter modeling approach, in which two models are superpositioned. One is a steady-state magnetic model outside the conductor, i.e., outside the element assembly of the SPS system, by considering the inductive character of the system. The other is a steady-state conduction model inside the conductor, which takes care of the resistive character of the system.

The conduction model is not different from the usual current density balance expression in mathematical form in terms of electric potential. The only difference is that this model is able to determine only the resistive portion of the electric potential. Also, the measurable voltage drop between the two rams represents both the resistive and the inductive contributions. This problem is solved by using a Fourier series development which evidences that the experimental counterpart of the modeled resistive portion of the RMS voltage can be obtained with $\varphi = \mathrm{R}I_{\mathrm{RMS}}$, where the resistance (R) is determined from the measured average voltage and electrical current by $\mathrm{R} = \bar{V}/\bar{I}$. As for the boundary conditions of the mechanical problem, the bottom surface of the lower ram is assumed to be subject to a uniformly distributed load, whereas a blocked displacement is applied at the top surface of upper ram in both the radial and the vertical directions. It is also assumed that all the exposed surfaces are free, which means that there is no radial and shear component of stress. Friction forces in horizontal contacts are neglected, so that continuity condition can be applied on stress axial component and vertical displacement between the two contacting sides of the interfaces.

Mathematically, the model has been coupled with both temperature and electrical current distributions. Therefore, as the thermal and electrical current balances are derived, the stress–strain distribution and the displacement temporal profile can be determined, because the temperature distribution represents the input for the mechanical problem. The assumption that the spatial domain of integration does not change substantially due to thermal expansion and mechanical load is reasonable for the SPS system without powder sample involved. The magnitude of the negative displacement due to the thermal expansion is about 1 mm, as compared with the height of the system which is 1 m. It is also reasonable to assume that the contributions of thermal expansion and mechanical loading are equally distributed among the different elements, i.e., the stainless steel electrodes. However, if powder samples are involved in the system, a positive displacement with a magnitude of about 3–5 mm will be observed in the shrinking powder sample. In this case, these assumptions may no longer be valid for the modeling. Instead, moving boundaries should be considered in modeling, especially when simulating the spatial–temporal profiles of density of the final products.

It is found that a pretty good agreement is observed, during both the heating and cooling processes. The problem is that the model prediction without including the contact resistances is not satisfactory. This conclusion is supported by independent simulation results [42], where axial temperature and RMS portion of electric potential profiles at steady state are assumed for the system assembly that is subject to a RMS current of 1200 A and a mechanical load of 10 MPa. Because the graphite elements have lower electrical resistance, the voltage drop is mainly at the contact interface with the stainless steel rams. Thermal contact resistances are less significant, because the conductive heat flux at the contact interface due to the Joule heat contribution is discontinuous. Even if, for the main part of their length, almost the whole length of the stainless steel rams is at ambient temperature, while only a very small portion at the end contacting with the graphite element is at high temperature, which however contributes significantly to the displacement measured experimentally. This is simply because the coefficient of thermal expansion of stainless steel is three times higher than that of graphite. Therefore, the water cooling system

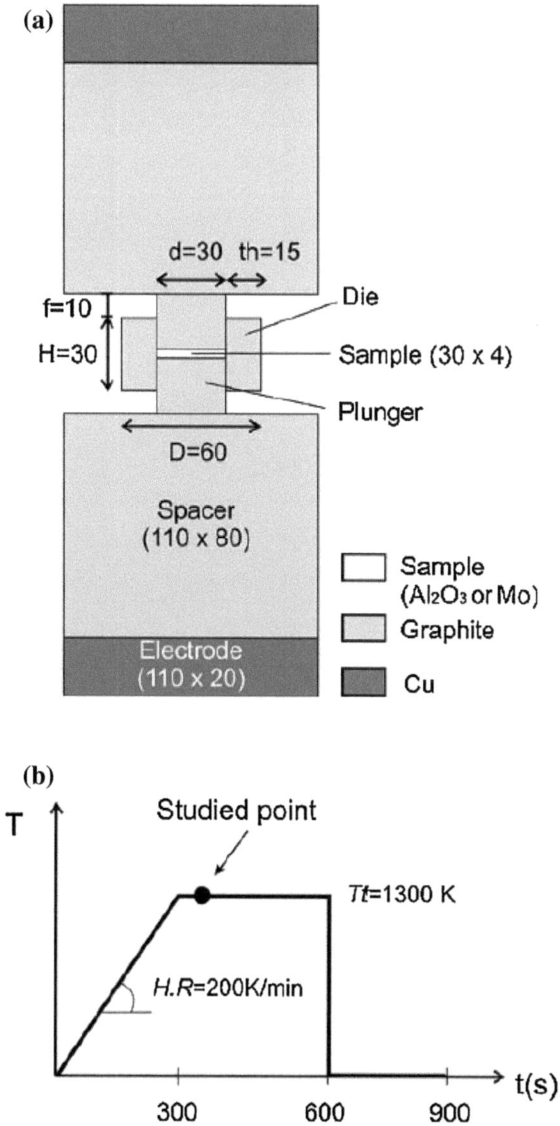

Fig. 6.31 **a** Reference geometry of the spark plasma sintering apparatus, with dimensions in millimeters. **b** Control temperature profile of the reference experiment. Reproduced with permission from [44]. Copyright © 2013, Springer

inside the full length of the rams must be modeled when simulating the displacement that is measured at bottom end of the lower ram [42].

More recently, there has been a report on systematic study of the ECAS behaviors, with Al_2O_3 (insulator) and Mo (conductor) as targeted materials [44]. The effects, including die geometry (thickness and height), heating rate, heat dissipation, material properties, and sintering temperature, on microstructural non-uniformities have been analyzed. Figure 6.31 shows reference geometry of the system for simulation. The parameters used in the analysis are divided into three

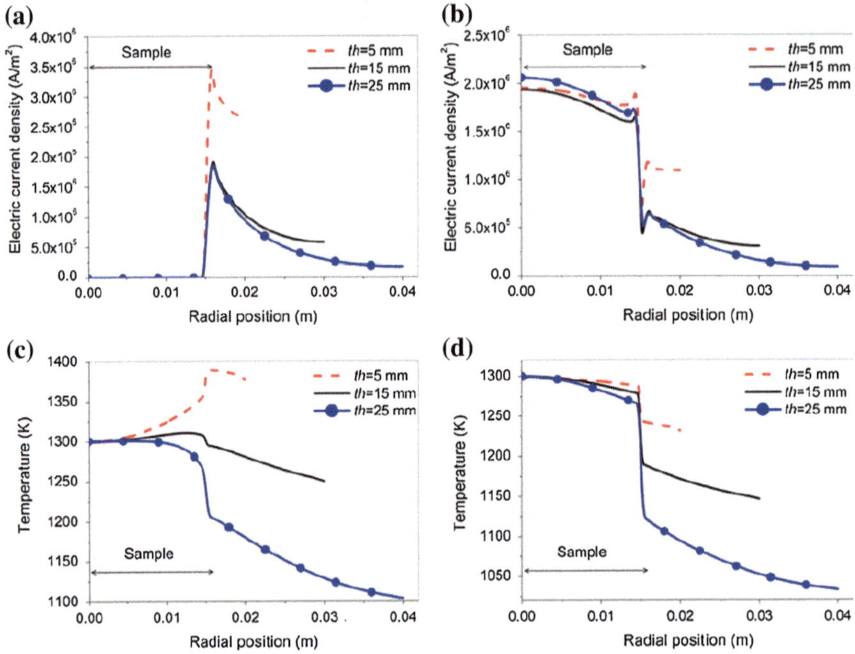

Fig. 6.32 The effect of the die thickness. Electrical current density distribution of Al_2O_3 (**a**) and Mo (**b**) samples. Temperature distribution of Al_2O_3 (**c**) and Mo (**d**) samples. All results are obtained with the reference geometry. Reproduced with permission from [44]. Copyright © 2013, Springer

groups. The first group defines the geometry of the plunger–die–sample assembly, including dimension of the sample, wall thickness and height of the die, and the exposed section of the plunger [44]. The second group includes the parameters controlling the thermal cycle, such as target temperature and heating rate. The final group includes the amount of radiative heat losses and the presence/absence of electrical and thermal contact resistances. Electrical and thermophysical properties of graphite used for the die are assumed to be constant.

Figure 6.32 shows simulation results on electrical current density and temperature distribution in radial direction for the two samples [44]. For insulative Al_2O_3 (Fig. 6.32a), the current cannot flow through the sample, but through the die which is around it instead. For different wall thicknesses, the current profiles are similar, which are characterized by the presence of a peak at near the edge of the sample and a marked drop toward the external surface of the die. The temperature profile within the sample (Fig. 6.32c) is determined by the temperature of the wall of the die. Therefore, a decrease in the wall thickness means an increase in the local current density, which thus produces higher localized Joule heating and higher temperatures at the edge of the sample. As a result, the temperature profile within the wall of the die decreases in the radial direction, due to both the current distribution and the heat losses by radiation. A discontinuity across the interface between the sample and the die is always observed, which is attributed to the presence of a thermal resistance at the interface. It is worth mentioning that the presence of both positive and negative radial temperature gradients for the insulative samples is of particular interest, because it offers a possibility to minimize the temperature gradients through the proper design of the die.

For the conductive sample (Mo), the electrical current flows through both the sample and the die (Fig. 6.32b). In this case, the current density distribution is only affected by the wall thickness of the die, resulting in the temperature distribution profiles as shown in Fig. 6.32d. Because a large fraction of the electrical current flows through the sample and consequently through the plungers that are adjacent to top and bottom surfaces of the sample, higher temperatures are always observed at the central region of the sample (Fig. 6.32d) in this case. As the thickness of the die wall is decreased, a more homogeneous temperature distribution within the sample is observed, but the effect is very marginal [44].

The effects of other factors, such as the free length of the plungers and contact resistances, have also been systematically studied [44]. From this systematic study, several conclusions have been arrived. First of all, there is possibility to optimize the experimental conditions so as to reduce the temperature inhomogeneities within the samples. As for the geometrical parameters, the temperature gradient in the sample is highly sensitive to the wall thickness and height of the graphite dies, while the exposed length has a negligible effect in this aspect. It is further found that the effects of various parameters on temperature distribution profiles are more significant for insulative samples than for conductive samples, due the difference in the flowing electrical current in the two cases.

The temperature distribution has no significant dependence on the heating rate if the analysis is carried out after the temperature is stabilized. The transient effects on the sintering of the materials can be safely neglected. The positive effect of insulation of the external surfaces of the die on homogeneous temperature distribution of the samples is only pronounced for conductive samples. For insulative samples, the actual result even could be opposite, depending on the size of the samples. The effect of contact resistances is also related to the type of the samples. The presence of contact resistances means an increase in the temperature inhomogeneities for the insulative samples, while for the conductive samples, the effect is negligible.

Due to the intrinsic complexity of the problem of the sintering technique and the strongly correlated relationship among the various variable parameters, it is still a challenge to predict the effect on the temperature distribution of the samples, by just adjusting one of the experimental parameters. In most cases, a thorough optimization of all the experimental conditions is necessary, especially for samples with large dimensions and for scale-up production for industrial applications [44].

Another example, which is necessarily mentioned, is the study on the scalability of SPS [45, 46], because the previous common sense is that the technique is not suitable for large-scale production [47]. Although efforts have been made in trying to scale-up the production, the progress is far away the requirement for real applications [48, 49].

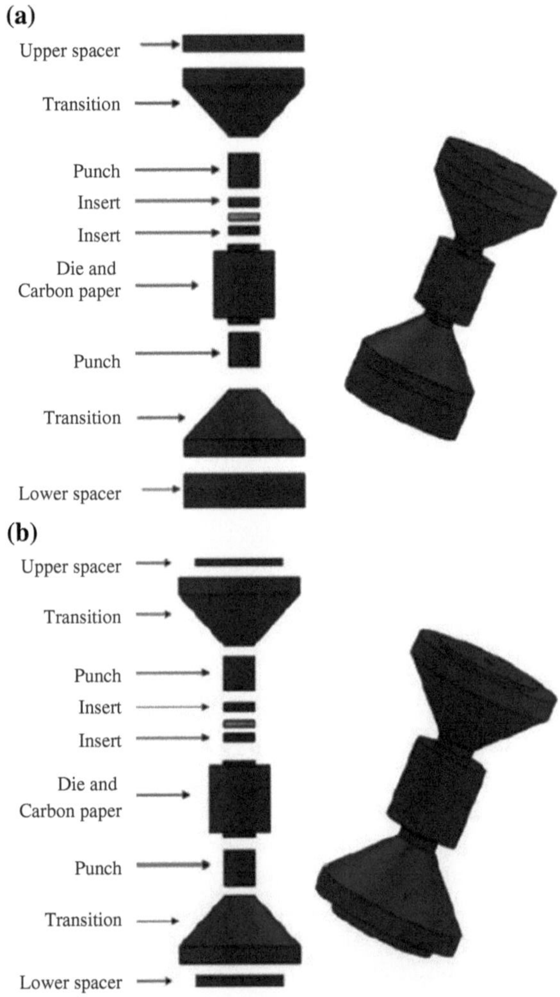

Fig. 6.33 Tooling assembly for the spark plasma sintering (SPS) experiments: **a** 15 mm and **b** 56 mm. Reproduced with permission from [45]. Copyright © 2012, John Wiley & Sons

Experimental studies are carried out first, with four dies fabricated to produce final products with diameters of 15, 40, 48, and 56 mm, as shown representatively in Fig. 6.33 [45]. The assembly components include external spacer, transition spacer, punch, insert, and sample. The die is used to encase the sample, as well as the two inserts and portions of the upper and lower punches. The lateral surfaces of the die and end faces of the punches are covered with carbon paper to prevent material adhesion to the graphite components. The inner and outer diameters and height of the die are maintained in an approximate aspect ratio of 1:2:2. A commercial Al_2O_3 powder is used for this study.

According to the experimental results, in terms of relative density and grain-pore structure, it is found that, despite certain deviations in some cases, the samples with different tooling sizes and temperature regimes are quite similar, when they are processing with the same SPS facility (FCT). The samples also have relatively uniform density and grain size spatial distributions, which means that SPS is potential to be scaled up. At the same time, a significant size impact in case of high heating rates and large specimen sizes has been observed. In addition, when different SPS systems are used, the outcomes could be possibly different.

By using a thermo-electromechanical finite element framework, shrinkage and grain growth kinetics, as well as the temperature, porosity, and grain size gradients, of the four samples with different sizes have been analyzed. The thermal and electrical conduction problems are strongly coupled, which has been taken into account by expressing the electrical properties of the materials as functions of temperature. The thermophysical parameters, such as heat capacity and thermal conductivities, are also expressed as functions of temperature. The thermal, electrical, and porosity components can be linked because the electrical and heat transfer properties can be expressed as functions of porosity. The consolidated system of partial differential equations has been solved with all the boundary conditions, including electrical, thermal, and mechanical, which are assigned simultaneously at the same time step.

As boundary and initial conditions, for the electrical component of the modeling framework, the end of the lower external spacer is considered to be grounded and thus its potential is set to be zero. A potential is applied at the end of the upper external spacer, which is specified to generate a preset time–temperature profile on the outer die surface. The free surfaces are regarded to be electrically isolated.

For the temperature component of the modeling framework, the initial temperature is set to be the ambient temperature (300 K). Thermal boundary conditions on the tooling surfaces are considered in the forms of a specified boundary temperature or heat flux. The temperatures of the two extreme upper and lower surfaces of the external spacers are assumed to be constant of 300 K. Heat loss due to radiation is considered at the lateral surfaces to be: $f = v\varepsilon\left(T_{\mathrm{W}}^4 - W_0^4\right)$, where f is the heat flux per unit area, m is the emissivity, which is assumed to be 0.8, $\varepsilon = 5.6704 \times 10^{-8}$ W m^2 K^4 called the Stefan–Boltzmann constant, T_{W} is the temperature of at surface of the die, and T_0 is the ambient temperature [23]. Contacts are modeled as perfect interfaces without losses. All simulation results are compared with experimental results [46].

Figure 6.34 shows numerical results of the finite element modeling code in terms of longitudinal and transversal slices for the final temperature distributions of the 56 mm–100 °C min^{-1} simulation (a) and the 15 mm–100 °C min^{-1} simulation (b). The transversal slices are taken at the middle of the die, so as to expose the temperature distribution inside the specimen.

Temperature evolution profiles at the center of the samples for the three heating rates for the 56 mm–100 °C min^{-1} simulation and for the 15 mm–100 °C min^{-1} simulation are shown in Fig. 6.35a, b, as a comparison. It is observed that the temperature at the center of the samples reaches the holding temperature earlier than for the 300 °C min^{-1} case, but with similar stabilized value to the other two cases. The 100 °C min^{-1} heating regime is slower for the 56 mm sample, but reaches a

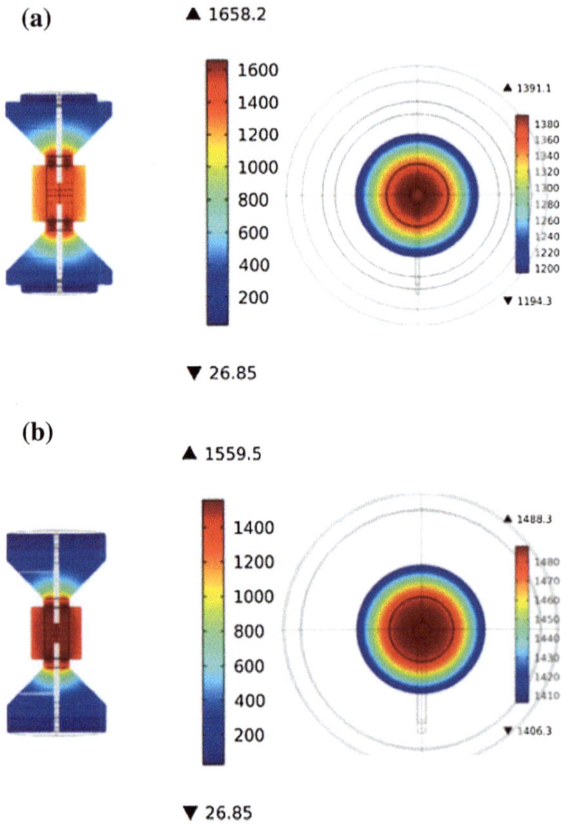

Fig. 6.34 **a** 56 mm configuration–100 °C min^{-1} regime: temperature distribution. *Left* longitudinal view. *Right* transversal view, with unit of °C. **b** 15 mm configuration-100 °C min^{-1} regime: temperature distribution. *Left* longitudinal view. *Right* transversal view, with unit of °C. Reproduced with permission from [46]. Copyright © 2012, John Wiley & Sons

slightly higher temperature. This trend is opposite for 15 mm sample. Therefore, the two samples of different size should experience different evolution in material structures, such as relative density and average grain size [46].

Figure 6.35c, d demonstrates the evolution of relative density of the two samples, both of which can achieve a high level of densification of about 95 % [46]. The 56 mm sample exhibits a much slower densification progress, delayed for about 300 s, for the 225 °C min^{-1} simulation, as compared with the 300 °C min^{-1} simulation. The densification is slower than for the 100 °C min^{-1} simulation, which is readily attributed to the lower holding temperature that it can achieve and the shorter holding time. For the 15 mm sample, the higher temperatures developed at 100 and 200 °C min^{-1}, as compared with at 300 °C min^{-1} regime, are ascribed to their higher shrinkage rates.

Figure 6.35e, f shows evolutions of average grain size of the two samples for the three regimes. For both the 56 and 15 mm samples, the average grain size is larger for the 300 °C min^{-1}, probably due to the earlier significant levels of densification, on the one hand, and larger temperature gradients, on the other hand. However, for the two lower heating rates, the increase of the heating rate shows an opposite

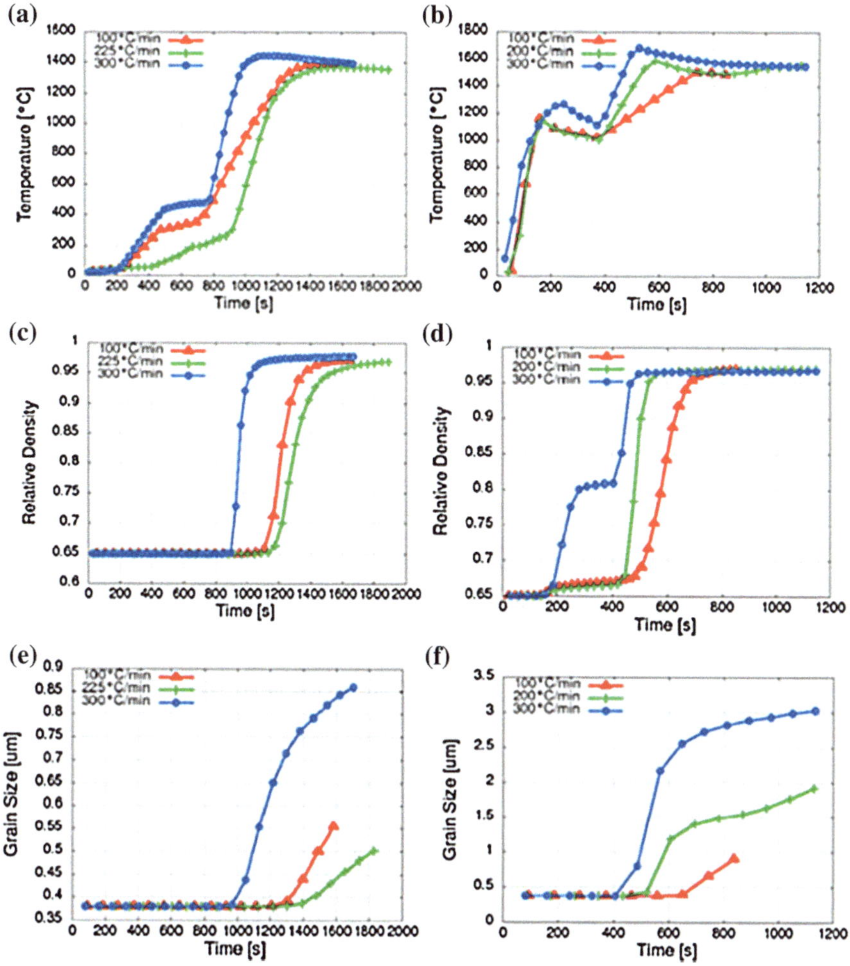

Fig. 6.35 Modeling results for 56 mm (*left*) versus 15 mm (*right*) specimens: **a** and **b** temperature evolution in the center of the specimen; **c** and **d** relative density; **e** and **f** grain size. The calculation results are given for different heating rates of 100, 225 °C min^{-1} (200 °C min^{-1} for the 15 mm specimen), and 300 °C min^{-1}. Reproduced with permission from [46]. Copyright © 2012, John Wiley & Sons

impact in 56 and 15 mm samples. This implies that there has been a complex interplay of the time of processing maximum temperature and temperature/relative density gradients [46].

Gradients of temperature, relative density, and grain size, as functions of sample size, are shown in Fig. 6.36 for two heating rates: 100 and 300 °C min^{-1} [46]. The results are obtained for two time durations: an intermediate and the final time. The intermediate time corresponds approximately to the point at which the holding period starts for each combination of the sample size—heating rate. Similarly, the

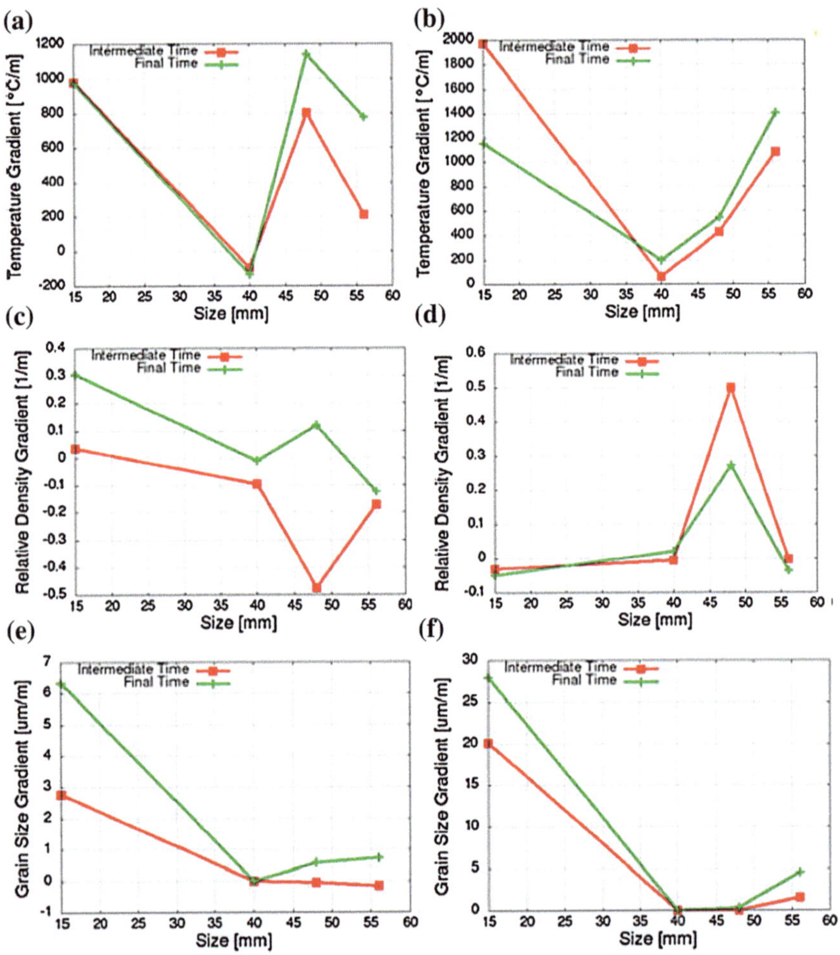

Fig. 6.36 Computed dependence on the spark plasma sintering (*SPS*) tooling size for 100 °C min^{-1} (*left*) and 300 °C min^{-1} (*right*) of: **a** and **b** temperature gradient; **c** and **d** relative density gradient; **e** and **f** grain size gradient. The calculation results are given for an intermediate and final time of SPS. Reproduced with permission from [46]. Copyright © 2012, John Wiley & Sons

final time corresponds to the end of the holding time, which also represents the end of the simulation. The simulated gradients depending on the sample size are a reflection of the heating patterns of the samples with different sizes. Therefore, establishment of reliable modeling approaches is an important requirement for large-scale industrial production of ceramics by using the SPS techniques, which could be a subject of future studies in this area.

Distributions of temperature and stress in a functionally graded material (FGM) based on Ti and TiB during a SPS process have been simulated, with the development of a coupled electrical–thermal–mechanical finite element model [32]. It is

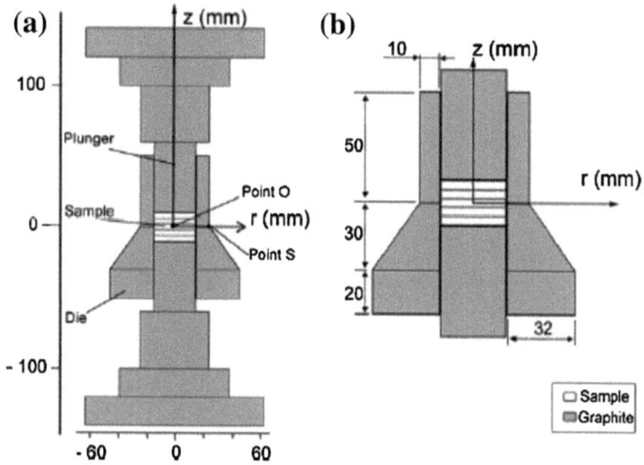

Fig. 6.37 Configuration of the SPS system: **a** schematic diagram of the SPS equipment and **b** die dimensions used in the simulation. Reproduced with permission from [32]. Copyright © 2012, Elsevier

found a stable axial temperature gradient in the sample is observed if a die with an area-changing cross section is used in the SPS process, which can be used as a strategy to ensure homogeneity of FGM during the sintering with SPS. High stress and the stress gradient are observed at bottom of the specimen, which could result in microstructural heterogeneity and cracks between adjacent layers. The simulation results also indicate that both the temperature and stress gradients increase with increasing heating rate, suggesting a principle to use heating rate as low as possible.

A schematic diagram of the SPS equipment is shown in Fig. 6.37a, which consists of a graphite die, a specimen, two plungers, and six protection spacers [32]. The die, plungers, and spacers are all made of dense graphite. Figure 6.37b shows schematic of the die used in the simulation. The sintered FGM sample has five layers of equal thickness, with a diameter of 30 mm and a total thickness (height) of 20 mm. Weight percentages of Ti in the Ti–TiB layers are 45, 55, 65, 75, and 85 %, respectively. The sample is considered to be isotropic in the simulation in order to establish a simple model. The properties of the different layers in the sample are calculated by using the mixture rule, as discussed earlier.

During SPS process, heat is generated due to the application of an electrical current; therefore, the temperature distribution in the die is determined by the distribution of the electrical current [32]. Also, the density of the electrical current is inversely proportional to the cross-sectional area of the die. As a result, the current density decreases gradually from top to bottom of the die, so that Joule heating is more significant at top of the die, leading to a temperature gradient in the axial direction. The highest temperature at Point A is 1481 K, while the lowest temperature at Point L is 1385 K. The maximum temperature difference in this direction is 88 K. This axial temperature gradient is crucial to fabricate FGMs with desired homogeneity.

Because all the three stresses, radial, axial, and circumferential, are compression stress, it is less likely to form cracks and fractures in the sample during the stage of heating [32]. It is also found that both the radial stress and the circumferential stress are maximized at the bottom of the sample, whereas the maximum axial stress is

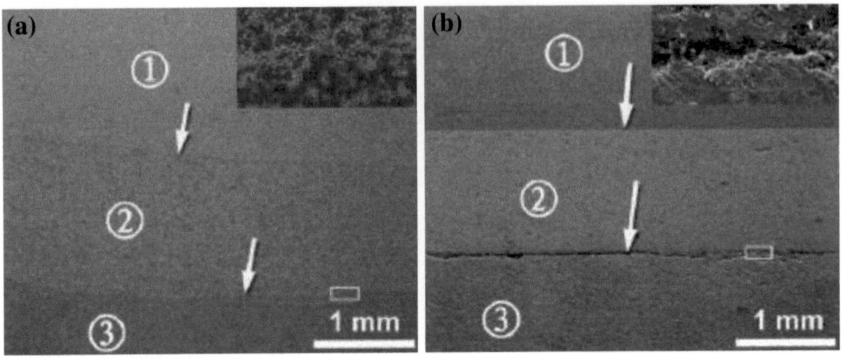

Fig. 6.38 SEM images of the FGM samples sintered by using the SPS process with dies of different structures: **a** the die with an area-changing cross section and **b** a hollow cylinder die. The *inset* shows a magnified image of the *squared* area. Reproduced with permission from [32]. Copyright © 2012, Elsevier

observed in the center of the sample. The hydrostatic stress along Lines CD, EF, GH, and IJ decrease at an increasing rate, while the hydrostatic stress at top and bottom of the sample increases in the radial direction. In addition, the hydrostatic stress at the bottom of the sample is significantly higher than that at all the other horizontal planes. The hydrostatic stress for Line AK is in the range of 74–100 MPa, whereas that for Line BL is from 45 to 109 MPa, which means that the difference in stress for Line BL is about 2.5 times larger than that for Line AK. The difference in hydrostatic stress between adjacent layers in the inner part of the sample is generally less than 10 MPa, while that in the rest part is higher. The maximum difference in stress at the bottom portion is 35 MPa. Highest stress and largest stress gradient are located at the edges of the sample.

Figure 6.38 shows SEM images of two samples, A and B, which are processed by using dies with area-changing cross section and hollow cylinder, respectively [32]. It is observed that the sample A has a denser microstructure than the sample B. Also, the sample B has macro-crack at the interface between the second layer and the third layer. There are also many pores at that interface, which is responsible for the low bonding strength of the interface. Therefore, the use of the die with area-changing cross section is a solution to ensure the homogeneity of the FGM during the processing with SPS.

According to this study, serval conclusions can be arrived as follows [32]. (i) A stable temperature gradient can be achieved in the axial direction of the sample by using a die with an area-changing cross section. (ii) The axial temperature gradient increases with increasing heating rate, due to the increase in the rate of Joule heat generation and the decrease in heat diffusion at top of the die. (iii) Both the stress and the stress gradient at the bottom portion of the specimen are significantly higher than those at the samples' center. (iv) The maximum hydrostatic stress is observed at the final stage of heating, and both the hydrostatic stress and the stress gradient increase with increasing heating rate. (v) The sample processed by using the die with an area-changing cross section has much better microstructure than the sample processed with the normal die, which would ensure desired mechanical properties of the final products.

Although the studies on simulations of densification behaviors of transparent ceramics by using ECAS (SPS) process have not been widely reported, the results that have been discussed above can be used as a reference. This is especially true for those of oxide materials, because they share similar thermophysical properties with transparent ceramics.

6.3 Microwave Sintering

6.3.1 Brief Introduction

Microwave sintering is a special sintering process that is fundamentally different from the conventional sintering techniques. Microwave energy belongs to electromagnetic energy, with frequencies ranging from 300 MHz to 300 GHz, corresponding to wavelengths between 1 mm and 1 m, as shown in Fig. 6.39 [50]. The fundamental difference between microwave sinterings is essentially different from the conventional sintering in the heating mechanism. The temperature profile for both methods is shown in Fig. 6.40 [50]. In microwave process, the processed materials themselves absorb microwave power and then microwave is converted into heat within the samples rapidly [34, 35, 50]. Because the power produced by the microwave source can be used only to heat the materials, while the furnace

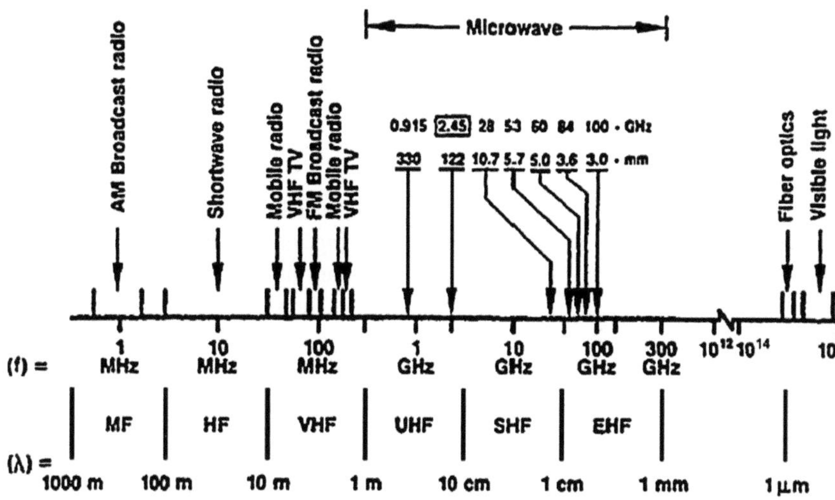

Fig. 6.39 Electromagnetic spectrum and frequencies used for microwave processing of materials. Reproduced with permission from [50]. Copyright © 2010, Elsevier

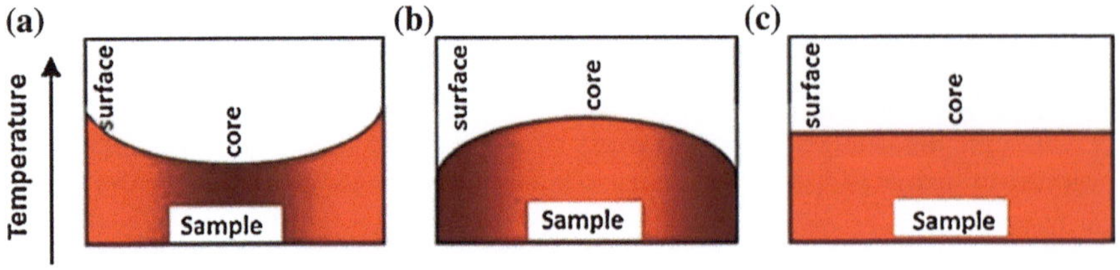

Fig. 6.40 Temperature profiles within the materials to be sintered by using different techniques: **a** conventional heating, **b** microwave heating, and **c** microwave hybrid heating. Reproduced with permission from [50]. Copyright © 2010, Elsevier

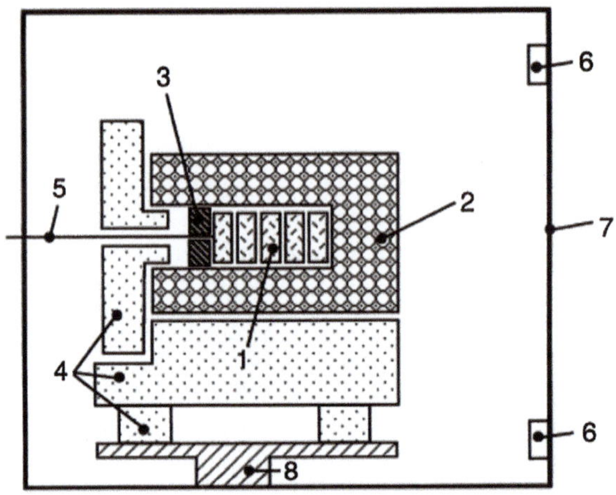

Fig. 6.41 Configuration of heating chamber of a microwave sintering setup: *1* samples, *2* bubble alumina thermal insulation in sintered alumina container, *3* sintered alumina disk, *4* porous alumina parts providing thermal insulation, *5* Pt-sheathed S-type thermocouple, *6* MW insertion points, *7* applicator wall, and *8* metal stand. Reproduced with permission from [51]. Copyright © 1999, Elsevier

elements are not heated, much higher heating rates can be achieved when using microwave sintering. In addition, a high sintering rate is benefit to bring out products with finer and less defective microstructures and thus enhanced functional properties [36]. Therefore, microwave sintering to produce ceramics has various advantages, such as higher energy efficiency, enhanced reaction and sintering rate, time-consuming and cost-effectiveness. Specifically, transparent ceramics can be fabricated at lower sintering temperatures for much shorter sintering times by using microwave sintering [43].

As compared with the traditional furnaces, MW furnaces have relatively more complicated configurations, due to the involvement of MWs. Two examples are demonstrated here. The first configuration is able to sinter samples batch by batch

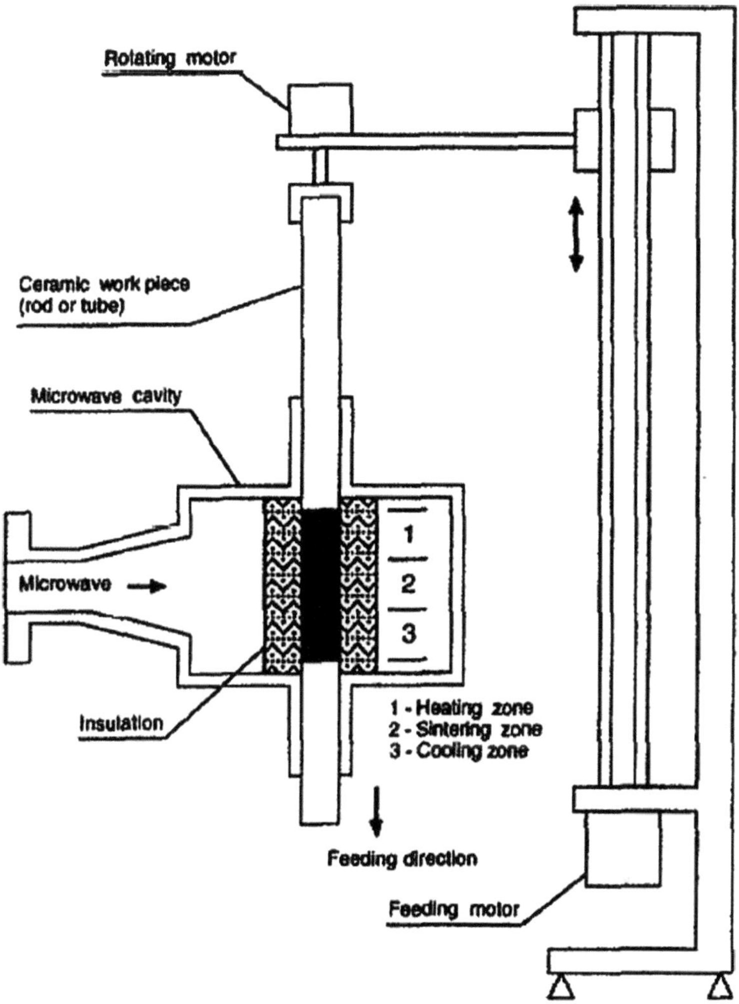

Fig. 6.42 Schematic drawing of apparatus for continuous microwave sintering. Reproduced with permission from [52]. Copyright © 1998, Elsevier

[51], while the second type can support continuous experiments and thus is suitable for large-scale production [52], as shown in Figs. 6.41 and 6.42, respectively.

6.3.2 Theoretical Aspects

Theoretical understanding is important for the development and analysis of microwave sintering technique. For example, optimization of processing parameters is tedious and very time-consuming. By using modeling and numerical simulations, this work could be significantly simplified. Microwave technique is an interdisciplinary subject that involves several areas, including electromagnetics, physics, and materials science. A model of microwave sintering process is to obtain the temporal evolution and spatial distribution of (i) electromagnetic field in the materials to be processed, (ii) temperature, and (iii) other physical properties of the materials, such as stresses, mass fluxes, porosity, and grain size. These properties are dependent on the evolution and distribution of temperature within the body of the materials, while they could also be influenced by the nonthermal effect of electromagnetic field.

The distribution of electromagnetic field within the materials to be processed is determined by various factors, such as effective dielectric and magnetic parameters of the materials that are dependent on temperature and microstructure, dimensions and properties of the applicator, and the degree of matching between the microwave source, transmission line, and applicator. Consequently, the distribution of the electromagnetic field will affect the distribution of temperature within the sintered object. Of course, other factors, such as effective absorption properties, heat capacity, and thermal conductivity of the materials, as well as heat transportation characteristics of the materials and the furnaces, will also have their specific effects. Ultimately, all these will be reflected by the properties of the final sintered products. While detailed theoretical elaboration of microwave sintering can be found in Ref. [39], a brief description will be presented in this section.

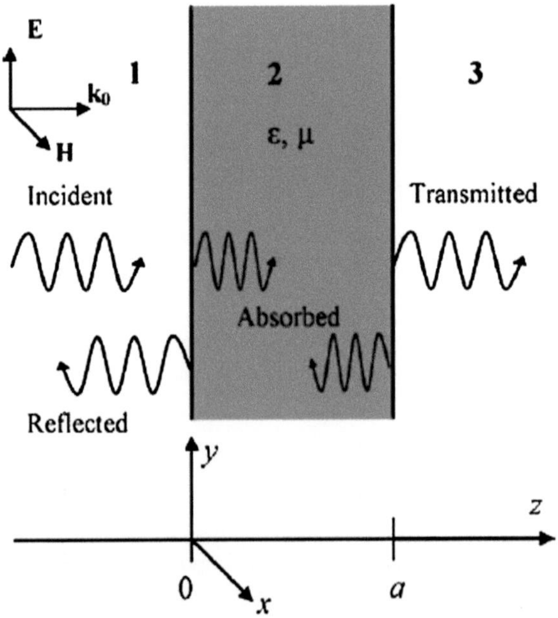

Fig. 6.43 Normal incidence of a plane electromagnetic wave on a slab of material from vacuum. Reproduced with permission from [39]. Copyright © 2013, John Wiley & Sons

6.3.2.1 Theoretical Modeling

For simplicity, normal incidence of a plane electromagnetic wave on a slab of material, with a thickness of a, from vacuum, is considered, as shown in Fig. 6.43 [39]. In the regions (**1**) and (**2**), there are waves traveling in both forward and backward directions, because of the reflections at the front and back surfaces of the material slab. However, in the region (**3**), there is only wave traveling in the forward direction. Complex amplitudes, E_x and H_y, of the electric and magnetic fields, of the electromagnetic wave, are given by the following equations [53–55]:

$$E(z,t) - x_0 \mathrm{Re}\left[E_x(z)\mathrm{e}^{i\omega t}\right], \tag{6.36}$$

$$H(z,t) - y_0 \mathrm{Re}\left[H_y(z)\mathrm{e}^{i\omega t}\right]. \tag{6.37}$$

Then, the electric field, in the three regions, can be written in the form:

$$E_x = E_0\left(\mathrm{e}^{ik_0 z} + R\cdot \mathrm{e}^{-ik_0 z}\right), \text{ region } (\mathbf{1}), \tag{6.38}$$

$$E_x = E_0\left(A\cdot \mathrm{e}^{ikz} + B\cdot \mathrm{e}^{-ikz}\right), \text{ region } (\mathbf{2}), \tag{6.39}$$

$$E_x = E_0 T\cdot \mathrm{e}^{ik_0 z}, \text{ region } (\mathbf{3}), \tag{6.40}$$

where E_0 is the amplitude of electric field in the incident wave, ω is frequency, $k_0 = \omega/c$ is the wave number of vacuum, c is the velocity of light, $k = k_0\sqrt{\varepsilon\mu}$ is the wave number in the material, ε is relative dielectric permittivity, and μ is relative magnetic permeability of the material.

Similarly, expressions for the magnetic field in the three regions are given by:

$$H_y = H_0\left(\mathrm{e}^{ik_0 z} - R\cdot \mathrm{e}^{-ik_0 z}\right), \text{ region } (\mathbf{1}), \tag{6.41}$$

$$E_y = \frac{H_0}{\zeta}\left(A\cdot \mathrm{e}^{ikz} - B\cdot \mathrm{e}^{-ikz}\right), \text{ region } (\mathbf{2}), \tag{6.42}$$

$$H_y = H_0 T\cdot \mathrm{e}^{ik_0 z}, \text{ region } (\mathbf{3}), \tag{6.43}$$

where H_0 is the amplitude of magnetic field in the incident wave and $\zeta = \sqrt{\mu/\varepsilon}$ is the impedance of the material. For a plane wave in vacuum, the amplitudes of the electric and magnetic field, E_0 and H_0, can be related by the following equation:

$$E_0 = \zeta_0 H_0, \tag{6.44}$$

where $\zeta_0 = \sqrt{\mu_0/\varepsilon_0}$ is the impedance of free space, ε_0 is the relative dielectric constant, and μ_0 is the relative magnetic permeability of vacuum, with $c = 1/\sqrt{\mu_0\varepsilon_0}$.

In Eqs. (6.38–6.40) and (6.41–6.43), A, B, R, and T are unknown coefficients, which are determined the boundary conditions that require continuity of the electric and magnetic field components, E_x and H_y, at the front and back surfaces, i.e., $z = 0$ and $z = a$. By imposing four boundary conditions and solving four algebraic equations, the coefficients can be obtained as follows:

$$A = \frac{2\left(\frac{1}{\zeta} + 1\right)(\cos ka - i \sin ka)}{\frac{2}{\zeta}\cos ka - i\left(\frac{1}{\zeta^2} + 1\right)\sin ka}, \tag{6.45}$$

$$B = \frac{2\left(\frac{1}{\zeta} - 1\right)(\cos ka + i \sin ka)}{\frac{2}{\zeta}\cos ka - i\left(\frac{1}{\zeta^2} + 1\right)\sin ka}, \tag{6.46}$$

$$R = \frac{i\left(\frac{1}{\zeta^2} - 1\right)\sin ka}{\frac{2}{\zeta}\cos ka - i\left(\frac{1}{\zeta^2} + 1\right)\sin ka}, \tag{6.47}$$

$$T = \frac{\frac{2}{\zeta}(\cos ka - i \sin ka)}{\frac{2}{\zeta}\cos ka - i\left(\frac{1}{\zeta^2} + 1\right)\sin ka}, \tag{6.48}$$

The energy flux density in the plane electromagnetic wave is equal to the average value of the Poynting vector, given by:

$$\bar{\Xi} = |\bar{E} \times \bar{H}| = \frac{1}{2} Re\left(E_x H_y^*\right), \tag{6.49}$$

where the overline denotes averaging over the field period, while the asterisk denotes complex conjugation. Therefore, the energy flux densities, in the incident wave, in the wave reflected from the slab and that transmitted through the slab, are $(E_0H_0)/2$, $(E_0H_0|R|^2)/2$ and $(E_0H_0|T|^2)/2$, respectively.

If the materials have dielectrically and magnetically lossless, both ε and μ are their real parts. In this case, the wave number k also has only real part and there is:

$$|R|^2 + |T|^2 = 1, \tag{6.50}$$

which means that the total energy flux density, carried away by the reflected and transmitted waves, is equal to the energy flux density in the incident wave.

However, if the material has dielectric and magnetic losses, both ε and/or μ will have imaginary parts, ε'' and μ''. As a result, the wave number k will be complex. The power density, w, with a unit of per unit square, which is the energy absorbed by the slab of the material, can be calculated by using following equation:

$$w = \frac{\omega}{2} \int_{0}^{a} \left(\varepsilon_0 \varepsilon'' |E_x|^2 + \mu_0 \mu'' |H_y|^2 \right) \mathrm{d}z, \tag{6.51}$$

where the components, E_x and H_y, within the material slab, are given by the Eqs. (6.38–6.40) and (6.41–6.42) for the region (**2**), with coefficients A and B being derived from Eqs. (6.45) and (6.46). Therefore, the energy absorbed in the material slab is the difference between the energy flux density in the original incident wave and the total energy flux density due to the reflected and the transmitted waves, i.e.,

$$w = \frac{1}{2} E_0 H_0 (1 - |R|^2 - |T|^2). \tag{6.52}$$

For materials with nonuniform properties, the most general method is to divide the material layers, with each layer having a fixed value of complex dielectric permittivity and magnetic permeability. In this case, the incident wave can be calculated by using the boundary conditions at all boundaries between every adjacent layers and solving the linear algebraic equations of the field amplitudes. Therefore, the power absorbed in each layer can be calculated accordingly [39, 56]. Besides planar slabs, materials with other shapes, such as cylinders and spheres, have also been simulated, although the electromagnetic field distribution and structure in these configurations would be much more complicated [57–60].

In addition to the interaction between the electromagnetic wave and the materials, the heating effect of applicators is also an important issue in microwave sintering. This is simply because applicators are necessary to build the cavity resonators. Various methods, such as finite-difference time domain (FDTD) [61–63], finite element method (FEM) [64], transmission line matrix (TLM) [65, 66], method of moments (MOM), have been used for such purposes.

Most transparent ceramics are nonmagnetic materials; therefore, their dielectric properties, i.e., dielectric constant and loss tangent, will play very important roles in determining the effectiveness and efficiency of microwave sintering. Complex dielectric permittivity is expressed as $\varepsilon = \varepsilon' - j\varepsilon''$, where ε' and ε'' are real and imaginary parts, respectively. In this case, dielectric loss tangent is defined as $\tan\delta = \varepsilon''/\varepsilon'$. While dielectric permittivity of a given material is somehow a constant, its loss tangent could be varied over a wide range, depending on various factors, including purity, density, grain size, size distribution, and defects [67–69].

6.3.3 Heat Transfer and Sintering

As mentioned earlier, microwave heating is much faster than the conventional methods, because the microwave power can be directly delivered into the material to be sintered. In this respect, thermal conduction to dissipate and transport heat is not the only determining factor of microwave heating process. The temperature within the material during microwave heating also follows the heat conduction equation:

$$c_p\rho\frac{\partial T}{\partial t} - \nabla\cdot(\kappa\nabla T) = w, \tag{6.53}$$

where c_p is specific heat capacity, ρ is density, κ is thermal conductivity of the material, and w is local density of energy released in the material. For microwave heating, w is usually given by:

$$w = \frac{1}{2}\omega\left(\varepsilon_0\varepsilon''\mathrm{E}^2 + \mu_0\mu''\mathrm{H}^2\right). \tag{6.54}$$

It was found that the distribution of temperature in the materials processed by using microwave heating is generally non-uniform. The distribution is parabolic with a maximum at the center of the material, which is usually referred to as an inverse temperature profile and is a specific characteristic of microwave sintering. At short timescales, the temperature distribution follows the power distribution. Otherwise, it is determined by thermal conduction of the entire dimensions of the materials [70].

If the distribution of the microwave power is uniform or follows certain trend, e.g., exponentially decaying from the surface into the bulk of the materials, heat conduction equation alone is sufficient for the modeling of the process. This is especially true for highloss materials. Simulation methods, including finite elements and finite differences, have been used to analyze the microwave sintering process based on heat conduction equation [71, 72].

In practice, however, the distribution of electromagnetic field of a material is strongly affected by the variation in dielectric and magnetic properties with temperature and time. Therefore, it is necessary to simulate the microwave sintering process by considering the coupling of electromagnetic and thermal aspects. It is knows as self-consistent model, in which there is an electrodynamic part to describe the electric and magnetic fields and a thermal part to determine the temperature. At the same time, the temperature in turn has influence on all the material parameters, including ε, μ, c_p, and κ. If the temperature dependencies of these parameters can be expressed with certain describe forms, the modeling would have analytical solutions [73–76]. In addition, it is also very important to consider the electromagnetic field in the whole microwave cavity [77–79]. This is because the variation in material parameters in turn would alter the matching characteristics of the cavity [80]. For such purposes, various models have been developed and applied to microwave sintering [81–83].

The ultimate goal is to simulate the densification of the ceramics during microwave sintering. Microwave sintering is different from the conventional sintering in certain aspects, e.g., inhomogeneity of temperature distribution in the materials [84]. There have been only limited attempts to simulate the temperature-dependent densification by coupling the electromagnetic and thermal effects. A simple way is to fit the relation of density and temperature [85]. However, in this case, the evolution of dimension of the samples and the properties of the materials are not described. An alternative simple approach is the use of master sintering curves [86, 87]. More sophisticated simulations include continuum sintering models and finite-difference electromagnetic simulation, which also have their disadvantages [88].

6.3.4 *Nonthermal Effects*

Another important issue in microwave sintering is the nonthermal effects of electromagnetic field, which should be considered in the modelings and simulations [39]. As stated previously, there is an essential difference in principle between microwave and conventional sintering processes. Microwave sintering is to use coherent electromagnetic field to drive oscillatory motions of charged particles in the materials, while conventional sintering is realized through energy exchange between the furnace and materials, due to the quasi-equilibrium thermal electromagnetic radiation with a continuous spectrum predominantly in the infrared range. The thermal radiation is absorbed by the materials through the excitation of oscillations with an equilibrium spectrum that can be described by the Maxwell–Boltzmann statistics. This absorption is usually at the surface of the materials, which is then transferred to the whole materials through thermal conduction. In summary, accurate simulation and modeling of microwave sintering either for all materials in general or for transparent ceramics in particular are still a great challenge.

6.4 Summary

Various transparent ceramics have been fabricated by using SPS (ECAS) and MW sintering. Both sintering techniques have similar advantages, such as rapid heating rate, low sintering temperature, and short sintering time duration. However, comparatively, their application in fabrication of transparent ceramics is still much less popular, as compared with the conventional sintering techniques. Possible reasons include the requirement of sophisticated facility in general, as well as expensive setup for SPS and relative inaccuracy in measurement of temperature for microwave sintering in particular. In addition, theoretical understanding on mechanisms of the densification and grain growth of transparent ceramics is still lacking, as compared with that of the nontransparent materials. It is expected that more and more works will be reported to process transparent ceramics with these two new sintering technologies.

References

1. Orru R, Licheri R, Locci AM, Cincotti A, Cao G (2009) Consolidation/synthesis of materials by electric current activated/assisted sintering. Mater Sci Eng R-Rep 63:127–287
2. Weissler GA (1981) Resistance sintering with alumina dies. Int J Powder Metall 17:107–118
3. Groza JR, Zavaliangos A (2000) Sintering activation by external electrical field. Mater Sci Eng A-Struct Mater Prop Microstruct Process 287:171–177
4. Zavaliangos A, Zhang J, Krammer M, Groza JR (2004) Temperature evolution during field activated sintering. Mater Sci Eng A-Struct Mater Prop Microstruct Process 379:218–228
5. Wu X, Guo J (2007) Effect of liquid phase on densification in electric-discharge compaction. J Mater Sci 42:7787–7793
6. Alp T, Alhassani STS, Johnson W (1985) The electrical-discharge comaction of powder-mechanics and materials structure. J Eng Mater Technol-Trans ASME 107:186–195
7. An YB, Oh NH, Chun YW, Kim YH, Kim DK, Park JS et al (2005) Mechanical properties of environmental-electro-discharge-sintered porous Ti implants. Mater Lett 59:2178–2182
8. An YB, Oh NH, Chun YW, Kim YH, Park JS, Choi KO et al (2005) Surface characteristics of porous titanium implants fabricated by environmental electro-discharge sintering of spherical Ti powders in a vacuum atmosphere. Scripta Mater 53:905–908
9. Rajagopalan PK, Desai SV, Kalghatgi RS, Krishnan TS, Bose DK (2000) Studies on the electric discharge compaction of metal powders. Mater Sci Eng A-Struct Mater Prop Microstruct Process 280:289–293
10. Grigoriev EG, Rosliakov AV (2007) Electro-discharge compaction of WC-Co and W-Ni-Fe-Co composite materials. J Mater Process Technol 191:182–184
11. Williams DJ, Johnson W (1982) Neck formation and growth in high-voltage discharge forming of metal powders. Powder Metall 25:85–89
12. Wu XY, Zhang W, Li DX, Guo JD (2007) Microstructure of WC in WC-Co cemented carbides consolidated by electric discharge. Mater Sci Technol 23:627–629
13. Zavodov NN, Kozlov AV, Luzganov SN, Polishchuk VP, Shurupov AV (1999) Sintering of metal powders by a series of heavy current pulses. High Temp 37:130–135
14. Qiu J, Shibata T, Rock C, Okazaki K (1997) Electro-discharge consolidation of atomized high strength aluminum powders. Mater Trans JIM 38:226–231
15. Matsugi K, Ishibashi N, Hatayama T, Yanagisawa O (1996) Microstructure of spark sintered titanium-aluminide compacts. Intermetallics 4:457–467
16. Ervin DR, Bourell DL, Persad C, Rabenberg L (1988) Structure and properties of high-energy, high-rate consolidated molybdenum alloy TZM. Mater Sci Eng A-Struct Mater Prop Microstruct Process 102:25–30
17. Raghunathan SK, Persad C, Bourell DL, Marcus HL (1991) High-energy, high-rate consolidation of tungsten and tungsten-based composite powders. Mater Sci Eng A-Struct Mater Prop Microstruct Process 131:243–253
18. Song Z, Kishimoto S, Shinya N (2003) Fabrication of Ni-P alloy closed cellular solid containing polymer by the pulse current hot pressing technique. J Mater Sci 38:4211–4219
19. Xie GQ, Ohashi O, Chiba K, Yamaguchi N, Song MH, Furuya K et al (2003) Frequency effect on pulse electric current sintering process of pure aluminum powder. Mater Sci Eng A-Struct Mater Prop Microstruct Process 359:384–390
20. Vanmeensel K, Laptev A, Hennicke J, Vleugels J, Van der Biest O (2005) Modelling of the temperature distribution during field assisted sintering. Acta Mater 53:4379–4388
21. Chen W, Anselmi-Tamburini U, Garay JE, Groza JR, Munir ZA (2005) Fundamental investigations on the spark plasma sintering/synthesis process—I. Effect of dc pulsing on reactivity. Mater Sci Eng A-Struct Mater Prop Microstruct Process 394:132–138
22. Anselmi-Tamburini U, Garay JE, Munir ZA (2005) Fundamental investigations on the spark plasma sintering/synthesis process III. Current effect on reactivity. Mater Sci Eng A-Struct Mater Prop Microstruct Process 407:24–30

23. Anselmi-Tamburini U, Gennari S, Garay JE, Munir ZA (2005) Fundamental investigations on the spark plasma sintering/synthesis process—II. Modeling of current and temperature distributions. Mater Sci Eng A-Struct Mater Prop Microstruct Process 394:139–148
24. Raichenko AI, Chernikova ES (1989) A mathematical-model of electric-heating of the prorous-medium using current-supplying electrode punches. Sov Powder Metall Metal Ceram 28:365–371
25. Anselmi-Tamburini U, Garay JE, Munir ZA (2006) Fast low-temperature consolidation of bulk nanometric ceramic materials. Scripta Mater 54:823–828
26. Jaroszewicz J, Michalski A (2006) Preparation of a TiB_2 composite with a nickel matrix by pulse plasma sintering with combustion synthesis. J Eur Ceram Soc 26:2427–2430
27. Michalski A, Jaroszewicz J, Rosinski M, Siemiaszko D (2006) NiAl-Al2O3 composites produced by pulse plasma sintering with the participation of the SHS reaction. Intermetallics 14:603–606
28. Wang YC, Fu ZY (2002) Study of temperature field in spark plasma sintering. Mater Sci Eng B-Solid State Mater Adv Technol 90:34–37
29. Heian EM, Feng A, Munir ZA (2002) A kinetic model for the field-activated synthesis of $MoSi_2$/SiC composites: simulation of SPS conditions. Acta Mater 50:3331–3346
30. Gu XF, Zhang LM, Zhang DM, Yang MJ, Wang ZZ (2006) Spark effect on the densification of SiCp/Al composites by SPS. J Wuhan Univ Technol-Mater Sci Ed 21:72–75
31. Matsugi K, Kuramoto H, Hatayama T, Yanagisawa O (2003) Temperature distribution at steady state under constant current discharge in spark sintering process of Ti and Al_2O_3 powders. J Mater Process Technol 134:225–232
32. Wei S, Zhang Z-H, Shen X-B, Wang F-C, Sun M-Y, Yang R et al (2012) Simulation of temperature and stress distributions in functionally graded materials synthesized by a spark plasma sintering process. Comput Mater Sci 60:168–175
33. Matsugi K, Kuramoto H, Yanagisawa O, Kiritani M (2003) A case study for production of perfectly sintered complex compacts in rapid consolidation by spark sintering. Mater Sci Eng A-Struct Mater Prop Microstruct Process 354:234–242
34. Katz JD (1992) Microwave sintering of ceramics. Annu Rev Mater Sci 22:153–170
35. Thostenson ET, Chou TW (1999) Microwave processing: fundamentals and applications. Compos Part A-Appl Sci Manuf 30:1055–1071
36. Bykov YV, Rybakov KI, Semenov VE (2001) High-temperature microwave processing of materials. J Phys D-Appl Phys 34:R55–R75
37. Vanmeensel K, Laptev A, Van der Biest O, Vleugels J (2007) The influence of percolation during pulsed electric current sintering of ZrO_2-TiN powder compacts with varying TiN content. Acta Mater 55:1801–1811
38. Vanmeensel K, Laptev A, Van der Biest O, Vleugels J (2007) Field assisted sintering of electro-conductive ZrO_2-based composites. J Eur Ceram Soc 27:979–985
39. Rybakov KI, Olevsky EA, Krikun EV (2013) Microwave sintering: Fundamentals and modeling. J Am Ceram Soc 96:1003–1020
40. Olevsky E, Froyen L (2006) Constitutive modeling of spark-plasma sintering of conductive materials. Scripta Mater 55:1175–1178
41. Wang X, Casolco SR, Xu G, Garay JE (2007) Finite element modeling of electric current-activated sintering: the effect of coupled electrical potential, temperature and stress. Acta Mater 55:3611–3622
42. Cincotti A, Locci AM, Orru R, Cao G (2007) Modeling of SPS apparatus: temperature, current and strain distribution with no powders. AIChE J 53:703–719
43. Bykov Y, Egorov S, Eremeev A (2013) Fabrication of transparent ceramics by millimeter-wave sintering. Phys Status Solidi C 10:945–951
44. Munoz S, Anselmi-Tamburini U (2013) Parametric investigation of temperature distribution in field activated sintering apparatus. Int J Adv Manuf Technol 65:127–140
45. Olevsky EA, Bradbury WL, Haines CD, Martin DG, Kapoor D (2012) Fundamental aspects of spark plasma sintering: I. Experimental analysis of scalability. J Am Ceram Soc 95:2406–2413

46. Olevsky EA, Garcia-Cardona C, Bradbury WL, Haines CD, Martin DG, Kapoor D (2012) Fundamental aspects of spark plasma sintering: II. Finite element analysis of scalability. J Am Ceram Soc 95:2414–2422
47. Lenel FV (1955) Resistance sintering under pressure. Trans Am Inst Min Metall Eng 203:158–167
48. Graeve OA, Carrillo-Heian EM, Feng A, Munir ZA (2001) Modeling of wave configuration during electrically ignited combustion synthesis. J Mater Res 16:93–100
49. Tokita M (2003) Large-size WC/Co functionally graded materials fabricated by spark plasma sintering (SPS) method. In: Pan W, Gong J, Zhang L, Chen L (eds) Functionally graded materials VII. Trans Tech Publication, Durnten, pp 39–44
50. Oghbaei M, Mirzaee O (2010) Microwave versus conventional sintering: a review of fundamentals, advantages and applications. J Alloy Compd 494:175–189
51. Goldstein A, Travitzky N, Singurindy A, Kravchik M (1999) Direct microwave sintering of yttria-stabilized zirconia at 2 center dot 45 GHz. J Eur Ceram Soc 19:2067–2072
52. Agrawal DK (1998) Microwave processing of ceramics. Curr Opin Solid State Mater Sci 3:480–485
53. Rothwell EJ, Cloud MJ (2009) Electromagnetics. Taylor & Francis Group, New York
54. Kong JA (1990) Electromagnetic wave theory, 2nd edn. Wiley, New York
55. Shevgaonkar RK (2006) Electromagnetic waves. Tata McGraw-Hill, New Delhi
56. Vriezinga CA (1996) Thermal runaway and bistability in microwave heated isothermal slabs. J Appl Phys 79:1779–1783
57. Vriezinga CA (1998) Thermal runaway in microwave heated isothermal slabs, cylinders, and spheres. J Appl Phys 83:438–442
58. Fliflet AW (2008) Self-consistent electromagnetic-thermal model for calculating the temperature of a ceramic cylinder irradiated by a high-power millimeter-wave beam. IEEE Trans Plasma Sci 36:582–590
59. Kulumbaev EB, Semenov VE, Rybakov KI (2007) Stability of microwave heating of ceramic materials in a cylindrical cavity. J Phys D-Appl Phys 40:6809–6817
60. Kozlov PV, Rafatov IR, Kulumbaev EB, Lelevkin VM (2007) On modelling of microwave heating of a ceramic material. J Phys D-Appl Phys 40:2927–2935
61. Subirats M, Iskander MF, White MJ, Kiggans JO (1997) FDTD simulation of microwave sintering in large (500/4000 liter) multimode cavities. J Microw Power Electromagn Energy 32:161–170
62. White MJ, Iskander MF, Huang ZL (1997) Development of a multigrid FDTD code for three-dimensional applications. IEEE Trans Antennas Propag 45:1512–1517
63. Iskander MF, Smith RL, Andrade AOM, Kimrey H, Walsh LM (1994) FDTD simulation of microwave sintering of ceramics in multimode cavities. IEEE Trans Microw Theory Tech 42:793–800
64. Georghiou GE, Ehlers RA, Hallac A, Malan H, Papadakis AP, Metaxas AC (2006) Finite elements in the simulation of dielectric heating systems. Springer, Berlin
65. Amri A, Saidane A (2003) TLM simulation of microwave hybrid sintering of multiple samples in a multimode cavity. Int J Numer Model-Electron Netw Devices Fields 16:271–285
66. Amri A, Saidane A (2004) TLM modelling of microwave sintering of multiple alumina samples. IEE Proc-Sci Measure Technol 151:259–266
67. Meng BS, Klein BDB, Booske JH, Cooper RF (1996) Microwave absorption in insulating dielectric ionic crystals including the role of point defects. Phys Rev B 53:12777–12785
68. Penn SJ, Alford NM, Templeton A, Wang XR, Xu MS, Reece M et al (1997) Effect of porosity and grain size on the microwave dielectric properties of sintered alumina. J Am Ceram Soc 80:1885–1888
69. Templeton A, Wang XR, Penn SJ, Webb SJ, Cohen LF, Alford NM (2000) Microwave dielectric loss of titanium oxide. J Am Ceram Soc 83:95–100
70. Bhattacharya M, Basak T (2008) Generalized scaling on forecasting heating patterns for microwave processing. AIChE J 54:56–73

71. Peng Z, Hwang J-Y, Andriese M, Bell W, Huang X, Wang X (2011) Numerical simulation of heat transfer during microwave heating of magnetite. ISIJ Int 51:884–888
72. Shukla AK, Mondal A, Upadhyaya A (2010) Numerical modeling of microwave heating. Sci Sinter 42:99–124
73. Mishra P, Sethi G, Upadhyaya A (2006) Modeling of microwave heating of particulate metals. Metall Mater Trans B-Process Metall Mater Process Sci 37:839–845
74. Lasri J, Ramesh PD, Schachter L (2000) Energy conversion during microwave sintering of a multiphase ceramic surrounded by a susceptor. J Am Ceram Soc 83:1465–1468
75. Santos T, Valente MA, Monteiro J, Sousa J, Costa LC (2011) Electromagnetic and thermal history during microwave heating. Appl Therm Eng 31:3255–3261
76. Spotz MS, Skamser DJ, Johnson DL (1995) Thermal-stability of ceramic materials in microwave-heating. J Am Ceram Soc 78:1041–1048
77. Gupta N, Midha V, Balakotaiah V, Economou DJ (1999) Bifurcation analysis of thermal runaway in microwave heating of ceramics. J Electrochem Soc 146:4659–4665
78. Alpert Y, Jerby E (1999) Coupled thermal-electromagnetic model for microwave heating of temperature-dependent dielectric media. IEEE Trans Plasma Sci 27:555–562
79. Wu X, Thomas JR, Davis WA (2002) Control of thermal runaway in microwave resonant cavities. J Appl Phys 92:3374–3380
80. Kriegsmann GA (1997) Cavity effects in microwave heating of ceramics. Siam J Appl Mathe 57:382–400
81. Alliouat M, Lecluse Y, Massieu J, Mazo L (1990) Control algorithm for microwave sintering in a resonant system. J Microw Power Electromagn Energy 25:25–31
82. Beale GO, Arteaga FJ, Black WM (1992) Design and evaluation of a controller for the process of microwave joinning of ceramics. IEEE Trans Industr Electron 39:301–312
83. Liu B, Marchant TR (2002) The occurrence of limit-cycles during feedback control of microwave heating. Math Comput Model 35:1095–1118
84. Rybakov KI, Semenov VE (1996) Densification of powder materials in non-uniform temperature fields. Philos Mag A-Phys Condens Matter Struct Defects Mech Prop 73:295–307
85. Birnboim A, Carmel Y (1999) Simulation of microwave sintering of ceramic bodies with complex geometry. J Am Ceram Soc 82:3024–3030
86. Su HH, Johnson DL (1996) Master sintering curve: a practical approach to sintering. J Am Ceram Soc 79:3211–3217
87. Su HH, Johnson DL (1997) A practical approach to sintering. Am Ceram Soc Bull 76:72–76
88. Chatterjee A, Basak T, Ayappa KG (1998) Analysis of microwave sintering of ceramics. AIChE J 44:2302–2311

Chapter 7
Sintering and Densification of Transparent Ceramics

7.1 Introduction

There are various sintering techniques that can be used to fabricate transparent ceramics. Conventional sintering techniques include vacuum sintering, hot pressing (HP), and hot isostatic pressing (HIP), while spark plasma sintering (SPS) is more popular than microwave sintering in the new sintering technique category. Every method has its own advantages and disadvantages. Different methods can be combined to offer higher sintering efficiency. The selection of sintering technique is also dependent on materials.

7.2 Vacuum Sintering

Vacuum sintering refers to a sintering process technique carried out with a vacuum equipment at a high vacuum level to achieve desired densification. Vacuum sintering has been widely used to fabricate various transparent ceramics, including garnet [1–7], spinel [8–11], alumina [12, 13], and rare earth sesquioxides ceramics, such as Y_2O_3 [14–18], Sc_2O_3 [19], pyrochlore oxides [20, 21], and so on.

When using vacuum sintering to fabricate YAG ceramics, the procedure usually consists of vacuum sintering followed by air annealing [22–38]. For example, transparent polycrystalline Nd:YAG ceramics can be synthesized by using solid-state reactive sintering the mixture of Al_2O_3, Y_2O_3, and Nd_2O_3 in a tungsten mesh-heated vacuum furnace (M60, Centorr Vacuum) at 5×10^{-6} Torr during holding [35]. Fully dense transparent samples with average grain sizes of ~ 50 μm were obtained at 1800 °C. The sintering temperature was slightly affected by the concentration of Nd. A small amount of SiO_2 doping lowered the sintering temperature by ~ 100 °C. SiO_2 is usually introduced by using tetraethyl orthosilicate (TEOS).

L.B. Kong et al., *Transparent Ceramics*, Topics in Mining, Metallurgy and Materials Engineering, DOI 10.1007/978-3-319-18956-7_7

After sintering, the specimens were annealed at 1400 °C for 2 h in air for decoloring.

The first YAG transparent ceramics for laser applications was prepared by using vacuum sintering [5]. High-purity fine Al_2O_3 and Y_2O_3 powders presynthesized by using wet chemical methods, with the composition of YAG, were mixed with ethyl silicate as a solution of 0.5 wt% in ethyl alcohol, by using ball milling. After milling, the dried powder mixture was compacted by using isostatic pressing at 140 MPa. The pressed pellets were sintered at 1600–1850 °C at 1.3×10^{-3} Pa. Optical properties, microhardness, and thermal conductivity of the YAG ceramics sintered at 1800 °C were comparable with those of YAG single crystals.

Highly transparent polycrystalline Er^{3+}:$Y_3Al_5O_{12}$ (Er:YAG) ceramics with Er^{3+} ion concentrations of 1–90 % were prepared by using solid-state reaction combined with vacuum sintering technique [22]. Commercial powders of α-Al_2O_3, Y_2O_3, and Er_2O_3 with purity of >99.99 % were thoroughly mixed by using ball milling, according to Er:YAG compositions with different contents of Er, together with a trace of TEOS as sintering aid. The powder mixtures were dry-pressed at 3T to form pellets with a diameter of 25 mm, which were isostatically cold-pressed at 250 MPa. The pellets were vacuum sintered at temperatures of up to 1800 °C for different time durations at 10^{-3} Pa, after which they were annealed at 1450 °C for 20 h in air.

Figure 7.1 shows photographs of the mirror-polished Er:YAG transparent ceramics with different Er contents of from 1 to 90 %. With increasing content of Er, the specimens appeared pinker and pinker, due to the light absorption of Er^{3+} at the visible band. All the Er:YAG ceramics exhibited excellent optical properties.

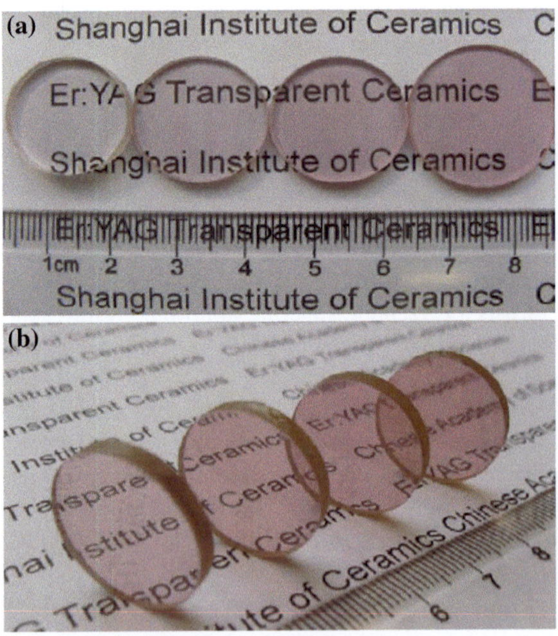

Fig. 7.1 Photographs of the mirror-polished Er:YAG ceramics with different contents of Er (from *left to right*): **a** 1, 5, 10, and 15 at.% and **b** 30, 50, 70, and 90 at.%. Reproduced with permission from [22]. Copyright © 2011, Elsevier

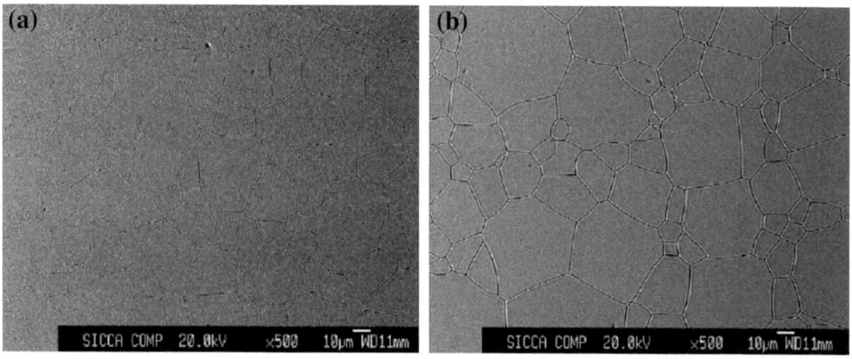

Fig. 7.2 Electron probe micro-analyzer (EPMA) images of the Er:YAG ceramics: **a** 1 % and **b** 90 %. Reproduced with permission from [22]. Copyright © 2011, Elsevier

EPMA images of the ceramics with 1 and 90 % Er are shown in Fig. 7.2. Both samples were very compact with very few pores. No secondary phases were observed both at the grain boundaries and in the inner grains. Average grain size of the 1 % Er:YAG ceramics was about 30 μm. The 90 % Er:YAG sample had a larger average grain size, probably due to the large particle size of starting Er_2O_3 powder. The high-resolution TEM (HRTEM) image indicated that the grain boundaries of the ceramics were clean and narrow, with a width of about 1 nm, thus leading to much negligible optical scattering.

It was found that there was a strong dependence of grain sizes on the location of the samples with green bodies formed by using uniaxial pressing with die, which actually indicated the presence of distributions in density of the green bodies. As mentioned before, cold isostatic pressing can only increase the density of the green body by <10 % and the density distribution of the green bodies would not be affected significantly. For two-directional pressing, the middle part of the green body possessed the lowest density, thus leading to largest grains after sintering. The grain size increased from the center part to the ends of the cylinder sample. For one-directional pressing, the density of the green body decreased from the top end to the bottom end. Figure 7.3 shows the scenario of one-directional pressing [22]. This observation is an agreement with previous discussion on ceramic powder packing.

Microstructural evolution during vacuum sintering of Nd:YAG transparent ceramics has been systematically studied [1]. Through microstructural observations, the microstructural maps and grain size density and grain size–pore size sintering trajectories of Nd:YAG ceramics, as a function of silica content, were established. For densities higher than 99.7 %, the occurrence of intragranular porosity was correlated to a critical pore radius of 0.16 μm. The presence of silica favored the formation of intragranular pores, which was attributed to the increase in the grain growth rate, as compared with the densification. An analytical model was established by coupling the analytical laws derived from sintering trajectories and the classical theory of light diffusion, which could be used to correlate the microstructural features of the transparent Nd:YAG ceramics to their optical properties.

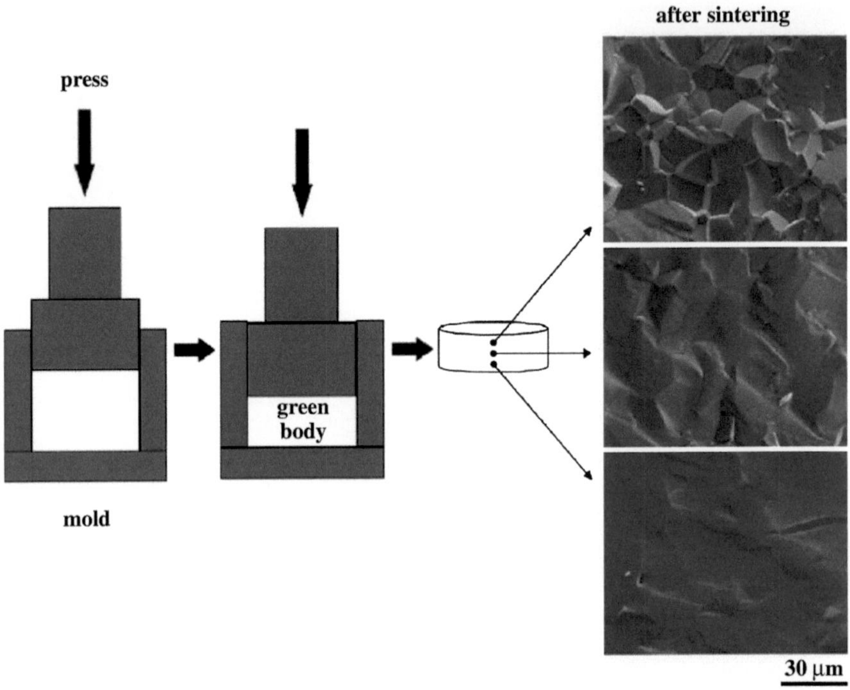

Fig. 7.3 Demonstration of the relationship between density and grain size distribution in a representative sample with green body formed by using one-directional axial press. Reproduced with permission from [22]. Copyright © 2011, Elsevier

Submicrometer α-Al_2O_3 (purity >99.99 %), Y_2O_3 (purity >99.99 %), and Nd_2O_3 (purity >99.99 %) powders were mixed to form the $Y_{2.94}Nd_{0.06}Al_5O_{12}$ (2 at.% Nd: YAG), by using ball milling in water with an organic dispersant and SiO_2 (purity >98 %) as sintering aid with contents of up to 0.3 wt%. Reactive sintering was conducted in a tungsten mesh-heated furnace at vacuum of $\leq 10^{-2}$ Pa and temperatures of 1000–1800 °C for different times durations, at heating/cooling rate of 5 °C min^{-1}.

Figure 7.4 shows SEM images of the undoped and 0.1 wt% SiO_2-doped Nd: YAG ceramics sintered at vacuum for 5 h at temperatures in the range of 1580–1730 °C [1]. Three types of microstructures could be observed, depending on the experimental parameters and the doping. Type one was porous microstructure, with the presence of open pores, having relative density (ρ) ranging from 60 to 90 % and submicrometer grain sizes of about 500 nm, as shown in Fig. 7.4a. Type two was dense microstructure, containing some closed intergranular pores, with relative density ranging from 90 to 99 % and micrometer grain sizes of about 1 μm, as illustrated in Fig. 7.4b. Type three was fully dense microstructure, with an average grain size of >1 μm, as seen in Fig. 7.4c–f. In the third-type microstructure, intragranular pores could occasionally be observed, as demonstrated in Fig. 7.4d. It

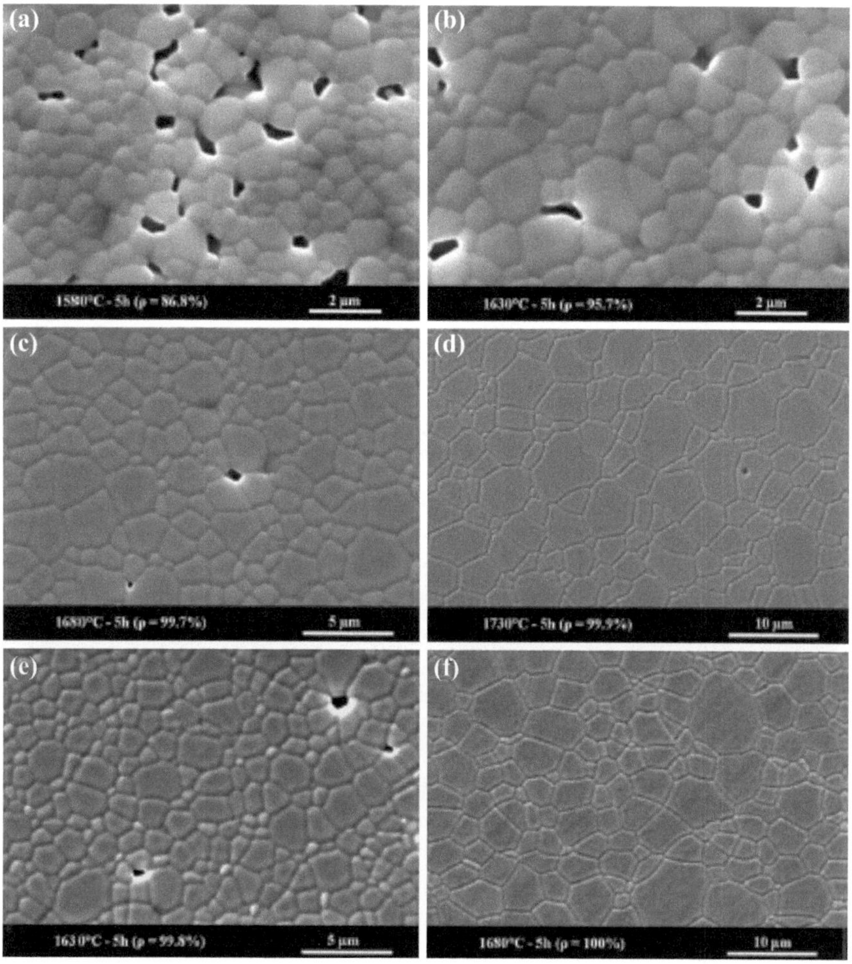

Fig. 7.4 SEM images of the undoped Nd:YAG ceramics sintered at vacuum for 5 h at temperatures of 1580 °C (**a**), 1630 °C (**b**), 1680 °C (**c**), and 1730 °C (**d**), as well as the 0.1 wt% SiO_2-doped Nd:YAG ceramics sintered for 5 h at 1630 °C (**e**) and 1680 °C (**f**). Reproduced with permission from [1]. Copyright © 2013, John Wiley & Sons

was shown that significant grain growth occurred at the final stage of the densification, i.e., with $\rho > 85$ %. The comparisons, between Fig. 7.4c–e, and d–f, indicated that the doping of 0.1 wt% of silica had no significant effect on overall microstructural profiles of the Nd:YAG ceramics, but could decrease the sintering temperature by about 50 °C correspondingly.

Sintering trajectories, i.e., grain size (G) versus relative density (ρ), of the undoped and silica-doped samples sintered for 2 h at 1450–1800 °C, are shown in Fig. 7.5 [11]. It was observed that grain size increased at a very slow rate until the

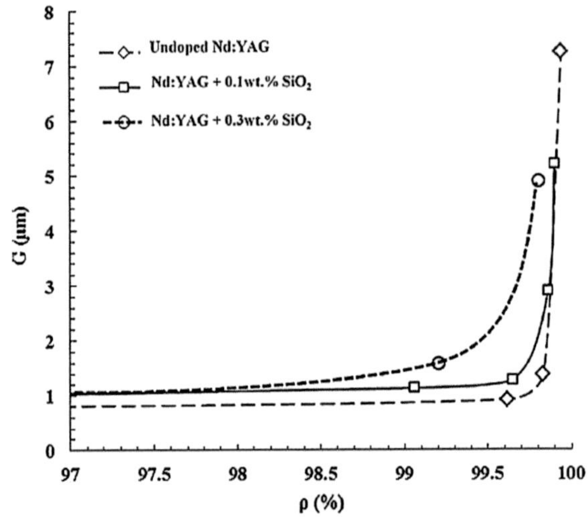

Fig. 7.5 *G–ρ* sintering trajectories of the undoped and silica-doped Nd:YAG ceramics sintered at vacuum for 2 h at temperatures ranging from 1450 to 1800 °C. Reproduced with permission from [1]. Copyright © 2013, John Wiley & Sons

relative density approached 99 %. However, after that, there was a sharp increase in grain size. The presence of silica could promote grain growth and densification.

Transmittance spectra of the samples with a thickness of 2.5 mm, labeled with their corresponding pore volume densities, are shown in Fig. 7.6a [11]. Obviously, the baseline of transmittance increased with decreasing volume density of the pores, demonstrating the optical scattering effect of pores in the Nd:YAG ceramics. Photographs of the corresponding Nd:YAG ceramic samples are shown in Fig. 7.6b. Analytical model indicated that the attenuation coefficient could be linked to the volume density of pores. It was concluded that once having similar values in pore volume density, e.g., $<3 \times 10^3$ cm^3, Nd:YAG ceramics could have similar optical properties of their single-crystal counterparts. This study provided a useful reference to other transparent ceramic materials.

Transparent alumina ceramics with high transmission have been developed by using a rapid vacuum sintering processing at a lower sintering temperature and shorter time duration, as compared with the conventional pressureless sintering [13]. The alumina sintered at 1670 °C for 5 min exhibited a residual porosity of as low as 0.002 %, thus showing a real in-line transmission of 64 % at of 1100 nm. The sintering time had an effect on the densification and transparency of the alumina ceramics sintered at 1650 °C.

A high-purity commercial α-Al_2O_3 (99.99 %) was mixed with 0.1 wt% MgO (99.99 %, 30 nm) as sintering aid by using ball milling with high-purity alumina balls. The milled and dried mixture was made into pellets with diameter of 15 mm and thickness of 2 mm at a uniaxial pressure of 200 MPa. The green pellets were presintered at 1200 °C for 2 h in muffle furnace and finally sintered in vacuum at 1500–1700 °C for 0–30 min, at a heating rate of 100 °C min^{-1}, which was followed by a natural cooling. All sintered samples were mirror-polished on both surfaces and thermally etched at 1400 °C for 1 h in air.

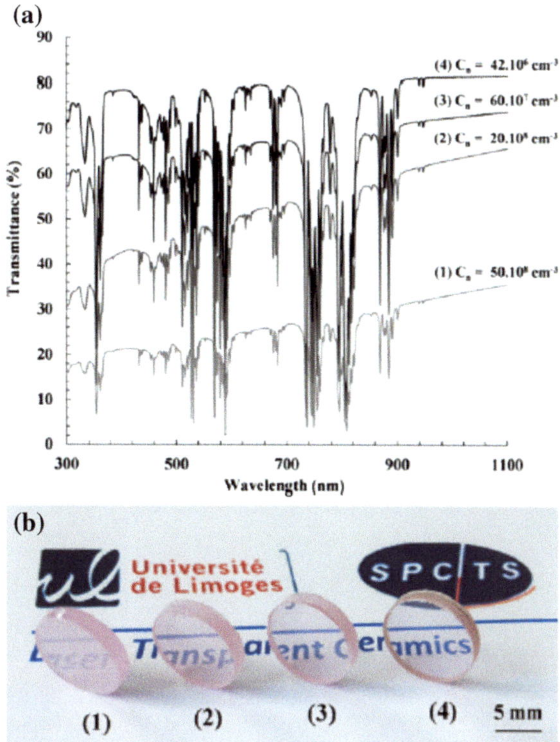

Fig. 7.6 **a** Optical transmittance spectra of the 0.3 wt% SiO_2-doped Nd:YAG transparent ceramics with different pore volume densities (C_n) after sintering at vacuum (samples thickness was 2.5 mm). **b** Photographs of the Nd:YAG ceramics corresponding to the samples in (**a**). Reproduced with permission from [1]. Copyright © 2013, John Wiley & Sons

SEM images of the ceramics sintered at different temperatures are shown in Fig. 7.7 [13]. Porosities of the samples sintered at 1570, 1600, 1650, and 1670 °C for 5 min were 1.256, 0.705, 0.075, and 0.002 %, with corresponding average grain sizes of 3.54, 5.47, 8.55, and 11.14 μm, respectively. Both the grain size and density of the samples were increased with increasing sintering temperature. A total of 1600 °C was the critical temperature for complete densification, above which all the samples demonstrated a homogeneous microstructure with clear grain boundaries and without abnormal grain growth.

Figure 7.8 shows SEM images of the samples sintered at 1650 °C for different time durations [13]. All samples exhibited very high density, homogeneous microstructure, and clear grain boundary, although residual pores at triple junction points of grain boundary were occasionally observed. It was found that prolonged sintering time resulted in an increase in pore size.

During the rapid vacuum sintering, the densification process of the transparent alumina can be attributed to the presence of a temperature gradient in the sample and the formation of nonequilibrium grain boundaries [39, 40]. The nonequilibrium grain boundaries had higher energy, thus providing a stronger driving force for grain growth. At the same time, due to the large thickness of the nonequilibrium grain boundaries, the grain-boundary mobility was decreased. Therefore, densification was maintained, while grain growth was suppressed.

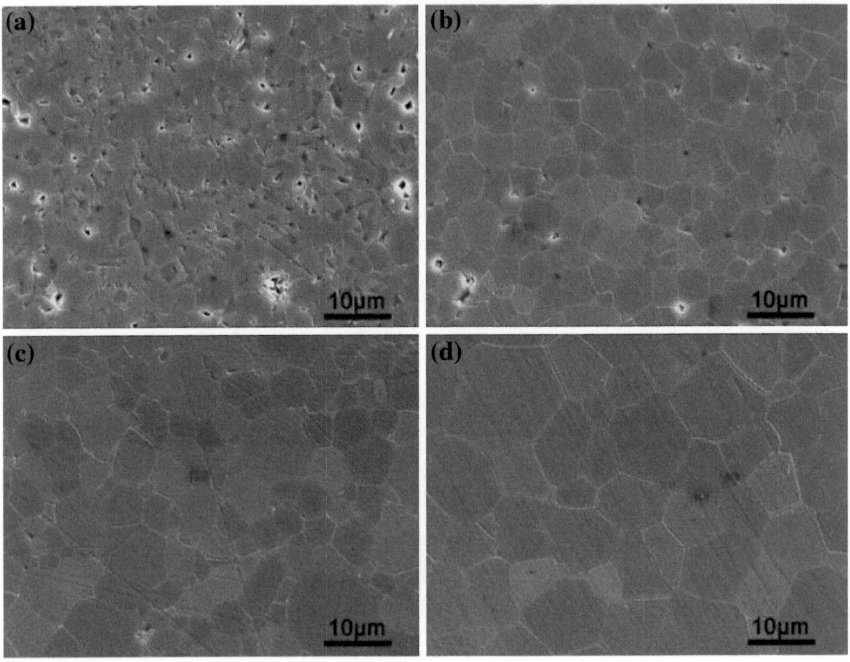

Fig. 7.7 SEM images of the alumina ceramics sintered for 5 min at different temperatures: **a** 1570 °C, **b** 1600 °C, **c** 1650 °C, and **d** 1670 °C. Reproduced with permission from [13]. Copyright © 2012, John Wiley & Sons

The mechanism for the improvement in transparency of vacuum sintered Y_2O_3 ceramics due to the addition of ZrO_2 has been analyzed [41]. Y_2O_3 (99.999 %) and ZrO_2 (99.99 %,) with an average particle size of 1 μm were mixed, according to the formula of $(Y_{1-x}Zr_x)_2O_3$ (x = 0–0.10), by using a ball milled in absolute ethyl alcohol with agate balls. After milling and drying, pellets with 20 mm in diameter were formed with stainless steel at 15 MPa and then isostatically pressed at 250 MPa. Sintering was conducted at 1800 °C for 15 h in vacuum.

Combined with sintering kinetics, thermodynamics, and experimental results, the mechanism of ZrO_2 in improving the transparency of Y_2O_3 ceramic was clarified. The addition of ZrO_2 led to the formation of ZrO_2–Y_2O_3 binary system with decreased melting point, thus reducing the sintering temperature. The radius of Zr^{4+} is 0.80 Å, which is close to that of Y^{3+} (0.90 Å), so that there was only very small lattice distortion when Y^{3+} was substituted with Zr^{4+}.

At the final stage of sintering, grains with large size and low surface energy tend to grow at the expense of the surrounding smaller grains, which is known as discontinuous or exaggerated grain growth. During this process, due to the high mobility of grain boundaries, pores are easily trapped in grains, which is harmful to

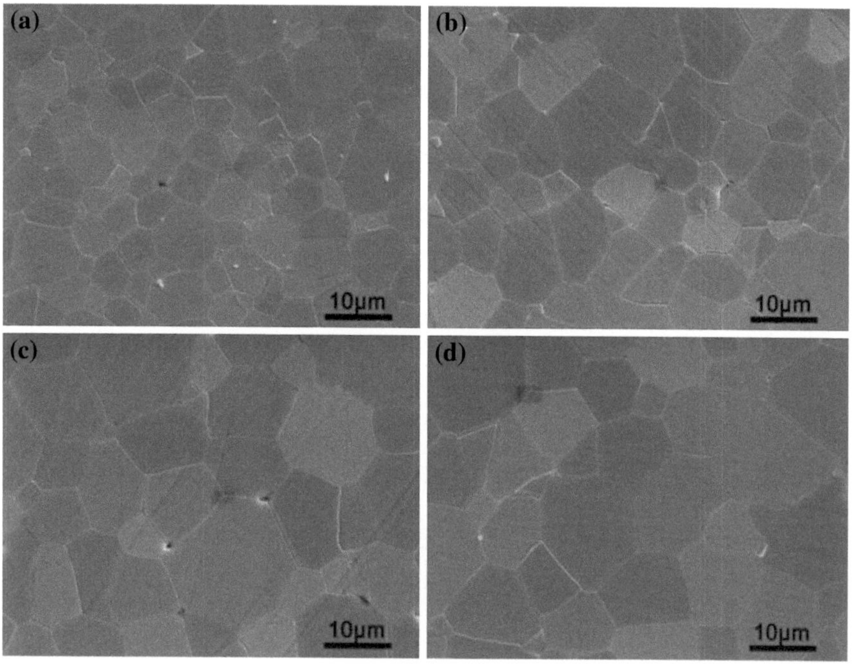

Fig. 7.8 SEM images of the alumina ceramics sintered at 1650 °C for different time durations: **a** 0 min, **b** 10 min, **c** 20 min, and **d** 30 min. Reproduced with permission from [13]. Copyright © 2012, John Wiley & Sons

the transparency. Due to its high melting point (2715 °C), ZrO_2 is prone to form solid solution at the grain boundaries, so as to reduce the grain-boundary mobility. At the same time, secondary recrystallization could occur, due to the solute drag mechanism, which would be beneficial to the elimination of pores and the homogenization of grain sizes, two critical factors to ensure high optical transmittance of the Y_2O_3 ceramics.

It is well known that the diffusion coefficient of O^{2-} is much higher than that of Y^{3+} in pure Y_2O_3. Therefore, cation diffusion is the rate-controlling step of grain-boundary migration, i.e., the diffusion of $[Y_i^{\cdot\cdot\cdot}]$ dominates the grain-boundary mobility [42]. There are following defects and defect reactions in Y_2O_3.

For Schottky defects, there is:

$$Y_2O_3 \Leftrightarrow 2V_Y^{\prime\prime\prime} + 3V_O^{\cdot\cdot}. \quad (7.1)$$

For Frenkel defects, there are:

$$O_O \Leftrightarrow O_i^{\prime\prime} + V_O^{\cdot\cdot}, \quad (7.2)$$

$$YY \Leftrightarrow V_Y''' + Y_i^{\cdots}. \tag{7.3}$$

After doping with ZrO_2, there is:

$$2ZrO_2 \longrightarrow Y_2O_3 2Zr_Y^{\cdot} + O_i'' + 3O_O. \tag{7.4}$$

As a result, when every two Zr^{4+} ions were substituted, one O_i'' was produced, so that $[V_Y''']$ was increased, while $[Y_i^{\cdots}]$ was decreased. The decrease in $[Y_i^{\cdots}]$ resulted in a reduction in grain-boundary mobility and sintering rate, which ensured the escape of pores and densification of the ceramics. For example, average grain size of the ZrO_2-doped ceramics was about 15 μm, which was much smaller than those widely reported in the open literature. In terms of optical performance of the transparent ceramics, 3 at.% ZrO_2 was the optimized doping concentration.

7.3 Hot Pressure (HP) Sintering

Hot pressure (HP) sintering is a high-pressure low-strain-rate powder metallurgy process, which is used to form a powder or powder compact at a temperature that is sufficiently high to induce sintering and creep processes. This is achieved by the simultaneous application of heat and pressure. Densification at high pressure proceeds through the rearrangement of particles and plastic flow at the particle contacts. HP is mainly used to fabricate hard and brittle materials. As a result, HP has become an important technique to obtain transparent ceramics.

In 1977, transparent spinel with small grain size (1–2 μm) was prepared by using low-temperature (1400 °C) and high-pressure (70 MPa) hot pressing [43]. Transparent magnesia–alumina spinel ceramics were prepared from equimolar mixture of the oxides, which were derived from $Mg(OH)_2$ and $Al(OH)_3$ after calcining at 900 °C for 2 h and 1400 °C for 2 h, respectively. Formation of spinel from the oxide mixture was accompanied with a volume expansion of about 7.9 %, which was suppressed by applying a pressure of >330 kg cm^{-2} at 950 °C, so as to promote the densification of the spinel ceramics. A step heating schedule was used, i.e., hot-pressing technique at 770 kg cm^{-2} at 1300 and 1350 °C for 0.5 h and 1400 °C for 1 h. The transparent spinel ceramics exhibited optical transmittance of 35–75 % in the visible range and 75–85 % in the IR range.

Since then, HP became a widely used technique to prepare transparent spinel ceramics. For example, HP process was used to fabricate $MgAl_2O_4$ transparent ceramics, with a nearly 100 % densification [44]. Spinel powders for the production of transparent polycrystalline ceramic windows were produced by using both the traditional ceramic solid-state reaction and solgel methods. Among them, the powders produced from the reaction of organomagnesium compounds with surface modified boehmite precursors were best to produce high-quality transparent spinel ceramics. The powder synthesis method allowed fine control over the particle size,

size distribution, purity, and stoichiometry. The process involved the formation of a boehmite solgel from the hydrolysis of aluminum alkoxides. The boehmite nanoparticles were treated through a surface modification with carboxylic acids. The surface modified boehmite nanoparticles were then used to make a precursor powder of pure phase spinel through metal exchange at room temperature with magnesium acetylacetonate.

Different from the conventional sintering process which usually requires high sintering temperatures (>1600 °C), HP process is able to achieve desired densification at relatively low temperature, which is therefore also called low-temperature high-pressure (LTHP) process. Due to its low-temperature requirement, it can be used to develop nanosized ceramics [45–48]. The sintering mechanism at high pressures is entirely different from that at ambient pressure [45]. High pressure can suppress the grain growth and initiate plastic deformation to eliminate pores and/or additional phases existing in triple junctions of the grains. However, the conventional sintering process is controlled by grain growth to avoid imperfections between grain boundaries when preparing transparent ceramics.

High-resolution transmission electron microscopy (HRTEM) and selected area electron diffraction (SAED) of two selected samples, sintered at 4.0 GPa at 600 °C and 4.0 GPa at 1100 °C, are shown in Fig. 7.9 [45]. The average grain size observed from Fig. 7.9a was about 40 nm. The nanostructured profile of the sample was confirmed by the corresponding SAED pattern. The HRTEM image showed a clear grain boundary without any secondary phase and/or in-between phases, as demonstrated in Fig. 7.9b. The clean grain boundaries had no contribution to scattering, which was responsible for the high transparency of the $MgAl_2O_4$ ceramics sintered at 4.0 GPa at 600 °C.

As the sintering temperature was increased to 1100 °C, the grain/crystallite size increased significantly, so that the nanostructured profile was lost due to the grain growth, as illustrated in Fig. 7.9c. It is also found that the crystals have a wider particle size distribution, probably due to the presence of a temperature gradient in the sample. Consequently, imperfections, nonuniform crystallites and irregular grain boundaries, were present, which would pose negative effects on optical transparency of the ceramics. The HRTEM image of $MgAl_2O_4$ ceramics sintered at 4.0 GPa at 1100 °C taken along the [332] zone axis, shown in Fig. 7.9d, together with the corresponding SAED pattern, indicated a highly crystallized feature of the sample, demonstrating the grain growth of the materials at 1100 °C.

Therefore, the formation temperature of the transparent $MgAl_2O_4$ spinel ceramics could be significantly reduced, if sufficiently high pressure is available. As the pressure is in the range of 2–5 GPa, very low sintering temperatures of 500–700 °C can lead to nanosized transparent spinel ceramics [46]. The nanosized ceramics were highly transparent even though their relative densities are less than 99 %, due to the low or negligible light scattering from the nanosized grains and pores. The LTHP process is able to densify nanosized powders without significant grain growth, which becomes a key advantage for the development of transparent ceramics, as discussed above. It is of special interest when mechanical strength is an important requirement.

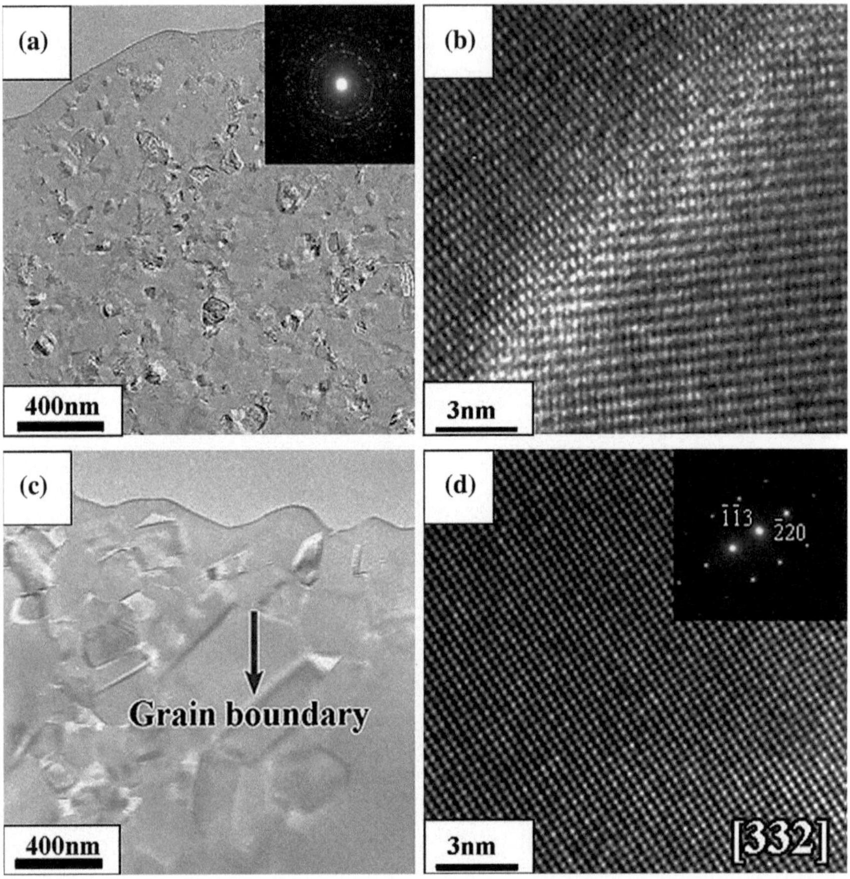

Fig. 7.9 **a** TEM image showing microstructure of the $MgAl_2O_4$ nanocrystalline transparent ceramics sintered at 4.0 GPa and 600 °C for 30 min. The corresponding SAED pattern in the inset proved the characteristics of the polycrystalline nanograins. **b** HRTEM image of the $MgAl_2O_4$ nanoceramics. **c** TEM image showing microstructure of the $MgAl_2O_4$ ceramics sintered at 4.0 GPa at 1100 °C for 30 min, with an average grain size of the smaller crystallites to be ~500 nm. **d** HRTEM image of the $MgAl_2O_4$ ceramics sintered at 4.0 GPa/1100 °C taken along the [332] direction, with the corresponding SAED pattern shown as the inset. Reproduced with permission from [45]. Copyright © 2010, Elsevier

Another example is the development of a nanocrystalline spinel ceramics with 50 % increase in hardness by using hot pressing [49]. It is well accepted that mechanical strength and hardness of a material can be increased by decreasing its grain size, as described by the empirical Hall–Petch relationship [50, 51]. Based on this principle, an integrated approach was developed by using nanosized spinel powder, combined with high-pressure and low-temperature sintering to fabricate fully dense and high-purity nanocrystalline ceramics with nanometer-sized grains. The Hall–Petch relationship was confirmed by the hardness of the transparent spinel ceramics with grain sizes down 28 nm. As a result, the nanosized spinel ceramics exhibited a 50 % increase in hardness without a decline in fracture resistance. More importantly, the nanocrystalline spinel ceramics had near theoretical optical transparency.

Figure 7.10 shows a schematic diagram of the high-pressure high-temperature sample holder [49]. Spinel powders were pressed into green compacts without binder within glove box in an evacuable die at 1900 kg cm^{-2}. The ejected green compacts were mechanically sealed in a metal capsule. The sealed sample capsule was inserted into the high-pressure assembly, which consisted of a partially sintered ZrO_2 container saturated with CsCl in the shape of a rectangular cuboid with truncated edges, a graphite heater, mica, and molybdenum heater contacts. K-type thermocouples were inserted through both high-pressure cell assembly lids. A pressed boron nitride tablet isolated the exposed thermocouple junction from the metal sample capsule. High-pressure experiments were performed in a pressless split-sphere apparatus (BARS) equipped with an 8–6-type multi-anvil system. After about 25 min pressurization to 2 GPa, samples were sintered at temperatures from 740 to 845 °C for 15 min. The nanosized spinel ceramics are transparent in the visible spectrum, without major absorption bands from ultraviolet to short-infrared wavelengths. The maximum transmission was about 81 % at infrared wavelengths of 1200–1350 nm.

The fact that the ceramics were fully dense and of high quality, without residual porosity or obvious grain growth, was further confirmed by the SEM and TEM images, as shown in Figs. 7.11 and 7.12 [49]. SEM images of the polished surface of the ceramics derived from the 200 nm powder and the same 30.8 nm nanocrystalline ceramics are shown in Fig. 7.11a, b, respectively. Any potential variability in the sintering environment has been eliminated, because the samples were sintered in the same sample capsule at 2 GPa and 795 °C. Fine pores were present in the submicron sample, while very few features, which included a surface scratch and dust particles, were present at similar magnification for the nanocrystalline

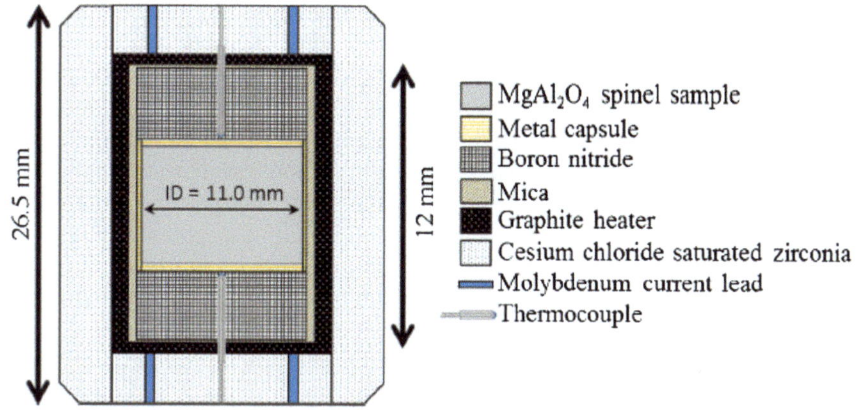

Fig. 7.10 Schematic of the high-pressure cell assembly. Reproduced with permission from [49]. Copyright © 2014, Elsevier

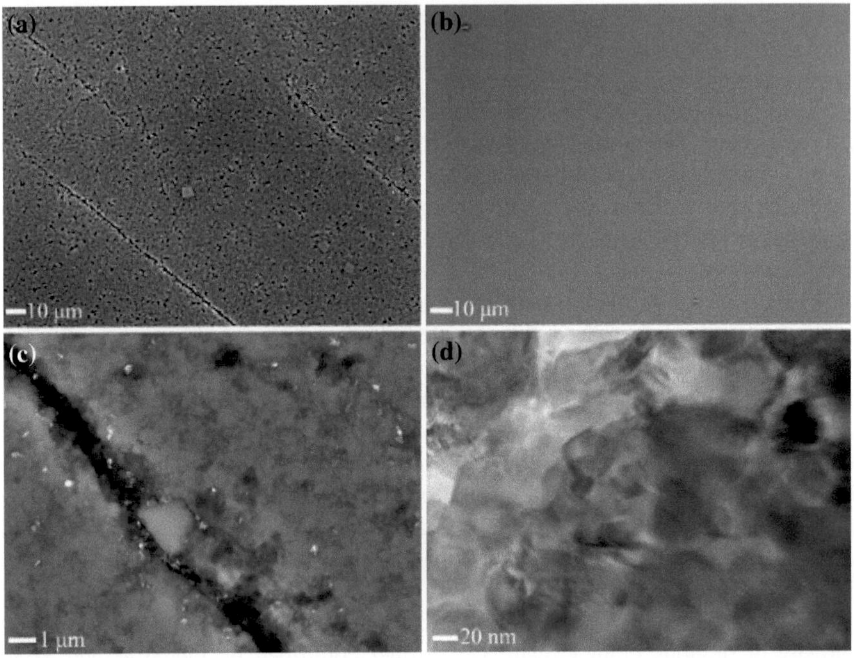

Fig. 7.11 Microstructures of the porous and fully dense spinel ceramics produced by sintering at 2 GPa and 795 °C. **a** and **c** SEM micrographs of the porous submicron structure of spinel ceramic derived from 200-nm-sized spinel powder. **b** SEM and **d** TEM images of the microstructure of the fully dense nanosized spinel ceramics. Reproduced with permission from [49]. Copyright © 2014, Elsevier

ceramics. At higher magnifications, a submicron grain structure was observed in the coarse ceramics and abnormally large grains were present along with the porous structure. The grain structures of the nanocrystalline ceramics are clearly demonstrated by the TEM images, as shown in Figs. 7.11d and 7.12.

Figure 7.12 shows TEM images of different grain-boundary regions of the nanosized spinel ceramics [49]. There were nanoscale grains where the crystalline lattice extended into the boundary, as shown in Fig. 7.12a, d. Also, grain boundaries decorated with a small, ~ 1 nm, amorphous phase region were observed, as shown in Fig. 7.12b. Such amorphous phases were also present at the junctions of three grains, as shown in Fig. 7.11c. Noticing the highly transparent characteristics of the nanosized sample, the slight fraction of amorphous phase at the grain boundaries had negligible influence on its optical properties and density.

Similar process was also used to develop yttrium aluminum garnet ($Y_3Al_5O_{12}$, YAG) transparent ceramics with nanosized grains [52]. In this study, a high-pressure cell was used, where disk samples were placed in a capsule made of NaCl, both to ensure the quasi-hydrostatic compression of the sample and prevent its contact with the graphite heater. The whole structure was then contained in the pyrophyllite high-pressure cell. The sample temperature in the high-pressure cell

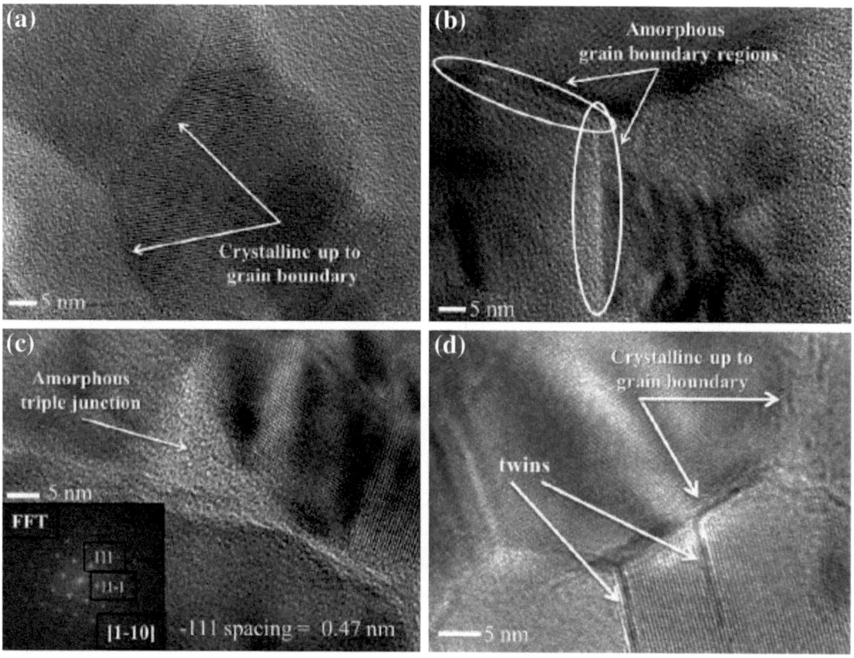

Fig. 7.12 TEM images of the fully dense nanosized spinel ceramics produced by sintering at 2 GPa and 795 °C. **a** and **b** Different areas revealing that some grain boundaries contain ~1-nm amorphous phase regions. **c** An amorphous triple junction with the inset showing the fast Fourier transform of the main figure. **d** Image from a region containing nanotwins. Reproduced with permission from [49]. Copyright © 2014, Elsevier

was measured directly by using a standard Ni–Cr/Ni–Si thermocouple. The cell pressure was monitored by using a calibrated oil pressure meter. Desired pressures were first applied to the samples before raising the temperature. The pressure range was 2.0–5.0 GPa, and temperature range was 300–500 °C, with holding time of 30 min. After that, the pressure was released first, and then, the samples were cooled down to room temperature at a rate of 15 °C min^{-1}. By doing in this way, the residual stress of the samples could be released in order to avoid the happening of cracking. Optimized processing parameters for transparent YAG ceramics were 5 GPa and 450 °C.

One of the characteristics of HP is freedom to optimize the sintering parameters, which has been demonstrated when HP sintering was used to fabricate transparent Y_2O_3 ceramics [53, 54]. For instance, stepwise process has been reported to fabricate Eu:Y_2O_3 transparent ceramics [54]. The program included a high-temperature ramp at constant pressure and a pressure ramp at constant high temperature. After a dwell at 1150 °C to outgas the samples, a pressure of 10 MPa was applied, followed by a temperature ramp to 1580 °C at a heating rate of 8 °C min^{-1}. After a 30 min dwell at constant temperature and pressure, the pressure was raised to 40 MPa at a rate of 0.4 MPa min^{-1}. With this stepwise application of pressure, optimized temperature was 1580 °C.

HP sintering process is a very unique and feasible technique to prepare transparent electro-optic ceramics, including PLZT [55–57], PZN–PLZT [58] and PMN–PT [56, 59–61], and other ferroelectric ceramics [62], which contain volatile element Pb. In the early 1970s, (Pb, La)(Zr, Ti)O_3 (PLZT) transparent ferroelectric ceramics were synthesized by using HP [43]. The transparent PLZT ferroelectric ceramic materials have found a variety of electrooptic applications. They were prepared from mixed oxides by using hot-pressing sintering at 1100 °C for 16 h at 2000 psi. Transmission measurements in the visible and infrared showed that these materials exhibited a nearly constant response from the absorption edge of 0.37 μm to ~6 μm. The highest transmission values were observed for compositions containing >8 at.% La. Specific compositions exhibited electro-optic memory, conventional linear, and quadratic electrooptic effects, with performances comparable with those of single crystals.

Highly transparent ceramics of La-doped $0.75Pb(Mg_{1/3}Nb_{2/3})O_3$–$0.25PbTiO_3$ (PMN–PT) were fabricated by using a two-stage sintering method [59]. Green pellets were first sintered in an oxygen atmosphere (OA) at different temperatures, which were then hot pressed (HP) at temperatures of >1000 °C for more than 8 h at pressures of 50–100 MPa. The ceramics prepared in this way exhibited a transparency of as high as 65 % at infrared wavelength. Large quadratic electro-optic coefficient of 66×10^{-16} $(m/v)^2$ was obtained, which was the highest value reported in the literature for the ceramics with similar compositions.

The precursor powder with a composition of La-doped PMN–PT 3/75/25 was synthesized by using the columbite precursor method, starting with PbO, MgO, Nb_2O_5, TiO_2, and La_2O_3, in which small amount of excessive PbO was used to promote the densification through liquid-phase sintering. The powder was made into pellets by using uniaxial pressing. The pellets were first sintered in an OA at different temperatures. The sintered pellets were then hot pressed (HP) at a temperature of >1000 °C for more than 8 h at pressures of 50–100 MPa. In the OA–HP two-stage sintering method, the OA sintering temperatures were different: (A) 1150 °C, OA–HP; (B) 1200 °C, OA–HP; and (C) 1230 °C, OA–HP.

The PMN–PT ceramics exhibited a transparency of as high as 65 % in infrared region, as shown in Fig. 7.13a. SEM images of PMN–PT 3/75/25 ceramics sintered with different schedules are shown in Fig. 7.13b, c. All samples showed transgranular fracture character, which was attributed to the increased homogeneity between grain boundary and grain, as well as firm conjunct strength of grains caused by the liquid phase that was formed during the sintering. It was found that the sample with more transgranular fracture had higher transparency, which suggested that grain boundary with an increased homogeneity and firm strength would have reduced reflection and scattering of light. This observation once again indicates that the quality of grain boundary plays a more important role in determining optical properties of transparent ceramics, as discussed previously.

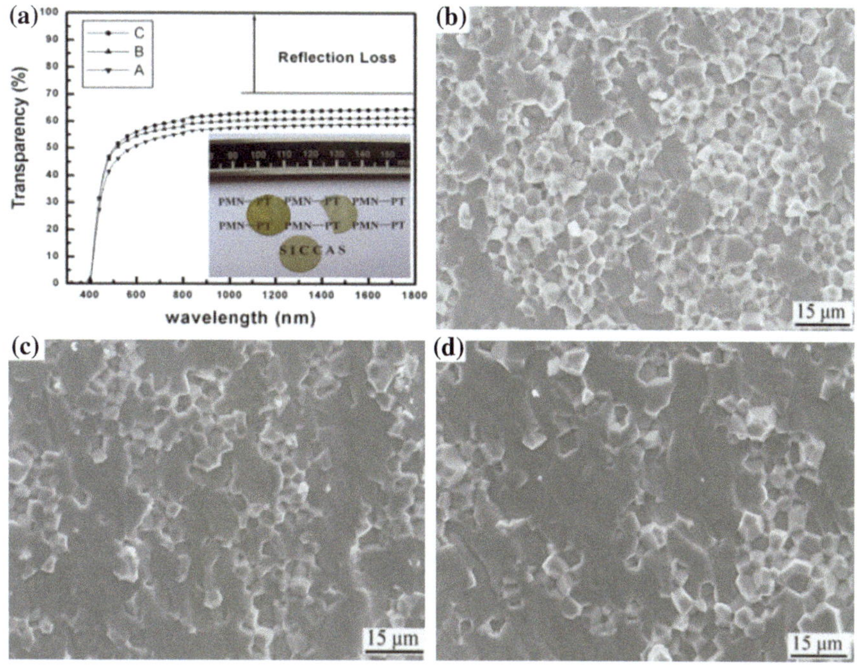

Fig. 7.13 Transmission curves of the PMN–PT 3/75/25 ceramics with a thickness of 0.5 mm (**a**) and SEM images of the samples sintered with different schedules: **b** sample A, **c** sample B, and **d** sample C. Reproduced with permission from [59]. Copyright © 2010, John Wiley & Sons

HP process was found to be the most feasible method to prepare transparent or translucent α-sialon ceramics [63, 64], which could not be achieved by using the conventional pressless sintering. Single-phase α-sialon ceramics with high optical transmittance have been prepared by using hot pressing [63]. HP was also used to tailor grain morphologies and preferential texturing microstructures of α-sialon ceramics [64–66]. Other examples include translucent MgO ceramics fabricated by using hot-pressing, with nanopowder of MgO containing 2–4 % LiF as a sintering aid [67]. HP process has even been used to prepare fluoride ceramics (CaF_2), such as transparent Yb:CaF_2 ceramics [68, 69].

7.4 Hot Isostatic Pressure (HIP) Sintering

Hot isostatic pressing (HIP) is a fabrication process that is used to reduce the porosity of metals and increase the density of various ceramic materials [70–74]. The HIP process subjects a component to both elevated temperature and isostatic gas pressure in a high-pressure containment vessel. The chamber is heated, causing the pressure inside the vessel to increase. Many systems use associated gas pumping to achieve the desired pressure levels. Pressure is applied to the material from all directions.

Materials are hot isostatically pressed (HIPed) in order to achieve the maximum possible densification, which has become a key technique to achieve high optical transmittance. To reduce the fabrication cost, hot isostatic pressing is usually used as the last step, although it has been proved to be a critical step to prepare high-quality transparent ceramics. This is simply because HIP involves sophisticates equipment and critical experimental conditions.

HIP process has been widely used to synthesize transparent armor ceramics, such as alumina (Al_2O_3) [75–79], YAG [80–83], and spinel ($MgAl_2O_4$) [84–88]. HIP process can be combined with HP. Hot pressing followed by hot isostatic pressing (HIP) proved to be more feasible to fabricate transparent $MgAl_2O_4$ ceramics was widely used [84, 85, 88]. Nowadays, HIP process is also used to synthesize cubic sesquioxide ceramics, including Y_2O_3 [89], Sc_2O_3 [90], and Lu_2O_3 [91]. In these cases, vacuum sintering to remove closed pores with a subsequent HIP step was usually employed, which provided an alternate processing route to the fabrication of fully dense sesquioxide ceramics with less possibility for contamination and reduction than hot pressing.

A simple method has been established to achieve nearly 100 % relative density of α–alumina ceramics by using hot isostatic pressing (HIP) [75]. In this study, transparencies of the samples were correlated to their grain size and residual porosity. A commercially available α-alumina powder was wet milled in order to realize de-agglomeration. After that, the average particle size of the powder was 100–150 nm. Suspensions with ~75 wt% solid with any organic additive were cast with a nylon filter (pore diameter = 0.2 μm) to form green bodies. The wet bodies were dried at 65 °C for 15 h and then calcined at 600 °C for 30 min in vacuum of <1 mbar. Then, the samples were sintered naturally in vacuum of $\sim 10^{-6}$ mbar at temperatures of about 1200 °C for 2–17 h. The samples with porosities of <10 % were subject to HIP at about 1200 °C at 170 MPa in argon for 3–5 h. The HIP furnace was made of molybdenum. The densities of the HIP processed samples were measured with a very precise method, by measuring mass of the ceramics outside and inside water, which were then compared with the density of alumina single crystal to obtain relative densities.

Figure 7.14 shows porosity values of the samples after natural sintering in vacuum versus the treatment temperature and time [75]. With a constant treatment time, the porosity always increases as the sintering temperature is decreased. It was found in this work that if the porosity was higher than 8 %, the HIP treatment was not workable. The samples sintered at 1200 and 1220 °C; time duration of 2 h was sufficient. However, the samples sintered at 1170 and 1185 °C should be kept for 17 h to decrease the porosity to be <4 %.

In addition, the treatment temperature and time also showed their effects on grain size of the samples, as shown in Fig. 7.15 [75]. As stated earlier, the grain size of the starting powder was 100–150 nm. After the treatments at 1200 and 1220 °C for 2 h, the grain size was <400 nm, which was close to that of the sample sintered at 1170 °C for 17 h. However, after sintering at 1185 °C for 17 h, the grain size was

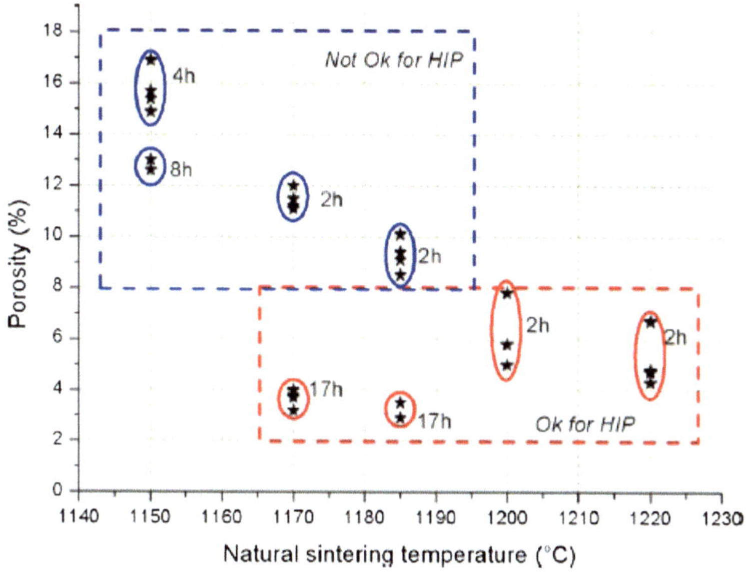

Fig. 7.14 Total porosity versus the natural sintering temperature, for different treatment times, at the pressure of ~10^{-6} mbar. Reproduced with permission from [75]. Copyright © 2011, Elsevier

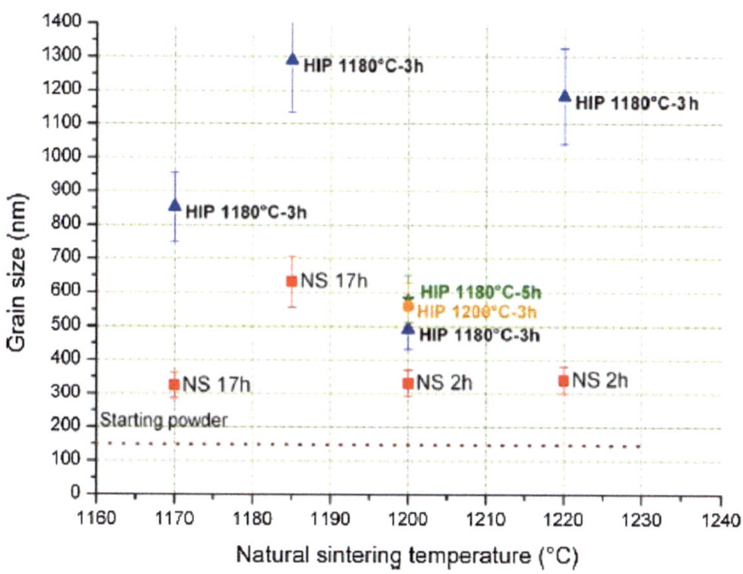

Fig. 7.15 Grain sizes of the ceramics versus the natural sintering (NS) temperature: (*square*) after natural sintering, (*triangle*) after HIP at 1180 °C and 170 MPa for 3 h, (*disc*) after HIP at 1200 °C and 170 MPa for 3 h, and (*star*) after HIP at 1180 °C and 170 MPa for 5 h. The HIPed ceramic values are placed versus the temperature of the corresponding natural sintering. Reproduced with permission from [75]. Copyright © 2011, Elsevier

>600 nm. The HIP treatment led to an increase in grain size. The increase in grain size after the HIP processing was simply based on the grain size of the samples after the natural sintering. For a given natural sintering temperature, the variation in the HIP condition had relatively less effect on the grain size as compared with that of the natural sintering.

Figure 7.16 shows relative densities of the samples [75]. It was observed that the effect of the natural sintering temperature was also quite obvious. At the same HIP condition, the higher the natural sintering temperature, the weaker the effect of HIP on the relative density would be. The samples naturally sintered at 1170 and 1185 °C followed by HIP at 1180 °C for 3 h and the sample sintered at 1200 °C followed by HIP at 1180 °C for 5 h exhibited almost the same relative density of 0.9985, while the different in their grain sizes was quite large, i.e., between 580 and 1300 nm. Therefore, this achievement implies that high relative densities can be realized, while grain growth can be well controlled at the same time. It is believed to be also applicable to other transparent ceramic materials.

Highly transparent 8 mol% Y_2O_3–ZrO_2 (8YSZ) ceramics were fabricated by an improved hot isostatic pressing method [92]. Combined with presintering treatment, microstructures, such as positional difference in residual pores, inter-, or intra-granular pores, of the 8YSZ ceramics could be well controlled. Commercially available 8 mol% Y_2O_3–ZrO_2 powder was used as a starting material. Green bodies were formed by using cold isostatic pressing method. Pellets with diameter of 20 mm and thickness of 3 mm were made by die pressing at 50 MPa and then isostatically cold-pressed at 200 MPa. The green bodies were presintered in an

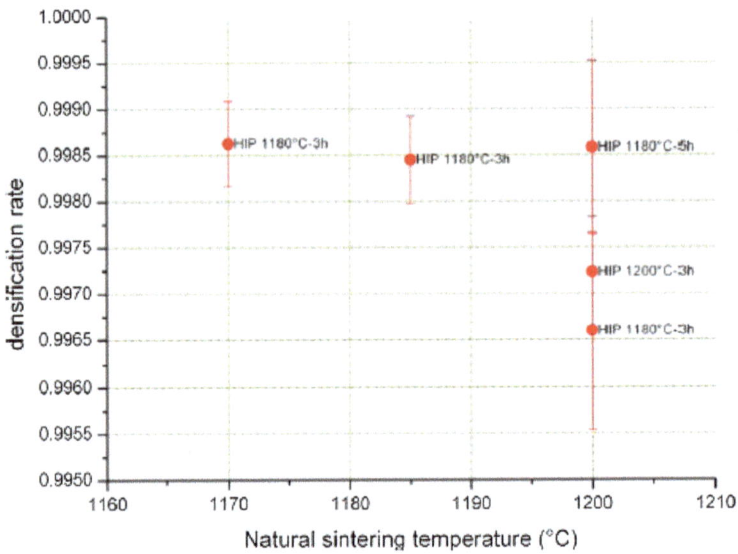

Fig. 7.16 Relative density (τ) after HIP treatment versus the natural sintering temperature. The HIP parameters are indicated for each data. Reproduced with permission from [75]. Copyright © 2011, Elsevier

alumina tube furnace in air. The heating rate was 100 °C h^{-1}, and the holding time was 2 h at temperatures of <1650 °C. The presintered samples were finally treated by using hot isostatic pressing, at temperatures of 1350–1750 °C for 1 h in argon gas at pressure of 150 MPa. The temperature and pressure were simultaneously raised at a heating rate of 500 °C h^{-1}.

According to the experimental results, a microstructure model for pore elimination during the hot isostatic pressing has been proposed to illustrate microstructural evolution of a green body, as shown in Fig. 7.17 [92]. At low presintering temperatures, small intergranular pores are surrounded with fine grains. The intergranular pores transform to intragranular ones when the presintering temperatures are sufficiently high to facilitate grain growth. It is suggested that such intergranular pores are easily removed during the hot isostatic pressing. On the one hand, fine grains have large plastic deformability, which is favorite for pores to shrink at the hot isostatic pressures [93]. On the other hand, it is well known that high-temperature plastic deformability is increased with decreasing grain size [94, 95]. Also, the rapid migration closed gas is through grain boundaries. Small grain size means more grain boundaries, thus being beneficial to the shrinkage of pores. The model has been found to be applicable to many other oxide ceramics [96, 97].

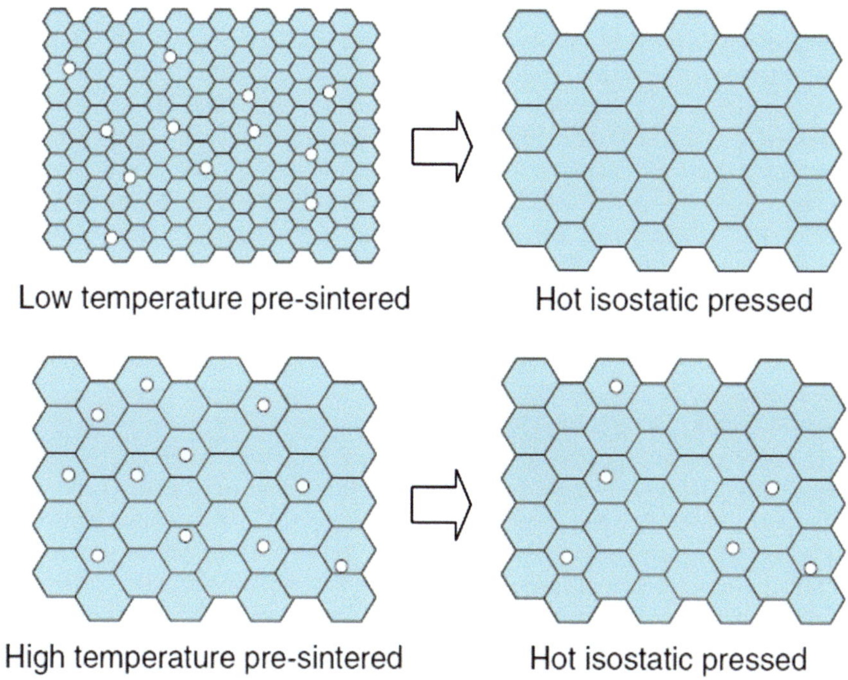

Fig. 7.17 Microstructure model for pore elimination during the sintering of hot isostatic pressing. Reproduced with permission from [92]. Copyright © 2008, John Wiley & Sons

The feasibility of HIP can be further demonstrated by the following examples. Fully dense transparent garnet ceramics were derived cold-pressed green bodies that were subsequently vacuum sintered, with residual porosity being removed by hot isostatic pressing [98]. It was reported that hot isostatic pressing (HIPing) could be used to enhance the electrical and optical properties of PNNZT (50.0/15.5/34.5) [99]. After hot isostatic pressing at 1100 °C, the density was increased to >99 %. Transparent mullite ceramics could be obtained by pressureless sintered combined with hot isostatically pressed (HIP) [100]. In the HIPed mullite, with porosity of <1 %, a transmittance of 40 % was observed in the VIS range and up to 80 % in NIR. Transparent Yb:CaF_2 ceramics could be fabricated by sintering and hot-pressing powders that were derived by a soft chemistry process [101]. There are also reports to use HIP process to consolidate hydroxyapatite (HAp) filter-cakes [102]. HAp ceramics were fully densified at 800 °C after 2 h when a hot isostatic pressure of 100 MPa was used. Besides, transparent α-sialon ceramics [103], transparent magnesium oxide (MgO) ceramics [104], and transparent $La_2Hf_2O_7$ (LHO) ceramics [105] were also fabricated by using HIP process.

7.5 SPS Processed Transparent Ceramics

Various transparent ceramics, such as alumina (Al_2O_3) [106–112], yttrium aluminum garnet (YAG) [113, 114], sesquioxides [115–120], ZrO_2 [121–123], MgO [124], $MgAl_2O_4$ [125–127], mullite [128, 129], hydroxyapatite (HAp) [130, 131], other complex oxide compunds [132], and nonoxides (e.g., AlN) [133], have been processed by using SPS. Also, SPS has been demonstrated to be an effect method to fabricate electro-optic ceramics, including lanthanum-doped lead zirconate stannate titanate (PLZSnT) antiferroelectric ceramics, $PbZrO_3$–$PbTiO_3$–$Pb(Zn_{1/3}Nb_{2/3})O_3$ (PZ–PT–PZN) ferroelectric ceramics [134], and lanthanum-doped lead zirconate titanate ($Pb_{0.92}La_{0.08}(Zr_{0.65}Ti_{0.35})_{0.98}O_3$, PLZT) ceramics [135]. Processing of these transparent ceramics by using the conventional sintering is a significant challenge, due to the content of volatile element (Pb). More recently, transparent $BaTiO_3$ and $SrTiO_3$ ceramics, as well as their solid solutions, $Ba_{1-x}Sr_xTiO_3$ (or BST in short), which can never be obtained by using the conventional sintering techniques, have been achieved by using SPS [136, 137].

The effects of SPS processing parameters on microstructure and optical performance of Al_2O_3 ceramics have been studied [138]. Commercial α-Al_2O_3 powder, with a purity of 99.99 % and an average particle size of 0.15 μm, was used directly in this study. Compacted samples were heated to 1150 °C at a uniaxial pressure of 80 MPa by using a SPS machine (SPS-1050, Sumitomo) with a pulse duration of 3.4 ms. Heating was conducted in a sequent way, with twelve DC pulses (40.8 ms) followed by zero current for 6.8 ms. The heating rate from 600 to 1150 °C was varied between 2 and 100 °C min^{-1}. Temperature was measured with an optical pyrometer focused on a nonthrough hole (1 mm in diameter and 2 mm in depth) of the graphite die. After holding for 20 min at the sintering temperature and subsequent annealing at

1000 °C for 10 min, the sintered disk was about 30 mm in diameter and 3 mm in thickness. In addition, for the heating rates of 8 and 50 °C min^{-1}, sintering time was varied between 0 and 5 h, in order to examine the grain growth behavior. The mechanical pressure was unloaded before annealing.

It was found that, with the sintering parameters stated above, grain size of the final products increased with increasing heating rate. For example, the sample heated at 10 °C min^{-1} had an average grain size of 0.29 μm, while the one heated at 100 °C min^{-1} showed a grain size of 0.55 μm. SEM images of representative samples are shown in Fig. 7.18 [138]. This observation is in contrast to common trend in conventional furnace sintering. It was attributed to the high defect concentration produced by rapid heating and the associated rapid deformation during densification. The defects produced by both rapid heating and deformation would be responsible for the accelerated grain growth, which was further demonstrated in Fig. 7.19. As the temperature just reached 1150 °C, the two heating rates led to samples with almost the same grain size. However, as the sintering time was increased, the grain size of the sample sintered at 50 °C min^{-1} became much larger

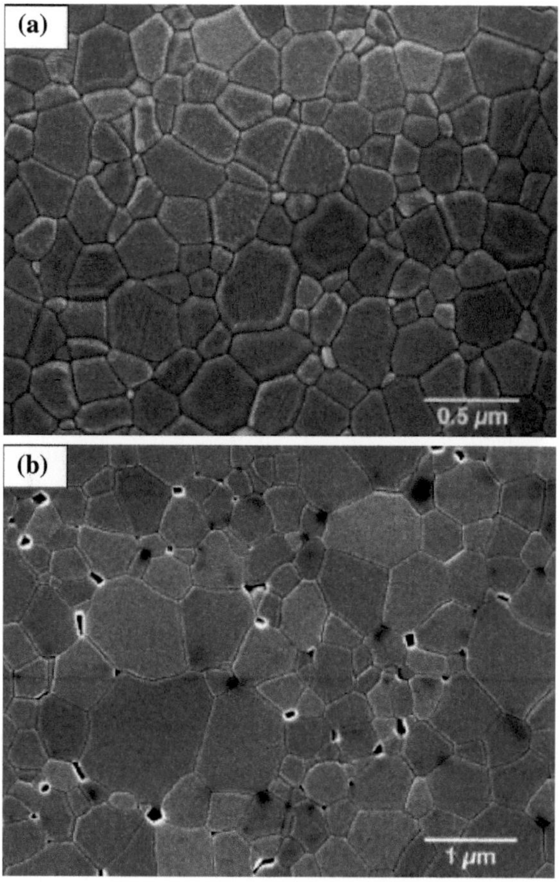

Fig. 7.18 Microstructures of the alumina sintered at a heating rate of **a** 2 °C min^{-1} and **b** 100 °C min^{-1}. Reproduced with permission from [138]. Copyright © 2009, Elsevier

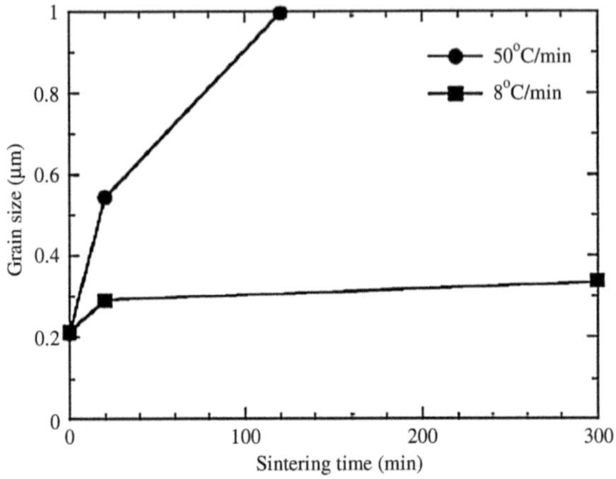

Fig. 7.19 Grain growth behaviors at 1150 °C for two different heating rates. Reproduced with permission from [138]. Copyright © 2009, Elsevier

than that of the one sintered at 8 °C min^{-1}. The effect of heating rate on porosity was normal, i.e., porosity increased with increasing heating rate.

Figure 7.20 shows photographs of the Al_2O_3 ceramics sintered at 1150 °C for 20 m with different heating rates, while tier in-line transmission curves are shown in Fig. 7.21 [138]. It is clearly demonstrated that the heating rate should be as low as possible in order to achieve high optical transparency. However, in practice, a trade-off is necessary, because low heating rate means high energy consumption.

SPS technique even can be used to fabricate samples with special shapes, such as hemispherical domes. For instance, transparent polycrystalline alumina domes have been obtained by combining sintering and forming into one step in minutes [139]. In contrast, fabrication of such items by using the convention sintering technology requires hours. Therefore, SPS provides an unprecedented opportunity to make optically transparent domes at much lower cost, in this respect. In addition, alumina transparent ceramics processed by using SPS have relatively finer grains, due to the rapid heating rate and short sintering time duration.

It is possible to further optimize the sintering effectiveness and efficiency when using SPS. For example, a two-step pressure profile was used to demonstrate the feasibility of SPS in processing $MgAl_2O_4$ [140]. At a low preload pressure, 5 MPa, with a normal fast heating rate of 100 °C min^{-1}, samples could have high in-line transmittance of 51 % at 550 nm and 85 % at 2000 nm. Sintering was carried out in

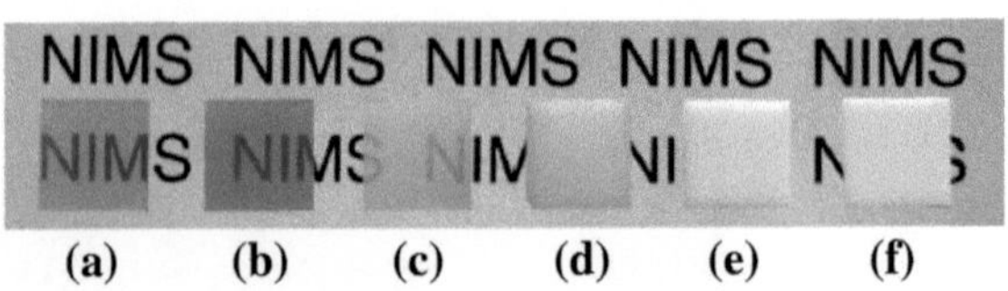

Fig. 7.20 Alumina ceramics sintered by SPS at a heating rate of **a** 2 °C min^{-1}, **b** 5 °C min^{-1}, **c** 10 °C min^{-1}, **d** 25 °C min^{-1}, **e** 50 °C min^{-1}, and **f** 100 °C min^{-1}. All samples are 0.9 mm thick and are put on a transparent polystyrene plate 24 mm above the text. Reproduced with permission from [138]. Copyright © 2009, Elsevier

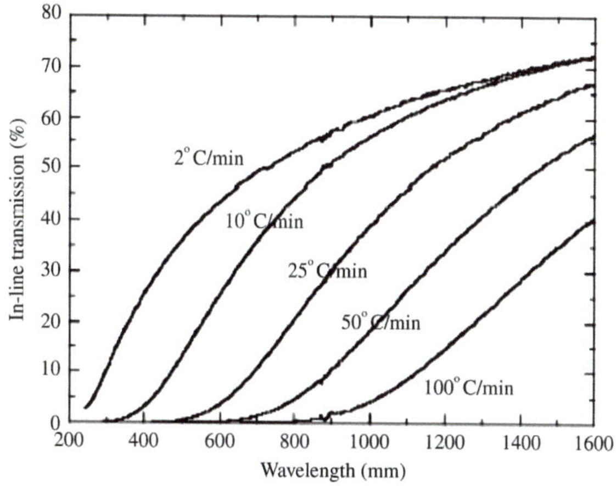

Fig. 7.21 In-line transmission of the alumina sintered at 1150 °C for 20 min. The sample thickness is about 0.9 mm. Reproduced with permission from [138]. Copyright © 2009, Elsevier

vacuum (4–6 Pa) with a Dr. Sinter 2050 SPS apparatus (Sumitomo Coal Mining Company Ltd., Japan). The temperature was increased to 600 °C within 3 min and then further increased to 1250 °C in 6 min, while the final sintering temperature of 1300 °C was reached in 1 min. The sintering time at the final stage for all experiments was 3 min. Five preload pressures, i.e., $P_1 = 5$, 10, 20, 30, 50, and 100 MPa, were used in the experiment. Photographs of the samples are shown in Fig. 7.22. Obviously, the samples prepared at P_1 of 30, 50, and 100 MPa have a darker discoloration core of about 8 mm in size at their center [140]. The discoloration was attributed to the dislocations formed during the fast densification. Figure 7.23 shows in-line transmittance curves of the samples. It was concluded that a preload

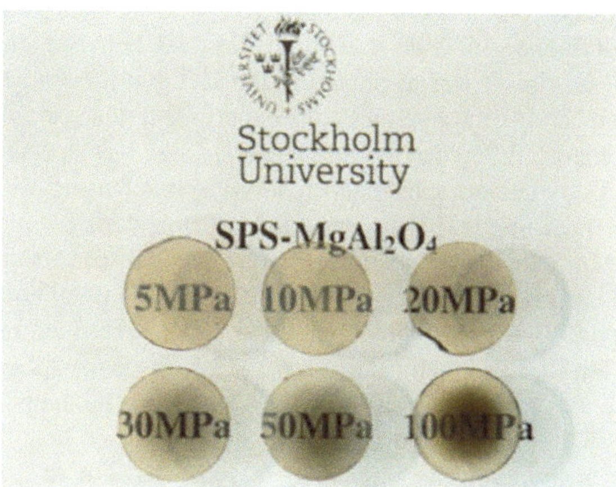

Fig. 7.22 Photographs of the transparent $MgAl_2O_4$ ceramics sintered at 1300 °C for 3 min at different preloaded pressures (P_1): 5 MPa (1.79 mm), 10 MPa (1.99 mm), 20 MPa (1.82 mm), 30 MPa (1.89 mm), 50 MPa (1.82 mm), and 100 MPa (1.60 mm). The samples were 15 mm above the paper, and thicknesses of the samples are indicated in parentheses. Reproduced with permission from [140]. Copyright © 2009, Elsevier

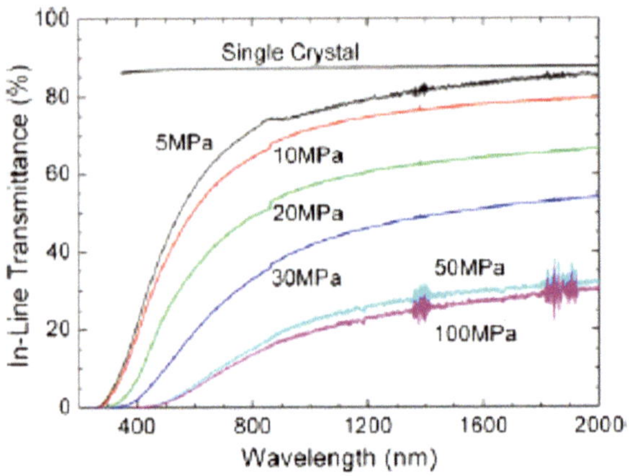

Fig. 7.23 In-line transmittance spectra of the polycrystalline $MgAl_2O_4$ sintered by SPS at different preload pressures P1. The theoretical transmittance of a single-crystal $MgAl_2O_4$ was calculated based on the refraction index as a function of wavelength. Reproduced with permission from [140]. Copyright © 2009, Elsevier

pressure P_1 should not be >20 MPa, so as to obtain acceptable optical transmittance approaching the theoretical limit in the wavelength range of >1000 nm.

Figure 7.24 shows SEM image of the samples processed at 5 MPa and 100 MPa [140]. It was found that both samples exhibited uniform microstructures, which implied that the application of a prepressure had no significant effect on microstructure of the final product. No residual pores were observed on surface the thermally etched sample, while only a very few pores were found by using TEM observation. Average grain sizes of the 5 and 100 MPa samples were 600–700 nm. No significant difference in grain size was observed between the discoloration center and the clean rim. It was attributed to the fact that the temperature gradient was <30 °C in the 12 mm graphite die in the experiment, which was not sufficiently large to lead to energy difference, so that the grain size was not affected.

Transparent Y_2O_3 ceramics have been fabricated by using SPS at moderate temperature and pressure [119]. It was found that sintering temperature had a significant effect on densification, microstructure, optical, and mechanical properties of the Y_2O_3 ceramics. Very high-purity Y_2O_3 powder (99.999 %) was used as the starting material. Before SPS processing, the powder was ball milled in ethanol with ZrO_2 balls for 12 h and then calcined at 1273 K in air for 7.2 ks. SPS-210 LX model SPS setup was used to densify the Y_2O_3 powder. The temperature was first increased to 873 K in 180 s and to 1373 K in 300 s and then held for 300 s. After that, the temperature was further increased to 1373–1823 K at a heating rate of 0.17 K s^{-1} and maintained at each temperature for 2.7 ks. A pressure of 10 MPa was preloaded between room temperature and 1373 K, which was increased to 100 MPa at >1373 K for final sintering. Post-annealing was carried out at 1123–1423 K for 21.6 ks in air.

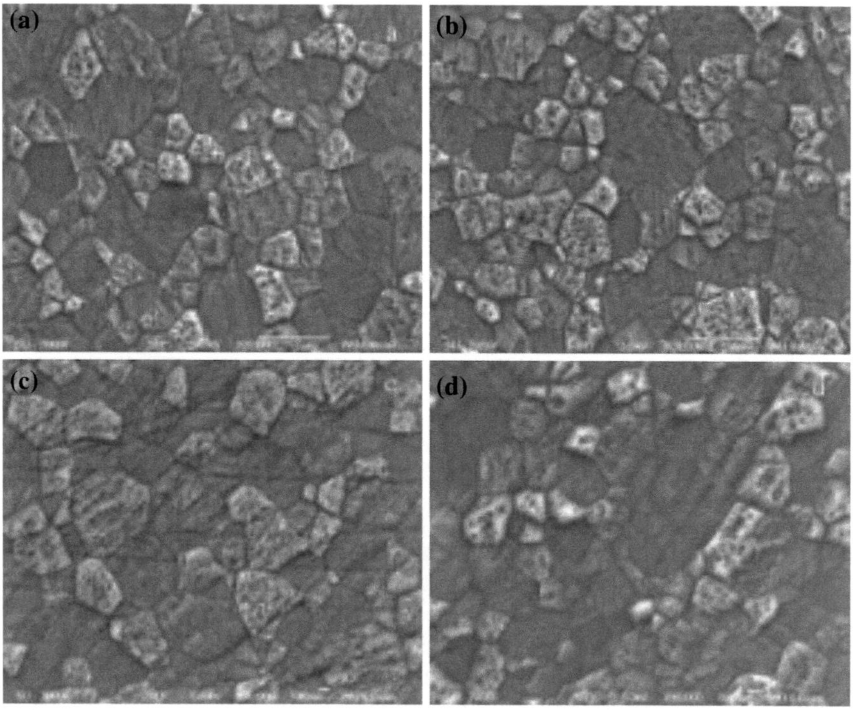

Fig. 7.24 The microstructure of transparent $MgAl_2O_4$ sintered by SPS at 1300 °C for 3 min at different pressures of P1: **a** and **b** 5 MPa and **c** and **d** 100 MPa. **a** and **c** SEM images taken near edge of the samples. **b** and **d** SEM images taken at the discolored center. Reproduced with permission from [140]. Copyright © 2009, Elsevier

Relative density of the sample was 98 % after sintering at 1373 K, while fully dense ceramics with relative density of above 99 % were achieved after sintering at 1473–1823 K. The average grain size was slightly increased from 0.24 to 0.32 μm as the sintering temperature was increased from 1473 to 1573 K. After sintering at 1823 K, the grain size was increased to 1.97 μm [119].

Transmittance spectra of the Y_2O_3 ceramics SPS sintered at 1573 K before and after annealing at 1123–1423 K for 21.6 ks are shown in Fig. 7.25 [119]. As the annealing temperature was increased, transmittance of the samples was increased gradually. The optimized annealing temperature was 1323 K, as shown in Fig. 7.25b. The ultraviolet absorption edge was blue-shifted from 308 to 256 nm, as the annealing temperature was increased from 1123 to 1223 K. No significant change was observed after annealing at higher temperatures of up to 1423 K.

Figure 7.26 shows transmittance spectra of Y_2O_3 bodies sintered at 1473–1773 K after annealing at 1323 K for 21.6 ks in air, together with those that were calculated with the refractive index of Y_2O_3 single crystal [119]. Transmittance of the Y_2O_3 ceramics was increased greatly after annealing. The sample sintered at 1573 K exhibited highest transmittance of 55.0 and 81.7 % at $\lambda = 550$ and 2000 nm.

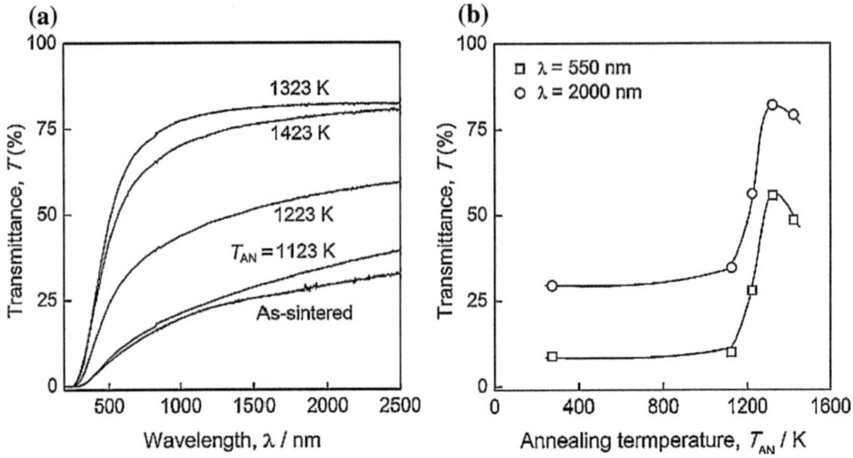

Fig. 7.25 **a** Transmittance spectra of the Y_2O_3 ceramics before and after annealing at 1123–1423 K for 21.6 ks using the samples sintered at 1573 K. **b** Transmittance at λ = 550 and 2000 nm as a function of annealing temperature. Reproduced with permission from [119]. Copyright © 2012, Elsevier

The one sintered at 1573 K after annealing reached nearly 99 % of the theoretical value in the infrared region, as shown in Fig. 7.26a. This temperature was much lower than those required by other sintering techniques.

Vickers hardness (H_V) and fracture toughness of the annealed Y_2O_3 ceramics as a function of sintering temperature are shown in Fig. 7.27a, while relationship between H_V and grain size is illustrated in Fig. 7.27b [119]. H_V of the Y_2O_3 ceramics sintered

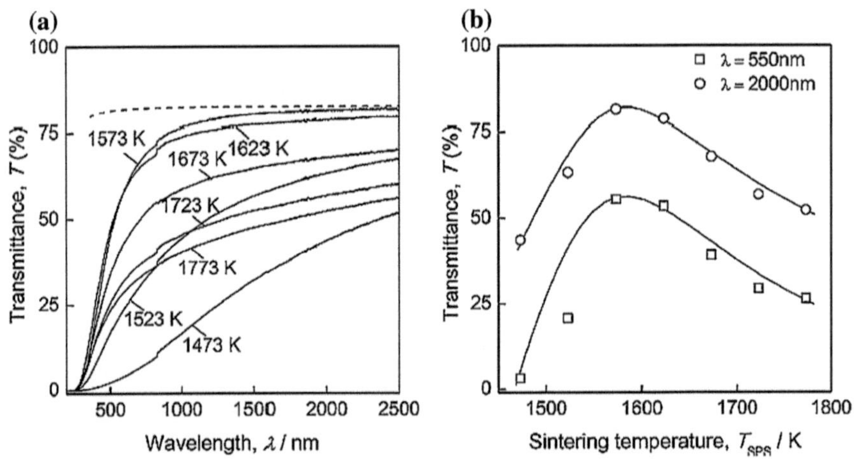

Fig. 7.26 **a** Transmittance spectra of the Y_2O_3 ceramics sintered at 1473–1773 K after annealing at 1323 K. **b** Transmittance at λ = 550 and 2000 nm as a function of sintering temperature. The dashed line indicates transmittance calculated from the refractive index of Y_2O_3 single crystal. Reproduced with permission from [119]. Copyright © 2012, Elsevier

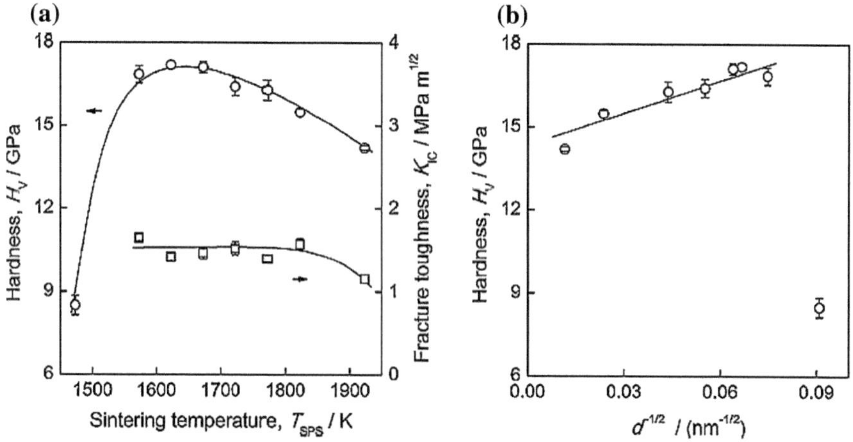

Fig. 7.27 **a** Vickers hardness and fracture toughness of the Y_2O_3 ceramics annealed at 1323 K as a function of sintering temperature. **b** Vickers hardness as a function of the inverse square root of grain size. Reproduced with permission from [119]. Copyright © 2012, Elsevier

at 1373 K was 8.3 GPa. It was sharply increased to ~17.2 GPa after sintering at 1573 K. A maximum value of ~17.5 GPa was achieved at the sintering temperature of 1623 K. After that, the value of H_V steadily decreased to ~14 GPa after sintering at 1923 K. Fracture toughness was in the range of 1.0–1.5 MPa m$^{1/2}$, which was not significantly influenced by sintering temperature, especially over 1573–1823 K. H_V of the Y_2O_3 ceramics showed a linear relationship with $d^{-1/2}$, i.e., the Hall–Petch relation. The fact that H_V increases with $d^{-1/2}$ in ceramics has been explained in terms of the reduced free path for dislocations as the grain size is decreased [141].

Transparent YAG ceramics have been fabricated by using SPS (Dr. Sinter 2050) and commercial YAG powder as starting material [114]. The YAG powder was of nonagglomerated spherical shaped particles with average particle diameter of 34 ± 17 nm and lognormal particle size distribution. Graphite foils (Grafoil) were used as spacers between the sample to be sintered and the graphite die and the punches. A preloaded pressure of 100 MPa was applied before the heating was started. The pressure was maintained during the whole heating process and at the SPS temperatures during the sintering. The temperature was increased to 600 °C within 3 min, after which it was raised to sintering temperatures at a heating rate of 100 °C min^{-1}.

During the densification of nanocrystalline powder compacts by using cold pressing, it is very common that large pores could be entrapped, due to the agglomeration of the fine particles. Remove of such pores during pressureless sintering and densification is quite difficult, due to the presence of the junctions with high dihedral angles at the pore–grain boundaries. However, the SPS compact of YAG at 900 °C exhibited particles with homogeneous distribution and pores with comparable size with the particle sizes, as shown in Fig. 7.28 [114]. This was mainly because the YAG powder had no agglomeration. Such homogeneous and small pore size could not result in full densification when using pressureless sintering, because many pores had high coordination numbers. However, this problem could be addressed by using SPS, due to the application of pressures. Pressures would assist the closure of such pores through grain sliding.

Figure 7.29 shows SEM images of the samples sintered at higher temperatures [114]. The sample sintered at 1250 °C was still opaque, while the one sintered at 1400 °C had been transparent. Both samples showed homogeneous microstructure. Relative density of the YAG samples was increased significantly from 83 to 99.6 %, although the temperature range was as narrow as 150 °C. Grain size in this temperature range was increased in a parabolic manner. Significant grain growth was observed at the temperatures of above 1400 °C. The grain growth behaviors below and above 1400 °C were different [114].

SPS time duration had effect on densification and grain growth of YAG ceramics. For example, as the SPS duration was increased from 3 to 6 min, the relative density was increased tremendously, especially at lower temperatures, e.g., <1300 °C, as shown in Fig. 7.30 [114]. Therefore, if the sintering duration was

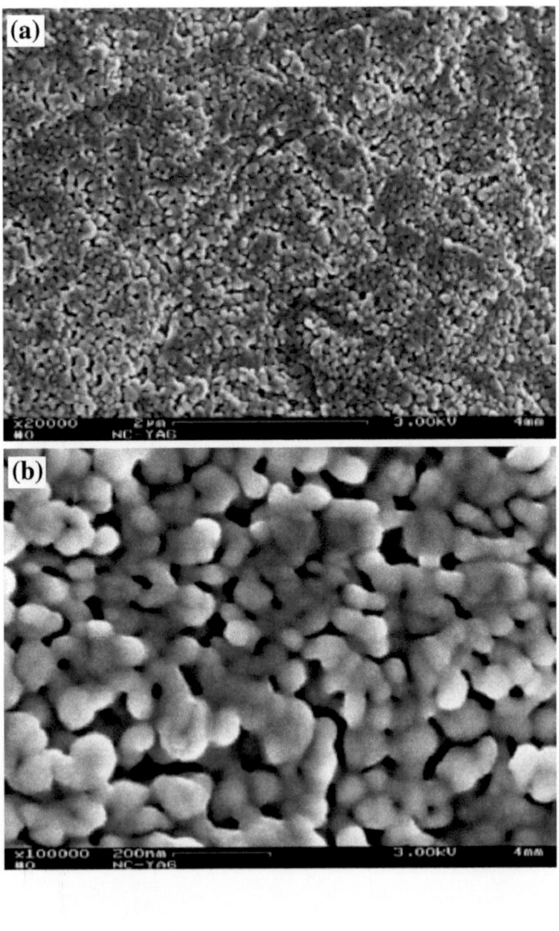

Fig. 7.28 Low (**a**) and high (**b**) magnification SEM images of the samples derived from the YAG compact subjected to SPS for 3 min at 900 °C and 100 MPa, which were further annealed at 900 °C for 5 h. Homogeneous distribution of the particles and pores with comparable sizes could be observed. Reproduced with permission from [114]. Copyright © 2007, Elsevier

Fig. 7.29 SEM images of fracture surfaces of YAG ceramics after SPS for 3 min at 100 MPa at different temperatures: **a** 1250 °C and **b** 1450 °C. Reproduced with permission from [114]. Copyright © 2007, Elsevier

increased from 3 to 6 min, the sintering temperature could be reduced by about 75 °C, for a given relative density. After 6 min sintering at 1350 °C, transparent YAG ceramics could be obtained. It was worth mentioning that the sample sintered for 6 min at 1300 °C had higher density than the one sintered to 3 min at 1350 °C, which indicated the negative effect of grain growth on densification during SPS. In this case, the densification was governed by the atomic transport processes at surface of the particles and particle–particle interfaces. Densification and grain growth by grain rotation and sliding would convert the originally high-angle grain boundaries to low-angle grain boundaries that in turn are annealed out by short-circuit diffusion. TEM observations verified the presence of such low-angle grain boundaries in dense ceramics was confirmed by TEM analysis, as shown in Fig. 7.31 [114]. In addition, Fig. 7.30 also indicated the effect of pressure.

In another study, it was found that the way to treat the precursor powder had significant effect on performance of YAG ceramics by using SPS [142]. When sintering YAG synthesised by using reverse-strike co-precipitation with SPS, the precursor powder was calcined at 1000 °C and dispersed by using ball milling with

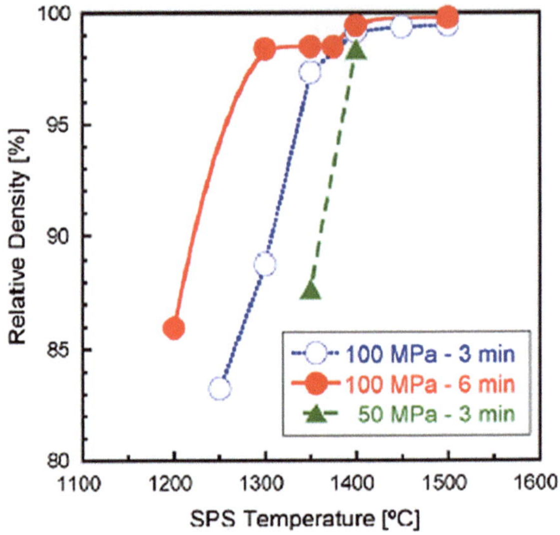

Fig. 7.30 Relative density of the YAG powder compacts after SPS for 3 and 6 min at 50 and 100 MPa, as a function sintering temperature. Reproduced with permission from [114]. Copyright © 2007, Elsevier

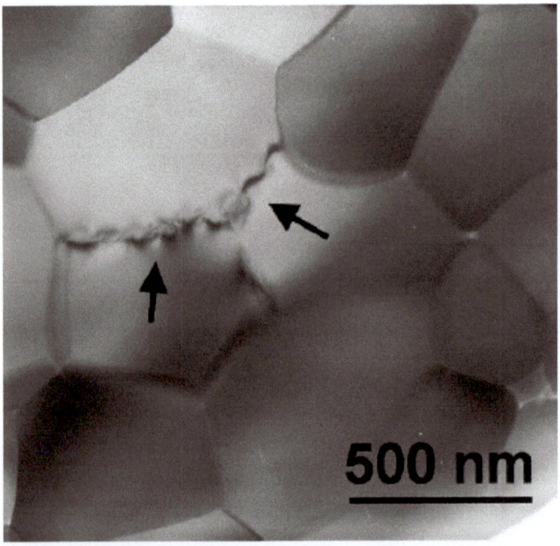

Fig. 7.31 TEM image of the YAG ceramics after SPS for 3 min at 1400 °C and 100 MPa. The high-angle grain boundaries appeared as straight lines. The low-angle grain boundaries (*arrowed*) appeared to be curvy due to the strain field caused by dislocations. Nanometer-sized residual pores were present at the grain-boundary junctions. Reproduced with permission from [114]. Copyright © 2007, Elsevier

α-alumina (BM_A) or zirconia (BMz) balls or ultrasonication (US). All the three dispersed powders could be consolidated by using SPS to nearly theoretical density, but only the US powder was optically transparent, as shown in Fig. 7.32.

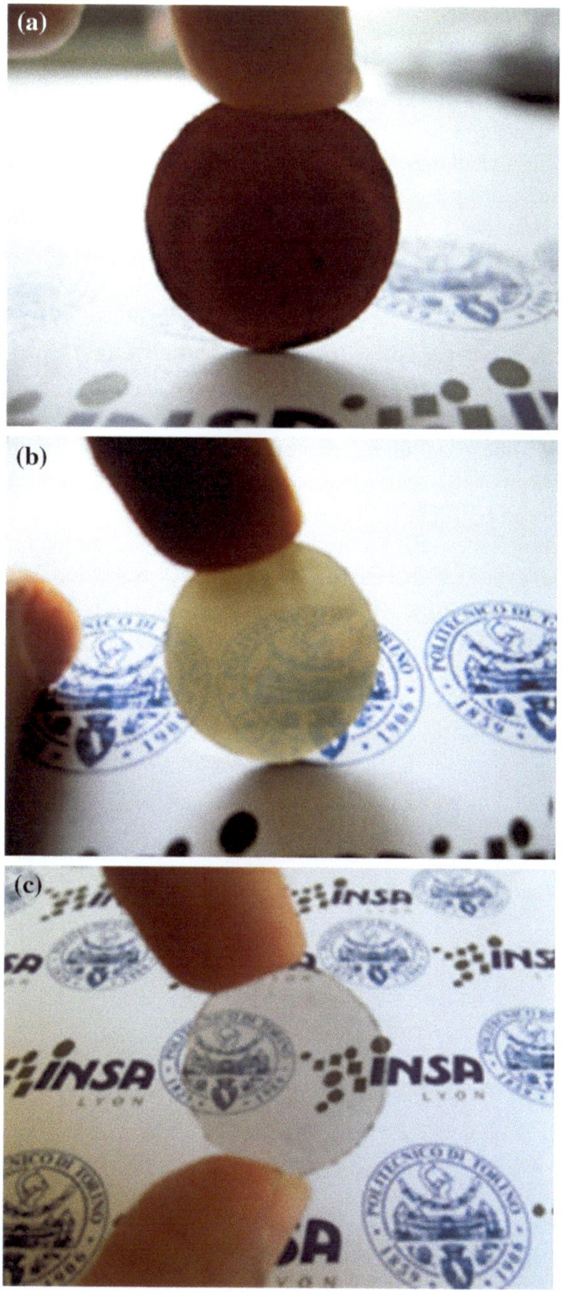

Fig. 7.32 Photographs of the samples derived from powders of BM_Z (**a**), BM_A (**b**), and US (**c**) after thermal annealing. Reproduced with permission from [142]. Copyright © 2013, Elsevier

The 1-mm-thick transparent sample possessed a transmittance of about 60 % at 600 nm. According to Raman spectroscopy, it was found that there were some phonon vibrational shifts, which implied that minor secondary phases could be have introduced due to the contamination from the milling media. Such minor secondary phases could not be detectable by using XRD, because of their very small quantity. This result further indicates that special carefulness is important when processing transparent ceramics.

Recently, SPS has been used to prepare pyrochlore transparent ceramics, including $Lu_2Ti_2O_7$ [143, 144], $Lu_2Hf_2O_7$ [145], and Lu_3NbO_7 [146, 147], which is still a challenge by using other sintering techniques. Figure 7.33 shows photograph and transmittance curves of transparent $Lu_2Ti_2O_7$ pyrochlore ceramics, which was synthesized by using SPS through reactive sintering directly from Lu_2O_3 and TiO_2 oxide precursors at 1723 K for 45 min [143]. The annealed ceramics demonstrated a transmittance of 72 % at wavelength of 2000 nm and 40 % transmittance at 550 nm. The average grain size was 14.5 μm with uniform microstructure. Representative results of Lu_3NbO_7 transparent ceramics, fabricated by using SPS, are shown in Figs. 7.34 and 7.35 [147].

Another distinctive advantage of SPS is its ability to processing the transparent ceramics containing elements that are unstable in air, such as nitrogen (N), phosphorus (P), and fluorine (F). Examples include AlON [148] and Yb^{3+}-doped $Sr_5(PO_4)_3F$ [149]. More significantly, $Sr_5(PO_4)_3F$ is a hexagonal crystal, instead of cubic ones as required according to the general understanding on transparent ceramics, as discussed previously. 20 mol% Yb^{3+}-doped $Sr_5(PO_4)_3F$ nanopowder was synthesized by using a co-precipitation method, with $Sr(NO_3)_2$, $Yb(NO_3)_3$, $(NH4)_2HPO_4$, and NH_4F as starting materials [149]. Solution containing 0.03 M $Sr(NO_3)_2$ and 0.0013 M $Yb(NO_3)_3$ was mixed with solution of 0.013 M NH_4F and 0.04 M $(NH_4)_2HPO_4$ to precipitate the precursor particles. The powder was ball milled for 20 h using ethanol as a solvent first and then dried at 70 °C and sieved

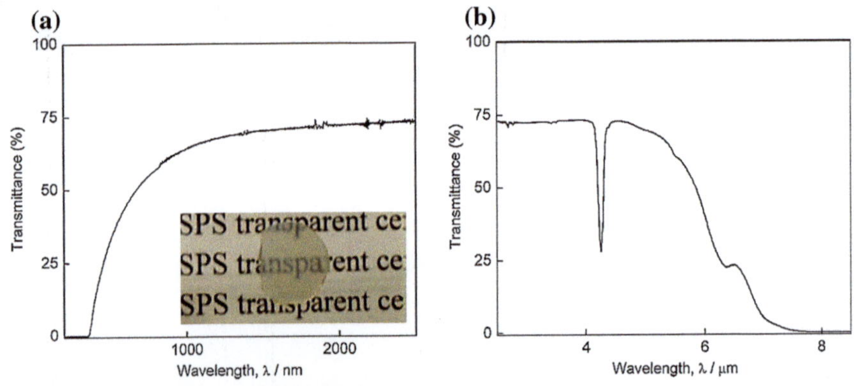

Fig. 7.33 Transmittance spectra of $Lu_2Ti_2O_7$ sintered at 1723 K for 45 min in different ranges: **a** 190–2500 nm and **b** 2.5–8.5 μm. The inset is photograph of the transparent ceramics. Reproduced with permission from [143]. Copyright © 2011, Elsevier

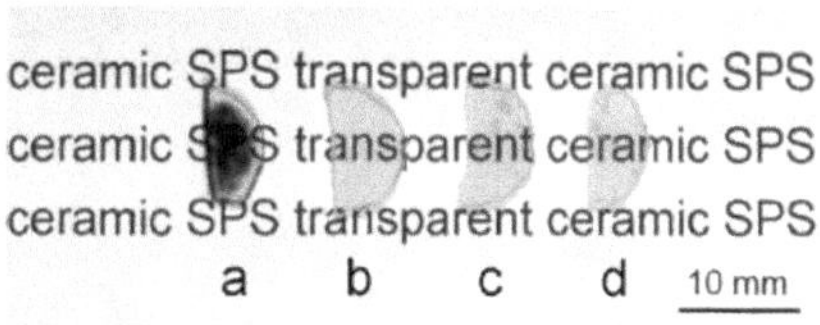

Fig. 7.34 Photographs of the Lu_3NbO_7 ceramics SPS sintered at 1500 °C and annealed at various temperatures in air for 6 h: **a** 750 °C, **b** 850 °C, **c** 950 °C, and **d** 1050 °C. Reproduced with permission from [147]. Copyright © 2013, Elsevier

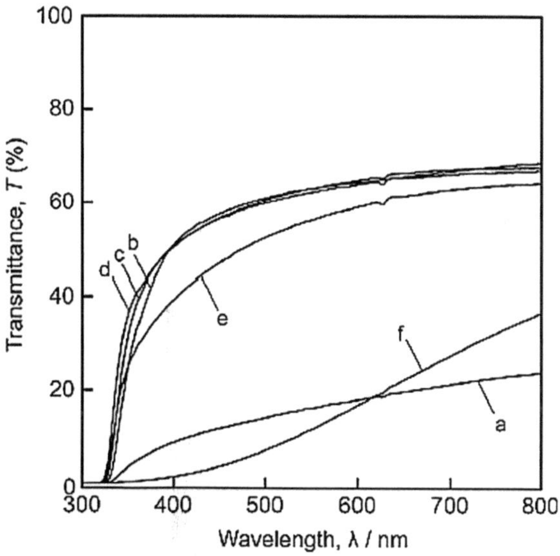

Fig. 7.35 Transmittance curves of the Lu_3NbO_7 ceramics SPS sintered at 1450 °C and annealed at various temperatures in air for 6 h: **a** 750 °C, **b** 850 °C, **c** 950 °C, **d** 1050 °C, **e** 1150 °C, and **f** 1250 °C. Reproduced with permission from [147]. Copyright © 2013, Elsevier

with a 200 mesh screen. The powder was finally put into graphite mold and then separated by foil and BN spraying coatings from the mold. The sample was sintered in a FCT System SPS facility, at a heating rate of 50 °C min^{-1} from RT to 1050 °C at a loading pressure of 100 MPa for 8 min. A trance of second phase, $Sr_9Yb(PO_4)_7$, was found in the sintered ceramics, which could be suppressed by using a small amount of SrF_2.

Photographs of representative Yb^{3+}-doped $Sr_5(PO_4)_3F$ transparent ceramics are shown in Fig. 7.36 [149]. A TEM image of the ceramics is also included in the figure. It was found that almost no porosity could be observed in the SPS-sintered sample. The grain sizes were ranged from 40 to 200 nm, with an average value of about 150 nm. In-line optical transmittance spectrum indicated that the $Sr_5(PO_4)_3F$:Yb ceramics had a strong absorption in the range of 850–980 nm, showing high potential to be high-power laser material. The transmittance near the emission wavelength was above 74 % at 1000 nm.

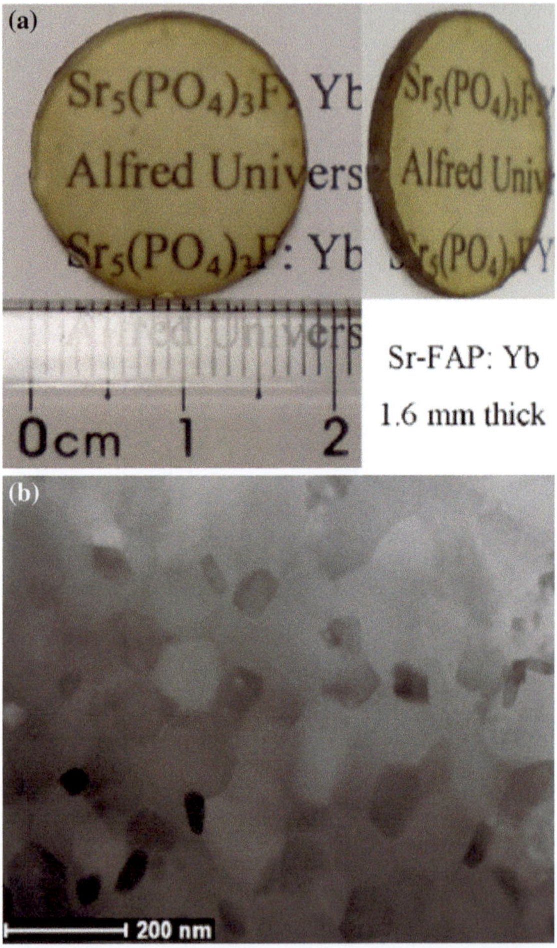

Fig. 7.36 Photographs (*top*) and TEM image (*bottom*) of the Yb^{3+}-doped $Sr_5(PO_4)_3F$ transparent ceramics processed by using SPS. Reproduced with permission from [149]. Copyright © 2012, John Wiley & Sons

7.6 MW-Processed Transparent Ceramics

The use of MW sintering to fabricate transparent ceramics is relatively less popular currently. Transparent ceramics that have been processed by using MW sintering include Al_2O_3 [150–152], YAG [153], AlON [154], and mullite [155]. MW sintering has been shown to be an effective technique to produce transparent ceramics as compared with the conventional sintering technologies.

First example discussed in this section is MW-processed Al_2O_3 transparent ceramics reported by Cheng et al. [150]. High-purity (99.99 %) commercial Al_2O_3 powder with average particle size of 0.15 μm was used as precursor. A total of 0.05 wt% of MgO was incorporated through a wet-chemical route with $Mg(NO_3)_2{\cdot}5H_2O$ as sintering promoter. Green pellets, with relative densities of

52–54 %, were prepared by dry pressing uniaxially, followed by cold isostatic pressing at 280 MPa. The compacted pellets were preheated at 1100 °C for 2 h in a resistance furnace to burn out the binder. Microwave sintering was carried out using a TE103 single-mode microwave applicator that was coupled with a 1.5-kW microwave generator operating at 2.45 GHz for small samples with diameters of <12.7 mm and a multi-mode microwave applicator with a 6-kW microwave power source for large samples with diameters of up to 25.4 mm. Ultrahigh-purity hydrogen under ambient pressure was used as sintering atmosphere.

Figure 7.37 shows photograph of a representative transparent Al_2O_3 ceramics that was microwave sintered at 1750 °C for 45 min [150]. In the reported experimental range, optical transmittance of the samples was slightly increased with increasing sintering time. Figure 7.38 shows SEM images of the samples microwave sintered at 1750 °C for different time durations. All samples exhibited uniform microstructure and grain size, with almost 100 % relative density, i.e., 3.97 g cm^{-3}. As the sintering time was increased from 15 to 45 min, the average grain size was increased from 20 to 40 μm.

Some microwave sintered transparent Al_2O_3 samples were post-sintered in microwave field at 1850–1880 °C for 2 h under ultrahigh-purity hydrogen (H_2) atmosphere, in order to convert the polycrystalline ceramics to single crystals. The processing is schematically shown in Fig. 7.39 [150]. A 9.5-mm-diameter Al_2O_3 disk, which was an as-sintered transparent sample by using MW sintering at 1750 °C for 30 min and supported by a high-purity Al_2O_3 tube, was placed in the single-mode microwave cavity to apply microwave post-sintering treatment. It was found that the temperatures at the center and periphery of the Al_2O_3 disk could be slightly different, due to the cooling effect of the flowing H_2 on the peripheral area.

As shown in Fig. 7.40, significant change in microstructure was observed in the post-sintered Al_2O_3 sample that was microwave heated at 1880 °C for 30 min [150]. It was demonstrated that the conversion of the polycrystalline ceramics to single crystal could be achieved by using the microwave post-sintering process. The peripheral area of the sample was still polycrystalline with average grain size of

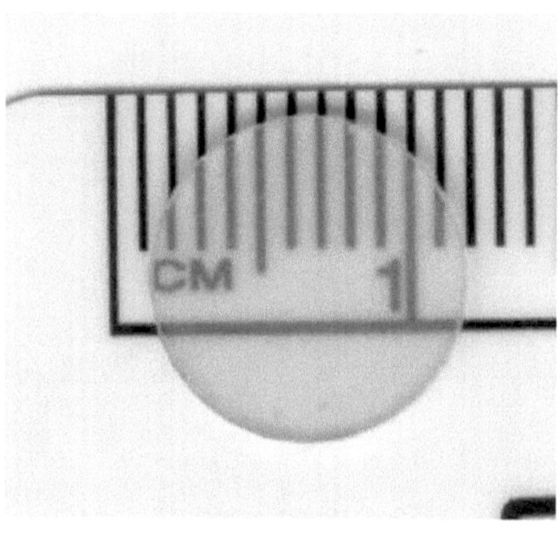

Fig. 7.37 Photograph of a representative highly transparent Al_2O_3 ceramics microwave sintered at 1750 °C for 45 min. Reproduced with permission from [150]. Copyright © 2002, Elsevier

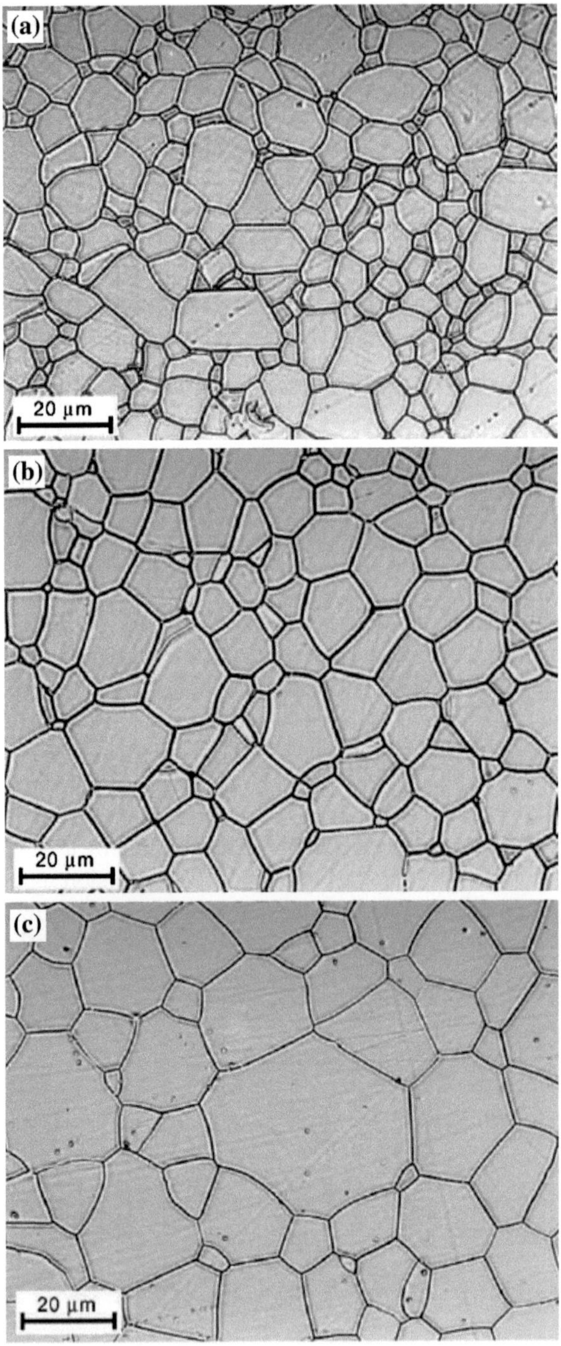

Fig. 7.38 SEM images of the Al_2O_3 ceramics with 0.05 wt% MgO microwave sintered at 1750 °C at a heating rate of 100 °C min^{-1} for different time durations: **a** 15, **b** 30, and **c** 45 min. Reproduced with permission from [150]. Copyright © 2002, Elsevier

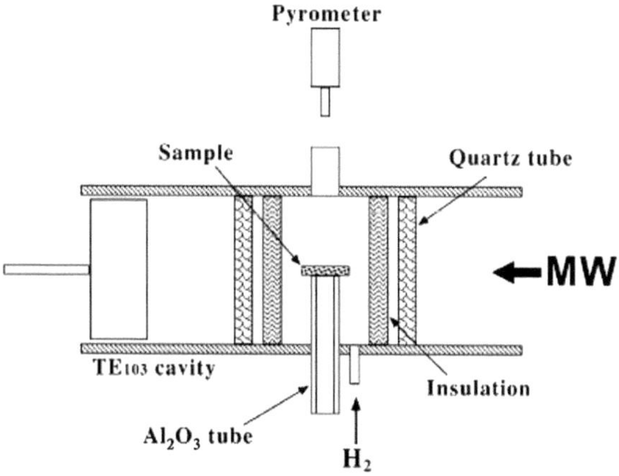

Fig. 7.39 Schematic of the microwave setup for solid-state conversion of polycrystalline Al_2O_3 sample to single-crystal sapphire. Reproduced with permission from [150]. Copyright © 2002, Elsevier

30–40 μm, as shown at the top panel in Fig. 7.40, whereas the center part of the post-sintered sample was converted into single crystal without the presence of grain boundaries, i.e., the bottom panel of the figure. The converted sample exhibited an enhanced transmittance by about 20 % as compared with the as-sintered ones, showing the feasibility and advantage of MW sintering technique.

Another example of transparent Al_2O_3 was fabricated by using microwave sintering at 2.54 GHz [151]. Figure 7.41 shows a schematic diagram of the microwave sintering with a special designed insulating casket. The hole on the top cover was used for infrared pyrometer. SiC powder was used as a microwave absorber, acting as a low-temperature microwave susceptor to preheat the Al_2O_3 samples, because Al_2O_3 is a poor microwave absorber at room temperature.

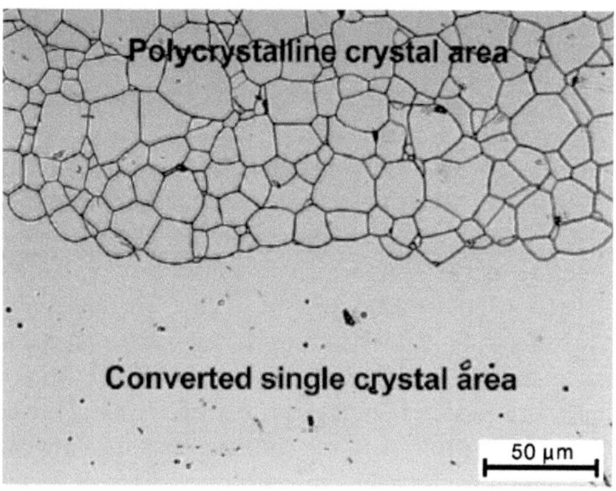

Fig. 7.40 SEM image of the Al_2O_3 sample microwave post-sintered at 1880 °C for 120 min, showing microstructure development due to the conversion of polycrystalline to single-crystal structure. Reproduced with permission from [150]. Copyright © 2002, Elsevier

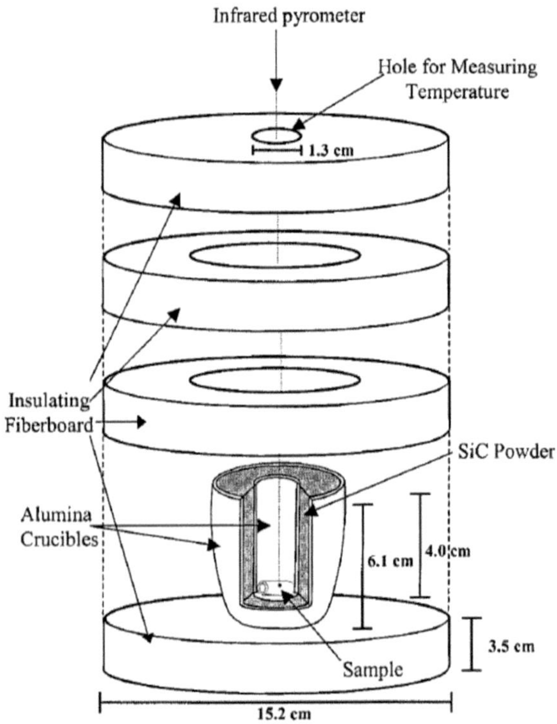

Fig. 7.41 Schematic diagram of a microwave sintering with specially designed insulating casket. Reproduced with permission from [151]. Copyright © 2003, John Wiley & Sons

The precursor alumina powder was doped with 350 ppm Y_2O_3 and 500 ppm MgO. Binder was removed by heating to 650 °C at 5 °C min^{-1} with by using a conventional furnace. The average green density of the tubes after binder removal was 1.63 ± 0.01 g cm^{-3}, i.e., 40.85 ± 0.35 % relative density. A multi-mode 6-kW, 2.45-GHz microwave furnace was used to sintering the samples. The sintering temperature and heating rate of microwave sintering were controlled by adjusting the input power. The sintering temperatures in the microwave furnace were measured by using digital infrared pyrometer.

Figure 7.42 shows density of the samples, for microwave and conventional sintering, as a function of sintering temperature [151]. Without dwelling at the sintering temperatures, microwave sintered Al_2O_3 samples reached relative density of 95 % after sintering at 1350 °C, which was much lower than 1600 °C for the conventionally heated samples. Also, the microwave sintered sample was almost fully densified after sintering at 1400 °C, while the density of the sample sintered through the conventional sintering was only about 50 %. Microwave sintering reduced the temperature for the samples start to densify by about 300 °C.

It was found that both the microwave sintering and the conventional sintering exhibited a similar grain growth behavior, as shown in Fig. 7.43 [151]. Therefore, the enhancement in densification of Al_2O_3 by the microwave sintering was not due to the fast heating rate. However, the microwave densification corresponded to apparent activation energy of 85 ± 10 kJ mol^{-1}, much lower than the value of

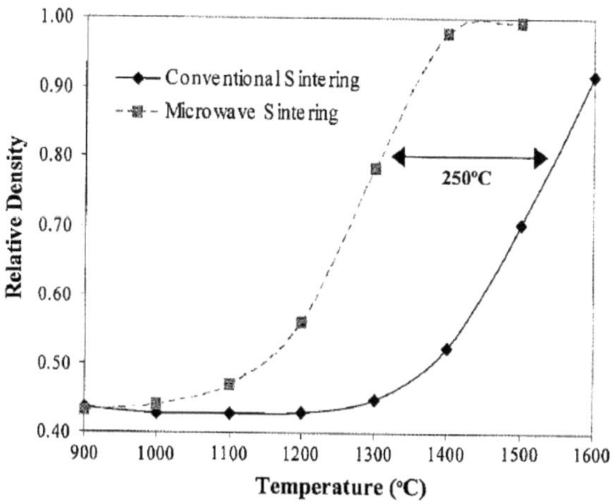

Fig. 7.42 Density of the samples, sintered by using microwave sintering and conventional sintering without holding at the sintering temperature, as a function of sintering temperature. Reproduced with permission from [151]. Copyright © 2003, John Wiley & Sons

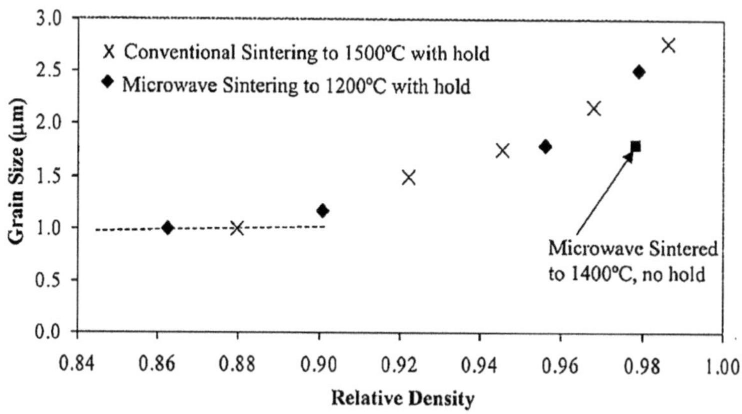

Fig. 7.43 Grain growth trajectories of the microwave sintered and conventionally sintered samples. Reproduced with permission from [151]. Copyright © 2003, John Wiley & Sons

520 ± 14 kJ mol^{-1} for the conventional sintering, meaning that the there was a difference in densification mechanism between the two sintering techniques, which deserves for further investigation.

A gyrotron-based system operating at a frequency of 24 GHz with microwave power up to 6 kW has been used to develop YAG transparent ceramics through reactive sintering from oxide precursors [153]. TEOS with contents of 0.3 and 0.5 wt% was used to provide SiO_2 as sintering aid.

Figure 7.44 shows photograph and microstructure of representative sample [153]. The sample contained 1.0 % Yb and 0.5 wt% TEOS and sintered at 1730 °C

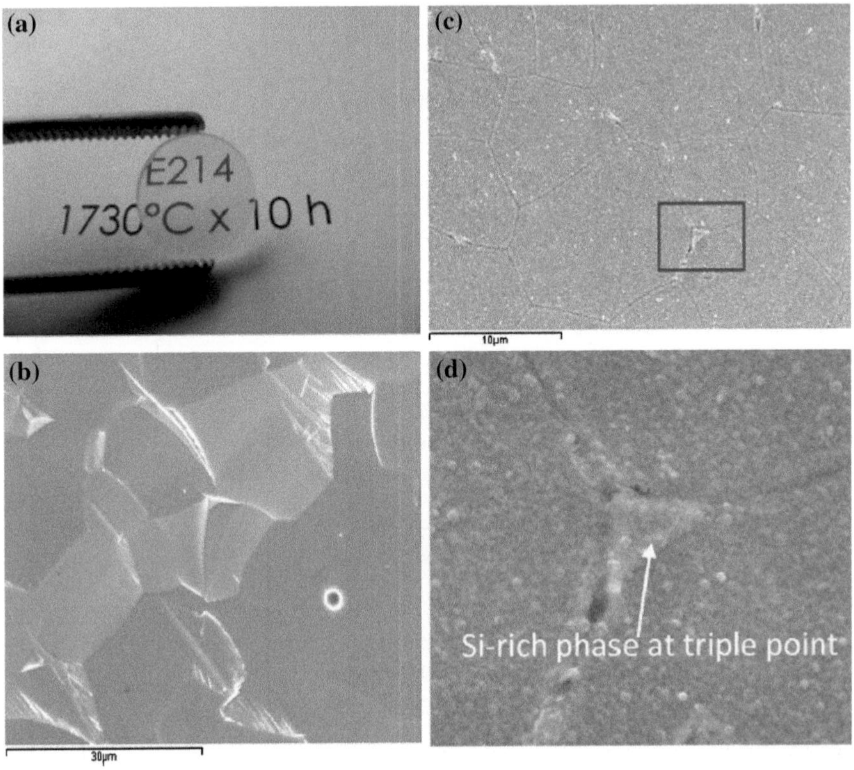

Fig. 7.44 *Photograph* **a** and SEM images of the sample 1.0 % Yb:YAG with 0.5 wt% TOES sintered at 1730 °C for 10 h: **b** fracture surface and **b**, **c** polished and thermally etched surface. Reproduced with permission from [153]. Copyright © 2013, Elsevier

for 10 h. As shown in Fig. 7.44b, on the fracture surface, few pores were observed, most of which were preferentially within the grains. The sample exhibited grain sizes ranging from 5 to 20 μm, as shown by the polished and thermally etched surface in Fig. 7.44c. It was also found that the content of TEOS had no significant effect on grain size of the YAG transparent ceramics. Figure 7.44d indicated that amorphous or crystalline phase enriched with SiO_2 could be found at some triple junctions or grain boundaries. Such phases might have been formed during the thermal etching process, due to the diffusion of silicon through the grains toward the grain boundaries.

As stated previously, the sintering behavior of various transparent ceramics, including YAG, can be improved by using silica-containing sintering aid. In conventional sintering, silica functions to form a liquid phase at about 1390 °C. The liquid phase makes it easier for the particles to rearrange and enhances mass transport by grain-boundary diffusion, so as to promote densification. During cooling from the sintering temperature, secondary phases of Al_2O_3–Y_2O_3–SiO_2, either amorphous or crystalline, would be formed. The presence of such phases is harmful to optical properties due to their light scattering. Therefore, a trade-off is necessary to optimize the content of silica, i.e., promoting densification yet without compromising optical performances.

One of the most distinctive features of the microwave sintering is that there is a special behavior in microwave power in the initial stage of heating. For example, in that study, there was a drop in microwave power as the temperature was above 700 °C [153]. It was attributed to the increase in microwave absorption of the sample, due to the increase in their ionic conductivity. The microwave power was also influenced by the exothermal reactions of the intermediate $Y_4Al_2O_9$ and $YAlO_3$.

The formation of YAG through solid-state reaction is controlled by the diffusion of Al^{3+} [156]. Such diffusion of Al^{3+} was enhanced by the microwave field, so that the reaction temperature of YAG and its sintering temperature were decreased, as evidenced by the pores entrapped in grains, as shown in Fig. 7.44b. Similarly, the formation of the liquid phase was also promoted. In other words, the content of sintering aids could be reduced when using microwave sintering, which however needs further clarification.

The enhancement in mass transport of polycrystalline materials has been explained by the so-called ponderomotive action of electromagnetic field on charged particles [157, 158]. The theory suggested that the ponderomotive effect is more effective in ionic crystalline materials with high microwave absorption and it is maximized when the imaginary part and real parts of the complex dielectric permittivity of the materials have very close values. It can be expected the microwave processing parameters could be further optimized in such a way that all diffusion processes would be operated concurrently during the densification of powder compacts. In this case, the amount of sintering aids could be greatly reduced or even completely avoided.

Besides, Roy et al. found that single-phase ALON could be synthesized from Al_2O_3 and AlN by using microwave heating at 1650 °C for 1 h, while fully dense and transparent ALON ceramics were obtained through microwave sintering at 1800 °C for 1 h [154]. A total transmission of 60 % was achieved for the polished sample with a thickness of 0.6 mm. The microwave sintering was carried out by using a homemade 1.5-kW, 2.45-GHz single-mode microwave applicator in nitrogen (N_2) flowing at ambient pressure. The heating rate was about 100 °C min^{-1} by controlling the incident microwave power.

Transparent mullite ceramics were developed by using microwave sintering at about 1300 °C [155]. The microwave sintering was conducted by using a 2.45-GHz, 900-W microwave furnace. The samples were heated in the microwave field directly from room temperature to 1300 °C and held at the temperature for 10 min, then allowed to cool down naturally. The results indicated that aerogel of mullite stoichiometry, free of agglomeration and highly active, was key factor to achieving transparent mullite ceramics.

7.7 Concluding Remarks

Vacuum sintering is the most widely used technique to fabricate transparent ceramics, due to its various advantages, such as simple processing, large size product, and relatively cost-effectiveness. HP and HIP offer lower sintering temperatures, so that transparent ceramics with fine-grained microstructures could be obtained, which make them attractive in some applications requiring strong mechanical performances.

References

1. Boulesteix R, Maitre A, Chretien L, Rabinovitch Y, Salle C (2013) Microstructural evolution during vacuum sintering of yttrium aluminum garnet transparent ceramics: toward the origin of residual porosity affecting the transparency. J Am Ceram Soc 96:1724–1731
2. Stevenson AJ, Bittel BC, Leh CG, Li X, Dickey EC, Lenahan PM et al (2011) Color center formation in vacuum sintered $Nd_{3x}Y_{3-3x}Al_5O_{12}$ transparent ceramics. Appl Phys Lett 98:051906
3. Tang F, Cao YG, Huang JQ, Guo W, Liu HG, Huang QF et al (2012) Multilayer YAG/RE: YAG/YAG laser ceramic prepared by tape casting and vacuum sintering method. J Eur Ceram Soc 32:3995–4002
4. Xu J, Shi Y, Xie JJ, Lei F (2013) Fabrication, microstructure, and luminescent properties of Ce^{3+}-Doped $Lu_3Al_5O_{12}$ (Ce:LuAG) transparent ceramics by low-temperature vacuum sintering. J Am Ceram Soc 96:1930–1936
5. Ikesue A, Furusato I, Kamata K (1995) Fabrication of polycrystalline transparent YAG ceramics by a solid-state reaction method. J Am Ceram Soc 78:225–228
6. Ikesue A, Kamata K, Yoshida K (1995) Synthesis of Nd^{3+}, Cr^{3+}-coped YAG ceramics for high-efficiency solid-state lasers. J Am Ceram Soc 78:2545–2547
7. Ikesue A, Kinoshita T, Kamata K, Yoshida K (1995) Fabrication and optical properties of high-performance polycrystalline Nd-YAG ceramics for solid-state lasers. J Am Ceram Soc 78:1033–1040
8. Chang XH, He DW, Zhang J, Zou YT, Wang JH, Lu TC (2010) Effects of vacuum heat treatment of nano powder on sintering of transparent nano-crystalline $MgAl_2O_4$ ceramics. In: Pan W, Gong J, (eds) High-performance ceramics VI. p 657–660
9. Li JG, Ikegami T, Lee JH, Mori T, Yajima Y (2001) A wet-chemical process yielding reactive magnesium aluminate spinel ($MgAl_2O_4$) powder. Ceram Int 27:481–489
10. Sun YF, Lu TC, Wang XJ (2007) Preparation and optical properties of transparent Mn: $MgAl_2O_4$ ceramics. Rare Metal Mater Eng 36:438–441
11. Rozenburg K, Reimanis IE, Kleebe HJ, Cook RL (2008) Sintering kinetics of a $MgAl_2O_4$ spinel doped with LiF. J Am Ceram Soc 91:444–450
12. Mata-Osoro G, Moya JS, Pecharroman C (2012) Transparent alumina by vacuum sintering. J Eur Ceram Soc 32:2925–2933
13. Zhang X, Liang S, Zhang P, Zhao T, Bai Y, Bao CG et al (2012) Fabrication of transparent alumina by rapid vacuum pressureless sintering technology. J Am Ceram Soc 95:2116–2119
14. Li WJ, Zhou SM, Liu N, Lin H, Teng H, Li YK et al (2010) Synthesis and spectral properties of Yb^{3+}/Ho^{3+} co-doped yttria 2 mu m transparent ceramics. Mater Lett 64:1344–1346
15. Wen L, Sun XD, Lu Q, Xu GX, Hu XZ (2006) Synthesis of yttria nanopowders for transparent yttria ceramics. Opt Mater 29:239–245

16. Jin LL, Zhou GH, Shimai S, Zhang J, Wang SW (2010) ZrO_2-doped Y_2O_3 transparent ceramics via slip casting and vacuum sintering. J Eur Ceram Soc 30:2139–2143
17. Huang YH, Jiang DL, Zhang JX, Lin QL (2009) Fabrication of transparent lanthanum-doped yttria ceramics by combination of two-step sintering and vacuum sintering. J Am Ceram Soc 92:2883–2887
18. Hou XR, Zhou SM, Jia TT, Lin H, Teng H (2011) Effect of Nd concentration on structural and optical properties of Nd:Y_2O_3 transparent ceramic. J Lumin 131:1953–1958
19. Wang Y, Lu B, Sun X, Sun T, Xu H (2011) Synthesis of nanocrystalline Sc_2O_3 powder and fabrication of transparent Sc_2O_3 ceramics. Adv Appl Ceram 110:95–98
20. Wang ZJ, Zhou GH, Qin XP, Yang Y, Zhang GJ, Menke Y et al (2014) Transparent $La_{2-x}Gd_xZr_2O_7$ ceramics obtained by combustion method and vacuum sintering. J Alloy Compd 585:497–502
21. Zhou GH, Wang ZJ, Zhou BZ, Zhao Y, Zhang GJ, Wang SW (2013) Fabrication of transparent $Y_2Hf_2O_7$ ceramics via vacuum sintering. Opt Mater 35:774–777
22. Zhou J, Zhang WX, Wang LA, Shen YQ, Li J, Liu WB et al (2011) Fabrication, microstructure and optical properties of polycrystalline Er^{3+}:$Y_3Al_5O_{12}$ ceramics. Ceram Int 37:119–125
23. Zhou J, Zhang WX, Huang TD, Wang LA, Li J, Liu WB et al (2011) Optical properties of Er, Yb co-doped YAG transparent ceramics. Ceram Int 37:513–519
24. Liu WB, Li J, Jiang BX, Zhang D, Pan YB (2012) Effect of La_2O_3 on microstructures and laser properties of Nd:YAG ceramics. J Alloy Compd 512:1–4
25. Li J, Wu YS, Pan YB, Liu WB, Huang LP, Guo JK (2008) Laminar-structured YAG/Nd: YAG/YAG transparent ceramics for solid-state lasers. Int J Appl Ceram Technol 5:360–364
26. Appiagyei KA, Messing GL, Dumm JQ (2008) Aqueous slip casting of transparent yttrium aluminum garnet (YAG) ceramics. Ceram Int 34:1309–1313
27. Li YK, Zhou SM, Lin H, Hou XR, Li WJ, Teng H et al (2010) Fabrication of Nd:YAG transparent ceramics with TEOS, MgO and compound additives as sintering aids. J Alloy Compd 502:225–230
28. Qin XP, Yang H, Zhou GH, Luo DW, Yang Y, Zhang J et al (2012) Fabrication and properties of highly transparent Er:YAG ceramics. Opt Mater 34:973–976
29. Gong H, Tang DY, Huang H, Ma J (2009) Fabrication of yttrium aluminum garnet transparent ceramics from yttria nanopowders synthesized by carbonate precipitation. J Electroceram 23:89–93
30. Qin XP, Zhou GH, Yang H, Wong JI, Zhang J, Luo DW et al (2012) Fabrication and plasma resistance properties of transparent YAG ceramics. Ceram Int 38:2529–2535
31. Gong H, Zhang J, Tang DY, Xie GQ, Huang H, Ma J (2011) Fabrication and laser performance of highly transparent Nd:YAG ceramics from well-dispersed Nd:Y_2O_3 nanopowders by freeze-drying. J Nanopart Res 13:3853–3860
32. Yang H, Qin XP, Zhang J, Ma J, Tang DY, Wang SW et al (2012) The effect of MgO and SiO_2 codoping on the properties of Nd:YAG transparent ceramic. Opt Mater 34:940–943
33. Zhang W, Lu TC, Wei NA, Wang YZ, Ma BY, Li F et al (2012) Assessment of light scattering by pores in Nd:YAG transparent ceramics. J Alloy Compd 520:36–41
34. Ramirez MO, Wisdom J, Li H, Aung YL, Stitt J, Messing GL et al (2008) Three-dimensional grain boundary spectroscopy in transparent high power ceramic laser materials. Opt Express 16:5965–5973
35. Lee SH, Kochawattana S, Messing GL, Dumm JQ, Quarles G, Castillo V (2006) Solid-state reactive sintering of transparent polycrystalline Nd:YAG ceramics. J Am Ceram Soc 89:1945–1950
36. Yagi H, Takaichi K, Ueda K, Yanagitani T, Karninskii AA (2006) Influence of annealing conditions on the optical properties of chromium-doped ceramic $Y_3Al_5O_{12}$. Opt Mater 29:392–396
37. Yagi H, Yanagitani T, Takaichi K, Ueda K, Kaminskii AA (2007) Characterizations and laser performances of highly transparent Nd^{3+}:$Y_3Al_5O_{12}$ laser ceramics. Opt Mater 29:1258–1262

38. Serantoni M, Piancastelli A, Costa AL, Esposito L (2012) Improvements in the production of Yb:YAG transparent ceramic materials: spray drying optimisation. Opt Mater 34:995–1001
39. Chen DJ, Mayo MJ (1996) Rapid rate sintering of nanocrystalline ZrO_2-3mol% Y_2O_3. J Am Ceram Soc 79:906–912
40. Zhang J, Meng F, Todd RI, Fu Z (2010) The nature of grain boundaries in alumina fabricated by fast sintering. Scripta Mater 62:658–661
41. Hou XR, Zhou SM, Li WJ, Li YK (2010) Study on the effect and mechanism of zirconia on the sinterability of yttria transparent ceramic. J Eur Ceram Soc 30:3125–3129
42. Chen PL, Chen IW (1996) Grain boundary mobility in Y_2O_3: defect mechanism and dopant effects. J Am Ceram Soc 79:1801–1809
43. Hamano K, Kanzaki S (1977) Fabrication of transparent spinel ceramics by reactive hot-pressing. J Ceram Soc Jpn 85:225–230
44. Cook R, Kochis M, Reimanis I, Kleebe HJ (2005) A new powder production route for transparent spinel windows: powder synthesis and window properties. In: Tustison RW (ed) Window and dome technologies and materials IX. SPIE-The International Society for Optical Engineering, Bellingham, pp 41–47
45. Zou YT, He DW, Wei XK, Yu RC, Lu TC, Chang XH et al (2010) Nanosintering mechanism of $MgAl_2O_4$ transparent ceramics under high pressure. Mater Chem Phys 123:529–533
46. Lu TC, Chang XH, Qi JQ, Luo XJ, Wei QM, Zhu S, et al (2006) Low-temperature high-pressure preparation of transparent nanocrystalline $MgAl_2O_4$ ceramics. Appl Phys Lett 88 (21):213120–213120
47. Zhang J, Lu TC, Chang XH, Wei N, Xu W (2009) Related mechanism of transparency in $MgAl_2O_4$ nano-ceramics prepared by sintering under high pressure and low temperature. J Phys D-Appl Phys 42(5):052002
48. Chen QY, Meng CM, Lu TC, Chang XH, Ji GF, Zhang L et al (2010) Enhancement of sintering ability of magnesium aluminate spinel ($MgAl_2O_4$) ceramic nanopowders by shock compression. Powder Technol 200:91–95
49. Wollmershauser JA, Feigelson BN, Gorzkowski EP, Ellis CT, Goswami R, Qadri SB et al (2014) An extended hardness limit in bulk nanoceramics. Acta Mater 69:9–16
50. Hall EO (1951) The deformation and aging of mild steel, 3. Discussion of results. Proc Phys Soc Lond Sect B 64:747–753
51. Petch NJ (1953) The cleavage strength of polycrystals. J Iron Steel Inst 174:25–28
52. Liu J, Liu K, Wang HS, Gao F, Liao R (2012) Preparation of silicon nitride porous ceramics. In: Pan W, Gong JH (eds) High-performance ceramics III, parts 1 and 2. Trans Tech Publications Ltd., Stafa-Zurich, pp 824–827
53. Hou XR, Zhou SM, Jia TT, Lin H, Teng H (2011) White light emission in $Tm^{3+}/Er^{3+}/Yb^{3+}$ tri-doped Y_2O_3 transparent ceramic. J Alloy Compd 509:2793–2796
54. Podowitz SR, Gaume R, Feigelson RS (2010) Effect of europium concentration on densification of transparent Eu:Y_2O_3 scintillator ceramics using hot pressing. J Am Ceram Soc 93:82–88
55. Haertling GH, Land CE (1971) Hot-pressed (Pb, La)(Zr, Ti)O_3 ferroelectric ceramics for electrooptic applications. J Am Ceram Soc 54:1–11
56. Jiang H, Zou YK, Chen Q, Li KK, Zhang R, Wang Y et al (2005) Transparent electro-optic ceramics and devices. In: Ming H, Zhang XP, Chen MY (eds) Optoelectronic devices and integration, pts 1 and 2. SPIE-The International Society for Optical Engineering, Bellingham, pp 380–394
57. Sun P, Xu CN, Akiyama M, Watanabe T (1999) Controlled oxygen partial pressure sintering of (Pb, La)(Zr, Ti)O_3 ceramics. J Am Ceram Soc 82:1447–1450
58. Yin QR, Ding AL, Zheng XS, Qiu PS, Shen MR, Cao WW (2004) Preparation and characterization of transparent PZN–PLZT ceramics. J Mater Res 19:729–732
59. Ruan W, Li GR, Zeng JT, Bian JJ, Kamzina LS, Zeng HR et al (2010) Large electro-optic effect in La-doped $0.75Pb(Mg_{1/3}Nb_{2/3})O_3$-$0.25PbTiO_3$ transparent ceramic by two-stage sintering. J Am Ceram Soc 93:2128–2131

60. Zeng X, He XY, Cheng WX, Qiu PS, Xia B (2014) Effect of Dy substitution on ferroelectric, optical and electro-optic properties of transparent $Pb_{0.90}La_{0.10}(Zr_{0.65}Ti_{0.35})O_3$ ceramics. Ceram Int 40:6197–6202
61. Ji WL, He XY, Cheng WX, Qiu PS, Zeng X, Xia B et al (2015) Effect of La content on dielectric, ferroelectric and electro-optic properties of $Pb(Mg_{1/3}Nb_{2/3})O_3$–$PbTiO_3$ transparent ceramics. Ceram Int 41:1950–1956
62. Giniewicz JR, McHenry DA, Shrout TR, Jang SJ, Bhalla AS, Ainger FW (1990) Characterization of $(1-x)Pb(Mg_{1/3}Nb_{2/3})O_3$-$(x)PbTiO_3$ and $Pb(Sc_{1/2}Ta_{1/2})O_3$ transparent ceramics prepared by uniaxial hot-pressing. Ferroelectrics 109:167–172
63. Su XL, Wang PL, Chen WW, Zhu B, Cheng YB, Yan DS (2004) Translucent α-sialon ceramics by hot pressing. J Am Ceram Soc 87:730–732
64. Xie ZH, Hoffman M, Moon RJ, Munroe PR, Cheng YB (2004) Subsurface indentation damage and mechanical characterization of α-sialon ceramics. J Am Ceram Soc 87:2114–2124
65. Xie ZH, Hoffman M, Cheng YB (2002) Microstructural tailoring and characterization of a calcium alpha-SiAlON composition. J Am Ceram Soc 85:812–818
66. Jones MI, Hyuga H, Hirao K (2003) Optical and mechanical properties of α/β composite sialons. J Am Ceram Soc 86:520–522
67. Fang Y, Agrawal D, Skandan G, Jain M (2004) Fabrication of translucent MgO ceramics using nanopowders. Mater Lett 58:551–554
68. Lyberis A, Stevenson AJ, Suganuma A, Ricaud S, Druon F, Herbst F et al (2012) Effect of Yb^{3+} concentration on optical properties of $Yb:CaF_2$ transparent ceramics. Opt Mater 34:965–968
69. Lyberis A, Patriarche G, Gredin P, Vivien D, Mortier M (2011) Origin of light scattering in ytterbium doped calcium fluoride transparent ceramic for high power lasers. J Eur Ceram Soc 31:1619–1630
70. Swinkels FB, Wilkinson DS, Arzt E, Ashby MF (1983) Mechanism of hot-isostatic pressing. Acta Metall 31:1829–1840
71. Li WB, Ashby MF, Easterling KE (1987) On densification and shape change during hot isostatic pressing. Acta Metall 35:2831–2842
72. Atkinson HV, Davies S (2000) Fundamental aspects of hot isostatic pressing: an overview. Metall Mater Trans A-Phys Metall Mater Sci 31:2981–3000
73. Loh NL, Sia KY (1992) An overview of hot isostatic pressing. J Mater Process Technol 30:45–65
74. Bocanegra-Bernal MH (2004) Hot isostatic pressing (HIP) technology and its applications to metals and ceramics. J Mater Sci 39:6399–6420
75. Petit J, Dethare P, Sergent A, Marino R, Ritti MH, Landais S et al (2011) Sintering of α-alumina for highly transparent ceramic applications. J Eur Ceram Soc 31:1957–1963
76. Krell A, Klimke J (2006) Effects of the homogeneity of particle coordination on solid-state sintering of transparent alumina. J Am Ceram Soc 89:1985–1992
77. Krell A, Klimke J, Hutzler T (2009) Advanced spinel and sub-μm Al_2O_3 for transparent armour applications. J Eur Ceram Soc 29:275–281
78. Krell A, Blank P, Ma HW, Hutzler T, Nebelung M (2003) Processing of high-density submicrometer Al_2O_3 for new applications. J Am Ceram Soc 86:546–553
79. Krell A, Blank P, Ma HW, Hutzler T, van Bruggen MPB, Apetz R (2003) Transparent sintered corundum with high hardness and strength. J Am Ceram Soc 86:12–18
80. Ikesue A, Kamata K (1996) Microstructure and optical properties of hot isostatically pressed Nd:YAC ceramics. J Am Ceram Soc 79:1927–1933
81. Lee SH, Kupp ER, Stevenson AJ, Anderson JM, Messing GL, Li X et al (2009) Hot isostatic pressing of transparent Nd:YAG ceramics. J Am Ceram Soc 92:1456–1463
82. Suarez M, Fernandez A, Menendez JL, Nygren M, Torrecillas R, Zhao Z (2010) Hot isostatic pressing of optically active Nd:YAG powders doped by a colloidal processing route. J Eur Ceram Soc 30:1489–1494

83. Zhang W, Lu TC, Ma BY, Wei N, Lu ZW, Li F et al (2013) Improvement of optical properties of Nd:YAG transparent ceramics by post-annealing and post hot isostatic pressing. Opt Mater 35:2405–2410
84. Dericioglu AF, Kagawa Y (2003) Effect of grain boundary microcracking on the light transmittance of sintered transparent $MgAl_2O_4$. J Eur Ceram Soc 23:951–959
85. Dericioglu AF, Boccaccini AR, Dlouhy I, Kagawa Y (2005) Effect of chemical composition on the optical properties and fracture toughness of transparent magnesium aluminate spinel ceramics. Mater Trans 46:996–1003
86. Esposito L, Piancastelli A, Miceli P, Martelli S (2015) A thermodynamic approach to obtaining transparent spinel ($MgAl_2O_4$) by hot pressing. J Eur Ceram Soc 35:651–661
87. Sutorik AC, Gilde G, Cooper C, Wright J, Hilton C (2012) The effect of varied amounts of LiF sintering aid on the transparency of alumina rich spinel ceramic with the composition $MgO\text{-}1.5Al_2O_3$. J Am Ceram Soc 95:1807–1810
88. Sutorik AC, Gilde G, Swab JJ, Cooper C, Gamble R, Shanholtz E (2012) Transparent solid solution magnesium aluminate spinel polycrystalline ceramic with the alumina-rich composition $MgO\text{-}1.2Al_2O_3$. J Am Ceram Soc 95:636–643
89. Mouzon J, Maitre A, Frisk L, Lehto N, Oden M (2009) Fabrication of transparent yttria by HIP and the glass-encapsulation method. J Eur Ceram Soc 29:311–316
90. Serivalsatit K, Ballato J (2010) Submicrometer grain-sized transparent erbium-doped scandia ceramics. J Am Ceram Soc 93:3657–3662
91. Seeley ZM, Kuntz JD, Cherepy NJ, Payne SA (2011) Transparent Lu_2O_3:Eu ceramics by sinter and HIP optimization. Opt Mater 33:1721–1726
92. Tsukuma K, Yamashita I, Kusunose T (2008) Transparent 8 mol% $Y_2O_3\text{-}ZrO_2$ (8Y) ceramics. J Am Ceram Soc 91:813–818
93. Oh KS, Kim DY, Cho SJ (1995) Shrinkage of large isolated pores during hot isostatic pressing of presintered alumina ceramics. J Am Ceram Soc 78:2537–2540
94. Sakuma T, Ikuhara Y, Takigawa Y, Thavorniti P (1997) Importance of grain boundary chemistry on the high-temperature plastic flow in oxide ceramics. Mater Sci Eng A-Struct Mater Prop Microstruct Process 234:226–229
95. Yoshida H, Kuwabara A, Yamamoto T, Ikuhara Y, Sakuma T (2005) High temperature plastic flow and grain boundary chemistry in oxide ceramics. J Mater Sci 40:3129–3135
96. Hayashi K, Kobayashi O, Toyoda S, Morinaga K (1991) Transmission optical-properties of polycrystalline alumina with submicro grains. Mater Trans JIM 32:1024–1029
97. Tsukuma K (2006) Transparent $MgAl_2O_4$ spinel ceramics produced by HIP post-sintering. J Ceram Soc Jpn 114:802–806
98. Cherepy NJ, Kuntz JD, Roberts JJ, Hurst TA, Drury OB, Sanner RD et al (2008) Transparent ceramic scintillator fabrication, properties and applications. In: Burger A, Franks LA, James RB (eds) Hard X-ray, gamma-ray, and neutron detector physics X. SPIE-The International Society for Optical Engineering, Bellingham
99. Alberta EF, Bhalla AS (2001) Piezoelectric and dielectric properties of transparent $Pb(Ni_{1/3}Nb_{2/3})_{1-x-y}Zr_xTi_yO_3$ ceramics prepared by hot isostatic pressing. Int J Inorg Mater 3:987–995
100. Schneider H, Schmucker M, Ikeda K, Kaysser WA (1993) Optically translucent mullite ceramics. J Am Ceram Soc 76:2912–2914
101. Aubry P, Bensalah A, Gredin P, Patriarche G, Vivien D, Mortier M (2009) Synthesis and optical characterizations of Yb-doped CaF_2 ceramics. Opt Mater 31:750–753
102. Uematsu K, Takagi M, Honda T, Uchida N, Saito K (1989) Transparent hydroxyapatite prepared by hog isostatic pressing of filter cake. J Am Ceram Soc 72:1476–1478
103. Mandal H (1999) New developments in α-SiAlON ceramics. J Eur Ceram Soc 19:2349–2357
104. Itatani K, Tsujimoto T, Kishimoto A (2006) Thermal and optical properties of transparent magnesium oxide ceramics fabricated by post hot-isostatic pressing. J Eur Ceram Soc 26:639–645
105. Ji YM, Jiang DY, Fen T, Shi JL (2005) Fabrication of transparent $La_2Hf_2O_7$ ceramics from combustion synthesized powders. Mater Res Bull 40:553–559

106. Roussel N, Lallemant L, Durand B, Guillemet S, Ching JYC, Fantozzi G et al (2011) Effects of the nature of the doping salt and of the thermal pre-treatment and sintering temperature on spark plasma sintering of transparent alumina. Ceram Int 37:3565–3573
107. Kim BN, Hiraga K, Morita K, Yoshida H (2007) Spark plasma sintering of transparent alumina. Scripta Mater 57:607–610
108. Wang C, Zhao Z (2010) Transparent polycrystalline ruby ceramic by spark plasma sintering. Mater Res Bull 45:1127–1131
109. Jin XH, Gao L, Sun J (2010) Highly transparent alumina spark plasma sintered from common-grade commercial powder: the effect of powder treatment. J Am Ceram Soc 93:1232–1236
110. Aman Y, Garnier V, Djurado E (2009) Influence of green state processes on the sintering behaviour and the subsequent optical properties of spark plasma sintered alumina. J Eur Ceram Soc 29:3363–3370
111. Dang KQ, Takei S, Kawahara M, Nanko M (2011) Pulsed electric current sintering of transparent Cr-doped Al_2O_3. Ceram Int 37:957–963
112. Nanko M, Dang KQ (2014) Two-step pulsed electric current sintering of transparent Al_2O_3 ceramics. Adv Appl Ceram 113:80–84
113. Chaim R, Marder-Jaeckel R, Shen JZ (2006) Transparent YAG ceramics by surface softening of nanoparticles in spark plasma sintering. Mater Sci Eng A-Struct Mater Prop Microstruct Process 429:74–78
114. Chaim R, Kalina M, Shen JZ (2007) Transparent yttrium aluminum garnet (YAG) ceramics by spark plasma sintering. J Eur Ceram Soc 27:3331–3337
115. An LQ, Ito A, Goto T (2011) Two-step pressure sintering of transparent lutetium oxide by spark plasma sintering. J Eur Ceram Soc 31:1597–1602
116. Yoshida H, Morita K, Kim BN, Hiraga K, Yamanaka K, Soga K et al (2011) Low-temperature spark plasma sintering of yttria ceramics with ultrafine grain size. J Am Ceram Soc 94:3301–3307
117. Yoshida H, Morita K, Kim BN, Hiraga K, Kodo M, Soga K et al (2008) Densification of nanocrystalline yttria by low temperature spark plasma sintering. J Am Ceram Soc 91:1707–1710
118. Zhang HB, Kim BN, Morita K, Yoshida H, Hiraga K, Sakka Y (2011) Fabrication of transparent yttria by high-pressure spark plasma sintering. J Am Ceram Soc 94:3206–3210
119. An LQ, Ito A, Goto T (2012) Transparent yttria produced by spark plasma sintering at moderate temperature and pressure profiles. J Eur Ceram Soc 32:1035–1040
120. An LQ, Ito A, Goto T (2011) Fabrication of transparent lutetium oxide by spark plasma sintering. J Am Ceram Soc 94:695–698
121. Anselmi-Tamburini U, Woolman JN, Munir ZA (2007) Transparent nanometric cubic and tetragonal zirconia obtained by high-pressure pulsed electric current sintering. Adv Funct Mater 17:3267–3273
122. Zhang HB, Kim BN, Morita K, Hiraga HYK, Sakka Y (2011) Effect of sintering temperature on optical properties and microstructure of translucent zirconia prepared by high-pressure spark plasma sintering. Sci Technol Adv Mater 12:055003
123. Lei LW, Fu ZY, Wang H, Lee SW, Niihara K (2012) Transparent yttria stabilized zirconia from glycine-nitrate process by spark plasma sintering. Ceram Int 38:23–28
124. Chaim R, Shen ZJ, Nygren M (2004) Transparent nanocrystalline MgO by rapid and low-temperature spark plasma sintering. J Mater Res 19:2527–2531
125. Frage N, Cohen S, Meir S, Kalabukhov S, Dariel MP (2007) Spark plasma sintering (SPS) of transparent magnesium-aluminate spinel. J Mater Sci 42:3273–3275
126. Meir S, Kalabukhov S, Froumin N, Dariel MP, Frage N (2009) Synthesis and densification of transparent magnesium aluminate spinel by SPS processing. J Am Ceram Soc 92:358–364
127. Morita K, Kim BN, Hiraga K, Yoshida H (2008) Fabrication of transparent $MgAl_2O_4$ spinel polycrystal by spark plasma sintering processing. Scripta Mater 58:1114–1117

128. Zhang GM, Wang YC, Fu ZY, Wang H, Wang WM, Zhang JY et al (2009) Transparent mullite ceramic from single-phase gel by spark plasma sintering. J Eur Ceram Soc 29:2705–2711
129. Zhang GM, Fu ZY, Wang YC, Wang H, Wang WM, Zhang JY (2009) Effect of precursors on transmittance and microstructure of mullite ceramics. In: Kim H, Yang JF, Sekino T, Lee SW (eds) Eco-materials processing and design X. p 429–432
130. Watanabe Y, Ikoma T, Monkawa A, Suetsugu Y, Yamada H, Tanaka J et al (2005) Fabrication of transparent hydroxyapatite sintered body with high crystal orientation by pulse electric current sintering. J Am Ceram Soc 88:243–245
131. Eriksson M, Liu Y, Hu JF, Gao L, Nygren M, Shen ZJ (2011) Transparent hydroxyapatite ceramics with nanograin structure prepared by high pressure spark plasma sintering at the minimized sintering temperature. J Eur Ceram Soc 31:1533–1540
132. An L, Ito A, Goto T (2011) Fabrication of transparent Lu_3NbO_7 by spark plasma sintering. Mater Lett 65:3167–3169
133. Xiong Y, Fu ZY, Wang YC, Quan F (2006) Fabrication of transparent AIN ceramics. J Mater Sci 41:2537–2539
134. Wu YJ, Kimura R, Uekawa N, Kakegawa K, Sasaki Y (2002) Spark plasma sintering of transparent $PbZrO_3$–$PbTiO_3$–$Pb(Zn_{1/3}Nb_{2/3})O_3$ ceramics. Japan J Appl Phys Part 2-Lett 41: L219–L221
135. Wu YJ, Li J, Kimura R, Uekawa N, Kakegawa K (2005) Effects of preparation conditions on the structural and optical properties of spark plasma-sintered PLZT (8/65/35) ceramics. J Am Ceram Soc 88:3327–3331
136. Liu J, Shen ZJ, Yao WL, Zhao YH, Mukherjee AK (2010) Visible and infrared transparency in lead-free bulk $BaTiO_3$ and $SrTiO_3$ nanoceramics. Nanotechnology 21:075706
137. Wu YJ, Wang N, Wu SY, Chen XM (2011) Transparent barium strontium titanate ceramics prepared by spark plasma sintering. J Am Ceram Soc 94:1343–1345
138. Kim B-N, Hiraga K, Morita K, Yoshida H (2009) Effects of heating rate on microstructure and transparency of spark-plasma-sintered alumina. J Eur Ceram Soc 29:323–327
139. Jiang D, Hulbert DM, Anselmi-Tamburini U, Ng T, Land D, Mukherjee AK (2007) Spark plasma sintering and forming of transparent polycrystalline Al_2O_3 windows and domes. In: Tustison RW (ed) Window and dome technologies and materials X. SPIE-The International Society for Optical Engineering, Bellingham, p 54509
140. Wang C, Zhao Z (2009) Transparent $MgAl_2O_4$ ceramic produced by spark plasma sintering. Scripta Mater 61:193–196
141. Krell A, Blank P (1995) Grain-size dependence of hardness in dense submicrometer alumina. J Am Ceram Soc 78:1118–1120
142. Palmero P, Bonelli B, Fantozzi G, Spina G, Bonnefont G, Montanaro L et al (2013) Surface and mechanical properties of transparent polycrystalline YAG fabricated by SPS. Mater Res Bull 48:2589–2597
143. An LQ, Ito A, Goto T (2011) Highly transparent lutetium titanium oxide produced by spark plasma sintenng. J Eur Ceram Soc 31:237–240
144. An LQ, Ito A, Goto T (2012) Effect of calcination temperature on the fabrication of transparent lutetium titanate by spark plasma sintering. Ceram Int 38:4973–4977
145. An LQ, Ito A, Goto T (2013) Fabrication of transparent $Lu_2Hf_2O_7$ by reactive spark plasma sintering. Opt Mater 35:817–819
146. An LQ, Ito A, Goto T (2011) Fabrication of transparent Lu_3NbO_7 by spark plasma sintering. Mater Lett 65:3167–3169
147. An LQ, Ito A, Goto T (2013) Transparent Lu_3NbO_7 bodies prepared by reactive spark plasma sintering and their optical and mechanical properties. Ceram Int 39:383–387
148. Sahin FC, Kanbur HE, Apak B (2012) Preparation of AlON ceramics via reactive spark plasma sintering. J Eur Ceram Soc 32:925–929
149. Chen S, Wu YQ, Yang Y (2013) Spark plasma sintering of hexagonal structure Yb^{3+}-doped $Sr_5(PO_4)_3F$ transparent ceramics. J Am Ceram Soc 96:1694–1697

150. Cheng JP, Agrawal D, Zhang YJ, Roy R (2002) Microwave sintering of transparent alumina. Mater Lett 56:587–592
151. Brosnan KH, Messing GL, Agrawal DK (2003) Microwave sintering of alumina at 2.45 GHz. J Am Ceram Soc 86:1307–1312
152. Fang Y, Cheng JP, Agrawal DK (2004) Effect of powder reactivity on microwave sintering of alumina. Mater Lett 58:498–501
153. Esposito L, Piancastelli A, Bykov Y, Egorov S, Eremeev A (2013) Microwave sintering of Yb:YAG transparent laser ceramics. Opt Mater 35:761–765
154. Cheng JP, Agrawal D, Zhang YJ, Roy R (2001) Microwave reactive sintering to fully transparent aluminum oxynitride (AlON) ceramics. J Materi Sci Lett 20:77–79
155. Fang Y, Roy R, Agrawal DK, Roy DM (1996) Transparent mullite ceramics from diphasic aerogels by microwave and conventional processings. Mater Lett 28:11–15
156. Li MJ, Wu YC, Yen FS, Huang CY (2011) Influence of ionic mobility on the phase transformation route in $Y_3Al_5O_{12}$ (YAG) stoichiometry. J Eur Ceram Soc 31:2099–2106
157. Rybakov KI, Semenov VE (1994) Possibility of plastic-deformation of an ionic-crystal due to the nonthermal influence of a high-frequency electric-field. Phys Rev B 49:64–68
158. Rybakov KI, Semenov VE (1995) Mass-transport in ionic-crystals induced by the ponderomotive action of a high-frequency electric-field. Phys Rev B 52:3030–3033

Chapter 8
Grain Growth and Microstructure Development

8.1 Introduction

The properties of a ceramics are determined by its chemical composition intrinsically and microstructure extrinsically. For ceramics with a given composition, microstructure means the overall feature, which reflects its grain size and morphology, grain size distribution, porosity, pore size and distribution, type and quality of grain boundaries, as well as the nature and distribution of second-phases, and so on. For most applications, especially optical transparency, microstructural control means to achieve full densification, narrow distribution of grain size, least contamination, and so on [1, 2]. The microstructure of a ceramics is directly related the quality of the green compact, which is dependent on the properties of the precursor powder, as well as the consolidation method used to form the green body. Therefore, the control of microstructure of ceramics is a systematic engineering. When all other conditions have been fixed, controlling microstructure during the sintering process becomes especially important. Densification is almost inevitably accompanied by the growth of both grains and pores, which is known as microstructure coarsening. Densification process has been discussed previously; this chapter focuses on coarsening process and the interrelationship between densification and coarsening, in terms of microstructural control.

Although various factors have their influence on microstructure of a ceramics, the effect of grain growth is the most pronounced. There are two types of grain growth: normal and abnormal grain growth (AGG). Normal grain growth (NGG) means that the increase in grain size follows a trend of expected magnification, while the grain shapes have no significant change. In contrast, AGG is characterized by the presence of a limited number of extremely large grains, which are formed at the expense of the smaller ones surrounding them. AGG is usually thought to have a destructive effect on microstructure of ceramics, unless it is a specific requirement. Theoretical and experimental results indicate that the key to control the microstructure or avoid the AGG is to decrease the grain growth rate, but increase the

L.B. Kong et al., *Transparent Ceramics*, Topics in Mining,
Metallurgy and Materials Engineering, DOI 10.1007/978-3-319-18956-7_8

densification rate. Strategies to achieve this include homogeneous packing of fine particles with a narrow size distribution, doping with secondary phase particles, use of liquid-phase sintering, application of external pressure and employment of new sintering techniques, e.g., spark plasma sintering or SPS.

8.2 General Concepts

8.2.1 Features of Grain Growth

Grain growth is to mean the dynamic increase in average grain size of a given ceramic system, which usually takes place at high temperatures during calcining and sintering processes. It is understood that an increase in grain size should be at the expense of smaller grains that surround relatively large growing ones. In other words, the consequence of grain growth is the decrease in the total number of grains. Statistical data indicated that only one eighth of the grains could survive when the average grain size of the system doubles every time. Therefore, grain growth is also called coarsening.

8.2.2 Microscopic Features of Grain Growth

Ceramics consist of crystalline grains and noncrystalline or disordered grain boundaries, which are defined as the regions between adjacent grains. Figure 8.1 shows schematically a section across two adjacent grains [3]. In most ceramics, the thickness of the grain boundary is 0.5–1 nm. Grain growth takes place through the diffusion of atoms or ions over less than an interatomic distance from one side of the boundary to the other side. As a result, one grain grows at the expense of other smaller ones. Generally, the atoms migrate from the convex surface to the concave surface, because the chemical potential of the atoms at the convex surface is higher than that of the atoms at the concave surface, as discussed in previous chapters. Due to the flux, the boundary moves toward its center of curvature, as shown in Fig. 8.2 [4].

8.2.3 Driving Force of Grain Growth

Because the atoms or ions in disordered grain boundaries have higher energy than those in the bulk of crystalline grains. A specific energy, known as grain-boundary energy, γ_{gb}, is usually used to characterize the grain boundaries. The decrease in the energy through the reduction of the total area of the grain boundaries is the driving force of grain growth. Therefore, grain growth is also accompanied by the reduction in volume fraction of grain boundaries.

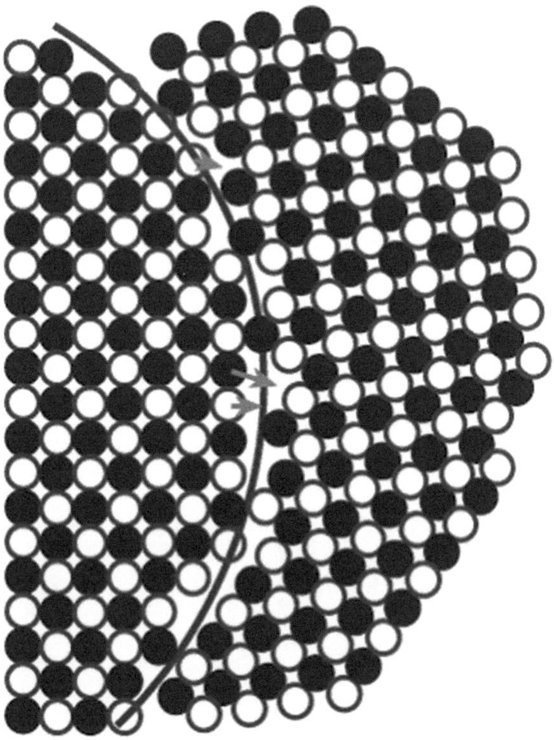

Fig. 8.1 Classic picture of a grain boundary and its migration. The boundary migrates from *left* to *right* as the atoms diffuse from the convex side of the boundary to the concave side. Reproduced with permission from [3]. Copyright © 2007, Springer

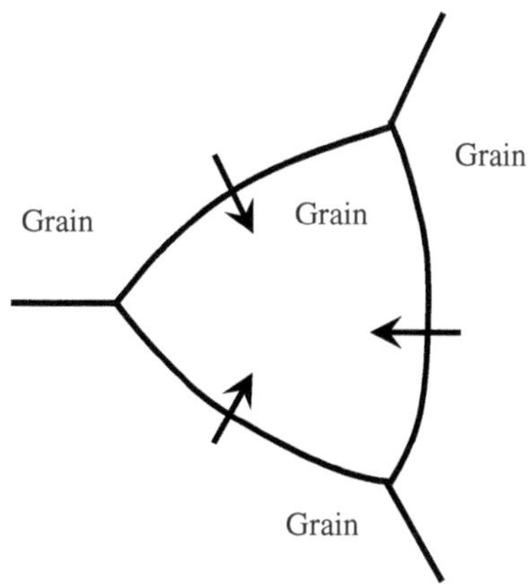

Fig. 8.2 Movement of the grain boundaries toward their center of curvature. Reproduced with permission from [4]. Copyright © 2003, CRC Press

8.2.4 Abnormal Grain Growth

As mentioned above, there are two types of grain growth experienced by ceramics: (i) normal grain growth and (ii) abnormal grain growth. AGG is also known as several other names, including exaggerated grain growth, discontinuous grain growth, or secondary crystallization. NGG is more like a simple magnification in average grain size, without significant change in grain size distribution, so that it is generally known as scaling or self-similarity property. However, when AGG occurs, the ceramics would have a distinctive variation in microstructure, with the presence of extremely large grains that are embedded in a matrix of smaller ones. Therefore, a bimodal distribution is usually observed. The microstructures of the ceramics are generally characterized by nearly equiaxial grains. In some cases, grains with an elongated or plate-like morphology are observed, which usually is called anisotropic grain growth. Anisotropic grain growth in most cases is only observed in ceramics with special crystal structures. Representative SEM images and photograph of NGG and AGG of $BaTiO_3$ ceramics, which could be controlled through the variation of oxygen partial pressure during sintering, are shown in Fig. 8.3 [5].

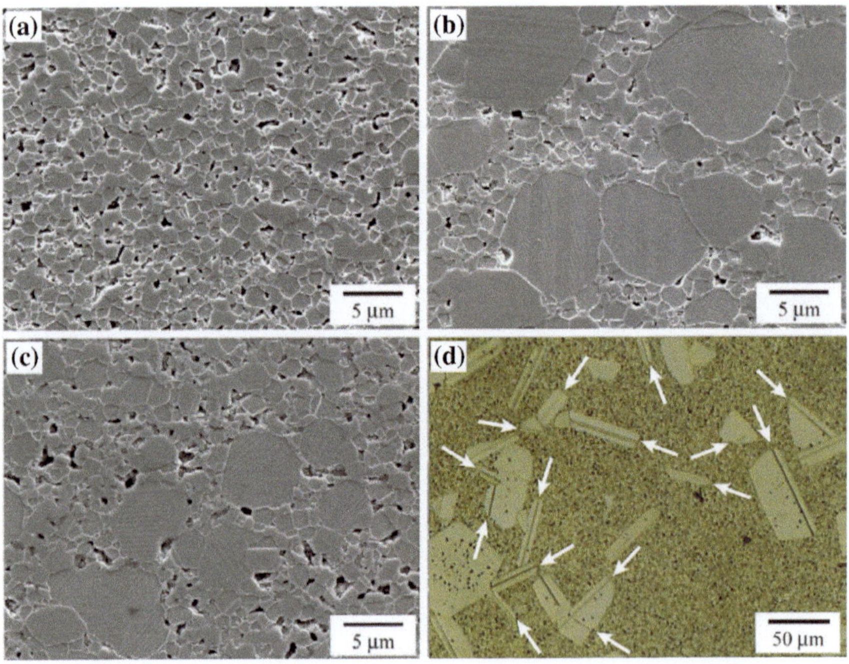

Fig. 8.3 SEM images and photograph of the $BaTiO_3$ ceramics: **a** normal grain growth, **b**, **c** abnormal grain growth, and **d** abnormal anisotropic grain growth. Reproduced with permission from [5]. Copyright © 2006, Elsevier

8.2.5 Grain Growth Control

At least two reasons can be used to support the control of grain growth during sintering of ceramics. On the one hand, as the grain size is increased, the diffusion distance of matter transport is increased. Therefore, grain growth has a negative effect on densification. On the other hand, many properties of ceramics are directly determined by their grain size, especially mechanical strengths [1]. For instance, fracture strength of ceramics generally increases with decreasing grain size, G, following the relation of $K_{1c} \infty 1/G^{1/2}$ [6–8]. In this respect, control of grain growth while not sacrificing optical performance is still a challenge in the fabrication of transparent ceramics. Similarly, magnetic and electrical properties of ceramics are determined and also have close relation with their grain sizes.

As discussed previously, densification of ceramics is attributed to the flux of matter from the grain boundaries or source to the pores or sinks. When the sintering is governed by diffusion mechanisms, the rate of densification is dependent on grain size G, through the following equation:

$$\frac{1}{\rho}\frac{\mathrm{d}\rho}{\mathrm{d}t} = \frac{K}{G^m}, \tag{8.1}$$

where K is a temperature-dependent constant, while the exponent has different values for different densification mechanisms, e.g., with $m = 3$ for lattice diffusion and $m = 4$ for grain-boundary diffusion.

Therefore, to have a high densification rate, there should be a short diffusion distance between the source of matter and the sink, i.e., the grain size must remain as small as possible. This is the reason why nanosized powders have high sinterability. Equation (8.1) also indicates that as the grain growth rate is increased, the densification rate would be decreased. As a result, prolonged sintering time is required to achieve the desired density, which in turn could trigger the occurrence of AGG. Once AGG takes place, the pores would be trapped inside the grains and thus cannot be removed. Therefore, grain growth control is an effective technique to achieve high densification rate. Generally, AGG should be avoided, but it is also useful in certain occasion, i.e., growth of single crystals from ceramic matrix, as discussed later.

8.3 Ostwald Ripening and LSW Theory

Ostwald ripening is defined as the coarsening of particles in a solid or liquid. Almost all grain growth and pore growth during sintering of ceramics can be described by the using Ostwald ripening. Figure 8.4 shows a system with spherical particles with different radii dispersed in a medium, in which the particles have certain level of solubility [4].

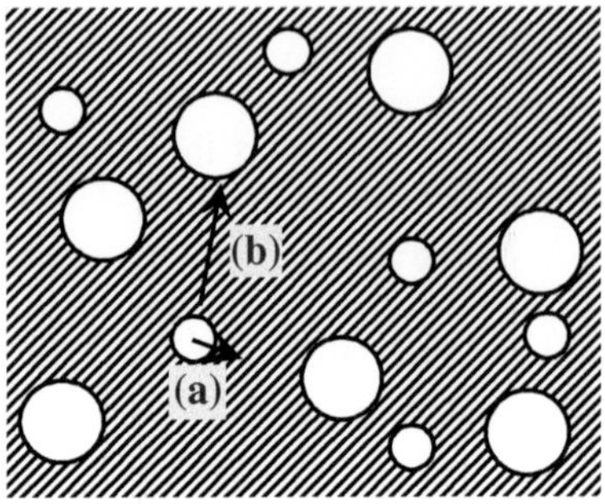

Fig. 8.4 Coarsening of particles in a medium by matter transport from the smaller particles to the larger ones. The rate of coarsening is controlled by two mechanisms: **a** reaction at the interface between the particles and the medium and **b** diffusion through the medium. Reproduced with permission from [4]. Copyright © 2003, CRC Press

As stated before, chemical potential of the atoms under the surface of a sphere with a radius a is given by:

$$\mu = \mu_0 + \frac{2\gamma\Omega}{a}, \tag{8.2}$$

where μ_0 is the chemical potential of the atoms under a flat surface, γ is the specific energy of the interface between the sphere and the medium and Ω is the atomic volume. The atoms under the surface of the sphere have higher chemical potential and thus higher solubility in the surrounding medium than the atoms under a flat surface. If solutions are assumed to be ideal, chemical potential and concentration can be related by the following expression:

$$kT \ln\left(\frac{C}{C_0}\right) = \mu - \mu_0 = \frac{2\gamma\Omega}{a}, \tag{8.3}$$

where C is concentration of the solute surrounding a particle with radius of a, C_0 is the concentration over a flat surface, k is the Boltzmann constant, and T is the absolute temperature. If $\Delta C = C - C_0$ is small, $\ln(C/C_0) \approx \Delta C/C_0$, so that Eq. (8.3) becomes

$$\frac{\Delta C}{C_0} = \frac{2\gamma\Omega}{kTa}. \tag{8.4}$$

The smaller the particles, the higher the solute concentration around the particles will be. As a result, there is always a net flux of matter from the smaller particles to the larger ones, i.e., the smaller particles are continuously dissolved, while the particles keep growing, which is driven by the reduction in the interfacial area between the particles and the medium. This process is known as Ostwald ripening.

8.3.1 LSW Theory

The theory of Ostwald ripening is also known as the LSW theory [9–12]. The LSW theory is used to describe the coarsening process of idealized systems with three conditions. Firstly, a particle grows at a rate that is the same as the atomic flux at its surface. Secondly, the particles have a continuous size distribution. Lastly, the total volume or mass of the particles follows mass conservation.

Besides the above three basic conditions, there are also other assumptions: (i) the precipitate and the medium are isotropic fluids, (ii) the precipitates are spherical particles, (iii) the number of precipitates is sufficiently large to ensure continuous distribution of radius of the precipitates, (iv) radius of the precipitate is the only factor to determine the solute concentration at the surface of the precipitate, (v) nucleation and precipitate coalescence are neglected, and (vi) the total volume of the system is infinite.

The rate of coarsening is controlled by two mechanisms, i.e., interface reaction mechanism and diffusion-controlled mechanism, as shown in Fig. 8.4. For interface reaction mechanism, coarsening rate is controlled either by the solubility of the particles into the medium or the deposition of the solute onto the particle surfaces. For diffusion-controlled mechanism, the diffusion of atoms through the medium to combine with larger particles is the controlling step.

8.3.2 Interface Reaction Mechanism

In this case, it is assumed that the rate of transfer of atoms is proportional to the difference between the solute concentration around a precipitate with radius, a, which is given by Eq. (8.4) and the average concentration of the solute, C^*, which is defined as the concentration that is in equilibrium with a precipitate with radius, a^*, which is kept unchanged. Also, the change in C^* with increasing a^* is neglected. Therefore, the rate of change in the radius of the precipitate is given by [4]:

$$\frac{\mathrm{d}a}{\mathrm{d}t} = -\alpha_{\mathrm{T}}\Omega(C_{\mathrm{a}} - C^*), \tag{8.5}$$

where α_{T} is a transfer constant, while the negative sign in the equation is used, because $C_{\mathrm{a}} - C^*$ is positive for smaller precipitates, whereas the radius, a, decreases with time. If the total volume of the precipitates is constant, there is

$$4\pi \sum_i a_i^2 \frac{\mathrm{d}a_i}{\mathrm{d}t} = 0. \tag{8.6}$$

The summation means that all the precipitates in the system are included. Equation (8.5) can be rewritten as follows:

$$\frac{\mathrm{d}a}{\mathrm{d}t} = \alpha_{\mathrm{T}}\Omega[(C^* - C_0) - (C_{\mathrm{a}} - C_0)], \tag{8.7}$$

By putting Eq. (8.7) into Eq. (8.6), there is

$$\sum_i a_i^2(C^* - C_0) = \sum_i a_i^2(C_{\mathrm{a}_i} - C_0). \tag{8.8}$$

By putting $C_{\mathrm{a}_i} - C_0 = \Delta C$ and ΔC from Eq. (8.4) into Eq. (8.8), there is

$$C^* - C_0 = \frac{2\gamma\Omega C_0}{kT}\frac{\sum a_i}{\sum a_i^2}. \tag{8.9}$$

By putting Eq. (8.9) into Eq. (8.7), there is

$$\frac{\mathrm{d}a}{\mathrm{d}t} = \frac{2\alpha_{\mathrm{T}}\gamma\Omega^2 C_0}{kT}\left(\frac{\sum a_i}{\sum a_i^2} - \frac{1}{a}\right). \tag{8.10}$$

Let $a^* = \sum a_i^2 / \sum a_i$, Eq. (8.10) becomes:

$$\frac{\mathrm{d}a}{\mathrm{d}t} = \frac{2\alpha_{\mathrm{T}}\gamma\Omega^2 C_0}{kT}\left(\frac{1}{a^*} - \frac{1}{a}\right). \tag{8.11}$$

Equation (8.11) indicates that the rate of change in the radius of the precipitate is proportional to the difference between the critical precipitate curvature and the actual precipitate curvature.

The evolution of the system of precipitates can be described by using a distribution function $f(a, t)$, with which $f(a, t)\,\mathrm{d}a$ represents the fractional number of precipitates in the radium range of $(a, a + \mathrm{d}a)$. The distribution function must satisfy the continuity equation:

$$\frac{\mathrm{d}f}{\mathrm{d}t} + \frac{\partial}{\partial a}\left[f\left(\frac{\mathrm{d}a}{\mathrm{d}t}\right)\right] = 0. \tag{8.12}$$

From the solution of the coupled differential Eqs. (8.11) and (8.12), it is found that the rate of the coarsening reaches a steady state, so that the precipitate size distribution remains stationary against time, after a prolonged duration of coarsening process. When a reduced size, $s = a/a^*$, is used to represent the radius of the precipitate, the distribution function will have the following forms [4]:

$$f(s,t)\begin{cases} \approx s\left(\dfrac{2}{2-s}\right)^5 \exp\left(-\dfrac{3s}{2-s}\right), & \text{for} \quad 0<s<2, \\ = 0, & \text{for} \quad s>2. \end{cases} \tag{8.13}$$

This asymptotic distribution is independent on the initial distribution of the system before the coarsening starts. In this case, the average radius, $\bar{a}$, which is the arithmetic mean radius, is given by $\bar{a} = (8/9)a^*$, while the maximum particle radius is $2a^*$. The critical radius in this steady-state coarsening regime increases parabolically through the following equation:

$$(a^*)^2 - (a_0^*)^2 = \left(\frac{\alpha_T C_0 \gamma \Omega^2}{kT}\right)t. \tag{8.14}$$

8.3.3 *Diffusion-Controlled Mechanism*

Similar procedure can be used to derive the coarsening rate governed by the diffusion mechanism. The rate of change in the particle radius, which is related to the diffusive flux at the precipitate surface, is given by [4, 13]:

$$\frac{\mathrm{d}a}{\mathrm{d}t} = -D\Omega\frac{\mathrm{d}C}{\mathrm{d}a}, \tag{8.15}$$

where D is the diffusion coefficient of the solute atoms in the medium.

For a dilute dispersion of particles, Eq. (8.15) can be rewritten as follows:

$$\frac{\mathrm{d}a}{\mathrm{d}t} = D\Omega\left(\frac{C^* - C_a}{a}\right) = \frac{2D\gamma\Omega^2 C_0}{kTa}\left(\frac{1}{a^*} - \frac{1}{a}\right). \tag{8.16}$$

From the solution of the coupled differential Eqs. (8.12) and (8.16), the distribution function for steady-state coarsening has the following forms [4, 13]:

$$f(s,t)\begin{cases} \approx s^2\left(\dfrac{3}{3+s}\right)^{7/3}\left(\dfrac{3/2}{3/2-s}\right)^{11/3} \exp\left(\dfrac{-s}{3/2-s}\right), & \text{for} \quad 0<s<3/2, \\ = 0, & \text{for} \quad s>3/2. \end{cases} \tag{8.17}$$

The distribution function is also independent of the initial size distribution, but the average radius $\bar{a}$ should be equal to a^*, while the maximum particle radius is equal to $3a^*/2$. For the mechanism of diffusion control, the critical or average radius in the steady-state coarsening regime increases with time following a cubic law, which is given by:

$$(a^*)^3 - (a_0^*)^3 = \left(\frac{8DC_0\gamma\Omega^2}{9kT}\right)t. \tag{8.18}$$

8.3.4 Deviation and Modification of LSW Theory

Deviations have been found in the predictions of the LSW theory. For instance, a practical system could have broader and more symmetrical steady-state distribution in diffusion-controlled coarsening. Also, the rate constant predicted by the LSW theory is often different from the experimental observations. In addition, volume fraction of the precipitates sometimes should be included, which is assumed to have no effect in the LSW theory [14–16]. It is found that as the volume fraction increases, the rate constant is increased while the size distribution function is broadened with increasing volume fraction of the precipitates [17]. Further modifications have made the LSW theory to be more agreement with experimental results [15, 18].

8.3.5 Time-Dependent Ostwald Ripening

Before the scaling regime mentioned above is reached, the distribution function varies with time. The precipitates could have a modified Gaussian distribution of sizes, which is given by [19]:

$$Y(a) = a^2 \exp\left[-\frac{(a - a_{\mathrm{m}})^2}{2a_{\mathrm{w}}^2}\right], \tag{8.19}$$

where the parameters a_{m} and a_{w} are the position of the maximum and width of the distribution, respectively.

By using the distributions with varying a_{m} and a_{w}, the variation in the properties of the distribution with time can be analyzed. For the distributions with a broad width, the standard deviation of the distribution decreases to the characteristic steady-state value, whereas for the distributions with a narrow width, the standard deviation increases to the steady-state value. We may interpret this behavior during the transient regime to mean that the scaling regime acts as a strong attractor for the evolution of the precipitate size distribution.

8.4 Topological and Interfacial Tensions

In most theories, grains and grain boundaries are treated as isolated items. However, a real dense polycrystalline solid consists of a space-filling array of grains, where have certain topological requirements of space filling and local requirements for equilibrium of the interfacial tensions. Figure 8.5 shows a two-dimensional section through a dense polycrystalline solid [3]. The structure consists of vertices joined by edges, also called sides, which surround faces.

It is assumed that the face at infinity is not counted, the numbers of faces *F*, edges *E*, and vertices *V* follow the Euler's equation:

$$F - E + V = 1. \tag{8.20}$$

For stable topological structures, i.e., those in which the topological features are not changed by small deformations, the number of edges that intersect at a vertex is 3. For isotropic grain-boundary energies, i.e., grain boundaries have the same value of γ_{gb}, if the grain-boundary tensions are balanced, the edges meet at an angle of 120°. In this case, if *N* is the number of sides, a hexagon with $N = 6$ has plane sides, while a polygon with $N > 6$ will have concave sides, whereas those with $N < 6$ have convex sides. Since the grain boundary migrates toward its center of curvature, grains with $N > 6$ tend to grow, while those with $N < 6$ tend to shrink.

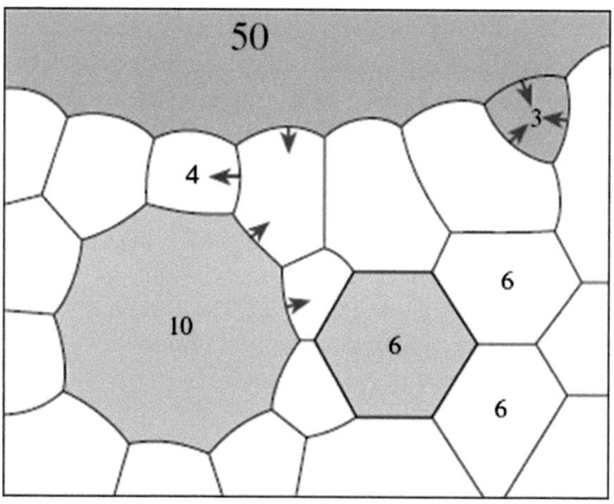

Fig. 8.5 Sketch of a section through a dense polycrystalline solid. The sign of the curvature changes as the number of sides increases from less than six to more than six. The *arrows* indicate the direction in which the boundaries migrate. Reproduced with permission from [3]. Copyright © 2007, Springer

In three dimension, the faces surround cells, if the cell at infinity is not counted, there is

$$F - E + V - C = 1, \tag{8.21}$$

where C is the number of cells. If the interfacial energies are assumed to be isotropic, the surfaces meet in groups of three at angles of 120° for interfacial tension to be balanced, along lines which themselves meet in groups of four mutually at an angle of 109.5°. This is the angle subtended by straight lines joining the corners of a regular tetrahedron. No regular polyhedron with plane sides has exactly this angle between its edges. The nearest approach to space filling by a regular plane-sided polyhedron is obtained with tetrakaidecahedra arranged on a body-centered cubic lattice, but even with this structure, the angles are not exactly those required and the boundaries must become curved to satisfy local equilibrium at the vertices. In general, real grains are arranged randomly and also have different sizes.

8.5 Normal Grain Growth in Dense Solids

8.5.1 The Burke and Turnbull Model

NGG in dense polycrystalline solids has been extensively studied in the open literature [20, 21]. The driving force for grain growth process in earlier models is considered to be the pressure gradient across the grain boundary that facilitates the transport of mass, whereas chemical potential gradient is used as the driving force in the later ones [22]. In the Burke and Turnbull model, an isolated part of the grain boundary is considered, with the assumption that the grain growth equation derived from the analysis describes the average behavior of the system. It is also assumed that the grain-boundary energy γ_{gb} is isotropic and independent on the crystallographic direction, with a constant grain-boundary width γ_{gb}.

If the instantaneous rate of grain growth is used to represent the average rate of grain-boundary migration v_b, also known as grain-boundary velocity, there is

$$v_b \approx \frac{dG}{dt}, \tag{8.22}$$

where G is the average grain size. The average rate of grain-boundary migration v_b can be represented by the product of the driving force for the grain-boundary migration F_b and the grain-boundary mobility M_b, so that there is

$$v_b = M_b F_b, \tag{8.23}$$

where M_b includes all effects related to the mechanism of migration. The pressure difference across the grain boundary is given by the equation of Young and Laplace:

$$\Delta p = \gamma_{gb}\left(\frac{1}{r_1} + \frac{1}{r_2}\right), \quad (8.24)$$

where γ_{gb} is the specific grain-boundary energy, i.e., energy per unit area and r_1 and r_2 are the principal radii of curvature of the boundaries. If the radius of the boundary is assumed to be proportional to the average grain size G, there is

$$\frac{1}{r_1} + \frac{1}{r_2} = \frac{\alpha}{G}, \quad (8.25)$$

where α is a geometrical constant that is determined by the shape of the boundary. If the driving force for the atomic diffusion across the grain boundary is equal to the gradient in the chemical potential, there is

$$F_b = \frac{d\mu}{dx} = \frac{d}{dx}(\Omega\Delta p) = \frac{1}{\delta_{gb}}\frac{\Omega\gamma_{gb}\alpha}{G}, \quad (8.26)$$

where Ω is the atomic volume and $dx = \delta_{gb}$ is the width of the grain boundary. From Eq. (4.92), the atomic flux across the boundary is given by

$$J = \frac{D_a}{\Omega kT}\frac{d\mu}{dx} = \frac{D_a}{\Omega kT}\frac{\Omega\gamma_{gb}\alpha}{\delta_{gb}G}, \quad (8.27)$$

where D_a is the atomic diffusion coefficient across the grain boundary. Therefore, the grain-boundary velocity becomes

$$v_b \approx \frac{dG}{dt} = \Omega J = \frac{D_a}{kT}\frac{\Omega}{\delta_{gb}}\frac{\alpha\gamma_{gb}}{G}. \quad (8.28)$$

Pressure difference across the grain boundary, $\alpha\gamma_{gb}/G$, can also be used as the driving force, so that v_b is expressed as follows:

$$v_b \approx \frac{dG}{dt} = M_b\left(\frac{\alpha\gamma_{gb}}{G}\right). \quad (8.29)$$

By comparing Eqs. (8.28) and (8.29), M_b can be defined by the following expression:

$$M_b = \frac{D_a}{kT}\left(\frac{\Omega}{\delta_{gb}}\right). \quad (8.30)$$

Integrating Eq. (8.29) yields:

$$G^2 - G_0^2 = Kt, \tag{8.31}$$

where G_0 is the initial grain size, i.e., $G = G_0$, at $t = 0$, while K is a temperature-dependent growth factor, which is given by

$$K = 2\alpha\gamma_{gb}M_b. \tag{8.32}$$

Equation (8.31) is the parabolic grain growth law, with the same form as the LSW equation for interface reaction-controlled Ostwald ripening given in Eq. (8.14). The growth factor K as a function of temperature T can be described by using the Arrhenius equation, i.e., $K = K_0 \exp(-Q/RT)$, where K_0 is a constant, R is the gas constant and Q is the activation energy of the grain growth.

The boundary mobility M_b is determined by the diffusion coefficient D_a for the atomic migrates across the grain boundary of the pure material, which is called intrinsic boundary mobility. In ionic solids, because both cations and anions could diffuse, D_a is the diffusion coefficient of the species that are rate-limiting or lowest. In real ceramics, various drag forces, such as segregated solutes, inclusions, pores, and second-phase films, can be applied to the grain boundary. As a result, experimental boundary mobility is lower than the M_b given by Eq. (8.30) in most cases.

8.5.2 Mean-Field Theory

In mean-field theory, the change in size of an isolated grain is considered, which is embedded in an environment that represents the average effect of all the other grains [23]. In this theory, the rate of grain growth is related to the radius a through the following equation:

$$\frac{da}{dt} = \alpha_1\gamma_{gb}M_b\left(\frac{1}{a^*} - \frac{1}{a}\right), \tag{8.33}$$

where α_1 is a geometrical factor that is equal to 1/2 for two dimension and 1 for three dimension, a is the radius of circle or sphere that has the same area or volume as the grain, and a^* is a critical grain size, i.e., if $a > a^*$, the grain grows, while if $a < a^*$, the grain shrinks.

Equation (8.33) is similar to the LSW equation of Eq. (8.11), where the rate of change in the radius of the grain in Ostwald ripening is controlled by the interface reaction, so that the critical radius for the rate of change is given by [23]:

$$\frac{d(a^*)}{dt} = \frac{1}{2}\alpha_1\gamma_{gb}M_b. \tag{8.34}$$

If it is assumed that the terms on the right-hand side of Eq. (8.34) are not varied with time, integration of the equation gives rise to parabolic grain growth kinetics, which is analogous to the Eq. (8.31). In addition, the size distribution during the steady-state grain growth can be expressed by the following equation [23]:

$$f(s,t) = (2e)^{\beta} \frac{\beta s}{(2-s)^{2+\beta}} \exp\left(\frac{-2\beta}{2-s}\right), \tag{8.35}$$

where s is called reduced size, that is equal to a/a^*; e is the base of the natural logarithm, i.e., $e = 2.718$, while $\beta = 2$ for two dimension and 3 for three dimension. The distribution predicted by using this equation has a sharp peak [24].

The mean-field approach has been further developed [25, 26]. If the grain size distribution is assumed to be lognormal and time-independent when plotted as a function of the reduced size $s = a/\bar{a}$, where $\bar{a}$ is the average grain size, similar parabolic growth kinetics could be obtained [25]. Alternatively, a random walk process is used to model the motion of grain boundary, by attributing the drift to larger average sizes to the fluctuation of the grain size. In this case, the grain growth kinetics is also parabolic, while a time invariant grain size distribution function can be obtained as follows [26]:

$$f(s,t) = As \exp\left(-\frac{s^2}{2}\right), \tag{8.36}$$

where A is a constant, with certain distribution functions [24, 27].

Therefore, all the theories indicate that the NGG in polycrystalline solids can be treated as the Ostwald ripening governed by the interface reaction mechanism. However, in practice, the grain growth data cannot be always described by the parabolic law, so that a general grain growth equation is used:

$$G^m - G_0^m = Kt, \tag{8.37}$$

where the exponent m is usually in the range of 2–4, which is dependent on temperature to a certain degree. Although $m = 3$ has been observed in most ceramics, it could be attributed to different mechanisms.

8.5.3 Topological Analysis

Topological requirements of space filling, which are neglected by the theories of grain growth, have an argument that the volume of a shrinking grain must be shared with all the grains in the structure [28]. Therefore, the changes in topological parameters, such as the numbers of faces, edges and vertices, must also be shared with the other grains. Two constants of grain growth, i.e., sweep constant θ and structure gradient σ, have been introduced. The sweep constant is defined as the number of grains that are lost when the grain boundaries throughout the structure sweep through an equivalent unit volume of the solid. There is some doubt about whether the sweep constant is indeed constant [29]. The structure gradient is defined as follows:

$$\sigma = \frac{M_V S_V}{N_V}, \tag{8.38}$$

where M_V, S_V, and N_V are the curvature, surface area, and number of grains, respectively, per unit volume of the solid. Available experimental data have confirmed that σ is indeed constant.

In the kinetics of grain growth, the mean boundary is equal to the product of the mobility of the grain boundary and the force applied to it, which is given by

$$v_b = M_b \gamma_{gb} M_V. \tag{8.39}$$

With a unit volume of the solid, the volume sweeps out per unit time is $v_b S_V$, while the number of grains that are lost per unit time is $\theta v_b S_V$, according to the definition of the sweep constant. The total volume transferred from the grains that are going to disappear to the grains that are remained per unit time is $\theta v_b S_V G$, where V_G is the average volume of the grains. The rate of increase in the average volume per grain can be derived by:

$$\frac{dV_G}{dt} = \frac{\theta v_b S_V V_G}{N_V}, \tag{8.40}$$

where N_V, the number of grains per unit volume, is equal to $1/V_G$. When S_V and v_b are substituted with those derived from Eqs. (8.38) and (8.39), Eq. (8.40) becomes

$$\frac{dV_G}{dt} = \frac{M_b \gamma_{gb} \theta \sigma}{N_V}. \tag{8.41}$$

If it is assumed that the terms on the right-hand side of Eq. (8.41) are constant, integrating Eq. (8.41) yields:

$$V_G - V_{G_0} = \left(\frac{M_b \gamma_{gb} \theta \sigma}{N_V} \right) t, \tag{8.42}$$

where V_{G_0} and N_{V_0} are the average grain volume and the average number of grains at $t = 0$. Experimental data indicate that V_G indeed has a linear dependence on time.

8.5.4 Simulation of Normal Grain Growth

Computer simulations have been used to explore the complexities of grain growth [21]. Two-dimensional models are usually used to simulate grain growth, either for direct simulation where a boundary network is constructed [30–32] or for statistical approach in which ensemble probabilities for a collection of grains are determined from a series of interrelated equations to find the behaviors of typical grain types [21, 22]. The probabilistic models can include both the topological requirements of space filling and the detailed local effects of grain-boundary curvature in the simulations. The microstructure is mapped out onto a discrete lattice, in which each lattice site is assigned a number between 1 and Q, corresponding to the orientation of the grain in it [27]. Generally, a sufficiently large value of Q (>30) is selected to limit the impingement of grains of the same orientation. The grain-boundary segment is defined in between two sites with different orientations.

For two dimension, the average area per grain $\bar{A}$ is related to the average grain size G. After an initial simulation of a shrinking circular grain embedded in an infinite matrix, i.e., $Q = 2$, which is similar to the mean-field approach, it is found that the growth of the grain size with time can be described by the following equation:

$$\bar{A} - \bar{A}_0 = Kt, \tag{8.43}$$

where $\bar{A}_0$ is the initial area of the grain and K is a constant. Therefore, the grain growth kinetics of the isolated grains is parabolic, which is in agreement with the prediction by the mean-field theories and the Burke–Turnbull theory.

For an interconnected network of polycrystalline grains, the kinetics is on longer parabolic. Grain growth exponent obtained in the simulations is $m = 2.44$, which is different from that value predicted by the mean-field theories, i.e., $m = 2$ [27]. In the mean-field theories, the driving force for grain growth is the reduction of the curvature or area of the boundary, whereas in the lattice models used in the simulations, the curvature is discretely allocated as kinks on the boundary, which can be eliminated by two mechanisms. First mechanism is the meeting and annihilation of two kinks with same orientation but opposite signs, which means that the grain growth is driven only by curvature. It is similar to the simulation of a circular grain embedded in an infinite matrix. The second mechanism is related to the adsorption of a kink at a vertex where more than two grains meet, which requires the presence of vertices. Vertices can reduce the curvature without causing grain growth by absorbing the kinks. Therefore, the growth is slower as compared to that of the circular grain in the infinite matrix, so that the value of m is relatively high.

8.6 Abnormal Grain Growth

As stated previously, the presence of extremely large grains is detrimental to full densification and thus some properties of ceramics. Therefore, in order to control AGG, it is necessary to understand its mechanism.

8.6.1 Origins of Abnormal Grain Growth

The occurrence of AGG is often attributed to nonuniform the distribution of particle size in the starting powders. Equation (8.35) indicates that a steady-state normal grain size distribution has a maximum or cut-off grain size, which means that AGG would take place if there are grains that are larger than twice the average size in two dimension. However, theoretical models can hardly predict AGG. For example, it is found in a Monte Carlo simulation that although there are large grains at certain stages, they would not grow abnormally, if the system is isotropic with uniform grain-boundary energy and mobility, as shown in a systematic study [33–35]. This is because the normal grains will grow at higher rates, so that the large or abnormal grains are eventually included in the normal size distribution.

When there is a large grain with a radius a is embedded in a matrix of fine normal grains, similar to Eq. (8.11), its growth rate can be expressed as [36]

$$\frac{\mathrm{d}a}{\mathrm{d}t} = 2\gamma_{\mathrm{gb}} M_{\mathrm{b}} \left(\frac{1}{a^*} - \frac{1}{a} \right), \tag{8.44}$$

where a^* is the critical radius at which a grain neither grows nor shrinks. Therefore, the relative growth rate of the large grain is given by

$$\frac{\mathrm{d}}{\mathrm{d}t}\left(\frac{a}{a^*} \right) = \frac{1}{(a^*)^2} \left(a^* \frac{\mathrm{d}a}{\mathrm{d}t} - a \frac{\mathrm{d}a^*}{\mathrm{d}t} \right). \tag{8.45}$$

Because the number of the fine normal grains is much larger than that of the abnormal grain, the time dependence of a^* can be described in a similar way to that given by Eq. (8.14):

$$\frac{\mathrm{d}a^*}{\mathrm{d}t} = \frac{\gamma_{\mathrm{gb}} M_{\mathrm{b}}}{2a^*}. \tag{8.46}$$

Putting Eqs. (8.44) and (8.46) into Eq. (8.45) yields

$$\frac{\mathrm{d}}{\mathrm{d}t}\left(\frac{a}{a^*} \right) = -\frac{\gamma_{\mathrm{gb}} M_{\mathrm{b}}}{2aa^*} \left(\frac{a}{a^*} - 2 \right)^2. \tag{8.47}$$

Equation (8.47) indicates that the relative growth rate of the large grain is always negative, except for the case of $a = a^*$, for which it is zero. Therefore, those abnormal grains, i.e., with grain sizes of $a > 2a^*$, will not grow to a size that is significantly outside the normal grain size range, due to the upper limit of $2a^*$. Moreover, due to their irregularities in shape and fluctuations in size, they cannot remain unchanged in size at exactly $2a^*$, after they are included in the normal grain size distribution. Instead, they tend to shrink. As a result, size difference is not a sufficient condition to initiate AGG in isotropic systems.

Both computer simulations and theoretical analysis have demonstrated that AGG most likely occurs in anisotropic systems, i.e., the grain-boundary energy and mobility are variable [37–40]. Once the boundaries have a higher mobility and/or a lower energy than the surrounding matrix grains, AGG is triggered. It has been observed that the abnormal grain embedded in a fine-grained matrix could be 6 times larger than the average size, which has a mobility 7.5 times that of the fine grains [37]. According to prediction, the abnormal grain grows faster than the average-sized grains in the surrounding matrix.

The properties of grain boundaries can be varied in several ways in practical ceramics. One of the characteristics is the structure and mis-orientation of the grain boundary. Special or low-angle grain boundaries, which are formed when two grains have only a slight mis-orientation relative to one another, possess a lower energy than general boundaries that have high mis-orientation angles. Transfer of matter from the surrounding grains to the low-energy boundaries promotes AGG. Alternatively, low-energy grain boundaries are generally believed to have low mobility, so growth is slow normal to and rapid parallel to the low-energy boundaries, often resulting in faceting and anisotropic abnormal growth.

During the sintering process of ceramics, there could be happenings, such as the release of solutes, formation of secondary phases, and the presence of pores from moving grain boundaries, which bring out a sudden increase in the boundary mobility, thus leading to AGG. The formation of liquid phases has also been acknowledged to be a cause of AGG, because the boundary mobility can be significantly increased due to the presence of the liquid films at the grain boundaries. This is the reason why AGG is easily observed in liquid-phase sintering [41]. Inhomogeneities in physical properties and chemical compositions, e.g., inhomogeneous packing and nonuniform distribution of dopants and secondary phases, result in inhomogeneous microstructures, which is another major cause of AGG, because the local microstructural heterogeneity creates differences in the boundary mobility and energy.

8.6.2 Applications of Abnormal Grain Growth

As mentioned above, it is desirable to prevent AGG. However, if AGG can be well controlled, it is also useful for several ceramic materials. For example, in situ growth of anisotropic abnormal grains in a fine-grained matrix, with controlled size distributions, has been used to increase the fracture toughness of SiC, Si_3N_4, Al_2O_3, and mullite ceramics [42–48]. Preferential alignment of growing anisotropic grains has been employed to fabricate ceramic materials with textured microstructure, which is also called template grain growth (TGG) [49–56].

Another application of AGG is to obtain single crystals through ceramic processing. Previously, various single crystals, such as ferrites and ferroelectrics, have been obtained by using this technique [57–60]. More importantly, this method has also been used to convert ceramics to solid-state laser single crystals [61–63].

Desirably, monodispersed starting powders should be used, so as to form a fine-grained matrix with a narrow size distribution. At the same time, a large seed crystal with high quality is embedded into the fine-grained matrix. The system is then annealed at high temperatures, so that the seed crystal will grow into the powder matrix through the migration of its boundary. Under proper conditions, the entire sample could be grown into a single grain, i.e., single crystal.

The derivation of single crystals from polycrystalline ceramics by sintering is known as solid-state crystal growth (SSCG) [64]. A schematic diagram of the single-crystal sintering method is shown in Fig. 8.6. First, a polycrystalline Nd: YAG ceramic material with a relative density of over 95 % was prepared. Then, one face was polished and contacted with a seed crystal, which could be single-crystal YAG with any crystal orientation of ⟨111⟩, ⟨100⟩ and ⟨110⟩. When the seeded ceramics was heated at a high temperature (below the melting point) of over 1700 °C, single crystallization occurred in the solid state due to continuous grain growth. When the surface energy of the fine grains of polycrystalline ceramic (E_p) is much greater than that of the seed crystal (E_s) (i.e., $E_p \gg E_s$), the fine grains tend to change to a thermodynamically stable condition at higher temperature. As a result, fine grains are absorbed into the seed crystal, and grain growth occurs continuously. Finally, the polycrystalline materials change to single-crystal materials in a solid-state condition.

A trace amount of solid SiO_2, which in most cases is in the form of colloidal silica, was added to prepare Nd:YAG sintered bodies. Then, they were heat treated at 1780 °C for various soaking times: 0, 2, and 5 h. Reflected microscopic images are shown in Fig. 8.7. Upon reaching the sintering temperature, the microstructure was still characterized by grains with a uniform size distribution. However, as the sintering time was increased to 2 h, a few of abnormal grains with diameters of up to 1 mm were observed. As the sintering was further increased to 5 h, the abnormal

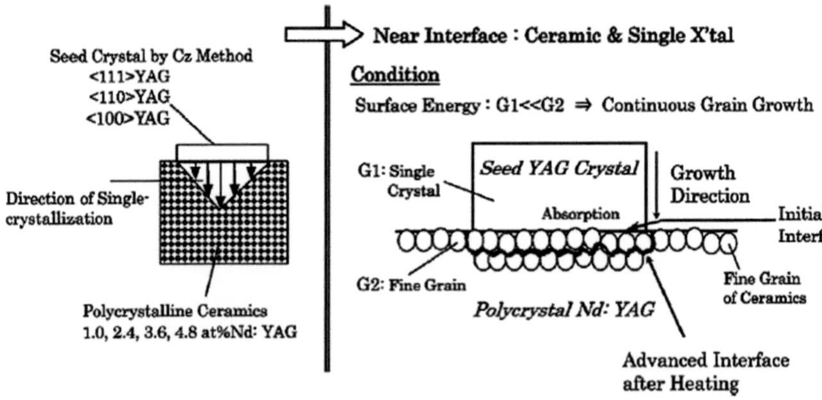

Fig. 8.6 Illustration of the production of single-crystal ceramic material by the sintering (nonmelting) conversion method. Reproduced with permission from [61]. Copyright © 2007, Elsevier

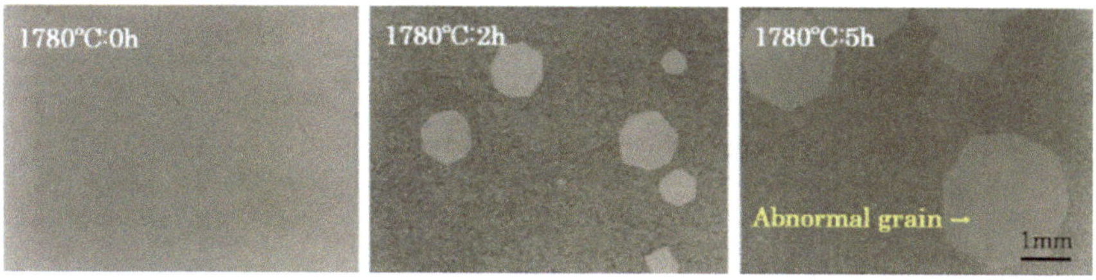

Fig. 8.7 Variation in microstructure of Nd:YAG ceramics doped with a small amount of SiO_2 sintered at 1780 °C for 0, 2, and 5 h. Reproduced with permission from [61]. Copyright © 2007, Elsevier

grain sizes were largely increased. In this case, the silica acted as both a seed crystal and an accelerator of the grain growth. In solid-state crystal growth, the formation and growth of abnormal grains provides the driving force for continuous crystallization.

The relationship between the heat treatment temperature and rate of crystal growth has been studied. Crystallization started at about 1700 °C, and the crystal growth rate increased with an increase in heat treatment temperature. In this experiment, a maximum growth rate of 1.7 mm h^{-1} was reached. Using the CZ method, the rate of crystal growth is generally about 0.2 mm h^{-1}. Therefore, greater or equivalent growth rate can be realized using the SSCG method.

The appearance of heavily doped Nd:YAG single crystals, with doping concentrations of 2.4, 3.6, and 4.8 %, prepared by the SSCG method is shown in Fig. 8.8a. A typical reflected (polarized) microscopic image of a near growth interface of a 2.4 % Nd:YAG sample is shown in Fig. 8.8b. It was confirmed that grain growth occurred from a seed crystal toward polycrystalline directions. It

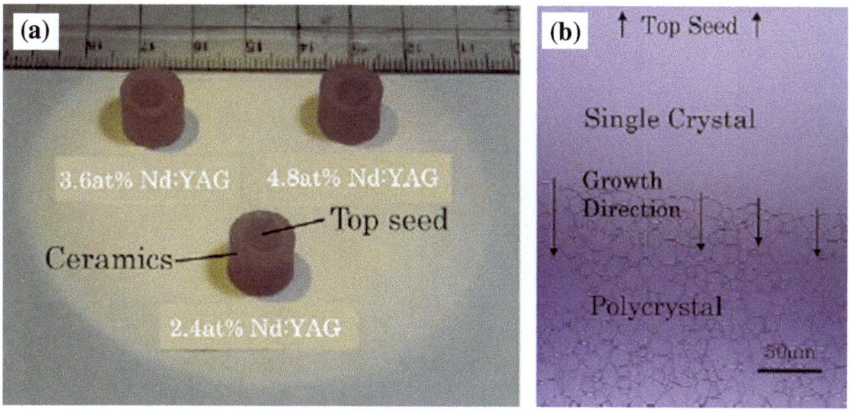

Fig. 8.8 **a** Photographs of the 2.4, 3.6, and 4.8 at % Nd:YAG ceramics after heat treatment for single crystallization. **b** Microstructure of the 2.4 at.% Nd:YAG ceramic sample at the interface between the single crystal and polycrystal. Reproduced with permission from [61]. Copyright © 2007, Elsevier

should be noted that heavily doped Nd:YAG single crystal can be produced by the SSCG method. Using the CZ (melt-growth) method, the segregation coefficient of Nd ions in the YAG host crystal is very small, and it is difficult to dope more than 1 at % Nd ions homogeneously into the YAG host crystal. In the case of the SSCG method, since there is no solid–liquid interface, the concept of segregation is not relevant for this solid process. In this process, polycrystalline grains are gradually absorbed into the seed single crystal.

Therefore, if the composition of each polycrystalline Nd:YAG grain is very homogeneous in the ceramic, then the homogeneity of the grown single-crystal ceramics will also be very high. Transmission electron microscopy (TEM) image and energy dispersive spectroscopy (EDS) analysis have been used to study the growth interface between single crystal and polycrystalline part. There was no boundary phase or secondary phase near the growth interface, and the growth interface was very similar to the grain boundary of the host polycrystalline ceramics. The EDS analysis results, however, showed a trace amount of Si (SiO_2) at the growth interface and at the grain boundaries, but not in the crystal grains. Studies on the crystal growth mechanisms are still under investigation, but it is certain that the Si component is closely involved in the growth mechanism.

Figure 8.9 shows XRD patterns of normal polycrystalline Nd:YAG ceramic, and the SSCG-grown single-crystal ceramics with ⟨111⟩ and ⟨110⟩ crystal orientations. In the case of normal sintered Nd:YAG ceramic without the seed crystal, random crystal orientations were observed. But in the case of the SSCG-grown single-crystal ceramics, the crystal orientations were the same as the seed crystal. Therefore, it was confirmed that single crystallization by sintering is technologically possible.

SSCG-grown Cr^{4+}:YAG single-crystal ceramic could be well bonded with a ⟨100⟩-oriented seed crystal. Using this technology, it is possible to dope various types of laser active ions. This material was grown at 1700 °C, and the crystal growth rate was 4 mm h^{-1}. The crystal growth rate can be controlled mainly using

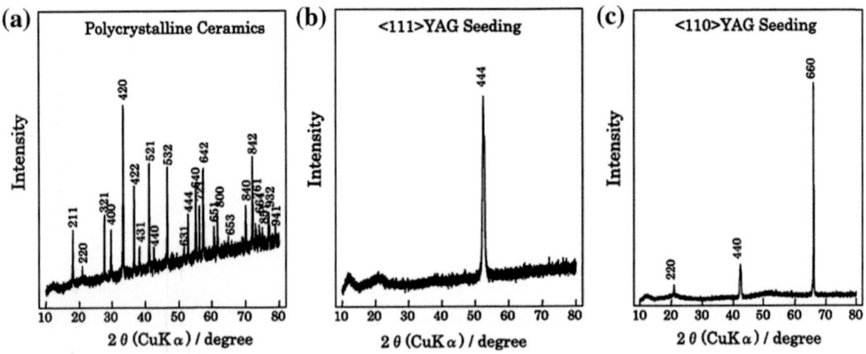

Fig. 8.9 XRD patterns of the polycrystalline (**a**), ⟨111⟩ (**b**) and ⟨110⟩ (**c**) seeded Nd:YAG samples. The characteristic peaks of the ⟨111⟩ and ⟨110⟩ seeded crystals are confirmed. Reproduced with permission from [61]. Copyright © 2007, Elsevier

(i) the seed crystal orientation, (ii) the grain size of the polycrystalline ceramic, and (iii) the heat treatment temperature. The maximum growth rate achieved was 5 mm h^{-1}.

Laser performance of the SSCG-grown Nd:YAG single-crystal ceramic and that of pore-free normal Nd:YAG polycrystalline ceramic, with Nd concentration of 2.4 at.%, have been compared. A Ti:sapphire laser was used as an excitation source. A 99.9 % reflection mirror and an output mirror with 5 % transmission were used. The laser gain media samples were AR coated, but the laser oscillation test was performed without optimizing the laser resonator. Both samples reached a slope efficiency of almost 60 %, but the single-crystal ceramic showed about 6 % higher efficiency than the polycrystalline ceramic. This improvement was probably because scattering due to grain boundaries does not occur in the single-crystal ceramic material. Since the single-crystal ceramic materials produced by the SSCG method have no grain boundaries, it is considered that this kind of material is the best quality for laser materials.

Various garnet crystals of Nd-doped $(Y,Gd,La,Lu)_3(Sc,Al)_5O_{12}$ were grown from Nd-doped $(Y,Gd,La,Lu)_3(Sc,Al)_5O_{12}$ polycrystalline precursors, by using controlled AGG at temperatures between 1750 and 1850 °C [65]. The composition of the expected Nd-doped $(Y,Gd,La,Lu)_3(Sc,Al)_5O_{12}$ single crystal was $Nd_{0.01}$:$(Y_{0.85}La_{0.05}Lu_{0.05}Gd_{0.05})_{2.99}$:$(Sc_{0.3}Al_{0.7})_5O_{12}$. The Nd-doped $(Y,Gd,La,Lu)_3(Sc,Al)_5O_{12}$ precursor materials were prepared using stoichiometric amounts of the following analytical grade reagents: yttrium oxide (Y_2O_3), scandium oxide (Sc_2O_3), lanthanum oxide (La_2O_3), lutetium oxide (Lu_2O_3), gadolinium oxide (Gd_2O_3), neodymium oxide (Nd_2O_3), aluminum nitrate hydrate ($Al(NO_3)_3{\cdot}9H_2O$), nitric acid (HNO_3), and ammonia ($NH_3{\cdot}4H_2O$).

Y_2O_3, Sc_2O_3, La_2O_3, Lu_2O_3, Gd_2O_3, and Nd_2O_3 were dissolved in nitric acid. Aluminum nitrate was dissolved in distilled water. The two solutions were then mixed together. An appropriate amount of the mixture was stirred for 3 h at the same temperature. In the following step, ammonia ($NH_3{\cdot}4H_2O$) was added drop-wise to the above-described solutions as a precipitator with continuous stirring. The sediment was washed several times using deionized water and then filtered. The sediment was then dried in an oven for 24 h at 100 °C.

The sediment was ground in an agate mortar, sieved, placed in alumina crucibles, and burned for 2 h at 850 °C in air (heating rate of 1 °C min^{-1}). The powder was compacted into pellets with diameter of 20 mm at 20–30 MPa and was then cold-isostatically pressed at 200 MPa. The localized codoping of the disks was achieved by applying colloidal SiO_2 with 15 wt% in H_2O to one point of each disk. The codoped disks were allowed to dry for at least 5 min prior to sintering. After sintering at 1750–1850 °C for 20 h under a H_2 atmosphere without pressure in a tungsten wire furnace, the specimens were cut and double-polished to 1 mm in thickness for spectral analysis.

The samples obtained were transparent. Figure 8.10 shows a photograph of a Nd-doped $(Y,Gd,La,Lu)_3(Sc,Al)_5O_{12}$ single crystal produced using the solid-state single-crystal conversion (SCC) process [65]. The size of the single crystal of Nd-doped $(Y,Gd,La,Lu)_3(Sc,Al)_5O_{12}$ was approximately 4 mm and thickness was

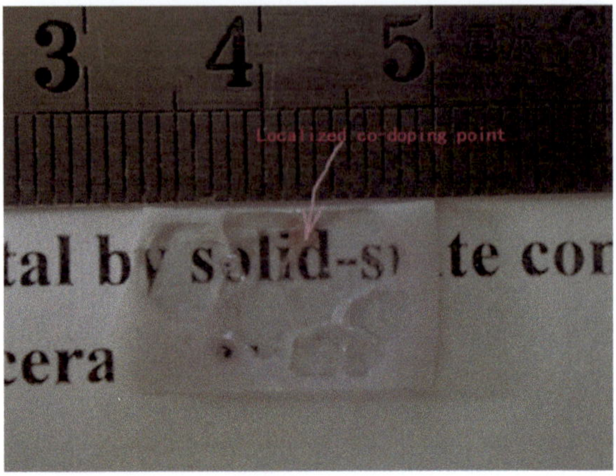

Fig. 8.10 Photograph of the mixed-metal garnet single crystal by using solid-state ceramic conversion process (polished sample with thickness = 0.8 mm). Reproduced with permission from [65]. Copyright © 2014, Elsevier

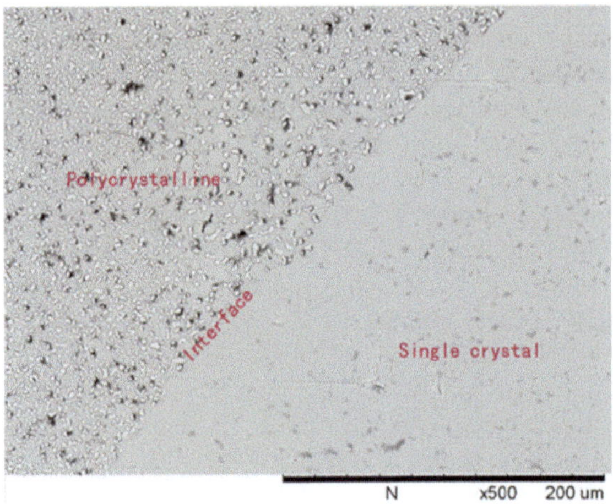

Fig. 8.11 SEM image of interface between the polycrystalline and the as-prepared single-crystal part. Reproduced with permission from [65]. Copyright © 2014, Elsevier

0.8 mm. The relative density was 99.8 %. Figure 8.11 shows an SEM image of interface between the grown single-crystal and the polycrystalline ceramics [65]. The grains of the polycrystalline ceramics were equiaxed, with an average grain size of ~30 μm. Many residual pores with small sizes at the grain boundaries and within the grain interiors could be observed. Single crystals with a maximum size of 4 mm and promising optical performances could be obtained by using this method.

Yttrium aluminum garnet (YAG) ceramics with stoichiometric Y_2O_3:Al_2O_3 = 3:5 ratio, as well as with excessive either Y_2O_3 or Al_2O_3 of up to 5 mol%, were compared, in order to study the composition effect of YAG on crystal growth behavior during the SSCG process [62]. This work was carried out, because previous results doped yttrium aluminum garnet $Y_3Al_5O_{12}$ (YAG) ceramics were contradictory, in terms of crystal growth rate with respect to compositions of the YAG studies.

All the ceramics were prepared through solid-state reaction of nanopowders of Y_2O_3 and Al_2O_3, which were synthesized by using chemical precipitation from yttrium nitrate and ammonium–aluminum sulfate with ammonium bicarbonate solution as precipitant. All mixed powders were compacted at uniaxial pressure 20 MPa first and further consolidated by using CIP at 200 MPa into pellets with diameter of 10 mm and thickness of 2 mm. The samples were then sintered in vacuum at 1730–1750 °C for 5–10 h. Besides the stoichiometric 3:5 composition, excessive Y_2O_3, or Al_2O_3 were from 0.2 to 5 mol%. 0.8 mol% of Nd_2O_3 was included in some samples.

(111) YAG-polished crystal seed plates with 200-nm SiO_x layers deposited on them by magnetron sputtering were pressed onto polished ceramic disks [66]. Thermal annealing was conducted at 1800–1850 °C or 1890 °C for 10–12 in, in N_2 + 5 % O_2, at heating/cooling rates of 150 °C h^{-1}. Slices of 1 mm thick, perpendicular to the seed–ceramics contacting surfaces, were cut from the samples after thermal annealing, with both sides to be polished.

Figure 8.12 shows photographs of representative samples. Stoichiometric ceramics samples were transparent, samples with 1 % excess Y_2O_3 or Al_2O_3 were translucent, and those with 3 % and more excess oxides were opaque. Grain size in all samples was about 20 μm. Porosity increased slightly with increasing contents of the excessive oxides.

Figure 8.13 shows SEM image of the grown crystal after thermal etching. Grain boundaries were clearly observed in etched ceramic part, while newly grown single-

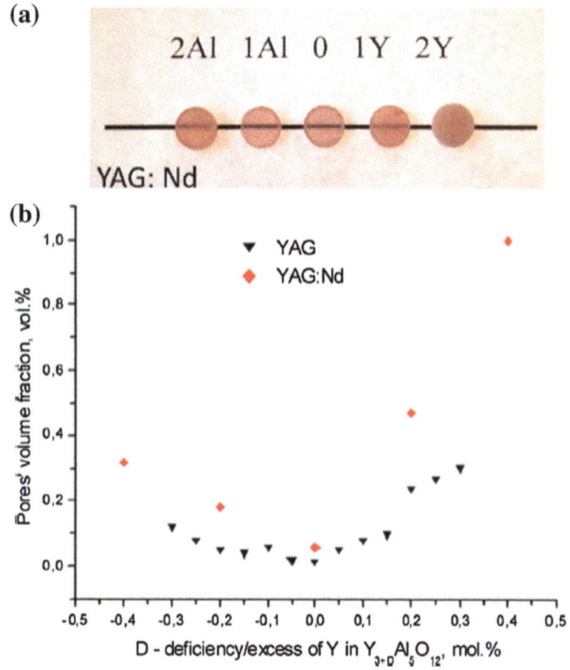

Fig. 8.12 **a** Photographs of representative YAG:Nd and YAG ceramics. Every sample differs from its neighbors by 1 mol% excessive Y_2O_3 (1Y, 2Y) or Al_2O_3 (1A, 2A), while "0" means stoichiometric Y/Al ratio of 3:5. **b** Porosity (vol.%) as a function of Y/Al ratio for pure and Nd-doped YAG ceramics with different levels of excessive oxides. Reproduced with permission from [62]. Copyright © 2013, Elsevier

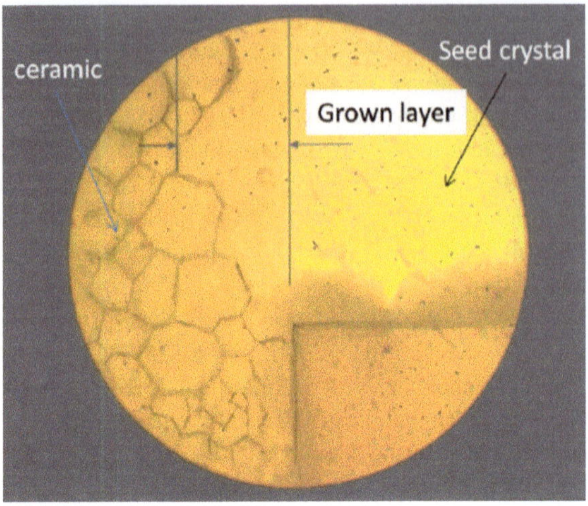

Fig. 8.13 SEM image of the grown crystal after thermal etching to show the grain boundaries in the ceramics, without boundary being observed in the grown crystal. Reproduced with permission from [62]. Copyright © 2013, Elsevier

crystal part has no grain boundaries. Experimental results indicated that the growth rates of the samples with excessive Al_2O_3 were higher that of the stoichiometric sample. The sample with 5 % excessive Al_2O_3 at 1890 °C exhibited the highest growth rate of 0.15 mm h^{-1}, which was about higher than that of stoichiometric sample by nearly 20 times.

Nonuniform growth was observed in central part of the sample with 1.5 mol% excessive Y_2O_3, as shown in Fig. 8.14. The central part had a higher growth rate

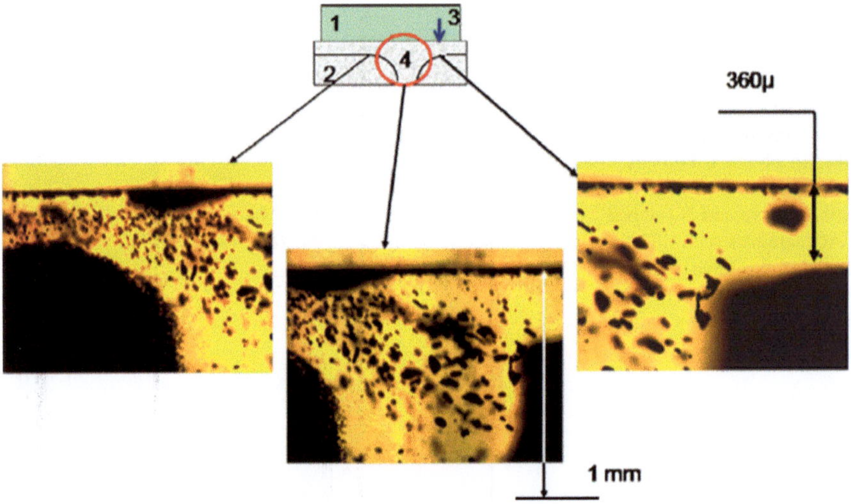

Fig. 8.14 Nonuniform growth of the crystal layer in the center part of the YAG ceramics with 1.5 mol% excessive Y_2O_3. Diagram of the layered structure is shown at the *top panel 1* seed YAG crystal, *2* ceramics, *3* SiO_x layer, and *4* grown crystal layer. Reproduced with permission from [62]. Copyright © 2013, Elsevier

than the rest area by about three times, which was different from normal growth behavior, i.e., such growths are from outer parts toward central part. The anomaly was attributed to possible nonuniform contact at the seed crystal–ceramics interface. In addition, there is no difference in SSCG rate between undoped and 0.8 mol% Nd-doped samples.

The effect of contact between the seed crystal and the ceramics on the SSCG growth efficiency has been examined. One example is shown in Fig. 8.15, where the two edges of same seed had different contacting qualities, i.e., one was better than the other. Poor contact led to slow growth rate and the formation of pores, as shown Fig. 8.15a. In contrast, high-quality contact resulted in high growth rate and reduced number of pores, as demonstrated in Fig. 8.15b.

EDS measurement results indicated that, in direction perpendicular to the seed-ceramic interface, some local regions possessed enrichment in Y or Al after annealing at temperatures of >1800 °C. The number and dimension of such regions increased with increasing content of excessive Y_2O_3 and Al_2O_3 in the initial compositions. Meanwhile, the Y–Al ratio was very close to the stoichiometric ratio of 3:5 in other areas. These either Y or Al enriched regions could participate the formation of liquid phase with SiO_x, which promoted crystal conversion and pore elimination. This interesting issue deserves further investigation.

Abnormal grain growth (AGG) can also be used to directly convert polycrystalline Al_2O_3 (PCA) ceramics into single crystals, without using seeds. For

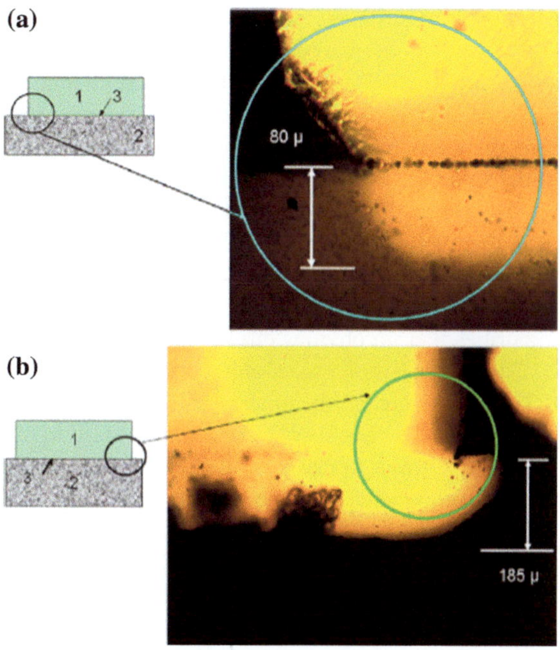

Fig. 8.15 Lateral growth of the crystals in the ceramics with 4.7 mol% excessive Y_2O_3: **a** low rate of lateral growth and **b** high rate of lateral growth. (*1* seed YAG crystal, *2* ceramic, and *3* SiO_x layer). Reproduced with permission from [62]. Copyright © 2013, Elsevier

example, more than 30–40 % of polycrystalline, MgO-doped Al_2O_3 tubes were converted into single crystals of sapphire by AGG in the solid state at a high temperature of 1880 °C [67]. Most crystals were 4–10 cm in length in the tubes with wall thicknesses of 1/2 and 3/4 mm and outer diameters of 5 and 7 mm, respectively, with their *c*-axes being oriented ~90° and 45° to the tube axis. The AGG was initiated, due to the low values of MgO concentration of about 50 ppm in the bulk. The unconverted parts had grains with millimeter sizes. The difference in grain structures between the converted and the unconverted tubes was attributed to nonuniform concentration of MgO in the extruded tubes.

The growth front of the migrating crystal boundary possessed nonuniform shapes, while the interface between the single crystal and the polycrystalline matrix contained "curved" boundary segments, characterizing typical AGG in a single-phase material. The average velocity of most migrating crystal boundaries was as high as 1.5 cm h^{-1}. The average grain-boundary mobility at 1880 °C was estimated to be 2×10^{-10} m^3 $(N\ s)^{-1}$, which was the highest value for Al_2O_3 and was higher than the calculated intrinsic mobility by a factor of 2.5. Such a conversion could on take place, when a codopant of CaO, La_2O_3, or ZrO_2 was used at concentrations of several hundred ppm.

A commercial alumina powder, mixed with 620 ppm of MgO using magnesium nitrate as a precursor and appropriate amounts of organic binder and lubricant, was used to extrude tubes. The alumina powder had a specific surface area of 8.8 m^2 g^{-1} and median particle size of 0.31 μm. Tubes of 100 cm in length were extruded, which were dried for 16 h at 100 °C and prefired at 1050 °C in air to remove organic binders. Finally, the tubes were sintered at 1880 °C in H_2 for different time durations. The average concentrations of MgO (C_{MgO}) and impurities were measured in sintered PCA tubes by using wet chemical analysis.

Figure 8.16 shows C_{MgO} and grain size the samples as a function of sintering time duration [67]. After sintering 3 h, the C_{MgO} in the PCA decreased from the initial value of 660 ppm to about 380 ppm. The MgO content continued to decrease to 40–55 ppm after sintering form 9 h, which showed no further significant decrease thereafter when tubes were converted to single crystals. The solid-state crystal conversion (SSCC) event was started at about 9 h, which was found to be related with very low C_{MgO} of 50 ppm. Due to their very low levels, the concentrations of impurities had no obvious effects on the SSCC event.

The average grain size of the polycrystalline matrix increased rapidly from 0.3 μm in the green tube to about 30 μm in the sample sintered at 1880 °C for 3 h. After that, the grains slowly increased from 30 to 60 μm, as the sintering time was increased from 3 to 18 h. The sizes of the largest grains in the polycrystalline regions were about twice the average grain size, indicating that the event was still NGG. The average size of the "surface" grains on the outer diameter (OD) surface of the sample sintered for 9 h was 65 μm, which was larger than the interior grain size of 50 μm by about 30 %. In this sample, the largest surface grains were 200–250 μm, 3–4 times larger than the average grain size. It means that the largest surface grains started to grow abnormally, while those in the sample interior and on the inner diameter (I)D surface of the sintered tube still experienced NGG.

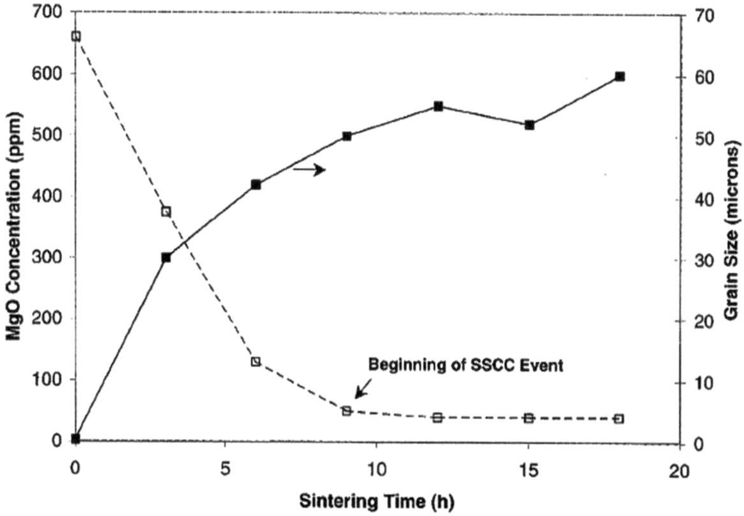

Fig. 8.16 Average MgO concentration and grain size in PCA tubes with 5 mm OD as a function of sintering time at 1880 °C. Reproduced with permission from [67]. Copyright © 2002, John Wiley & Sons

Figure 8.17a shows a detailed image of the single crystal/PCA interface [67]. The interfacial boundary between the single crystal and the polycrystalline matrix with an average grain size of ~30 μm consisted of many "curved" boundary segments, indicative of typical AGG. There were two distinctive types of surface features in newly grown single crystals. The surface sapphire crystal undulations

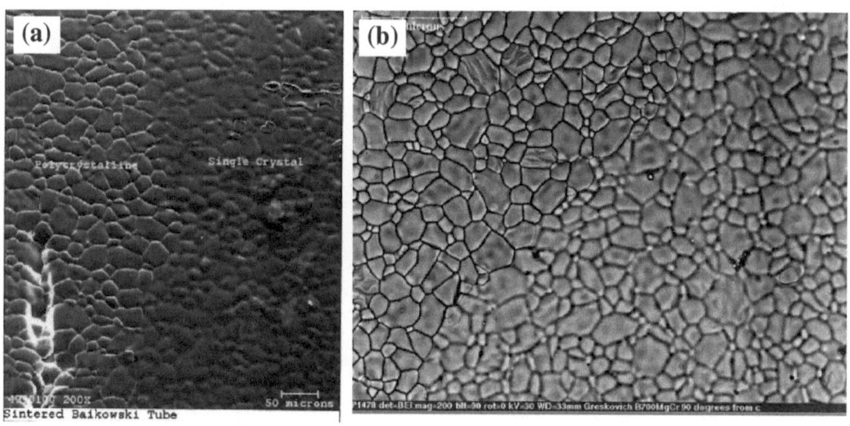

Fig. 8.17 **a** Typical interface between a growing single crystal and the PCA matrix on the OD surface. **b** Optical micrograph of the interface showing the "ghost" structure in the single-crystal region. Reproduced with permission from [67]. Copyright © 2002, John Wiley & Sons

appeared to be preexisting on the PCA material before conversion, which was even more visible in Fig. 8.17b. Also, single or groups of isolated PCA grains were observed the grown crystal. These isolated grains were Al_2O_3 grains that were misoriented with the grown crystal. This is a low-cost process of SSCC that could be used to convert complex shapes of polycrystalline ceramics into single crystals through AGG, which is believed to be applicable to other materials and thus deserves further investigation.

Another example is on the conversion of polycrystalline MgO-doped Al_2O_3 tubes to single-crystal sapphire with the codoping of SiO_2 [68]. The intentional codoping with SiO_2 before sintering was used to trigger the occurrence of AGG, leading to full conversion of tube surfaces to single crystal without adversely affecting the densification. The degree of surface conversion was strongly dependent on experimental parameters, such as sintering temperature and codoping level. Surface-converted tubes had excellent physical properties, including good thermal cycling resistance and optical properties.

Figure 8.18a shows microstructure of an MgO-doped PCA tube without codoping with SiO_2, which was sintered at an optically calibrated temperature of 1886 °C for 4 h [68]. The grains were equiaxed with an average grain size of ~30 μm, while small-sized residual pores both at the grain boundaries and in grain interiors could be observed. The tube exhibited good translucency. Microstructure of an $MgO–SiO_2$-codoped Al_2O_3 tube sintered at 1886 °C for 4 h is shown in Fig. 8.18b. It was found that the localized codoping of one end of prefired MgO-doped Al_2O_3 tubes with a very small amount of colloidal SiO_2 before sintering led

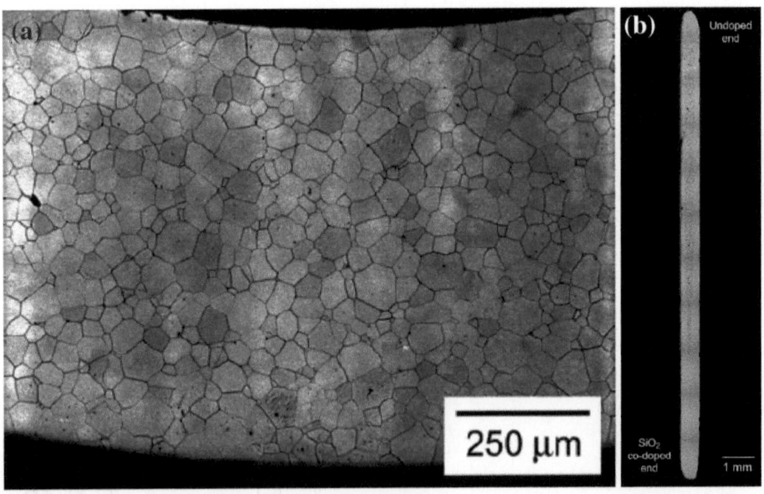

Fig. 8.18 Optical micrographs of the microstructure: **a** an MgO-doped PCA tube (not codoped) sintered at 1886 °C for 4 h and **b** an $MgO–SiO_2$-codoped Al_2O_3 tube (10 μL SiO_2) sintered at 1886 °C for 4 h, in which codoping of one end of the tube with colloidal SiO_2 before sintering resulted in full densification followed by the full conversion of the surface to single crystal. Reproduced with permission from [68]. Copyright © 2004, John Wiley & Sons

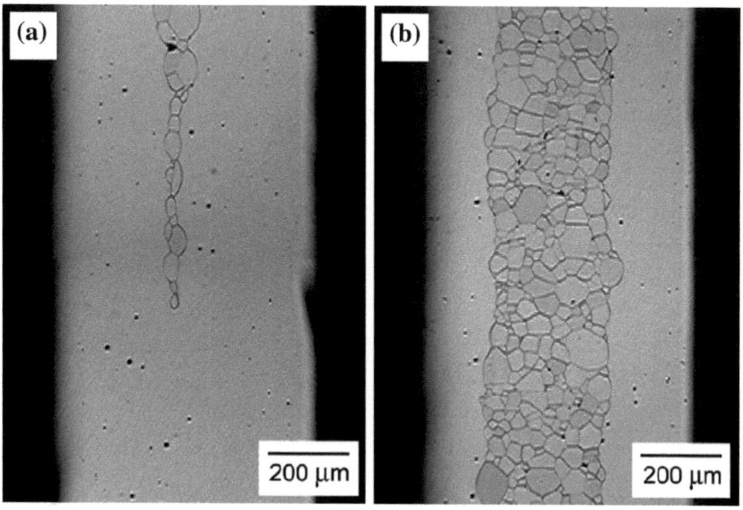

Fig. 8.19 Optical micrographs that show magnified views of the $MgO–SiO_2$-codoped Al_2O_3 tube in Fig. 3, **a** near to and **b** away from the codoped end. Reproduced with permission from [68]. Copyright © 2004, John Wiley & Sons

to the full conversion of the tube surface to single-crystal sapphire. The remaining core was still dense PCA. Nearly 1/4 of the length of the tube (~4 mm) were fully converted to single crystal across the entire thickness of the tube wall. The remaining length of the tube at the other end was converted in the surface regions of ~200 μm in thickness, a central of PCA core region of 300 μm in thickness was enveloped by the single-crystal layers. Figure 8.19 shows high magnification graphs of the $MgO–SiO_2$-codoped Al_2O_3 tube of Fig. 8.18b, near to (a) and away from the codoped end (b), respectively.

Mechanism of the conversion to single crystals from an MgO-doped PCA precursor has been studied [69]. The single crystals were grown through controlled AGG at temperatures between 1670 and 1945 °C. It was observed that CaO impurities segregated at the boundary between the single crystal and the polycrystalline matrix, at which a thin layer of wetting intergranular film was formed. The presence of this film facilitated to produce highly mobile grain boundaries that were required to trigger the single-crystal conversion. The measured grain-boundary mobilities agreed well with the values calculated from the data for a grain boundary containing a film with properties of bulk glass. It was found that the highest grain-boundary mobility was an extrinsic effect, rather than intrinsic behavior of the materials.

The samples were prefired Al_2O_3 tubes doped with 500 ppm MgO, with an inner diameter (ID) of 1.25 mm, an outer diameter (OD) of 3.30 mm, and a length of 30 mm. Sintering was conducted in a refractory metal furnace, in which the samples were placed vertically on the end of a 0.1-mm-diameter molybdenum wire. Sintering conditions were N_2–5 %H_2 at 1945 °C, N_2–7 %H_2 at 1840 °C, N_2–10 % H_2 at 1800 °C, and pure H_2 at 1670 °C for 0–2 h.

Single crystals with sizes of up to 30 mm could be grown reproducibly. The size of the single crystals was only limited by the size of the hot zone of the furnace. Figure 8.20 shows an image of the interface between a grown single crystal and the polycrystalline matrix [69]. Such boundaries were found to be curved at all scales and rough on the atomic scale. AGG took place after a certain while at the sintering temperatures. An intergranular film with a thickness of 10–20 nm was observed at the grain boundaries in quenched samples. Figure 8.21 shows a diffuse dark-field image of the boundary, indicating that the film was amorphous. The thickness of

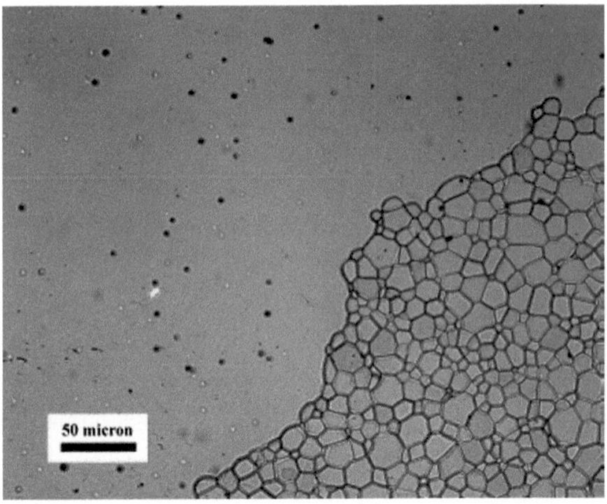

Fig. 8.20 Polished and etched interface between a growing single crystal and the polycrystalline region. Reproduced with permission from [69]. Copyright © 2007, John Wiley & Sons

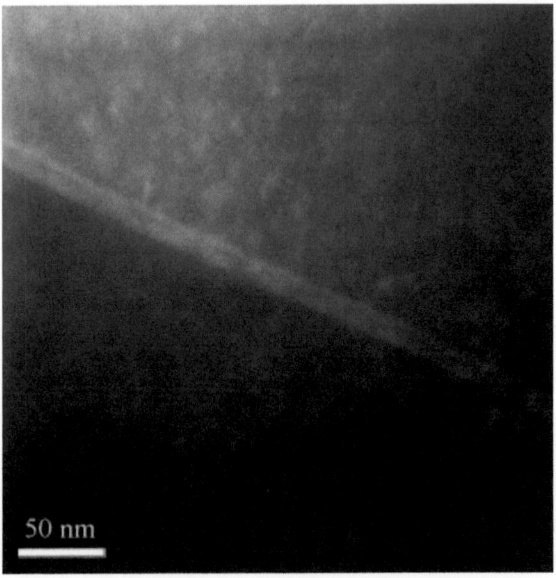

Fig. 8.21 Diffuse *dark-field* image of an intergranular film on the boundary of a growing single crystal that was quenched during growth. Reproduced with permission from [69]. Copyright © 2007, John Wiley & Sons

film varied only very slightly from individual boundary to the other. Chemical analysis of EDS suggested that calcium is the only detectable segregant at the grain boundary. The normal grains in front of the growing single crystal did not contain such a film. Therefore, the mechanism of the SSCC in polycrystalline alumina is due to the rapid diffusion through an intergranular film that behaves in a manner similar to bulk glass, rather than intrinsic grain growth.

The principle and the current status of solid-state conversion of single crystals from polycrystalline precursors have been well summarized recently [70]. As discussed before, NGG has a linear dependence on boundary migration with respect to the driving force. Figure 8.22 shows schematically the free energy state of an atom across a boundary. When there is an energy barrier Δg^*, i.e., activation energy for diffusion, and an energy difference of the atom, Δg, i.e., the driving force, between the two states on grain I and grain II, a net atomic flux between them results in the migration of the boundary. For a system in which the energy difference between the two adjacent grains comes from a nonzero curvature of the boundary, i.e., curvature-driven boundary migration is present, the driving force is expressed as the capillary energy, so that the mobility varies as an exponential function of the activation energy. In this diffusion-controlled boundary migration, it is assumed that the energy state of the atom is only position-dependent, g_{I} on grain I and g_{II} on grain II.

It is acknowledged that boundary migration involves at least two steps: (i) jump of atoms from one grain to its adjacent one—diffusion across the boundary and then (ii) attachment onto the adjacent grain—interface reaction. In this case, the migration is governed by the slower step, either diffusion or interface reaction [71, 72]. In all the conventional mechanisms, the migration has been assumed to be controlled only by the diffusion. However, if the free energy state for the attachment

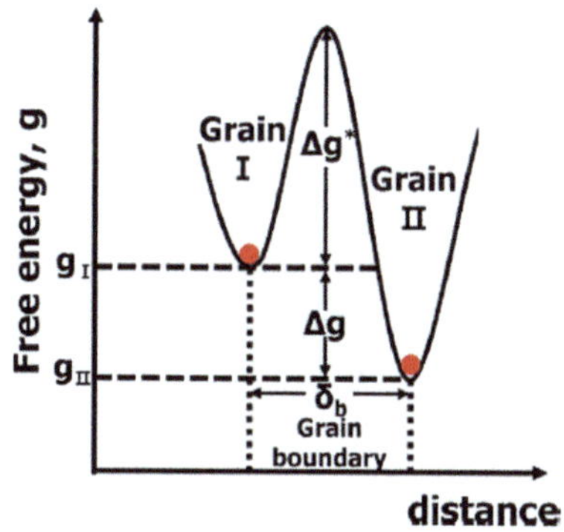

Fig. 8.22 Schematic demonstrating the energy state of atoms, g_{I} and g_{II}, on the surfaces of two adjacent grains, grain I and grain II, with an atomically disordered boundary. The energy states of an atom before and after jumping across the boundary are the same as those on grain I and grain II, g_{I} and g_{II}. Δg^* is the activation energy for atom diffusion across the boundary and Δg is the driving force for atom diffusion. Reproduced with permission from [70]. Copyright © 2015, John Wiley & Sons

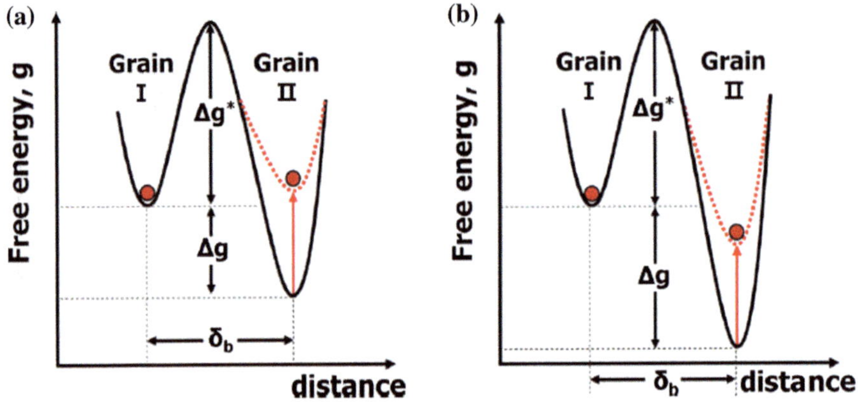

Fig. 8.23 Schematic of the energy state of an atom before and after jumping across a boundary with an atomically ordered structure, for the cases of the energy increase of the attached atom **a** larger and **b** smaller than the driving force for atom diffusion, Δg. Reproduced with permission from [70]. Copyright © 2015, John Wiley & Sons

of the atom is increased to a level that is higher than the original state, the atom will detach and jump back to the original grain, with the free energy diagram shown in Fig. 8.23a. The migration of the boundary is thus governed by a stable attachment, which means that the interface reaction is similar to the case of the two-dimensional nucleation on a single crystal. If the energy increase for the attachment of an atom on a flat surface is lower than the driving force, as shown in Fig. 8.23b, the attached atom will be stable and will not return to the original grain. In this case, the diffusion of atoms governs the overall kinetics.

According to the mean-field concept for boundary migration, every individual grain has its own driving force, depending on its size relative to the average size, so that the driving force for growth of grains in a polycrystalline system has a distribution, with the largest grain having the maximum value, Δg_{max} [71, 72]. Grain growth behavior is then governed by the value of Δg_{max} relative to Δg_c. The principle of microstructural evolution is deduced for solid/liquid two-phase systems with the coupling effect of Δg_{max} and Δg_c. Totally, there are four types of grain growth behavior: (i) normal grain growth (NGG) if $\Delta g_c = 0$, (ii) pseudo-normal grain growth (PNGG) if $0 < \Delta g_c \ll \Delta g_{max}$, (iii) abnormal grain growth (AGG) if $0 < \Delta g_c \sim \Delta g_{max}$, and (iv) stagnant grain growth (SGG) if $\Delta g_{max} \ll \Delta g_c$.

In order for solid-state conversion of single crystals, grain growth in the matrix should be suppressed. Preferably, the condition of $\Delta g_{max} < \Delta g_c$ should be maintained during the conversion, while the driving force for the growth of the seed crystal, Δg_{seed}, should be larger than Δg_c. Therefore, it is necessary to adjust Δg_{max} and Δg_c. The maximum driving force is determined by the average grain size and the grain size distribution. As a result, the properties of the initial powder determine the magnitude of Δg_{max} at the beginning of the sintering. The critical driving force Δg_c is dependent on temperature, doping, and atmosphere, e.g., oxygen partial pressure. In solid-state conversion experiments, the seed crystal can be placed either

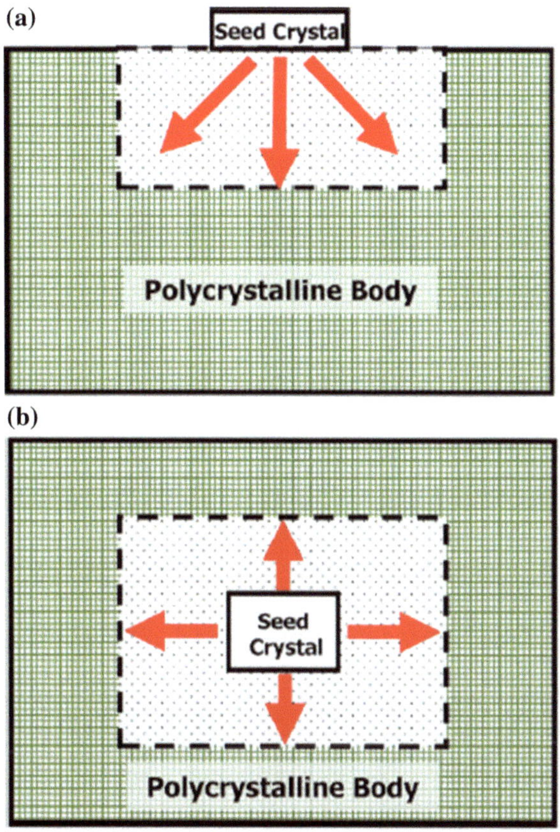

Fig. 8.24 Schematics illustrating the solid-state conversion of single crystals by **a** diffusion bonding and **b** embedding of a seed crystal. Reproduced with permission from [70]. Copyright © 2015, John Wiley & Sons

on top of the sintered polycrystalline ceramics or embedded within the powder compact, as schematically shown in Fig. 8.24 [70]. The seed crystal can also be created through nucleation from the polycrystalline matrix by using temperature gradient or dopant.

8.7 Grain-Boundary Mobility

For single-phase solids, the grain-boundary mobility M_b is called intrinsic boundary mobility, which is defined by Eq. (8.30), so that the grain growth is controlled by the diffusion of the atoms in the grains across the grain boundary. In this case, the effects of various other factors, such as dopants, impurities, secondary phases and pore, which are almost always present in ceramics, on the rate of atom diffusion across the grain boundary, are neglected. In practice, boundary mobilities obtained from experimental grain growth data are usually significantly lower than the intrinsic mobility predicted by the theories.

8.7.1 Effect of Fine Second-Phase Particles

It is assumed that a polycrystalline solid contains randomly dispersed fine second-phase particles, also known as fine inclusions or precipitates, which are insoluble and immobile in the system. When a grain boundary encounters a fine particle during its movement driven due to its curvature, it will be pinned by the particle, while the rest portion of the grain boundary continues to move. The grain boundary will break away only as the rest portion has moved for a sufficiently far distance. Therefore, if the grain boundary encounters a sufficient number of particles, it could be entirely pinned, i.e., the motion of the grain boundary is completely inhibited. This scenario can be described by using two models.

If the pinning particles are assumed to be monosized, spherical, insoluble, immobile and randomly distributed in the polycrystalline solid, for a grain boundary with principal radii of curvature, a_1 and a_2, the driving force per unit area for the motion of the grain boundary is given by [73]:

$$F_b = \gamma_{gb}\left(\frac{1}{a_1} + \frac{1}{a_2}\right). \tag{8.48}$$

If both a_1 and a_2 are proportional to the grain size G, there is

$$F_b = \frac{\alpha\gamma_{gb}}{G}, \tag{8.49}$$

where α is a geometrical shape factor, e.g., $\alpha = 2$ for a spherical grain. When the grain boundary intersects a particle, it is dragged by the particle, as shown in Fig. 8.25a, b. Due to the dragging effect of the particle, the grain boundary needs additional work to migrate, as compared to a particle-free boundary, i.e., there is a

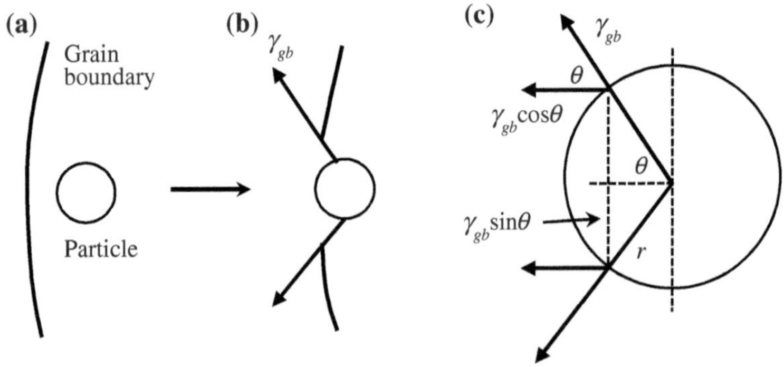

Fig. 8.25 Interaction of a grain boundary with an immobile particle. **a** Approach of the boundary toward the particle. **b** Interaction between the grain boundary and the particle leading to a retarding force on the boundary. **c** Detailed geometry of the particle–grain boundary interaction. Reproduced with permission from [4]. Copyright © 2003, CRC Press

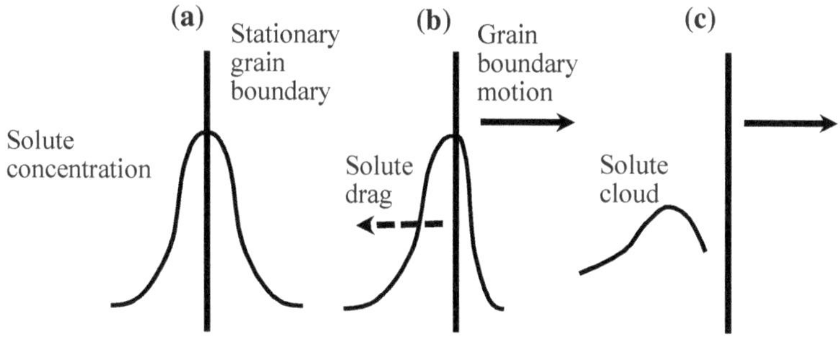

Fig. 8.26 Sketch of the solute drag effect produced by the segregation of dopants to the grain boundaries. **a** Symmetrical distribution of the dopant in the region of a stationary grain boundary. **b** For a moving boundary, the dopant distribution becomes asymmetrical if the diffusion coefficient of the dopant atoms across the boundary is different from that of the host atoms. The asymmetrical distribution produces a drag on the boundary. **c** Breakaway of the boundary from the dopant leaving a solute cloud behind. Reproduced with permission from [4]. Copyright © 2003, CRC Press

retarding force applied to the grain boundary. If r is the radius of the inclusion, as shown in Fig. 8.26c, the retarding force of the particle applied to the boundary is given by

$$F_{\mathrm{r}} = \gamma_{\mathrm{gb}} \cos\theta(2\pi r \sin\theta). \tag{8.50}$$

Therefore, the retarding force is the product of the perimeter of contact and the grain-boundary tension, which is in an opposite direction to that of the grain-boundary migration. At $\theta = 45°$, $\sin\theta\cos\theta = 1/2$, so that the retarding force is maximized, which is given by

$$F_{\mathrm{r}}^{\max} = \pi r \gamma_{\mathrm{gb}}. \tag{8.51}$$

If there are N_{A} inclusion particles per unit area in the grain boundary, the maximum retarding force per unit area of the boundary is given by

$$F_{\mathrm{d}}^{\max} = N_{\mathrm{A}} \pi r \gamma_{\mathrm{gb}}. \tag{8.52}$$

Although it is difficult to determine N_{A}, it is related to N_{V}, the number of inclusions per unit volume, which is given by

$$N_{\mathrm{A}} = 2rN_{\mathrm{V}}. \tag{8.53}$$

If the volume fraction of the inclusions in the solid is f, there is

$$N_{\mathrm{V}} = \frac{f}{(4/3)\pi r^3}. \tag{8.54}$$

When N_A in Eq. (8.52) is substituted with N_V determined by Eqs. (8.53) and (8.54), the retarding force becomes:

$$F_d^{max} = \frac{3f\gamma_{gb}}{2r}. \tag{8.55}$$

The net driving force per unit area of the grain boundary is thus given by

$$F_{net} = F_b - F_d^{max} = \gamma_{gb}\left(\frac{\alpha}{G} - \frac{3f}{2r}\right). \tag{8.56}$$

When $F_{net} = 0$, the grain boundary stops migrating, which occurs if the following equation is valid:

$$G = G_L = \frac{2\alpha r}{3f}, \tag{8.57}$$

where G_L is called the limiting grain size. Equation (8.57) is also known as the Zener relationship [73], which means that the limiting grain size is proportional to the size of the inclusion particle and inversely proportional to the volume fraction of inclusion particle. Although there have been various modifications to the Zener model, the dependence of G_L on f is the same [74–77].

Monte Carlo computer simulation has been modified to analyze NGG of single-phase solids that are dispersed with fine secondary phase particles that are assumed to be monosized, spherical, insoluble, immobile, and randomly distributed [34, 78–80]. It is found that the grain growth is finally limited at a specific size. During the growing of the grains, the microstructures have the characteristics of NGG, with the same grain size distribution and growth exponent as those of the systems without the secondary phase inclusions. Comparatively, the simulated microstructures for NGG with the presence of inclusions possess grains with irregular shapes.

It is also revealed that both the time used to reach the limiting grain size and the value of the limiting grain size decrease with increasing volume fraction of the secondary phase particles, f. Furthermore, when the total number of inclusions is constant, the fraction of inclusions ϕ_p that are located at the grain boundaries decreases with time. This occurrence becomes more and more pronounced, with decreasing volume fraction of the secondary phase particles, f. The limiting grain size G_L is given by

$$G_L = K_S \frac{r}{(\varphi_p f)^{1/3}}, \tag{8.58}$$

where K_S is a constant and r is radius of the inclusions. Comparatively, the Zener model overestimates the driving force for grain growth, because the effect of the inclusions on the curvature of the boundary is neglected, while only the work to drag the inclusion is considered. Disagreement is occasionally observed between the simulation results and experimental data, demonstrating the complicated situations of real materials [81–88].

8.7.2 Effect of Dopants–Solute Drag

Dopant is also called solute, which is dissolved as solid solution in polycrystalline solids. If there is an interaction potential for the solute to be attracted to or repelled from the grain boundary, the solute atoms or ions will have a nonuniform distribution at the grain boundaries. The interaction could be due to lattice strain energy caused by size mismatch between the solute and host ions and/or electrostatic potential energy for aliovalent solutes.

When the grain boundary is at a stationary state, the concentration profiles of the solute ions at both sides of the boundary are symmetrical, as shown in Fig. 8.26a. Because the forces of interaction at the two sides are balanced, the net force of interaction is zero. Once the boundary starts to migrate, the profiles of the dopant concentration at both sides will become asymmetrical, because the diffusion rate of the solute ions across the boundary is different from that of the host diffusion, as shown in Fig. 8.26b. This asymmetry of dopant concentration profile produces a retarding force or drag applied to the grain boundary, so as to decrease the driving force for migration of the boundary. When the driving force for the migration of the grain boundary is sufficiently high, the boundary will break away from the solute, which also known as solute cloud, as shown in Fig. 8.26c. In this case, mobility of the grain boundary is close to its intrinsic value. Various models have been developed to describe grain-boundary migration with solute drag [89–93].

The chemical potential of the solute atoms in the near grain-boundary region is given by

$$\mu = kT \ln C(x) + U(x) + U_0, \tag{8.59}$$

where $C(x)$ and $U(x)$ are concentration and energy functions of x, while U_0 is a constant, such that $U(\infty) = 0$. At steady-state conditions, the composition profile of the solute can be described by the following equation:

$$\frac{\mathrm{d}C}{\mathrm{d}t} = -v_\mathrm{b}\frac{\mathrm{d}C}{\mathrm{d}x}. \tag{8.60}$$

According to diffusion equation, the flux of the solute atoms is given by

$$J = -\frac{D_\mathrm{b} C}{\Omega kT}\frac{\mathrm{d}\mu}{\mathrm{d}x}, \tag{8.61}$$

where D_b is the diffusion coefficient for the solute atoms across the grain boundary. According to Eq. (8.59), Eq. (8.61) can be rewritten as follows:

$$J = -\frac{D_b}{\Omega}\frac{dC}{dx} + \frac{D_b C}{\Omega kT}\frac{dU}{dx}. \tag{8.62}$$

The concentration profile of the solute can now be calculated from the continuity, which is given by

$$\frac{dJ}{dx} + \frac{1}{\Omega}\frac{dC}{dt} = 0, \tag{8.63}$$

where the boundary conditions are $dC/dx = 0$, $dU/dx = 0$, and $C(x) = C_\infty$ at $x = \infty$. The concentration C_∞ can be taken as that in the interior of the grain. Therefore, $C(x)$ must satisfy the equation:

$$D_b\frac{dC}{dx} + \frac{D_b C}{kT}\frac{dU}{dx} + v_b(C - C_\infty) = 0. \tag{8.64}$$

The solute atom applies a force $-(dU/dx)$ to the grain boundary, so that the net force applied by all the solute atoms is given by

$$F_s = -N_V\int_{-\infty}^{\infty}[C(x) - C_\infty]\frac{dU}{dx}dx, \tag{8.65}$$

where N_V is the number of host atoms per unit volume. In the analysis, the $C(x)$ derived from Eq. (8.64) is used to calculate F_s with Eq. (8.65). It is found that there is an approximate solution that is valid for both low and high boundary velocities, which is given by:

$$F_s = \frac{\alpha C_\infty v_b}{1 + \beta^2 v_b^2}, \tag{8.66}$$

where the parameters α and β are given by the following expressions:

$$\alpha = 4N_V kT\int_{-\infty}^{\infty}\frac{\sinh^2[U(x)/2kT]}{D_b(x)}dx, \tag{8.67}$$

$$\frac{\alpha}{\beta^2} = \frac{N_V}{kT}\int_{-\infty}^{\infty}\left(\frac{dU}{dx}\right)D_b(x)dx. \tag{8.68}$$

In this case, α is the solute drag per unit velocity and per unit dopant concentration in the low boundary velocity limit, while $1/\beta$ is the drift velocity of impurity atom moving across the grain boundary. According to Eq. (8.67), solutes with either attractive or repulsive interaction energy, with the same magnitude, will apply a similar drag force to the grain boundary.

The total drag force on the boundary is the sum of the intrinsic drag F_b and the drag due to the dopant atoms F_s, which is given by

$$F = F_b + F_s = \frac{v_b}{M_b} + \frac{\alpha C_\infty v_b}{1 + \beta^2 v_b^2}, \tag{8.69}$$

where M_b is the intrinsic boundary mobility defined by Eq. (8.30). With the low boundary velocity limit, because the term $\beta^2 v_b^2$ in Eq. (8.69) can be neglected, there is

$$v_b = \frac{F}{1/M_b + \alpha C_\infty}. \tag{8.70}$$

At the beginning, the mobility due to solute drag is constant. However, as the velocity increases, the grain boundary continually separates from the solute. When the velocity becomes sufficiently high, the boundary will migrate at its intrinsic velocity. Therefore, a transition is observed from the solute drag-limited velocity to the intrinsic velocity, over a range of driving forces [94–96].

According to Eq. (8.70), the boundary mobility M'_b is equal to v_b/F, which can be expressed in terms of the intrinsic component M_b and the solute drag component M_s:

$$M'_b = \left(\frac{1}{M_b} + \frac{1}{M_s}\right)^{-1}, \tag{8.71}$$

where $M_s = 1/\alpha C_\infty$. If the solute segregates at the grain boundary and the center of the boundary contributes most significantly to the drag effect, α can be expressed by the following equation:

$$\alpha = \frac{4 N_V kT \delta_{gb} Q}{D_b}, \tag{8.72}$$

where Q is a partition coefficient (>1) for the dopant to distribute between the boundary region and the interior of the grain, such that the solute concentration in the boundary region is QC_∞. Therefore, mobility due to the solute drag is given by

$$M_s = \frac{D_b}{4 N_V kT \delta_{gb} Q C_\infty}. \tag{8.73}$$

This equation indicates that the presence of dopants is the most effective way to reduce the boundary mobility if the diffusion coefficient of the rate-limiting species D_b is low and the segregated solute concentration QC_∞ is high.

8.8 Grain Growth and Pore Evolution

Densification is accompanied by grain growth, while migration of the boundaries during the grain growth leads to coalescence of the pores, so that the average size of the pores also increases, as shown in Fig. 8.27 [3]. Both the grain growth and pore coalescence contribute to the coarsening at the later stages of sintering of ceramics.

8.8.1 Thermodynamics of Pore–Boundary Interactions

Whether a pore shrinks or not is determined by the free energy change that accompanies the change in pore size. If the pore shrinks, there will be a decrease in the free energy due to the decrease in the pore surface area, while there is also an increase due to the increase in the grain-boundary area. If the pore shrinks, the former (decrease) must be larger than the latter (increase). The equilibrium shape of the pore is determined by the dihedral angle ψ, which is defined by

$$\cos\frac{\psi}{2} = \frac{\gamma_{gb}}{2\gamma_{SV}}, \tag{8.74}$$

where γ_{SV} and γ_{gb} are the interfacial tensions at the pore surface and in the grain-boundary interface, respectively.

In two dimensions, a pore can have a dihedral angle $\psi = 120°$, which is surrounded by N other grains. The number N is called the pore coordination number (CN). Similar to the case of a grain surrounded by other grains, if $N = 6$, the pore has straight sides; otherwise, it has convex sides for $N < 6$ and concave sides for $N > 6$, as shown in Fig. 8.28. The surface of the pore will move toward its center of curvature, so the pore with $N < 6$ will shrink, whereas the one with $N > 6$ will expand. The pore is metastable for N = 6, so that the number is called critical pore CN, or N_c. Accordingly, if the pores have convex sides, i.e., N < 6, the decrease in the pore

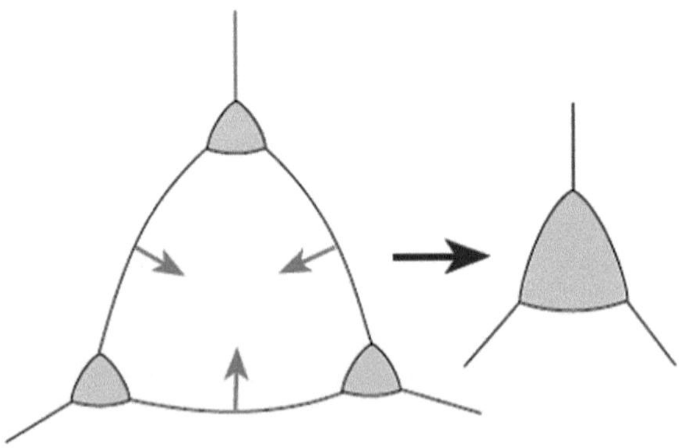

Fig. 8.27 Schematic illustration of grain growth accompanied by pore coalescence. Reproduced with permission from [3]. Copyright © 2007, Springer

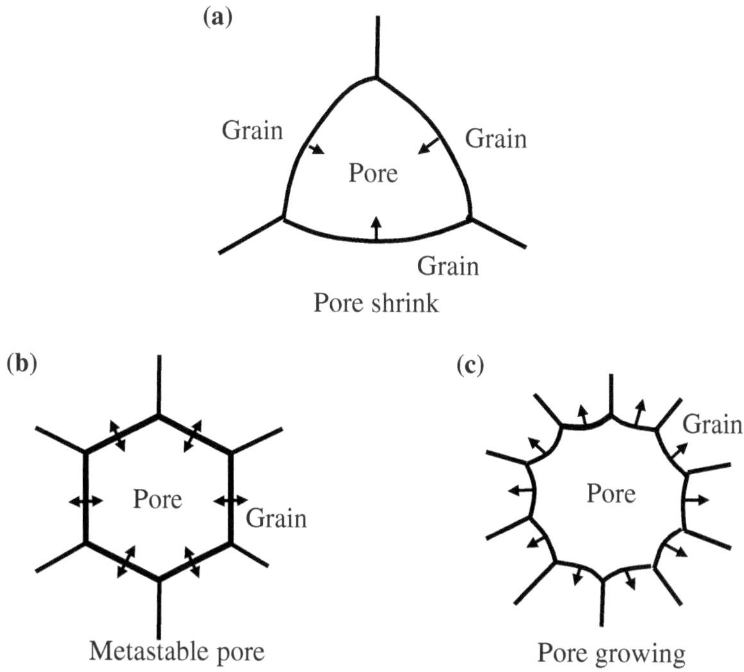

Fig. 8.28 Pore stability in two dimensions for a dihedral angle of 120° : **a** shrinking pore, **b** metastable pore and **c** growing pore. Reproduced with permission from [4]. Copyright © 2003, CRC Press

surface energy is larger than the increase in the grain-boundary energy. In contrast, if the pores have concave sides ($N > 6$), the decrease in pore surface energy is smaller than the increase in grain-boundary energy. For the metastable pores ($N = 6$), the increase in grain-boundary energy is balanced by the decrease in pore surface energy.

The geometrical considerations can be extended to three dimensions, in which the pore is a polyhedron [97]. Taking r_s as the radius of curvature of the circumscribed sphere around a polyhedral pore surrounded by grains, the ratio of the radius of curvature of the pore, r to r_s, depends both on the dihedral angle and the pore CN. When the surfaces of the pore become flat ($r = \infty$), the pore is metastable. As a result, the pore neither grows nor shrinks. In this case, the ratio r_s/r is zero, with which N_c is defined. For example, when a pore has a dihedral angle of 120°, $N_c = 12$. Therefore, a pore with $N < 12$ will shrink, while the one with $N > 12$ will expand. However, theoretical analysis indicated that, in practice, the pore with $N > N_c$ will grow to a limited size and then stop, instead of grow continuously [98, 99].

Therefore, when powder is poorly compacted, the compact will contain pores that are larger than the grain size. In this case, it is very difficult to densify the compact, especially when the pores have low dihedral angles, due to the large CN. In order to avoid this problem, it is necessary to prepare compacts with high green density and uniform pore size distribution, by using advanced techniques, such as cold isostatic pressing and colloidal methods for compaction.

8.8.2 Kinetics of Pore-Boundary Interactions

With the definition of the pore mobility, it is possible to analyze the influence of the interactions between the pores and the grain boundaries on the kinetics of grain growth. There are two cases that can be considered: (i) the pore becomes separated from the boundary and (ii) the pore remains attached to the boundary.

When $v_p < v_b$, pore separation will be observed, and this condition can also be written as follows:

$$F_pM_p < FM_b, \tag{8.75}$$

where F is the effective driving force on the boundary. If F_d is the drag force exerted by a pore, then the balance of the forces requires that F_d is equal and opposite to F_p. Considering unit area of the boundary in which there are N_A pores, Eq. (8.75) can be written as follows:

$$F_pM_p < (F_b - N_AF_p)M_b, \tag{8.76}$$

where F_b is the driving force on the pore-free boundary due to its curvature. Rearranging Eq. (8.76), the condition for pore separation can be expressed as follows:

$$F_b > N_AF_p + \frac{F_pM_p}{M_b}. \tag{8.77}$$

The condition for pore attachment to the boundary is $v_p = v_b$, which can also be written as follows:

$$F_pM_p = (F_b - N_AF_p)M_b. \tag{8.78}$$

Putting $v_p = F_pM_p = v_b$ Eq. (8.78) and rearranging, there is

$$v_b = F_b\frac{M_pM_b}{N_AM_b + M_p}. \tag{8.79}$$

Two limiting conditions can be defined. When $N_AM_b \gg M_p$, there is

$$v_b = \frac{F_bM_p}{N_A}. \tag{8.80}$$

The effective driving force on the boundary is $F = F_b - N_AF_p$, and from Eq. (8.78), there is $F = v_p/M_b$. From $v_p = v_b = F_bM_p/N_A$, there is $F = F_bM_p/N_AM_b \ll F_b$. The driving force on the boundary is almost balanced by the drag of the pores, while the boundary migration is limited by the pore mobility. This condition is referred to as pore control. The other limiting condition is $N_AM_b \ll M_p$, in which case:

$$v_b = F_b M_b. \tag{8.81}$$

The drag exerted by the pores is $N_A F_p = N_A(F_b M_b/M_p) \ll F_b$. The presence of the pores has almost no effect on the boundary velocity, which is a condition known as boundary control.

8.8.3 Grain Growth Kinetics

With the simplified model for nearly spherical isolated pores on the grain boundary, the equations for the grain growth kinetics can be derived, if it is assumed that the grain growth is controlled by the pore mobility, i.e., pore control. If pore migration occurs through surface diffusion, there is

$$v_b \approx \frac{dG}{dt} = \frac{F_b}{N_A} \frac{D_s \delta_s \Omega}{\pi k T r^4}. \tag{8.82}$$

Taking $F_b \approx 1/G$, $N_A \approx 1/X^2 \approx 1/G^2$, where X is the interpore distance, and assuming that the coarsening is due to the grain growth and pore coalescence, so that $r \approx G$. Rearranging Eq. (8.82) gives

$$\frac{dG}{dt} = \frac{K_1}{G^3}, \tag{8.83}$$

where K_1 is a constant at a given temperature. Integration of Eq. (8.83) yields

$$G^4 = G_0^4 + K_2 t, \tag{8.84}$$

where G_0 is the grain size at $t = 0$ and K_2 is a constant. Similarly, grain growth equations can be derived by the same procedure for the other mechanisms.

Using the general form of the grain growth equation, $G^m = G_0^m + Kt$, the exponent m for the various mechanisms can be obtained. Except for the mechanism involving solution of second-phase particles ($m = 1$), the m values lie in the range 2–4. In many ceramics, $m = 3$ has been reported. Therefore, this value can correspond to at least five mechanisms. In other words, the m values determined from grain growth data cannot be reliably used to imply physical significance.

8.9 Simultaneous Densification and Grain Growth

In solid-state sintering, the processes of densification and grain growth (coarsening) are usually discussed separately. However, at the final stage of sintering, the two processes take place simultaneously with their rates have mutual influence. To fully understand and control the sintering process, it is necessary to consider both densification and coarsening and their interaction. Because each process by itself is already, it is unrealistic to have a theoretical model that is able to predict the coupling of densification and coarsening. Therefore, in practice, simpler approaches are generally employed to have a qualitative understanding of the interaction. As a reference, two models are discussed in this part: (i) Brook model and (ii) Yan, Cannon and Chowdhry model.

Various theoretical models to describe simultaneous densification and grain growth, as well as the transition to AGG, have been well established [100, 101]. It is assumed that an idealized final-stage microstructure consists of a nearly spherical pore on at isolated grain boundary, the considerations in this case include the transition from pore drag-controlled boundary migration to intrinsic or solute drag boundary migration, with the pores either attached or separated from the boundary, which are as functions of grain size G and pore size $2r$. As mentioned earlier, there are two limiting cases: pore control and boundary control. The conditions that separate these two cases are represented by a curve, called the equal mobility curve, and defined by the condition that the pore mobility is equal to the boundary mobility, which is,

$$N_{\mathrm{A}}M_{\mathrm{b}} = M_{\mathrm{p}}. \tag{8.85}$$

If the pore migration occurs through surface diffusion, the appropriate relations for M_{b} and M_{p} can be derived. If $N_{\mathrm{A}} \approx 1/X^2$, where X is the interpore distance, and $X \approx G$, Eq. (8.82) gives

$$G_{\mathrm{em}} = \left(\frac{D_{\mathrm{a}}\pi}{D_{\mathrm{s}}\delta_{\mathrm{s}}\delta_{\mathrm{gb}}}\right)^{1/2} r^2, \tag{8.86}$$

where G_{em} is the grain size at the condition of the equal mobility. Using logarithmic axes plot G versus $2r$, the equal mobility condition is represented by a straight line with a slope of 2.

In determining the conditions for separation of the boundary from the pore, the maximum force exerted by the grain boundary on a pore to drag it along is given by Eq. (8.51), which is

$$F_{\mathrm{p}}^{\max} = \pi r\gamma_{\mathrm{gb}}. \tag{8.87}$$

The maximum velocity for the pore is therefore given by:

$$v_{\mathrm{p}}^{\max} = M_{\mathrm{p}}F_{\mathrm{p}}^{\max}. \tag{8.88}$$

If the velocity of the boundary with the attached pore exceeds $v_{\mathrm{p}}^{\max}$, separation will occur. The limiting condition for separation can therefore be written as follows:

$$v_{\mathrm{b}} = v_{\mathrm{p}}^{\max}. \tag{8.89}$$

Substituting for v_{b} in Eq. (8.79), there is

$$\frac{M_{\mathrm{b}}M_{\mathrm{p}}}{N_{\mathrm{A}}M_{\mathrm{b}} + M_{\mathrm{p}}}F_{\mathrm{b}} = M_{\mathrm{p}}F_{\mathrm{b}}^{\max}. \tag{8.90}$$

When $N_{\mathrm{A}} \approx 1/X^2$, there is

$$G_{\mathrm{sep}} = \left(\frac{\pi r}{X^2} + \frac{D_{\mathrm{s}}\delta_{\mathrm{s}}\delta_{\mathrm{gb}}}{D_{aa}r^3}\right)^{-1}, \tag{8.91}$$

where G_{sep} is the grain size when the boundary separates from the pore. If $X \approx G$, Eq. (8.91) can be written as

$$\left(\frac{D_{\mathrm{s}}\delta_{\mathrm{s}}\delta_{\mathrm{gb}}}{D_{aa}r^3}\right)G_{\mathrm{sep}}^2 - G_{\mathrm{sep}} + \pi r = 0. \tag{8.92}$$

The solution to this quadratic equation determines the separation curve.

It is to be expected that factors such as grain size distribution, the number of pores at the grain boundaries, and the dihedral angle all have effect on the simple relationships between the grain-boundary curvature and pore size for the separation condition [98, 99]. In a refined model, it is demonstrated that when pore separation occurs, the assumption of $X \approx G$ is not valid [102–105]. The effect of dihedral angle has also been considered [106–108]. As shown in Fig. 8.29, the effective area of the boundary intersected by a pore with a constant volume increases as the dihedral angle decreases, resulting in greater pore drag and less boundary separating from the pore [106].

The effects of a grain size distribution and solute drag have also been incorporated into the basic model [109, 110]. The critical density at which pore separation occurs is significantly lower for the powder compacts with a wide distribution of particle size than for the powder with a narrow size distribution. Separation of only a fraction of the pores from the boundary is sufficient to cause AGG, which is different from the assumption in the simple analysis that all pores should separate from the boundaries.

Another theoretical analysis of simultaneous densification and grain growth is the Yan, Cannon and Chowdhry model [110]. In this model, the final-stage microstructure is assumed to consist of tetrakaidecahedral grains with spherical pores at the grain corners, the dependences of the achievable final density and the extent of coarsening on the ratio of the coarsening rate to the densification rate can be determined. The instantaneous rate of change in the pore radius r is given by:

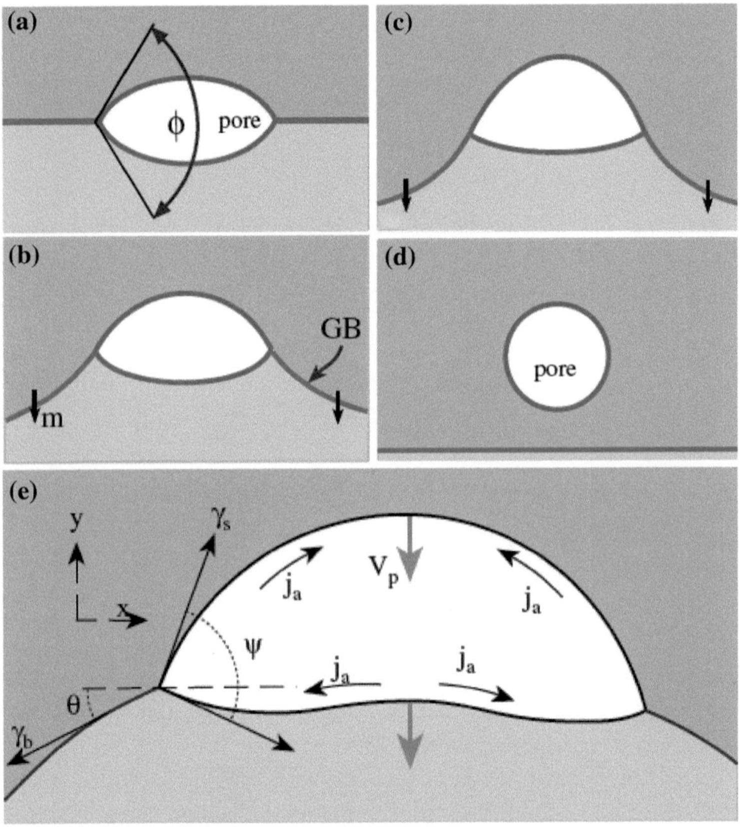

Fig. 8.29 Schematic of the distortion that accompanies the motion of pores attached to grain boundaries and the atomic flux from the leading surface to the trailing surface. Reproduced with permission from [106]. Copyright © 1983, John Wiley & Sons. Adopted from permission from [3]. Copyright © 2007, Springer

$$\frac{\mathrm{d}r}{\mathrm{d}t} = \left(\frac{\mathrm{d}r}{\mathrm{d}t}\right)_P + \left(\frac{\mathrm{d}r}{\mathrm{d}t}\right)_G, \tag{8.93}$$

where the first term on the right-hand side is the rate of change in the pore size at constant porosity, due to the coarsening process, which is a positive value, while the second term is the rate of change at constant grain size, due to the densification process, which is a negative value. For a coarsening process involving grain growth and pore coalescence, there is

$$\frac{\mathrm{d}r}{\mathrm{d}t} = \frac{r}{G}\left(\frac{\mathrm{d}G}{\mathrm{d}t}\right)_P + \left(\frac{\mathrm{d}r}{\mathrm{d}t}\right)_G. \tag{8.94}$$

Since $dG/dr = (dG/dt)/(dr/dt)$, there is

$$\frac{dG}{dr} = \frac{dG/dt}{(r/G)(dG/dt) + (dr/dt)_G}. \tag{8.95}$$

This equation can be written as follows:

$$\frac{d\ln G}{d\ln r} = \frac{\Gamma}{\Gamma - 1}, \tag{8.96}$$

where Γ is the ratio of the coarsening rate to the densification rate, which is given by:

$$\Gamma = -\frac{(r/G)(dG/dt)}{(dr/dt)_G}. \tag{8.97}$$

In this case, Γ is calculated by assuming specific models for densification and grain growth, where only one densification mechanism and one coarsening mechanism are dominant.

The Yan, Cannon and Chowdhry model has been modified by including the conditions to separate the boundary from the pore [111, 112]. The calculated results

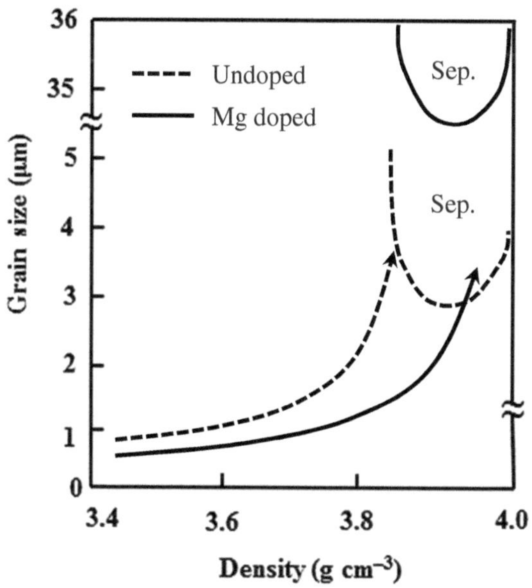

Fig. 8.30 Grain size–density map for Al_2O_3, illustrating the effect of raising the surface diffusion coefficient by a factor of 4, reducing the lattice diffusion coefficient by a factor of 2, and reducing the grain-boundary mobility by a factor of 34. This has the effect of flattening the grain size–density trajectory and raising the separation region to larger grain sizes, thereby making it possible to sinter to full density. Reproduced with permission from [112]. Copyright © 1990, John Wiley & Sons

for Al_2O_3, where the densification is controlled by lattice diffusion, while the grain growth or coarsening is controlled by surface diffusion-controlled pore drag, are shown in Fig. 8.30 [112]. Due to the combined features of grain size versus density trajectory and the conditions for separation, the treatment is effective to discuss the influence of dopants, temperature, and other variables on microstructural evolution. It is observed that the separation region is shifted to larger grain sizes, when MgO is used as a dopant. This is because the presence of MgO decreases the grain-boundary mobility of Al_2O_3. As a result, the grain size versus density trajectory bypasses the separation region. Flattening of the trajectory means that densification rate is increased, while the grain growth rate is suppressed. In addition, the trajectory can also be flattened by controlling the heating steps during the sintering, e.g., two-step sintering, as discussed later.

8.10 Strategies to Control Microstructure of Ceramics

Most applications of ceramics require products with high-density and controlled (small) grain size [110, 113–117]. The discussion of the previous section indicates that when suitable processing procedures are employed, such an end point is achievable through fabrication routes that have the effect of increasing the ratio of the densification rate to the grain growth (coarsening) rate or avoiding the separation region (abnormal grain growth). The principles governing these fabrication routes are discussed as follows.

8.10.1 Sintering at External Pressures

When compared to sintering, hot pressing produces an increase in the driving force for densification. For an equivalent microstructure, the dependence of the densification rate on the driving force can be written as follows:

$$\dot{\rho}_{\mathrm{hp}} \sim (\Sigma + p_{\mathrm{a}}), \quad \dot{\rho}_{\mathrm{s}} \sim \Sigma, \tag{8.98}$$

where the subscripts in $\dot{\rho}_{\mathrm{hp}}$ and $\dot{\rho}_{\mathrm{s}}$ stand for hot pressing and normal sintering, respectively, Σ is the sintering stress and p_{a} is the applied pressure. Since the grain-boundary mobility is not changed, there is

$$\dot{G}_{\mathrm{hp}} \approx \dot{G}_{\mathrm{s}}. \tag{8.99}$$

When $p_{\mathrm{a}} \gg \Sigma$, there is $\left(\dot{\rho}/\dot{G}\right)_{\mathrm{hp}} \gg \left(\dot{\rho}/\dot{G}\right)_{\mathrm{s}}$. Therefore, by using hot pressing, it is possible to produce ceramics with high density coupled with small grain size, due to the ability to increase the value of $\dot{\rho}/\dot{G}$. This is the simple reason why HP, HIP, and SPS could be used to produce ceramics with high densification and small grain sizes.

8.10.2 Use of Dopants and Inclusions

The effects of dopants on microstructure of ceramics are important but complex. If the dopants segregate at the grain boundary, they will reduce the boundary mobility by the so-called solute drag effect. In this case, there is $\dot{G}_{\text{doped}} < \dot{G}_{\text{undoped}}$. Dopants can also influence the densification process, although this effect is usually relatively weaker compared to the effect on grain growth rate. The effectiveness of dopants is reflected by the ability to reduce the value of $\dot{G}$ or that to increase the value of $\dot{\rho}/\dot{G}$.

The role of inclusions sometimes is similar to that of dopants. If the drag on the grain boundary is sufficiently strong, the boundary could be pinned. Similarly, their effect on densification is not very significant. The effectiveness of inclusions is almost the same as that of dopants, in terms of $\dot{G}$ or $\dot{\rho}/\dot{G}$.

It has been that similar dopants could have different and even opposite effects on grain growth. One example has been shown for AlON transparent ceramics doped with Y_2O_3 and La_2O_3 [118]. In this study, γ-AlON powder was mixed with different concentrations of Y_2O_3 (99.99 % purity) and La_2O_3 (99.99 % purity) by using high-energy ball milling. The mixtures were then dried and passed through a 200-mesh sieve. Cylindrical pellets with a diameter of 20 mm and a thick of 4 mm were pressed at a uniaxial pressure of 5 MPa and then isostatically pressed at a pressure of 200 MPa. The pellets with a relative density of about 50 % were pressureless sintered at 1820–1950 °C for different durations in nitrogen.

Figure 8.31 shows optical transmittance and grain size of the AlON transparent ceramics, as a function of concentration of Y_2O_3 and La_2O_3 [118]. The grain size increased gradually with increasing concentration of Y_2O_3 from 0.06 to 0.19 wt%, which means that the presence of Y^{3+} enhanced the mobility of grain boundary. This was probably attributed to the formation of Y_2O_3–Al_2O_3 liquid phase at the grain boundaries, which significantly enhanced the mass transfer of the samples during the sintering process, because of the increase in the liquid diffusion coefficient. In contrast, the grain size decreased with increasing content of La_2O_3, as shown in Fig. 8.31b, i.e., La^{3+} acted as a grain-boundary inhibitor. Although two possible reasons, i.e., pore drag and particle pinning mechanisms, could be employed to explain the observation, given the fact that the doping level was relatively low, the particle pinning was not likely to be the key mechanism. Therefore, dragging effect of the solute ions La^{3+} along the pore-free sections of the boundaries was the grain growth inhibition mechanism.

The different effects of the two dopants could be combined to form a codopant for AlON. On the one hand, a small amount of La^{3+} could largely decelerate the mass transfer and hence suppressed the grain growth. At the same time, a low grain-boundary mobility would be more effective for the pores to escape along the grain boundaries instead of from interior of the grains. On the other hand, a large quantity of Y^{3+} accelerated the mass transfer during the sintering process, which thus facilitated the migration atoms toward pores and enhanced the elimination of pores. An optimized codoping was 0.12 wt% Y_2O_3 plus 0.09 wt% La_2O_3.

Fig. 8.31 Reproduced with permission from [118]. Copyright © 2015, Elsevier

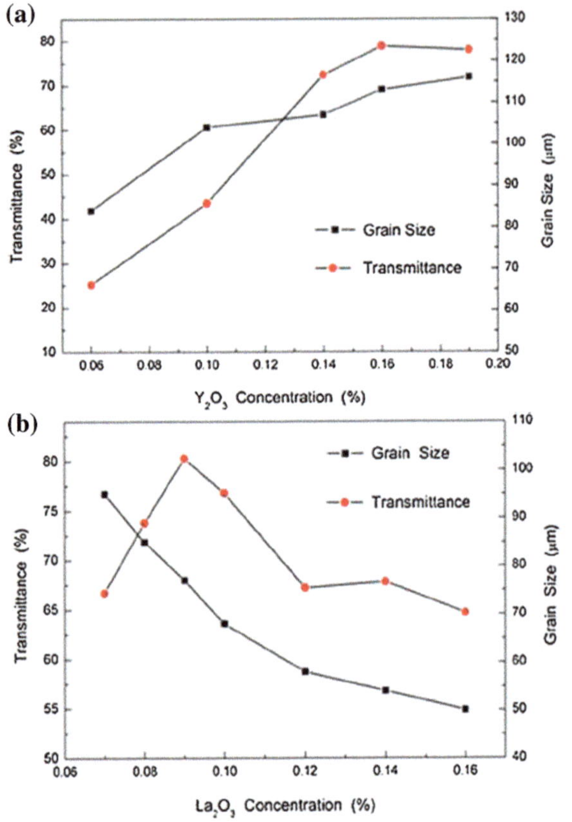

8.10.3 *Use of Fine Particles with Uniform Packing*

Uniformly compacted fine powders have small pores with low CN (i.e., $N < N_c$). The densification rate for such a system is higher than for a similar system with heterogeneous packing. Furthermore, if the particle size distribution is narrow, the driving force for grain growth due to the curvature of the boundary is small. The effectiveness of this route can therefore be interpreted in terms of an increase in the value of $\dot{\rho}/\dot{G}$.

8.10.4 *Control of Firing Schedule*

When a powder system has a higher activation energy for densification than that for grain growth, it is possible to heat it up quickly to a sufficiently high temperature to achieve a high densification without significant grain growth, i.e., $\dot{\rho}/\dot{G}$. With this principle, new sintering strategies to control grain size and microstructure have emerged, including two-step [116, 117] and multiple-step sintering [119, 120]. Especially, two-step sintering has been widely used to process transparent ceramics [121–127].

Figure 8.32 shows representatively schematic diagrams of normal single-step and two-step sintering [128]. The heating and cooling rates were 100 °C h^{-1}, which can be varied according to the properties of materials or specific requirements. Usually, the first-step temperature (T_1) is higher than the second-step temperature (T_2). There are various combinations of the two temperatures, thus providing high flexibility and feasibility.

A successful instance has been demonstrated for Y_2O_3 ceramics [116]. The green body was first heated to a higher temperature to achieve an intermediate density and then cooled down a lower temperature, at which the sample was held until it was fully densified. For pure nanosized Y_2O_3, the sample was first heated to 1310 °C and then cooled to 1150 °C, full density was achieved after holding for 20 h, during which there was no grain growth. For different compositions, the scheme should be different.

The key to the two-step sintering is to get a sufficiently high starting density during the first step. It has been shown that, when the density of a ceramic compact is above 70 %, all pores become subcritical and unstable against shrinkage. These pores can be readily filled, so long as it is allowed by the grain-boundary diffusion, even though the particle network is frozen. Generally, densities of >75 % are sufficient to maintain the second-step sintering. The absence of grain growth in second-step sintering can

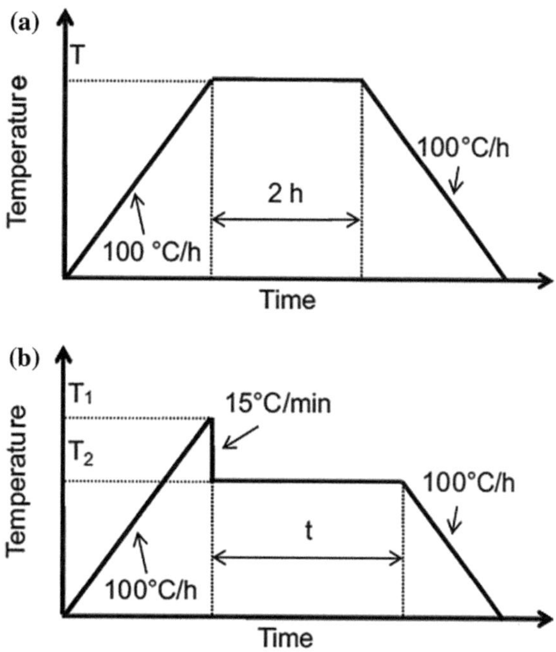

Fig. 8.32 Heating programs of **a** single and **b** two-step sintering methods. Reproduced with permission from [128]. Copyright © 2012, Elsevier

be attributed to kinetics. Grain growth is significantly dynamic, so that the microstructure of the sintered body constantly refreshes. As mentioned earlier, one eighth of all grains would survive when the size of the grains doubles every time. Such an evolution could be an origin of enhanced kinetics. This is the reason why the final-stage sintering always is accompanied by a rapid grain growth. In the two-step sintering, the microstructure form in the first-step sintering at the higher temperature is "frozen" and in the second-step sintering at a lower temperature. In this case, the kinetics is slowed down, but it is still sufficient for densification.

A two-step sintering method, in which the sample was first heat to 1800 °C without dwelling and cooled 1600 °C by holding for 8 h, was used to fabricate transparent YAG ceramics in vacuum (10^{-3} Pa) [125]. The YAG powder was synthesized, from $Y(NO_3)_3{\cdot}6H_2O$ (>99.9 % purity) and $Al(NO_3)_3{\cdot}9H_2O$ (>99.9 % purity), by using a coprecipitation method with ammonium hydrogen carbonate as the precipitant. The as-synthesized YAG powders were ball milled with 0.5 wt% sintering aid tetraethoxysilane (TEOS) for 12 h. The milled powders were dry-pressed at 10 MPa into Φ20 mm pellets and then cold-isostatically pressed at 200 MPa. The pressed samples were presintered at 1000 °C for 10 h in air. Finally, the powder compacts were sintered in a molybdenum wire-heated vacuum furnace by using a two-step sintering schedule at a vacuum of 10^{-3} Pa.

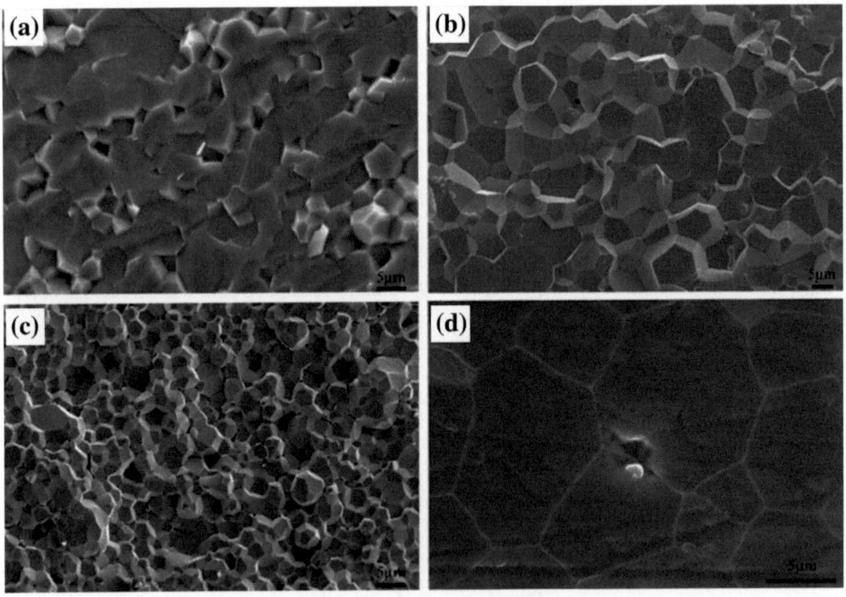

Fig. 8.33 SEM images of the ceramics sintered with different schemes: **a** two-step sintering (1800 °C for 0 h + 1600 °C for 8 h), **b** single-step sintering at 1800 °C for 0 h, **c** single-step sintering at 1600 °C for 8 h and **d** single-step sintering at 1800 °C for 0 h. The images of **a–c** were from fractured surfaces, while that of **d** was from polished surface. Reproduced with permission from [125]. Copyright © 2012, Elsevier

During the two-step sintering, the samples were first heated to 1800 °C without soaking, but immediately cooled down and held at 1600 °C for 8 h, i.e., 1800 °C for 0 h + 1600 °C for 8 h. The sintered samples were annealed at 1450 °C for 6 h in air. Two single-step sintering schemes, 1800 and 1600 °C, were included for comparison. For the 1800 °C scheme, the samples were rapidly heated to the sintering temperature and immediately cooled down to room temperature. For the 1600 °C scheme, the samples were sintering at the temperature for 8 h.

Figure 8.33 shows microstructures of samples sintered with different schemes [125]. As shown in Fig. 8.33a, the two-step sintered sample exhibited a dense and pore-free microstructure, consisting of uniform grains of about 4 μm without AGG. However, the single-step sintered samples at 1800 °C for 0 h and 1600 °C for 8 h, as shown in Fig. 8.33b, c, respectively, possessed obvious pores. After reaching 1800 °C, a sufficiently high density was achieved so sustain the two-step sintering, as shown in Fig. 8.33d. The second-step sintering at 1600 °C for 8 h was sufficient to promote the grain-boundary diffusion through the continuous framework formed during the first-step sintering, while effectively suppressing the grain-boundary migration.

8.10.5 Use of Liquid-Phase Sintering

A second-phase that forms a liquid at the firing temperature can provide a fast diffusion path for densification but grain growth by the Ostwald ripening process may also be enhanced. In this case, high density is normally accompanied by appreciable grain growth. This commonly used fabrication approach is the subject of the next chapter.

8.11 Concluding Remarks

Grain and microstructural development during solid-state sintering of polycrystalline ceramics are governed by basic principles. With simple models, equations for kinetics of NGG could be derived, which, however, are usually for the analysis of an isolated grain boundary or a single grain, without considering the topological requirements of space filling. Computer simulations are playing an increasing role in this aspect. The reduction of uniform grain-boundary energy cannot be used to explain AGG. Two possible reasons have been acknowledged: (i) there is a local driving force that is higher than that due to the geometry or (ii) the boundary mobility of a growing grain is higher than that of ordinary boundaries. AGG should be avoided in general ceramic processing, but it can be utilized specifically to realize solid-state crystal conversion from polycrystalline ceramics. Various strategies could be explored to obtain transparent ceramics with high density, together with controlled grain size and desired microstructure.

References

1. Kingery WD, Bowen HK, Uhlmann DR (1976) Introduction to ceramics, 2nd edn. Wiley, New York
2. German RM (2010) Coarsening in sintering: grain shape distribution, grain size distribution, and grain growth kinetics in solid-pore systems. Crit Rev Solid State Mater Sci 35:263–305
3. Carter CB, Norton MG (2007) Ceramics materials: science and engineering. Springer, Berlin
4. Rahaman MN (2003) Ceramic processing and sintering, 2nd edn. CRC Press, New York
5. Jung YI, Choi SY, Kang SJL (2006) Effect of oxygen partial pressure on grain boundary structure and grain growth behavior in $BaTiO_3$. Acta Mater 54:2849–2855
6. Krell A (2014) Comments on "an extended hardness limit in bulk nanoceramics". In: JA Wollmershauser, BN Feigelson, EP Gorzkowski, CT Ellis, R Goswami, SB Qadri, JG Tischler, FJ Kub, RK Everett (eds) Acta Mater, vol 69, 9–16. (Scripta Mater. 2014;92:63–64)
7. Wollmershauser JA, Feigelson BN, Gorzkowski EP, Ellis CT, Goswami R, Qadri SB et al (2014) Reply to comments on "an extended hardness limit in bulk nanoceramics. Acta Mater 69:9–16. (Scripta Mater. 2014;92:65–68)
8. Wollmershauser JA, Feigelson BN, Gorzkowski EP, Ellis CT, Goswami R, Qadri SB et al (2014) An extended hardness limit in bulk nanoceramics. Acta Mater 69:9–16
9. Greenwood GW (1956) The growth of disersed precipitates in solutions. Acta Metall 4:243–248
10. Lifshitz IM, Slyozov VV (1961) The kinetics of precipitation from supersaturated solid solutions. J Phys Chem Solids 19:35–50
11. Enomoto Y, Kawasaki K, Tokuyama M (1987) Computer modeling of Ostwald ripening. Acta Metall 35:907–913
12. Enomoto Y, Tokuyama M, Kawasaki K (1986) Finite volume fraction effects on Ostwald ripening. Acta Metall 34:2119–2128
13. Wagner C (1961) Theorie der alterung von niederschlagen durch umlosen (Ostwald-reifung). Zeitschrift Fur Elektrochemie 65:581–591
14. Brailsford AD, Wynblatt P (1979) Dependence of Ostwald ripening kinetics on particle—volume fraction. Acta Metall 27:489–497
15. Davies CKL, Nash P, Stevens RN (1980) Effect of volume fraction of precipitate on Ostwald ripening. Acta Metall 28:179–189
16. Marqusee JA, Ross J (1984) Theory of Ostwald ripening—competitive growth and its dependence on volume fraction. J Chem Phys 80:536–543
17. Ardell AJ (1972) Effect of volume fraction on particle coarsening—theoretical considereations. Acta Metall 20:61–71
18. Davies CKL, Nash P, Stevens RN (1980) Precipitation in Ni-Co alloys, 1. Continuous precipitation. J Mater Sci 15:1521–1532
19. Enomoto Y, Kawasaki K, Tokuyama M (1987) The time-dependent behavior of the Ostwald ripening for the finite volume fraction. Acta Metall 35:915–922
20. Atkinson HV (1988) Development of grain-structure in nickel-oxide scale. Mater Sci Technol 4:1052–1063
21. Atkinson HV (1988) Theories of normal grain-growth in pure single-phase systems. Acta Metall 36:469–491
22. Burke JE, Turnbull D (1952) Recrystallinzatin and grain growth. Prog Metal Phys 3:220–292
23. Hillert M (1965) On theory of normal and abnormal grain growth. Acta Metall 13:227–238
24. Srolovitz DJ, Anderson MP, Sahni PS, Grest GS (1984) Computer simulation of grain growth, 2. Grain size distribution, topology and dynamics. Acta Metall 32:793–802
25. Feltham P (1957) Grain growth in metals. Acta Metall 5:97–105
26. Louat NP (1974) Theory of normal grain growth. Acta Metall 22:721–724
27. Anderson MP, Srolovitz DJ, Grest GS, Sahni PS (1984) Computer simulation of grain growth, 1. Kinetics. Acta Metall 32:783–791

28. Rhines FN, Craig KR, Dehoff RT (1974) Mechanism of steady-state grain growth in aluminum. Metall Trans 5:413–425
29. Doherty RD (1975) Mechanism of steady-state grain growth in aluminum—comment. Metall Trans A6:588–590
30. Weaire D, Kermode JP (1983) Computer simulation of a two-dimensional soap froth, 1. Mechod and motivation. Phil Mag B Phys Condens Matter Stat Mech Electron Opt Magn Prop 48:245–259
31. Weaire D, Kermode JP (1983) The evolution of the structure of a two-dimensional soap froth. Phil Mag B Phys Condens Matter Stat Mech Electron Opt Magn Prop 47:L29–L31
32. Weaire D, Kermode JP (1984) Computer simulation of a two-dimensional soap froth, 2. Analysis of results. Phil Mag B Phys Condens Matter Stat Mech Electron Opt Magn Prop 50:379–395
33. Grest GS, Srolovitz DJ, Anderson MP (1985) Computer simulation of grain growth, 4. Anisotropic grain-boundary energies. Acta Metall 33:509–520
34. Srolovitz DJ, Anderson MP, Grest GS, Sahni PS (1984) Computer simulation of grain growth, 3. Influence of a particle dispersion. Acta Metall 32:1429–1438
35. Srolovitz DJ, Grest GS, Anderson MP (1985) Computer simulation of grain growth, 5. Abnormal grain growth. Acta Metall 33:2233–2247
36. Thompson CV, Frost HJ, Spaepen F (1987) The relative rates of secondary and normal grain growth. Acta Metall 35:887–890
37. Rollett AD, Srolovitz DJ, Anderson MP (1989) Simulation and theory of abnormal grain growth: anisotropic grain-boundary energies and mobilities. Acta Metall 37:1227–1240
38. Yang W, Chen LQ, Messing GL (1995) Computer-simulation of anisotropic grain growth. Mater Sci Eng A Struct Mater Prop Microstruct Process 195:179–187
39. Kunaver U, Kolar D (1993) Computer-simulation of anisotropic grain-growth in ceramics. Acta Metall Mater 41:2255–2263
40. Kunaver U, Kolar D (1998) Three-dimensional computer simulation of anisotropic grain growth in ceramics. Acta Mater 46:4629–4640
41. Bateman CA, Bennison SJ, Harmer MP (1989) Mechanism for the role of magensia in the sintering of alumina containing small amounts of a liquid-phase. J Am Ceram Soc 72:1241–1244
42. Cao JJ, MoberlyChan WJ, DeJonghe LC, Gilbert CJ, Ritchie RO (1996) In situ toughened silicon carbide with Al-B-C additions. J Am Ceram Soc 79:461–469
43. Becher PF, Sun EY, Plucknett KP, Alexander KB, Hsueh CH, Lin HT et al (1998) Microstructural design of silicon nitride with improved fracture toughness: I, effects of grain shape and size. J Am Ceram Soc 81:2821–2830
44. Sun EY, Becher PF, Plucknett KP, Hsueh CH, Alexander KB, Waters SB et al (1998) Microstructural design of silicon nitride with improved fracture toughness: II, effects of yttria and alumina additives. J Am Ceram Soc 81:2831–2840
45. Huang T, Rahaman MN, Mah TI, Parthasarathay TA (2000) Effect of SiO_2 and Y_2O_3 additives on the anisotropic grain growth of dense mullite. J Mater Res 15:718–726
46. Gonenli IE, Messing GL (2001) Texturing of mullite by templated grain growth with aluminum borate whiskers. J Eur Ceram Soc 21:2495–2501
47. Hong SH, Messing GL (1999) Development of textured mullite by templated grain growth. J Am Ceram Soc 82:867–872
48. MacLaren I, Cannon RM, Gulgun MA, Voytovych R, Popescu-Pogrion N, Scheu C et al (2003) Abnormal grain growth in alumina: synergistic effects of yttria and silica. J Am Ceram Soc 86:650–659
49. Seabaugh MM, Kerscht IH, Messing GL (1997) Texture development by templated grain growth in liquid-phase-sintered alpha-alumina. J Am Ceram Soc 80:1181–1188
50. Amorin H, Ursic H, Ramos P, Holc J, Moreno R, Chateigner D et al (2014) $Pb(Mg_{1/3}Nb_{2/3})O_3$-$PbTiO_3$ textured ceramics with high piezoelectric response by a novel templated grain growth approach. J Am Ceram Soc 97:420–426

51. Chang YF, Poterala S, Yener D, Messing GL (2013) Fabrication of highly textured fine-grained α-alumina by templated grain growth of nanoscale precursors. J Am Ceram Soc 96:1390–1397
52. Snel MD, van Hoolst J, de Wilde AM, Mertens M, Snijkers F, Luyten J (2009) Influence of tape cast parameters on texture formation in alumina by templated grain growth. J Eur Ceram Soc 29:2757–2763
53. Kimura T, Yi Y, Sakurai F (2010) Mechanisms of texture development in lead-free piezoelectric ceramics with perovskite structure made by the templated grain growth process. Materials 3:4965–4978
54. Ma S, Fuh JYH, Zhang YF, Lu L (2010) Synthesis of anisotropic lead titanate powders for templated grain growth of textured piezoelectric ceramics. Surf Rev Lett 17:159–164
55. Messing GL, Trolier-McKinstry S, Sabolsky EM, Duran C, Kwon S, Brahmaroutu B et al (2004) Templated grain growth of textured piezoelectric ceramics. Crit Rev Solid State Mater Sci 29:45–96
56. Richter T, Schuh C, Moos R, Suvaci E (2008) Single crystal growth and texturing of lead-based piezoelectric ceramics via templated grain growth process. Funct Mater Lett 1:127–132
57. Khan A, Gorzkowski EP, Scotch AM, Leite ER, Li T, Chan HM et al (2003) Influence of excess PbO additions on 111 single-crystal growth of $Pb(Mg_1/3Nb_{2/3})O_3$-35 mol% $PbTiO_3$ by seeded polycrystal conversion. J Am Ceram Soc 86:2176–2181
58. Khan A, Meschke FA, Li T, Scotch AM, Chan HM, Harmer MP (1999) Growth of $Pb(Mg_{1/3}Nb_{2/3})O_3$-35 mol% $PbTiO_3$ single crystals from (111) substrates by seeded polycrystal conversion. J Am Ceram Soc 82:2958–2962
59. Li T, Scotch AM, Chan HM, Harmer MP, Park SE, Shrout TR et al (1998) Single crystals of $Pb(Mg_{1/3}Nb_{2/3})O_3$-35 mol% $PbTiO_3$ from polycrystalline precursors. J Am Ceram Soc 81:244–248
60. Yamamoto T, Sakuma T (1994) Fabrication of barium titanate single-crystals by solid-state grain-growth. J Am Ceram Soc 77:1107–1109
61. Ikesue A, Aung YL, Yoda T, Nakayama S, Kamimura T (2007) Fabrication and laser performance of polycrystal and single crystal Nd:YAG by advanced ceramic processing. Opt Mater 29:1289–1294
62. Bagayev SN, Kaminskii AA, Kopylov YL, Kotelyanskii IM, Kravchenko VB, Luzanov VA (2013) Single crystal growth in YAG ceramics of different stoichiometry. Opt Mater 35:757–760
63. Maxim I, Irina V, Vladimir K (2012) Growth of optical grade yttrium oxide single crystal via ceramic technology. Opt Mater 34:955–958
64. Ikesue A, Aung YL, Lupei V (2014) Ceramic lasers. Cambridge University Press, Cambridge
65. Li Z, Nian H, Feng T, Jiang D (2014) Solid-state grain growth of garnet single crystals of complex composition. Ceram Int 40:10193–10196
66. Bagayev SN, Kaminskii AA, Kopylov YL, Kotelyanskii IM, Kravchenko VB (2012) Simple method to join YAG ceramics and crystals. Opt Mater 34:951–954
67. Scott C, Kaliszewski M, Greskovich C, Levinson L (2002) Conversion of polycrystalline Al_2O_3 into single-crystal sapphire by abnormal grain growth. J Am Ceram Soc 85:1275–1280
68. Thompson GS, Henderson PA, Harmer MP, Wei GC, Rhodes WH (2004) Conversion of polycrystalline alumina to single-crystal sapphire by localized codoping with silica. J Am Ceram Soc 87:1879–1882
69. Dillon SJ, Harmer MP (2007) Mechanism of "solid-state" single-crystal conversion in alumina. J Am Ceram Soc 90:993–995
70. Kang SJL, Park JH, Ko SY, Lee HY (2015) Solid-state conversion of single crystals: the principle and the state-of-the-art. J Am Ceram Soc 98:347–360
71. Kang SJL (2013) Boundary structure-dependent grain growth behavior in polycrystals: model and principle. In: Barnett M (ed) Recrystallization and grain growth, pp 377–382

72. Kang SJL, Lee MG, An SM (2009) Microstructural evolution during sintering with control of the interface structure. J Am Ceram Soc 92:1464–1471
73. Smith CS (1948) Grains, phases and interfaces—an interpretation of microstructure. Trans Am Inst Min Metall Eng 175:15–51
74. Louat N (1983) The inhibition of grain-boundary motion by a dispersion of particles. Phil Mag A Phys Condens Matter Struct Defects Mech Prop 47:903–912
75. Haroun NA (1980) Theory of inclusion controlled grain-growth. J Mater Sci 15:2816–2822
76. Hunderi O, Nes E, Ryum N (1989) On the Zener drag—Addendum. Acta Metall 37:129–133
77. Nes E, Ryum N, Hunderi O (1985) On the Zener drag. Acta Metall 33:11–22
78. Anderson MP, Grest GS, Doherty RD, Li K, Srolovitz DJ (1989) Inhibition of grain-growth by 2nd phase particles—3 dimensional Monte-Carlo computer simulations. Scr Metall 23:753–758
79. Anderson MP, Grest GS, Srolovitz DJ (1989) Computation simulation of normal grain-growth in 3 dimensions. Phil Mag B Phys Condens Matter Stat Mech Electron Opt Magn Prop 59:293–329
80. Doherty RD, Srolovitz DJ, Rollett AD, Anderson MP (1987) On the volume fraction dependence of particle limited grain-growth. Scr Metall 21:675–679
81. Lange FF, Yamaguchi T, Davis BI, Morgan PED (1988) Effect of ZrO_2 inclusions on the sintering of Al_2O_3. J Am Ceram Soc 71:446–448
82. Lange FF, Hirlinger MM (1984) Hinderance of grain-growth in Al_2O_3 by ZrO_2 inclusions. J Am Ceram Soc 67:164–168
83. Lange FF, Hirlinger MM (1987) Grain-growth in 2-phase ceramics—Al_2O_3 inclusions in ZrO_2. J Am Ceram Soc 70:827–830
84. Kibbel B, Heuer AH (1986) Exaggerated grain-growth in ZrO_2-toughened Al_2O_3. J Am Ceram Soc 69:231–236
85. Green DJ (1982) Critical microstructures for microcracking in Al_2O_3-ZrO_2 composites. J Am Ceram Soc 65:610–614
86. Stearns LC, Harmer MP (1996) Particle-inhibited grain growth in Al_2O_3-SiC, 1. Experimental results. J Am Ceram Soc 79:3013–3019
87. Steams LC, Harmer MP (1996) Particle-inhibited grain growth in Al_2O_3-SiC, 2. Equilibrium and kinetic analysis. J Am Ceram Soc 79:3020–3028
88. Cantwell PR, Tang M, Dillon SJ, Luo J, Rohrer GS, Harmer MP (2014) Grain boundary complexions. Acta Mater 62:1–48
89. Dillon SJ, Behera SK, Harmer MP (2008) An experimentally quantifiable solute drag factor. Acta Mater 56:1374–1379
90. Hillert M, Sundman B (1976) Treatment of solute drag on moving grain-boundaries and phase interfaces in binary-alloys. Acta Metall 24:731–743
91. Hillert M, Sundman B (1977) Solute-drag treatment of transition from diffusion-controlled to diffusion less solidification. Acta Metall 25:11–18
92. Cahn JW (1962) Impurity-drag effect in grain boundary motion. Acta Metall 10:789–798
93. Drolet JP, Galibois A (1968) Impurity-drag effect on grain growth. Acta Metall 16:1387–1399
94. Glaeser AM, Bowen HK, Cannon RM (1986) Grain-boundary migration in LiF, 1. Mobility measurements. J Am Ceram Soc 69:119–126
95. Glaeser AM, Bowen HK, Cannon RM (1986) Grain-boundary migration in LiF, 2. Microstructural characteristics. J Am Ceram Soc 69:299–309
96. Glaeser AM, Bowen HK, Cannon RM (1986) Background impurity effects on grain-boundary migration in LiF. Mater Sci Eng 79:111–117
97. Kingery WD, Francois B (1965) Grain growth in porous compacts. J Am Ceram Soc 48:546–547
98. Handwerker CA, Dynys JM, Cannon RM, Coble RL (1990) Metal reference line technique for obtaining dihedral angles from surface grooves. J Am Ceram Soc 73:1365–1370
99. Handwerker CA, Dynys JM, Cannon RM, Coble RL (1990) Dihedral angles in magnesia and alumina—distributions from surface thermal grooves. J Am Ceram Soc 73:1371–1377

100. Brook RJ (1969) Pore-grain boundary interactions and grain growth. J Am Ceram Soc 52:56–57
101. Brook RJ (1969) Pores and grain growth kinetics. J Am Ceram Soc 52:339–340
102. Kurtz SK, Carpay FMA (1980) Microstructure and normal grain-growth in metals and ceramics, 1. Theory. J Appl Phys 51:5725–5744
103. Kurtz SK, Carpay FMA (1980) Microstructure and normal grain-growth in metals and ceramics, 2. Experiment. J Appl Phys 51:5745–5754
104. Carpay FMA (1977) Discontinuous grain-growth and pore drag. J Am Ceram Soc 60:82–83
105. Carpay FMA (1978) Normal grain-growth. Berichte Der Bunsen-Gesellschaft-Phys Chem Chem Phys 82:306–308
106. Sakarcan M, Hsueh CH, Evans AG (1983) Experimental assessment of pore breakaway during sintering. J Am Ceram Soc 66:456–461
107. Hsueh CH, Evans AG, Coble RL (1982) Microstructure development during final intermediate stage sintering, 1. Pore grain-boundary separation. Acta Metall 30:1269–1279
108. Hsueh CH, Evans AG (1983) Microstructure evolution during sintering—the role of evaporation condensation. Acta Metall 31:189–198
109. Yan MF, Cannon RM, Bowen HK, Chowdhry U (1983) Effect of grain-size distribution on sintered density. Mater Sci Eng 60:275–281
110. Yan MF (1981) Microstructural control in the processing of electronic ceramics. Mater Sci Eng 48:53–72
111. Bennison SJ, Harmer MP (1983) Effect of MgO solute on the kinetics of grain-growth in Al_2O_3. J Am Ceram Soc 66:C90–C92
112. Bennison SJ, Harmer MP (1990) Effect of magnesia solute on surface-diffusion in sapphire and the role of magensia in the sintering of alumina. J Am Ceram Soc 73:833–837
113. Brook RJ (1985) Processing technology for high-performance ceramics. Mater Sci Eng 71:305–312
114. Mostaghaci H, Brook RJ (1983) Production of dense and fine-grain size $BaTiO_3$ by fast firing. Trans J Brit Ceram Soc 82:167–170
115. Shaw NJ, Brook RJ (1986) Structure and grain coarsening during the sintering of alumina. J Am Ceram Soc 69:107–110
116. Chen IW, Wang XH (2000) Sintering dense nanocrystalline ceramics without final-stage grain growth. Nature 404:168–171
117. Wang XH, Chen PL, Chen IW (2006) Two-step sintering of ceramics with constant grain-size, I. Y_2O_3. J Am Ceram Soc 89:431–437
118. Wang J, Zhang F, Chen F, Zhang J, Zhang HL, Tian R et al (2015) Effect of Y_2O_3 and La_2O_3 on the sinterability of gamma-AlON transparent ceramics. J Eur Ceram Soc 35:23–28
119. Kong LB, Ma J, Zhang TS, Zhang RF (2002) Transparent lead lanthanum zirconate titanate ceramics derived from oxide mixture via a repeated annealing process. J Mater Res 17:929–932
120. Kong LB, Ma J, Zhu W, Tan OK (2002) Transparent PLZT8/65/35 ceramics from constituent oxides mechanically modified by high-energy ball milling. J Mater Sci Lett 21:197–199
121. Chen ZH, Li JT, Xu JJ, Hu ZG (2008) Fabrication of YAG transparent ceramics by two-step sintering. Ceram Int 34:1709–1712
122. Huang YH, Jiang DL, Zhang JX, Lin QL (2009) Fabrication of transparent lanthanum-doped yttria ceramics by combination of two-step sintering and vacuum sintering. J Am Ceram Soc 92:2883–2887
123. Kim DS, Lee JH, Sung RJ, Kim SW, Kim HS, Park JS (2007) Improvement of translucency in Al_2O_3 ceramics by two-step sintering technique. J Eur Ceram Soc 27:3629–3632
124. Li J, Chen Q, Feng GY, Wu WJ, Xiao DQ, Zhu JG (2012) Optical properties of the polycrystalline transparent Nd:YAG ceramics prepared by two-step sintering. Ceram Int 38: S649–S652
125. Li XX, Zheng BY, Odoom-Wubah T, Huang JL (2013) Co-precipitation synthesis and two-step sintering of YAG powders for transparent ceramics. Ceram Int 39:7983–7988

126. Seeley Z, Cherepy N, Payne S (2013) Two-step sintering of $Gd_{0.3}Lu_{1.6}Eu_{0.1}O_3$ transparent ceramic scintillator. Opt Mater Express 3
127. Nanko M, Dang KQ (2014) Two-step pulsed electric current sintering of transparent Al_2O_3 ceramics. Adv Appl Ceram 113:80–84
128. Isobe T, Ooyama A, Shimizu M, Nakajima A (2012) Pore size control of Al_2O_3 ceramics using two-step sintering. Ceram Int 38:787–793

Chapter 9
Laser Applications

9.1 Introduction

Transparent ceramics have found various applications, including lasers, scintillator, electro-optical, lighting, and armors, with their applications in solid-state lasers being the most important and most extensively and intensively studied. Due to the richness in content, transparent ceramics for the applications in solid-state lasers will be presented in detail in this chapter, while the rest of the applications will be discussed in the next chapter. Ceramic lasers have been detailed and well documented in Ref. [1]; thus, a brief description will be given to representative examples in this chapter. Noting the wide range of materials and laser devices that have been available in the open literature, examples will be given in an arbitrary way, without specific ordering or intentional classifying arrangement. Effort has been made to provide a description as complete as possible, but there is no intention to cover all the openly published works on solid-state ceramic lasers. The order of the ceramic laser materials is arranged according to the rare earth elements, while within each group of the rare earth elements, discussions are according to the types of laser devices.

9.2 Ceramic Solid-State Lasers

Ceramic lasers have been extensively studied in all aspects, i.e., laser-related materials, their processing and characterization, and devices and applications [2–18]. As discussed before, the performance of a solid-state laser is a synergistic effect of the laser materials, the pumping process, and the design of the whole laser system. The key to high performance is to minimize the nonradiative de-excitation processes, so as to (i) reduce the wastage of excitation and the generation of heat and (ii) control the thermal field inside the laser material and the thermomechanical and thermo-optical processes.

L.B. Kong et al., *Transparent Ceramics*, Topics in Mining, Metallurgy and Materials Engineering, DOI 10.1007/978-3-319-18956-7_9

9.2.1 Pumping Schemes

In principle, the pumping schemes must fulfill the two requirements: (i) high absorption of pump radiation and (ii) optimized utilization of the absorbed radiation for laser emission. At the same time, the manifestation of de-excitation processes that have no contribution to the laser emission scheme should be prevented. Besides the absorption coefficient of the laser materials, the superposition of the pump radiation spectrum with the absorption spectrum of the laser materials could also influence the pump absorption. Therefore, when selecting the pumping sources, it is necessary to consider the following: (i) absorption properties of the laser materials, including spectrum peak position and FWHM and cross sections, (ii) chemical and physical properties of the laser materials, such as concentration of dopants and dimensions, and (iii) design of the laser devices, in order to have a complete absorption and proper distribution of the absorbed power.

The efficiency of utilization of the absorbed excitation in the laser process and the limitation of the parasitic heat generation are determined by the quantum defect between the pump and laser emission. For a given laser emission scheme, the quantum of the laser emission is determined by the position of the emitting, the terminal levels, and their positions relative to the ground state, which are all fixed. As a result, the most effective way to reduce the quantum defect is to decrease or eliminate the upper quantum defect between the pump energy level and the emitting laser level. Optimization of thermalization of the energy levels with their neighboring levels could be an additional way to further decrease the quantum defect.

Usually, pumping sources with broadband or multiple emissions, e.g., flashlamps and solar radiations, exhibit less superposition with the absorption lines of RE^{3+} ions, so that a large portion of the pump radiation would oscillate inside the laser materials, without being absorbed by the laser-active ions. Furthermore, this type of pumping has a variety of absorption transitions, thus leading to various quantum defects. In this case, an average quantum defect is used.

The width of the emission line of diode lasers is usually several nanometers, which is too broad for pumping some intense but narrow absorption bands of the RE^{3+} ions in many laser materials, especially at low temperatures. As a result, the pump radiation will not be entirely absorbed. Also, diode lasers often have thermal shift in both the emission lines and the absorption lines of the RE^{3+} ions. Therefore, when using laser materials with multicenter structure to have a broad emission spectrum, it should be especially careful that there could be broad and sometimes multiple absorption lines, due to the multi-pole structural centers. In this case, the emission spectrum of the pump source should have a complete superposition on the whole inhomogeneously broadened absorption line, so as to grant excitation for all the centers, in order to make the full use of the broad emission to generate short-pulse lasers.

When the resonant pumping is close to the emitting laser level, or preferably the pumping is directly into this level, low quantum defect laser schemes could be realized, thus leading to increased laser performances and reduced heat generation. The quantum defect can be further reduced, when the pumping is on the transitions that are from the thermally populated upper Stark levels of the ground manifold. With accidental spectral degeneracy of several such transitions, the pumping efficiency can be increased. Due to its effect on the intracavity laser intensity I_ω at the fundamental frequency ω, direct pumping into the emitting level can facilitate a great enhancement in the efficiency of the nonlinear $n\omega$ processes, because of the relation of $I_{n\omega} \propto (I_\omega)^n$ [19, 20]. Also, due to the reduction in thermal effects on the quality of the beam, an additional increment in the global efficiency is available.

9.2.1.1 Nd-Doped Ceramic Materials

Because of the presence of strong absorption levels in the visible and infrared regions, Nd-doped laser ceramics can be pumped with a large variety of pumping sources, including lamps, diode lasers, and solar radiation. However, when using the lamp pumping, the efficiency of the pump emission spectrum is relatively low and serious quantum defect is present. As a consequence, resonant pumping with diode lasers at the levels close to the emitting level is generally used to achieve high-performance laser emission. In the initial experiments on diode-pumped Nd:YAG lasers, direct pumping into the level $^4F_{3/2}$ in the 870 nm range was used [21–23]. The layout of the end-pumping experiment is shown in Fig. 9.1 [21].

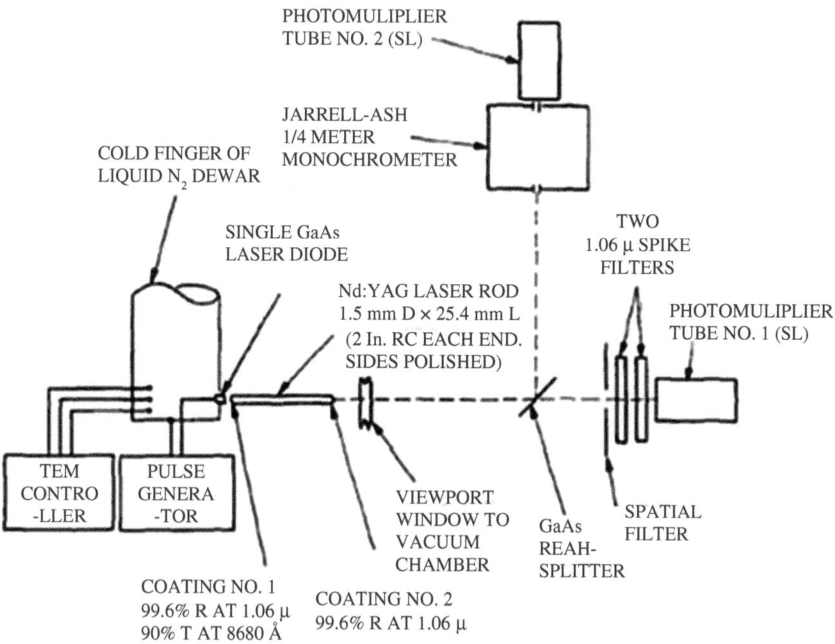

Fig. 9.1 Layout of the GaAs diode-pumped Nd:YAG laser experiment. Reproduced with permission from [21]. Copyright © 1968, American Institute of Physics

However, the power of the diodes was too low and the absorption of the 1 at.% Nd:YAG at this level was too weak, with an absorption coefficient of about 4 cm^{-1}. Therefore, it was later changed to pumping into the strong absorption of $^4I_{9/2} \rightarrow {}^4F_{5/2}$ at about 809 nm, having an absorption coefficient of about 11.4 cm^{-1}, by using more efficient and powerful AlGaAs diode lasers. In this case, a relative quantum defect of ~0.24 was observed for the laser emission at 1.06 μm, about 30 % of which represented the upper part, without having any effect on the laser emission.

As the concentration of Nd is increased, the absorption coefficients for various transitions to $^4F_{3/2}$ could be largely increased, so that nearly all the transitions that are originated from the Stark components of the ground manifold can be used for pumping at this level. It is obvious that the higher the Stark component is, the less the quantum defect will be, although the small fractional thermal population could have a negative effect on the absorption. Accordingly, the relative quantum defect for the 1064 nm laser emission was decreased from 18.3 % for the transition of $^4I_{9/2}(1) \rightarrow {}^4F_{3/2}(2)$ to 11 % for the transition of $^4I_{9/2}(5) \rightarrow {}^4F_{3/2}(1)$.

Besides the largest room temperature effective absorption coefficient at level $^4F_{3/2}$, corresponding to the <1 nm sharp 869 nm transition of $^4I_{9/2}(1) \rightarrow {}^4F_{3/2}(2)$, the thermally activated or hot-band transitions of $^4I_{9/2}(2) \rightarrow {}^4F_{3/2}(1)$ and $^4I_{9/2}(3) \rightarrow {}^4F_{3/2}(2)$ of Nd:YAG, with quasi-degeneracy, leading to a ~3 nm broad two-peak (885.7 and 884.4 nm) band centered at 885 nm, can be used for diode laser pumping. The room temperature average absorption coefficient of this band of 1 at.% Nd is about 1.65 cm^{-1}. The fractional thermal coefficients (f) of the Stark levels of the ground manifold $^4I_{9/2}$ have a dependence on temperature in the range of 300–400 K, leading to a difference in the effective absorption cross section at different temperatures.

With increasing temperature in this range, f is decreased by ~15.4 % from 0.465 to 0.405 for the first Stark level Z_1, kept nearly unchanged for Z_2, and increased slightly for the upper Stark levels. The sum of the f coefficients for levels Z_2 and Z_3, corresponding to the 885-nm band, is increased by 5.2 % from 0.424 to 0.446, which partially compensates the thermal decrease in the peak absorption and emission cross sections. Therefore, the thermal stability of pump absorption and laser emission of the 885 nm pumping is higher than that of the traditional 809 nm pumping. Also, the relative quantum defect could be reduced by 30 and 55.5 %, i.e., from 0.24 to 0.168 and from 0.145 to 0.065, for the 1064 and 946 nm laser emissions, respectively.

The quasi-degeneracy of two thermally activated transitions has also been observed in other Nd-doped laser materials. One example is Nd:GSGG, in which the hot bands of $^4I_{9/2}(2) \rightarrow {}^4F_{3/2}(1)$ and $^4I_{9/2}(3) \rightarrow {}^4F_{3/2}(2)$ are coincident [24]. In addition, the sum of the thermal population coefficients, $f_2 + f_3 = 0.444$, is larger than $f_1 = 0.425$ at room temperature, which becomes $f_2 + f_3 = 0.458$ and $f_1 = 0.374$ at 400 K, leading to an increase in the difference. The spectroscopic and emission decay measurements indicated that the transparent Nd-doped GSGG ceramics exhibited similar properties to those of single crystals.

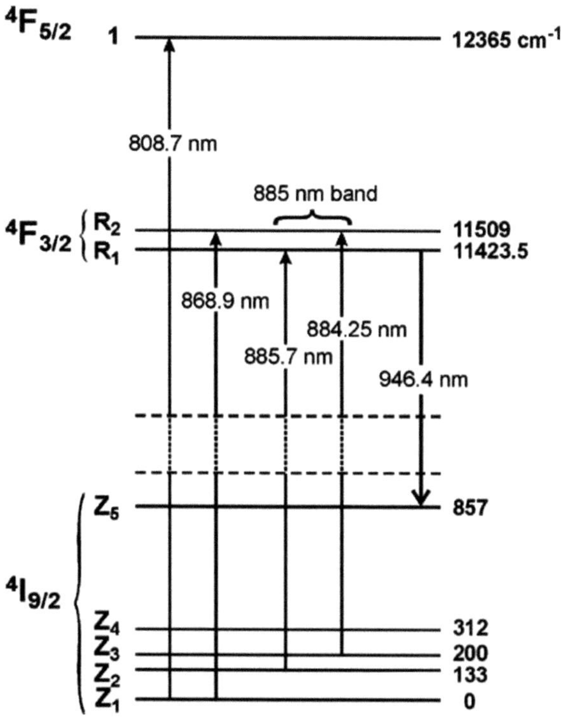

Fig. 9.2 Transitions involved in the pumping and emission of the room temperature 946 nm Nd:YAG laser. Reproduced with permission from [33]. Copyright © 2002, Elsevier

Due to the availability of diode lasers with extended wavelength range and increased power, it is possible to pump the Nd lasers into the emitting level. In this case, the effect of low absorption cross sections of the 869- and 885-nm bands in the $^4I_{9/2} \rightarrow {}^4F_{3/2}$ spectrum of Nd:YAG on its pump absorption efficiency can be manipulated by controlling the path of the pump radiation inside the laser materials. Various strategies have been developed to do so, such as laser rods with end-pumped configurations [25–28], zigzag configurations or multi-pass pumping [19] and high doping concentrations [29–32]. However, the long laser rods lead to increased residual optical losses, complicated multi-pass pumping, and possible self-quenching of the emission in case of high Nd concentrations.

Laser emission with pumping at 946 nm due to the $^4I_{9/2}(5) \rightarrow {}^4F_{3/2}(1)$ transition, as shown in Fig. 9.2 [33], with the lowest quantum defect, was observed in Nd: YAG crystal [34]. However, the low thermal population of the Z_5 level at room temperature led to a low efficiency with respect to the incident pump power. Although this problem could be addressed by increasing temperature, there is another concern of thermalization of the emitting level of $^4F_{5/2}$.

Pump diode lasers are usually operated in CW or QCW mode, which cannot be used for the generation of high-energy pulses. To generate high-energy pulses, is it still necessary to use pumping of Nd ceramic lasers with flashlamps. One example is the pumping of Nd:YAG ceramics with lamps, in which the ultraviolet part was cut by using special Sm-doped filters to prevent the absorption due to the stable or transient color centers [35]. It was found that these materials could be used to replace Nd:glass materials for high-energy lasers, due to their superior thermo-mechanical properties. Because the lamp emission cannot be completely absorbed by the Nd^{3+} emission, codoping with strong absorbing ions, such as Cr^{3+}, has been used to transfer the excitation to Nd^{3+} [36, 37]. This approach is especially useful in the case of solar-pumped and white light-pumped Nd:YAG lasers [38–42].

For diode laser pumping, Nd-doped sesquioxides offer several lines in the near-infrared region, with the strongest one being in the region of 820–825 nm, corresponding to the transition of $^4I_{9/2} \rightarrow {}^4F_{5/2}$. It showed room temperature absorption cross sections of 5–7 $\times$ 10^{-20} cm^2, followed by absorption at about 808 nm with an absorption cross section of about 4 $\times$ 10^{-20} cm^2. Pumping in the transition of $^4I_{9/2} \rightarrow {}^4F_{3/2}$ can be realized on the line of $Z_1 \rightarrow R_2$ at about 877 nm or on the hot-band transition line of $Z_2 \rightarrow R_1$, both with cross sections 2 $\times$ 10^{-20} cm^2, without spectrally degenerated hot bands. These ceramics can also be pumped with flashlamps.

9.2.1.2 Yb-Doped Ceramics

Yb-doped laser materials have simple electronic structures, with an excited manifold at about 10,000 cm^{-1}, which can be pumped only by infrared diode lasers. Due to the weak absorption of the pump radiation, high concentrations of Yb have to be used or a long pump radiation path inside the laser materials should be adopted. One of the problems at very high concentrations of Yb is the presence of the onset of parasitic de-excitation due to the cooperative processes within the system of the Yb ions or the migration-assisted energy transfer to accidental impurities. Due to the quasi-three-level characteristic of Yb laser emission, large laser rods are generally not used, so that the long absorption path is realized through multi-pass pump radiation by using thin materials or long fibers doped at very low concentrations.

The transition of $^2F_{7/2}(1) \rightarrow {}^2F_{5/2}(1)$ in the range of 965–980 nm gives the strongest absorption line of Yb^{3+}. However, because its linewidth is decreased with decreasing temperature, it cannot be used for pumping process. Therefore, the broader but much less intense transition of $^2F_{7/2}(1) \rightarrow {}^2F_{5/2}(2)$ at about 940 nm is usually used. Sensitization of Yb^{3+} emission through energy transfer from Nd^{3+} with high transfer efficiency has been observed in sesquioxides and YAG [43–46] and heavily doped (Nd,Yb):CaF_2 single crystals [47–49] and ceramics [50].

9.2.1.3 Other RE^{3+}-Doped Ceramics

Diode lasers in the range of 800 nm and 940–980 nm enable excitation of infrared (1.5–3 μm) emitting levels for other RE^{3+} ions. For example, the 3-μm emitting level of $^4I_{11/2}$ and the 1.5-μm emitting level of $^4I_{13/2}$ of Er^{3+} can be excited by 940–980 nm pumping into the level $^4I_{11/2}$, either directly or through sensitization with Yb^{3+}, while the 2 μm emission of the level $^3F_4$ of Tm^{3+} can be excited by 790–800 nm pumping into the level $^3H_4$, followed by the quantum cutting ($^3H_4$, $^3H_6$) → ($^3F_4$, $^3F_4$) cross-relaxation. Due to the development of efficient diode lasers in the infrared range of 1.5–2 μm, the resonant pumping into the emitting level of quasi-three-level infrared emission of RE^{3+} has attracted much interest. This is because the pumping offers a low quantum defect laser emission and thus the high efficiency in using the absorbed radiation, which has been demonstrated by Ho^{3+}-, Er^{3+}-, and Tm^{3+}-doped ceramics.

9.2.2 Radiative and Nonradiative Processes

The excited ions of the pumped laser materials in a laser resonator can be de-excited by various radiative (either laser or luminescence) and nonradiative (electron–phonon interaction or energy transfer) processes. Also, the amount of ions participating in laser emission is dependent on the laser emission efficiency (η_l). The laser emission is produced by the excited ions inside the laser mode volume and pumped above the threshold. The excited ions inside the pumped volume but outside the laser mode volume and those that form the inversion of population at the laser threshold can be de-excited by luminescence and nonradiative processes.

The relevance of the superposition integral of the laser mode and the pumped volume η_v is determined by the regime of laser emission. The material factors that influence the laser threshold and thus the laser emission efficiency η_l, in the case of CW emission, have been evaluated. Several of these factors, such as the emission quantum efficiency and the effective lifetime, can be influenced by the conditions of the experiment (concentration of the doping ions, temperature), whereas the quantum defect ratio $\eta_{qd}^{(l)}$ is influenced by the pump wavelength.

9.2.2.1 De-excitation of Emitting Levels

At concentrations of Nd^{3+}, self-quenching of the emission due to the downconversion or upconversion within the system of the doping ions can reduce the emission quantum efficiency (η_{qe}), which influences the emission threshold and the generation of heat. The effects of C_{Nd} and the pump intensity on η_{qe} in Nd-doped laser materials have been well documented. The calculated concentration dependences of effective lifetime ($\tau_{eff} = \tau_f \eta_{qe}$) for Nd-doped YAG and GSGG have been calculated and compared [24].

The increased self-quenching of emission, as shown in Fig. 9.3, has been widely acknowledged as a disadvantage of Nd laser materials with increasing doping concentration [30]. Actually, the effect of C_{Nd} should be evaluated by considering both the pump absorption efficiency (η_a) and the emission quantum efficiency (η_{qe}) that have effects on the laser parameter, i.e., the incident pump power [29, 30, 32].

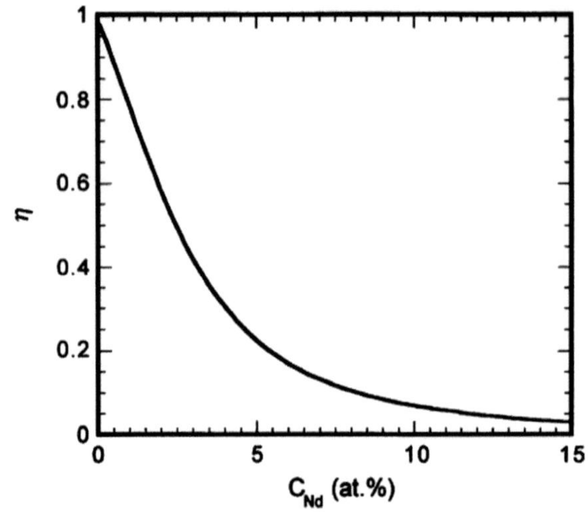

Fig. 9.3 Calculated concentration dependence of emission quantum efficiency η_{qe} (η). Reproduced with permission from [30]. Copyright © 2001, Elsevier

The efficiency is not influenced by η_{qe} but also dependent on η_a, while the laser threshold is inversely proportional to the product $\eta_a\eta_{qe}$, which can be used as a figure of merit to measure the effect of concentration on threshold. Therefore, although the decrease in η_{qe} with increasing C_{Nd} enhances the threshold, represented by absorbed power $P_{th}^{(a)}$, this effect can be compensated by the increased absorption in certain concentration ranges, which results in a lowered laser threshold, represented by incident power $P_{th}^{(in)}$. Noting that for a given pumping source, the absorption efficiency of a laser material is also influenced by its geometry and dimension, and the product $C_{Nd}\eta_{qe}$ can be used to evaluate the effect of C_{Nd} on the laser threshold $P_{th}^{(in)}$, in terms of intrinsic properties of materials.

The calculated product $C_{Nd}\eta_{qe}$ for Nd:YAG is shown in Fig. 9.4, indicating that the overall effect of C_{Nd} in certain concentration ranges, e.g., the range of 3–4 at.%, on threshold could be positive [30]. Furthermore, the increased η_a at high C_{Nd} offers a larger slope efficiency $\eta_{sl}^{(i)}$. Therefore, the high value of C_{Nd} required to have efficient pump absorption in the 885 nm pumping of Nd:YAG is not harmful to the overall laser emission performance. In addition, there is also a positive effect on the quantum effect (Stokes) ratio $\eta_{qd}^{(l)}$. For more accurate evaluation of the effect of C_{Nd}, it is necessary to use the actual value of η_a, because it provides information on the optimal concentration and size of the laser materials.

The concentration of Nd can also affect the lineshape and width of the absorption and emission lines in a selective manner, through the crystal field effect related to the ensembles of the Nd^{3+} ions. For instance, the linewidth of the $^4I_{9/2} \rightarrow {^4F_{3/2}}$ absorption of Nd:YAG ceramics can be increased by up to 10 %, in the C_{Nd} range of 1–9 at.%. This increase in linewidth inevitably leads to reduction in the peak cross sections. In addition, the asymmetric perturbation effects and the variation in the crystal lattice, related to the variation in C_{Nd}, also slightly shift the position of

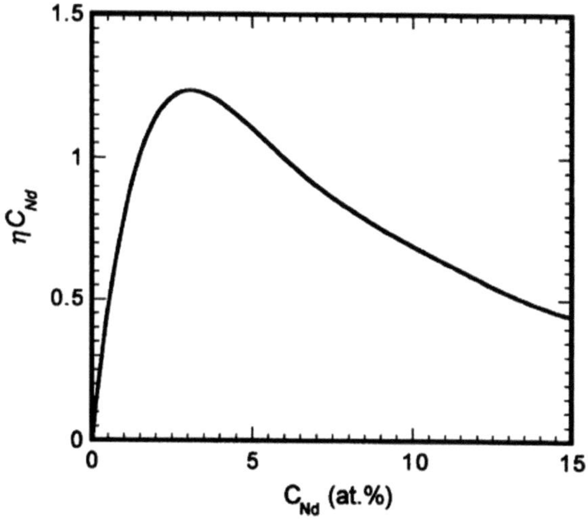

Fig. 9.4 Calculated concentration dependence of $\eta_{qe}C_{Nd}$ (ηC_{Nd}). Reproduced with permission from [30]. Copyright © 2001, Elsevier

the lines. Because the broadening induced by the variation in concentration is through the crystal field perturbation effects, the emission quantum efficiency varies nonuniformly across the emission band, so that the emission lineshape of a transition can be different from that of the absorption.

The emission properties of Nd-doped ceramic lasers are also affected by temperature, because a variation in temperature could change the fractional coefficients of the two Stark levels (R_1 and R_2) of the emitting manifold $^4F_{3/2}$ and the emission cross sections (σ_e), thus leading to the change in the effective emission cross sections through $\sigma_{eff} = \sigma_e f_l$. The crystal field splitting of this manifold could be in the range of 0–250 cm^{-1}, depending on the host materials that are used. For example, Nd:YAG has a splitting of ~84 cm^{-1}, leading to strong emission lines from the two Stark components, which are linearly dependent on temperature. The main emission line at 1064 nm is contributed by transitions from both levels. At room temperature, the fractional thermal coefficients of levels, R_1 and R_2, are 0.6 and 0.4, respectively. Although R_2 has a lower value, the emission is dominant, due to the larger absolute emission cross section σ_e. With decreasing temperature, there is a crossover, between the effective emission cross sections, $\sigma_{eff} = \sigma_e f_l$, of the two transitions, below which the emission from level R_1 becomes dominant. The effective emission cross sections from these two Stark levels have been systematically studied, over a wide temperature range of 77–650 K, as shown in Fig. 9.5 [51].

9.2.2.2 Global Balance of De-excitation Processes

Dependences, of global fractional laser emission efficiency $\eta^{(l)}$, luminescence emission efficiency $\eta^{(f)}$, and heat load parameter η_h, on operating point P/P_{th} in absorbed power for 1 and 2 at.% Nd:YAG, have been systematically studied, for

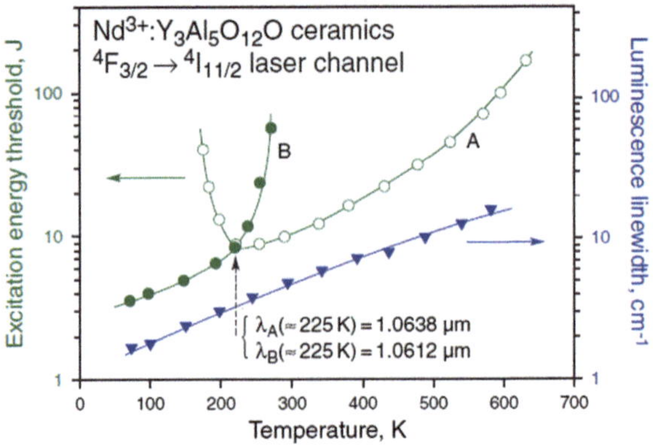

Fig. 9.5 Temperature dependences of threshold excitation energy for the A- and B-lasing lines, together with luminescence linewidth for the A inter-Stark transition of fine-grained garnet ceramics Nd^{3+}:$Y_3Al_5O_{12}$. Reproduced with permission from [51]. Copyright © 2009, John Wiley & Sons

pumping at 809 and 885 nm, with the assumption of $\eta_v = 1$. As the unit of power for all curves is the threshold for 1 at.% Nd:YAG pumped at 809 nm, it was found that 885 nm pumping determines higher $\eta^{(l)}$ and lower η_h than 809 nm pumping and above the specific threshold for each pump wavelength.

According to the powers dissipated in each de-excitation process, $P_{Las} = \eta^{(l)}P^{(a)}$, $P_{Fluo} = \eta^{(f)}P^{(a)}$, and $P_{Heat} = \eta_h P^{(a)}$, it is observed that P_{Fluo} increases first linearly with the absorbed power up to the specific threshold of all situations and then keeps unchanged, with the same value for all cases. At the same time, P_{Heat} increases up to the threshold value for each case and then changes to a slope which depends on the pumping wavelength, but is independent on Nd concentration.

The variation in CW laser emission efficiency η_l of Nd:YAG with the pump-to-laser volume superposition efficiency η_v describes the dependence of the efficiencies of de-excitation processes on η_v. It has been demonstrated that a decreased volume superposition means that the higher the laser threshold, the lower the laser slope efficiency. At the same time, there is an increase in the heat load parameter η_h.

The variation of η_h with η_v determines the slope of the change in the thermal power P_{Heat} above the laser threshold, $\eta_{h,ath}$. For instance, the heat load parameter $\eta_h^{(f)}$ of 1 at.% Nd:YAG in the absence of laser emission, which determines the slope of P_{Heat}, i.e., $\eta_{h,bth}$, below the threshold, calculated with the emission quantum efficiency $\eta_{qe} = 0.8$ is ~0.377. Above the threshold, for CW 1064 nm laser emission pumped at 809 nm, the values of the slope $\eta_{h,ath}$ are 0.24, 0.254, 0.267, and 0.281, for $\eta_v = 1$, 0.9, 0.8, and 0.7, respectively. If the slope efficiency is lower than that predicted by the measured residual losses, the outcoupling losses, and the quantum defect, it means that the pump-to-laser volume superposition is not sufficiently good.

9.2.2.3 Spatial Distribution of De-excitation Processes

The spatial distributions of the absorbed power and the CW laser mode determine the spatial distribution of the de-excitation processes. In most cases, the limits of pump power are determined by the maximum local efficiencies of the de-excitation processes, instead of global values. For example, thermal stress fracture could be observed when the heat generated is beyond the fracture limit in certain regions of the pumped laser materials, while the averaged value over the whole volume is pretty low.

It has been clearly demonstrated that the distribution of heat generation in CW Nd-doped lasers during the laser emission is usually different from the distribution of the absorbed power, which is attributed to both the pump-mode volume superposition and the operating point. Cooling the laser rods could alter the thermal field inside the pumped laser materials. According to theoretical calculation, if heat is uniformly generated in a cylindrical rod, uniform lateral cooling leads to a parabolic thermal field, with the maximum at central axis of the rod, which results in a spherical thermal lens effect [52]. The distribution of luminescence emission is important in limiting the amplified spontaneous emission (ASE). For example, cladding laser rods with materials that absorb the luminescence from the rods and transform it into heat introduces an additional heat generation in the annular direction.

Laser emission efficiency η_l has a strong dependence on the distribution of the de-excitation processes. It has been proved that proper management of the pump and laser mode volumes is crucial to achieve high CW laser efficiency, reduce the parasitic luminescence, and decrease the heat generation, which can result in the loss of excitation by ASE, distortion of the beam, and mechanical fracture. These properties are also important for the design of cooling systems and evaluation of material characteristics, e.g., emission quantum efficiency from thermal lensing data [1, 11].

For end-pumped uniformly doped laser rods, the strongest absorption at the entrance face can lead to excessive heat generation, which limits the maximum power of the laser severely and can result in reduction in the doping concentration and lengthening of the rod, which leads to negative effect on the residual optical losses. Theoretical calculation results indicated that in order to mitigate such effects and optimize the cooling process of the rods, it is necessary to create a (quasi-) uniform distribution of heat generation along the rod and avoid the presence of regions with excessive absorption and heating, which can be achieved by using rods with a profiled distribution of Nd concentration along the rod or multi-segmented rods consisting of parts with different uniform Nd concentrations [53–55]. However, this approach is a strong challenge from the material processing point of view.

9.2.2.4 Heat Generation and Extension of Power Scaling

Because heat generation is the main limiting factor for power scaling of Nd lasers, a generalized figure of merit has been proposed to measure the effects of pump wavelength, Nd concentration, and operating point on scaling, which is the ratio of the fractional laser efficiency to the heat load parameter, $(\eta^{(1)}/\eta_h)$ [11]. Pumping on transitions with quantum defects will offer an enhancement in laser efficiency and a reduction in heat generation simultaneously. For Nd-doped ceramic lasers, this can be fulfilled by direct pumping into the emitting manifold. Based on the dependence of the generalized figure of merit $(\eta^{(1)}/\eta_h)$ on these factors for Nd:YAG, for pumping with 809- and 885-nm diode lasers, the highest figure of merit in terms of absorbed power is observed in the 885-nm lightly pumped doped material.

9.2.3 Ceramic Laser Materials and Components

When using melting techniques to grow single crystals, the presence of segregation coefficient causes the variation in the concentration of doping ions in the crystal growth direction, which limits the useful length of the crystals. Furthermore, when two types of ions are used for codoping, e.g., sensitized laser materials, the difference in the segregation coefficients of the different ions will result in different distributions, which will affect the sensitization process. However, as discussed before, this problem can be readily addressed by using ceramic techniques [56–65]. Various garnet and sesquioxide ceramic laser rods with uniform RE^{3+} distribution have been developed. For 1 at.% Nd:YAG, high-quality rods with a diameter of up to ~10 mm and a length of up to 150–200 mm, and plates with a dimension of up to 100 cm^2, have been available for various laser designs.

One of the thermomechanical effects, leading to thermal lensing, is the bulging of the rods, due to the nonuniform distribution of the thermal field inside the pumped and externally cooled laser materials. One strategy to address this problem is to bond nondoped segments to both ends of the doped rod of the laser material, which enhances the dissipation of heat and mechanical restriction of bulging. The end caps are either thermally bonded or mechanically clamped to the rod, which makes it difficult to have a perfect contact. To solve this problem, a simple strategy has been proposed, in which a thin layer of SiO_2 was deposited on the surfaces to be joined. In this case, perfect bonding of crystalline and ceramic YAG could be realized by heating at 1200 °C without applying any pressure [66].

Monolithic composite laser materials with undoped caps can be produced directly from ceramics [67–74]. An alternative way is the use of undoped material to clad the laser-active materials with any shape, in which the undoped material can serve to enhance dissipation of heat and even guide the laser radiation. In this case, there is diffusion of doping ions into the undoped part, which has been studied by various methods [75–79]. It is found that in case of Nd:YAG, a diffusion layer with the thickness of up to tens of micrometers can be formed, depending on the fabrication process.

Very complex laser materials for multiple total-reflection active mirrors (TRAM) have also been produced by using ceramic techniques [80–82]. Composite multifunctional monolithic components consisting of the active material and a proper absorber, e.g., Nd:YAG/Cr^{4+}:YAG, acting as a *Q*-switch, were prepared by using ceramic technology, without involving bounding but with acceptable mechanical strength. Mixed ceramic materials doped uniformly with laser ions and SA ions have been reported. However, the presence of SA ions can cause losses and thus affect the threshold. In addition, although the doping concentration of Cr^{4+} in this case is very low, energy transfer from the laser ions can be observed, especially when the ions have high concentrations, due to the efficient migration-assisted energy transfer. Ceramic techniques also help to fabricate monolithic composite multi-functional materials, consisting of an active laser part and a part that acts as a suppressor for ASE, such as Nd:YAG/Sm:YAG.

Ceramic techniques have been proved to be useful in the fabrication of high-performance Faraday rotators for isolation of laser amplifiers and birefringence compensation in two-pass high-power laser systems. The main requirements for such components are a high Verdet constant, high optical quality, high laser damage threshold, high thermal strength, and feasible size scalability. Terbium-doped garnet crystals have very good functional parameters, for which ceramic techniques are useful for size scaling and improving the thermomechanical properties. The Verdet constants for ceramic terbium gallium garnet (TGG) [83] and terbium aluminum garnet (TAG) [84, 85] are similar to those of their single-crystal counterparts. In addition, thermochemical methods have been employed to strengthen the surfaces of ceramic rods [86, 87].

9.2.4 Practical Ceramic Lasers

9.2.4.1 Nd-Doped Ceramic Lasers

Due to its unique spectroscopic and emission dynamic properties, Nd^{3+} ion can be used for laser emission over a wide range of temporal regimes, from CW to very short pulses of below picosecond. The transitions from the two Stark levels of the manifold $^4F_{3/2}$ to the Stark levels of the manifolds $^4I_{9/2}$, $^4I_{11/2}$, and $^4I_{13/2}$, corresponding to the ranges of 0.9, 1.0, and 1.3 μm, respectively, showed pretty large emission cross sections. Various cubic crystalline materials, including garnets and sesquioxides, have been used as hosts for Nd laser emission, which formed the major part of ceramic laser materials.

Free-Generation Nd Ceramic Lasers

Nd-Doped Garnets

Nd-doped garnet ceramic lasers have been intensively studied, with numerous reports being available in the open literature. This part will cover (i) Nd-doped garnet lasers at 1 μm, (ii) laser emission at 1.3 μm, (iii) laser emission at 0.9 μm,

(iv) multi-wavelength Nd-doped ceramic laser emission, and (v) sensitized CW laser emission.

Nd:YAG is still the most popular Nd-doped garnet ceramic laser materials. At room temperature, the transitions of $^4F_{3/2}$ (R_i) → $^4I_{11/2}$ (Y_j) have several lines with sufficiently large emission cross sections, which can be used for laser emission. The most intense and widely used emission line is at 1064.15 nm, with emission cross sections of 2.8–3 × 10^{-19} cm^2, which is based mainly on the transition of 2 → 3, with a minor contribution from the transition of 1 → 2. In this transition, there are various other lines for potential laser emission, which include 2 → 1 at 1052.1 nm (room temperature effective emission cross section of 0.94 × 10^{-19} cm^2), 1 → 1 at 1061.5 nm (0.94 × 10^{-19} cm^2), 2 → 4 at 1068.2 nm (0.6 × 10^{-19} cm^2), 1 → 3 at 1073.7 nm (1.64 × 10^{-19} cm^2), 1 → 4 at 1077.9 nm (0.76 × 10^{-19} cm^2), 2 → 6 at 1111.9 nm (0.36 × 10^{-19} cm^2), 1 → 5 at 1115.8 nm (0.41 × 10^{-19} cm^2), and 1 → 1 at 1122.5 nm (0.4 × 10^{-19} cm^2) [51].

The first Nd:YAG ceramic laser showing CW laser emission was made with 1.1 at.% Nd-doped-YAG ceramics which were prepared by using solid-state reactive vacuum sintering method, with an efficiency compatible to the corresponding crystal laser under diode laser pumping at 809 nm [2]. Significant improvement has been made for this composition by using new synthetic technologies, such as wet chemical methods, so that the doping concentration of Nd can be largely increased up to 6.9 at.% [34, 88, 89]. 1.5-kW high-power CW laser with diode laser pumping has been realized, due to the availability of large-sized fine-grained ceramic laser rods [90].

The Nd:YAG ceramics were prepared from precursor powder synthesized by using chemical coprecipitation method, with mixed aqueous solutions of aluminum, yttrium, and neodymium chlorides as sources of the cations and solution of ammonium hydrogen carbonate as precipitant. The precursor was calcinated at 1200 °C to obtain oxide raw powders. The powders were compacted and finally sintered at 1750 °C in vacuum for 5–20 h [90].

A high-power virtual point source (VPS) laser diode (LD) system was used as the CW pumping source to characterize laser performances of the Nd:YAG ceramics, with the schematic diagram of the VPS setup shown in Fig. 9.6 [90]. In this system, 32 sets of LDs with a maximum output of each of 10 W at 807 nm were used to form

Fig. 9.6 Photographs of the polycrystalline Nd:YAG ceramic laser rods. Reproduced with permission from [90]. Copyright © 2002, Elsevier

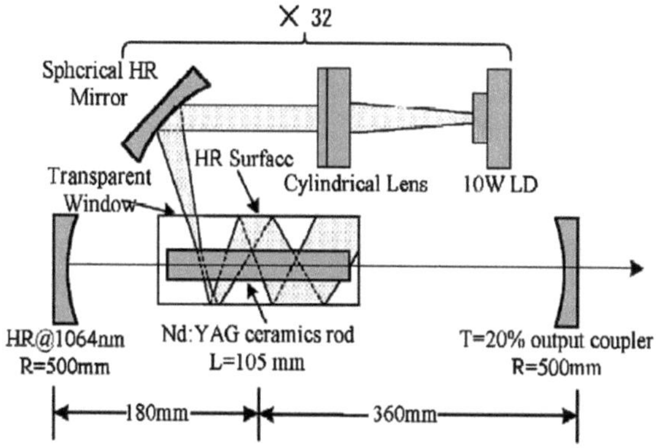

Fig. 9.7 Schematic diagram of the laser cavity and VPS laser diode pumping system. Reproduced with permission from [90]. Copyright © 2002, Elsevier

a symmetrical ring-shaped pumping source, so as to ensure the angular uniformity. The LD output was focused on a point or line at the rod axis by 32 sets of optics, each composed of a cylindrical lens and a high-reflectivity spherical mirror. The axial illuminations were completed by one coaxial cylinder with an Ag-coated side surface and Au-coated end surfaces, which functioned as a huge "double-clad fiber." The point of focus, acting as a VPS, was imaged and reimaged along the rod axis and reflected by the end surfaces. Therefore, after multiple passes, the entire lasing rod was illuminated. By rotating the spherical mirrors, the pump radiation distribution could be controlled. In the experiment, the centrally depressed pump radiation distribution was adopted to decrease the thermal lensing effect.

Fig. 9.7 shows schematic diagram of the laser cavity and VPS laser diode pumping system [90]. Nd:YAG ceramic rods, with a dimension of Ø4 mm × 105 mm and 0.6 at.% Nd^{3+}, were used for the experiments, with flat end faces that were antireflection-coated at 1064 nm. Two concave mirrors, both with curvature of 500 mm, were used to form the laser cavity. One was high-reflectivity-coated, and the other had a reflectivity of about 80 % at the laser wavelength. Figure 9.8 show the laser outputs at 1064 nm as a function of pump power [90]. At the pump power of

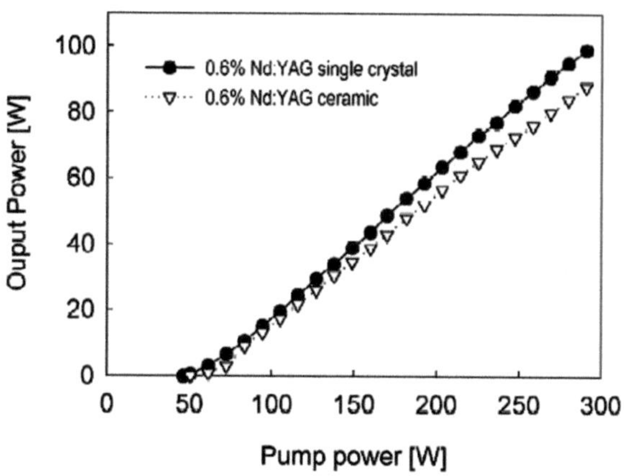

Fig. 9.8 Laser outputs of the Nd:YAG ceramic and single-crystal lasers as a function of VPS pump power. Reproduced with permission from [90]. Copyright © 2002, Elsevier

290 W, a multi-mode CW laser output of 88 W was achieved. The corresponding optical-to-optical efficiency was about 30 %. In order to compare with a single-crystal rod laser, an Nd:YAG single-crystal rod of Ø4 mm × 104 mm 0.6 % was used as a comparison. The output power was 99 W with an optical-to-optical efficiency of 34 %, which was only 4 % higher than that of the Nd:YAG ceramic rod. It has been observed that in most cases, Nd:YAG ceramic lasers exhibited performances comparable to the single-crystal counterparts [57, 91–93].

It was found that composite ceramic rods consisting of Nd-doped YAG core cladded with Sm:YAG demonstrated higher performances as QCW lasers than the uniformly doped rods and the rods with undoped YAG cladding [94]. The power of diode laser-pumped Nd:YAG lasers could be as high as 100 kW. One example is the system of seven coherently coupled master oscillator power amplifier (MOPA) lasers [95], while another example is a system with low-duty regime, in which six modules were connected in pairs to thin Nd:YAG ceramic plates in a single-aperture power oscillator with the laser beam traveling in a zigzag manner, i.e., the so-called thin ZAG concept [96]. The individual ceramic rods, in single-rod oscillator configuration, offered an average power of 426 W, with optical–optical efficiency similar to that of single-crystal rods. Even higher power, i.e., 67 kW, has been observed in a burst-mode high aperture heat capacity laser, with a cascade of five Nd:YAG ceramic plates with a dimension of 10 × 10 × 2 cm^3, which were face-pumped with 808-nm diode lasers [97].

A QCW (1 kHz) laser with 808.5-nm transverse diode-pumped MOPA architecture, consisting of three 1 at.% Nd:YAG ceramic rods, showed a high power of 1.02 kW [98]. The module consisted of four parts, i.e., Nd:YAG ceramic rod, a flow tube, a reflector, and sixty diode bars mounted to brass heat sinks. The Nd:YAG ceramic rods had a dimension of Ø6 mm × 100 mm, which were made by using the conventional ceramic processing method. The Nd:YAG rod was surrounded by a diffusive reflector and pumped by laser diodes from five directions. The LD arrays were arranged in a two-line fivefold symmetry. The beam coupling method could achieve efficient pumping of the Nd:YAG rod and produce a uniform pumping distribution within the entire cross section of the rod. The diode bars and ceramic rod were both cooled to 21 °C with deionized water flowing through the flow tube. The diode bar array was operated in quasi-CW mode with a fixed repetition rate of 1 kHz at central wavelength of 808.5 nm. The maximum pump power was 1150 W for each laser module.

Experimental setup of the MOPA laser system is shown in Fig. 9.9 [98]. The master oscillator contained two laser modules, with a plano–plano symmetrical structure. One of the laser modules served as the amplifier stage. The internal time sequence was controlled by using a synchronous controller. In the setup, L1 is the distance from the mirror M1 to the left end surface of the Nd:YAG rod, L2 is the distance between the Nd:YAG rods, and L3 is the distance from the mirror M2 to the right end surface of the Nd:YAG rod. L1 and L3 were 8 mm, which formed a symmetric resonator. L2 was 6 mm that was limited by the mechanical size of laser modules. The distance between the amplifier stage and the output mirror M2, i.e., L4, was 15 mm. M1 was coated to have 99.8 % high reflectance at 1064 nm. M2

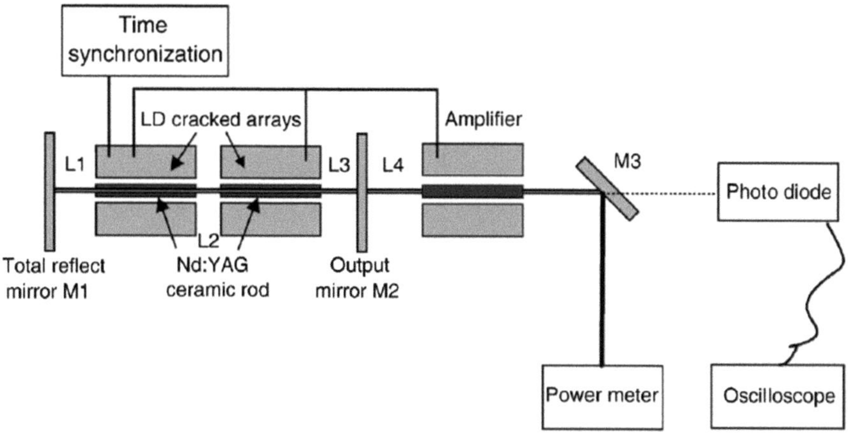

Fig. 9.9 Schematic diagram of the MOPA system for the Nd:YAG ceramic laser. Reproduced with permission from [98]. Copyright © 2010, Elsevier

was the output coupler (OC), which was coated with 20 % transmittance at 1064 nm. M3 was coated at 1064 nm to have 99.8 % high reflectance at the incidence angle of 45°. The output power was monitored by using an Ophir 5-kW power meter.

Figure 9.10 shows the average output power of the laser as a function of the LD pump power. The output power was increased almost linearly with increasing incident pump power. The threshold power was about 640 W. The maximum output power of 1.02 kW was achieved at the pump power of 3433 W, corresponding to an optical–optical conversion efficiency of 29.7 %. The slope efficiency was estimated to be 40.2 %. The MOPA scheme offered high beam quality.

The Nd:YAG ceramic lasers directly pumped at 885 nm with Ti:sapphire or diode lasers have exhibited enhanced laser performances, e.g., lower threshold and higher slope efficiency, by using uncoated laser rods [29, 99, 100]. The performances were further increased to close to the limit of the quantum defect, by using

Fig. 9.10 Output power versus the total LD pump power for the Nd:YAG ceramic MOPA system. Reproduced with permission from [98]. Copyright © 2010, Elsevier

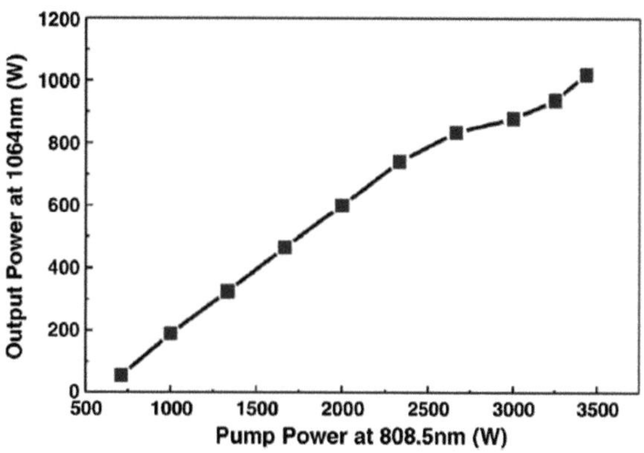

AR-coated higher-quality ceramics. The improvement in the CW laser parameters was nearly 10 %, according to the prediction by the quantum defect ratio, which was attributed to the additional contribution of the reduced heat generation and the increased beam quality. In addition, a highly efficient 250-W laser under diode laser pumping has been developed [101]. Therefore, development of 885-nm laser diodes indicated that the scaling of power at this pump wavelength is highly potential.

A flashlamp-pumped 1064 nm Nd:YAG ceramic laser has been reported to be operated in the QCW free-generation regime under 5 ms pulse excitation at 20 Hz by using a Xe lamp, with the ultraviolet radiation being filtered by using a Sm-doped flow glass tube. An average power of 387 W was achieved by using 1.1 at.% Nd:YAG ceramic rod, with a diameter of 8.3 mm and a length of 152 mm, with their performances slightly better than those of the single-crystal counterparts [102]. The Nd:YAG ceramic rods showed no solarization, which means that they can be used to generate high-energy pulses with lamp pumping.

Various other emission transitions in the 1 μm range have been reported by using Nd:YAG ceramics. Laser emission on such transitions can be selected by using wavelength selection optics, e.g., mirrors and intracavity elements for gratings, prisms, etalons, and selective absorbers, so as to avoid the competition from other transitions. It is also possible by operating under such conditions that the competing transitions cannot reach the laser threshold. With this concept, highly efficient QCW laser of 180 μs emission with 1.1 kHz repetition rate of the optical slope efficiency of 24.7 % has been achieved at 1123 nm by using Nd:YAG ceramics under 1000 W diode laser pumping, in which a strictly controlled coating was used to avoid the competition from nearby transitions [103–106]. Efficient CW emission of Nd:YAG ceramic laser at 1052 nm has also been demonstrated [107].

Laser emissions of Nd:YAG ceramic lasers in the 1.3 and 0.9 μm ranges are relatively less overwhelming. Laser emission at 1.3 μm of Nd:YAG ceramics with a high slope efficiency of ~35 % has been achieved by using the intense transition $^4F_{3/2}(2) \rightarrow {}^4I_{13/2}(1)$ at 1.319 μm, which has a room temperature emission cross section of ~0.8×10^{-19} cm^2 [7]. CW laser based on Nd:YAG ceramics at 946 nm due to the transition of $^4F_{3/2}(2) \rightarrow {}^4I_{9/2}(5)$ exhibited performances that are almost the same as those of crystals [108].

Multiple laser emissions at several discrete wavelengths has been realized by using broadband or multiple dichroic resonator optics, at conditions of pumping that allow for the threshold to be reached by the transitions. For example, multi-wavelength laser emission can be achieved by using unicenter laser materials on the emission lines inside the same intermanifold transition or between different manifolds. An alternative is to use multicenter materials with emission lines from various structural centers that are excited simultaneously by pumping. Laser emission at two or more wavelengths due to the transitions inside the same or different inter-manifold transitions has been demonstrated in both Nd-doped crystals and ceramics. For instance, for the 1.3 μm emission range, simultaneous emission could be achieved on the 1.319 and 1.338 μm lines by using Nd:YAG ceramics [109]. Similarly, a further tri-wavelength laser emission including these transition lines and the line of 1.064 μm has been realized [110]. These multi-wavelength lasers

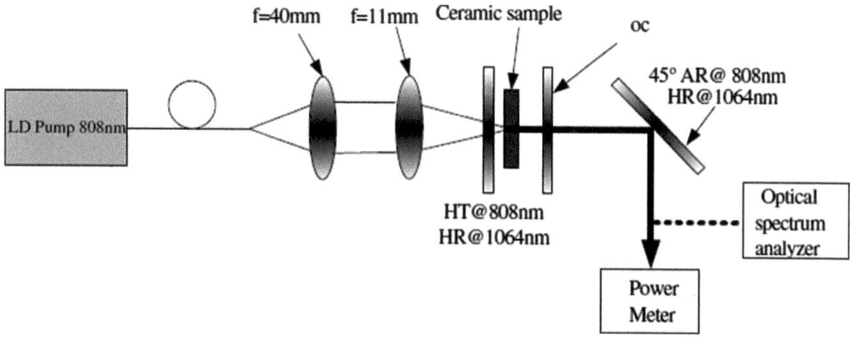

Fig. 9.11 Schematic diagram of the experimental setup. Reproduced with permission from [111]. Copyright © 2008, Elsevier

could be found applications in nonlinear processes, including mixing and frequency multiplication, which would enable the generation of radiations from visible to terahertz region.

Multi-wavelength laser emission can also be achieved by using disordered systems, in which the multicenter structure can provide different emission wavelengths for the same transition from the different centers. For example, dual-wavelength emission has been observed in a ceramic thin-disk laser made with Nd-doped disordered solid solution of $Y_3Sc_{1.5}Al_{3.5}O_{12}$ (YSAG) [111]. A schematic diagram of the experiment setup is shown in Fig. 9.11. The Nd:$Y_3Sc_{1.5}Al_{3.5}O_{12}$ ceramic disk was doped with 4 at.% Nd^{3+}, with a dimension of Ø12 mm × 1 mm. Both surfaces of the sample were coated with antireflection layers at 1064 nm. The laser experiments were carried out at room temperature without using active cooling. A fiber-coupled LD working at the central wavelength of 808 nm was used as the pump source. The fiber core diameter was 200 μm with the numerical aperture of 0.22. With a two-lense coupling system, the pump beam was focused on the surface of the ceramic disk to produce a pump light footprint of about 60 μm in diameter. A plano–plano cavity with an overall length of 7 mm was used. The front mirror was coated with antireflection layer at 808 nm and highly reflective at 1064 nm. The plane-parallel rear mirror was used as the OC. A dichroic beam splitter (45°) was used to reflect the laser and filter the pump light.

It was found that the ceramics absorbed about 64 % of the pump power irradiated on its surface. The OCs with transmissions of 3, 3.9, 6, and 10 % at 1064 nm were used for laser output experiments. The laser threshold was increased from 0.345 to 1.03 W as the transmission of the OC was increased. The maximum output power of 0.356 W was obtained at an absorbed pumping power of 1.96 W, as the OC had T_{oc} = 10 %. An optical–optical efficiency of 18.2 % and a slope efficiency of 23.2 % were observed. The emission spectra of the laser output are shown in Fig. 9.12, in which the absorbed pump power was 1.52 W with the OC having a transmission of 10 %. Obviously, two wavelengths oscillated simultaneously [111].

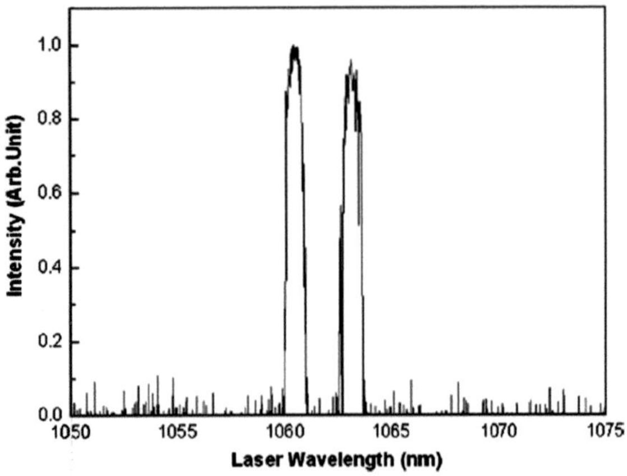

Fig. 9.12 Emission spectra of the YSAG ceramic laser, with two wavelengths oscillating simultaneously. Reproduced with permission from [111]. Copyright © 2008, Elsevier

Therefore, compositionally disordered garnets with intrinsic disordering, such as the CNGG–CLNGG family, with multicenter structure of spectra, would be very useful for multi-wavelength emission. Furthermore, the possibility of high concentration and uniform distribution of doping ions in ceramics make it possible to tailor the ceramic laser materials with sensitized emissions. For example, because Cr^{3+} has strong and broad absorption bands in the visible, (Cr^{3+}, Nd):YAG ceramics can be used for lamp or solar pumping. Comparatively, due to the difference in segregation coefficients between the two ions, the conditions of sensitization in single crystals are nonuniform.

With the presence of highly transparent (Cr^{3+},Nd^{3+}):YAG ceramics [37], a quite number of highly efficient laser emissions and amplifications have been developed by the ceramics under lamp pumping [36, 39–41, 112, 113]. Under solar pumping using a 4 m^2 Fresnel lens collector, 80 W CW laser emission has been demonstrated, with solar laser power conversion efficiency of 4.3 % [42, 114, 115]. A 120-watt continuous-wave solar-pumped laser has been realized by an Nd:YAG rod with a liquid light guide lens [116]. As a result, (Cr^{3+}, Nd):YAG ceramics are potential candidates that can be used to construct highly efficient lamp-pumped Nd-doped ceramic lasers and solar-pumped lasers. It is expected to further increase the energy transfer from Cr^{3+} to Nd^{3+} through the optimization of the laser materials.

Nd-Doped Sesquioxide Ceramic Lasers

As discussed earlier, the first Nd-doped oxide ceramic laser was Y_2O_3-doped with 10 % ThO_2 under lamp pumping, which was operated in free generation of 150-μs pulses, but with a low efficiency of 0.14 % at 40 J pumping, which was attributed to high residual optical losses [3, 117, 118]. Higher performance of 1-μm laser for CW emission under diode laser pumping was later achieved by using fine-grained ceramics of Nd:Y_2O_3 [119] and Nd:Lu_2O_3 [5]. Due to the large crystal field splitting in these materials, the transitions with the largest emission cross sections at

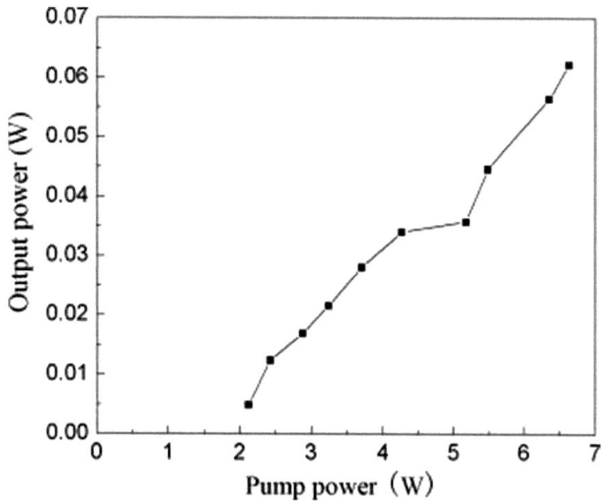

Fig. 9.13 Output power versus the absorbed power of the Nd:$Y_{1.8}La_{0.2}O_3$ ceramic laser. Reproduced with permission from [120]. Copyright © 2011, Elsevier

room temperature were from the lowest Stark component of the emitting level. In addition, in both ceramic materials, the strongest transitions, with close emission cross sections, are terminated at the lowest two Stark components of the manifold $^4I_{11/2}$ with a difference of only 35 cm^{-1}, and a dual CW laser emission was observed at 1074.6 and 1078.6 nm for Y_2O_3 and 1075.9 and 1080 nm for Lu_2O_3, respectively. Nd-doped solid solution $Y_{1.8}La_{0.2}O_3$ ceramics, showing a strong broadening of the emission line of up to 5 nm, provided a laser emission at 1079 nm under diode laser pumping [120]. Figure 9.13 shows the dependence of output power of ceramic laser, as a function of the absorbed power, indicating that the laser had a threshold of about 2.2 W. The output power of the laser was very high, because the Nd:$Y_{1.8}La_{0.2}O_3$ ceramics still had absorption and scattering losses which imposed significant negative effect on the laser performance. It is expected that through optimizing concentration of Nd^{3+}, thickness of a gain medium, and output coupling efficiency, more efficient lasers could be achieved.

Other Nd-Doped Ceramic Lasers

Nd^{3+} ions in intrinsically disordered ceramics have broad absorption and emission bands, which could be used for various applications, such as multi-wavelength emission, tunable laser emission, and generation of ultrashort pulses. For example, under diode laser pumping at 809 nm, cubic perovskite disordered Ba(Zr,Mg,Ta)O_3 transparent ceramics can show single (1075 nm) or multi-wavelength emissions in the range of 1062–1075 nm, with a value of FWHM that is potential for short-pulse mode-locked emission [121]. The transparent Nd^{3+}:BZMT ceramics were prepared by using an optimized ceramic fabrication process, from high-purity $BaCO_3$, ZrO_2, $MgCO_3$, Ta_2O_5, and Nd_2O_3 powders. The Nd^{3+} ions could take any position of the cations at the B-site in the perovskite structure. Disordered nature was observed, according to the characteristic broadened fluorescence spectra.

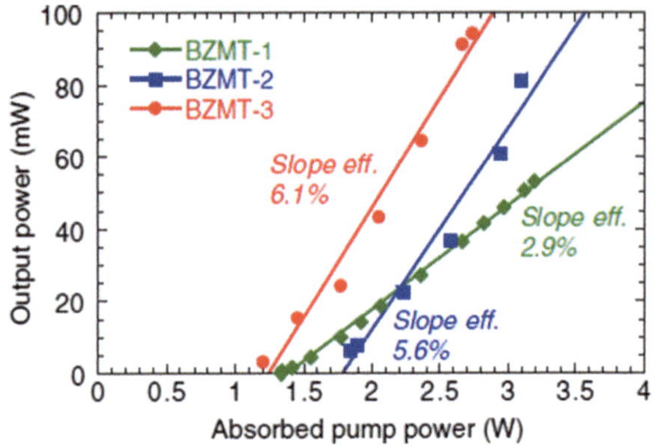

Fig. 9.14 Laser properties of the Nd^{3+}:BZMT ceramics. Reproduced with permission from [121]. Copyright © 2011, Elsevier

Three samples were studied for comparison, where BZMT-1 was the ceramics with a bandwidth of 26 nm, while BZMT-2 and BZMT-3 had a slightly broader bandwidth of ~30 nm with different peak positions. The thickness of BZMT-1 was 4 mm, while that of BZMT-2 and BZMT-3 was 2 mm. All ceramics were uncoated. A simple two-mirror resonator pumped by a fiber-coupled LD was used for laser measurement. Output powers of as a function of absorbed pump power are shown in Fig. 9.14 [121]. Because no coating was used, the high-reflection (HR) loss of ~12 % for each surface due to the high indices of >2.0 limited the efficiency, but the slope efficiencies of the BZMT-2 and BZMT-3 ceramics were increased by about two times in comparison with slope efficiency of the BZMT-1 ceramics. Because BZMT-3 had the lowest laser threshold and highest slope efficiency of 6.1 %, it was used in the mode-locked laser. BZMT-1 showed single-wavelength operation at 1075 nm, while BZMT-2 and BZMT-3 showed multi-wavelength broadband operations.

Figure 9.15 shows schematic diagram of the mode-locked Nd^{3+}:BZMT ceramic laser, in which a z-shaped astigmatically compensated cavity was used [121]. A 3.3-mm-thick 1 mol% Nd^{3+}:BZMT ceramic (BZMT-3) was arranged at the Brewster angle and pumped by a broad-stripe LD, with an emission area of 1 × 95 μm and a maximum power of 5 W at 809 nm. The pump beam was focused into the ceramic to $1/e^2$ diameters of 25 × 134 μm through the beam shaping optics.

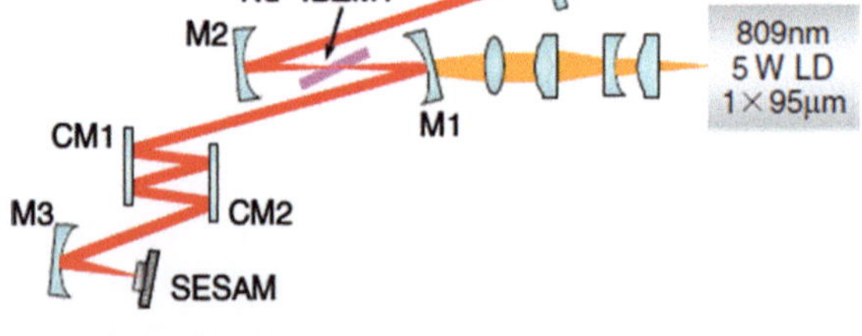

Fig. 9.15 Experimental setup of the mode-locked Nd^{3+}:BZMT ceramic laser. Reproduced with permission from [121]. Copyright © 2011, Elsevier

Fig. 9.16 Output power of the mode-locked Nd^{3+}:BZMT ceramic laser. Reproduced with permission from [121]. Copyright © 2011, Elsevier

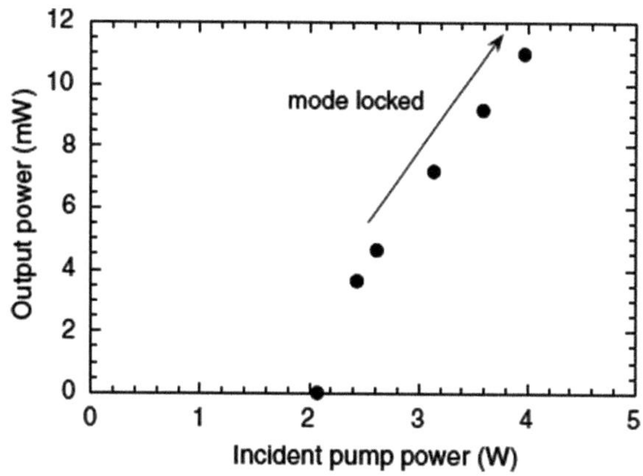

The maximum incident pump power was 3.97 W. The folding mirrors, i.e., M1 and M2, had a 100 mm radius of curvature (ROC) and AR-coated below 1000 nm and HR-coated above 1020 nm. The laser beam was focused onto a semiconductor saturable absorption mirror (SESAM) by a concave mirror M3 with ROC = 50 mm. The chirped mirror pair, i.e., CM1 and CM2, had a group-delay dispersion of $-2000\ \text{fs}^2$ per one round trip. The OC had a 1 % transmittance.

Figure 9.16 shows power evolution of the ceramic laser [121]. At the incident pump power of 2.6 W, mode locking was self-started. The maximum average power was 11 mW. According to the spectrum and autocorrelation trace of the output, it was observed that the center wavelength was 1062 nm, with a spectral bandwidth of 1.5 nm. The sech^2-fit pulse width was 1.4 ps. It is expected that shorter pulse could be generated by optimizing the setup and improving the quality of the transparent ceramics.

Electro-optic lead zirconate titanate PLZT (10/65/35) ferroelectric ceramics doped with 1 at.% Nd offered CW laser emission at 1064.4 nm under 100 ms pulsed CW diode laser pumping at 805 nm, with potential for multi-functional laser emission [122, 123]. A high single-pass gain of 13.0 dB was obtained at 1064.4 nm in the thin electro-optic ceramic plate of Nd:PLZT. By using the samples doped with 1.0 and 0.5 mol% Nd^{3+}, pumping with a diode laser, gains were achieved over a broad seed laser power, ranging from tens of nW to tens of mW at room temperature. Due to the broad absorption spectra of Nd:PLZT, the laser was insensitive to the pump wavelength. The Nd:PLZT ceramics had a total scattering loss coefficient of $<0.06\ \text{cm}^{-1}$ and exhibited a considerable quadratic electro-optic coefficient of $>0.30 \times 10^{-16}\ \text{m}^2\ \text{V}^{-2}$ [122].

Nonoxide transparent ceramics can be used as host for Nd lasers. For instance, quasi-CW pulsed diode laser pumping of 0.5 at.% Nd:SrF_2 at 790 nm, corresponding to the maximum of the $^4I_{9/2} \rightarrow {}^4F_{5/2}$ absorption of the isolated tetragonal centers L, provided a double-wavelength laser emission, with a main line at 1037 nm and a weaker line at 1044 nm. The slope efficiency of absorbed power was 19 %, slightly lower than that of single crystals. Pumping at 796 nm, corresponding to the maximum of the composite absorption band for the center L and the Nd-pair center M, there was an increase in the laser threshold and a reduction in the slope efficiency for the L centers, while no emission was observed from the M centers [124].

Besides the cubic laser materials, there have also been reports on anisotropic ceramic materials, e.g., strontium fluorapatite S-FAP, with grains aligned by using a magnetic field prior to sintering, has demonstrated QCW laser emission at 1063.1 nm under 807.5 nm diode laser pumping, although efficiency is still relatively low [125, 126].

Q-Switched Nd-Doped Ceramic Lasers

Nd:YAG ceramic laser can be *Q*-switched by using either of the active methods, such as electro-optic and acousto-optic switching, or passive methods, depending on the saturable absorptions.

Active *Q*-switching of Nd:YAG ceramic lasers has been achieved by using electro-optic and acousto-optic switches. For instance, a high repetition rate (1000 Hz) laser emission at 1064 nm, with pulses of 23 mJ within the duration of 12 ns, has been achieved, by using electro-optic *Q*-switching under CW diode laser pumping at 808 nm [127]. A QCW pumping at 808 nm with a repetition rate of 100 Hz has facilitated the generation of pulses of 50 mJ with duration of 6 ns [128]. A laser emission at 1064 nm with an average power of >40 W and a high spatial quality have been realized, with an 808-nm diode-pumped and acousto-optically *Q*-switched MOPA, with a bounce configuration based on 2 at.% Nd:YAG ceramics, showing pulses of 35–80 ns in the repetition range of 10–150 kHz [129]. Diode laser-pumped and acousto-optically *Q*-switched laser emission with a repetition rate of tens of kHz, by using Nd:YAG ceramics, has been available at 1319 [130] and 946 nm [131]. Currently, the performances of actively *Q*-switched emissions of Nd: YAG ceramic lasers are comparable with those of the best single-crystal lasers.

One example is diode-pumped *Q*-switched Nd:YAG ceramic laser operating at 946 nm [131]. Figure 9.17 shows experimental schematic diagram of the diode-pumped actively *Q*-switched 946-nm Nd:YAG ceramic laser. A Nd:YAG ceramic sample doped with 0.6 at.% Nd with a dimension of Ø4 mm × 5 mm was used. When using a fiber-coupled laser diode with the fiber diameter of 600 μm as the

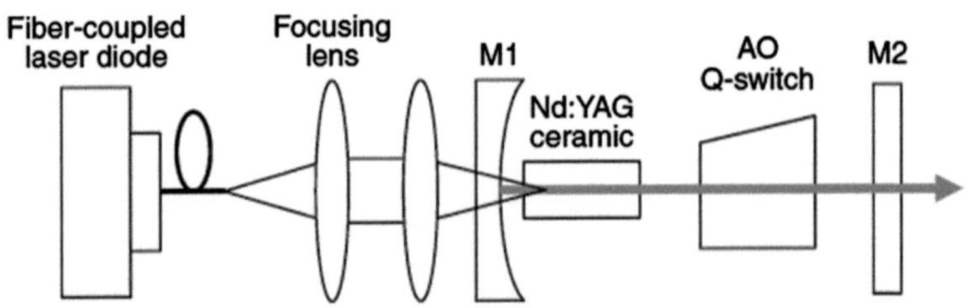

Fig. 9.17 Schematic of the diode-pumped actively *Q*-switched 946 nm Nd:YAG ceramic laser. Reproduced with permission from [131]. Copyright © 2009, John Wiley & Sons

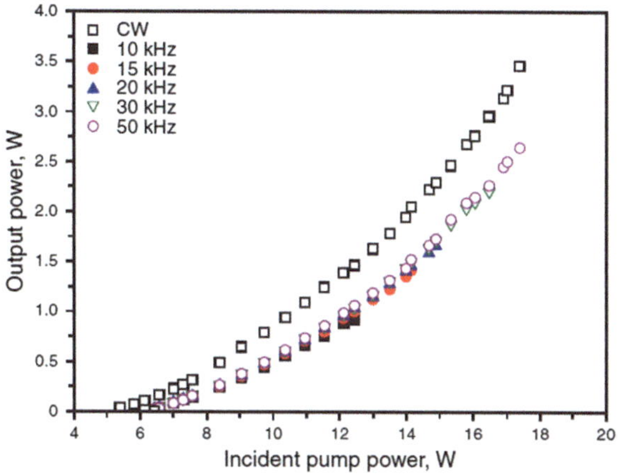

Fig. 9.18 Average output powers at 946 nm from the *Q*-switched Nd:YAG ceramic laser as a function of the incident pump power, with the coupling fiber of the pumping LD having a diameter of 600 μm. Reproduced with permission from [131]. Copyright © 2009, John Wiley & Sons

pump source, an average output power of 2.64 W at a repetition rate of 50 kHz was observed, at the incident pump power of 17.39 W, as shown in Fig. 9.18. When the fiber diameter was 400 μm, at the incident pump power of 11.56 W, an average output power of 1.79 W was obtained at a repetition rate of 50 kHz, corresponding to an optical conversion efficiency of 15.5 %. A maximum pulse energy of 120.89 μJ and a minimum pulse width of 56 ns were observed at a repetition rate of 10 kHz.

Although semiconductors, e.g., GaAs, have been attempted, the saturable absorbers (SAs) used for *Q*-switching Nd-doped lasers are mainly based on transparent materials doped with transition metal (TM) ions. The main parameters of SAs are saturation fluence, $E_s = h\nu/\sigma_{gs}$, initial transmission, and residual absorption. Saturation fluence is a material parameter of the SAs, while initial transmission can be controlled through the concentration of the doping ions and the size of the laser components, whereas the residual absorption is also technologically controllable.

The TM ions usually used for passive *Q*-switching of infrared solid-state lasers include Cr^{4+} (940–1100 nm range), V^{3+} (1.30–1.35 μm), and Co^{2+} (1.3–1.6 μm). Although all these ions can be doped in cubic materials, i.e., Cr^{4+} and V^{3+} in garnets and Co^{2+} in spinels, only Cr^{4+}:YAG ceramics have been reported for ceramic SA *Q*-switches [37, 42, 132, 133]. The SA materials are usually used as components together with the Nd laser-active materials in the cavity. In this case, if the laser materials and the *Q*-switch materials have a same host, such as Nd:YAG lasers passively *Q*-switched with Cr^{4+}:YAG, the difference in refractive index between them should be very small, so that there is no reflection at the interface when the two materials are intimately joined. For single crystals, they can be either diffusively bonded or mechanically clamped. In contrast, when one or both of them

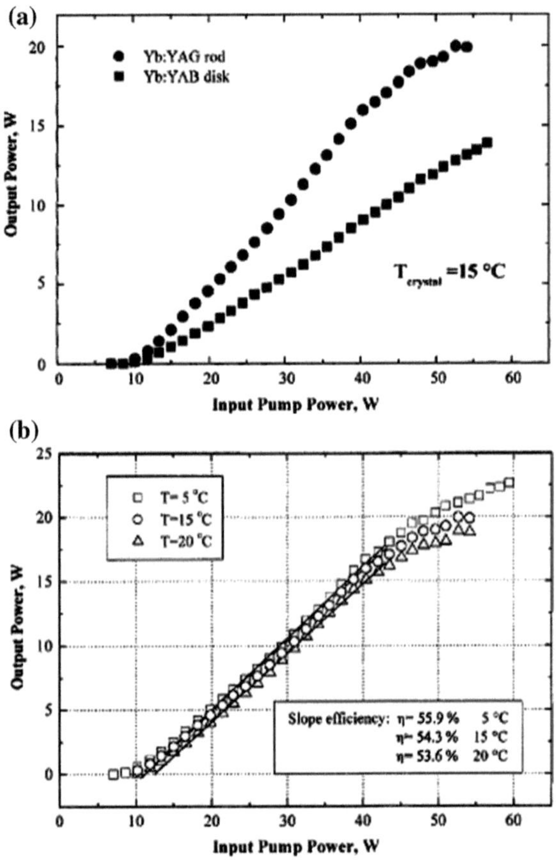

Fig. 9.19 **a** Laser output powers of the free-running CW Yb:YAG laser rod with C_{Yb} = 10 at% and disk with C_{Yb} = 8.5 at%, as a function of incident pumping power at 940 nm, at heat sink temperature of 15 °C. **b** Laser output powers of the free-running CW Yb:YAG laser rod with C_{Yb} = 10 at%, as a function of incident pumping power at 940 nm, at different heat sink temperatures. Reproduced with permission from [138]. Copyright © 2002, Elsevier

are ceramics, monolithic composites with uniform codoping of Cr^{4+} and Nd^{3+} can be readily fabricated.

Theoretical and experimental studies indicated that when using SAs for passive *Q*-switching of solid-state lasers, in order to increase the energy and decrease the duration of the pulse, it is necessary to use SAs with low initial transmission, increase the transmission of the outcoupling mirror, and shorten the laser resonator [134–139]. For instance, the repetitive modulation in kHz frequency domain of a passively *Q*-switched, diode-pumped Yb:YAG laser, by Cr^{4+}:YAG, Cr^{4+}:LuAG, and Cr^{4+}:GSGG SAs, was investigated in a report [138]. With an aim toward the design of passively *Q*-switched Yb:YAG microlaser, free-running performances of a rod and a disk Yb:YAG were characterized and evaluated. Representative laser output power results are shown in Fig. 9.19. The experimental results, such as gain and loss, together with the levels of the stimulated emission cross section, e.g., $\sigma_{em} = 3.3 \times 10^{-20}$ cm^2, were in a good agreement with the predictions of numerical models. *Q*-switched pulses with maximum peak power of ~10.4 kW, with energy level of ~0.5 mJ pulse^{-1}, have been extracted, with an extraction efficiency 30 %.

It was acknowledged that pump intensity has negligible effect on energy and duration of pulses, but is able to influence the repetition rate. Most Cr^{4+}:YAG SA passive Q-switching of Nd lasers that have been reported in the open literature are CW diode-pumped lasers, with performances that are constantly increased. For example, the most widely reported 808-nm diode laser-pumped and Cr^{4+}:YAG Q-switched 1064-nm Nd:YAG ceramic lasers have pulse energies ranging from several mJ to as high as 200 μJ, peak powers ranging from 100 W to tens of kW and duration of tens of ns [140–143]. Passively Q-switched ceramic Nd^{3+}:YAG lasers with ceramic Cr^{4+}:YAG SAs have been demonstrated [143]. When the lasers were pumped by a 1-W CW laser diode, optical–optical efficiency of as high as 22 % could be achieved by using Cr^{4+}:YAG with an initial transmission of 94–79 %. The results were comparable to those of their single-crystal counterparts.

The instability of passive Cr^{4+}:YAG Q-switched emission of end-pumped Nd:YAG ceramic rod lasers has been suppressed by cladding with Sm:YAG ceramic layers [94]. Furthermore, for SAs with low T_0, the Sm:YAG-cladded ceramic rods lead to pulse energies that were higher than or at least close to those realized by using undoped YAG-cladded Nd:YAG rods or uniformly doped Nd:YAG rods. Efficient Cr^{4+}:YAG passively Q-switched emissions of Nd:YAG ceramic lasers, with peak powers of >20 kW, at wavelengths of 1052 and 1064 nm, have been achieved [102, 144]. Figure 9.20 shows average laser output powers at 1064 nm as a function of the pump power [102]. For the ceramic rod, a maximum average output power of 387 W was achieved, at the pump power of 16.4 kW, which was even slightly higher than the 370 W obtained for the single-crystal rod in the same pumping condition. The optical quality of the ceramic YAG rod was comparable to that of the single-crystal YAG rod under the flashlamp pumping scheme.

High-energy pulses, ranging from several mJ to tens of mJ, have been generated, by pulsed pumping of passively Q-switched 1-μm Nd:YAG lasers. 5-Hz flashlamp pumping of passively Q-switched ceramic YAG has been achieved by using a 1 at.%

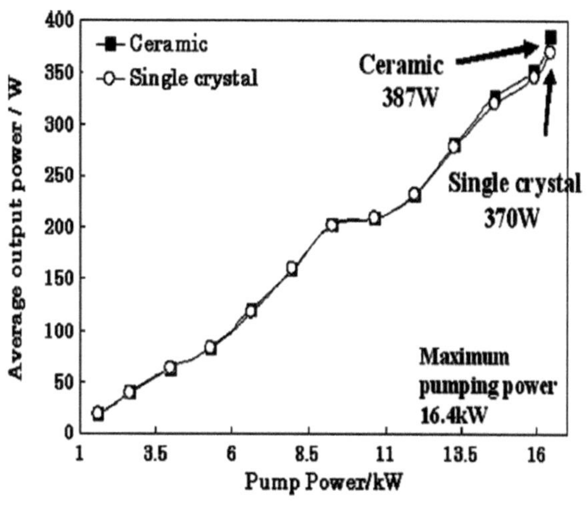

Fig. 9.20 Average laser output powers at 1064 nm as a function of the pump power. Reproduced with permission from [102]. Copyright © 2007, Elsevier

Nd-doped YAG ceramic rod codoped with 0.1 at.% Cr^{3+} to increase absorption [145]. In the free-generation regime, this laser could generate pulses of 2.1 J. By inserting a high T_0 (90 %) SA Cr^{4+}:YAG ceramics in the cavity, Q-switched pulses, with each pulse of ~300 ns and 1.1 J burst of 130 μs, have been generated for each pump pulse. If T_0 was decreased to 30 %, the multiple pulses became to be a single pulse of 130 ns. The laser threshold and the slope efficiency were also decreased, while the pulse energy was decreased to only 80 mJ. The threshold of the Nd/Cr: YAG ceramic laser oscillator was decreased to be <10 % of that of an Nd:YAG oscillator. Calculation results indicated that the output laser energy of the Nd/Cr: YAG ceramic laser was higher than that of Nd:YAG by about 3 times.

It was found that if a combined active–passive Q-switching was used for high-energy-pulse pumped Nd:YAG crystal lasers, the multiple pulse emission disappeared; instead, a single pulse was observed, with much higher energy, shorter duration, and controlled jitter [146]. In that case, a combined mechanical LiF:F_2^- color center passive Q-switch was used for the YAG:Nd lasers. This approach is also useful to CW pumped Nd:YAG single-crystal lasers [147, 148]. A continuous-wave (CW) flashlamp-pumped Nd:YAG laser, passively Q-switched by Cr^{4+}:YAG and LiF:F_2^- crystals, by using active–passive method, was demonstrated in terms of output characteristics [149]. In this study, a new method was used to control the small-signal transmission of the passive Q-switch, which was realized by using a miniature laser placed outside of the resonator. The feasibility was described theoretically and confirmed experimentally. Moreover, the combination of a Cr^{4+}: YAG passive Q-switch and an acousto-optic Q-switch that worked together in an Nd:YAG CW pumped laser was exploited. With a two-rod resonator and a Cr^{4+}: YAG crystal as saturable absorber, an average laser output power of >180 W was achieved, with a pulse-to-pulse stability of 1 %.

It is expected that such an approach can be used to generate giant pulses with flashlamp-pumped (Cr,Nd):YAG ceramics. For instance, pulse emission with energy of ~2.4 mJ and peak power of 2.8 MW at repetition rates of 5–100 Hz has been achieved by using a passively Q-switched composite ceramic of 1.1 at.% Nd: YAG/Cr^{4+}:YAG under QCW 807 nm diode laser pumping [71]. As the laser was pumped by three independent beams, it could generate three laser beams, which can be focused for penetration at three points, with positions that can be controlled. This function can be used in car engines to provide optimized ignition. This configuration has several advantages. For example, the heat deposited in the unit volume was effectively decreased, the alignment was easily realized, while the laser medium could be ceramics. Moreover, the Nd:YAG/Cr^{4+}:YAG monolithic structure ensured robustness and compactness of the laser system, which is especially important if it is mounted on engine. If one pump line configuration was employed, the high-energy laser beam should be split into three fascicles, which had to be directed at certain angles and then focused. As a result, the probability of damaging the laser media would be increased, because the laser beam had higher intensity, due to thermal effects. At the same time, the guiding line would become more complicated.

(Nd,Cr):YAG ceramics have been also used as an active-mirror amplifier under flashlamp, arc lamp, and solar pumping, for free-generation (160 μs) or *Q*-switched pulses with a low repetition rate, in the form of lamp-pumped lasers. The characteristics of amplification for short and long pulses were similar, while the output energy was limited by the saturation of the amplified input laser at low fluences [39]. In contrast, this amplifier is pumped with a CW arc lamp, and it would be significantly efficient to provide high optical conversion with high repetition rates [38].

Passive *Q*-switching of a 1.3-μm bounce Nd:YAG ceramic laser with crystalline V^{3+}:YAG saturable absorber, with ground-state absorption cross section 7.2×10^{-18} cm^2 and negligible excited-state absorption, was able to generate pulses with repetition rates of 10–114 kHz and duration of 55–140 ns [150]. Figure 9.21 shows

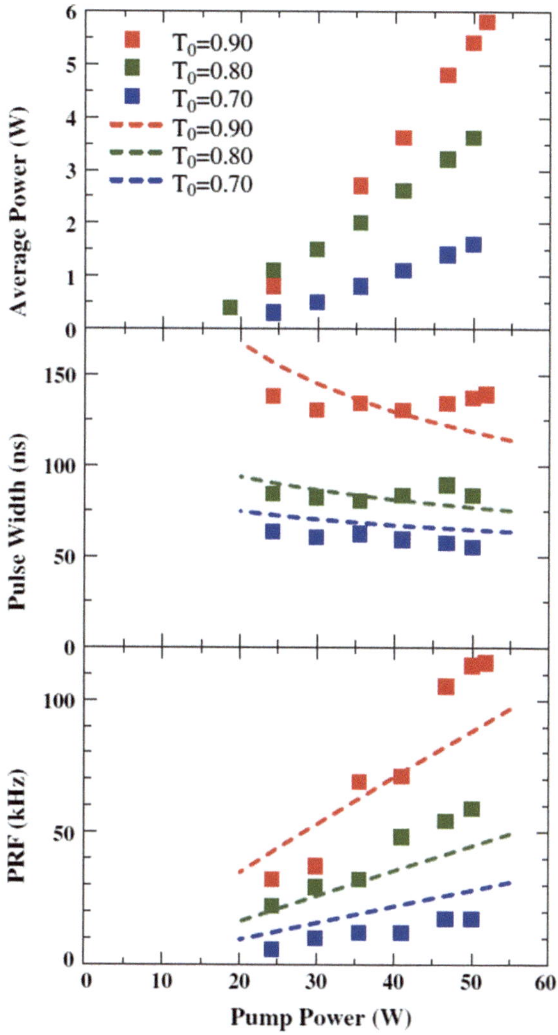

Fig. 9.21 Experimental data of the *Q*-switching operation with the V:YAG crystals of various initial transmittances. *Red*, *green*, and *blue squares* represent experimental results, while *red*, *green*, and *blue broken curves* stand for the simulated results. Reproduced with permission from [150]. Copyright © 2009, Elsevier

experimental results of the passive Q-switching operation of the laser. Best performance was obtained for the V:YAG sample with an unsaturated transmission of 0.9 and 80 % reflectivity OC. Q-switched output power of 5.8 W was observed for pump power of 52 W, corresponding to an optical-to-optical conversion efficiency of 11.5 %. Threshold for laser operation was ~20 W, with a slope efficiency of 19 %. The output was typically multi-mode in the horizontal, but single mode in the vertical direction. Pulse duration of the laser output was ~135 ns at all pump levels, whereas the pulse repetition frequency (PRF) was 32 kHz near the threshold and increased to 114 kHz at the maximum pump power of 52 W. Q-switching performance of the system was affected by the initial transmissions of the V:YAG crystals. With dense V:YAG crystals, Q-switching output had pulse duration of <100 ns, with PRF being limited to <100 kHz, even at the maximum pumping power. The corresponding peak power of the output was up to 1.7 kW for the V:YAG crystal with an initial transmission of 0.70.

A similar laser, with diode end-pumping, could produce pulses of 128 ns at a repetition rate of 230 kHz, with a slope efficiency of 9 % [151, 152]. One of the examples was a LD-end-pumped passively Q-switched Nd:YAG ceramic laser, operated at 1319 nm with a V^{3+}:YAG saturable absorber [151]. An average output power of 1.8 W was achieved at the pump power of 23.7 W, which corresponded to an optical conversion efficiency of 7.8 % and a slope efficiency of 9 %. The minimum pulse width of 128 ns at a pulse repetition rate of 230 kHz was obtained with a $T = 2.8$ % OC at the pump power of 23.7 W. Another example was a side-pumped Nd-doped Gd(0.6)Y(0.4)VO(4) bounce laser, which was combined with a V:YAG saturable absorber crystal [152]. It offered a passively Q-switched output of 6.5 W at 1.3 μm. Output powers of 6.5 and 6 W were observed at a maximum pump level for the multi- and TEM(00)-mode operations, demonstrating optical-to-optical efficiencies of 17.5 and 16.2 %, respectively.

The self-Q-switching properties of codoped (Cr^{4+},Nd/Yb):YAG crystals have been intensively studied, with the attempt to develop single-mode microchip lasers [153, 154]. Similarly, (Cr^{4+},Nd):YAG ceramics have also been available for potential microchip laser applications [155, 156]. In one study, optical properties of Cr,Yb:YAG and Cr,Nd:YAG crystals and composite Yb:YAG/Cr:YAG ceramics self-Q-switched solid-state laser materials were presented, with their merits being evaluated [153]. Although tens of kilowatts of peak power was observed for monolithic microchip Cr,Yb:YAG laser, its performance was limited, due to the absorption of Cr^{4+} at 940 nm and the strong fluorescence quenching at high concentrations of Cr^{4+}. Comparatively, composite Yb:YAG/Cr:YAG ceramics were more promising to achieve high pulse energy and peak power, with optimized lasing and Q-switching.

Mode-Locked Nd-Doped Ceramic Lasers

Three types of mode locking are discussed in the subsection: (i) passive saturable absorption mode locking, (ii) effective saturable absorption mode locking, and (iii) mode locking by controlled mirror reflection. Mode locking of solid-state lasers is an efficient approach to reduce the pulse duration. It has been shown that mode locking can be realized by using either active or passive methods. Active methods include the use of acousto-optic modulators, whereas passive methods are based on intracavity intensity-dependent modification of the losses in the resonator with real or effective SAs or modification of the reflectivity of mirrors through nonlinear processes.

Generation of stable regular trains of picosecond pulses by using active mode locking has been achieved in various CW pumped Nd-doped single-crystal lasers, but it has not been so essentially available by using Nd-doped ceramic lasers [157–160]. For SA mode locking, the pulse duration (τ_p) is limited by the product $\tau_p \Delta\nu$, where $\Delta\nu$ is the full width at half maximum (FWHM) of the emission of the laser-active material in the frequency domain, which is 0.44 and 0.315 for Gaussian and $sech^2$ temporal dependences, respectively. Passive mode locking requires relaxation times from the upper state of the absorber, which is populated by absorption of the laser emission, at an order of picoseconds.

As mentioned before, Cr^{4+} ion in YAG shows saturable ground-state absorption (GSA) to its first excited state in the 1 μm range of Nd lasers, with a lifetime for this excited state of about 4.2 μs. Therefore, mode locking is supported at the intracavity intensities due to the bleach of the absorber, although efficient *Q*-switching can be facilitated. If the intracavity intensity can be increased, strong excited-state absorption (ESA) of the 1 μm laser emission from the first excited state of Cr to upper levels will be available. Because the lifetime of these upper levels is very short, which is usually <0.1 ns, this ESA could be used for mode locking, although its saturation intensity could be significantly larger than that of *Q*-switching, e.g., by as large as five orders of magnitude. In this case, the train of mode-locked pulses is generally modulated in intensity within the shape of the *Q*-switched temporal pulse, i.e., *Q*-switched mode-locking (QML) regime. The QML regime can produce mode-locking (ML) pulses with intensity higher than that of normal ML. This regime of emission using Cr^{4+}:YAG as SA has been intensively investigated for Nd-doped single-crystal lasers, from which criteria for optimization have been well established [161]. However, such studies have rarely reported for ceramics. Furthermore, for (Cr^{4+}, Nd^{3+}) codoped YAG, the saturation intensity for ESA could be significantly reduced to a value that is larger than that of *Q*-switching by only two orders of magnitude [162, 163].

It has been found that nanosized carbons, such as single-walled carbon nanotubes and graphenes, can be used as SAs for ML of solid-state lasers, as long as these carbon materials are semiconductive. For example, semiconducting SWCNTs with a high content of 80 % have been produced by using a polymer extraction technique. These SWCNTs exhibited a saturation intensity of 1.7 MW cm^{-2}, modulation depth of 6 %, and nonresidual loss of 5.5 %, which ensured to generate stable pulses of 8.3 ps, with a repetition rate of 90 MHz, when combined with a 0.5 at.% Nd:YAG ceramic laser pumped with a CW 808-nm diode [164].

Figure 9.22 shows the experimental setup, in which M4 was the SWCNT mirror [164]. An X-fold cavity was used, with the fundamental pulse repetition rate being

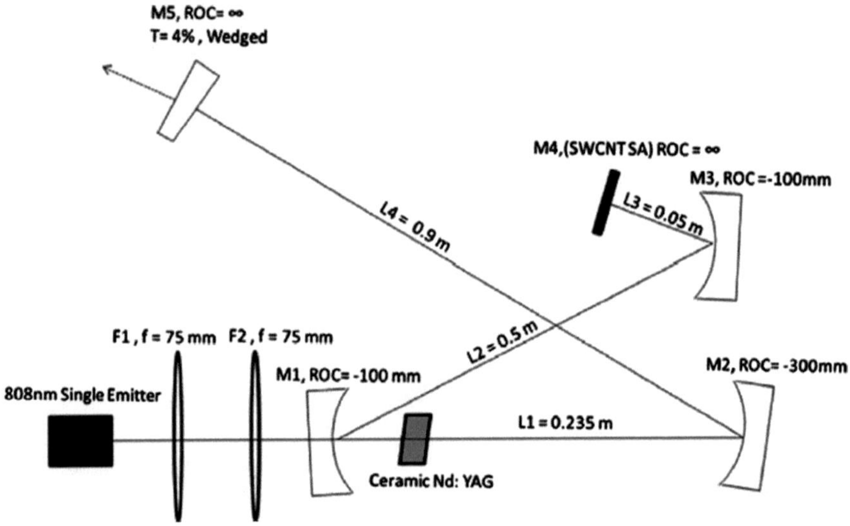

Fig. 9.22 Schematic of the ceramic Nd:YAG mode-locked laser. The SWCNT SA mirror/SESAM was placed at *M4*. *F1* and *F2* are beam reshaping lens. *M1*, *M2*, and *M3* are concave mirrors, while *M5* is the output coupler. Reproduced with permission from [164]. Copyright © 2011, Elsevier

90 MHz. The laser crystal used in the setup was a piece of Nd:YAG ceramics, with an Nd doping concentration of 0.5 at.% and a physical dimension of $3 \times 3 \times 4$ mm^3. The crystal had antireflection coating at the pump and laser wavelengths on both surfaces. It was wrapped in a piece of indium foiled and placed within a water-cooled housing that was maintained at 18 °C. The Nd: YAG ceramic was pumped by using a single emitter at 808 nm. The OC used in it had a transmission of 4 % at 1064 nm, with a 30° wedge to reduce back-reflections into the cavity. It was found that the laser spot size within the crystal had a radius of about 50 μm, while the spot size of the SWCNT SA was about 30 μm.

Although this laser had an average power of only 130 mW for stable mode-locked emission, the results are of interests, due to the special properties of graphene. For instance, a single layer of graphene can absorb ~2.3 % of the incident light. It is insensitive to the wavelength and has very fast recovery time. Because the number of layers of graphenes on the transparent substrates can be readily controlled, it is possible to control the transparency of the ML SAs. By using graphene SA, a diode laser-pumped Nd:YAG ceramic could generate 4-ps pulses with a repetition rate of 88 MHz and average powers of up to 100 mW [165].

If the reflectivity of the rear mirror of lasers can be rapidly increased from a low to a higher value through certain modification, it can be used to achieve ML, which is known as ML by controlled mirror reflection. One way to control the reflectivity is to insert a nonlinear component, whose transmission is changed under the action of the intracavity intensity, in front of the 100 % reflective mirror. This component can be either a conventional nonlinear optical material or a semiconductor saturable absorber. To achieve the high stability and performance, the semiconductor SA can be coated with a layer of another material with slightly lower reflectivity, thus forming a Fabry–Perot resonator that ensures precise control of the SA bleaching. Such a device is also known as semiconductor saturable absorption mirror (SESAM).

The SESAMs have very low residual losses, which could be as low as 0.2 %, as well as short recovery time of ~20 ps, with modulation depth of about 1 %. Under low CW diode laser pumping at 808 nm, the 1064-nm Nd:YAG ceramic laser was worked in the QML regime. When pumping at higher power, stable ML was observed, with pulses of 8.3 ps at a repetition rate of 130 MHz, together with an average power of 1.59 W and optical efficiency of 21 % [166]. In a separate study, higher output power of 8.6 W was observed, with pulses of 26 ps and repetition rate of 78 MHz [167]. In this case, the 1.2 at.% Nd:YAG ceramic was pumped with an 808-nm laser diode, placed in a 1.92-m cavity, and passively mode-locked at 1064 nm with a 1 % modulation depth SESAM. At a pump power of 11.1 W, the laser created an average power of 2.6 W, with a slope efficiency of 27 %. The ceramic exhibited no peak power degradation during a 20-h test of doubling efficiency with periodically poled lithium tantalate with a near stoichiometric composition.

It has also been found that *Q*-switching instability of the SESAM mode-locked Nd:YAG ceramic laser could be suppressed by optimizing the modulation depth of SESAM and the length of the laser resonator [168]. High-quality Nd:YAG ceramic, having the concentration of Nd^{3+} to be $C_{Nd} = 2$ at.%, with a physical dimension of $2 \times 2 \times 2$ mm^3 was used in the experiment. The Nd:YAG ceramics had an average size of 3 μm. The sample faces were not antireflection-coated for the pump and the laser radiations. The residual absorption coefficient at 1.06 μm was <0.02 cm^{-1}, implying that the scattering was negligible. With optimum conditions, continuous-wave ML could be achieved for pump powers of >0.4 W. At the maximum absorbed pump power of 3 W, an average output power of about 0.75 W at a repetition rate of 52 MHz was observed. Figure 9.23 shows output laser power as a function of the absorbed pump power for the cavity length of 3 m.

The inhomogeneously broadened emission lines in some solid solutions or intrinsically disordered garnet materials provided possibilities of realizing shorter mode-locked pulses. To make the full use of the inhomogeneously broadened emission lines, pumping over the whole width of the absorption bands is required. For example, an Nd-doped $Y_3ScAl_4O_{12}$ ceramic laser, pumped with CW Ti:sapphire laser, using a nonlinear LBO-controlled mirror, has been reported to show passive ML, with a slope efficiency of 49.5 % and duration of 10 ps at 77 MHz [169]. Highly transparent Nd^{3+}-doped $Y_3ScAl_4O_{12}$ (YSAG) ceramics with concentrations of Nd^{3+} (C_{Nd}) up to 5.0 at.%, synthesized from high-purity (>99.99 %) commercial oxide powders by using solid-state reaction method, were used in the experiment. Experimental results clearly demonstrated that the Nd:YSAG ceramics should be a promising candidate for applications in microchip lasers or mode-locked lasers.

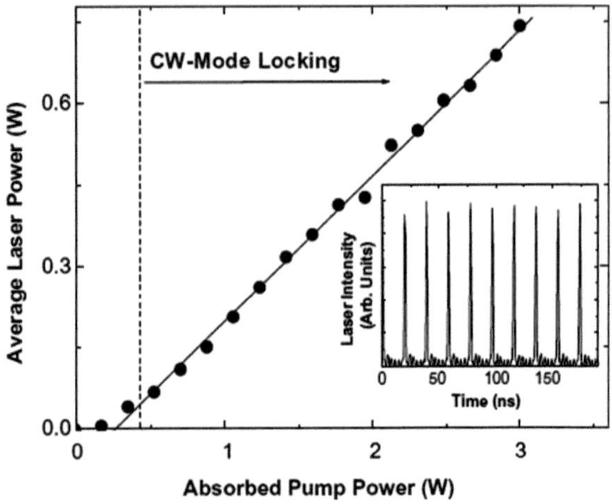

Fig. 9.23 Average output power as a function of the absorbed pump power. Cavity length was 3 m. The *inset* shows a detail of the pulse train obtained under the maximum absorbed pump power. Reproduced with permission from [168]. Copyright © 2009, Elsevier

It has been shown that subpicosecond solid-state lasers, with emission linewidth of 5.3 nm, could be achieved by using Nd^{3+}-doped mixed scandium garnets, i.e., solid solution GSAG–YSGG ceramics with a 2:1 composition [170]. The ceramics used in the experiment were 12 mm in diameter and 1 mm in thickness. An inhomogeneous broadening of Nd^{3+} fluorescence lines in a bandwidth above 5 nm was observed, which was significantly broader than that of Nd:YAG, thus leading to the subpicosecond mode-locked pulse duration. It was also found that the emission cross section of 7.8×10^{-20} cm^2 was sufficient for the efficient energy extraction. The new Nd-doped laser ceramics had a thermal conductivity of 4.7 W mK^{-1}. The GSAG–YSGG ceramics can be good candidates as laser host materials for diode-pumped subpicosecond laser systems with high efficiencies and high repetition rates.

Nd-doped garnets with intrinsically disordered structure based on CNGG could be used for short-pulse generation [171, 172]. For instance, 1061-nm SESAM mode-locked pulses with duration of 900 fs, corresponding to a FWHM of 1.8 nm, with a repetition rate of 88 MHz, have been observed by using Nd:CLNGG crystals [171]. Passive ML of diode-pumped Nd:CLNGG laser has also been demonstrated [172]. The laser was operated at a central wavelength of 1061.2 nm, exhibiting a maximum average output power of 101 mW. The mode-locked pulses had pulse duration of 2.0 ps, a spectral bandwidth of 1.2 nm, and a repetition rate of 88.4 MHz. The mode locking of the laser was enabled by a SESAM. Therefore, such Nd:CLNGG crystals could be used as gain media for ultrashort-pulse generation at 1 μm. Similarly, a passively mode-locked composite Nd:CLNGG/Nd: CNGG crystal hybrid laser has been developed to generate pulses of 534 fs with an effective FWHM of 2.2–2.3 nm [173].

Nd-doped CLNGG single crystals and ceramics have been studied and compared [174]. Single crystals of RE^{3+}-doped CLNGG (RE = Nd, Yb) were grown by using the Czochralski method with T_{melt} of ~1548 °C) from the congruent melt composition, $(Ca_{1-x}RE_x)_3Li_{0.275}Nb_zGa_{(4.725-z)}O_{12}$ with $z = 1.775 - 1.5x$, which was prepared by mixing appropriate amounts of RELNGG ($RE_3Li_{0.275}Nb_{0.275}Ga_{4.45}O_{12}$) and CLNGG ($Ca_3Li_{0.275}Nb_{1.775}Ga_{2.95}O_{12}$) powders, synthesized by using solid-state reaction method. The values of x were 0.01 of Nd and 0.03 and 0.05 for Yb, while z was adjusted according to the requirement of charge neutrality. Ceramic samples were obtained by using solid-state reaction method at 1200 °C and vacuum sintering at 1400 °C. It was found that ceramics had similar performances as compared with their single-crystal counterparts in terms of short-pulse generation. Furthermore, the additional disordering of CLNTGG offered possibility to further broaden the emission lines. In addition, SESAM mode-locked 1062 nm emission was also observed in disordered Nd-doped $Ba(Zr,Mg,Ta)O_3$ ceramics, which had a short pulse of 1.4 ps with FWHM of 1.5 nm [121].

The achievable energy of pulses in passive ML with GaAs SA has been limited by the strong two-photon absorption, due to its effect on the linear absorption at high intracavity intensities. Fortunately, the reflectivity of the thick GaAs wafers can be altered through optical interference modulation [175]. By using a GaAs wafer coated with a high reflectivity layer as the end mirror of the laser, the reflectivity can be modulated, due to the combination of Kerr effect and optical interference, which acted as an effective saturable absorber, thus resulting in ML of the laser beam. For example, a 0.5 % Nd:YAG ceramic laser under CW diode pumping at 806 nm has been reported to demonstrate mode-locked pulses of 4.1 ps at a repetition rate of 129 MHz, with an average power of 2.84 W and a slope efficiency of 48 % [175]. The Nd:YAG ceramic used in the experiment had an Nd^{3+} concentration of 0.5 at.% and a physical dimension of 3×3 mm^2 in cross section and 4 mm in length. Both sides of the ceramic were AR-coated at the laser wavelength.

9.2.4.2 Ho-Doped Ceramic Lasers

Direct pumping in the emitting manifold $^5I_7$ of Ho^{3+} exhibited efficient quasi-three-level CW laser emission in the range of 2 μm. For example, 1.91 μm pumping with a Tm:YLF laser of 1 at.% Ho:YAG ceramic laser provided a 2.09 μm laser emission. [176]. Ho:YAG ceramics, with doping concentration of 1 at.% and an average grain size of ~15 μm, were synthesized by using solid-state reaction method and sintered at 1760 °C for 20 h in vacuum, followed by thermal annealing at 1400 °C for 1 h in air. Figure 9.24 shows the laser output power versus the absorbed pump power and laser spectrum of the 1 at.% Ho:YAG ceramics. A maximum laser output power of 1.95 W was obtained at only 6.7 W of absorbed Tm pump power. A linear fit to the data yielded a slope efficiency of 44.19 %, with a threshold of about 2.3 W. Although the laser performance of the Ho:YAG ceramics was still lower than that of single crystals, it is believed that further improvement could be achieved through materials and processing optimizations.

1907 nm Tm fiber laser pumping of 1.5 at.% Ho:YAG ceramic laser created 2097 nm CW laser radiation, with a power of 21 W, a slope efficiency of 63.3 %,

Fig. 9.24 Laser output power versus the absorbed pump power (a) and laser spectrum (b) of the 1 at.% Ho:YAG ceramics. Reproduced with permission from [176]. Copyright © 2010, Elsevier

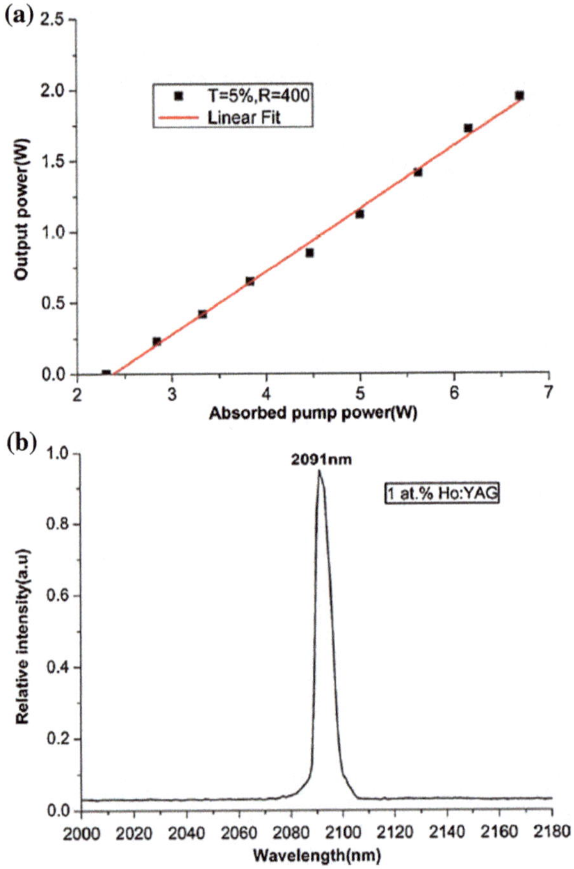

and an optical efficiency of 61 % in incident power [177]. Polycrystalline Ho:YAG ceramics with Ho^{3+} concentrations of 1.5 and 2.0 at.% were used for the experiment, which were fabricated by using solid-state reaction method and vacuum sintering. They had dimensions of 2 × 3 mm^2 in cross section and 14 mm in thickness, with both end faces being antireflection-coated at 1800–2100 nm with $T > 99.7$ %. The Ho:YAG ceramic laser showed no rollover in output power even at the highest pump power, suggesting that potential power scaling of the laser by increasing the pump power is possible.

In a separate study, lasing characteristics of the Ho:YAG ceramics, doped with 0.8, 1.0, 1.5, 2.0, and 4.0 at.% Ho^{3+}, were evaluated and compared [178]. The laser arrangement used in the experiment is illustrated schematically in Fig. 9.25. A simple two-mirror resonator was employed, which consisted of a plane pump input coupler (IC) with high reflectivity of $R > 99.8$ % at the lasing wavelength of 2050–2250 nm and high transmission of $T > 95$ % at 1850–1960 nm, as well as a concave OC of 100 mm ROC, with ~6 % transmission at 2000–2250 nm and high reflectivity at the pump wavelength that enabled a double-pass pump absorption.

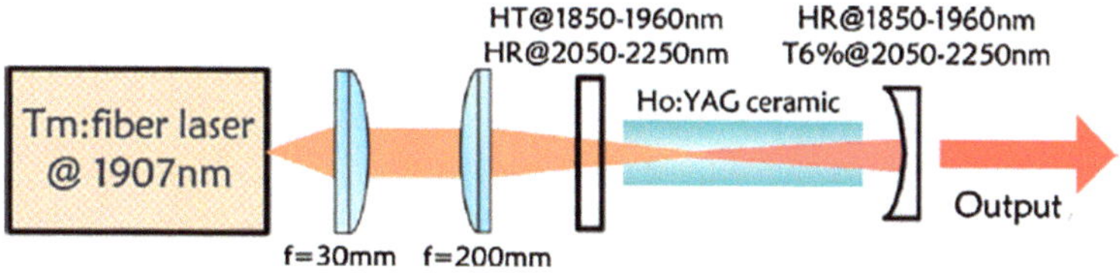

Fig. 9.25 Schematic diagram of the Ho:YAG ceramic laser measurement setup. Reproduced with permission from [178]. Copyright © 2013, Elsevier

Figure 9.26 shows output powers of the samples, as a function of incident pump power at 1907 nm [178]. With an OC of 6 % transmission, 9.4 W of output power at 2097 nm was achieved, by using the 0.8 at.% Ho:YAG ceramics at 15.1 W of pump power, which corresponded to a slope efficiency of 64.0 % with respect to incident pump power. The prospects for improvement in output power and lasing efficiency via further optimization in Ho^{3+} doping concentration were considered. The significant increase in threshold pump power and the decrease in the slope efficiency of the Ho-doped YAG ceramics with high doping concentrations indicated that energy transfer upconversion process played an important role in the Ho:YAG laser systems. Therefore, further improvement in laser performances should focus on lower Ho^{3+} doping concentrations, especially for lasers in *Q*-switched mode of operation, in which excitation density in the upper manifold is relatively high.

There have been reports on graphene passively *Q*-switched Ho:YAG lasers [179, 180]. The first graphene passively *Q*-switched Ho:YAG ceramic laser was operated at the central wavelength of 2097 nm [180]. It generated stable pulses with repetition rates of 28–64 kHz and pulse widths of 2.6–9 μs, under 1907 nm thulium

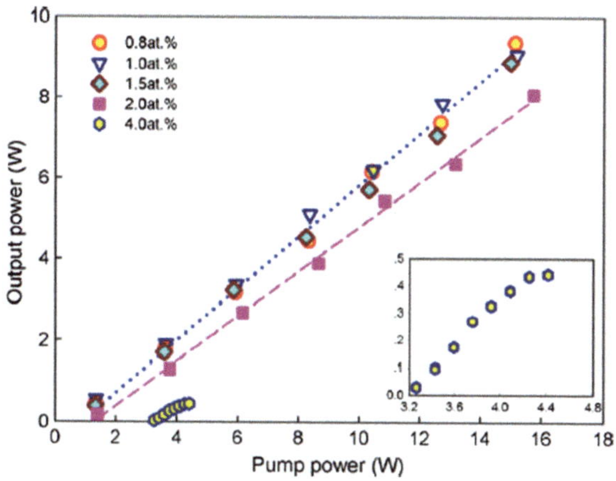

Fig. 9.26 Output powers versus incident pump power of the Ho:YAG ceramic samples doped with 0.8, 1.0, 1.5, 2.0, and 4.0 at.% Ho^{3+}, while an output coupler of 6 % transmission was used. Reproduced with permission from [178]. Copyright © 2013, Elsevier

fiber laser pumping. Under pump power of 3.27 W, maximum average power of 264 mW with 9.3 μJ pulse energy has been achieved.

A 2.09-μm in-band pumped graphene passively *Q*-switched Ho: YAG laser has been demonstrated [179]. A single layer of graphene deposited on a quartz substrate was used as the saturable absorber for the *Q*-switching operation. A minimum pulse width of 2.11 μs was obtained at an average output power of 100 mW, corresponding to a pulse repetition frequency of 57.1 kHz and a pulse energy of 1.75 μJ. The beam quality factor M^2 of the *Q*-switched laser was 1.18 and 1.22 in the horizontal and longitudinal directions, respectively. The optical-to-optical conversion efficiency of the passively *Q*-switched laser was 4.3 %, which was the highest conversion efficiency in the 2 μm wavelength. Therefore, the application of graphene could make the Ho:YAG crystal to be a potential gain medium in the 2 μm range.

Resonant 1.93 μm diode laser pumping of Ho:Y_2O_3 ceramic laser could produce CW emission at 2.12 μm with a slope efficiency of 35 % and a beam propagation factor of M^2 of 1.1 [181]. The transparent Ho^{3+}-doped Y_2O_3 ceramics used in the experiments were fabricated by using solid-state reaction method. Cold isostatic pressing (CIP), sintering, and hot isostatic pressing at 1600 °C were combined to obtain high-quality ceramics, which showed an average grain size of ~5 μm. The absorption and stimulated emission cross sections of the Ho^{3+}:Y_2O_3 ceramics at 77 and 300 K were measured. The 3 at.% Ho:Y_2O_s ceramics exhibited a scattering loss of ~0.07 cm^{-1} and an overall intracavity loss of 10 % per pass. Based on these loss data, it is expected that a laser efficiency of >70 % could be obtained, if the single-pass cavity loss could be reduced to a reasonable value, e.g., 1 %.

9.2.4.3 Er-Doped Ceramic Lasers

Because of their low quantum defect, Er-doped transparent ceramics have interesting laser emissions, due to the infrared transitions of 1.5 μm $^4I_{13/2} \rightarrow {}^4I_{15/2}$ and 3 μm $^4I_{11/2} \rightarrow {}^4I_{13/2}$, under resonant pumping into the emitting manifold. The strong absorption band of Er:YAG at 1532.2 nm can be used for InP diode laser pumping of the $^4I_{13/2}$ level. This band collects the hot-band absorption transitions of $^4I_{15/2}(2) \rightarrow {}^4I_{13/2}(1)$ at 1532.1 nm and $^4I_{15/2}(4) \rightarrow {}^4I_{13/2}(3)$ at 1532.3 nm. With quasi-continuous-wave fiber laser pumping at 1532 nm, with 10-ms pulses and 5 % duty cycle, 7 W laser emission at 1647 nm, with a slope efficiency of 56.9 % in absorbed power at room temperature, has been produced, by using a Er:YAG composite ceramic rod, which was made of three segments by using tape casting, with 0, 0.25 and 0.5 at.% Er [182].

Similarly, with 1532 nm volume Bragg grating stabilized (Er, Yb) fiber laser CW pumping, 13.8 W 1645 nm radiation with a slope efficiency of 54.5 % in incident power could be obtained by using a 1 at.% Er:YAG ceramic rod [183]. In this study, polycrystalline Er:YAG ceramics with 0.5 and 1.0 at.% Er^{3+} doping concentrations, prepared by using solid-state reaction method, were cut and polished to have a dimension of 2 × 3 mm^2 in cross section and 14 mm in length/

thickness, with both end faces being antireflection (AR)-coated at 1500–1700 nm. More importantly, the linearity of the output power as a function of pump power implied that a potential power scaling of the lasers can be expected.

CW 1617 nm emission with a slope efficiency of 51.7 % has been observed in a 0.5 at.% Er:YAG ceramic laser [184]. Efficient operation of the high-power erbium-doped polycrystalline Er:YAG ceramic laser was resonantly pumped by using a high-power 1532-nm Er,Yb fiber laser. Lasing characteristics of Er:YAG ceramics with different Er^{3+} concentrations were evaluated and compared. With an OC of 15 % transmission, the 0.5 at.% Er^{3+}-doped YAG ceramic laser generated 14 W of output power at 1617 nm for 28.8 W of incident pump power at 1532 nm.

Er:Sc_2O_3 ceramics pumped with an Er-doped fiber laser at 1535 nm in a laser resonator with volume Bragg grating sharp-edge dichroic input mirror could offer 3.3 W of CW laser emission at 1558 nm with a slope efficiency of 45 % [185]. Due to the less efficient electron–phonon interaction determined by the lower phonon quanta, the lifetime of luminescence of Er-doped Y_2O_3 ceramics is significantly longer than that of YAG and Sc_2O_3 ceramics. As a result, laser emission in the 3 μm range has been mainly achieved by using Er-doped Y_2O_3 ceramics. For example, with a spectrally narrowed laser diode CW pumping at 974 nm, laser emission dominated by the transition of $^4I_{11/2}(1) \rightarrow {}^4I_{13/2}(1)$ at 2707 nm, containing unstable admixture from other transitions, including 2715, 2725, and 2740 nm, with a slope efficiency of 15 % at room temperature, has been created by using a 2 at.% Er:Y_2O_3 ceramic [186]. Commercially available Er^{3+}:Y_2O_3 ceramics with 0.5 and 2 at.% doping concentrations were used in the experiment.

When operating at cryogenic temperatures, both the maximum extractable power and the slope efficiency could be increased. For instance, when pumping with single-emitter narrow-band (FWHM of ~0.3 nm) surface-emitting distributed-feedback laser diodes at 974 nm, the slope efficiency was increased to 27.5 %, which was very close to the quantum defect limit. In addition, with spatially combined and fiber-coupled two-dimensional arrays of such laser diodes, high power of 14 W laser emission with a slope efficiency of ~26 % could be generated by using a 2 at.% Er:Y_2O_3 ceramic [187, 188]. In this case, the laser generated stable CW and QCW radiations at 2715 nm, with unstable contributions from the emissions at 2707 and 2740 nm.

9.2.4.4 Tm-Doped Ceramic Lasers

Laser emission in the 2 μm range of Tm^{3+} in YAG ceramics was investigated under Ti:sapphire or diode laser pumping into the manifold $^3H_4$, using the downconversion process ($^3H_4$, $^3H_6$) → ($^3F_4$, $^3F_4$). In order to produce the efficient energy transfer downconversion, Tm concentrations of 6 at.% were typically used. Although the quantum defect inherent in this excitation process together with the efficiency of downconversion and with the quasi-three-level scheme of laser emission limits the performance of the laser, 2012 or 2015 nm CW emission slope efficiencies in the range 20–42 % have been reported [189–191].

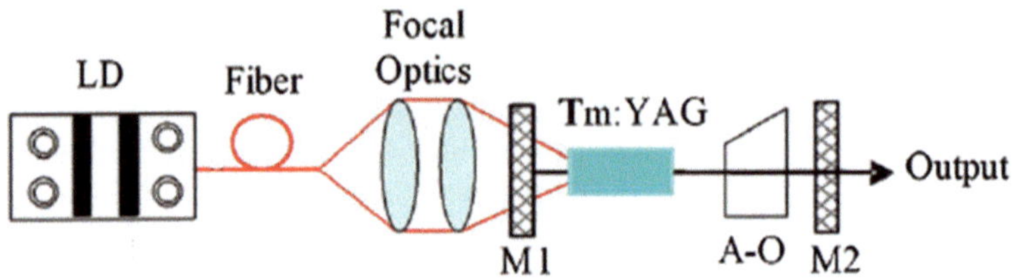

Fig. 9.27 Experimental setup of the Tm:YAG ceramic slab laser. Reproduced with permission from [192]. Copyright © 2013, Elsevier

A highly efficient continuous-wave (CW) and actively Q-switched laser was realized by using Tm:YAG ceramics [192]. Figure 9.27 shows experimental setup of the ceramic laser. Pumping with fiber-coupled laser diode at 792 nm, a maximum CW output power of 5.21 W with improved beam quality was achieved with a slope efficiency of 46.1 %. A significantly short-pulse width of 107 ns was observed, at a pulse repetition frequency (PRF) of 0.5 kHz, thus leading to the largest pulse energy of 1.44 mJ and the highest peak power of 13.42 kW. The results of the CW output powers of the Tm:YAG ceramic slab with different OCs, T = 2, 5 and 7 %, as a function of the absorbed pump power, are shown in Fig. 9.28a. The laser spectrum of the Tm:YAG ceramic in such condition is illustrated in Fig. 9.28b, with the central laser wavelength at 2019.5 nm, together with a full width at half maximum (FWHM) of about 3 nm and an M^2 factor of about 1.6, demonstrating the potential applications in the 2 μm Q-switched laser operation.

With increased pump power of 782-nm diode laser of >50 W, dual-wavelength CW emission at 2016 and 2006 nm was obtained, with a slope efficiency of >36 %, and Q-switching with an acousto-optic modulator produced pulses in the repetition range of 500 Hz–5 kHz, at 500 Hz duration of pulses of 69 ns, having an energy level of 20.4 mJ [193]. The Tm:YAG ceramics had a Tm doping concentration of 6 at.%. Furthermore, it was found that lifetime of the Tm:YAG ceramics was 10.5 ms, which was longer than that of its single-crystal counterparts by 5 %. For laser materials, longer lifetime means higher energy storage capability, while longer lifetime is benefit to generate high pulse energy at low repetition rates.

When compared with single crystals, transparent ceramics have unique advantages for efficient laser operation in the 2 μm wavelength range [193]. The scattering loss in ceramic has been acknowledged to the main limiting factor of highly efficient laser operation [88]. As scattering sources in ceramics, the sizes of pores and color centers are generally at micrometer scale. According to Rayleigh scattering law, the scattering losses are inversely proportional to four orders of the laser wavelength. As a result, the losses corresponding to these scattering sources at the 2 μm wavelength range are 1/16 times those at the 1 μm wavelength range. In other words, the scattering problem of ceramics is effectively avoided, when they are used for laser applications in the 2 μm wavelength range [194]. In contrast, no scattering is caused by the grain boundaries of ceramics for both the 1 and 2 μm wavelength ranges, because they are at nanometer scales [195–198]. In this case, the grain

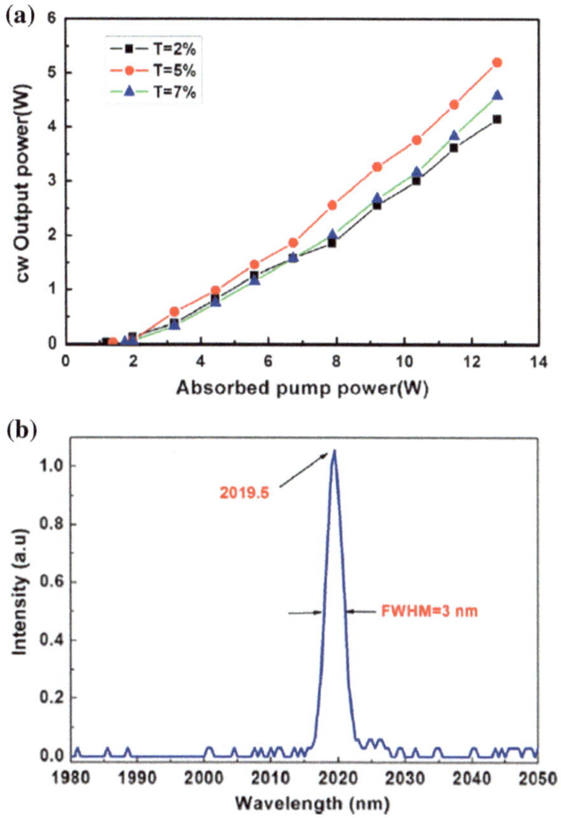

Fig. 9.28 **a** CW output powers with different output couplers, i.e., $T = 2$, 5, and 7 %, as a function of the absorbed pump power. **b** Laser spectrum of the Tm:YAG ceramics. Reproduced with permission from [192]. Copyright © 2013, Elsevier

boundaries can cause scattering of UV light. UV light could come from the upconversion process in 2 μm solid-state lasers, in which the population is excited to upper levels, emitting photon at the UV band. The upconversion process consumes the population inversion and degrades the laser efficiency. When the UV light is scattered, the upconversion rate could be decreased, thus leading to potential increase in high efficiencies of the 2-μm ceramic lasers.

Transparent polycrystalline ceramics, $(Y_{3-x}Tm_x)Al_5O_{12}$, with $x = 0$, 0.18, 0.9, 1.5, and 3.0, were fabricated by a solid-state reaction method with high-purity micrometer-sized precursor powders [199]. The Tm-doped ceramics made in this way had an almost perfect pore-free microstructure and high transparency. The average grain size and grain boundary width were 15 μm and 1 nm, respectively. When a ceramic slab of $(Y_{2.82}Tm_{0.18})Al_5O_{12}$ doped with 6 at.% Tm, with a dimension of $5 \times 5 \times 3.5$ mm^3, was end-pumped by a Ti:sapphire laser at 785 nm, a maximum output power of 725 mW was obtained, with a slope efficiency of 36.2 % and an optical-to-optical efficiency of 19.9 % at 2012 nm.

Similarly, resonant pumping of 4 and 6 at.% Tm:YAG ceramic with a 1617 nm Er:YAG laser could produce 2015 nm CW laser emission [200]. With an OC of

10 % transmission, a maximum output power of 7.3 W was obtained at 2015 nm under 12.8 W of incident pump power, corresponding to a slope efficiency of 62.3 %, for the 4 at.% Tm:YAG ceramics. According to estimation, scattering loss of the ceramic sample was <0.36 % cm^{-1}. It is expected that laser performances, such as the maximum output power and lasing efficiency, can be further improved, by using Tm^{3+}-doped Tm:YAG ceramics with lower doping concentrations, in which more uniform heat distribution and reduced upconversion could be more effectively achieved.

Diode laser pumping at 796 and 811 nm of 2 at.% Tm:Lu_2O_3 fine-grained ceramic laser, with an average grain size of only 550 nm, has produced 2066 nm CW emission, with a slope efficiency of ~40 % [201]. The average size of the ceramics was in the range of 540–560 nm. According to the absorption spectrum measured in the near-IR region, it was suggested that Tm:Lu_2O_3 ceramics could be pumped by commercially available laser diodes at wavelengths of 796 and 811 nm. The high transmission region of the ceramics in the mid-IR region extended to 7 μm. The luminescence spectrum in the region of 1.75–2.2 mm upon laser diode pumping indicated that there were strong lines peaked at wavelengths of 1942, 1965, and 2066 nm. According to ellipsometric measurements, refractive indices of the transparent Tm:Lu_2O_3 ceramics were $n \approx 1.92$ at the pump wavelength of 796 nm and $n \approx 1.94$ at the laser wavelength of 2066 nm.

9.2.4.5 Yb-Doped Ceramic Lasers

It is widely accepted that the room temperature spectroscopic properties of Yb-doped ceramic materials could offer nearly pure four-level emission on the transition of $^2F_{5/2}(1) \rightarrow {}^2F_{7/2}(4)$ and quasi-three-level laser emission on the much more intense transition of $^2F_{5/2}(1) \rightarrow {}^2F_{7/2}(3)$. The competition between the laser emissions on the two transitions can be influenced by ambient temperature and the temperature rise inside the pumped laser materials, which is determined by the generation of heat. The heat generation is related to the absorbed pump power and the transmission of the output mirror that could influence the intracavity intensity. At cryogenic temperatures, the coefficient $f_t(3)$ is significantly reduced, so that the laser can be generated in a four-level scheme on the transition of $^2F_{5/2}(1) \rightarrow {}^2F_{7/2}(3)$.

Yb-doped ceramic lasers have a problem of low absorption cross section for the broad absorption band of $^2F_{7/2}(1) \rightarrow {}^2F_{5/2}(2)$ near 940 nm, which is used usually for diode laser pumping. For ordered materials, the absorption line of $^2F_{7/2}(1) \rightarrow {}^2F_{5/2}(1)$, with a larger peak absorption cross section that could lead to low quantum defect laser emission, is very narrow, especially at low temperatures, so that diode laser pumping is difficult. However, the recent development of stabilized diode lasers has enabled the pumping in this transition in various Yb-doped ceramic lasers at room temperature. As a consequence, the small absorption cross sections require high concentrations of Yb or longer path of pump radiation inside the laser materials. According to spectroscopic studies, this absorption line can be broadened inhomogeneously by using disordered laser materials, although the peak absorption cross section could be reduced.

Due to the simple electronic structure of Yb^{3+}, Yb^{3+}-doped ceramic lasers have no problem of de-excitation processes like the concentration-dependent self-quenching by downconversion or upconversion cross-relaxation. However, a reduction in emission lifetime and emission quantum efficiency has been observed. In addition, cooperative processes could occur, due to the interaction between two excited Yb^{3+} ions, when the concentrations of Yb^{3+} are too high. The parasitic de-excitation processes are usually accompanied by the generation of heat that is up to twice as much as that due to the quantum defect [202]. Therefore, it is important to have a good matching between the pump and mode volume in the laser materials.

With the moderate emission cross section, long lifetime of the emission, and relatively broad emission lines, Yb-doped ceramics could have a wide range of laser applications, including CW, ultrashort pulses, and tunable emission. Yb-doped ceramic lasers, with a broad range of sizes and powers, from microchip lasers to high-power lasers, have been developed. Specifically, the high-power lasers and high-energy peak power amplifiers are of interest in nuclear physics for nuclear fusion applications.

Free-Generation Yb-Doped Ceramic Lasers

In this part, Yb-doped garnet and sesquioxide ceramic lasers are described first, at both room temperature and cryogenic temperatures. After that, Yb-doped fluoride ceramic lasers will be briefly discussed.

The first room temperature CW Yb:YAG ceramic laser was made of a less doped ceramics with Yb concentration of 1 at.% and thickness of 1.5 mm [203]. Due to the relatively low optical quality and the nonoptimized laser system, a relatively low slope efficiency of 26 % was achieved, under 940 nm diode laser pumping, with transmission T of the OC of ~5 % at 1030 nm laser wavelength, as shown in Fig. 9.29.

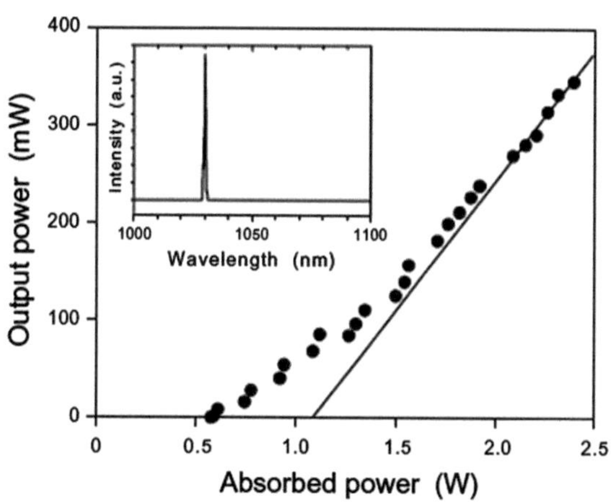

Fig. 9.29 Input–output dependence of the Yb^{3+}:YAG (C_{Yb} = 1 at.%) ceramic laser at 1030 nm. The *inset* shows the corresponding lasing spectrum. Reproduced with permission from [203]. Copyright © 2003, John Wiley & Sons

By using Yb:YAG ceramics with improved optical quality and higher Yb concentration, the lasing performances could be improved, so that a higher absorption efficiency and a reduction in thickness of the material were observed [204]. At $T = 5$ %, a low power pumping could facilitate laser emission at 1030 nm. At the same time, an increased pumping competition from the additional emission at 1049 nm was encountered, which became dominant at high powers. For $T = 10$ and 20 %, the 1030 nm emission remained dominant. Slope efficiencies of 79 % at 1030 nm and 67 % at 1049 nm were achieved for 1-mm-thick Yb:YAG ceramic plate with a doping concentration of $C_{Yb} = 9.8$ at.% under CW laser diode pumping. The results indicated that the Yb:YAG ceramic laser materials could be used for high-power solid-state lasers.

Further improvement in performances of such end-pumped Yb:YAG ceramics has been reported, with tens of watts and a slope efficiency of as high as >70 % [205–209]. An example is the use of diode laser at 968 nm [207]. In that study, the transparent Yb:YAG ceramics were fabricated by using solid-state reaction method at 1780 °C in vacuum, with commercial Al_2O_3, Y_2O_3 and Yb_2O_3 powders as raw materials and MgO and tetraethoxysilane (TEOS) as sintering aids. A high-power CW polycrystalline Yb:YAG ceramic laser was pumped by the fiber-pigtailed laser diode with a 400-μm fiber core. The Yb:YAG ceramic laser performance was evaluated for different concentrations of Yb^{3+} ions in the ceramics, by using a conventional end-pumped laser cavity consisting of two flat mirrors with OCs of different transmissions. A CW laser output of 40 W average power with M^2 factor of 5.8 was achieved, under 120 W incident pump power, by using the 5 mol% Yb-doped ceramics.

With a front face-pumped compact active-mirror laser (CAMIL) structure, an Yb:YAG/YAG composite ceramic disk laser with pumping wavelength at 970 run has been reported [210]. The laser was operated in both CW and *Q*-switching modes. Under CW operation, laser output power of 1.05 W with 2 % transmission OC was achieved at the wavelength of 1031 nm. *Q*-switched laser output was obtained by using an acousto-optic *Q*-switch, with repetition rates of 1–30 kHz and pulse widths of 166–700 ns.

The reabsorption that limits the 1030 nm emission of Yb:YAG lasers can be eliminated by reducing thickness of the ceramics, but the efficiency of the conventional end-pumping could be reduced. Two approaches have been used to address this problem. One approach is the use of side pumping of a thin laser ceramics, with a diameter sufficiently large to ensure efficient pump absorption, while the other is to use a thin disk based on face pumping, with repeated passes of the pump radiation inside the laser material through multiple reflection. In both of the approaches, the end mirror is coated on the laser material and the removal of heat or cooling is realized through this face.

The heat flow in this case is along the laser axis, and no thermal lensing is encountered. Increased power has been achieved by 940 nm diode laser side pumping of a 10 % Yb:YAG thin ceramic disk, with a diameter of 3.7 mm and a thickness of 0.2 mm, cladded by undoped YAG ceramics that acted as a pump light pipe [211–213]. The increased powder was attributed to the increased pump absorption efficiency and the simultaneous reduction of reabsorption, leading to 414 W CW with a slope efficiency of 47 % and 520 W QCW 1030 nm laser emission (10 ms, 10 Hz) with a slope efficiency of 52 %, corresponding to output power densities of 3.9 kW cm^{-2} and 0.19 MW cm^{-3}, respectively.

The thin-disk strategy has been successfully applied to various crystalline Yb: YAG lasers [214–216]. Principal ideas of thin-disk laser design and the advantages for operating different laser materials have been demonstrated [214]. The potential of the thin-disk laser has been supported by the results for CW and *Q*-switched operation, as well as the amplifications of short (nanosecond) and ultrashort (picosecond and femtosecond) pulses. Scaling laws for this laser design showed that the power limit for CW operation was far above 40 kW by using one single disk, with an energy limit of higher than 3 J from one disk in pulsed operation. In addition, the thin-disk laser concept could be applied to optically pumped semiconductor structures. When pumping directly into the quantum wells, the energy defect between the pump photon and the laser photon could be <5 %, thus reducing the amount of the waste heat generated inside the semiconductor structure. First results demonstrate the potential of this new laser concept. The thin-disk laser technology could find various potential industrial applications.

The aperture of these lasers is 1–2 cm^2, with scalability that can be limited by deleterious ASE and heating effect. It has been shown that bonding an undoped cap to the Yb-doped disk could reduce ASE [63]. A simple model was used to describe the active material. It was observed that the reduction of the loss was important for the power scaling, because this loss, together with the heat generation, determined the maximum power that could be achieved at a given geometry. Usually, the loss has been assumed to be sufficiently low. With this model, the power enhancement of a composite thin-disk laser made of an undoped layer bonded over a thin active layer to reduce ASE losses could be estimated analytically. Scaling laws for the parameters of a disk laser were thus suggested for the cases with and without an anti-ASE cap. The predicted maximum achievable powers for a given laser material were in a good agreement with the published experimental data. The use of the anti-ASE cap resulted in an increase in the maximum output power, which was proportional to the square of the logarithm of the round-trip loss.

By using a 9 at.% Yb:YAG ceramic 0.2-mm thin disk that was bonded with a 1-mm-thick undoped YAG ceramic cap, stable 1030 nm multi-mode laser emission, with a high power of 6.5 kW and a slope efficiency of 57 %, could be achieved in single thin-disk configuration [217]. An output density of 2.5 kW cm^{-2} was observed under a pump density of 5 kW cm^{-2}. Analysis results suggested that the laser was operated with inhomogeneous gain saturation, which was attributed to the enhanced spatial hole-burning effect when the gain element was adjacent to a mirror. The pump threshold and output intensities were independent of pump spot size, which validated the area scaling. Thermal lensing contributions, including thermal expansion-induced disk flexure, pump-edge-induced temperature profile, and a strong thermal imprint of the cooling nozzle due to the direct jet impingement on the HR-coated side, were observed. Weak absorption of the 1030 nm intracavity intensity in the undoped cap and/or the antireflection (AR) coating led to additional

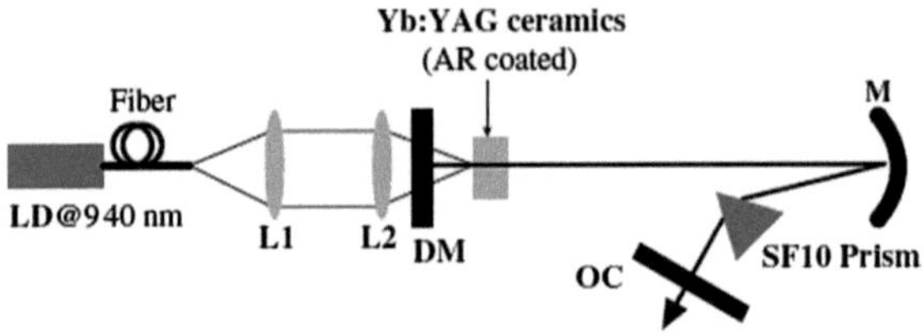

Fig. 9.30 Experimental setup for the tunable Yb:YAG ceramic laser. *LD* fiber-coupled diode laser, *L1*, *L2* focusing lenses, *DM* flat dichroic mirror, *M* a concave mirror (ROC = 250 mm), and *OC* output coupler. Reproduced with permission from [209]. Copyright © 2008, Elsevier

heating, which limited the extracted power intensity. Therefore, it was suggested that the Yb:YAG ceramics can be scaled to higher powers, by using optimized thin-disk elements, together with improved disk thermal management strategies.

Broadly tunable laser emission with wavelength range that is dependent on the transmission of the output mirror has been reported by using 9.8 at.% Yb:YAG ceramics under diode laser pumping at 940 nm [218]. For $T = 1$ %, the laser was continuously tunable between 992.5 nm and 1098 nm, which could be extended to a range of 992.5–1110.83 nm for $T = 0.1$ %, although a decrease in the maximum extractable power was encountered. With these wide tuning ranges, it is possible to use Yb:YAG ceramics to generate ultrashort pulses.

A similar high-power efficient ceramic Yb:YAG laser was demonstrated at a room temperature of 20 °C, with an Yb concentration of 9.8 at.%, a gain medium of 1 mm, a pumping power of 13.8 W, an OC of $T = 10$ %, and a cavity length of 20 mm [209]. Experimental setup for the Yb:YAG ceramic laser is shown in Fig. 9.30. Tunability of the 1.5 W Yb:YAG ceramic laser was characterized by using a 1 % OC and a prism. Figure 9.31 shows dependences of the output power versus the absorbed pump power, which were determined by considering the absorption efficiency difference between nonlasing and lasing cases, i.e., absorption efficiency remains constant with increasing pump intensity. A 6.8 W CW output power was achieved, with a slope efficiency of 72 %. The ceramic Yb:YAG laser exhibited a continuous tunability within a spectral range of 63.5 nm, from 1020.1 to 1083.6 nm, for $T = 1$ % at a maximum output power of 1.6 W. Such a tunable laser could be very attractive for femtosecond applications and suitable for a laser amplifier, due to the maximum single-pass gain of 1.2 at a gain medium thickness of 1 mm.

According to the performances achieved by using the Yb:YAG ceramics with C_{Yb} of up to 10 at.%, the power scaling is limited mainly by the capability of dissipating the heat that is generated in the pumped laser materials. It is well known that in the face-cooled Yb laser materials, the thinner the laser elements, the more effective the heat dissipation will be. Therefore, in order to optimize the pump absorption and power scaling, it is necessary to use Yb:YAG ceramics with high C_{Yb} and high excitation densities. Laser emissions at 1030 nm of 20 % Yb:YAG ceramics and single crystals have been investigated as a comparison, with a

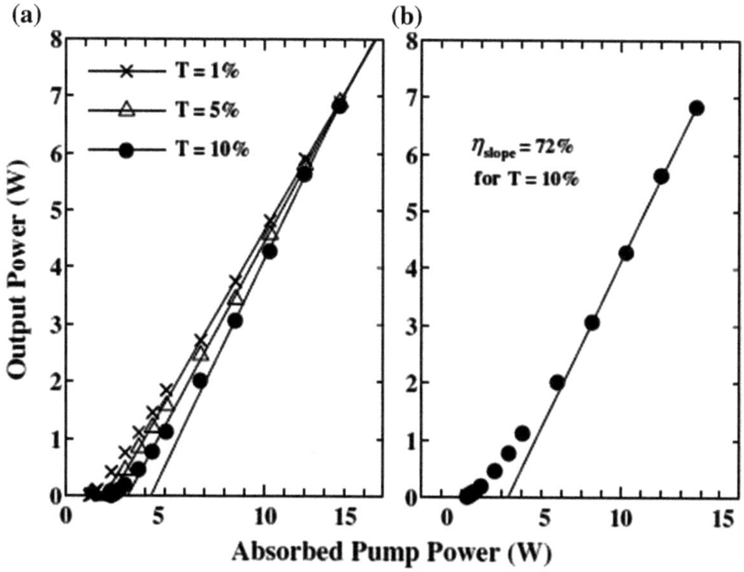

Fig. 9.31 Input–output power dependences of the efficient Yb:YAG (C_{Yb} = 9.8 at.%) ceramic laser: **a** with output couplers of T = 1, 5, and 10 % and **b** with only T = 10 %. Reproduced with permission from [209]. Copyright © 2008, Elsevier

minilaser double-pump-pass configuration, consisting of 1-mm plates mounted on a copper heat sink, with diode laser pumping at 940 nm [219]. A slope efficiency of as high as 52 % and an optical-to-optical efficiency of >48 % have been obtained. It was found that the ceramic lasers exhibited higher maximum extractable power and linearity with the absorbed 940 nm pump power than the single-crystal counterparts. In other words, in case of heavy doping, Yb:YAG ceramics have better laser performance than their single-crystal counterparts, which is one of the advantages of ceramic laser technologies, as mentioned previously.

Similar comparative study has been carried out on Yb:YAG ceramics and crystals with higher doping concentrations C_{Yb}, with absorbed pump power at 940 nm in the range 5–5.6 W, by using a similar two-pump-pass laser configuration, in which a spot of ~170 μm was focused on the laser materials [220]. At a concentration of 10 at.%, single crystals exhibited a slightly higher laser performance. With increase in Yb concentration, the performances of both types of materials were decreased, with the effect on single crystals being stronger. For instance, at 20 at.% C_{Yb}, ceramics demonstrated higher performances than single crystals.

Laser parameters can be influenced by various processes. Some processes, which have strong effect on the slope efficiency, such as optical losses, e.g., reabsorption of the laser emission and the pump-mode volume superposition, should be specifically takin into account. A study has been conducted to compare the performances of a 2-mm-thick 10 at.% Yb:YAG and a 1-mm-thick 20 at.% Yb:YAG ceramics, with a higher pumping power at 940 nm, combined with the optimization of the output mirror, in a single-pass laser configuration [221]. For the 1030 nm laser emission, with $T = 6$, 12, and 18.5 %, the laser performances of the 10 and 20 at.% Yb:YAG ceramics were almost the same, with the output powers, at maximum pumping of 21 W, corresponding to ~13 W absorbed power, to be 8.7 and 8.2 W, respectively. For the 1050 nm laser emission, with $T = 1.5$ %, the 10 at.% Yb:YAG ceramics showed slightly higher performances, with the maximum output powers of about 8 W.

There have been qualitative models established to explain the drop in performance observed in the Yb:YAG ceramics with high concentrations of Yb at high excitation intensities. At the same time, it has been accepted that high doping concentration Yb:YAG should have high performances. No quantitative explanation of the experimental observations has been available until now. As a consequence, in reference, the drop in performance has once been attributed to the energy transfer to impurities, because their concentration was increased with increasing concentration of Yb [220]. However, it is obvious that such energy transfer processes could have influence on the lifetime of emission without the presence of laser emission, i.e., the emission quantum efficiency, which in turn should increase the emission threshold, rather than directly influence the slope efficiency of the CW lasers. The better performance of ceramics with high doping concentrations was attributed to the increased phonon transmission, due to the segregation of the Yb^{3+} ions at grain boundaries, but the fact is that the segregation of Yb at grain boundaries in YAG ceramics is very low [222]. In addition, heavily doped Yb:YAG ceramics exhibited strong photoconductivity, which increased significantly with increasing absorbed pump power.

Qualitative models involving upconversion of excitation to the charge transfer band and then to the conduction band of the materials, including the involvement of transient or permanent Yb^{2+} centers, have been reported [223, 224], whereas a quantitative explanation of the observed effects has not been available. According to the thermal effects of the Yb:YAG ceramics doped with 10 and 20 at.% Yb at different intracavity intensities, the presence of an additional nonlinear effect at high Yb concentrations and high excitations has been confirmed [221].

In this respect, $Lu_3Al_5O_{12}$ (LuAG) has been acknowledged to be a promising host for Yb, because its thermal conductivity is not strongly affected by the doping of Yb [225–227]. For example, diode laser pumping at 940 nm of a 5 at.% Yb: LuAG ceramic laser exhibited 1030 nm CW laser emission, with a slope efficiency of 63 %, which was influenced by the residual absorption [227]. The Yb:LuAG ceramics with high optical quality were prepared by using solid-state reactive vacuum sintering technique. With uncoated Yb:LuAG ceramic sample, a maximum output power of 7 W could be achieved under 13 W absorbed pump power. Because Yb:LuAG has higher thermal conductivity than Yb:YAG at high doping concentrations, Yb:LuAG ceramics could be an attractive laser gain medium for high-power solid-state laser applications.

According to the previous discussion, Yb:YAG lasers with thin-disk multi-pass configurations can generate very high CW powers, which is only limited by the heat generation of the laser materials. Furthermore, the multi-pass techniques require complicated advanced optical systems. Another disadvantage of these lasers is the requirement of a high-reflection coating on the disk, which can reduce heat transfer to the coolant. Fortunately, most of these problems of Yb-doped ceramic lasers encountered at room temperature can be readily addressed once they are operated at cryogenic temperatures.

Nowadays, sophisticated cooling techniques have been available to ensure the laser materials to be operated at cryogenic temperatures, in which the costs can be well compensated by the advantages that can be gained in laser performances. Spectroscopic advantages include increased emission cross section and reduced population of the terminal level that transforms the 1030 nm emission of the strong transition of $^2F_{5/2}(1) \rightarrow {}^2F_{7/2}(3)$ into a four-level scheme, while material properties are greatly improved, such as increased thermal conductivity, reduced thermal expansion, and decreased thermo-optic coefficient, which have been systematically discussed, especially for Yb:YAG ceramic lasers [80–82, 228]. Because of these advantages, thicker Yb:YAG plates can be used and the range of tailorable Yb concentration to optimize the laser performance is widened. For example, at cryogenic temperatures, Yb:YAG ceramics can be applied with the total-reflection active-mirror (TRAM) technique, which uses the total reflection on bottom of the laser material plate, instead of the use of a high-reflection coating [81].

An additional advantage is the possibility to separate the path of pump radiation from that of the laser beam. By using a composite laser material, consisting of a 180- or 400-μm-thick 9.8 at.% Yb:YAG ceramic plate, with 40×40 mm^2 bottom surface bonded to an undoped YAG ceramic and with the bottom surface of the system be cooled by using direct contact with liquid nitrogen, an output power of 273 W has been achieved, under 940 nm diode laser pumping, with a slope efficiency of 72 % and an optical–optical efficiency of 65 %. Another example of TRAM laser was constructed by using a composite ceramic material with three active Yb:YAG plates, which were assembled in a zigzag configuration on an undoped YAG ceramic [80]. It has been expected that with this approach, Yb:YAG ceramic lasers can be scaled up to 10 kW.

Similarly to LuAG, due to their high thermal conductivities, all the sesquioxides are potential host materials for laser applications. The only problem is that this advantage is weakened as the doping concentrations of laser-active ions are increased. Room temperature emission of Yb^{3+} can be achieved by using transparent ceramics of all cubic sesquioxides, i.e., Y_2O_3, Lu_2O_3, and Sc_2O_3. These ceramic lasers can be pumped with diode lasers at the broad $^2F_{7/2}(1) \rightarrow {}^2F_{5/2}(2)$ absorption band in the region 937–950 nm or at the more intense but significantly sharper $^2F_{7/2}(1) \rightarrow {}^2F_{5/2}(2)$ absorption band in the region 976–980 nm. Similar to garnets, sesquioxides can also allow concurrent room temperature emission in the region 1030–1040 nm due to the transition of $^2F_{5/2}(1) \rightarrow {}^2F_{7/2}(3)$ in a quasi-three-level scheme or in the region 1077–1080 nm due to the transition of $^2F_{5/2}(1) \rightarrow {}^2F_{7/2}(4)$ in a four-level scheme. The emission wavelength can be selected by controlling the reabsorption by controlling the size of the laser material and the concentration of Yb. Additional strategy is the use of selective optical components.

The performances of Yb-doped Y_2O_3 [8, 229–233], Lu_2O_3 [234–238], and Sc_2O_3 [6, 234, 235, 239–241] ceramic lasers have been continuously improved, due to the constant improvement in laser materials and laser designs. Slope efficiencies of >80 % and an optical–optical efficiencies of >70 % for the transition of $^2F_{5/2}(1) \rightarrow {}^2F_{7/2}(4)$ have been widely reported. The emission lines are at the nanometer scale, showing potential applications for short-pulse emission. By using solid solution $(Y_{0.9}La_{0.1})_2O_3$ ceramics, efficient emission at 1080 nm has been achieved under diode pumping at 976 nm. Smooth tunable emission in the range of 1018–1086 nm has been obtained under pumping at 940 nm, with broadband lasing of up to 30 nm [242]. The 1030 nm emission of 10 at.% Yb-doped Y_2O_3, Lu_2O_3, and Sc_2O_3 ceramics at cryogenic temperatures has been studied in the free-generation regime, under pulsed diode laser pumping at 940 nm [243]. It was found that the improvement in the 1030 nm emission was attributed to the reduction in reabsorption. In the temperature range of 80–320 K, the output energy could be increased by 3–6 times, by using 600-μs pump pulses with energies of up to 1 J at a repetition rate of 10 Hz, which has been attributed to the lowered emission threshold and the enhanced slope efficiency.

Besides oxides, fluoride ceramics doped with Yb have also been studied for laser emissions. In this case, material processing was found to play an important role. For example, laser emission of diode laser-pumped 1 at.% Yb:CaF_2 ceramics derived from nanopowders synthesized by using a wet chemical method showed high residual optical losses [244]. In contrast, a 3 at.% Yb:CaF_2 ceramics fabricated by using hot pressing of Bridgman crystalline material exhibited promising laser performance, with a slope efficiency ~30 % in CW and ~37 % in QCW regimes at room temperature [245]. A ternary system, 65 %CaF_2–30 %SrF_2–5 %YbF_3, prepared by using hot pressing, demonstrated even higher laser performance at 1052 nm, with a slope efficiency of 45 %, under 972 nm diode laser pumping, which was close to that of single crystals [246].

Q-Switched Emission of Yb-Doped Ceramic Lasers

Electro-optic *Q*-switching of Yb:YAG ceramics at cryogenic temperatures could exhibit high laser performances. Yb:YAG ceramic lasers in passive *Q*-switching with Cr^{4+}:YAG SAs could generate very short pulses. For example, by using separate ceramic Yb:YAG active material and ceramic Cr^{4+}:YAG SA with initial transmission of 80 %, *Q*-switched pulses of as short as 335 ps with peak power of >150 kW at a repetition rate of 5 kHz have been generated, with a high-quality beam that was close to the diffraction limit [247]. The absorption and emission spectra of the Yb:YAG ceramics were identical to those of the Yb:YAG single crystals. Two main emission peaks were centered at 1030 and 1049 nm, with an effective peak emission cross section of 2.2×10^{-20} cm^2 at 1030 nm, which was about six times that at 1049 nm. In the experiment for the passively *Q*-switched Yb:YAG ceramic miniature laser, a Cr^{4+}:YAG ceramic was used as the saturable absorber.

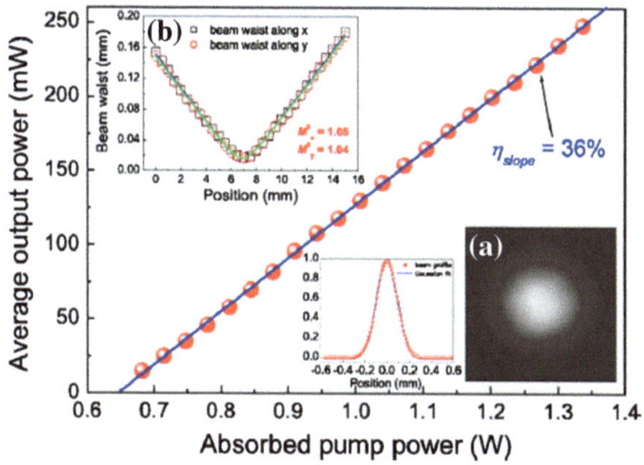

Fig. 9.32 Average output power as a function of the absorbed pump power for the passively *Q*-switched Yb:YAG/Cr^{4+}:YAG ceramic miniature laser. The *inset* **a** shows the output beam profile and transverse beam profile and **b** shows the measured beam quality factors. Reproduced with permission from [247]. Copyright © 2007, American Institute of Physics

Average output power as a function of the absorbed pump power and output beam profile are shown in Fig. 9.32 [247]. The absorbed pump power threshold was about 0.66 W, owing to the low initial transmission of Cr^{4+}:YAG and the high transmission of the OC. The average output power increased nearly linearly with the absorbed pump power, at a slope efficiency of about 36 %. Maximum average output power of 250 mW was achieved at the absorbed pump power of 1.34 W, corresponding to an optical-to-optical efficiency of 19 %. The transverse output beam profile is shown in the inset (a) of Fig. 9.32. The beam profile was close to fundamental transverse electromagnetic mode. Measured position-dependent beam radii near the focus are shown in the inset (b) of Fig. 9.32. Near-diffraction-limited output beam qualities, with M_x^2 of 1.05 and M_y^2 of 1.04, were achieved. The output beam diameter near the output mirror was about 100 μm. Furthermore, stable single-longitudinal-mode oscillation could be obtained by increasing the pump beam diameter incident on the laser ceramics at higher pump power.

Similarly, by using composite of 9.8 at.% Yb:YAG/0.1 at.% Cr^{4+}:YAG ceramics with 64 % initial SA transmission, pulse energies of 125 μJ in 1.2 ns duration could be generated, with peak power of 105 kW at a repetition rate of 3.8 kHz for 2.55 W absorbed 940 nm power [248]. The composite Yb:YAG/Cr:YAG ceramics were fabricated by using vacuum sintering technique and nanocrystalline synthetic technology. The diameter of the composite Yb:YAG/Cr:YAG ceramics was 8 mm, with the thicknesses of Yb:YAG and Cr:YAG to be 8 and 4 mm, respectively. Laser performance of the composite Yb:YAG/Cr:YAG ceramic laser was studied by using the plane-concave cavity method. The threshold absorbed pump power was 0.9 W, which was attributed to the low initial transmission of Cr^{4+}:YAG at 1030 nm. Average output power increased almost linearly with absorbed pump power above the absorbed pump power threshold, at a slope efficiency of 27 %. A near-diffraction-limited beam quality with M^2 of <1.35 was obtained. It is believed that the laser performance can be further improved by using high transmission OC or optimizing the ratio of thickness of Yb:YAG and Cr:YAG in the composite ceramics.

If the transmission of SA was increased to 70 %, combined with an OC with $T = 50$ %, pulses of 172 μJ with duration of 237 ps and peak power of 0.72 MW could be realized, at a repetition rate of 3.5 kHz for an absorbed power of 3.53 W. Moreover, because no sign of saturation was observed at the maximum power, potential scaling of such materials is present [153]. Specifically, the fluorescence lifetime of Cr,Yb:YAG was shortened due to the introduction of Cr, and there is a severe fluorescence quenching effect with increasing concentration of Cr. As a result, although laser diode-pumped Cr,Yb:YAG self-*Q*-switched microchip lasers with subnanosecond pulse width and peak power of several tens of kW have been realized, the laser pulse energy cannot be further scaled up by simply increasing the concentration of Cr, due to the strong fluorescence quenching effect. Stable multi-longitudinal-mode laser oscillation associated with the antiphase dynamics from the Cr,Yb:YAG microchip lasers with near-diffraction-limited beam quality still makes the Cr,Yb:YAG self-*Q*-switched lasers potential candidates for material processings and related applications. Composite Yb:YAG/Cr:YAG ceramics have showed very promising future for generating high pulse energy and high peak powers.

Mode-Locked Emission of Yb-Doped Ceramic Lasers

Mode-locked Yb-doped ceramic lasers discussed in the part include garnets and sesquioxides. Mode lockings of Yb:YAG ceramic lasers under CW 940 or 968 nm diode laser pumping by using SESAMs have been reported [249, 250]. In this case, the duration of the pulse was limited to several hundreds of femtoseconds, due to the relatively narrow emission lines. The Yb concentration was usually about 10 at %, while the emission wavelength, either 1030 or 1048 nm, was selected according to the thickness of the laser materials or the transmission of the output mirrors.

Dual-wavelength emission, with average powers of up to several watts and slope efficiencies of up to 76 %, has also reported [251]. The dual wavelength was realized in an independently mode-locked Yb:YAG ceramic laser in a single cavity. Dual-wavelength-mode locking, at 1033.6 and 1047.6 nm, was operated simultaneously in one beam. Each pulse width was measured to be about 380 fs, by using an autocorrelator. Their spectral widths were 4.50 nm centered at 1033.6 and 3.08 nm centered at 1047.6 nm. This was the first dual-wavelength-mode locking observed in Yb-doped solid-state lasers at the time when the work was published.

SESAM mode-locked pulses of shorter duration could be achieved, through the inhomogeneous broadening in the Yb-doped solid solution garnet systems. One example is Yb-doped $Y_3ScAl_4O_{12}$ ceramics, which exhibited an emission linewidth of 12.5 nm at 1035.8 nm, larger than that of YAG by about 50 %, so as to generate shorter mode-locked pulses of 280 fs [252]. Even shorter pulses of 69 fs were generated in a SESAM mode-locked laser, which was based on the more disordered solid solution $(YGd_2)(Sc_2Al_2Ga)O_{12}$ ceramics [253]. In this case, the laser wavelength was at 1042 nm, which demonstrated a considerable shift with respect to the value of 1031 nm in CW emission. The spectral linewidth of the mode-locked pulse was 22 nm, which was significantly larger than the luminescence FWHM of 16 nm, while the broadening was attributed to the participation of the Kerr lens ML.

Due to their unique emission cross section and linewidth, Yb-doped sesquioxide ceramics are able to generate short pulse by ML. Furthermore, because of the highly nonlinear refractive index, strong Kerr lensing and self-phase modulation could be observed in these types of materials. As a result, as long as the length of the laser material is sufficient long to support the high nonlinear effects, ML can be operated [254]. ML of Yb-doped sesquioxide ceramics by SESAM can generate pulses in the range of 200 fs to several ps [255, 256]. For example, a diode-pumped passively mode-locked Yb^{3+}:Lu_2O_3 ceramic laser has been available, by using a SESAM [255]. Almost transform-limited 357-fs pulses at the center wavelength of 1033.5 nm, with a maximum average power of 352 mW, were achieved in this case. The efficiency against the absorbed pump power was 32 %, by using a laser diode pumping at a repetition rate of 97 MHz.

ML of the 1078 nm emission was observed in a heavily doped Yb:Y_2O_3 ceramic, which was realized by the self-mode locking due to the diffraction loss induced by thermal lensing, in combination with the Kerr self-focusing [257]. Significantly shortened pulse duration, due to the Kerr effect in the presence of SESAM, has been observed in all Yb-doped cubic sesquioxide ceramics, with examples including 92 fs for Sc_2O_3 [258], 65 fs for Lu_2O_3 [259], and 68 fs for Y_2O_3 [260]. Furthermore, by using two ceramic gain elements, e.g., Yb:Y_2O_3 and Yb:Sc_2O_3, with carefully selected doping levels and thicknesses, pulses of as short as 53 fs could be generated [260]. A 66 fs pulse duration with a 1.5 W average power and a 53 fs pulse duration with a 1 W average power were achieved without using water cooling. A piece of Yb^{3+}:Sc_2O_3 ceramics with a thickness of 1 mm and C_{Yb} = 2.5 at.% and a piece of Yb^{3+}:Y_2O_3 ceramics with a thickness of 1.5 mm and C_{Yb} = 1.8 at.% were used simultaneously, corresponding to α of 0.46. They were physically contacted to each other without any coating and arranged at the Brewster angle. Because the two ceramics have very close refractive index values [254], their Brewster angles are almost the same, so that the additional loss of the contact was very small.

Yb-Doped Ceramic Amplifiers

Due to their spectroscopic properties, such as long emission lifetime and moderate emission cross section, Yb-doped ceramics can be used for storage of inversion and extraction of the stored energy in amplifiers. The amplification characteristics of Yb:YAG ceramics at cryogenic temperatures allowed for their transformation of the laser scheme to a four-level scheme and lead to an improvement in thermal properties of the materials. A MOPA laser system, consisting of a fiber oscillator delivering 10-ns pulses of energy below 15 pJ at a repetition rate of 100 Hz, was

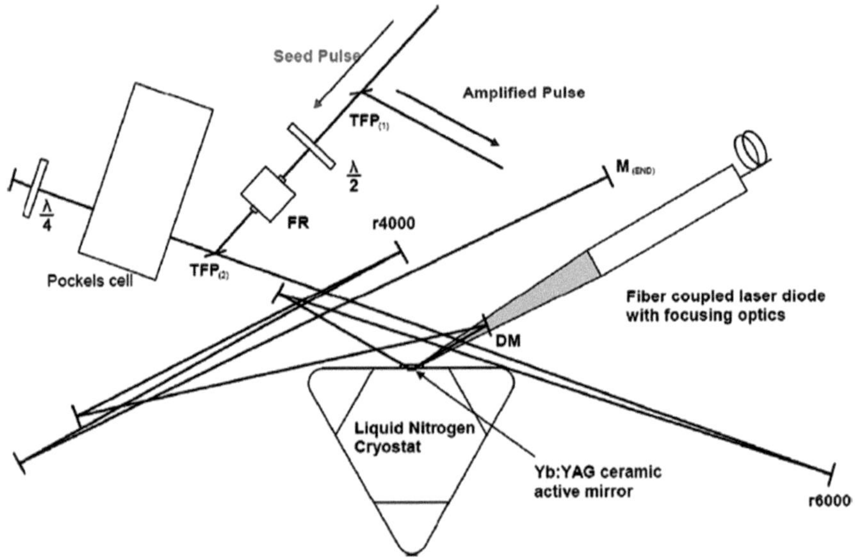

Fig. 9.33 Schematic layout of the regenerative amplifier based on an active-mirror cryogenic cooled Yb:YAG ceramic. $TFP_{(1)}$ and $TFP_{(2)}$ thin film polarizers, *FR* Faraday rotator, *DM* dichroic flat mirror (AR 750–1000 nm, HR 990–1170 nm), and *Pockels cell* Lasermetrics 5046E. Reproduced with permission from [261]. Copyright © 2009, Elsevier

first amplified, in a QCW of 1000 μs at 100 Hz 938-nm diode laser-pumped active-mirror regenerative amplifier, consisting of 9.8 at.% Yb:YAG ceramics to obtain pulses of 2.4 mJ and finally amplified to higher than 200 mJ, in a four-pass main amplifier based on a 700-μs diode laser-pumped 5.8 at.% Yb:YAG ceramic [261].

The configuration of the regenerative amplifier used in the experiments is shown in Fig. 9.33 [261]. The Yb:YAG ceramics had a dimension of $10 \times 10 \times 2\ mm^3$ in an active-mirror arrangement. One surface was HR-coated for 1030 and 940 nm, and the other was uncoated. The seed pulses enter the gain material from the noncoated surface at the Brewster angle, with a TEM00 mode size on the Yb:YAG of 1.8 mm ($1/e^2$). The amplified pulses from the regenerative amplifier were collimated and rotated to "p" polarization, which was used as the seed input for the four-pass main amplifier arrangement. Using the amplified output from the regenerative amplifier as the seed for the main amplifier, the pulse energy after four passes was measured for 700 μs pump duration, with experimental results shown in Fig. 9.34. For the initial experiment, a seed input energy of 2.4 mJ was used. Figure 9.35 shows output pulse energies, as a function of the laser diode peak power, for the same pulse duration used in the small-signal gain measurements. For a seed input energy of 1.8 mJ, the maximum pulse energy obtained was 150 mJ with pump duration between 200 and 700 μs. The optical–optical efficiency of the main amplifier was 19 %, and the slope efficiency was 30 %. At the same time, limitation of scaling due to the saturation and ASE was observed. However, above

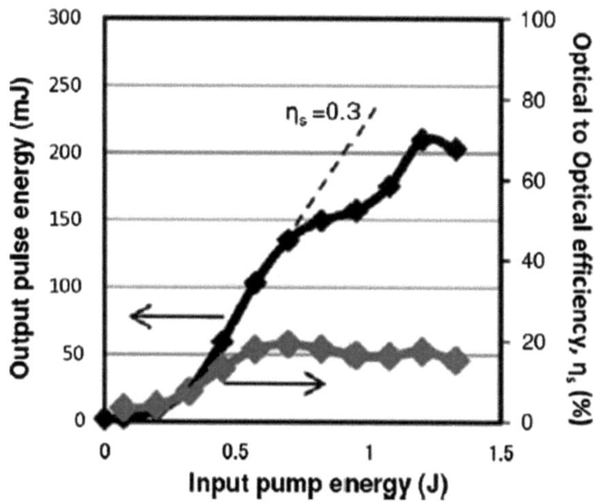

Fig. 9.34 Output performance of the main amplifier with a repetition rate of 100 Hz: output pulse energy and optical-to-optical efficiency, as a function of the diode pump energy, at an input seed pulse energy of 2.4 mJ. Reproduced with permission from [261]. Copyright © 2009, Elsevier

this energy, strong saturation effect was present, due to the enhanced ASE effects, which became even stronger if the repetition rates were increased.

Higher energy amplification has been made available by using regenerative amplifier with the total-reflection active-mirror (TRAM) design [262]. In this report, the TRAM consisted of an undoped YAG prism and a thin 20 at.% Yb:YAG ceramic layer that possessed a dimension of 30 × 30 × 0.40 mm^3. The cross section of the YAG prism was an isosceles triangle, with a base of 30 mm and a base angle of 31.2°. Output pulse energies of >5.0 mJ were achieved at repetition rates of 10–450 Hz, with an energy extraction of the amplifier of 41 %. A high energy fluence of 3.5 J cm^{-2} on the active media could be obtained by separating the input

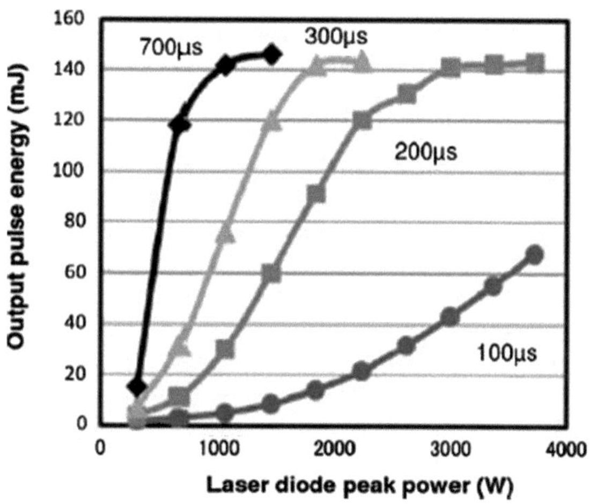

Fig. 9.35 Output performance of the main amplifier with image relay optics at a repetition rate of 10 Hz: output energy as a function of the laser diode peak power for different pump duration. Reproduced with permission from [261]. Copyright © 2009, Elsevier

and output surfaces of the TRAM. In addition, the absence of high-reflection (HR) coating can address the potential problem that the coating be damaged at high energies.

9.2.4.6 Waveguide Ceramic Lasers

It is well known that high-quality three-dimensional photonic structures, e.g., buried channel waveguides, which can be used for laser emissions, have been readily created on transparent laser ceramics, by using irradiations of short-pulse lasers [263–271] and particle beams [272–278]. This process is based on the controlled modification of the refractive index of the electronic processes and thermally induced transformations associated with laser irradiation. Laser writing to create such structures is a one-step process without the requirement of special preparation. It is a very fast, highly reproductive, and thus low-cost process. It has been employed to fabricate various integrated photonic devices.

Scanning confocal Raman scattering and luminescence imaging of femtosecond laser written double-filament Nd:YAG ceramic waveguide lasers have been used to elucidate the nature and extension of the microstructural modifications induced in the Nd:YAG ceramics by the laser inscription procedure [263]. It was found that femtosecond filament inscription caused a well-localized irreversible damage at the filament's core of the beam focus volume. The surroundings of the filaments were characterized by a high density of thermally removable defects and relevant strain-induced changes in the unit cell volume.

According to the confocal luminescence images, the spatial localization and extension of the refractive index modifications, i.e., the damage-induced refractive index reduction at filaments and strain-field-induced refractive index increment at the surroundings of filaments, were identified [263]. Net refractive index map of the waveguide could be established, by comparing the simulated and experimental beam propagations. The strain-field-induced refractive index increment was the main mechanism to form the waveguide, while the reduction in refractive index induced by damage at the filaments was responsible for the stronger mode confinement.

The fine-grained Nd:YAG ceramic sample used in this work was a $5 \times 5 \times 5\ \text{mm}^3$ cube with all its faces polished up to optical quality ($\lambda/4$), with a nominal Nd^{3+} concentration of 2 at.% [263]. Figure 9.36a shows a microscope optical transmission image of end face of the waveguide, where k is the fs laser beam propagation direction, E the linearly polarized electric vector of femtosecond light, and S the writing translation direction. Figure 9.36b shows the obtained near-field intensity distribution of the fundamental mode that was recorded with a charge-coupled device (CCD). The waveguide mode was highly elliptic, with a ratio of the mode size along the x- and y-directions to $w_y/w_x \approx 3$. Although the waveguide was multi-mode at 632.8 nm, most of the guided light propagated in the fundamental mode at these coupling conditions. In addition, both polarizations were confined by the waveguide.

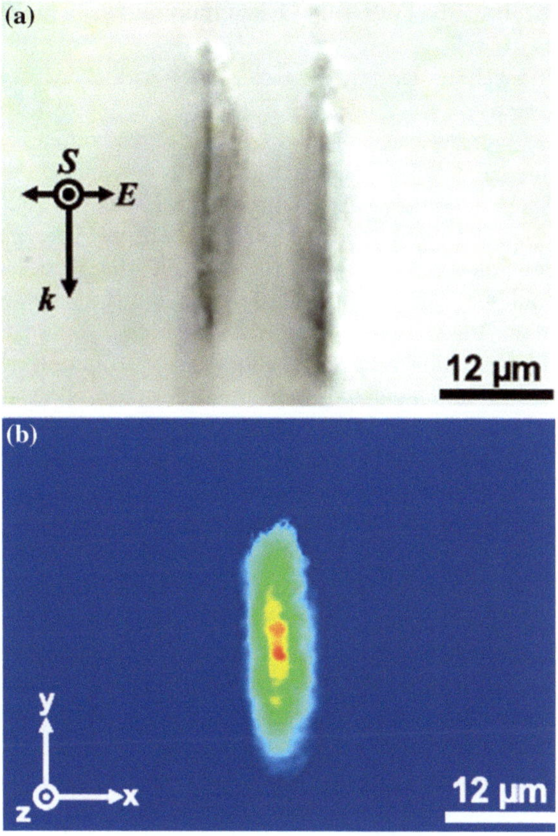

Fig. 9.36 **a** *xy*-plane optical transmission micrograph of the double-filament Nd:YAG ceramic waveguide, where *k* is the fs laser beam propagation direction, *E* the linearly polarized electric vector of femtosecond light, and *S* the writing translation direction. **b** Near-field intensity distribution of the fundamental mode at 632.8 nm, obtained by beam end coupling along the *z*-direction. Reproduced with permission from [263]. Copyright © 2009, Springer

9.2.4.7 Nonlinear Ceramic Lasers

The fundamental emission of ceramic lasers can be modified by various nonlinear processes. For instance, 104 W high-power 532 nm green radiation has been created by frequency doubling of *Q*-switched composite Nd:YAG ceramics [279], while lasing effects at wavelengths determined by the third-order susceptibility $\chi^{(3)}$ have been observed in various ceramics, including YAG, $(YGd_2)(Sc_2Al_2Ga)O_{12}$, and cubic sesquioxide [280–283].

For example, by using CW diode laser stacked arrays, side-pumping *Q*-switched composite ceramic Nd:YAG rod laser based on a type II KTP crystal intracavity frequency-doubling, a high-power high-stability green laser has been demonstrated [279]. An average output power of 104 W was achieved, at a repetition rate of 10.6 kHz, having a diode-to-green optical conversion efficiency of 10.9 %. At the average output power of about 100 W, the measured pulse width was 132 ns, with power fluctuation of <0.2 %. These experimental results indicated that the green laser system using the Nd:YAG transparent ceramics provided a higher laser performance and output stability than the conventional Nd:YAG single-crystal green laser system.

9.2.4.8 Effect of Grain Orientation on Performances of Ceramic Lasers

It has been accepted that in a laser with a rod of optically isotropic crystalline material, e.g., cubic garnets, the strain and stress caused by the thermal field inside the pumped laser material could lead to a difference between the radial and azimuthal refractive indices, which is usually called thermally induced birefringence. Such a thermally induced birefringence is closely related to the direction of the laser beam with respect to the cubic axes of the laser material, which determines orientation-dependent bifocal thermal lensing effect. Meanwhile, phase distortion could be induced, which thus results in orientation-dependent partial depolarization of polarized beam that travels inside the laser materials. Therefore, a complex depolarization behavior could be observed in polycrystalline transparent ceramics, due to the random orientation of grains. The global thermal depolarization behaviors of Nd:YAG ceramics have been extensively studied. It has been found that the Nd:YAG ceramics could have a similar behavior to that of (111)-oriented single crystals [284–288]. The effect becomes stronger and stronger as the absorbed pump power is increased [284, 285].

In the absence of lasing, the depolarization in Nd:YAG ceramics depends on the concentration of Nd^{3+}, because the thermal effect directly influences the emission quantum efficiency that is strongly dependent on the concentration of Nd^{3+} [284]. The measured depolarization of Nd:YAG ceramics with Nd concentrations of up to 3.4 at.% can be well linked to the calculated emission quantum efficiency. It was observed that due to the lower quantum defect, the depolarization under 885 nm diode laser pumping was smaller than that under 808 nm pumping. Also, under laser emission, the depolarization is smaller than that in the absence of lasing. In this case, there is a tendency that the heat load parameter is kept as a constant for each wavelength of pump, irrelevant to the concentration of Nd^{3+} in the ceramics. Both theoretical modeling and experimental data indicated that due to the thermal depolarization of Nd:YAG polycrystalline ceramics, the polarized and depolarized beams exhibited transverse modulation, with a spacing corresponding to the average grain size of the ceramics.

Thermally induced birefringence effects of representative Nd:YAG ceramic lasers have been accurately measured [285]. Depolarization for the ceramics was almost the same with that of (111)-cut single crystals, as the Nd^{3+} concentration was maintained at the same level. Under nonlasing condition, the thermally induced birefringence became more and more pronounced as the Nd^{3+} concentration was increased, due to the increased quenching effect. Under lasing condition, depolarization was found to be one-third of that under the nonlasing condition, which was attributed to the fact that the stimulated emission reduced the effect of the nonradiative process, so that the thermal load was effectively decreased. Furthermore, direct pumping into the emitting level by using an 885-nm pump source intrinsically reduced the heat generation, as compared with the conventional 808 nm pumping. It has been experimentally confirmed that the depolarization and the thermal load could be reduced by 30 % by using the direct pumping.

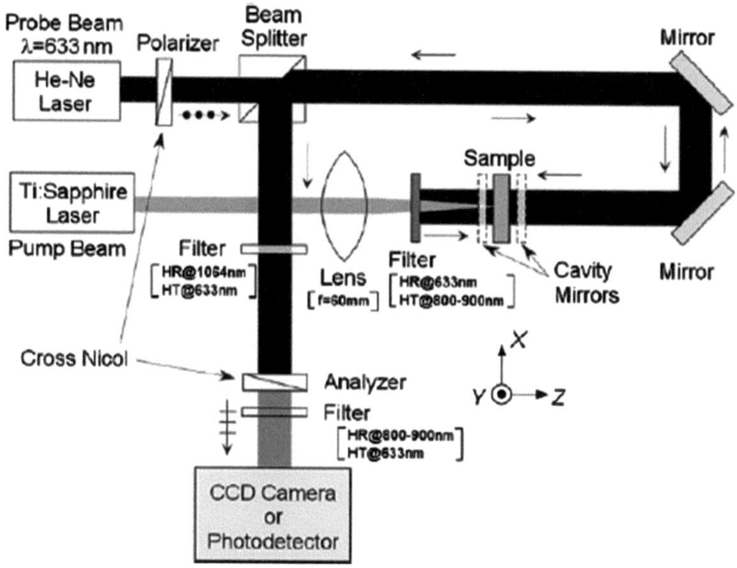

Fig. 9.37 Experimental setup to measure the depolarization induced by thermally induced birefringence. Reproduced with permission from [285]. Copyright © 2007, Elsevier

A pump–probe method for accurately measuring the depolarization caused by the thermally induced birefringence has been proposed, which is defined as the ratio of the depolarized power to the initially polarized power as $D_{pol} = P\perp/P_{initial}$. The experimental setup is shown in Fig. 9.37 [285]. Figure 9.38 shows depolarized beam patterns of the (111)-cut 1.0 at.% Nd:YAG single crystal and the 3.5 at.% Nd: YAG ceramics, which were recorded by using a CCD camera. Because the thermally induced birefringence in the (111) plane of Nd:YAG single crystal is circularly symmetric and occurs between the radial and the azimuthal directions, the depolarized beam formed a four-leaf-like pattern, as shown in Fig. 9.38a. Although a similar pattern was observed for the 3.5 at.% Nd:YAG ceramics, as illustrated in Fig. 9.38b, the degree of depolarization was higher as compared with that of the single crystal at the same absorbed pump power.

Figure 9.39 shows dependences of depolarizations of the Nd:YAG single crystal and ceramics, as a function of the Nd^{3+} concentration, at an absorbed pump power of 1 W [285]. It was observed that the Nd:YAG single crystal and the ceramics exhibited almost the same depolarization effect, if their Nd^{3+} concentrations were taken into account. Because the Nd:YAG ceramics consist of many single-crystal grains with an average grain size of tens of micrometers, which are randomly oriented, the thermally induced birefringence effect should be averaged among all the grains.

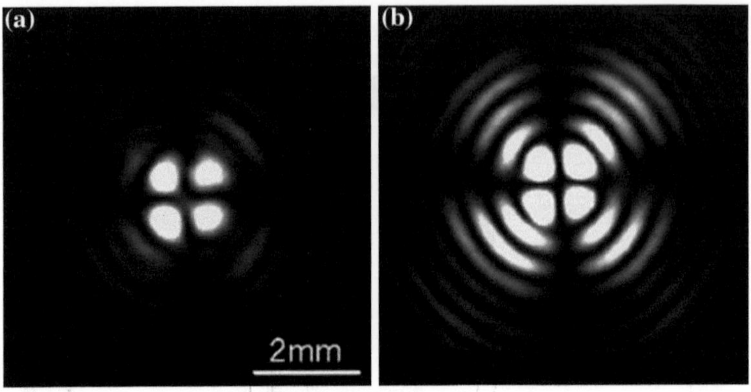

Fig. 9.38 Depolarized beam patterns under nonlasing: **a** (111)-cut 1.0 at.% Nd:YAG single crystal and **b** 3.5 at.% Nd:YAG ceramics, at an absorbed pump power of 1.7 W. Reproduced with permission from [285]. Copyright © 2007, Elsevier

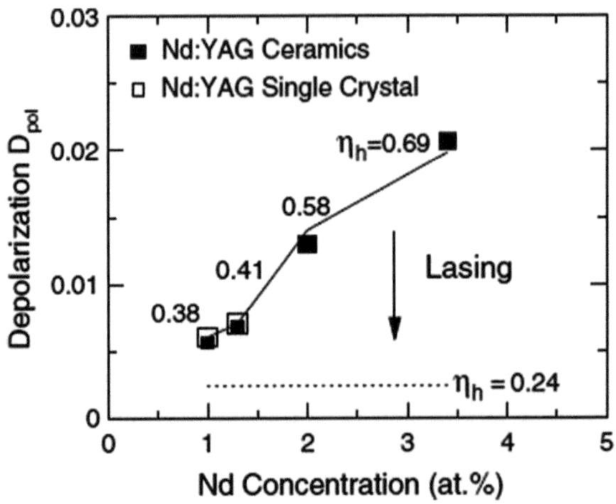

Fig. 9.39 Dependences of depolarization on Nd^{3+} concentration under nonlasing at an absorbed pump power of 1 W. The solid line represents the calculated depolarization, while the dotted line indicates the calculation under the ideal lasing condition. Reproduced with permission from [285]. Copyright © 2007, Elsevier

The depth of modulation is determined by the ratio of the thickness of the laser material to the grain size of the ceramics [287, 289–291]. Analytical expressions for eigen polarizations and phase delays in grains of thermally loaded Nd:YAG ceramic rods have been established [287]. It was suggested that the depolarization of radiation in polycrystalline ceramics led to beam modulation with a characteristic size of the order of the ceramic grain size. Therefore, increasing the ratio of rod length to grain size would be a reasonable approach to diminish the modulation depth and compensate for birefringence. Meanwhile, spatial modulation of a laser beam with a transverse size of the order of ceramic grain size has been experimentally identified corresponding to the thermal depolarization in Nd:YAG ceramics [289]. Depth of the modulation was increased with increasing heat release power and decreasing ratio of the laser sample length to the ceramic grain length. In addition, small-scale thermal modulation on phase distortion in ceramics has been confirmed theoretically and experimentally in undoped CaF_2 ceramics that were heated by using a CO_2 laser [292, 293].

The random distribution of phase distortions due to the random orientation of the ceramic grains led to local modes of oscillation in lasers, when the ratio of the thickness of the laser material to the grain size is sufficiently small, i.e., the laser material is sufficiently thin. Typical examples include the thin laser materials, which are used for microchip lasers. For instance, in a diode laser end-pumped highly doped Nd:YAG ceramic microchip laser with deposited mirrors, the global optical spectrum measured with a conventional wavemeter of 20 GHz resolution exhibited only a single-longitudinal-mode oscillation. However, when measured by using a 6-MHz high-resolution Fabry–Perot interferometer, closely spaced multiple components that fluctuate in time were observed in the laser emission modes, which implied that these laser modes contained the local modes emitted by the individual crystalline ceramic grains of the materials [294].

It was further demonstrated that the dynamics of the modulation by the coupled local modes could be very complex, in which additional quasi-periodic and chaotic relaxation oscillations have been observed. In addition, there were also periodic short-time quasi-Q-switched spikes, which were attributed to the presence of foreign inclusions with SA properties likely at the grain boundaries [295]. Experimental results indicated that such local modes are present in coarse-grained ceramics and absent in fine-grained samples.

Controlling the birefringence and the polarization of the laser emission is an important technical issue, relating to the applications of solid-state lasers. Various techniques have been explored to control the state of beam polarization, including configuration of resonator, insertion of intracavity elements, e.g., polarizers, grating mirrors and birefringent elements, direction and configuration of pump. Additionally, it is also possible to exploit the thermally induced modification of the refractive index, so as to increase the stability of the laser resonator to obtain the desired situation of beam polarization. These techniques have been successfully applied to Nd:YAG ceramic lasers, in order to control birefringence [296] and generate laser radiation with controlled polarization properties [297–300] and modal structures [301–303]. For example, arrays of vortex lasers have been obtained through beam controlling, which can be used as optical tweezers and cold atom trappers. Polarized laser emission was also realized in Nd:YAG ceramic lasers with passively Q-switching [304, 305] and ML [306].

9.3 Advanced Ceramic Lasers

9.3.1 Motivation and Overview

Laser gain medium made with single crystal and glasses can only have simple compositions and thus allow for uniform design. High laser performances, such as high power outputs, high beam qualities, and short-pulse generations, have been improved only by amplifying the emission generated from a high optical quality laser gain medium. The main objectives of research of laser gain media include improving optical quality, growing large-scale media, and synthesizing new materials. However, it has been shown that the conventional technologies have come to a situation that they are unable to bring out further improvement in laser performances. As a result, new laser elements and structures emerged, including composite lasers and fiber lasers, which will be discussed in the section.

Composite laser technology has pushed solid-state lasers to a new level of performance and functionality, which are much beyond those of simple element lasers [70–72, 154, 307–312]. Conventionally, a composite laser element is formed by bonding crystals with two or more compositions. Usually, three steps are involved: (i) polishing the bonding surfaces of the crystals, (ii) making close contact with the components, and (iii) heating at high temperatures to ensure strong bonding through thermal diffusion. Therefore, it is called diffusion bonding. This conventional diffusion bonding technology is limited to the bonding of flat surfaces only, while the bonding strength is not high as well. Comparatively, composite laser structures made of polycrystalline ceramics would offer more flexibility in designs, because ceramic technology starts with powders. Various composite lasers based on ceramics and single crystals will be elaborated, with emphasizing the advantages of ceramic technology [4].

Glass laser gain media can be produced at large scale without losing optical quality. One of their applications is the use as high-energy short-pulse laser sources for nuclear fusion. Although high-energy glass lasers can be readily developed, they cannot be used for CW operation and pulse operation with high repetition rates, due to the poor thermal conductivity of glasses.

On the other hand, glass optical fibers have been widely used in optical communications, because they can transmit information over long distances at high speed and have a large carrying capacity with low signal loss. Optical fibers used in communications generally consist of a core that carries the light pulses and cladding that reflects the light pulses back into the core. Such fibers are fabricated by using a so-called double crucible method. Two silica glasses with compositions of core shell are fused in a double crucible container. Then, the melted glasses are pulled into a long wire with the core–clad structure.

This fiber technology has also been used to fabricate glass fiber laser gain media. In this case, the core is doped with laser-active elements with a higher refractive index, whereas the cladding layer does not contain laser-active elements with a relatively lower refractive index. When a long and narrow glass fiber, with a diameter of tens of micrometers, is used as gain medium, it was possible to generate a high-quality laser beams, because the heat generated can be easily dissipated. Furthermore, light propagation is only inside the core, so that a highly concentrated laser beam can be generated without using focusing lens. The concept of glass fiber has been extended to crystalline materials.

Traditionally, materials produced by using the melt-growth process are single crystals, while those fabricated by using sintering are polycrystalline ceramics. Recent development has shown that the general understanding on this topic is slightly changed. For example, single crystal can be grown or converted from polycrystalline ceramic bodies, through a special process known as abnormal grain growth. This process is able to produce single crystals without melting. The initial success was mainly on Mn–Zn ferrite single crystals [313, 314], which has been extended to transparent ceramic laser materials later, as discussed in the previous chapter [315].

9.3.2 Composite Ceramic Lasers

9.3.2.1 Problems of Single-Crystal Composites

In the simplest case, a composite laser element is formed by bonding two single crystals with same crystal structure but with different compositions. Various advanced laser elements with complicated structures, such as end-cap and clad–core designs, which cannot be fabricated directly by using the melt-growth single-crystal technology, have been developed by using the composite laser technology [154, 307–311]. However, significant progress has been achieved with these advanced single-crystal composite lasers, and there are some critical issues that need to be tackled for practical applications.

Firstly, the directions of the two bonded crystals are different from each other, so that the component is difficult to produce. Secondly, although the bonding surfaces are polished to an optical grade before bonding, surface roughness at the level of nanoscale cannot be eliminated. As a result, interstices are formed at the interface, which leads to poor thermomechanical properties of the composites and thus laser damage especially at high powers. Thirdly, conventional diffusion bonding technology is limited to the bonding of flat surfaces, so that only simple composite structures can be produced. Also, the diffusion distance of laser-active ions is limited, because it is not possible to control the microstructure of the bonding interface in the crystal composite.

In crystal composites, there is no chemical bond between the two crystals at the interface, which leads to low mechanical strength, low thermal conductivity, and poor thermal diffusivity. Therefore, such crystal composites cannot be used for high-power lasers, because the defects at the interface will absorb thermal energy, thus leading to decrease in beam quality and lasing efficiency. Due to potential thermal and mechanical damage at the defect points, sample breakdown could happen with strong excitation at high powers.

One example is Ti:sapphire crystal composite prepared by using a diffusion bonding technique [316]. The crystal composite was fractured after a *Q*-switch laser characterization. Most parts of the fracture surface were flat, which means that these areas were not well bonded. When the bonding interface of the crystal composite was scanned by using a He–Ne laser, with the laser beam irradiating perpendicularly to the bonding interface, scattering beam lines along the bonding interface could be observed, indicating that interstices were formed at the bonding interface, i.e., the bonding was imperfect in this case.

In summary, although composite lasers have shown great potential in improving laser performances, the conventional bonding technology to fabricate single-crystal composites has hindered their applications in high-power and highly functional lasers. In this respect, ceramic technology has emerged to produce composite lasers, without the bonding problem of the conventional single-crystal composites.

9.3.2.2 Examples of Composite Ceramic Lasers

Composite laser materials fabricated by using ceramic processing technology are discussed in this part [10]. Representative ceramic laser composites include (i) layered structures, (ii) cylindrical structures, (iii) waveguide types, (iv) core–shell fibers, and (v) gradient types. Some of them have flat bonding interfaces, such as layered structures and waveguide types, which are similar to using conventional crystal bonding technology. Others are of complicated bonding interfaces, such as cylindrical core–shell structures, which cannot be fabricated by using the conventional crystal bonding technology.

Highly transparent YAG/1.0 at.% Nd:YAG/YAG composite ceramic slabs have been prepared by using solid-state reactive vacuum sintering method, from commercial Al_2O_3, Y_2O_3, and Nd_2O_3 powders, with tetraethyl orthosilicate (TEOS) and La_2O_3 as sintering aids [74]. By optimizing the types and concentrations of the sintering aids, pure YAG and 1.0 at.% Nd:YAG ceramics could be readily fabricated at the same sintering temperature, making it possible to construct the desired composite ceramics by using ceramic processing. Fully dense composite ceramic slabs were obtained at a temperature of 1750 °C for 50 h, with an average grain size of ~10 μm. In-line transmittance of the composite ceramics was >81 % near the visible wavelength of 400 nm. Output power of the composite ceramic rod with a dimension of Ø3 mm × 82 mm was 20.3 W at the pump power of 201 W, corresponding to an optical-to-optical conversion efficiency of 10.1 %.

The powders were weighted according to the chemical stoichiometry composition of YAG, without or with doping of 1.0 at.% Nd^{3+}. Ball milling with high-purity Al_2O_3 balls for 12 h in ethanol was used to mix the powders, together with the sintering aids of tetraethyl orthosilicate (TEOS, >99.99 %) and La_2O_3 (>99.99 %). The mixtures were calcined at 800 °C for 4 h to burn out the organic components. They were then compacted by using dry pressing at 100 MPa,

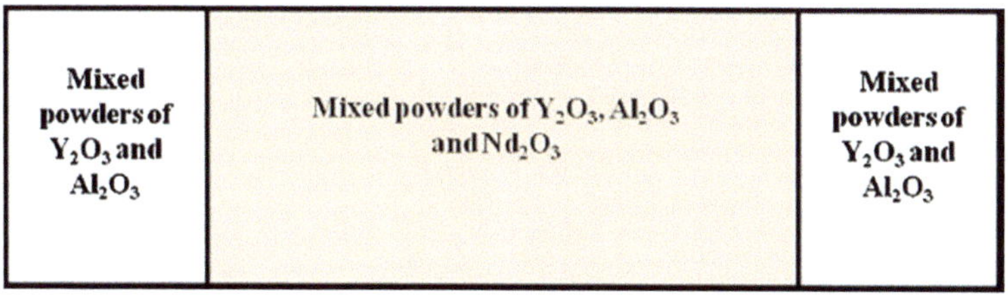

Fig. 9.40 Green body structure for the composite ceramics. Reproduced with permission from [74]. Copyright © 2012, Elsevier

followed by CIP at 250 MPa. Structure of the green body slab after pressing is shown in Fig. 9.40 [74]. The compacted pellets were sintered in vacuum of 10^{-5} Pa at 1750 and 1765 °C for up to 50 h. After sintering, the samples were annealed in air at 1450 °C for 20 h to completely remove internal stress and eliminate oxygen vacancies. Polished YAG/1.0 at.% Nd:YAG/YAG transparent composite ceramic slabs had a dimension of 110 × 55 × 5 mm^3.

In the composite ceramics, only 0.8 wt% TEOS was used to sinter the Nd:YAG region, while a combination of 1.0 wt% TEOS and 0.8 wt% La_2O_3 was used in the pure YAG region. The photograph of a representative composite ceramic slab sintered at 1750 °C for 50 h and annealed at 1450 °C for 20 h is shown in Fig. 9.41. According to EPMA micrograph of thermally etched surface, the sample exhibited no trace of abnormal grain growth and no inclusions present both at the grain boundaries and in the grain matrix. No clear interface was observed in the transition

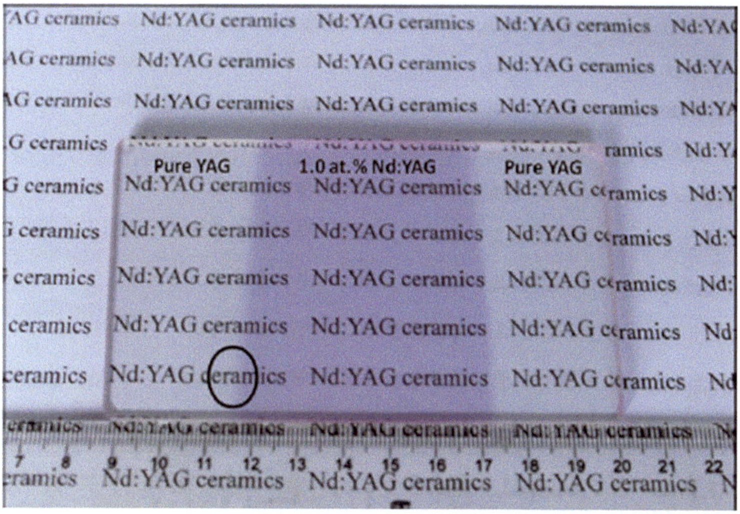

Fig. 9.41 Photograph of the composite ceramic slab sintered at 1750 °C for 50 h, with the pure YAG region having 1.0 wt% TEOS and 0.8 wt% La_2O_3 and the Nd:YAG region having 0.8 wt% TEOS, as sintering aids. Reproduced with permission from [74]. Copyright © 2012, Elsevier

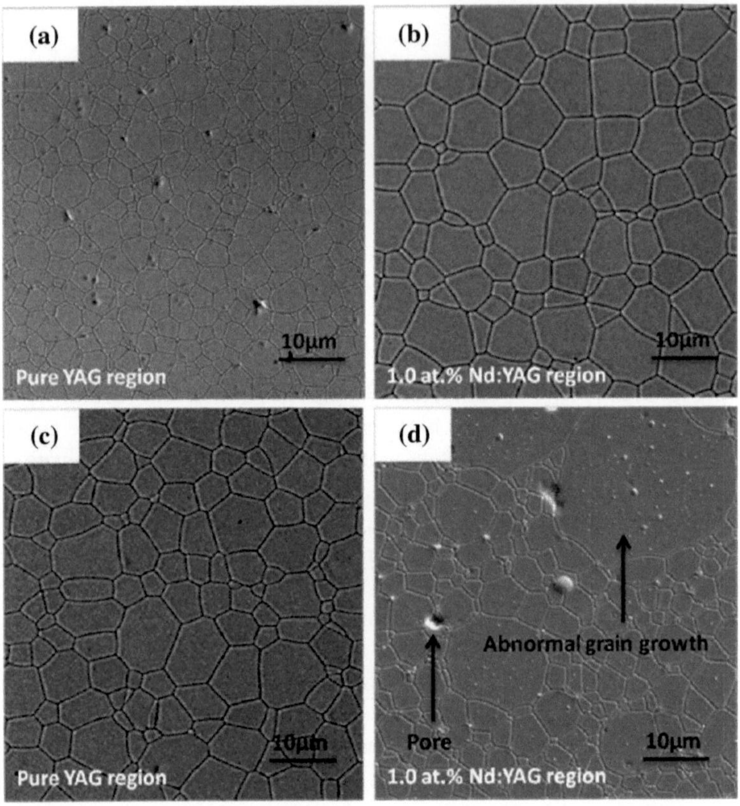

Fig. 9.42 EPMA micrographs of thermally etched surfaces of the composite ceramics sintered at 1750 °C (**a**, **b**) and 1765 °C for 50 h (**c**, **d**). Reproduced with permission from [74]. Copyright © 2012, Elsevier

region between the pure YAG and Nd:YAG region, which is advantageous to the conventional single-crystal bonding technology.

Figure 9.42 shows SEM images of the composite ceramics. In the pure YAG region, pores were observed inside the grains and at the grain boundaries, with an average grain size ~5 μm [74]. After sintering at 1765 °C, the grain size of the pure YAG region was increased to ~10 μm. The pores were removed completely, because the grain growth promoted the elimination of the large pores. However, abnormal grain growth was observed in the Nd:YAG region after sintering at 1765 °C. It could be attributed to the presence of Nd_2O_3, which caused lattice distortion of YAG due to the substitution of Y^{3+} by Nd^{3+}. Therefore, Nd:YAG had a lower sintering temperature than pure YAG. This problem has been addressed by the optimization of sintering aids.

A similar strategy has been used to develop multilayered YAG-Yb:YAG ceramics [317]. The layered structures with a tailored modulation of the doping level could facilitate to reduce the peak temperature, the temperature gradients, and also the thermally induced deformation of the laser materials, thus mitigating the overall thermal effects. The multilayered ceramics consisted two and three layers of pure YAG/10 at.% Yb:YAG and pure YAG/10 at.% Yb:YAG/pure YAG, with different compositions that were designed so as to control deformation and stresses and thus reduce the thermal lensing effect of the laser materials. The multilayered samples were assembled by linear and CIP and finally cosintered by using vacuum sintering. The Yb diffusion profile across the doped/undoped interface was identified, which has been related to output power of the lasers. An internal optical transmittance up to 96 % was achieved, while a laser output power up to 5 W, with a slope efficiency of as high as 74.3 %, was also obtained at the same time.

Tape casting is a very convenient method that can be used to fabricate transparent composite ceramic lasers [68, 73]. By using tape casting and vacuum sintering process, optical quality YAG/Re:YAG/YAG (Re = Yb or Nd) composite laser ceramics have been fabricated [68]. The influence of dispersant content on the rheological behavior of the slurry, microstructure, optical properties, and final laser performances of the composite ceramics was systematically studied. The laser output at 1030 nm was generated by using YAG/Yb:YAG/YAG composite ceramics with a threshold absorbed pump power of 4.33 W and slope efficiency of 12 %, as the transmission of OC (T_{oc}) was 2.3 %. With the YAG/Nd:YAG/YAG composite ceramics, 1064 nm laser output was observed. The slope efficiency was increased from 30 to 38 %, while T_{oc} was changed from 2.3 to 10 %.

Commercially available powders of high-purity oxides were used as starting materials, with chemical compositions of $(RE_xY_{1-x})_3Al_5O_{12}$ ($x = 0.02$ for RE = Nd and $x = 0.2$ for RE = Yb). The powders were mixed and milled for 24 h in a mixed solvent, consisting of 40 wt% ethanol and 60 wt% xylene and 4 wt% fish oil, which were used as dispersant to modify rheological behavior of the slurries. The powder-to-solvent weight ratio was 13:7. Then, binder and plasticizer were introduced into the slurry and milled for 16 h. 6 wt% polyvinyl butyral was used as binder, while 3 wt% butyl benzyl phthalate and 3 wt% polyalkylene glycol were used as plasticizers. Tape casting process was carried out with a gap height of 0.7 mm at a casting speed of 1.2 m min^{-1}. The thickness of the obtained tapes was 0.12 mm. Eighteen layers of the green tapes were stacked and laminated at 120 °C at 30 MPa for 10 min to obtain the green samples. The samples were calcined out in oxygen at 700 °C for 10 h to burn out the organic components and then sintered at 1720 °C for 20 h in vacuum of 10^{-6} Torr. After polishing and annealing treatment, optical-grade ceramics with a dimension of $10 \times 10 \times 2.1$ mm^3 were obtained.

Sandwich-like structure with three-segment composite rod was used to fabricate the composite ceramics, with 6 layers of YAG, 6 layers of 20 at.% Yb:YAG or 2 at.% Nd:YAG, and another 6 layers of YAG, as shown in Fig. 9.43a. A photograph of the composite laminated green bodies is shown in Fig. 9.43b. The design was based on the consideration of effective thermal management for laser operation. Higher doping concentration of rare earth ions leads to lower thermal conductivity and high heat generation when the Re:YAGs were used as laser gain medium. Therefore, undoped YAG layers are used to increase the heat dissipation effect.

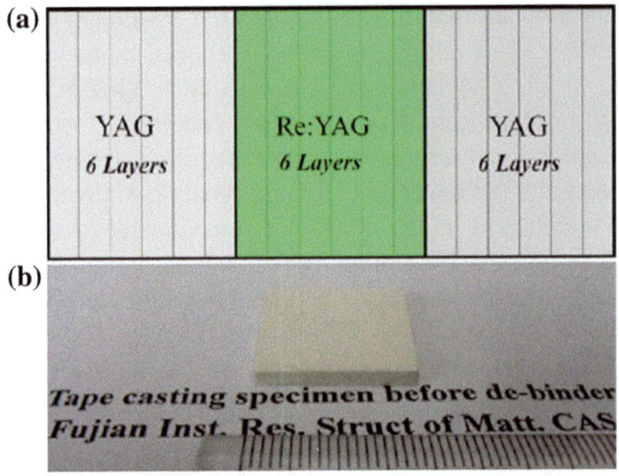

Fig. 9.43 Scheme of the composite ceramic along thickness direction (**a**) and a photograph of the laminated green body (**b**). Reproduced with permission from [68]. Copyright © 2012, Elsevier

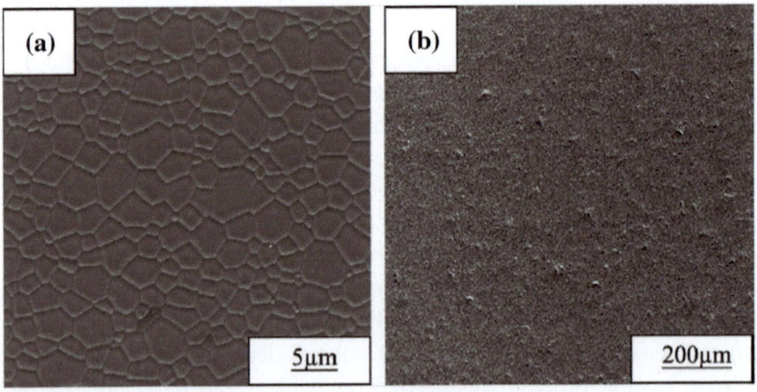

Fig. 9.44 SEM images of the Yb:YAG composite ceramics: **a** surface and **b** cross-sectional views. Reproduced with permission from [68]. Copyright © 2012, Elsevier

Figure 9.44a shows SEM image of the as-sintered YAG/Yb:YAG/YAG composite ceramics, showing a fully dense microstructure, with an average grain size of about 5 μm. Figure 9.44b shows cross-sectional view of the composite sample. It is found that the interfaces between the adjacent layers can be removed by the sintering process. Photographs of the two types of composite ceramics before and after the thermal annealing are shown in Fig. 9.45, with very nice physical appearances [68].

Yb:YAG ceramic slabs edge-cladded with Cr^{4+}:YAG have been reported by using vacuum sintering [318]. Commercial oxide powders were used as the raw materials and CaO was used as a charge compensator, keeping a Cr:Ca molar ratio of 1:3. After annealing, the Cr^{3+} ions were oxidized into Cr^{4+}. The lasing wavelength of 1030 nm of Yb^{3+} lies within the Cr^{4+} broad absorption range of 900–1100 nm.

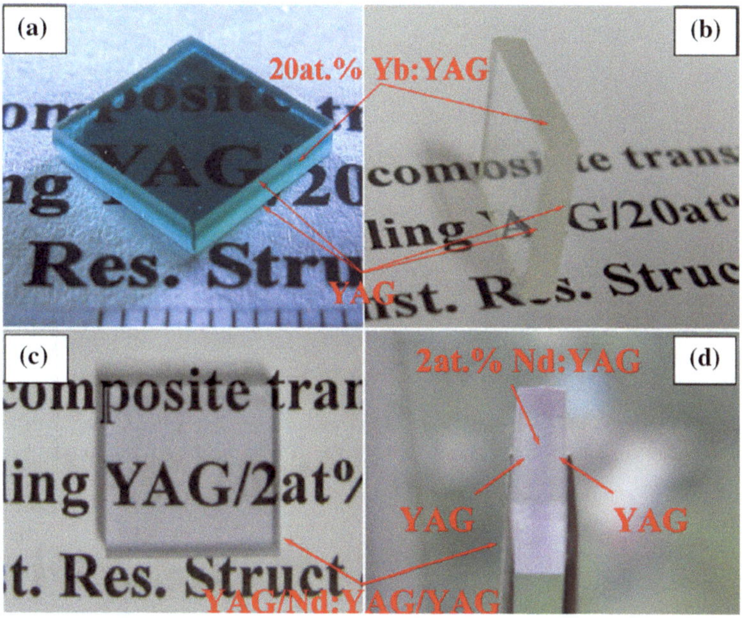

Fig. 9.45 Photographs of the YAG/Yb:YAG/YAG composite ceramics before (**a**) and after (**b**) annealing treatment and the YAG/Nd:YAG/YAG composite ceramics before (**c**) and after (**d**) annealing treatment. Reproduced with permission from [68]. Copyright © 2012, Elsevier

The intensity of the Cr^{4+} absorption peak was proportional to the concentration of Cr. The presence of excessive CaO led to abnormal grain growth and the formation of residual pores. Optimized concentration of Cr was 0.35 at.%. There was no interface effect present between the Cr:YAG and Yb:YAG regions in the composite ceramics. Transmission of the composite ceramics was comparable to that of single crystals.

High-purity oxide powders were used as starting materials, with compositions of 2 at.% Yb:YAG and Cr:YAG doped with 0.1, 0.35, and 0.5 at.% Cr, as well as TEOS as sintering aid. Typical ceramic processing was used to fabricate the composite ceramics. By using a custom-designed plexiglass mold, the mixed powders were dry-pressed at 50 MPa, followed by CIP at 250 MPa. Green compacts were sintered at 1770 °C for 50 h in vacuum of 10^{-5} Pa. After sintering, the samples were annealed at 1450 °C for 20 h in air to promote the transition of Cr^{3+} to Cr^{4+} and eliminate oxygen vacancies. Finally, highly transparent Yb:YAG ceramic slabs with Cr^{4+}:YAG edge cladding were obtained. Photographs of representative samples of the composite ceramics, before and after thermal annealing, are shown in Figs. 9.46 and 9.47, respectively [318]. The samples were fully transparent, while the edge cladding structures are clearly demonstrated, which are supported by spectroscopic experimental results.

Figure 9.48 shows fractured surface SEM images of representative samples. It is observed that the two samples both had uniform and fully dense microstructures.

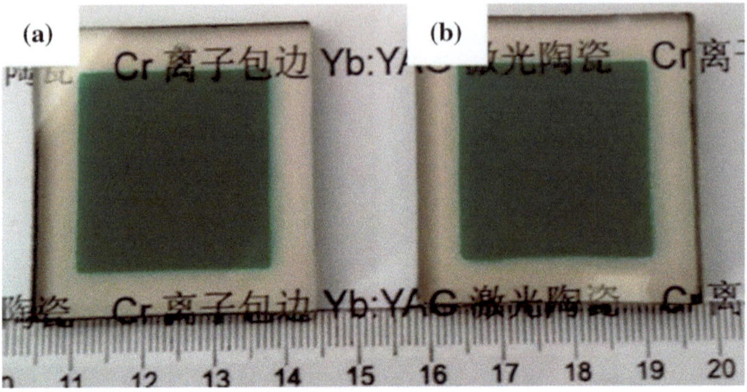

Fig. 9.46 Photographs of the 2 at.% Yb:YAG ceramic slabs with 0.1 at.% (**a**) and 0.35 at.% (**b**) Cr:YAG edge cladding before annealing. Reproduced with permission from [318]. Copyright © 2014, Elsevier

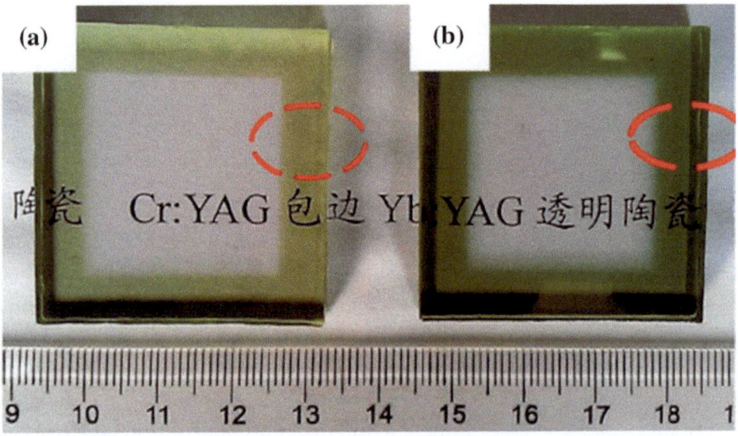

Fig. 9.47 Photographs of the 2 at.% Yb:YAG ceramic slabs with 0.1 at.% (**a**) and 0.35 at.% (**b**) Cr:YAG edge cladding after annealing. Reproduced with permission from [318]. Copyright © 2014, Elsevier

No obvious interfacial regions were observed. The 0.1 at.% Cr sample had an average grain size of ~20 μm, as shown in Fig. 9.48a, while the one with 0.35 at.% Cr exhibited an average grain size of ~35 μm, as illustrated in Fig. 9.48b. Because the Cr:Ca molar ratio was fixed at 1:3, high content of Cr means a high concentration of Ca, which had an additional function of promoting sintering. This is the reason why Cr concentration cannot be too high.

A colloidal cocasting process was reported to fabricate multi-segment composite ceramic laser gain materials [319]. The three-segment transparent composite rod of 62 mm long and 3 mm diameter consisted of undoped YAG, 0.25 % Er:YAG, and

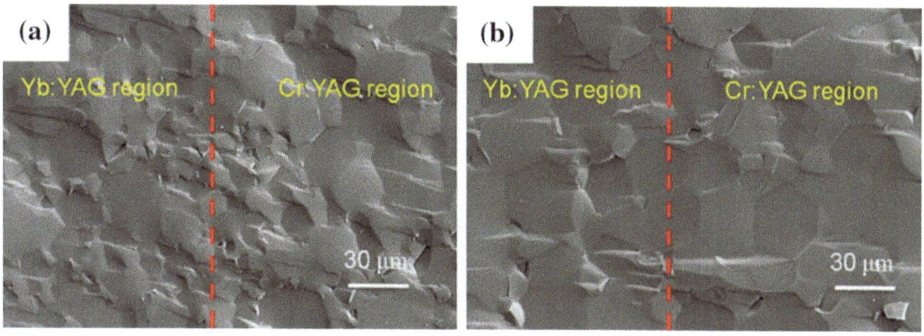

Fig. 9.48 Fracture surface SEM images of the 2 at.% Yb:YAG ceramic slabs with 0.1 at.% (**a**) and 0.35 at.% (**b**) Cr:YAG edge cladding after annealing. Reproduced with permission from [318]. Copyright © 2014, Elsevier

0.5 % Er:YAG. Schematic diagrams and photographs of the composite ceramics are shown in Fig. 9.49. The Er concentration profile in the composite had steep, controllable gradients at the segment interfaces, while maintaining constant dopant concentrations within each segment. The composite rod exhibited transmittance of 84 % at 1645 nm, which was the lasing wavelength, with a scattering loss of 0.4 % cm^{-1}. Laser operation of the composite Er:YAG ceramic rod was demonstrated, with almost the same lasing behavior as an Er:YAG single-crystal rod.

Colloidal slurries for cocasting were prepared by using oxide powders, with a composition of $Y_{3(1-x)}Er_{3x}Al_5O_{12}$ with x equal to 0.0025 and 0.005. Silica was used as sintering aid at a level of 0.14 wt% for all samples from TEOS. All the solvents,

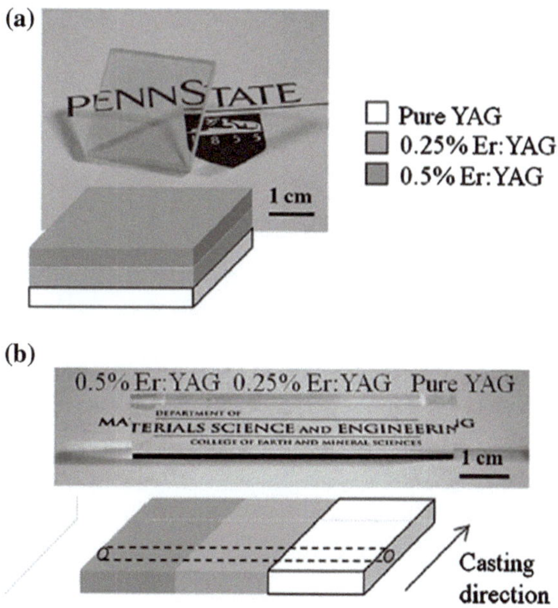

Fig. 9.49 Schematics and photographs of the Er:YAG/undoped YAG tape cast composites: **a** stacked and **b** cocast. The dotted lines indicate how a laser rod was cut from such composite. The composites were vacuum-sintered at 1650 °C for 4 h and HIPed in Ar at 1675 °C for 8 h. Reproduced with permission from [319]. Copyright © 2010, Cambridge University Press

dispersant, oxide powders, and TEOS were mixed by using ball milling to form slurries, followed by the addition of binder and plasticizers for an additional milling step to form pseudo-plastic suspensions. The colloidal suspensions were cocast with a three-reservoir doctor-blade fixture at a rate of 70 cm min^{-1} with a gap height of 0.356 mm onto silicone-coated Mylar.

Dried tapes were cut into pieces to form stacked composite parts. For example, a stacked composite, made of 30 layers of tape of 0.1 mm in thickness, was prepared by stacking 10 layers of each composition, i.e., undoped YAG, 0.25 % Er:YAG, and 0.5 % Er:YAG. The sections with different doping levels had the same thickness, while the interfaces in the composite were parallel to the tape layers. Tapes made by cocasting the three different slurries with the composite interfaces parallel to the length of the tape were formed into segmented composite slabs. Thirty-two layers were stacked with well alignment of the compositional interfaces. Densification was conducted by using vacuum sintering, combined with an optional hot isostatic pressing (HIP) step [319].

The three-step doping in the Er:YAG ceramic composite laser rod is clearly demonstrated in Fig. 9.50 [319]. The green glow was attributed to the Er^{3+} upconversion processes, when passing a 1532-nm laser beam through the length of the rod. In order for laser characterization, both ends of the rod were polished to the standard laser quality and parallelism, after which they were broadband-AR-coated, at wavelengths of 1520–1660 nm, in order to minimize the reflection at pump and lasing wavelengths. The laser output performance of the composite ceramic rod is shown in Fig. 9.51, together with that of a single-crystal laser sample for comparison.

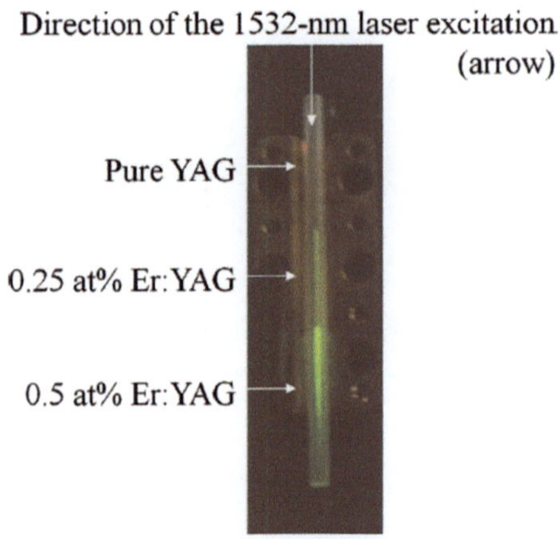

Fig. 9.50 Visual confirmation of the dopant concentration distributions in the Er:YAG composite rod with dimensions of 3 mm diameter × 62 mm. Upconversion of Er^{3+} ions excited using a 1532-nm laser caused the glow, which was progressively brighter as the dopant level was increased in the segments. Reproduced with permission from [319]. Copyright © 2010, Cambridge University Press

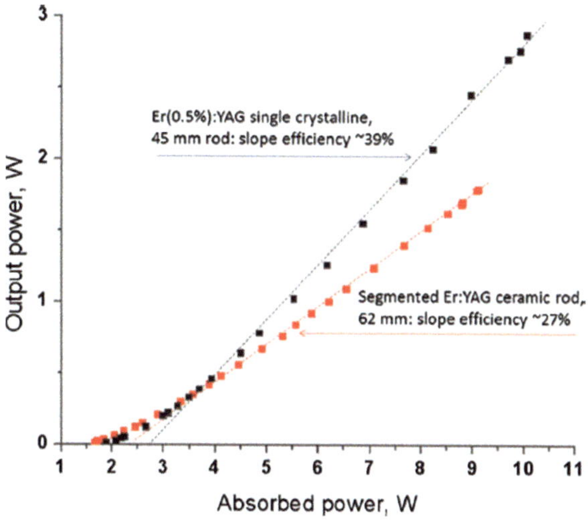

Fig. 9.51 Laser output as a function of absorbed pump power (quasi-CW) of the 62-mm (undoped YAG, 0.25 % Er:YAG, and 0.5 % Er:YAG) composite ceramic rod and the 45-mm (0.5 % Er:YAG) single-crystalline reference laser rod, with slope efficiencies being calculated from the dashed lines. Reproduced with permission from [319]. Copyright © 2010, Cambridge University Press

In summary, the advanced ceramic composite technology has offered more opportunity to fabricate complex structures of composite ceramic lasers, due to the availability of perfect inherent interface characteristics. It is expected that this technology will be a key subject of research and development of solid-state ceramic lasers.

9.3.3 Fiber Ceramic Lasers

Glasses could be excellent host materials for laser applications in terms of optical quality, while their poor thermal conductivity and mechanical strength have hindered their applications requiring high energy densities and high powers. Another advantage of glasses is that they can be easily made into wires or fibers [320]. Because the thickness of glass fibers could be as thin as tens of micrometers, the large surface-to-volume ratio is favorable to heat dissipation. As a result, glass fibers could be used for high-power lasers. Moreover, when using core–shell (clad) glass fibers, the refractive index of the cladding materials is lower than that of the core layer doped with laser-active ions, the generated laser beam can be confined in the core area. Therefore, it is possible to generate a high-quality laser beam without the requirement of focusing lens.

However, glass fibers have a serious shortcoming of low laser gain, because the doping concentrations of laser-active ions, such as Nd ions, are limited in glass media. As a result, glass fibers have a limited power generation per unit length, which is usually ≤ 1 W cm^{-1}. For example, to generate kilowatt laser emission, it is necessary to use a glass fiber with tens of meters in length. In this regard, it is quite obvious that single-crystal fibers could be promising candidates to address the problems of glass laser fibers.

Crystal fibers have been produced by the so-called micro-pulling-down or μ-PD method, which has principle similarities to the conventional melt-growth method [321]. After high-purity raw materials are melted in an iridium crucible, the melt is pulled down through a capillary tube at the bottom of the crucible and contacts with a small seed crystal. By pulling down the melt very slowly together with the seed crystal, a long and thin single crystal can thus be produced. The diameter of the grown crystals is about 1 mm, which is relatively larger than that of optical glass fibers. With the continuous development of this technique, the diameter of crystal fibers is continuously decreased. At the same time, the optical quality of the crystal fibers grown by using the μ-PD method is close to that of bulk crystals grown by using the CZ method. Crystal fibers developed at the initial stage often contained optically inhomogeneous regions, e.g., facets and core, which could not be used to achieve high-performance laser oscillations. Recent advancement has shown that laser oscillation has been produced by using μ-PD-grown crystal fibers, although the performance needs to be further improved [320, 322–325]. Examples of crystal fiber lasers are discussed in the following part.

Single-crystal fibers Yb^{3+}-doped $Y_3Al_5O_{12}$ (Yb:YAG), doped with 0.3 and 1 at.% Yb^{3+}, with lengths of up to 70 cm and diameters of 1 and 0.4 mm, have been grown by using micro-pulling-down (μ-PD) method [326]. According to their structural and optical characterizations, it was found that the fibers exhibited a perfect single-crystalline structure, with a homogeneous distribution of Yb^{3+} in the YAG host. In diode-pumped laser operation, a CW output power of 50 W at 1030 nm was observed for pump power of 180 W at 940 nm, by using the 1-mm-diameter fibers. For the 0.4-mm-diameter fibers, 28 W at 1030 nm was achieved for incident pump power of 100 W at 940 nm.

Micro-pulling-down is a melt-growth method, but different from the Czochralski or EFG method, because the crystal is grown by pulling the melt and the crystal in the downward direction, as shown in Fig. 9.52 [326]. Either radiative or conductive heating can be used to heat the crucible. In this study, a specially shaped iridium crucible was heated inductively in a neutral atmosphere to pull down the cylindrical Yb:YAG fiber crystals, as demonstrated in Fig. 9.52. The crucible was surrounded by ceramic protecting materials to ensure thermal homogeneity. Both <111> or <100> oriented seeds were used in the experiments. The radial and axial distributions of Yb^{3+} in the YAG fibers were characterized. Because internal stresses due to the temperature gradient could be produced during the crystallization process, defects within and at surface of the crystal fiber could be formed. The presence of defects will negatively influence light propagation, polarization state, beam quality, and hence laser mode [327, 328]. In this respect, the shape of the solid/liquid

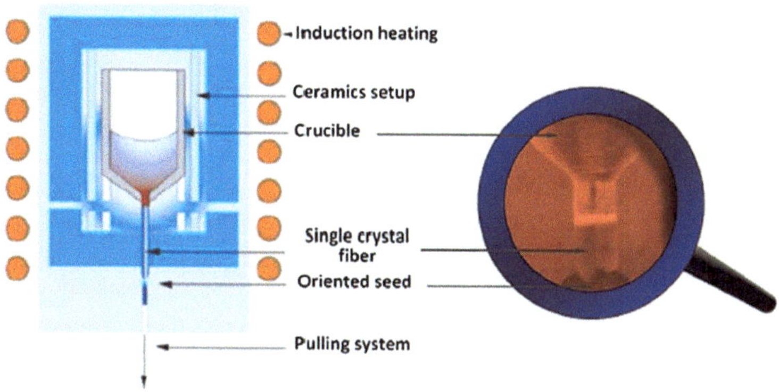

Fig. 9.52 Micro-pulling down setup and view of the growth of a single-crystal fiber. Reproduced with permission from [326]. Copyright © 2009, Elsevier

interface located below the nozzle of the crucible should be well controlled, so as to make sure that the grown fibers could have high crystallization and high optical quality, because these properties would have direct effect on the distribution of the dopants and thus laser efficiencies.

Crystallographic orientation, pulling rate, growth temperature, and geometries of the crucible and nozzle are all key parameters that can affect the quality of the crystal fibers. The seed orientation has a direct effect on features of cross section of the fibers, due to the formation of facets corresponding to crystallographic planes, as shown in Fig. 9.53. A growth rate of about 1 mm/min has been found to be propriate to avoid fast crystallization and the creation of colored centers. The Yb:YAG single-crystal fibers grown in this way could be as long as 1 m, with diameters of 0.4–3 mm.

Laser efficiencies of the single-crystal fibers are shown in Fig. 9.54 [326]. The 1-mm-diameter sample had a maximum CW output power of 50 W with a M^2 quality factor of 2.2 for an incident pump power of 180 W, corresponding to a slope efficiency of 35 %. This performance was the first report on 1 at.% Yb^{3+}-doped YAG single-crystal fibers grown by the micro-pulling-down technology. The 0.4-mm-diameter sample reached an output power of 28 W for incident pump power of 100 W. Due to the use of special cavity design, the emission was highly multi-mode with $M^2 > 10$.

Another example is LuAG fiber [329]. Undoped and Yb-doped $Lu_3Al_5O_{12}$ fiber single crystals were grown by using the micro-pulling-down (μ-PD) technique. The diameter of the crystal fibers is determined by the melt meniscus section formed by the crystallization surface. The longitudinal segregation coefficient (k) of the dopant (Yb^{3+}) was measured by using optical fluorescence and inductively coupled plasma (ICP). The crystal fiber was longitudinally pumped by a fiber-coupled laser diode emitting a maximum output power of 100 W at 940 nm. High-purity (5 N) Lu_2O_3, Al_2O_3, and Yb_2O_3 were used to synthesize $Lu_{3(1-x)}Yb_{3x}Al_5O_{12}$, by using the solid-state reaction method at 1400 °C for 30 h in air atmosphere.

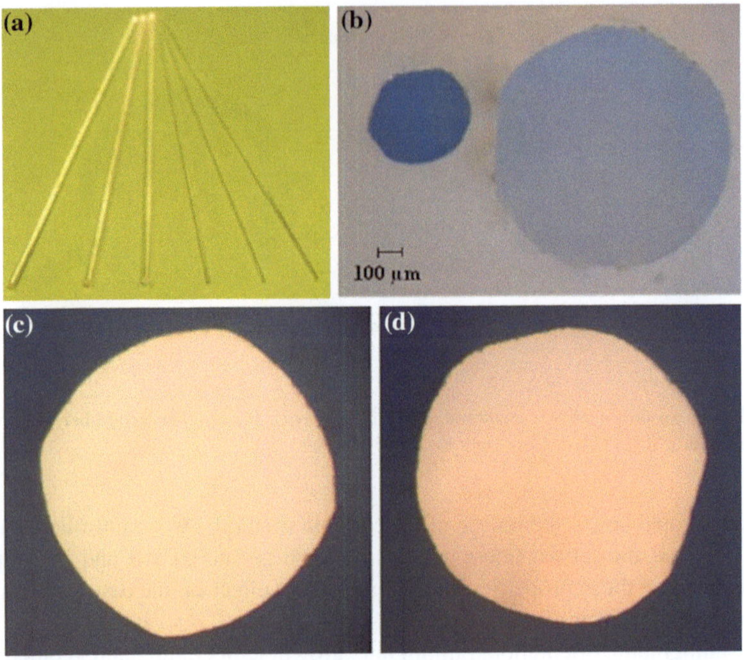

Fig. 9.53 **a** Photographs of the 50-mm-long, 1- and 0.4-mm-thick diameter samples cut from 1-m-long 1 at.% Yb-doped YAG single-crystal fibers grown by using the micro-pulling-down method. **b** Polished cross section of 0.4- and 1-mm-diameter <111> fibers and polished cross section of the 1-mm-diameter Yb:YAG single-crystal fiber with <100> orientation (**c**) and <111> orientation (**d**). Reproduced with permission from [326]. Copyright © 2009, Elsevier

Figure 9.55 shows the RF μ-PD setup that was used to grow the Yb^{3+}-doped LuAG fibers [329]. Induction heating generator, operated at 34 kHz and 25 kW, was used to heat the crucible and the materials. High-density and high-purity (99.7 %) alumina was used as ceramic mantle to protect the crucible for thermal insulation. Both heating and cooling were automatically controlled by using computer. Iridium crucible was used as the container. Crystal growth was carried out in argon atmosphere to prevent oxidation of iridium crucible. <111> LuAG seed crystal was dipped in the melt at the bottom of the crucible in the capillary die, while the melt temperature was adjusted so that a meniscus was supported. The molten zone and meniscus length were controlled by carefully adjusting the heating power. The whole assembly was enclosed in a quartz envelope, in which the gas flow could be well controlled and the growing process could be monitored by using a CCD camera.

Figure 9.56 shows photographs of representative samples of both the undoped and Yb-doped LuAG single-crystal fibers, while laser efficiency is shown in Fig. 9.57 [329]. The laser beam could be remained unguided inside the single-crystal fiber, with an estimated waist size of 120 μm in diameter, which was much

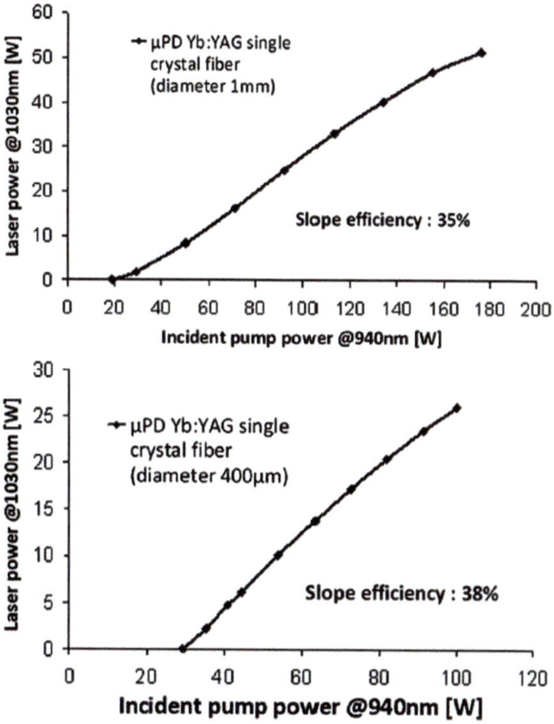

Fig. 9.54 Laser efficiencies in CW operation at 1030 nm under 940 nm pump power of the 1-mm-diameter Yb:YAG fiber (*above*) and of 0.4-mm-diameter Yb:YAG fiber (*below*). Reproduced with permission from [326]. Copyright © 2009, Elsevier

thinner than diameter of the whole pumped single-crystal fiber of about 800 μm. The efficiency was limited by the weak overlap between the pumped volume and the laser mode and the limited intrinsic transmission. In addition, the laser beam suffered diffraction losses at the crystal fiber extremities, which also affected the efficiency of the system. Nevertheless, the preliminary results based on the Yb: LuAG single-crystal fibers were very encouraging.

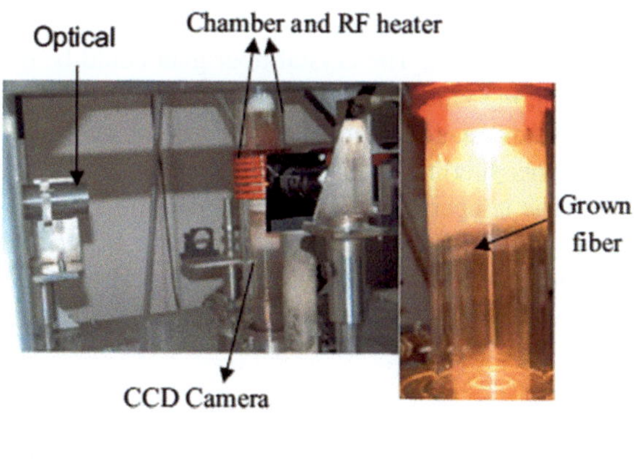

Fig. 9.55 Radio frequency (RF) micro-pulling-down facility for LuAG fiber crystal growth. Reproduced with permission from [329]. Copyright © 2009, Elsevier

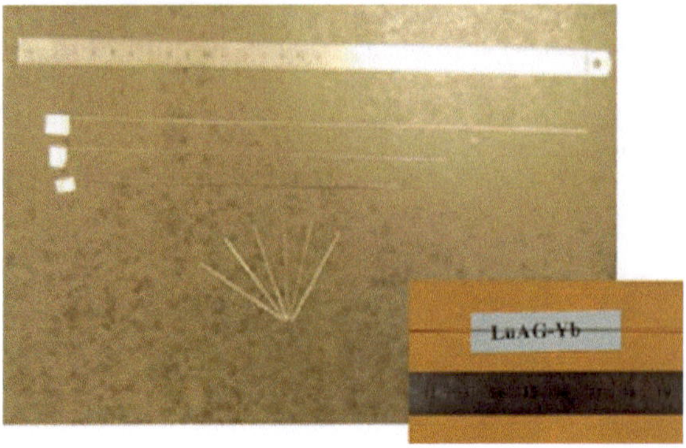

Fig. 9.56 Undoped and Yb-doped LuAG fibers grown under stationary state. Reproduced with permission from [329]. Copyright © 2009, Elsevier

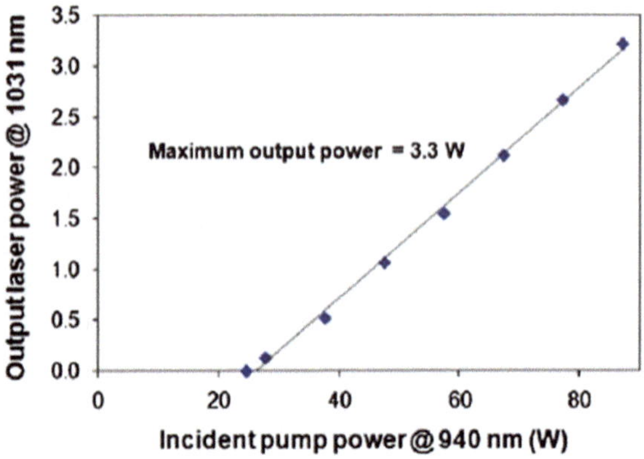

Fig. 9.57 Laser efficiency of the Yb:LuAG single-crystal fiber (0.8 mm diameter, 50 mm long, and 0.9 at.% doping level). Reproduced with permission from [329]. Copyright © 2009, Elsevier

A Cr^{4+}:YAG crystal fiber laser that was tunable from 1353 to 1509 nm, by using a birefringent filter, with a threshold pump power of 70 mW, operating at room temperature, has been reported [330]. The crystal fiber was 40 mm long, with a core diameter of 120 μm. The coupling loss between the crystal fiber and the collimator due to the lens aberration was a critical factor that limited the laser performance of the fiber, which, however, could be reduced by employing an aspheric lens with a short focal length. The crystal fiber gain could be optimized by aligning the pump polarization to the crystal axis. As compared with bulk crystals, the crystal fiber has various advantages, such as reduced pump threshold due to the light confinement and higher thermal dissipation efficiency due to the large surface area-to-volume ratio. The temperature rise of the crystal fiber with a 400 mW incident peak pump power with a duty cycle of 5 % was as low as 1.7 °C, in the case of passive cooling. Therefore, this low-threshold crystal fiber laser can be used as a tunable seed laser for power amplifier chains.

9.4 Summary

Ceramic lasers have formed an important part of solid-state lasers, which have been most extensively and intensively studied for various potential applications. Some of these lasers are already at a mature stage, but many others are still at the stage of investigation or development. The emergence of a wealth of new ceramic materials has promoted the fundamental understanding on lasing-related phenomena, such as quantum states, radiative and nonradiative de-excitation, interactions between the doping ions and energy transfer process, and various quantum electronic processes. Ceramic laser materials have also enriched the development of solid-state lasers, including extension of wavelength range, improvement of efficiency, generation of very short pulse, scaling of average, and peak power energy.

Ceramic processing is a flexible technology to develop laser materials, with controllable composition of host materials, doping concentration of active ions, and ion distributions. Therefore, it is possible to tailor the properties of laser materials to meet the requirements of specific applications. Thermomechanical and thermo-optic effects related to heat generation have been the main factors that limit the power or energy scaling of solid-state lasers. As a result, the management of intensity and distribution of thermal field inside the laser materials becomes a critical issue. Maximum utilization of radiation power from the excitation sources can be approached by improving pump absorption and laser emission efficiency and minimizing parasitic de-excitation. To achieve this, all the three parts of a laser, i.e., laser material, pumping system, and laser design, must be considered, from a system point of view.

Composite lasers and fiber lasers will continuously be hot topics of solid-state lasers. It is believed that more and more composite lasers will be developed. Ceramic fibers have high laser gains and thus allow high-power operation with short lengths. The main challenge of ceramic fiber lasers is the reduction in the core diameter. In addition, the solid-state crystal growth (SSCG) discussed in the last chapter also deserves further study. It is believed that transparent ceramics should have a bright future in research and development.

References

1. Ikesue A, Aung YL, Lupei V (2014) Ceramic lasers. Cambridge University Press, Cambridge
2. Ikesue A, Kinoshita T, Kamata K, Yoshida K (1995) Fabrication and optical properties of high-performance polycrystalline Nd-YAG ceramics for solid-state lasers. J Am Ceram Soc 78:1033–1040
3. Greskovich C, Chernoch JP (1973) Polycrystalline ceramic lasers. J Appl Phys 44:4599–4606
4. Ikesue A, Aung YL (2008) Ceramic laser materials. Nat Photonics 2:721–727
5. Lu J, Takaichi K, Uematsu T, Shirakawa A, Musha M, Ueda K et al (2002) Promising ceramic laser material: highly transparent Nd^{3+}:Lu_2O_3 ceramic. Appl Phys Lett 81:4324–4326

6. Lu J, Bisson JF, Takaichi K, Uematsu T, Shirakawa A, Musha M et al (2003) Yb^{3+}:Sc_2O_3 ceramic laser. Appl Phys Lett 83:1101–1103
7. Lu JH, Lu JR, Murai T, Takaichi K, Uematsu T, Xu JQ et al (2002) 36-W diode-pumped continuous-wave 1319-nm Nd:YAG ceramic laser. Opt Lett 27:1120–1122
8. Takaichi K, Yagi H, Lu JR, Bisson JF, Shirakawa A, Ueda K et al (2004) Highly efficient continuous-wave operation at 1030 and 1075 nm wavelengths of LD-pumped Yb^{3+}:Y_2O_3 ceramic lasers. Appl Phys Lett 84:317–319
9. Takaichi K, Yagi H, Shirakawa A, Ueda K, Yanagitani T, Kaminskii AA et al (2005) Highly transparent Yb-doped ceramics and the laser-diode-pumped ceramic lasers. IEEE, New York
10. Ikesue A, Aung YL (2006) Synthesis and performance of advanced ceramic lasers. J Am Ceram Soc 89:1936–1944
11. Lupei V (2009) Ceramic laser materials and the prospect for high power lasers. Opt Mater 31:701–706
12. Wang YB, Wang BS, Bo Y, Xu JL, Song S, Peng QJ et al (2010) High efficiency, high power QCW diode-side-pumped Nd:YAG ceramic laser at 1064 nm based on domestic ceramic. Chin Opt Lett 8:1144–1146
13. Gheorghe C, Lupei A, Lupei V, Ikesue A, Enculescu M (2011) Intensity parameters of Tm^{3+} doped Sc_2O_3 transparent ceramic laser material. Opt Mater 33:501–505
14. Liu WB, Zhang D, Zeng YP, Li J, Pan YB, Guo JK et al (2012) Diode-side-pumped 1123 nm Nd:YAG ceramic laser. Ceram Int 38:6969–6973
15. Wang Y, Chen H, Shen DY, Zhang J, Tang DY (2013) High power continuous-wave and graphene Q-switched operation of Er:YAG ceramic lasers at similar to 1.6 μm. J Opt Soc Korea 17:5–9
16. Liu Q, Liu J, Li J, Ivanov M, Medvedev A, Zeng Y et al (2014) Solid-state reactive sintering of YAG transparent ceramics for optical applications. J Alloy Compd 616:81–88
17. Wang J, Zhang F, Chen F, Zhang J, Zhang HL, Tian R et al (2015) Effect of Y_2O_3 and La_2O_3 on the sinterability of gamma-AlON transparent ceramics. J Eur Ceram Soc 35:23–28
18. Yang H, Zhang L, Luo D, Qiao X, Zhang J, Zhao T et al (2015) Optical properties of Ho: YAG and Ho:LuAG polycrystalline transparent ceramics. Opt Mater Expr 5:142–148
19. Lupei V (2003) Efficiency enhancement and power scaling of Nd lasers. Opt Mater 24:353–368
20. Lupei V, Pavel N, Taira T (2003) Basic enhancement of the overall optical efficiency of intracavity frequency-doubling devices for the 1 μm continuous-wave Nd: $Y_3Al_5O_{12}$ laser emission. Appl Phys Lett 83:3653–3655
21. Rosenkra LJ (1972) GaAs diode-pumped Nd-YAG laser. J Appl Phys 43:4603–4605
22. Rosenkra Lj, Rowley RS (1974) CW operation of a GaAs diode pumped Nd-YAG laser. IEEE J Quantum Electron QE10:710–711
23. Ross M (1968) YAG laser operation by semiconductor laser pumping. Proc Inst Electr Electron Eng 56:196–197
24. Lupei V, Lupei A, Gheorghe C, Ikesue A (2008) Comparative high-resolution spectroscopy and emission dynamics of Nd-doped GSGG crystals and transparent ceramics. J Lumin 128:885–887
25. Lavi R, Jackel S (2000) Thermally boosted pumping of neodymium lasers. Appl Opt 39:3093–3098
26. Lavi R, Jackel S, Tal A, Lebiush E, Tzuk Y, Apter M (2001) Thermally-boosted pumping of Nd:YAG using high power 885 nm diodes. In: Marshall C (ed) Advanced solid-state lasers, Proceedings 2001, pp 693–695
27. Lavi R, Jackel S, Tal A, Lebiush E, Tzuk Y, Goldring S (2001) 885 nm high-power diodes end-pumped Nd:YAG laser. Opt Commun 195:427–430
28. Lavi R, Jackel S, Tzuk Y, Winik M, Lebiush E, Katz M et al (1999) Efficient pumping scheme for neodymium-doped materials by direct excitation of the upper lasing level. Appl Opt 38:7382–7385

29. Lupei V, Lupei A, Georgescu S, Taira T, Sato Y, Ikesue A (2001) The effect of Nd concentration on the spectroscopic and emission decay properties of highly doped Nd:YAG ceramics. Phys Rev B 64:092102
30. Lupei V, Taira T, Lupei A, Pavel N, Shoji I, Ikesue A (2001) Spectroscopy and laser emission under hot band resonant pump in highly doped Nd:YAG ceramics. Opt Commun 195:225–232
31. Lupei V, Lupei A, Pavel N, Taira T, Shoji I, Ikesue A (2001) Laser emission under resonant pump in the emitting level of concentrated Nd:YAG ceramics. Appl Phys Lett 79:590–592
32. Lupei V, Lupei A, Pavel N, Taira T, Ikesue A (2001) Comparative investigation of spectroscopic and laser emission characteristics under direct 885-nm pump of concentrated Nd:YAG ceramics and crystals. Appl Phys B Lasers Opt 73:757–762
33. Lupei V, Aka G, Vivien D (2002) Quasi-three-level 946 nm CW laser emission of Nd:YAG under direct pumping at 885 nm into the emitting level. Opt Commun 204:399–405
34. Goldring S, Lavi R (2008) Nd:YAG laser pumped at 946 nm. Opt Lett 33:669–671
35. Yagi H, Yanagitani T, Ueda K-I (2006) Nd^{3+}:$Y_3Al_5O_{12}$ laser ceramics: Flashlamp pumped laser operation with a UV cut filter. J Alloy Compd 421:195–199
36. Yagi H, Yanagitani T, Yoshida H, Nakatsuka M, Ueda K (2006) Highly efficient flashlamp-pumped Cr^{3+} and Nd^{3+} codoped $Y_3Al_5O_{12}$ ceramic laser. Jpn J Appl Phys Part 1 Regul Pap Brief Commun Rev Pap 45:133–135
37. Ikesue A, Kamata K, Yoshida K (1995) Synthesis of Nd^{3+}, Cr^{3+}-coped YAG ceramics for high-efficiency solid-state lasers. J Am Ceram Soc 78:2545–2547
38. Saiki T, Motokoshi S, Imasaki K, Fujioka K, Yoshida H, Fujita H et al (2009) High repetition rate laser pulses amplified by Nd/Cr:YAG ceramic amplifier under CW arc-lamp-light pumping. Opt Commun 282:2556–2559
39. Saiki T, Motokoshi S, Imasaki K, Fujioka K, Yoshida H, Fujita H et al (2009) Laser pulses amplified by Nd/Cr:YAG ceramic amplifier using lamp and solar light sources. Opt Commun 282:1358–1362
40. Saiki T, Motokoshi S, Imasaki K, Nakatsuka M, Yamanaka C, Fujioka K et al (2009) Two-pass amplification of CW laser by Nd/Cr:YAG ceramic active mirror under lamp light pumping. Opt Commun 282:936–939
41. Saiki T, Nakatsuka M, Fujioka K, Motokoshi S, Imasaki K (2011) Cross-relaxation and spectral broadening of gain for Nd/Cr:YAG ceramic lasers with white-light pump source under high-temperature operation. Opt Commun 284:2980–2984
42. Saiki T, Nakatsuka M, Imasaki K (2010) Highly efficient lasing action of Nd^{3+}- and Cr^{3+}-doped yttrium aluminum garnet ceramics based on phonon-assisted cross-relaxation using solar light source. Jpn J Appl Phys 49:082702
43. Lupei V, Lupei A, Ikesue A (2005) Transparent Nd and (Nd, Yb)-doped Sc_2O_3 ceramics as potential new laser materials. Appl Phys Lett 86:111118
44. Lupei A, Lupei V, Ikesue A, Gheorghe C (2010) Spectroscopic and energy transfer investigation of Nd/Yb in Y2O3 transparent ceramics. J Opt Soc Am B Opt Phys 27:1002–1010
45. Lupei V, Lupei A, Gheorghe C, Hau S, Ikesue A (2009) Efficient sensitization of Yb^{3+} emission by Nd^{3+} in Y_2O_3 transparent ceramics and the prospect for high-energy Yb lasers. Opt Lett 34:2141–2143
46. Lupei V, Lupei A, Gheorghe C, Ikesue A (2010) Sensitized Yb^{3+} emission in (Nd, Yb): $Y_3Al_5O_{12}$ transparent ceramics. J Appl Phys 108:123112
47. Petit V, Camy P, Doualan JL, Moncorge R (2006) CW and tunable laser operation of Yb^{3+} in Nd: Yb:CaF_2. Appl Phys Lett 88:051111
48. Petit V, Camy P, Doualan JL, Portier X, Moncorge R (2008) Spectroscopy of Yb^{3+}: Ca_2: From isolated centers to clusters. Phys Rev B 78:085131
49. Petit V, Doualan JL, Camy P, Menard V, Moncorge R (2004) CW and tunable laser operation of Yb^{3+} doped CaF_2. Appl Phys B Lasers Opt 78:681–684
50. Kallel T, Hassairi MA, Dammak M, Lyberis A, Gredin P, Mortier M (2014) Spectra and energy levels of Yb^{3+} ions in CaF_2 transparent ceramics. J Alloy Compd 584:261–268

51. Kaminskii AA, Bagaev SN, Ueda K, Shirakawa A, Tokurakawa T, Yagi H et al (2009) Stimulated-emission spectroscopy of fine-grained "garnet" ceramics $Nd^{3+}:Y_3Al_5O_{12}$ in a wide temperature range between 77 and 650 K. Laser Phys Lett 6:682–687
52. Koechner W (1970) Absorbed pumb powder, thermal profile and stresses in a CW pumped Nd-YAG crystal. Appl Opt 9:1429–1434
53. Kracht D, Wilhelm R, Frede M, Dupre K, Ackermann L (2005) 407 W end-pumped multi-segmented Nd:YAG laser. Opt Express 13:10140–10144
54. Wilhelm R, Frede M, Kracht D (2008) Power scaling of end-pumped solid-state rod lasers by longitudinal dopant concentration gradients. IEEE J Quantum Electron 44:232–244
55. Denis T, Hahn S, Mebben S, Wilhelm R, Kolleck C, Neumann J et al (2010) Compact diode stack end pumped Nd:YAG amplifier using core doped ceramics. Appl Opt 49:811–816
56. Kracht D, Frede M, Wilhelm R, Fallnich C (2005) Comparison of crystalline and ceramic composite Nd:YAG for high power diode end-pumping. Opt Express 13:6212–6216
57. Kracht D, Freiburg D, Wilhelm R, Frede M, Fallnich C (2006) Core-doped ceramic Nd:YAG laser. Opt Express 14:2690–2694
58. Puncken O, Winkelmann L, Frede M, Wessels P, Neumann J, Kracht D (2012) Heat generation in Nd:YAG at different doping levels. Appl Opt 51:7586–7590
59. Wilhelm R, Freiburg D, Frede M, Kracht D, Fallnich C (2009) Design and comparison of composite rod crystals for power scaling of diode end-pumped Nd:YAG lasers. Opt Express 17:8229–8236
60. Straesser A, Ostermeyer M (2006) Improving the brightness of side pumped power amplifiers by using core doped ceramic rods. Opt Express 14:6687–6693
61. Huang T, Jiang B, Wu Y, Li J, Shi Y, Liu W et al (2009) Fabrication, microstructure and optical properties of titanium doped YAG transparent ceramic. J Alloy Compd 478:L16–L20
62. Wilhelm R, Freiburg D, Frede M, Kracht D (2008) End-pumped Nd:YAG laser with a longitudinal hyperbolic dopant concentration profile. Opt Express 16:20106–20116
63. Kouznetsov D, Bisson JF (2008) Role of undoped cap in the scaling of thin-disk lasers. J Opt Soc Am B Opt Phys 25:338–345
64. Kouznetsov D, Bisson JF, Dong J, Ueda KI (2006) Surface loss limit of the power scaling of a thin-disk laser. J Opt Soc Am B Opt Phys 23:1074–1082
65. Kouznetsov D, Bisson JF, Ueda K (2009) Scaling laws of disk lasers. Opt Mater 31:754–759
66. Bagayev SN, Kaminskii AA, Kopylov YL, Kotelyanskii IM, Kravchenko VB (2012) Simple method to join YAG ceramics and crystals. Opt Mater 34:951–954
67. Ge L, Li J, Zhou ZW, Qu HY, Dong MJ, Zhu Y et al (2014) Fabrication of composite YAG/Nd:YAG/YAG transparent ceramics for planar waveguide laser. Opt Mater Express 4:1042–1049
68. Tang F, Cao YG, Huang JQ, Guo W, Liu HG, Huang QF et al (2012) Multilayer YAG/RE:YAG/YAG laser ceramic prepared by tape casting and vacuum sintering method. J Eur Ceram Soc 32:3995–4002
69. Sandu O, Salamu G, Pavel N, Dascalu T, Chuchumishev D, Gaydardzhiev A et al (2012) High-peak power, passively Q-switched, composite, all-polycrystalline ceramic Nd:YAG/Cr^{4+}:YAG lasers. Quantum Electron 42:211–215
70. Ma J, Dong J, Ki Ueda, Kaminskii AA (2011) Optimization of Yb:YAG/Cr^{4+}:YAG composite ceramics passively Q-switched microchip lasers. Appl Phys B Lasers Opt 105:749–760
71. Pavel N, Tsunekane M, Taira T (2011) Composite, all-ceramics, high-peak power Nd:YAG/Cr^{4+}:YAG monolithic micro-laser with multiple-beam output for engine ignition. Opt Express 19:9378–9384
72. Salamu G, Ionescu A, Brandus CA, Sandu O, Pavel N, Dascalu T (2012) High-peak power, passively Q-switched, composite, all-poly-crystalline ceramics Nd:YAG/Cr^{4+}:YAG laser and generation of 532-nm green light. Laser Phys 22:68–73
73. Tang F, Cao Y, Huang J, Liu H, Guo W, Wang W (2012) Fabrication and laser behavior of composite Yb:YAG ceramic. J Am Ceram Soc 95:56–59

74. Liu WB, Zeng YP, Li J, Shen Y, Bo Y, Zong N et al (2012) Sintering and laser behavior of composite YAG/Nd:YAG/YAG transparent ceramics. J Alloy Compd 527:66–70
75. Marquardt K, Petrishcheva E, Abart R, Gardes E, Wirth R, Dohmen R et al (2010) Volume diffusion of Ytterbium in YAG: thin-film experiments and combined TEM-RBS analysis. Phys Chem Miner 37:751–760
76. Marquardt K, Petrishcheva E, Gardes E, Wirth R, Abart R, Heinrich W (2011) Grain boundary and volume diffusion experiments in yttrium aluminium garnet bicrystals at 1,723 K: aminiaturized study. Contrib Miner Petrol 162:739–749
77. Hollingsworth JP, Kuntz JD, Soules TF (2009) Neodymium ion diffusion during sintering of Nd:YAG transparent ceramics. J Phys D Appl Phys 42
78. Hollingsworth JP, Kuntz JD, Ryerson FJ, Soules TF (2011) Nd diffusion in YAG ceramics. Opt Mater 33:592–595
79. Wisdom JA, Gaume RM, Byer RL (2010) Laser-gain scanning microscopy: a new characterization technique for dopant engineered gain media. Opt Express 18:18912–18921
80. Furuse H, Kawanaka J, Miyanaga N, Saiki T, Imasaki K, Fujita M et al (2011) Zig-zag active-mirror laser with cryogenic Yb^{3+}:YAG/YAG composite ceramics. Opt Express 19:2448–2455
81. Furuse H, Kawanaka J, Takeshita K, Miyanaga N, Saiki T, Imasaki K et al (2009) Total-reflection active-mirror laser with cryogenic Yb:YAG ceramics. Opt Lett 34:3439–3441
82. Furuse H, Sakurai T, Chosrowjan H, Kawanaka J, Miyanaga N, Fujita M et al (2014) Amplification characteristics of a cryogenic Yb^{3+}:YAG total-reflection active-mirror laser. Appl Opt 53:1964–1969
83. Yoshida H, Tsubakimoto K, Fujimoto Y, Mikami K, Fujita H, Miyanaga N et al (2011) Optical properties and Faraday effect of ceramic terbium gallium garnet for a room temperature Faraday rotator. Opt Express 19:15181–15187
84. Lin H, Zhou S, Teng H (2011) Synthesis of $Tb_3Al_5O_{12}$ (TAG) transparent ceramics for potential magneto-optical applications. Opt Mater 33:1833–1836
85. Zheleznov D, Starobor A, Palashov O, Lin H, Zhou S (2014) Improving characteristics of Faraday isolators based on TAG ceramics by cerium doping. Opt Lett 39:2183–2186
86. Feldman R, Golan Y, Burshtein Z, Jackel S, Moshe I, Meir A et al (2011) Strengthening of poly-crystalline (ceramic) Nd:YAG elements for high-power laser applications. Opt Mater 33:695–701
87. Feldman R, Shimony Y, Lebiush E, Golan Y (2008) Effect of hot acid etching on the mechanical strength of ground YAG laser elements. J Phys Chem Solids 69:839–846
88. Lu J, Prabhu M, Song J, Li C, Xu J, Ueda K et al (2000) Optical properties and highly efficient laser oscillation of Nd:YAG ceramics. Appl Phys B Lasers Opt 71:469–473
89. Shoji I, Kurimura S, Sato Y, Taira T, Ikesue A, Yoshida K (2000) Optical properties and laser characteristics of highly $Nd3^{+}$-doped $Y_3Al_5O_{12}$ ceramics. Appl Phys Lett 77:939–941
90. Lu JR, Ueda K, Yagi H, Yanagitani T, Akiyama Y, Kaminskii AA (2002) Neodymium doped yttrium aluminum garnet ($Y_3Al_5O_{12}$) nanocrystalline ceramics—a new generation of solid state laser and optical materials. J Alloy Compd 341:220–225
91. Lupei V, Pavel N, Taira T (2002) Efficient laser emission in concentrated Nd laser materials under pumping into the emitting level. IEEE J Quantum Electron 38:240–245
92. Lupei V, Pavel N, Taira T (2002) Highly efficient laser emission in concentrated $Nd:YVO_4$ components under direct pumping into the emitting level. Opt Commun 201:431–435
93. Li HF, Xu DG, Yang Y, Wang YY, Zhou R, Zhang TL et al (2005) Experimental 511 W composite Nd:YAG ceramic laser. Chin Phys Lett 22:2565–2567
94. Huss R, Wilhelm R, Kolleck C, Neumann J, Kracht D (2010) Suppression of parasitic oscillations in a core-doped ceramic Nd:YAG laser by Sm:YAG cladding. Opt Express 18:13094–13101
95. McNaught SJ, Komine H, Weiss SB, Simpson R, Johnson AMF, Machan J et al (2009) 100 kW coherently combined sab MOPAs. In: 2009 IEEE conference on lasers and electro-optics and quantum electronics and laser science conference, pp 830–831

96. Mandl A, Klimek DE (2010) Textron's J-HPSSL 100 kW ThinZag (R) laser program. In: Conference on Lasers and Electro-Optics (CLEO)/Quantum Electronics and Laser Science Conference (QELS): WOS:000290513602227
97. Yamamoto RM, Parker JM, Allen KL, Allmon RW, Alviso KF, Barty CPJ et al (2007) Evolution of a solid state laser—art. no. 655205. In: Wood GL, Dubindkii MA (eds) Laser source technology for defense and security III, Proceedings of the Society of Photo-Optical Instrumentation Engineers (SPIE), 6552:55205
98. Li CY, Bo Y, Wang BS, Tian CY, Peng QJ, Cui DF et al (2010) A kilowatt level diode-side-pumped QCW Nd:YAG ceramic laser. Opt Commun 283:5145–5148
99. Lupei V, Pavel N, Taira T (2002) Highly efficient continuous-wave 946-nm Nd:YAG laser emission under direct 885-nm pumping. Appl Phys Lett 81:2677–2679
100. Lupei V, Pavel N, Taira T (2002) 1064 nm laser emission of highly doped Nd: Yttrium aluminum garnet under 885 nm diode laser pumping. Appl Phys Lett 80:4309–4311
101. Frede M, Wilhelm R, Kracht D (2006) 250 W end-pumped Nd:YAG laser with direct pumping into the upper laser level. Opt Lett 31:3618–3619
102. Yagi H, Yanagitani T, Takaichi K, Ueda K, Kaminskii AA (2007) Characterizations and laser performances of highly transparent Nd^{3+}:$Y_3Al_5O_{12}$ laser ceramics. Opt Mater 29:1258–1262
103. Zhang SS, Wang QP, Zhang XY, Cong ZH, Fan SZ, Liu ZJ et al (2009) Continuous-wave ceramic Nd:YAG laser at 1123 nm. Laser Phys Lett 6:864–867
104. Li C, Bo Y, Yang F, Wang Z, Xu Y, Wang Y et al (2010) 106.5 W high beam quality diode-side-pumped Nd:YAG laser at 1123 nm. Opt Express 18:7923–7928
105. Li CY, Bo Y, Xu JL, Tian CY, Peng QJ, Cui DF et al (2011) Simultaneous dual-wavelength oscillation at 1116 and 1123 nm of Nd:YAG laser. Opt Commun 284:4574–4576
106. Li CY, Bo Y, Xu YT, Yang F, Wang ZC, Wang BS et al (2010) 219.3 W CW diode-side-pumped 1123 nm Nd:YAG laser. Opt Commun 283:2885–2887
107. Wang Z, Liu H, Wang J, Lv Y, Sang Y, Lan R et al (2009) Passively Q-switched dual-wavelength laser output of LD-end-pumped ceramic Nd:YAG laser. Opt Express 17:12076–12081
108. Strohmaier SGP, Eichler HJ, Bisson JF, Yagi H, Takaichi K, Ueda K et al (2005) Ceramic Nd:YAG laser at 946 nm. Laser Phys Lett 2:383–386
109. Guo L, Lan R, Liu H, Yu H, Zhang H, Wang J et al (2010) 1319 nm and 1338 nm dual-wavelength operation of LD end-pumped Nd:YAG ceramic laser. Opt Express 18:9098–9106
110. Chen L, Wang Z, Liu H, Zhuang S, Yu H, Guo L et al (2010) Continuous-wave tri-wavelength operation at 1064, 1319 and 1338 nm of LD end-pumped Nd:YAG ceramic laser. Opt Express 18:22167–22173
111. Cai H, Zhou J, Feng T, Yao G, Qi Y, Lou Q et al (2008) Dual-wavelength competitive output in Nd:$Y_3Sc_{1.5}Al_{3.5}O_{12}$ ceramic disk laser. Opt Commun 281:4401–4405
112. Saiki T, Motokoshi S, Imasaki K, Fujioka K, Fujita H, Nakatsuka M et al (2008) Effective fluorescence lifetime and stimulated emission cross-section of Nd/Cr:YAG ceramics under CW lamplight pumping. Jpn J Appl Phys 47:7896–7902
113. Yagi H, Yanagitani T, Yoshida H, Nakatsuka M, Ueda K (2007) The optical properties and laser characteristics of Cr^{3+} and Nd^{3+} co-doped $Y_3Al_5O_{12}$ ceramics. Opt Laser Technol 39:1295–1300
114. Ohkubo T, Yabe T, Yoshida K, Uchida S, Funatsu T, Bagheri B et al (2009) Solar-pumped 80 W laser irradiated by a Fresnel lens. Opt Lett 34:175–177
115. Dinh TH, Ohkubo T, Yabe T (2014) Development of solar concentrators for high-power solar-pumped lasers. Appl Opt 53:2711–2719
116. Dinh TH, Ohkubo T, Yabe T, Kuboyama H (2012) 120 watt continuous wave solar-pumped laser with a liquid light-guide lens and an Nd:YAG rod. Opt Lett 37:2670–2672
117. Greskovich C, Chernoch JP (1974) Improved polycrysalline ceramic lasers. J Appl Phys 45:4495–4502
118. Greskovich C, Woods KN (1973) Fabrication of transparent ThO_2-doped Y_2O_3. Am Ceram Soc Bull 52:473–478
119. Lu JR, Lu JH, Murai T, Takaichi K, Uematsu T, Ueda K et al (2001) Nd^{3+}:Y_2O_3 ceramic laser. Jpn J Appl Phys Part 2 Lett 40:L1277–L1279

120. Yang QH, Lu SZ, Zhang B, Zhang HJ, Zhou J, Yuan ZJ et al (2011) Preparation and laser performance of Nd-doped yttrium lanthanum oxide transparent ceramic. Opt Mater 33:692–694
121. Kurokawa H, Shirakawa A, Tokurakawa M, Ueda K, Kuretake S, Tanaka N et al (2011) Broadband-gain Nd^{3+}-doped Ba(Zr, Mg, Ta)O_3 ceramic lasers for ultrashort pulse generation. Opt Mater 33:667–669
122. Zhang JWW, Zou YYK, Chen QS, Zhang R, Li KRK, Jiang H et al (2006) Optical amplification in Nd^{3+} doped electro-optic lanthanum lead zirconate titanate ceramics. Appl Phys Lett 89:061113
123. Zhao H, Sun XD, Zhang JW, Zou YYK, Li KWK, Wang YY et al (2011) Lasing action and optical amplification in Nd^{3+} doped electrooptic lanthanum lead zirconate titanate ceramics. Opt Express 19:2965–2971
124. Basiev TT, Doroshenko ME, Konyushkin VA, Osiko VV (2010) SrF_2:Nd^{3+} laser fluoride ceramics. Opt Lett 35:4009–4011
125. Akiyama J, Sato Y, Taira T (2011) Laser demonstration of diode-pumped Nd^{3+}-doped fluorapatite anisotropic ceramics. Appl Phys Express 4:022703
126. Akiyama J, Sato Y, Taira T (2010) Laser ceramics with rare-earth-doped anisotropic materials. Opt Lett 35:3598–3600
127. Wang CY, Ji JH, Qi YF, Lou QH, Zhu XL, Lu YT (2006) Kilohertz electro-optic Q-switched Nd:YAG ceramic laser. Chin Phys Lett 23:1797–1799
128. Qi YF, Zhu XL, Lou QH, Ji JH, Dong JX, Wei RR (2007) High-energy LDA switched side-pumped electro-optical Q-switched Nd:YAG ceramic laser. J Opt Soc Am B Opt Phys 24:1042–1045
129. Omatsu T, Nawata K, Sauder D, Minassian A, Damzen MJ (2006) Over 40-watt diffraction-limited Q-switched output from neodymium-doped YAG ceramic bounce amplifiers. Opt Express 14:8198–8204
130. Li P, Wang QP, Zhang XY, Wang YR, Chen XH (2010) Compact and efficient diode-pumped actively Q-switched 1319 nm Nd:YAG ceramic laser. Laser Phys 20:1603–1607
131. Zhang C, Zhang XY, Wang QP, Cong ZH, Fan SZ, Chen XH et al (2009) Diode-pumped Q-switched 946 nm Nd:YAG ceramic laser. Laser Phys Lett 6:521–525
132. Ikesue A, Yoshida K, Kamata K (1996) Transparent Cr^{4+}-doped YAG ceramics for tunable lasers. J Am Ceram Soc 79:507–509
133. Yagi H, Takaichi K, Ueda K, Yanagitani T, Karninskii AA (2006) Influence of annealing conditions on the optical properties of chromium-doped ceramic $Y_3Al_5O_{12}$. Opt Mater 29:392–396
134. Degnan JJ (1995) Optimization of passively Q-switched lasers. IEEE J Quantum Electron 31:1890–1901
135. Kalisky Y, Kalisky O, Rachum U, Boulon G, Brenier A (2007) Comparative performance of passively Q-switched diode-pumped Yb^{3+}-doped tungstate and garnet lasers using Cr^{4+}:YAG saturable absorber. IEEE J Sel Top Quantum Electron 13:502–510
136. Kalisky Y (2004) Cr^{4+}-doped crystals: their use as lasers and passive Q-switches. Prog Quantum Electron 28:249–303
137. Kalisky Y, Kravchik L, Kokta MR (2004) Performance of diode-end-pumped Cr^{4+}, Nd^{3+}: YAG self-Q-switched and Nd:YAG/Cr^{4+}:YAG diffusion bonded lasers. Opt Mater 24:607–614
138. Kalisky Y, Labbe C, Waichman K, Kravchik L, Rachum U, Deng P et al (2002) Passively Q-switched diode-pumped Yb:YAG laser using Cr^{4+}-doped garnets. Opt Mater 19:403–413
139. Sakai H, Kan H, Taira T (2008) >1 MW peak power single-mode high-brightness passively Q-switched Nd^{3+}:YAG microchip laser. Opt Express 16:19891–19899
140. Takaichi K, Lu JR, Murai T, Uematsu T, Shirakawa A, Ueda K et al (2002) Chromium doped $Y_3Al_5O_{12}$ ceramics—a novel saturable absorber for passively self-Q-switched one-micron solid state lasers. Jpn J Appl Phys Part 2 Lett 41:L96–L98
141. Lan R, Wang Z, Liu H, Yu H, Guo L, Chen L et al (2010) Passively Q-switched Nd:YAG ceramic laser towards large pulse energy and short pulse width. Laser Phys 20:187–191

142. Kong J, Zhang ZZ, Tang DY, Xie GQ, Chan CC, Shen YH (2008) Diode end-pumped passively Q-switched Nd:YAG ceramic laser with Cr^{4+}:YAG saturable absorber. Laser Phys 18:1508–1511
143. Feng Y, Lu JR, Takaichi K, Ueda K, Yagi H, Yanagitani T et al (2004) Passively Q-switched ceramic Nd^{3+}:YAG/Cr^{4+}:YAG lasers. Appl Opt 43:2944–2947
144. Yu HH, Zhang HJ, Wang ZP, Wang JY, Yu YG, Zhang XY et al (2009) Dual-wavelength neodymium-doped yttrium aluminum garnet laser with chromium-doped yttrium aluminum garnet as frequency selector. Appl Phys Lett 94:041126
145. Saiki T, Motokoshi S, Imasaki K, Fujioka K, Yoshida H, Fujita H et al (2009) Nd^{3+}- and Cr^{3+}-doped uttrium aluminum garnet ceramic pulse laser using Cr^{4+}-doped yttrium aluminum garnet crystal passive Q-switch. Jpn J Appl Phys 48:122501
146. Lupei V, Lupei A, Ionitamanzatu V, Georgescu S, Domsa F (1983) Combined mechanical-color center passive Q-switching of neodymium lasers. Opt Commun 48:203–206
147. Dascalu T, Pavel N, Lupei V, Philipps G, Beck T, Weber H (1996) Investigation of a passive Q-switched, externally controlled, quasicontinuous or continuous pumped Nd:YAG laser. Opt Eng 35:1247–1251
148. Dascalu T, Philipps G, Weber H (1997) Investigation of a Cr^{4+}:YAG passive Q-switch in CW pumped Nd:YAG lasers. Opt Laser Technol 29:145–149
149. Dascalu T, Acosta-Ortiz SE, Pavel N (2000) Nd : YAG laser continuous wave pumped, Q-switched by hybrid "passive-active" methods. Rev Mex Fis 46:320–328
150. Omatsu T, Minassian A, Damzen MJ (2009) Passive Q-switching of a diode-side-pumped Nd doped 1.3 μm ceramic YAG bounce laser. Opt Commun 282:4784–4788
151. Li P, Chen XH, Zhang HN, Wang QP (2011) Diode-end-pumped passively Q-switched 1319 nm Nd:YAG ceramic laser with a V^{3+}:YAG saturable absorber. Laser Phys 21:1708–1711
152. Omatsu T, Miyamoto K, Okida M, Minassian A, Damzen MJ (2010) 1.3-μm passive Q-switching of a Nd-doped mixed vanadate bounce laser in combination with a V:YAG saturable absorber. Appl Phys B Lasers Opt 101:65–70
153. Dong J, Ueda K, Yagi H, Kaminskii AA (2008) Laser-diode pumped self-Q-switched microchip lasers. Opt Rev 15:57–74
154. Ren Y, Dong J (2014) Passively Q-switched microchip lasers based on Yb:YAG/Cr^{4+}:YAG composite crystal. Opt Commun 312:163–167
155. Li J, Wu YS, Pan YB, Kou HM, Shi Y, Guo JK (2008) Densification and microstructure evolution of Cr^{4+}, Nd^{3+}:YAG transparent ceramics for self-Q-switched laser. Ceram Int 34:1675–1679
156. Li J, Wu YS, Pan YB, Guo JK (2006) Fabrication of Cr^{4+}, Nd^{3+}:YAG transparent ceramics for self-Q-switched laser. J Non-Cryst Solids 352:2404–2407
157. Ling WJ, Zhang SG, Zhang MX, Dong Z, Li K, Zuo YY et al (2010) High power continuous-wave actively mode-locked diode-pumped Nd:YAG laser. Chin Phys Lett 27:114202
158. Shcherbakov AS, Kosarsky AY, Moreno Zarate P, Campos Acosta J, Il'in YV, Tarasov IS (2011) Characterization of the train-average time-frequency parameters inherent in the low-power picosecond optical pulses generated by the actively mode-locked semiconductor laser with an external single-mode fiber cavity. Optik 122:136–141
159. Wojcik AK, Malara P, Blanchard R, Mansuripur TS, Capasso F, Belyanin A (2013) Generation of picosecond pulses and frequency combs in actively mode locked external ring cavity quantum cascade lasers. Appl Phys Lett 103
160. Yin K, Zhang BB, Yang WQ, Chen H, Chen SP, Hou J (2014) Flexible picosecond thulium-doped fiber laser using the active mode-locking technique. Opt Lett 39:4259–4262
161. Zhao J, Zhao SZ, Li K, Kong FM, Zhang G (2011) Optimization of passively Q-switched and mode-locked laser with Cr^{4+}:YAG saturable absorber. Opt Commun 284:1648–1651
162. Yang L, Feng B, Zhang Z, Gaebler V, Liu B, Eichler HJ (2003) Low mode-locking saturation intensity of co-doped Nd^{3+}, Cr^{4+}:YAG crystal as saturable absorber. Opt Mater 22:59–63

163. Yang L, Feng BH, Zhang ZG, Gaebler V, Liu BN, Eichler HJ et al (2002) Self-mode-locking in a diode-pumped self-Q-switched Nd^{3+}, Cr^{4+}:YAG laser. Chin Phys Lett 19:1450–1452
164. Tan WD, Chen FM, Knize RJ, Zhang J, Tang DY, Li LJ (2011) Passive mode locking of ceramic Nd: YAG using (7,5) semiconducting single walled carbon nanotubes. Opt Mater 33:679–683
165. Tan WD, Su CY, Knize RJ, Xie GQ, Li LJ, Tang DY (2010) Mode locking of ceramic Nd: yttrium aluminum garnet with graphene as a saturable absorber. Appl Phys Lett 96:031106
166. Guo L, Hou W, Zhang HB, Sun ZP, Cui DF, Xu ZY et al (2005) Diode-end-pumped passively mode-locked ceramic Nd:YAG Laser with a semiconductor saturable mirror. Opt Express 13:4085–4089
167. Wisdom JA, Hum DS, Digonnet MJF, Ikesue A, Fejer MM, Byer RL (2007) 2.6-watt average-power mode-locked ceramic Nd:YAG laser. In: Jiang S, Digonnet MJF (eds) Optical components and materials IV, Proceedings of the Society of Photo-Optical Instrumentation Engineers (SPIE), 6469:C4690
168. Martin-Rodriguez E, Molina P, Benayas A, Bausa LE, Garcia Sole J, Jaque D (2009) Suppression of Q-switching instabilities in a passively mode-locked $Nd:Y_3Al_5O_{12}$ ceramic laser. Opt Mater 31:725–728
169. Sato Y, Saikawa J, Taira T, Ikesue A (2007) Characteristics of Nd^{3+}-doped $Y_3ScAl_4O_{12}$ ceramic laser. Opt Mater 29:1277–1282
170. Okada H, Tanaka M, Kiriyama H, Nakai Y, Ochi Y, Sugiyama A et al (2010) Laser ceramic materials for subpicosecond solid-state lasers using Nd^{3+}-doped mixed scandium garnets. Opt Lett 35:3048–3050
171. Xie GQ, Tang DY, Tan WD, Luo H, Zhang HJ, Yu HH et al (2009) Subpicosecond pulse generation from a Nd:CLNGG disordered crystal laser. Opt Lett 34:103–105
172. Luo H, Tang DY, Xie GQ, Tan WD, Zhang HJ, Yu HH (2009) Diode-pumped passively mode-locked Nd:CLNGG laser. Opt Commun 282:291–293
173. Xie GQ, Qian LJ, Yuan P, Tang DY, Tan WD, Yu HH et al (2010) Generation of 534 fs pulses from a passively mode-locked Nd:CLNGG-CNGG disordered crystal hybrid laser. Laser Phys Lett 7:483–486
174. Lupei V, Lupei A, Gheorghe C, Gheorghe L, Achim A, Ikesue A (2012) Crystal field disorder effects in the optical spectra of Nd^{3+} and Yb^{3+}-doped calcium lithium niobium gallium garnets laser crystals and ceramics. J Appl Phys 112:063110
175. Xie GQ, Tang DY, Kong J, Qian LJ (2007) Passive mode-locking of a Nd:YAG ceramic laser by optical interference modulation in a GaAs wafer. Opt Express 15:5360–5365
176. Zhang WX, Zhou J, Liu WB, Li J, Wang L, Jiang BX et al (2010) Fabrication, properties and laser performance of Ho:YAG transparent ceramic. J Alloy Compd 506:745–748
177. Chen H, Shen DY, Zhang J, Yang H, Tang DY, Zhao T et al (2011) In-band pumped highly efficient Ho:YAG ceramic laser with 21 W output power at 2097 nm. Opt Lett 36:1575–1577
178. Zhao T, Chen H, Shen DY, Wang Y, Yang XF, Zhang J et al (2013) Effects of Ho^{3+}-doping concentration on the performances of resonantly pumped Ho:YAG ceramic lasers. Opt Mater 35:712–714
179. Yao BQ, Cui Z, Duan XM, Shen YJ, Wang J, Du YQ (2014) A graphene-based passively Q-switched Ho:YAG laser. Chin Phys Lett 31:074202
180. Zhao T, Wang Y, Chen H, Shen D (2014) Graphene passively Q-switched Ho:YAG ceramic laser. Appl Phys B Lasers Opt 116:947–950
181. Newburgh GA, Word-Daniels A, Michael A, Merkle LD, Ikesue A, Dubinskii M (2011) Resonantly diode-pumped $Ho^{3+}:Y_2O_3$ ceramic 2.1 μm laser. Opt Express 19:3604–3611
182. Ter-Gabrielyan N, Merkle LD, Kupp ER, Messing GL, Dubinskii M (2010) Efficient resonantly pumped tape cast composite ceramic Er:YAG laser at 1645 nm. Opt Lett 35:922–924
183. Shen DY, Chen H, Qin XP, Zhang J, Tang DY, Yang XF et al (2011) Polycrystalline ceramic Er:YAG laser in-band pumped. Appl Phys Express 4:052701
184. Zhang C, Shen DY, Wang Y, Qian LJ, Zhang J, Qin XP et al (2011) High-power polycrystalline Er:YAG ceramic laser at 1617 nm. Opt Lett 36:4767–4769

185. Ter-Gabrielyan N, Merkle LD, Ikesue A, Dubinskii M (2008) Ultralow quantum-defect eye-safe Er:Sc_2O_3 laser. Opt Lett 33:1524–1526
186. Sanamyan T, Simmons J, Dubinskii M (2010) Er^{3+}-doped Y_2O_3 ceramic laser at similar to 2.7 μm with direct diode pumping of the upper laser level. Laser Phys Lett 7:206–209
187. Sanamyan T, Simmons J, Dubinskii M (2010) Efficient cryo-cooled 2.7-μm Er^{3+}:Y_2O_3 ceramic laser with direct diode pumping of the upper laser level. Laser Phys Lett 7:569–572
188. Sanamyan T, Kanskar M, Xiao Y, Kedlaya D, Dubinskii M (2011) High power diode-pumped 2.7-μm Er^{3+}:Y_2O_3 laser with nearly quantum defect-limited efficiency. Opt Express 19:A1082–A1087
189. Cheng X-J, Xu J-Q, Zhang W-X, Jiang B-X, Pan Y-B (2009) End-pumped Tm:YAG ceramic slab lasers. Chin Phys Lett 26:074204
190. Zhang WX, Pan YB, Zhou J, Liu WB, Li J, Jiang BX et al (2009) Diode-pumped Tm:YAG ceramic laser. J Am Ceram Soc 92:2434–2437
191. Zou YW, Zhang YD, Zhong X, Wei ZY, Zhang WX, Jiang BX et al (2010) Efficient Tm: YAG ceramic laser at 2 μm. Chin Phys Lett 27:074213
192. Zhang SY, Wang X, Kong WJ, Yang QL, Xu JQ, Jiang BX et al (2013) Efficient Q-switched Tm:YAG ceramic slab laser pumped by a 792 nm fiber laser. Opt Commun 286:288–290
193. Zhang SY, Wang MJ, Xu L, Wang Y, Tang YL, Cheng XJ et al (2011) Efficient Q-switched Tm:YAG ceramic slab laser. Opt Express 19:727–732
194. Gallian A, Fedorov VV, Mirov SB, Badikov VV, Galkin SN, Voronkin EF et al (2006) Hot-pressed ceramic Cr^{2+}:ZnSe gain-switched laser. Opt Express 14:11694–11701
195. Cantwell PR, Tang M, Dillon SJ, Luo J, Rohrer GS, Harmer MP (2014) Grain boundary complexions. Acta Mater 62:1–48
196. Dericioglu AF, Kagawa Y (2003) Effect of grain boundary microcracking on the light transmittance of sintered transparent $MgAl_2O_4$. J Eur Ceram Soc 23:951–959
197. Ramirez MO, Wisdom J, Li H, Aung YL, Stitt J, Messing GL et al (2008) Three-dimensional grain boundary spectroscopy in transparent high power ceramic laser materials. Opt Express 16:5965–5973
198. Ikesue A, Yoshida K, Yamamoto T, Yamaga I (1997) Optical scattering centers in polycrystalline Nd:YAG laser. J Am Ceram Soc 80:1517–1522
199. Zhang WX, Pan YB, Zhou J, Liu WB, Li J, Zou YW et al (2011) Preparation and characterization of transparent Tm:YAG ceramics. Ceram Int 37:1133–1137
200. Wang Y, Shen DY, Chen H, Zhang J, Qin XP, Tang DY et al (2011) Highly efficient Tm: YAG ceramic laser resonantly pumped at 1617 nm. Opt Lett 36:4485–4487
201. Antipov OL, Golovkin SY, Gorshkov ON, Zakharov NG, Zinovev AP, Kasatkin AP et al (2011) Structural, optical, and spectroscopic properties and efficient two-micron lasing of new Tm^{3+}:Lu_2O_3 ceramics. Quantum Electron 41:863–868
202. Chenais S, Druon F, Forget S, Balembois F, Georges P (2006) On thermal effects in solid-state lasers: the case of ytterbium-doped materials. Prog Quantum Electron 30:89–153
203. Takaichi K, Yagi H, Lu J, Shirakawa A, Ueda K, Yanagitani T et al (2003) Yb^{3+}-doped $Y_3Al_5O_{12}$ ceramics—a new solid-state laser material. Phys Status Solidi A Appl Res 200: R5–R7
204. Dong J, Shirakawa A, Ueda K, Yagi H, Yanagitani T, Kaminskii AA (2006) Efficient Yb^{3+}: $Y_3Al_5O_{12}$ ceramic microchip lasers. Appl Phys Lett 89:091114
205. Pirri A, Alderighi D, Toci G, Vannini M (2009) High-efficiency, high-power and low threshold Yb^{3+}:YAG ceramic laser. Opt Express 17:23344–23349
206. Pirri A, Alderighi D, Toci G, Vannini M (2010) A Ceramic Based Yb^{3+}:YAG Laser. Laser Phys 20:931–935
207. Hao Q, Li WX, Pan HF, Zhang XY, Jiang BX, Pan YB et al (2009) Laser-diode pumped 40-W Yb:YAG ceramic laser. Opt Express 17:17734–17738
208. Nakamura S, Matsubara Y, Ogawa T, Wada S (2008) High-power high-efficiency Yb^{3+}-doped $Y_3Al_5O_{12}$ ceramic laser at room temperature. Jpn J Appl Phys 47:2149–2151
209. Nakamura S, Yoshioka H, Matsubara Y, Ogawa T, Wada S (2008) Efficient tunable Yb:YAG ceramic laser. Opt Commun 281:4411–4414

210. Cai H, Zhou J, Zhao HM, Qi YF, Lou QH, Dong JX et al (2008) Continuous-wave and Q-switched performance of an Yb:YAG/YAG composite thin disk ceramic laser pumped with 970-nm laser diode. Chin Opt Lett 6:852–854
211. Taira T (2007) RE^{3+}-ion-doped YAG ceramic lasers. IEEE J Sel Top Quantum Electron 13:798–809
212. Tsunekane M, Taira T (2005) High-power operation of diode edge-pumped, glue-bonded, composite Yb:$Y_3AI_5O_{12}$ microchip laser with ceramic, undoped YAG pump light-guide. Jpn J Appl Phys Part 2 Lett Expr Lett 44:L1164–L1167
213. Tsunekane M, Taira T (2007) High-power operation of diode edge-pumped, composite all-ceramic Yb:$Y_3Al_5O_{12}$ microchip laser. Appl Phys Lett 90:121101
214. Giesen A, Speiser J (2007) Fifteen years of work on thin-disk lasers: Results and scaling laws. IEEE J Sel Top Quantum Electron 13:598–609
215. Mende J, Spindler G, Speiser J, Giesen A (2009) Concept of neutral gain modules for power scaling of thin-disk lasers. Appl Phys B Lasers Opt 97:307–315
216. Petermann K, Fagundes-Peters D, Johannsen J, Mond M, Peters V, Romero JJ et al (2005) Highly Yb-doped oxides for thin-disc lasers. J Cryst Growth 275:135–140
217. Latham WP, Lobad A, Newell TC, Stalnaker D (2010) 6.5 kW, Yb:YAG ceramic thin disk laser. In: Phipps CR (ed) International symposium on high power laser ablation 2010, pp 758–764
218. Nakamura S, Yoshioka H, Ogawa T, Wada S (2009) Broadly tunable Yb^{3+}-doped $Y_3Al_5O_{12}$ ceramic laser at room temperature. Jpn J Appl Phys 48:060205
219. Dong J, Shirakawa A, Ueda K, Yagi H, Yanagitani T, Kaminskii AA (2007) Laser-diode pumped heavy-doped Yb:YAG ceramic lasers. Opt Lett 32:1890–1892
220. Dong J, Ueda K, Yagi H, Kaminskii AA, Cai Z (2009) Comparative study the effect of Yb concentrations on laser characteristics of Yb:YAG ceramics and crystals. Laser Phys Lett 6:282–289
221. Pirri A, Toci G, Alderighi D, Vannini M (2010) Effects of the excitation density on the laser output of two differently doped Yb:YAG ceramics. Opt Express 18:17262–17272
222. Chani VI, Boulon G, Zhao W, Yanagida T, Yoshikawa A (2010) Correlation between segregation of rare earth dopants in melt crystal growth and ceramic processing for optical applications. Jpn J Appl Phys 49:075601
223. Brandt C, Fredrich-Thornton ST, Petermann K, Huber G (2011) Photoconductivity in Yb-doped oxides at high excitation densities. Appl Phys B Lasers Opt 102:765–768
224. Bisson JF, Kouznetsov D, Ueda K, Fredrich-Thornton ST, Petermann K, Huber G (2007) Switching of emissivity and photoconductivity in highly doped Yb^{3+}:Y_2O_3 and Lu_2O_3 ceramics. Appl Phys Lett 90:201901
225. Luo D, Zhang J, Xu C, Lin H, Yang H, Zhu H et al (2013) Mode-locked Yb:LuAG ceramics laser. Phys Status Solidi C 10(6):967–968
226. Luo D, Zhang J, Xu C, Yang H, Lin H, Zhu H et al (2012) Yb:LuAG laser ceramics: a promising high power laser gain medium. Opt Mater Expr 2:1425–1431
227. Xu CW, Luo DW, Zhang J, Yang H, Qin XP, Tan WD et al (2012) Diode pumped highly efficient Yb:$Lu_3Al_5O_{12}$ ceramic laser. Laser Phys Lett 9:30–34
228. Furuse H, Yasuhara R, Hiraga K (2014) Thermo-optic properties of ceramic YAG at high temperatures. Opt Mater Expr 4:1794–1799
229. Lu JR, Takaichi K, Uematsu T, Shirakawa A, Musha M, Ueda K et al (2002) Yb^{3+}:Y_2O_3 ceramics—a novel solid-state laser material. Jpn J Appl Phys Part 2 Lett 41:L1373–L1375
230. Kong J, Tang DY, Chan CC, Lu J, Ueda K, Yagi H et al (2007) High-efficiency 1040 and 1078 nm laser emission of a Yb:Y_2O_3 ceramic laser with 976 nm diode pumping. Opt Lett 32:247–249
231. Kong J, Lu J, Takaichi K, Uematsu T, Ueda K, Tang DY et al (2003) Diode-pumped Yb:Y_2O_3 ceramic laser. Appl Phys Lett 82:2556–2558
232. Kong J, Tang DY, Lu J, Ueda K, Yagi H, Yanagitani T (2004) Passively mode-locked Yb:Y_2O_3 ceramic laser with a GaAs-saturable absorber mirror. Opt Commun 237:165–168

233. Kong J, Tang DY, Zhao B, Lu J, Ueda K, Yagi H et al (2005) Study on the diode-pumped Yb:Y_2O_3 ceramic laser. In: 2005 IEEE Pacific Rim conference on lasers and electro-optics, pp 142–143
234. Pirri A, Toci G, Ciofini M, Lapucci A, Gizzi LA, Labate L et al (2014) Thermal lens measurements in Yb-doped YAG, LuAG, Lu_2O_3, Sc_2O_3 ceramic lasers. In: 22nd international laser physics workshop (LPhys'13), vol 497, p 012013
235. Pirri A, Toci G, Vannini M (2012) Characterization and comparison of 1 % at Yb-doped Lu_2O_3 and Sc_2O_3 ceramics as laser gain media. Laser Phys 22:1851–1855
236. Kim WH, Baker C, Villalobos G, Frantz J, Shaw B, Lutz A et al (2011) Synthesis of high purity Yb^{3+}-doped Lu_2O_3 powder for high power solid-state lasers. J Am Ceram Soc 94:3001–3005
237. Sanghera J, Frantz J, Kim WH, Villalobos G, Baker C, Shaw B et al (2011) 10% Yb^{3+}-Lu_2O_3 ceramic laser with 74 % efficiency. Opt Lett 36:576–578
238. Sanghera J, Kim WH, Baker C, Villalobos G, Frantz J, Shaw B et al (2011) Laser oscillation in hot pressed 10 % Yb^{3+}:Lu_2O_3 ceramic. Opt Mater 33:670–674
239. Pirri A, Toci G, Nikl M, Vannini M (2012) High efficiency laser action of 1 % at. Yb^{3+}: Sc_2O_3 ceramic. Opt Express 20:22134–22142
240. Pirri A, Toci G, Vannini M (2011) First laser oscillation and broad tunability of 1 at. % Yb-doped Sc_2O_3 and Lu_2O_3 ceramics. Opt Lett 36:4284–4286
241. Lupei A, Lupei V, Gheorghe C, Ikesue A, Osiac E (2009) Upconversion emission of RE^{3+} in Sc_2O_3 ceramic under 800 nm pumping. Opt Mater 31:744–749
242. Hao Q, Li WX, Zeng HP, Yang QH, Dou CG, Zhou HX et al (2008) Low-threshold and broadly tunable lasers of Yb^{3+}-doped yttrium lanthanum oxide ceramic. Appl Phys Lett 92:211106
243. Bourdet GL, Casagrande O, Deguil-Robin N, Le Garrec B (2008) Performances of cryogenic cooled laser based on ytterbium doped sesquioxide ceramics. In: 5th international conference on inertial fusion sciences and applications (IFSA 2007), vol 112, p 032054
244. Lyberis A, Patriarche G, Gredin P, Vivien D, Mortier M (2011) Origin of light scattering in ytterbium doped calcium fluoride transparent ceramic for high power lasers. J Eur Ceram Soc 31:1619–1630
245. Akchurin MS, Basiev TT, Demidenko AA, Doroshenko ME, Fedorov PP, Garibin EA et al (2013) CaF^2: Yb laser ceramics. Opt Mater 35:444–450
246. Basiev TT, Doroshenko ME, Fedorov PP, Konyushkin VA, Kuznetsov SV, Osiko VV et al (2008) Efficient laser based on CaF_2-SrF_2-YbF_3 nanoceramics. Opt Lett 33:521–523
247. Dong J, Shirakawa A, Ueda K, Yagi H, Yanagitani T, Kaminskii AA (2007) Near-diffraction-limited passively Q-switched Yb:$Y_3Al_5O_{12}$ ceramic lasers with peak power >150 kW. Appl Phys Lett 90:131105
248. Dong J, Shirakawa A, Ueda K, Yagi H, Yanagitani T, Kaminskii AA (2007) Ytterbium and chromium doped composite $Y_3Al_5O_{12}$ ceramics self-Q-switched laser. Appl Phys Lett 90:191106
249. Yoshioka H, Nakamura S, Ogawa T, Wada S (2009) Diode-pumped mode-locked Yb:YAG ceramic laser. Opt Express 17:8919–8925
250. Zhou BB, Wei ZY, Zou YW, Zhang YD, Zhong X, Bourdet GL et al (2010) High-efficiency diode-pumped femtosecond Yb:YAG ceramic laser. Opt Lett 35:288–290
251. Yoshioka H, Nakamura S, Ogawa T, Wada S (2010) Dual-wavelength mode-locked Yb: YAG ceramic laser in single cavity. Opt Express 18:1479–1486
252. Saikawa J, Sato Y, Taira T, Ikesue A (2004) Passive mode locking of a mixed garnet Yb: $Y_3ScAl_4O_{12}$ ceramic laser. Appl Phys Lett 85:5845–5847
253. Tokurakawa M, Kurokawa H, Shirakawa A, Ueda K, Yagi H, Yanagitani T et al (2010) Continuous-wave and mode-locked lasers on the base of partially disordered crystalline Yb^{3+}:$(YGd_2)(Sc_2Al_2Ga)O_{12}$ ceramics. Opt Express 18:4390–4395
254. Senatsky Y, Shirakawa A, Sato Y, Hagiwara J, Lu J, Ueda K et al (2004) Nonlinear refractive index of ceramic laser media and perspectives of their usage in a high-power laser-driver. Laser Phys Lett 1:500–506

255. Tokurakawa M, Takaichi K, Shirakawa A, Ueda K, Yagi H, Hosokawa S et al (2006) Diode-pumped mode-locked Yb^{3+}:Lu_2O_3 ceramic laser. Opt Express 14:12832–12838
256. Tokurakawa M, Takaichi K, Shirakawa A, Ueda K, Yagi H, Yanagitani T et al (2007) Diode-pumped 188 fs mode-locked Yb^{3+}:Y_2O_3 ceramic laser. Appl Phys Lett 90:071101
257. Xie GQ, Tang DY, Zhao LM, Qian LJ, Ueda K (2007) High-power self-mode-locked Yb: Y_2O_3 ceramic laser. Opt Lett 32:2741–2743
258. Tokurakawa M, Shirakawa A, Ueda K-i, Yagi H, Yanagitani T, Kaminskii AA (2007) Diode-pumped sub-100 fs Kerr-lens mode-locked Yb^{3+}:Sc_2O_3 ceramic laser. Opt Lett 32:3382–3384
259. Tokurakawa M, Shirakawa A, K-i Ueda, Yagi H, Hosokawa S, Yanagitani T et al (2008) Diode-pumped 65 fs Kerr-lens mode-locked Yb^{3+}:Lu_2O_3 and nondoped Y_2O_3 combined ceramic laser. Opt Lett 33:1380–1382
260. Tokurakawa M, Shirakawa A, Ueda K, Yagi H, Noriyuki M, Yanagitani T et al (2009) Diode-pumped ultrashort-pulse generation based on Yb^{3+}:Sc_2O_3 and Yb^{3+}:Y_2O_3 ceramic multi-gain-media oscillator. Opt Express 17:3353–3361
261. Pearce S, Yasuhara R, Yoshida A, Kawanaka J, Kawashima T, Kan H (2009) Efficient generation of 200 mJ nanosecond pulses at 100 Hz repetition rate from a cryogenic cooled Yb:YAG MOPA system. Opt Commun 282:2199–2203
262. Takeuchi Y, Kawanaka J, Yoshida A, Yasuhara R, Kawashima T, Kan H et al (2011) Sub-kHz cryogenic Yb:YAG regenerative amplifier by using a total-reflection active mirror. Appl Phys B Lasers Opt 104:29–32
263. Rodenas A, Torchia GA, Lifante G, Cantelar E, Lamela J, Jaque F et al (2009) Refractive index change mechanisms in femtosecond laser written ceramic Nd:YAG waveguides: micro-spectroscopy experiments and beam propagation calculations. Appl Phys B Lasers Opt 95:85–96
264. Rodenas A, Zhou GY, Jaque D, Gu M (2008) Direct laser writing of three-dimensional photonic structures in Nd:yttrium aluminum garnet laser ceramics. Appl Phys Lett 93:151104
265. Torchia GA, Meilan PF, Rodenas A, Jaque D, Mendez C, Roso L (2007) Femtosecond laser written surface waveguides fabricated in Nd:YAG ceramics. Opt Express 15:13266–13271
266. Torchia GA, Rodenas A, Benayas A, Cantelar E, Roso L, Jaque D (2008) Highly efficient laser action in femtosecond-written Nd:yttrium aluminum garnet ceramic waveguides. Appl Phys Lett 92:111103
267. Benayas A, Silva WF, Rodenas A, Jacinto C, Vazquez de Aldana J, Chen F et al (2011) Ultrafast laser writing of optical waveguides in ceramic Yb:YAG: A study of thermal and non-thermal regimes. Appl Phys A Mater Sci Process 104:301–309
268. Choudhury D, Rodenas A, Paterson L, Diaz F, Jaque D, Kar AK (2013) Three-dimensional microstructuring of yttrium aluminum garnet crystals for laser active optofluidic applications. Appl Phys Lett 103:041101
269. Rodenas A, Benayas A, Macdonald JR, Zhang J, Tang DY, Jaque D et al (2011) Direct laser writing of near-IR step-index buried channel waveguides in rare earth doped YAG. Opt Lett 36:3395–3397
270. Benayas A, Dong NN, Yao YC, Chen F, Bettiol AA, Jaque D (2011) Thermal optimization and erasing of Nd:YAG proton beam written waveguides. Opt Lett 36:3278–3280
271. Benayas A, Silva WF, Jacinto C, Cantelar E, Lamela J, Jaque F et al (2010) Thermally resistant waveguides fabricated in Nd:YAG ceramics by crossing femtosecond damage filaments. Opt Lett 35:330–332
272. Dong NN, Benayas A, Jaque D, Tan Y, Chen F (2011) Fluorescence-quenching free channel waveguides in Yb:YAG ceramics by carbon ion implantation. J Lightwave Technol 29:1460–1464
273. Ren YY, Dong NN, Tan Y, Guan J, Chen F, Lu QM (2010) Continuous wave laser generation in proton implanted Nd:GGG planar waveguides. J Lightwave Technol 28:3578–3581

274. Ren YY, Tan Y, Chen F, Jaque D, Zhang HJ, Wang JY et al (2010) Optical channel waveguides in Nd:LGS laser crystals produced by proton implantation. Opt Express 18:16258–16263
275. Tan Y, Luan QF, Liu FQ, Akhmadaliev S, Zhou SQ, Chen F (2013) Swift carbon ion irradiated Nd:YAG ceramic optical waveguide amplifier. Opt Express 21:13992–13997
276. Tan Y, Zhang C, Chen F, Liu FQ, Jaque D, Lu QM (2011) Room-temperature continuous wave laser oscillations in Nd:YAG ceramic waveguides produced by carbon ion implantation. Appl Phys B Lasers Opt 103:837–840
277. Yao YC, Tan Y, Dong NN, Chen F, Bettiol AA (2010) Continuous wave Nd:YAG channel waveguide laser produced by focused proton beam writing. Opt Express 18:24516–24521
278. Tan Y, Chen F (2010) Proton-implanted optical channel waveguides in Nd:YAG laser ceramics. J Phys D Appl Phys 43:075105
279. Xu DY, Wang YY, Li HF, Yao JQ, Tsang YH (2007) 104 W high stability green laser generation by using diode laser pumped intracavity frequency-doubling Q-switched composite ceramic Nd:YAG laser. Opt Express 15:3991–3997
280. Kaminskii AA, Ueda K, Eichler HJ, Bagaev SN, Takaichi K, Lu J et al (2004) Observation of nonlinear lasing $\chi^{(3)}$-effects in highly transparent nanocrystalline Y_2O_3 and $Y_3Al_5O_{12}$ ceramics. Laser Phys Lett 1:6–11
281. Yanagida T, Fujimoto Y, Yagi H, Yanagitani T, Sugiyama M, Yamaji A et al (2013) Scintillation properties of transparent ceramics for Nd doped $(YGd_2)(Sc_2Al_2Ga)O_{12}$. Opt Mater 35:788–792
282. Kaminskii AA, Bagaev SN, Eichler HJ, Ueda K, Takaichi K, Shirakawa A et al (2006) Observation of high-order stokes and anti-stokes $\chi^{(3)}$-generation in highly transparent laser-host Lu_2O_3 ceramics. Laser Phys Lett 3:310–313
283. Kaminskii AA, Rhee H, Eichler HJ, Ueda K, Takaichi K, Shirakawa A et al (2008) New nonlinear-laser effects in crystalline fine-grained ceramics based on cubic Sc_2O_3 and Lu_2O_3 oxides: second and third harmonic generation, and cascaded self-sum-frequency mixing in UV spectral region. Laser Phys Lett 5:109–113
284. Shoji I, Sato Y, Kurimura S, Lupei V, Taira T, Ikesue A et al (2002) Thermal-birefringence-induced depolarization in Nd:YAG ceramics. Opt Lett 27:234–236
285. Shoji I, Taira T, Ikesue A (2007) Thermally-induced-birefringence effects of highly Nd^{3+}-doped $Y_3Al_5O_{12}$ ceramic lasers. Opt Mater 29:1271–1276
286. Kagan MA, Khazanov EA (2003) Compensation for thermally induced birefringence in polycrystalline ceramic active elements. Quantum Electron 33:876–882
287. Khazanov EA (2002) Thermally induced birefringence in Nd:YAG ceramics. Opt Lett 27:716–718
288. Ostermeyer M, Mudge D, Veitch PJ, Munch J (2006) Thermally induced birefringence in Nd:YAG slab lasers. Appl Opt 45:5368–5376
289. Mukhin IB, Palashov OV, Khazanov EA, Ikesue A, Aung YL (2005) Experimental study of thermally induced depolarization in Nd:YAG ceramics. Opt Express 13:5983–5987
290. Shin JS, Park S, Kong HJ (2010) Compensation of the thermally induced depolarization in a double-pass Nd:YAG rod amplifier with a stimulated Brillouin scattering phase conjugate mirror. Opt Commun 283:2402–2405
291. Yasuhara R, Furuse H (2013) Thermally induced depolarization in TGG ceramics. Opt Lett 38:1751–1753
292. Snetkov IL, Soloviev AA, Khazanov EA (2009) Study of a thermal lens in thin laser-ceramics discs. Quantum Electron 39:302–308
293. Soloviev A, Snetkov I, Zelenogorsky V, Kozhevatov I, Palashov O, Khazanov E (2008) Experimental study of thermal lens features in laser ceramics. Opt Express 16:21012–21021
294. Kawai R, Miyasaka Y, Otsuka K, Ohtomo T, Narita T, Ko JY et al (2004) Oscillation spectra and dynamic effects in a highly-doped microchip Nd:YAG ceramic laser. Opt Express 12:2293–2302

295. Otsuka K, Narita T, Miyasaka Y, Lin CC, Ko JY, Chu SC (2006) Nonlinear dynamics in thin-slice Nd:YAG ceramic lasers: coupled local-mode laser model. Appl Phys Lett 89:081117
296. Ma QL, Bo Y, Zong N, Peng QJ, Cui DF, Pan YB et al (2010) 108 W Nd YAG ceramic laser with birefringence compensation resonator. Opt Commun 283:5183–5186
297. Otsuka K, Ohtomo T (2008) Polarization properties of laser-diode-pumped micro-grained Nd:YAG ceramic lasers. Laser Phys Lett 5:659–663
298. Parriaux O, Bisson JF, Ueda K, Tonchev S, Gamet E, Pommier JC et al (2008) Polarization control of a Yb:YAG ceramic microchip laser by constructive-interference resonant grating mirror. J Mod Opt 55:1899–1912
299. Thirugnanasambandam MP, Senatsky Y, Ueda K-i (2011) Generation of radially and azimuthally polarized beams in Yb:YAG laser with intra-cavity lens and birefringent crystal. Opt Express 19:1905–1914
300. Moshe I, Jackel S, Lumer Y, Meir A, Feldman R, Shimony Y (2010) Use of polycrystalline Nd:YAG rods to achieve pure radially or azimuthally polarized beams from high-average-power lasers. Opt Lett 35:2511–2513
301. Thirugnanasambandam MP, Senatsky Y, Ueda K (2010) Generation of very-high order Laguerre-Gaussian modes in Yb:YAG ceramic laser. Laser Phys Lett 7:637–643
302. Tokunaga K, Chu SC, Hsiao HY, Ohtomo T, Otsuka K (2009) Spontaneous Mathieu-Gauss mode oscillation in micro-grained Nd:YAG ceramic lasers with azimuth laser-diode pumping. Laser Phys Lett 6:635–638
303. Thirugnanasambandam MP, Senatsky Y, Shirakawa A, Ueda K-i (2011) Multi-ring modes generation in Yb:YAG ceramic laser. Opt Mater 33:675–678
304. Li JL, Lin D, Zhong LX, Ueda K, Shirakawa A, Musha M et al (2009) Passively Q-switched Nd:YAG ceramic microchip laser with azimuthally polarized output. Laser Phys Lett 6:711–714
305. J-l Li, K-i Ueda, Musha M, Zhong L-x, Shirakawa A (2008) Radially polarized and pulsed output from passively Q-switched Nd:YAG ceramic microchip laser. Opt Lett 33:2686–2688
306. Huang J, Deng J, Cao Y, Weng W, Zheng H, Li J et al (2011) Passively mode-locked radially polarized laser based on ceramic Nd:YAG rod. Opt Express 19:2120–2125
307. Li SX, Li GQ, Zhao SZ, Wang XM, Yang KJ, Li DC et al (2014) Passively Q-switched laser performance of a composite Nd:YVO_4/Nd:YVO_4/Nd:YVO_4 crystal with GaAs saturable absorber. Opt Quant Electron 46:1179–1186
308. Zhu SQ, He Q, Wang SF, Chen ZQ, Li AM, Yin H et al (2014) High average power passively Q-switched laser diode side-pumped green laser by using Nd:YAG/Cr^{4+}:YAG/YAG composite crystal. J Laser Appl 26:032009
309. Jiang W, Zhu S, Chen ZQ, He Q, Chen Z, Wang S et al (2013) Green laser with V-shaped resonant cavity based on Nd:YAG/Cr^{4+}:YAG/YAG composite crystal rod. J Appl Spectrosc 80:694–697
310. Meng JY, Wang HX (2011) Single diode-pumped, 1.7 ns microchip laser by Nd:YAG/Cr^{4+}:YAG composite crystal. Laser Phys 21:79–81
311. Yin L, Li GQ, Zhao SZ, Li X, Cheng K, Zhang G (2011) Continuous wave and passively Q-switched laser performance of a composite crystal bonded with three Nd:YVO_4 single crystals. Laser Phys 21:1151–1154
312. Tang F, Cao YG, Huang JQ, Guo W, Liu HG, Wang WC et al (2012) Diode- pumped multilayer Yb:YAG composite ceramic laser. Laser Phys Lett 9:564–569
313. Hirota K, Satomi M, Matsuyama K, Kugimiya K (1990) The solid-state single-crystal growth from Mn-Zn ferrite polycrystal with additives. Mater Res Bull 25:1453–1459
314. Sugimoto M (1999) The past, present, and future of ferrites. J Am Ceram Soc 82:269–280
315. Kang SJL, Park JH, Ko SY, Lee HY (2015) Solid-state conversion of single crystals: the principle and the state-of-the-art. J Am Ceram Soc 98:347–360
316. Sugiyama A, Fukuyama H, Sasuga T, Arisawa T, Takuma H (1998) Direct bonding of Ti: sapphire laser crystals. Appl Opt 37:2407–2410

317. Esposito L, Hostasa J, Piancastelli A, Toci G, Alderighi D, Vannini M et al (2014) Multilayered YAG-Yb:YAG ceramics: manufacture and laser performance. J Mater Chem C 2:10138–10148
318. Liu WB, Li J, Liu J, Liu BL, Fu YL, Pan YB et al (2014) Study of Yb:YAG ceramic slab with Cr^{4+}:YAG edge cladding. Ceram Int 40:8879–8883
319. Kupp ER, Messing GL, Anderson JM, Gopalan V, Dumm JQ, Kraisinger C et al (2010) Co-casting and optical characteristics of transparent segmented composite Er:YAG laser ceramics. J Mater Res 25:476–483
320. Lebbou K, Perrodin D, Chani VI, Brenier A, Tillement O, Aloui O et al (2006) Fiber single-crystal growth from the melt for optical applications. J Am Ceram Soc 89:75–80
321. Yoshikawa A, Boulon G, Laversenne L, Canibano H, Lebbou K, Collombet A et al (2003) Growth and spectroscopic analysis of Yb^{3+}-doped $Y_3Al_5O_{12}$ fiber single crystals. J Appl Phys 94:5479–5488
322. Didierjean J, Castaing M, Balembois F, Georges P, Perrodin D, Fourmigue JM et al (2006) High-power laser with Nd:YAG single-crystal fiber grown by the micro-pulling-down technique. Opt Lett 31:3468–3470
323. Sangla D, Aubry N, Didierjean J, Perrodin D, Balembois F, Lebbou K et al (2009) Diode-pumped laser with Yb:YAG single-crystal fiber grown by the micro-pulling down technique. Appl Phys B Lasers Opt 94:203–207
324. Sangla D, Martial I, Aubry N, Didierjean J, Perrodin D, Balembois F et al (2009) High power laser operation with crystal fibers. Appl Phys B Lasers Opt 97:263–273
325. Ikesue A, Aung YL (2007) Progress in ceramic Nd:YAG laser In: Wood GL, Dubindkii MA (eds) Laser source technology for defense and security III, Proceedings of the Society of Photo-Optical Instrumentation Engineers (SPIE), p 55209
326. Aubry N, Sangla D, Mancini C, Didierjean J, Perrodin D, Fourmigue JM et al (2009) Characterizations of 0.4 and 1 mm diameter Yb:YAG single-crystal fibers grown by the micro-pulling-down method for laser applications. J Cryst Growth 311:4805–4811
327. Yang PZ, Deng PZ, Yin ZW, Tian YL (2000) The growth defects in Czochralski-grown Yb: YAG crystal. J Cryst Growth 218:87–92
328. Eakins DE, Held M, Norton MG, Bahr DF (2004) A study of fracture and defects in single crystal YAG. J Cryst Growth 267:502–509
329. Sangla D, Aubry N, Nehari A, Brenier A, Tillement O, Lebbou K et al (2009) Yb-doped $Lu_3Al_5O_{12}$ fibers single crystals grown under stationary stable state for laser application. J Cryst Growth 312:125–130
330. Jheng DY, Hsu KY, Liang YC, Huang SL (2015) Broadly tunable and low-threshold Cr^{4+}: YAG crystal fiber laser. IEEE J Sel Top Quantum Electron 21:0900608

Chapter 10
Other Applications

10.1 Brief Introduction

Besides their applications in solid-state lasers, transparent ceramics have been found to be applicable in many other areas, such as lighting, scintillators, opto-electric devices, optical systems and armors, which will be discussed in detail in this chapter. New applications of transparent ceramics, e.g., as biomaterials, will also be mentioned at the end of this chapter.

10.2 Lighting

With desirable thermal and mechanical properties, transparent ceramics have been used as envelopes for focused-beam short-arc lamps containing various fillers, operating at temperatures that are higher than the melting point of quartz. Translucent polycrystalline alumina (PCA) ceramics with coarse grains have been used as envelopes in discharge lamps for a quite long time. Translucent PCA doped with additives based on magnesia was discovered in the 1960s, which has become a key material in the lighting industry, especially for the widely used high-pressure sodium lamps [1, 2]. Initially, metal halide lamps with translucent alumina tubes, either cylindrical or spherical-like [3, 4], were used in applications for general illumination of white light. The lamps worked at the temperature of ~ 1200 °C, so that it was necessary to use Al_2O_3 [5–7].

Besides high optical transparency, there are other challenges when transparent ceramics are used for the applications in lighting, which include (i) strong bonding between the transparent ceramics and electrode system to complete the discharge enclosure, (ii) satisfactory thermomechanical properties in order to withstand the rapid heating and cooling cycles encountered in both the discharge tube and seal, (iii) high durability to resist the attack from lamp chemicals at high temperatures

L.B. Kong et al., *Transparent Ceramics*, Topics in Mining, Metallurgy and Materials Engineering, DOI 10.1007/978-3-319-18956-7_10

and (iv) stability in maintaining the optical quality throughout the life of use. Performance, efficiency, environmental sustainability, and cost effectiveness are the main considerations in the development of ceramic envelopes in lighting industry. Transparent ceramics offer a good opportunity to break the limits of envelope materials for lamps with continuously enhanced performances.

PCA ceramics with submicron grain size and high in-line transmittance were later developed by using HIP sintering, from nanosized alumina powders, which thus offered much improved illumination performances [5, 6, 8, 9]. Other transparent ceramics that have been used as envelope materials include AlON, YAG, Y_2O_3, and Dy_2O_3 [10–13]. Transparent ceramics with various shapes, e.g., cylindrical, bulgy, spherical and elliptical, can be fabricated by using the classical ceramic processing. A brief discussion is given to each of the transparent ceramics that have been used in lighting.

10.2.1 Sapphire

Straight sapphire tubes are used in special lamps, which can be grown through melting and solidification. Due to the dissolution of the Mo die in the molten alumina, a thin layer of fine Mo particles could be present. At the same time, pores are formed due to the shrinkage of solidification. These defects could affect the total transmittance of the sapphire tubes. In new designs of burners, the straight sapphire tubes were joined with PCA hats [14]. An interference was formed after the bonding, in which the initial sapphire–PCA boundary migrated toward the PCA side, because PCA was sintered and shrank sweeping across the original interface, thus leading to the formation of pores [15].

As discussed in previous chapters, PCA could be converted to single-crystal sapphire through the so-called solid-state crystal conversion (SSCC) process [16, 17]. When PCA doped with MgO was heated, an equiaxed grain structure of translucency was first achieved, which was followed by out-diffusion of the MgO dopant to <60 ppm, so that a high rate of intrinsic grain growth occurred, leading to the conversion of the PCA to transparent sapphire. The converted sapphire had similar microstructure to the PCA, in terms of surface roughness and residual pores. Complete conversion of large PCA ceramics to sapphire is still a great challenge, because the conversion usually takes place only near the surface region.

10.2.2 Alumina Ceramics with Fine Grains

PCA ceramics with submicron grain size and high in-line transmittance have found applications in metal halide lamps [18]. The in-line transmittance in the visible range could reach >70 % of that of sapphire, which is almost the theoretical limit for alumina ceramics with that range of grain sizes [19]. In addition, mechanical strength of PCA ceramics was nearly the same as that of sapphire, because of the small grain sizes [9]. Furthermore, such PCA ceramics can be made into complex shapes, as compared to the anisotropy and straight tubes of melt grown sapphire. Therefore, fine-grained PCA ceramics are promising candidates as envelopes in beamers. Figure 10.1 shows photographs of a fiber light-source placed inside the tubes, illustrating the definition of the submicron-grained ceramics is much better, as compared to that of the regular 15-μm-grained alumina [11]. The higher image resolution offered by the sub-μm alumina was attributed to its sharper curve of the angular dependence of the scattered light.

The transparent ceramics used for lamp envelope applications should have both high total and high in-line transmittance. Total transmittance, which is a measure of absorption of arc tubes, determines the total lumen output, while in-line transmittance that is a measure of scattering is critical to luminance or brightness of the arcs. However, a high in-line transmittance does not necessarily mean a high total transmittance [20]. The total transmittance is obtained by placing a small light pipe inside the body and measuring the total integrated flux of light passing through the tube wall.

It was found that the total transmittance of as-HIP-sintered discharge vessels from the sub-μm-grained alumina was 77.7 %, which was largely lower than that of normal sintered alumina parts of 98–99 %, with grain size of 15 μm. However, it can be improved by using thermal annealing. For example, the total transmittance of the sub-μm-grained alumina vessels was increased to 90.5 %, after annealing at 1025–1150 °C for 2–4 h in air. For commercial lighting applications, the total transmittance should be >92 %. Sometimes, the as-made tubes showed a gray color, which has been attributed to the presence of color centers in the PCA, including oxygen vacancy with one trapped electron and an associated magnesium solutes, double oxygen vacancy with four trapped electrons, and oxygen interstitial with two trapped holes.

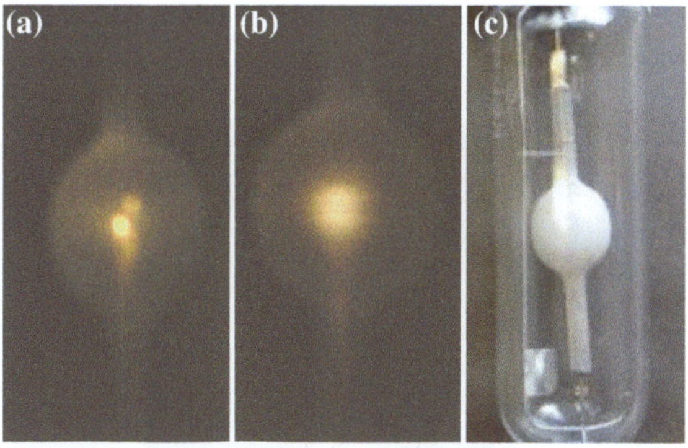

Fig. 10.1 Photographs of the polished tubes containing fiber-optical miniature light source illustrating the relatively sharp image of the source in sub-μm PCA tube (**a**) and the diffuse image in regular PCA tube (**b**). **c** One example of sub-μm PCA lamp. Reproduced with permission from [11]. Copyright © 2009, Elsevier

10.2.3 AlON (Aluminum Oxynitride)

Highly transparent AlON ($Al_{23}O_{27}N_5$) ceramics with cubic symmetry have been readily fabricated by various methods, as discussed earlier [21–28]. Because their mechanical strength and thermal expansion are close to those of PCA, AlON ceramics are also able to survive the stresses in ceramic metal halide lamps and thus can be used for the end enclosure structure of ceramic metal halide burners. AlON ceramics could have in-line transmittance in the visible that is close to that of sapphire [29]. However, thermodynamic analysis indicated that AlON was only stable in a small range of N_2 partial pressures at temperatures of <1640 °C, which seems to be a shortcoming of AlON ceramics for such applications. Fortunately, this problem could be addressed, since AlON can be stabilized using the doping of MgO [30–32]. Another issue for the use of AlON in projection lamps is its high oxidation behavior at temperatures of ≥1200 °C in air [31].

10.2.4 YAG (Yttrium Aluminate Garnet)

Transparent YAG ceramics (yttrium aluminate garnet, $Y_3Al_5O_{12}$) have excellent optical and mechanical properties. Besides their widespread applications in solid-state lasers, transparent YAG ceramics have also been used as envelopes in lamps [10, 11]. As expected, the corrosion resistance of YAG to rare-earth halide fills was less than that of PCA. For instance, the wall could be attacked by the rare-earth halide fills (ReI_3), such as TmI_3, after a prolonged duration of application, due to the following reaction:

$$Y_3Al_5O_{12} + ReI_3 \rightarrow Re_3Al_5O_{12} + YI_3. \qquad (10.1)$$

As a consequence, yttrium emission lines could be detected, while color shift and voltage drop in the YAG/ReI_3 lamps could be observed, as shown in Fig. 10.2. Therefore, it is necessary to find less corrosive fills for YAG lamps.

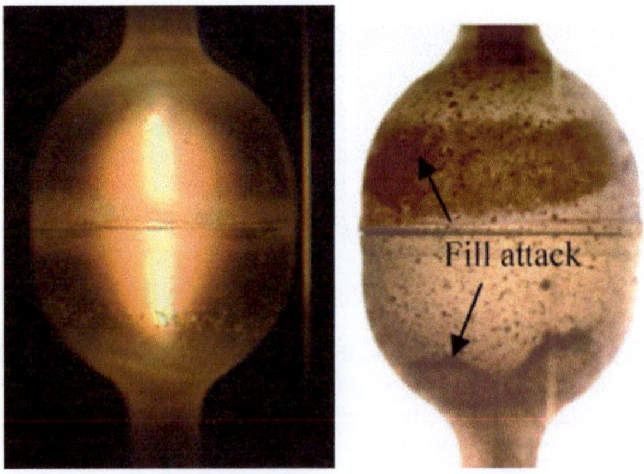

Fig. 10.2 Photographs of 2-piece YAG bulgy burner in operation (*left figure*) and after 1000 h showing "stains" or fill attack (*right figure*). Reproduced with permission from [11]. Copyright © 2009, Elsevier

10.2.5 Rare-Earth Oxide (Re_2O_3)

Polycrystalline rare-earth oxides are candidate tube materials for advanced beamers, due to their high optical transparency, low thermodynamic driving potentials for reactions with metal halide fills, high mechanical strength, and thermal shock resistance. One example is dysprosium oxide Dy_2O_3 [13]. The long infrared cutoff suggested that Dy_2O_3 has a lower emissivity and higher efficacy than PCA. Dy_2O_3 has several intrinsic absorption bands in the range of 275–450 nm. The free energy of formation (-1514 kJ mol^{-1}) at 1200 K for Dy_2O_3 is favorable, compared to those of Al_2O_3 (-1295 kJ mol^{-1}) and DyI_3 (-380 kJ mol^{-1}), and is nearly the same as that of Tm_2O_3 (-1526 kJ mol^{-1}), implying that Dy_2O_3 is stabile in the presence of the molten salts.

Transparent polycrystalline Dy_2O_3 ceramic parts were prepared by using the conventional ceramic process, from high-purity ultrafine precursor powders. The ceramics had an average grain size of ~3 μm. This small grain size is important for retaining a high mechanical strength, because fracture strength of ceramics typically increases with decreasing flaw size, which is closely related to grain size. Dy_2O_3 has a slight yellowish coloration due to the intrinsic transitions. Examples of the unpolished and polished polycrystalline Dy_2O_3 tubes are shown in Fig. 10.3a, b, respectively. The unpolished had a total transmittance of ~0.93 and an in-line transmittance of ~0.40, which was increased to ~0.75 after polishing. This value was higher than those of polished PCA with grain size of 20 μm, due to the cubic crystal structure of Dy_2O_3. Figure 10.3c shows a photograph of the polished Dy_2O_3 tube, with steel pin inserted to demonstrate the transparency. Dy_2O_3 tubes with light fiber (power on at two levels) placed inside the tube showed high degrees of in-line

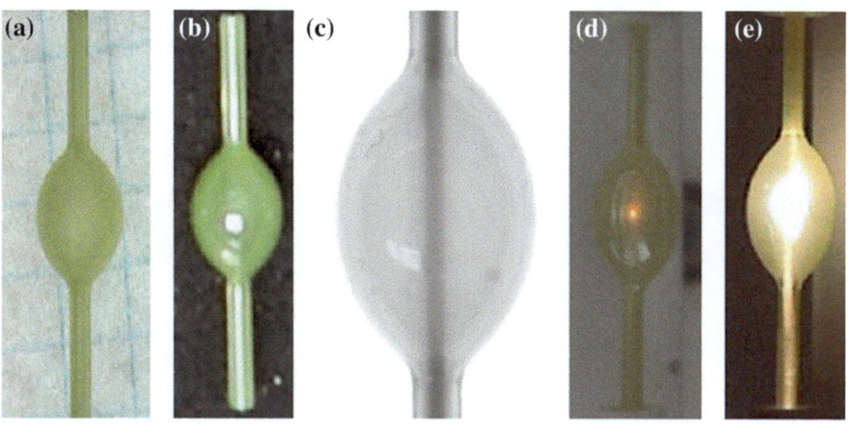

Fig. 10.3 Photographs showing Konoshima polycrystalline Dy_2O_3: **a** as-made, **b** polished, **c** polished with steel pin inserted to illustrate transparency, **d** with fiber light at low power placed inside the tube, and **e** with fiber light at high power placed inside the tube. Reproduced with permission from [13]. Copyright © 2008, IOP Publishing

transmittance, as shown in Fig. 10.3d, e. The infrared cutoff wavelength of Dy_2O_3 was ~9 μm, similar to ~8.5 μm of Y_2O_3, but longer than ~6 μm of PCA, due to the large atomic mass of Dy^{3+} as compared to that of Al^{3+}.

Figure 10.4 shows one of the Dy_2O_3 lamps in operation. The strong blue and UV absorption of the arc tube body material filters some of the plasma radiation and

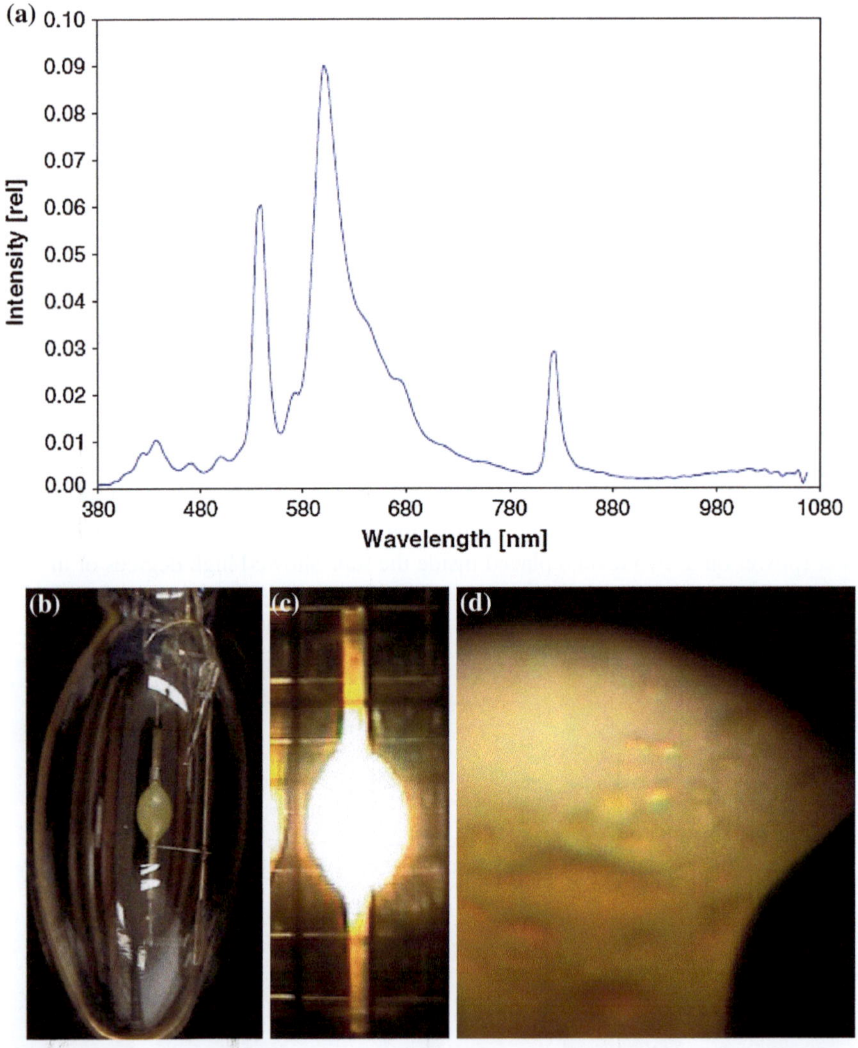

Fig. 10.4 Spectral power distribution from a burning lamp with coarse resolution (~5 nm), with attenuation at about 475 nm due to the absorption in the Dy_2O_3 (**a**), vacuum jacketed Dy_2O_3 lamp power off (**b**), Dy_2O_3 lamp in operation (**c**) and image of the lower funnel of the tube showing the condensed salt melt inside the burner (**d**). Reproduced with permission from [13]. Copyright © 2008, IOP Publishing

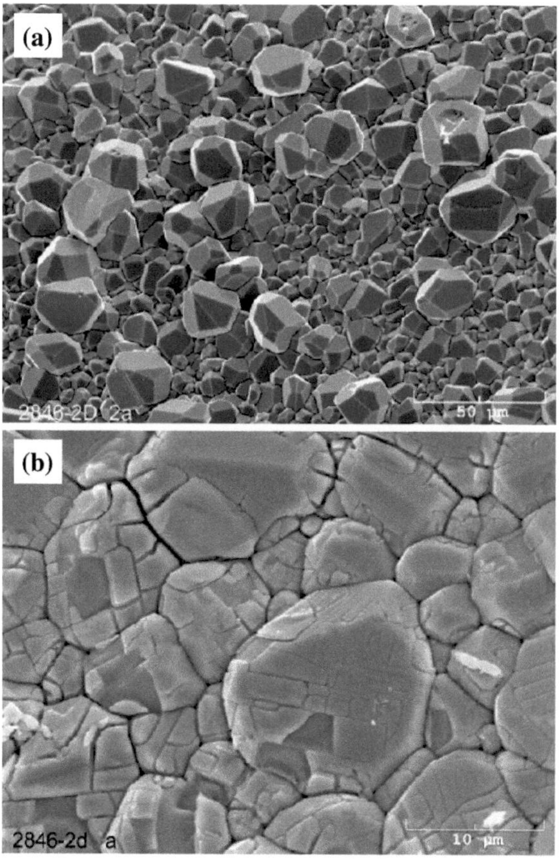

Fig. 10.5 SEM images showing grain-boundary etching (**a**) and redeposition (**b**) on the inside surface of Dy_2O_3 tube after ~500 h in lamps. Reproduced with permission from [13]. Copyright © 2008, IOP Publishing

alters the color of the lamps. Figure 10.4 also shows the spectral output of an experimental Dy_2O_3 lamp. Similar chemistries (nominally 3000 K) in quartz, sapphire, or PCA bodies, which exhibit no such absorption, yield spectral output with CCT ~3000 K. The Dy_2O_3 lamps were yellowish-white in color. The lamps properties were ~2500 K color temperature, ($x = 0.4815$, $y = 0.4193$) chromaticity, ~90 CRI, and ~84 lm W^{-1}. Lower efficacy and color temperature are expected since some of the plasma radiation is absorbed by the burner itself. PCA lamps with this chemistry typically operate above 90 lm W^{-1}.

In Dy_2O_3 lamps containing ReI_3 and Hg fills, the strong blue and UV absorption of the Dy_2O_3 tube body filters some of the plasma radiation and alters the lamp color. The lamps had color temperature of 2500 K, CRI of 90, and 84 lm W^{-1}. The performances were stable for up to 100 h. The hot spot was 1143 °C. Dy_2O_3 body absorbs the UV and blue radiation from Hg and rare-earth emission. This extra radiation, turned into heat, during run-up, led to a rapid warm-up behavior. Although calculated thermodynamic driving force for reactions is not very high, a certain degree of etching and corrosion occurred in the lamps. While the economic viability of the use of Dy_2O_3 ceramic in improved lamps remains open, the performances have demonstrated that developing new transparent ceramic envelope materials should be an effective way for advanced high-intensity discharge projection systems.

Facets of the Dy_2O_3 grains have been observed, as shown in Fig. 10.5a. The dome end of the vertically burnt discharge vessel showed more etching than the base end. Figure 10.5b shows the Dy_2O_3 grains of the arc tube with the presence of redepositions onto the grains, together with features of twin or cleavage-like lines in the redeposited layer. No Tm was detected in the Dy_2O_3 grains or deposits.

10.3 Scintillators

10.3.1 Brief Description

Inorganic scintillators are widely used as radiation detectors in a wide range of applications, such as high and medium energy physics and astrophysics detectors, spectroscopy, medical imaging, industrial nondestructive control systems, well oil logging, and homeland security [33–43]. Scintillators are optical materials that emit pulses of visible photons when excited with high-energy radiations. Gamma-ray spectrometers providing high sensitivity and effective isotope identification require high-energy resolution, high effective atomic number, and scintillators that can be produced with large sizes.

Scintillators are commonly used in radiation detectors for medical diagnostics, industrial inspection, dosimetry, nuclear medicine, and high-energy physics (HEP). In these applications, the scintillator is the primary radiation sensor that emits light or scintillates, when it is struck by high-energy photons, as shown schematically in Fig. 10.6. In this case, the scintillator is a luminescent material that absorbs the high-energy photons and then emits visible light. Solid scintillators are reflectively coated to direct the emitted light onto photodiodes or photomultiplier tubes for photoelectric conversion. There are various requirements for efficient scintillators, when they are used in radiation detectors for applications in imaging and photon counting.

Scintillators can be classified into six types of materials: (i) single crystals [44–50], (ii) polycrystalline ceramics [51, 52], (iii) glasses [53–56], (iv) powders [57–60], (v) plastics [61–66], and (vi) inert gases [67, 68]. The selection of scintillating materials is determined according to applications. For example, single-crystal scintillators, including NaI:Tl, CsI:Tl, $CdWO_4$, and $Bi_4Ge_3O_{12}$, are used in Anger (gamma) cameras, detectors for security and baggage inspection, X-ray detectors in computed tomography (CT), and block detectors in positron emission tomography (PET) and HEP, respectively. Ceramic scintillators, such as $(Y,Gd)_2O_3$:Eu, Lu_2O_3, and $Lu_3Ga_{5-x}In_xO_{12}$, have been employed in X-ray detectors for medical and industrial CT-scanning applications.

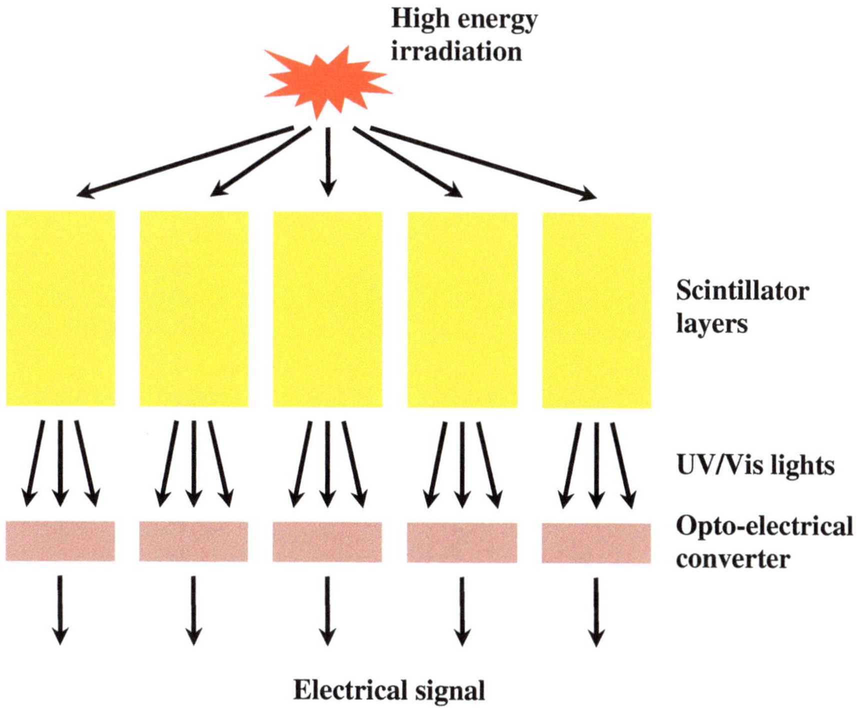

Fig. 10.6 Schematic of the working principle of scintillators

Glass scintillators, e.g., Tb-activated gadolinium silicate glass, are used in large-area and fiber-optic plate detectors in industrial X-ray inspection systems. Powder scintillators include LaOBr:Tm and Gd_2O_2S:Tb for X-ray intensifying screens and BaFBr:Eu powder for photostimulable storage applications. X-ray phosphor powders are used in medical radiography as conventional intensifying screens and photostimulable storage screens. In HEP experiments, plastic scintillators are used for triggering time-of-flight measurements and in sampling calorimeters. Pressurized Xe gas is used as an ionizing gas in both medical and industrial X-ray detectors for CT scanners. In this case, the Xe gas ionized by the X-rays is collected by a radial array of spaced tungsten plates with an imposed electric field, and the charge collected is proportional to the number of X-rays absorbed in the gas. Dense solid-state scintillators, single crystals and polycrystalline ceramics, provide largely higher detection efficiencies than other types of scintillators [69, 70].

Single-crystal scintillators currently used in detectors in commercial CT scanners are Tl-activated CsI and self-activated $CdWO_4$. CsI:Tl has high afterglow values of 0.3 % and radiation damage/hysteresis of 13.5 %, which would cause degradation of image quality, although it can be corrected with computer software. In addition, due to its low-light output of 30 %, CsI:Tl scintillator experienced problems of degraded signal-to-noise and reduced image quality. $CdWO_4$ is a monoclinic crystal that has poor cleavage properties, which thus makes it difficult to fabricate scintillator elements with precise dimensional control and high surface quality. Also, $CdWO_4$ has a high radiation damage of −2.9 %, which can lead to image artifacts if uncorrected. For single crystals, the issues in solid-state lasers are also remained in scintillators.

Transparent ceramic scintillators are a new class of materials developed for quantitative detection accuracy of approximately one part in a thousand. The first polycrystalline ceramic scintillator was developed by combining carefully controlled phosphor composition and doping technology with innovative technology on transparent, polycrystalline ceramics [35, 71, 72]. The major outcome of this development was the rare-earth ceramic scintillator $(Y,Gd)_2O_3$:Eu, which has been used specifically for high-performance medical detectors and commercial X-ray CT. The composition, $Y_{1.34}Gd_{0.60}Eu_{0.06}O_3$, is a cubic, single-phase solid solution of high-purity rare-earth oxides (≥99.99 %). A wet chemical coprecipitation process was used to synthesize the powder, which can be readily sintered into transparent luminescent ceramics. The chemical precipitation route allows homogeneous cod-oping at the molecular level. One of the dopants, Pr_2O_3, is used to reduce the X-ray-induced luminescent afterglow [35, 73–76]. As discussed before, high-quality transparent scintillators with such multiple compositions cannot be grown by using single-crystal methods, but can be readily synthesized by using the ceramic process. Furthermore, the polycrystalline characteristics of the luminescent ceramics, with an average grain size of 30 μm, make them to have high machinability. As a result, optical elements with dimensional accuracy of ±2.5 μm could be achieved. Therefore, ceramic scintillators are cost-effective yet highly efficient.

On the other hand, for scintillators, even the point defects and structural irregularities at atomic scale can seriously decrease the scintillating performance, because they would introduce trapping levels in the forbidden gap of the materials. For example, if a scintillator ceramics has an average grain size of tens of μm and a thickness of the grain-boundary layer of 1–2 nm, the concentration of the interface-related states or trapping states could be at the order of hundreds of ppm. The effect of the defects at such a concentration cannot be neglected, since they can influence the characteristics of the transport stage, due to the retrapping of the migrating carriers. Sometimes, even nonradiative traps might be introduced also, leading to degradation of the scintillation response and thus overall efficiency of the devices. Therefore, scintillating ceramics have even higher requirements, as compared to the transparent ceramics for other applications.

10.3.2 Properties of Scintillators

Properties that determine the performances of scintillators for their applications in high-energy detection systems include (i) optical transparency, (ii) X-ray stopping power, (iii) light output, (iv) primary speed and afterglow, and (v) radiation damage, which will be briefly elaborated in this part.

10.3.2.1 Optical Transparency

High transparency is a critical requirement of scintillator materials, because the visible scintillation photon must be transported efficiently to the photodetector. Applications normally require nearly complete absorption of the incident X-ray flux in the scintillator, and most of this absorption occurs in the entrance face of the scintillator, which is typically opposite to the photodetector. The isotropically emitted scintillation light must be transported to the photodetector by a combination of reflection and scattering at surfaces, as well as scattering within the material. The relative importance of these processes depends on the scintillator geometry as dictated by detector resolution (pixel width) and stopping power (pixel thickness). Minimization of the visible photon path length is critical for scintillators that have intrinsic optical absorption at the emission wavelength because this will result in efficiency losses, owing to the self-absorption. Increased path lengths also make the scintillator more susceptible to optical absorption and efficiency changes, because of the radiation damage.

Transparency of a polycrystalline ceramic scintillator, which is measured by the direct transmission of visible light at the scintillation wavelength, is determined by the microstructure of the ceramics that were developed during the sintering process. Residual pores, either at grain boundaries or entrapped within the grains, which have dimensions close to the emission wavelength, can act as scattering centers. Crystal structures that have nonisotropic optical properties can result in ceramics of reduced transparency, due to reflections at randomly oriented grain boundaries. Similarly, inclusions of second phases in the matrix composition or impurity phases can also decrease the transparency of scintillator ceramics. Sintering and processing techniques that have been discussed previously can be used to optimize the scintillator transparency for given geometries and applications.

10.3.2.2 X-Ray Stopping Power

Insufficient absorption of the X-ray flux incident on the scintillator could have various deleterious effects. First, the efficiency of the X-ray detection is diminished when X-ray photons are allowed to pass through the scintillator, without absorption and creation of excitons that excite luminescent centers. Second, in the common detector geometry where a photodiode is attached to the side of the scintillator opposite from where the X-rays enter, an X-ray that is not absorbed in the scintillator can be absorbed by the diode. This will cause the formation of electronic defect in the diode, so that an additional source of noise will be created, which in turn can degrade the performance of the whole detection system. Therefore, the scintillator material should be able to absorb all the incident X-rays ideally. Various detector designs have been proposed to minimize this effect, e.g., by placing the diode surface away from the direct path of the incoming X-ray beams [73, 77–79].

The fraction of incident X-ray flux of energy E that is transmitted through the scintillator is given by:

$$E_T = e^{-\alpha(E)x}, \tag{10.2}$$

where α(*E*) is the energy-dependent attenuation coefficient, and *x* is the thickness of the scintillator. Generally, α is the overall effect of photoelectric absorption, Compton scattering, and pair production processes. For a given element, each process is active over a certain X-ray energy range, with photoelectric and Compton scattering dominating at energy levels, which are used for medical diagnostics, usually in the range of 30–120 keV. The attenuation coefficient for a scintillator material is given by:

$$\alpha(E) = \rho\left(\sum_i \mu(E)M_i\right)\Big/\sum_i M_i, \tag{10.3}$$

where ρ is the scintillator density, *i* represents the elements in the scintillator, $\mu(E)$ is the mass attenuation coefficient of element *i* at energy *E*, and M_i is its molecular weight.

In order to minimize the transmitted flux given by Eq. 10.2, the scintillator thickness should be maximized, which is consistent with the requirement that the scintillation light be efficiently reflected out of the scintillator volume onto the diode surface. Furthermore, the scintillator density and $\mu(E)$ of the elements in the scintillator composition should also be maximized, which can realized by using scintillators containing elements with high atomic number. One example is the substitution of Y with Gd in the $(Y,Gd)_2O_3$:Eu ceramic scintillators.

The photoelectric absorption of an X-ray photon is rapidly followed by various processes that leave the luminescent centers in an excited state. High-energy mobile electrons release energy through the ionization and creation of secondary electrons. This relaxation continues until the process has insufficient energy to create further electron–hole pairs. On average, an X-ray of energy *E* creates $E/\beta E_g$ electron–hole pairs, where E_g is the band gap of the material and β has a value of 3–7 [80–83]. Further relaxation of electrons to the bottom of the conduction band and holes to the top of the valence band will lead to mobile excitons with energy of close to E_g. This energy can be relaxed radiatively, i.e., self-trapped exciton emission, or nonradiatively by trapping at defects in the lattice or luminescent sites. In scintillators doped with activator ions, recombination of the electron and hole at this site will make the activator in an excited state. Radiative decay from this excited state causes the emission of light photon and scintillation. Critical properties of a scintillator are determined by these relaxation processes, as well as their intermediate trapping states, including the scintillator light output, primary speed, afterglow, and radiation damage.

10.3.2.3 Relative Light Output

The relative light output of ceramic scintillators is an extrinsic property, which is usually measured as the current flowing out of a diode that is attached to the scintillators. To accurately measure the relative light output, it is necessary to maintain the energy and dose of the X-ray and the geometry of the scintillators to be at stable states. Light output is dependent on various factors, including (i) absorption coefficient of the scintillator at the energy level of the X-ray excitation, (ii) efficiency of the exciton creation and capture at the luminescent centers, (iii) intrinsic emission quantum efficiency of the centers, (iv) geometry of the scintillator (e.g., surface structure and reflectors), (v) transparency of the scintillator, and (vi) quantum efficiency of the diode at the emission wavelength. The intrinsic energy quantum efficiency of a scintillator is defined as the ratio of the photon energy emitted from the luminescent sites to the total energy trapped at those sites. Without the presence of any nonradiative relaxation processes, this ratio is given by [57–60]:

$$\eta = E_\lambda / \beta E_g, \tag{10.4}$$

where E_λ is the energy of the emitted photons. η is about 15 % for Gd_2O_2S:Pr and 10 % for CsI:Tl and $(Y,Gd)_2O_3$:Eu. In practice, however, nonradiative processes are always present at the luminescent centers, so that the real quantum efficiency is usually lower than that obtained according to Eq. (10.4).

For an X-ray detection system with scintillators, the detector efficiency is also affected by the coupling of the scintillator to the photodiode. Reflectors are generally used on surfaces of the scintillator to direct the scintillation light into the diode. Scintillator surface treatment and optical properties of the reflectors are important factors to determine the light to the diode. If the scintillators have high thickness-to-width ratios, internal light trapping and self-absorption at the scintillator emission wavelengths could diminish the efficiency of the detectors. Due to the polycrystalline characteristics, ceramic scintillators can be precisely machined, so that geometries of the scintillators can be readily optimized to maximize the efficiency of the reflector. Moreover, high-density scintillators generally have high indices of refraction, so that the light coupling to the Si photodiodes can be easily improved, by using optical couplers with an intermediate index of refraction.

Although the emission wavelength of the scintillators has also influence on the efficiency of the detectors, detection systems with high-quantum efficiency have been available over a wide range of spectrum, from the visible to the near-IR. In addition, matching the scintillator emission wavelength to the efficiency peak of a light detection system can be achieved by activating the scintillators with an activator that emits the desired wavelength. Also, the wide range of high-density ceramic hosts makes it possible to tune the internal crystal field of the activators, so that they will emit at desired wavelengths [84–89]. It is found that transition element activators, with 3d levels that interact with the surrounding crystal field, are good examples, such as $Gd_3Ga_5O_{12}$:Cr, in which the local environment of the Cr^{3+} activator ion has a sufficiently weak crystal field, so that its emission is from the $^4T_2 \rightarrow {}^4A_2$ electronic levels. This emission is centered at 730 nm in the host, which is well matched to the efficiency peak of Si photodiodes [90–92].

10.3.2.4 Primary Speed and Afterglow

Some applications have specific requirements on response time. If the incident X-ray beam is attenuated by the object to be imaged, the emission of the scintillator will be delayed. In this case, image quality will be degraded and image resolution will be reduced. It has been found that two mechanisms are responsible for such delayed emission. One is called primary speed, which is the intrinsic decay time of the emitting site or activator, and the other is known as afterglow, which is the delayed excitation and emission of activators, due to the trapping of electrons or holes by lattice defects and their delayed thermal release.

The primary speed is characterized by a decay time τ, which is obtained by fitting the following equation to the scintillator emission, after the excitation is turned off, i.e.,

$$I(t)/I_0 = e^{-t/\tau}, \tag{10.5}$$

The decay time is the inverse of the probability per unit time, A_r, of a radiative transition from excited state to ground state of the activator. This probability is determined by the symmetry and spin of the states, and the local crystal field experienced by the activator [84, 85, 88].

It is observed that the $5d \rightarrow 4f$ transition of Ce^{3+} ion has a large A_r, where the d and f states have opposite parity, so that it is an allowed electric dipole transition. As a result, this transition has a very short decay time, e.g., Ce^{3+} in $BaHfO_3$ has a decay time of $\tau = 20$ ns. Activators may occupy more than one symmetry site in a crystal lattice, such as Eu^{3+} in $(Y,Gd)_2O_3$:Eu, in which the Eu ion occupies both the symmetric S_6 site and the nonsymmetric C_2 site [93–95]. Due to the difference in crystal field between the two sites, they have different A_r and thus different τ. In this case, the decay time is the sum of two. Generally, the total probability A of a transition from the excited state to the relaxed state is given by:

$$A = A_\mathrm{r} + A_\mathrm{nr}, \tag{10.6}$$

where A_{nr} is the probability of the transition occurring nonradiatively due to the emission of phonons rather than photons. A_{nr} generally increases with increasing temperature, so that radiative decay time decreases with increasing temperature. This temperature dependence is fairly gradual, as compared to the temperature dependence of the second form of delayed emission, i.e., afterglow.

Visible emission from a scintillator with intensities that are higher than that predicted by using Eq. (10.5) is called afterglow. Afterglow emission can last $10^3\tau$ or even longer. When scintillators are used integrating technologies, such as medical CTs, this emission can become a significant contributor to the measured signals, especially when the detector experiences a sudden reduction in X-ray flux [96]. In photon-counting applications, afterglow can reduce the signal-to-noise ratio and thus increase the dead time of the detector [97].

Afterglow is caused by the trapping of mobile electrons or holes at the defect sites in the scintillator crystals. The defects could be intrinsic ones of the host lattice or extrinsic ones related to impurities. The binding energy of the trapping determines the amount of thermal energy required to release the carrier back to the conduction band for trapped electrons or valence band for trapped holes. For a first-order process, where the charge carrier is trapped and released by a single defect before it is trapped by the emitting ion, the resulting emission decay similarly is given by:

$$I(t,T)/I_0 = B(T)e^{-t/\tau_A}, \tag{10.7}$$

where the decay constant τ_A is related to the trap binding energy, which is strongly dependent on temperature T. At low temperatures, the traps remain filled, so as not to contribute to the emission. At high temperatures, the traps empty rapidly with a short decay constant. Similarly, there could be more than one trap at a given temperature [95], so that the afterglow decay is the sum of all of them, given by Eq. (10.7). The traps empty as they are filled, which could take times on the order of τ_A to come to an equilibrium population during the irradiation period. In this case, the magnitude and decay characteristics of the afterglow are closely related to the time duration for which the scintillator has been irradiated before the excitation is terminated [98]. To compare the afterglow effects between different scintillators, all the excitation conditions should be the same, such as time and temperature.

Afterglow is closely related to the distribution of doping ions. Advanced synthetic methods, such as chemical precipitation, have ensured the fabrication of transparent ceramic scintillators with homogeneous distribution of doping ions. Therefore, afterglow can be effectively decreased. It has been shown that afterglow of $(Y,Gd)_2O_3$:Eu scintillators can be greatly decreased by the doping with Pr^{3+} or Tb^{3+} [71]. This can be explained in terms of the multivalency of the dopants. The Eu^{3+} activators efficiently would trap electrons to form a transient state of Eu^{2+}, so that holes can be trapped at the afterglow defects, while Pr^{3+} and Tb^{3+} would trap holes to form Pr^{4+} and Tb^{4+}, which thus compete with the intrinsic traps that are responsible for the afterglow. The energy trapped on the Pr or Tb sites decays nonradiatively in the presence of Eu, so that the afterglow emission could be effectively reduced. Similarly, homogeneous distribution of the doping ions is also very important, because these nonradiative transitions could quench light output. In this respect, fabrication is a critical step to produce ceramic scintillators with reduced afterglow effects.

10.3.2.5 Radiation Damage

Scintillators are required to have high stability and reproducibility of light output for many applications. Radiation damage refers to the degradation of scintillation efficiency due to the formation of defects directly caused by the radiation. Because these defects are usually color centers, their electronic structures could impart the optical absorption bands at the emission wavelengths of the scintillators. Similar to that of afterglow, the binding energy of the color center determines the longevity of the damage. It has been found that this energy is generally quite large, so that the damage could last from seconds to even days at room temperature [97]. Usually, radiation damage would decrease the scintillation efficiency. However, radiation can be used to modify or neutralize the defect centers. In this case, the efficiency can be increased, due to the radiation.

By exposing the scintillator to a relevant dose of radiation and the measuring the change in light output before and after the damage dose, radiation damage can be characterized. The recovery is plotted as a function of time after the end of the damage pulse. To obtain accurate results, it is necessary to compensate for the variations in the source for the measurement. For scintillators with high afterglow, residual afterglow from the damage dose must be compensated. Because the absorption of the scintillation light is dependent on the path through the scintillator crystals, the damage is also dependent on the geometry and optical properties of the scintillator, including crystal thickness and transparency. The radiation damage mechanism can be clarified by monitoring the variation in optical transmission before and after the damage radiation.

Similar to afterglow, radiation damage can also be suppressed by materials processing of the scintillator materials, such as homogeneous doping. It is also found that oxygen partial pressure during the sintering has a significant effect on damage characteristics of $(Y,Gd)_2O_3$:Eu ceramic scintillators. This implies that anion defects could also influence the damage mechanism of ceramic scintillators.

10.3.3 Transparent Ceramic Scintillators

By using transparent ceramics, not only the production costs can be greatly reduced, but also the activator concentration and homogeneity can be significantly increased. Transparent ceramics have become alternatives to replace single crystals for scintillator applications, such as gamma-ray spectroscopy and radiography. The family of cerium-doped gadolinium garnet, GYGAG:Ce, scintillators has been found to meet these requirements. A new gamma spectroscopy scintillators, with high light yields and excellent phase stability, based on GYGAG:Ce and a radiography scintillator based on Lu_2O_3:Eu have been reported in the open literature [99, 100]. Various transparent ceramics, such as Lu_2O_3 [101–105], Ga_2O_3 [106], YAG:Ce [107], and LuAG [40, 108, 109], have formed a class of polycrystalline materials for promising scintillator applications. Representative samples will be discussed in a more detailed way in the following part.

It is important to mention that codoping is an important strategy to improve the performance of scintillator ceramics, e.g., $Gd_3Ga_5O_{12}$:Cr,Ce [35, 110]. In this case, the codoping of Ce^{3+} is used to reduce the afterglow through the capture of holes at the Ce ion sites, i.e., $Ce^{3+} \rightarrow Ce^{4+}$ conversion, and the subsequent nonradiative recombination with the migrating electrons and/or temporary Eu^{2+} centers.

Similarly, codoping of $(Y,Gd)_2O_3$:Eu with Pr^{3+} has become an efficient strategy to reduce the afterglow, as discussed earlier. Annealing in a controlled atmosphere can further diminish or passivate trapping states in the forbidden gap related to the grain interfaces. Eu^{3+}- or Tb^{3+}-doped Lu_2O_3 ceramics have been among the new optical ceramics that are suggested for scintillator applications, due to the slow 4f–4f transitions of the rare-earth ions [111].

Temporal dependence of X-ray excited scintillation from Eu^{3+}-activated Lu_2O_3 transparent ceramics has been reported both experimentally and theoretically [102]. The waveform of the decay has been measured on the timescales ranging from 100 ns to 5 min. Kinetic studies have been carried out and a computational model has been developed to describe the observed temporal behaviors. The model involved two overlapping Gaussian distributions of trap depths, with those ranging from 0.58 to 1.0 eV, which could have contributed to the afterglow. The ceramics has been studied by combined scintillation and thermoluminescence excited by prolonged gamma-ray irradiation [103]. The thermoluminescence glow curve partially confirmed and extended the model for afterglow following pulsed X-ray excitation.

It has been acknowledged that the extent to which the competing process of carrier trapping diminishes the efficiency of scintillator response cannot be satisfactorily assessed from the thermoluminescence data alone, while the absolute light output measurements are not available. Therefore, an experiment was established so as to determine the branching ratio of electron–hole pairs that contribute to thermoluminescence or scintillation [112, 113]. Energy storage by deep traps was probed by comparing thermoluminescence and scintillation light outputs. This scheme exploited a common activator with identical emission spectra for both thermoluminescence and scintillation.

A schematic diagram of the experimental arrangement is shown in Fig. 10.7 [103]. An electron Van de Graaff accelerator, operated at 1 MW and 1.0 μA, was employed as the primary radiation source, with the electron beam that was stopped by a thin copper target, which was used as an approximate point source of gamma rays. The sample was mounted on a copper pedestal enclosing a heating element. Sample temperature was monitored using a thermocouple that was attached to the pedestal, whereas luminescence was extracted using a shielded optical fiber and detected using a photomultiplier. Both the photomultiplier and thermocouple outputs were passed through electrometers to a computer, with data being recorded at an interval of 4 s for radiation times ranging from 15 s to 16 min. An unusually rapid temperature-ramp rate was employed to allow for the comparison of thermoluminescence and scintillation light outputs on comparable scales. The sample of the ceramic Lu_2O_3:Eu had a dimension of $2 \times 3 \times 8$ mm^3, with the activator concentration of 5 mol%.

Fast optical ceramics based on Ce-doped YAG have been reported [114]. Scintillation and luminescence properties of the highly dense transparent YAG:Ce ceramics were compared with those of a single crystal. Both the ceramics and the single crystal exhibited a dominant decay of ~85 ns, when they were excited with gamma rays. At the same time, the ceramics also showed a new rapid component of

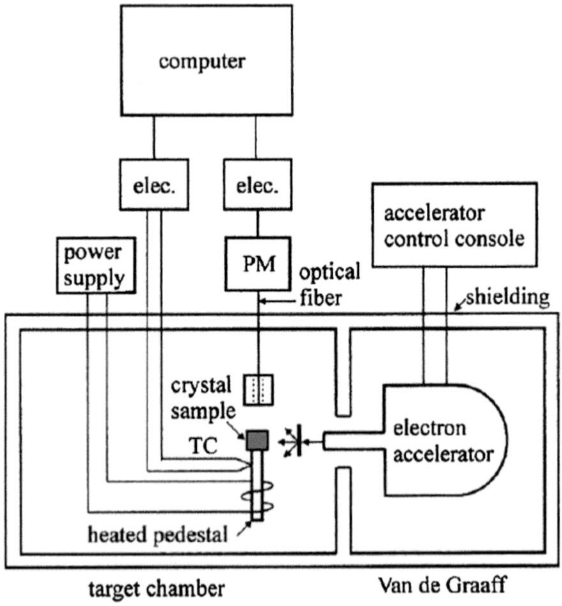

Fig. 10.7 Schematic diagram of the experimental setup. Reproduced with permission from [103]. Copyright © 2004, Elsevier

~20 ns, which was not observed in the single crystal. The scintillation output from the ceramic scintillator reached about 50 % of that from the single crystal, which was attributed to the diminishing effect of unusual loss processes caused by the deformed lattice at the grain boundaries. A phenomenological model involving distortion of the band structure was proposed, which could be used to well explain the results of kinetic measurements.

It was reported that the radio-luminescence intensity in the Ce^{3+} band at 550 nm was comparable for the YAG:Ce ceramics and single crystals, and the host 300–350 nm band was absent in the ceramics [115]. Optical quality YAG:Ce ceramics were prepared from YAG:Ce nanopowders. According to the experimental data in optical absorption, radioluminescence, scintillation decay, photoelectron yield, thermally stimulated luminescence and radiation-induced absorption, it was found that the antisite defects, Y_{Al}, i.e., Y at the Al site, were absence in the YAG:Ce ceramics. The presence of the antisite Y_{Al} defect was responsible for an UV emission, due to the luminescence center based on the defects [116]. Formation of this defect was dependent on temperature. Because the fabrication temperatures of the YAG-based ceramics are significantly lower that for the single-crystal growths, such Y_{Al} antisite defects are absent in ceramics [117]. The scintillation decay of the ceramics was faster than that of the single crystal, but its photoelectron yield measured with integration time of 1 μs was lower by 30–40 %. Except the photoelectron yield value, the overall performance of the YAG:Ce ceramics was found to be well comparable to the high-quality industrial YAG:Ce single crystals. Therefore, the YAG:Ce ceramics can be used as a promising candidate of scintillator materials.

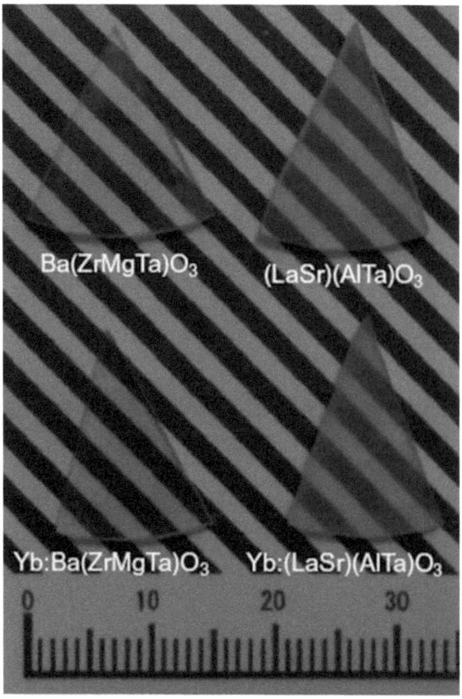

Fig. 10.8 Photographs of transparent ceramic specimens: pure and Yb 10 % doped Ba(ZrMgTa)O_3 and (LaSr)(AlTa)O_3. The sample thickness is 2 mm. Reproduced with permission from [123]. Copyright © 2011, Elsevier

Similarly, LuAG:Ce ceramics have been fabricated by using vacuum sintering from various precursor powders [118–121]. For example, nanosized highly sinterable Ce-doped lutetium aluminum garnet (LuAG:Ce) powders were synthesized by using a homogeneous coprecipitation route with urea as precipitant [120]. LuAG:Ce transparent ceramics with an average grain size of about 6 μm were obtained by using vacuum sintering at 1800 °C for 10 h without using sintering aids, from the powders calcined at 1000 °C for 2 h. The transmittance of the LuAG:Ce ceramics in the visible light region reached a value of 65 %. Radioluminescence spectrum of the ceramics displayed a broadband located at 480–650 nm, with two emissions due to the transitions from the lowest 5d excited state to the 4f ground state of Ce^{3+}, which were similar to those observed in LuAG:Ce single crystals [40, 122].

Optical and scintillation properties of complex perovskite transparent ceramic scintillators of pure and Yb^{3+}-doped Ba(ZrMgTa)O_3 [$Ba(Zr_{0.25}Mg_{0.25}Ta_{0.5})O_3$ and $Ba(Yb_{0.1}Zr_{0.2}Mg_{0.2}Ta_{0.5})O_3$] and (LaSr)(AlTa)$O_3$ [$(La_{0.5}Sr_{0.5})(Al_{0.8}Ta_{0.2})O_3$, and $(Yb_{0.1}La_{0.5}Sr_{0.4})(Al_{0.8}Ta_{0.2})O_3$] ceramics have been reported [123]. Their optical properties, including transmittance and photoluminescence spectra, were systematically evaluated. The ceramics demonstrated high transparency of 50–70 % in visible wavelength region. In addition, strong emission peaks were observed at 470 nm for Ba(ZrMgTa)O_3 under 284 nm excitation and at 500 nm for (LaSr)(AlTa)O_3 under 324 nm excitation. Photoluminescence decay times of Ba(ZrMgTa)

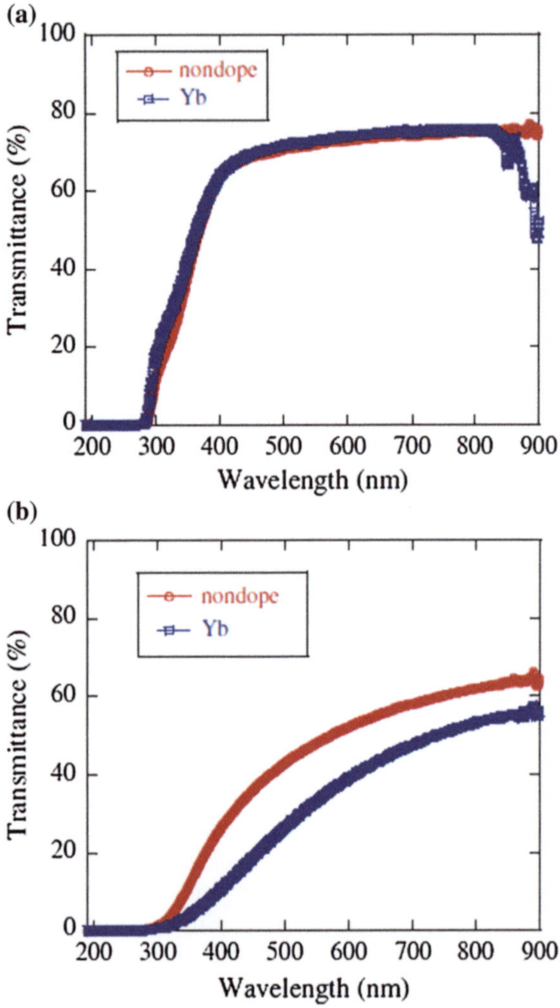

Fig. 10.9 Transmittance spectra of pure and Yb 10 % doped **a** Ba(ZrMgTa)O_3 and **b** (LaSr)(AlTa)O_3. Reproduced with permission from [123]. Copyright © 2011, Elsevier

O_3 and (LaSr)(AlTa)O_3 samples were 14 and 16 μs, respectively. According to optical properties, Yb^{3+} emission was not observed. In radioluminescence spectra, all specimens exhibited the same emission peaks with relatively higher emission intensity. ^{241}Am 5.5 MeV α-ray-induced pulse height spectra of the samples were also measured, where (LaSr)(AlTa)O_3 demonstrated ~500 ph/5.5 MeV α rays with a shaping time constant of 10 μs.

Although Ce^{3+} doping is usually used in scintillator ceramics, Yb^{3+} was used as the emission center in this study. Yb^{3+} charge transfer transition could facilitate fast and intense emission [124, 125]. Based on the experimental results, it was concluded that the strong thermal quenching in the scintillators was suppressed, so that fast and strong emission was achieved by using the materials at room temperature

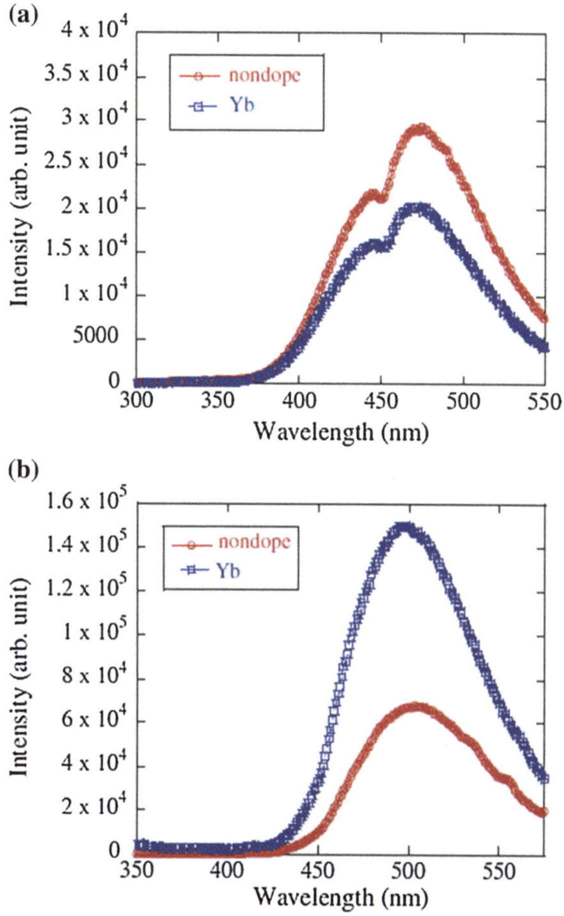

Fig. 10.10 The photoluminescence spectra of pure and Yb 10 % doped **a** Ba(ZrMgTa)O_3 (λ_{ex} = 284 nm) and **b** (LaSr)(AlTa)O_3 (λ_{ex} = 324 nm). Reproduced with permission from [123]. Copyright © 2011, Elsevier

[126–128]. However, it was observed that, as new scintillators, these complex perovskite ceramics have a different thermal quenching effect, as compared to the conventional scintillator materials, which deserves further investigation in the future. Figure 10.8 shows a photograph of the perovskite transparent ceramic samples, with a thickness of 2 mm [123]. Optical properties, including transmittance, photoluminescence, and photoluminescence decay time, have been characterized. Scintillation properties, such as radioluminescence and pulse height spectra, were studied in details.

Transmittance spectra of the perovskite ceramics are illustrated in Fig. 10.9 [123]. In Fig. 10.9a, absorption edge of the undoped Ba(ZrMgTa)O_3 was observed at about 300 nm and no other detectable absorption bands were found. Although the spectrum of the Yb 10 % doped Ba(ZrMgTa)O_3 had very similar feature, there was a difference between the two samples near 900 nm. The absorption bands were ascribed to the $^2F_{2/7}$ to $^2F_{2/5}$ transition of Yb^{3+}. On the other hand, the spectra in

Fig. 10.9b for the undoped and Yb 10 % doped $(LaSr)(AlTa)O_3$ were very similar in shape. However, the undoped sample had slightly higher transparency in the wavelength range.

Photoluminescence spectra of the ceramics are shown in Fig. 10.10 [123]. According to Fig. 10.10a, the undoped and Yb-doped $Ba(ZrMgTa)O_3$ had emission peaks at about 440 and 470 nm under the 286 nm excitation, with very similar profiles. The emission spectra of pure and Yb-doped $(LaSr)(AlTa)O_3$, shown in Fig. 10.10b, were recorded under 324 nm excitation. This group of ceramics had longer integration times than the $Ba(ZrMgTa)O_3$-based ceramics, due to low emission intensities. The shapes of the spectra of the two samples were quite similar. In these ceramics, only one intense peak at 500 nm was observed.

10.4 Ceramic Electro-Optic Devices

A variety of devices have been designed by using transparent electro-optical ceramic materials, including variable optical attenuators (VOA), polarization controllers (PC), sinusoidal filters, dynamic gain flattening filters, tunable optical filters, and *Q*-switches, which have been described in detail in Ref. [129]. A brief description is given for some devices as follows.

VOAs is a device used to reduce the power level of an optical signal, either in free space or in an optical fiber. The basic types of optical attenuators are fixed, step-wise variable and continuously variable. The type of devices has played an important role in controlling the optical signal levels throughout the network. Representative applications of VOA include pre-emphasis, channel balancing, and optical automatic gain control. In dense wavelength-division multiplexing (DWDM) transmission system, the optical powder between every two channels should be equalized before they are combined and guided into a single optical fiber. At add/drop network nodes, because the optical signals arrive individually from different points in a network, all channels must be equalized. In addition, VOA has found applications beyond the usual networks [129].

It is well known that light is in nature an electromagnetic wave, with electric and magnetic field components being perpendicular to each other and to its propagation direction. Therefore, light is actually a polarized electromagnetic field. As a result, light has distinct orientation in space and its propagation in a media is dependent on the polarization direction, which has exerted negative influence on the transmission rates of DWDM systems, such as polarization dependent loss (PDL) and polarization mode dispersion (PMD). To address these problems, it is necessary to have a good capability of polarization management. A dynamic PC is used to overcome such polarization-related problems. Therefore, PC can be used for PMD compensation, polarization scrambling, polarization multiplexing, polarization generation, and other polarization management functions [129].

The electro-optic properties of PLZT materials are intimately related to their ferroelectric properties. As a result, altering the ferroelectric polarization with an electric field, such as in a hysteresis loop, also produces a variation in the optical properties of the ceramics. Moreover, the magnitude of the observed electro-optic effect is dependent on both the strength and the direction of the electric field.

PLZT ceramics display optically uniaxial properties at microscopic scale and at macroscopic scale when they are polarized or activated with an electric field. There is one unique symmetry axis in uniaxial crystals, i.e., the optic axis that is colinear with the ferroelectric polarization vector in ceramic PLZT, which possesses optical properties different from the other two orthogonal axes. That is, light traveling in the direction along the optical axis and vibrating in the direction perpendicular to it possesses an index of refraction (n_O) that is different from the light traveling in the direction 90° to the optic axis and vibrating parallel to it (n_e). The absolute difference between the two indexes is defined as the birefringence; i.e., $n_e - n_O = \Delta n$. At macroscopic scale, Δn is equal to zero before poling and has a finite value after poling, depending on the composition of the materials and the degree of poling. For relaxor ferroelectric materials, Δn is not permanent, which is present only when an electric field is applied [129, 130].

A typical setup for determining the behavior of electro-optic ceramics is given in Fig. 10.11. Linearly polarized white light, on incident into the electrically energized ceramic, will be split into two orthogonally vibrating white light components, represented by red and green waves in the figure, with vibration directions that are defined by the crystallographic axes of the crystallites acting as one optical entity. Therefore, the axes are defined by the direction of the electric field. Because of the different refractive indices, n_e and n_O, the propagation velocities of the two components are different within the materials, which results in a phase shift that has been known as optical retardation. The total retardation (Γ) is a function of both Δn and optical path length (t), according to the following relationship:

$$\Gamma = \Delta nt, \tag{10.8}$$

where t is the ceramic thickness along the optical path. In the case of white light, e.g., average wavelength of 0.55 μm, as the applied voltage is sufficiently high, a

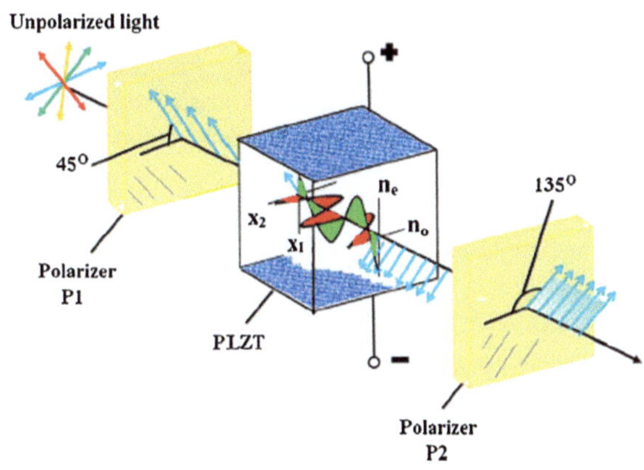

Fig. 10.11 Basic setup for evaluating electro-optic shutter/modulator characteristics (open condition shown). Reproduced with permission from [131]. Copyright © 1999, John Wiley & Sons

half-wave retardation could be achieved for one component relative to the other. The net result is that the vibration direction of the polarized light is rotated by 90°, so that it is transmitted by the second (crossed) polarizer. Switching of the ceramics from a state of zero retardation (no voltage) to half-wave (full voltage) creates an ON/OFF light shutter, whereas an analog modulator can be created by applying intermediate voltages. Color generation, including yellow, red, blue, and green, from the incoming white light can be realized by increasing the voltage beyond half-wavelength. If the monochromatic light is used, extending the voltage beyond the half-wavelength can only lead to a progression of repeating light and dark bands, thus acting as an interferometer.

There are two common types of electro-optic birefringent effects within the PLZT compositional phase diagram, i.e., (i) nonmemory quadratic (Kerr effect) and (ii) memory linear (Pockel effect). The respective electro-optic coefficients for these effects are calculated by using the following relationships:

$$R = -\frac{2\Gamma}{n^3 t E^2}, \tag{10.9}$$

$$r_c = -\frac{2\Gamma}{n^3 t E}, \tag{10.10}$$

where R is the quadratic coefficient and r_c is the linear coefficient, while n *is* the index of refraction, i.e., $n = 2.5$ for PLZT.

For thin plates and films deposited on substrate, interdigital electrodes (IDEs) are usually used, which are made on the free surface of the materials. In this case, the coefficients obtained are only a close approximation to the true coefficients, because the electric field lines are not truly transverse, instead, they are constrained to penetrate the materials in a nonlinear manner.

Highly transparent Dy-codoped lead lanthanum zirconate titanate (PLZT) ceramics, $[Pb_{0.9}(La_{1-x}Dy_x)_{0.1}](Zr_{0.65}Ti_{0.35})_{0.975}O_3$ or 10/65/35, 0.02Dy, 0.04Dy, and 0.06Dy for $x = 0$, 0.02, 0.04 and 0.06 mol% of Dy, respectively, have been studied for electro-optic properties [132]. The ceramics were prepared from oxide powders by using hot pressing sintering. The ceramics exhibited a transparency as high as 67 % at 632.8 nm, which was close to their theoretical value. The effects of substituting La by Dy on the phase structure, microstructure, dielectric, ferroelectric, optical, and electro-optic properties of PLZT were investigated. The results showed that Dy substitution decreased the coercive field and improved the optical transmittance and electro-optic properties of the PLZT ceramics. Moreover, the high half-wave voltage, determining the working voltage of devices, of the PLZT was reduced from 850 to 510 V, as a result of the Dy substitution. These Dy-modified PLZT transparent ceramics are the promising candidates for optical communication applications. A VOA based on the PLDZT exhibited excellent performances.

Figure 10.12 shows birefringence values of the perovskite transparent ceramics, as a function of electric field [123]. The quadratic EO coefficient R is proportional to the slope of Δn versus E^2 curve. The curves in Fig. 10.12 revealed that Dy substitution enhanced the quadratic EO coefficient R of the PLZT (10/65/35). As the electric voltage E is up to $V\lambda_{/2}$, EO coefficient R can be calculated according to Eq. (10.9). The quadratic EO coefficients for 10/65/35, 0.02Dy-, 0.04Dy-, and 0.06Dy-substituted PLZT were 2.566×10^{-16}, 2.83×10^{-16}, 3.30×10^{-16}, and 4.67×10^{-16} $(\text{m/V})^2$, respectively. Moreover, the half-wave voltage for 10/65/35, 0.02Dy-, 0.04Dy-, and 0.06Dy-substituted PLZT were 850, 610, 580, and 510 V, respectively. According to the dependence of quadratic EO coefficient and the half-wave voltage, as a function of the Dy dopant content in the PLDZT ceramics, it was observed that substituting La with a small amount of Dy increased the quadratic EO coefficient and reduced the half-wave voltage of the PLZT transparent ceramics.

As stated before, electro-optic properties of PLZT materials are related to their ferroelectric properties [131]. For PLDZT, the reduced coercive field (E_c) and the enhanced optical transmittance (T) induced by the Dy substitution have led to improved electro-optic properties. Because the working voltage of a device is mainly dependent on the half-wave voltage of the materials, the lowered $V\lambda/_2$ of materials should be favorable to the improvement in performances of the devices.

Optical characteristics of transparent lead magnesium niobate titanate (PMN-PT) electro-optic ceramics, including electro-optic phase modulation, electric hysteresis property and thermo-optic coefficient, have been investigated and reported [133]. Based on this transparent ceramics, a polarization independent electro-optic switch by using fiber Sagnac interferometer (FSI) structure has been realized. An initial π-shift was introduced into the Sagnac loop to eliminate the effect of the polarization orientation of the incident light on the switch performance. An electrically controllable PMNT phase retarder was used to switch the optical signal between the reflection and transmission ports. Theoretical analyses were provided to describe the switch performances of the devices, including the thermal characteristic and switching frequency response.

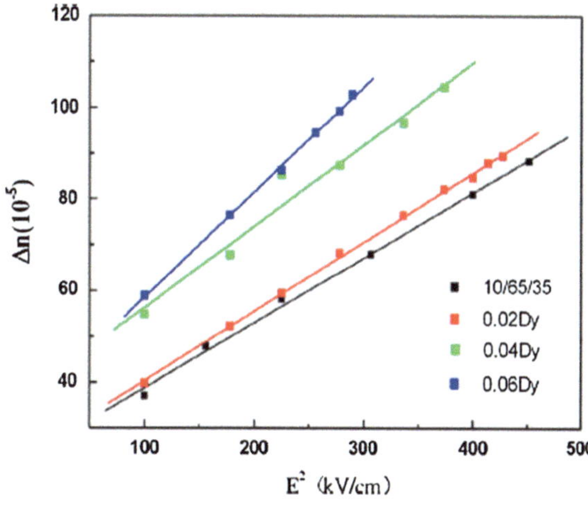

Fig. 10.12 Birefringences versus the quadratic electric field for the PLDZT ceramics with different concentrations of Dy. Reproduced with permission from [132]. Copyright © 2014, Elsevier

Figure 10.13 shows experimental setup for the optical characteristic measurement of PMNT ceramics [133]. The size of PMN-PT ceramic sample was 5 mm × 2 mm × 1 mm for length × width × thickness. Ti/Pt/Au layers were sputtered on both surfaces of the ceramics as electrodes. Two collimators were used to collimate the incident beam and receive the transmission beam. The output beam was detected by using an optical spectrometer and phase demodulation. Because the PMN-PT electro-optic ceramics have a large refractive index, i.e., n = 2.465, the ceramic samples could be considered as a Fabry–Perot (FP) resonator, which can be used to measure the electric hysteresis and thermo-optic coefficient. The applied voltage generated a transverse electro-optic effect for the transmission light beam.

Figure 10.14 shows the variation of phase shift of incident beam of the PMN-PT ceramic samples, as a function of the applied voltage [133]. A quadratic curve

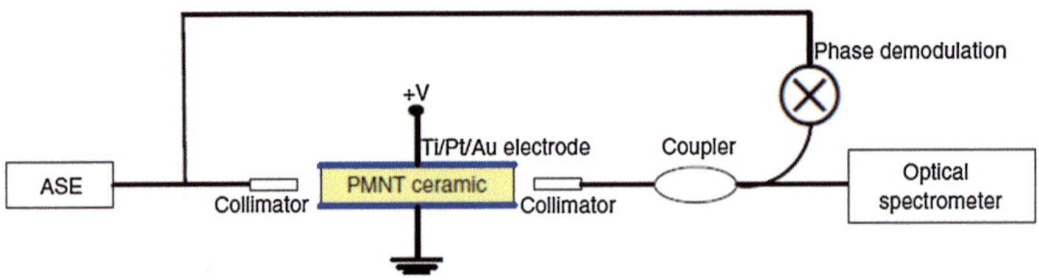

Fig. 10.13 Experimental setup for the optical characteristics measurement of PMNT ceramic. Reproduced with permission from [133]. Copyright © 2011, Elsevier

Fig. 10.14 Phase shift of the PMNT electro-optic ceramic sample with different applied voltage. Reproduced with permission from [133]. Copyright © 2011, Elsevier

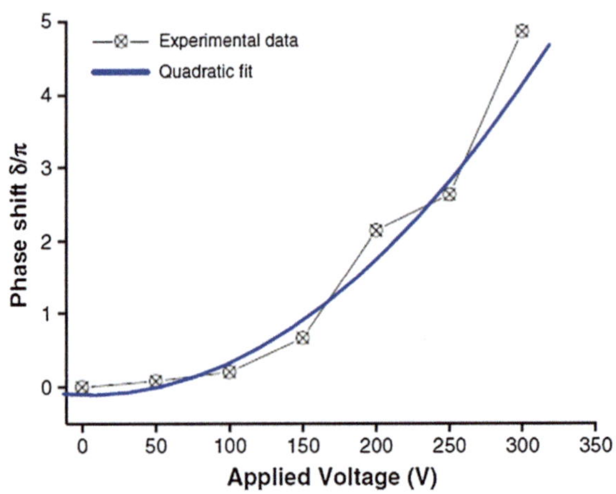

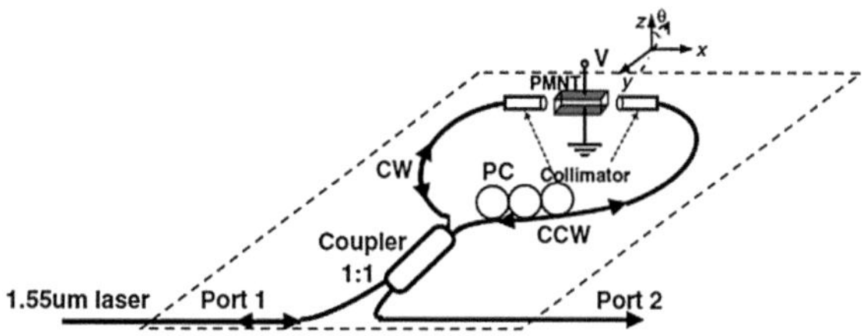

Fig. 10.15 Schematic diagram of the proposed electro-optic switch. Reproduced with permission from [133]. Copyright © 2011, Elsevier

relation was observed by fitting the experimental data. As the applied voltage was 158 V, phase shift as obtained, with the corresponding 2π, 3π, and 4π phase-shifted voltages to be 212, 256, and 294 V, respectively. Using this curve relation, the electro-optic coefficient of the PMNT ceramic sample was estimated to be about 3×10^{-15} m^2 V^{-2}, which is high than that of PLZT by about 15 times [134]. A resonant wavelength shift was used to estimate the hysteresis of the PMN-PT sample by taking it as a FP resonator, which exhibited a quite obvious FP interference effect. With increasing applied voltage, the resonant wavelength shifted to short wavelength gradually. When the applied voltage was reduced from high voltage to low voltage, the resonant wavelength went back to long wavelength, which was consistent with the voltage applying process. Therefore, the electric hysteresis of the PMN-PT ceramics was eliminated essentially, as compared to the PLZT ceramics.

Based on the properties of the PMN-PT electro-optic ceramic, a polarization independent electro-optic switch with an all single-mode FSI structure was proposed and designed, as shown in Fig. 10.15 [133]. The PMN-PT electro-optic ceramics was used as an electrically tunable arbitrary phase retarder, which can be used to switch the optical signal between the reflection port (i.e., Port 1) and transmission port (i.e., Port 2). When a linear polarized incident light was input into the interferometer from Port 1, it would be split into two counter propagation waves, (i) a clockwise wave (CW) and (ii) a counterclockwise wave (CCW). Assuming θ to the angle to one principal axis of the PMNT sample for the CW, the orientation for the CCW was $\pi - \theta$ or $-\theta$, due to the 180° direction change induced by the loop structure.

10.5 Optical Systems (Lens)

Markets for digital cameras and mobile phones with optic equipment are increasing rapidly. Frequently, a compact and small design of the lens systems is imperative for use and gives rise to the search for new transparent materials with high refractive index and high Abbe number. Optical glasses exhibit limits at $n = 1.50 - 1.85$, whereas higher values are observed for a number of crystalline ceramics [135–137]. Furthermore, ceramics exhibit a higher scratch and impact resistances [138].

Optical transparent polycrystalline ZrO_2 ceramics were fabricated by using solid-state sintering process, with vacuum sintering followed by hot isostatic pressing [139]. In the visible wavelength range of 400–800 nm, the in-line transmittance of the 5.6-mm-thick samples reaches 68 % at wavelength 600 nm, corresponding to an in-line absorbance based on 10 of $A_{10} = 0.08\ \text{cm}^{-1}$, which was about 90 % of theoretical limit. The refractive indices of the ZrO_2 optoceramics at 630 nm (n_d) were in the range of 2.10–2.20, depending on the content of TiO_2, which was used as a sintering aid. The appearance of birefringence was strongly correlated with the content of TiO_2. It was found that addition of TiO_2 and simultaneous increase in Y_2O_3 resulted in a decrease in birefringence.

In cubic transparent ceramics, the optical properties are isotropic in all directions. However, sinter aids and/or phase stabilizers could segregate at grain boundaries, which would cause fluctuations in optical properties. The birefringence properties of the ZrO_2–Y_2O_3–TiO_2 ceramic samples are shown in Fig. 10.16 [139]. It was found that the content of TiO_2 exhibited a significant effect on the feature of birefringence. With lower content of TiO_2, the birefringence was lower, but the transmission was decreased at the same time. It implies that there was a competition of transmission and homogeneity. This competition can be attributed to structural features and/or process-related effects. Calculation of reliable dispersion data requires high accurate refractive index data (5th digit). A value of 39 ± 1 for the best ZrO_2 optoceramic sample, i.e., Z10Y–5 wt% TiO_2–30 wt% Y_2O_3, has been calculated, as shown in Fig. 10.17 [139].

In order to release the residual strain, the samples were post-annealed at 1600 °C for 3 h in air, with low heating and cooling rates. After the post-annealing treatment, the high intensity of birefringence at the outer rim of the samples was significantly reduced. Therefore, it was confirmed that the birefringence was attributed to the presence of residual stress, and it typically appeared at the rim edge of the samples. The birefringence with intensity of >150 nm cm^{-1} could be readily eliminated by the post-annealing, as shown in 2D and 1D mappings of Fig. 10.18 [139]. However, the birefringence at center of the samples was still observed, which meant that the oxygen-related defects cannot be eliminated by using the post-annealing. On the other hand, the transmission was degraded by the post-annealing, which could be attributed to the possible formation of pores caused by the escape of Ar that was dissolved in the crystal lattice during the HIPing. In this respect, further studies are necessary to improve the properties of transparent ceramics for optical lens applications.

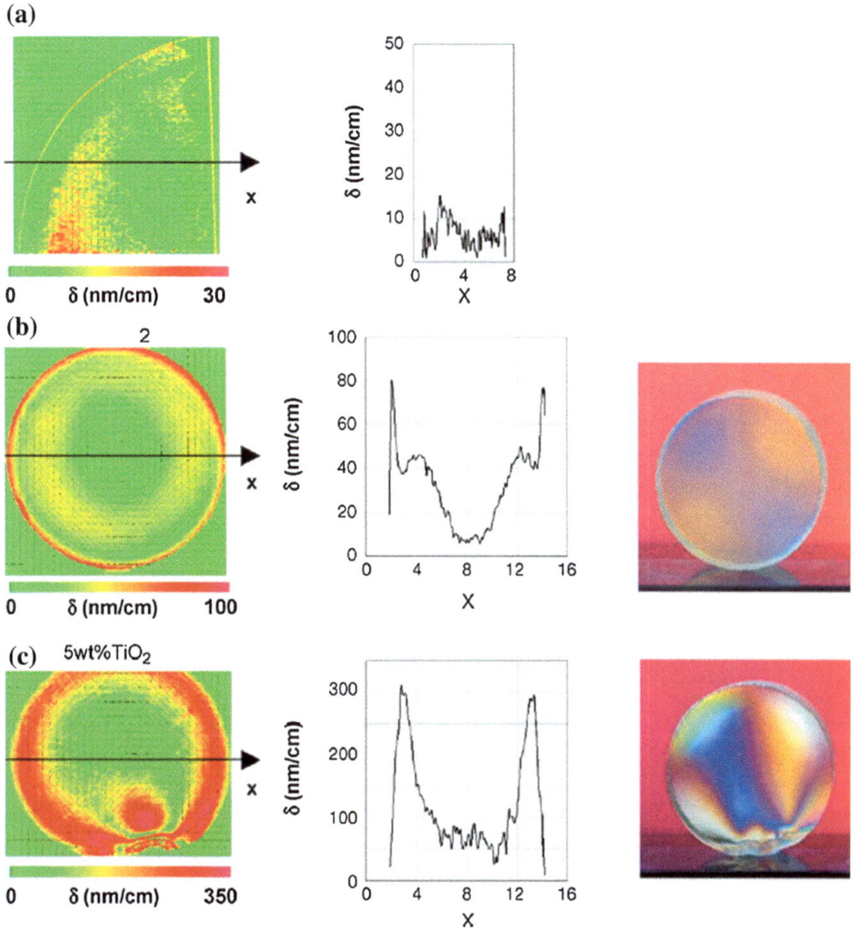

Fig. 10.16 Birefringences of the Z10Y–TiO_2 ceramics: **a** Z10Y–0 wt% TiO_2, **b** Z10Y–2 wt% TiO_2, and **c** Z10Y–5 wt% TiO_2. Schematic diagram of the proposed electro-optic switch. Reproduced with permission from [139]. Copyright © 2009, Elsevier

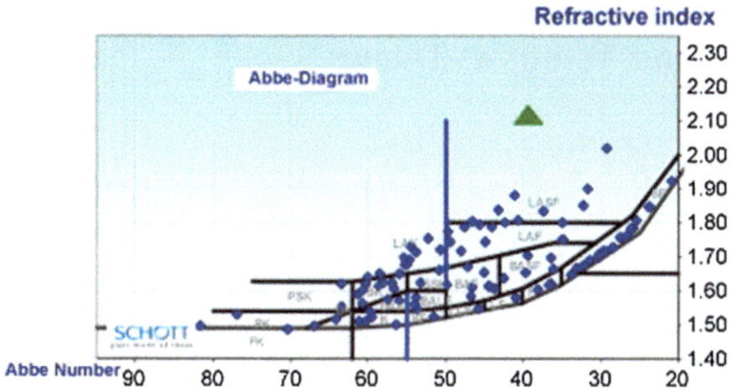

Fig. 10.17 Position of the sample Z10Y–5 wt% TiO_2–30 wt% Y_2O_3 in Abbe diagram (*triangle*). Schematic diagram of the proposed electro-optic switch. Reproduced with permission from [139]. Copyright © 2009, Elsevier

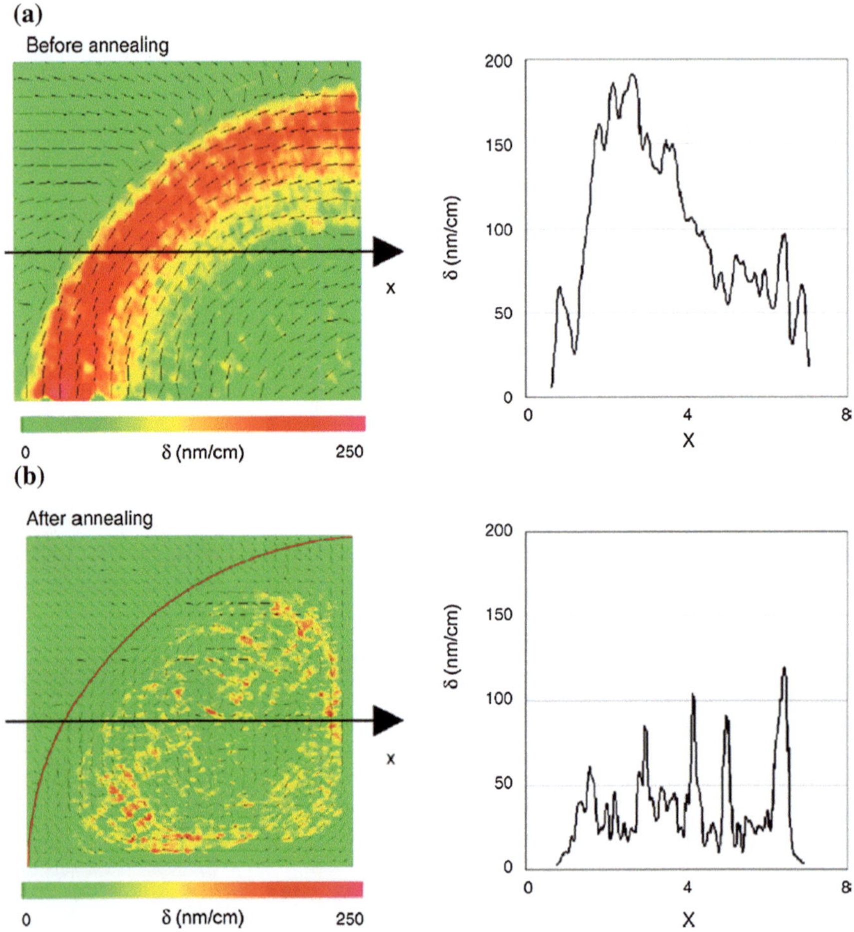

Fig. 10.18 Birefrengences of the Z10Y–5 wt% TiO_2 ceramics before (**a**) and after (**b**) post-annealing at 1600 °C for 3 h. Schematic diagram of the proposed electro-optic switch. Reproduced with permission from [139]. Copyright © 2009, Elsevier

10.6 Armors and Windows/Domes

10.6.1 Brief Description

Common applications for transparent armor include personnel protection in the form of face shields and visors, windows for ground vehicles and armored cars and windshields and lookdown windows for aircrafts. The widely used glass-based transparent armor windows are typically greater than five inches in thickness in order to provide the required multi-hit ballistic protection. Not only does this increase the wear cost for tactical vehicles which are already straining with their weight limits, but also the relatively soft glass facing can be degraded in the field by rock damage and blowing sand, leading to significant replacement cost. Therefore, there is a compelling need to reduce the weight of the transparent armor systems while simultaneously increase the ballistic protection capabilities.

There is also a need on large windows and domes for multi-mode weapon systems that are exposed to harsh environmental conditions [140–142]. When both strength and hardness are the critical performance metrics, the choices become very limited. For instance, fluorides which can be used as laser windows are unable to meet these requirements [143, 144]. Only a few materials can be considered, mainly including single crystals of sapphire [145–154], as well as ceramics of aluminum oxynitride (AlON) [155–159] and magnesium aluminate spinel ($MgAl_2O_4$) [21–23, 160–164]. Due to their high strength and hardness, these three materials have long been considered as a potential replacement for the ballistic glasses in transparent armor applications. Recent ballistic testings have shown that spinel, AlON, and sapphire possess similar performance. Significant progress has been made in this area. However, the development of windows based on transparent ceramics is still lagging behind the sapphire single crystals [145–154].

10.6.2 Transparent Ceramics Armors

As stated earlier, when compared to the sapphire single crystals, polycrystalline $MgAl_2O_4$ spinel [155–159] and AlON [21–23, 160] are superior for armors or armor-related applications. As the window materials, spinel is superior to sapphire and AlON due to its transparency up to 5 μm wavelength. Its high optical and MIR transparency in the range of 0.25–5.5 μm makes it an ideal candidate as windows and domes for EO and IR sensors on various defence systems, which are usually operated in harsh or extreme environmental conditions.

Mechanic strength is an important requirement for armor applications, which is closely related to grain size. Effects of grain size have been investigated into the lubricated sliding wear of three transparent magnesium aluminate ($MgAl_2O_4$) spinel ceramics with different grain sizes: nano, fine, and coarse [156]. Only fine spinel showed classical wear behavior, which was characterized by an initial mild wear, followed by a sharp transition to severe fracture-controlled wear. Worn surfaces of fine spinel showed extensive grain pullout, consistent with intergranular mode of fracture found in spinel. Both the nano- and coarse spinels exhibited gradual transition from mild wear to severe wear, with significantly lower overall wear rates as compared to the fine spinel sample. Worn surfaces in both nano- and coarse spinels demonstrated transgranular fracture and material removal, indicative of lateral-crack-induced chipping. The transgranular fracture mode in the nanospinel could be attributed to its stronger grain boundaries, due to the presence of Y_2O_3 sintering aid used for grain refinement. In contrast, the large grains of the coarse spinel were responsible for the transgranular fracture.

In the application of laser ignition to large caliber cannons, a critical element is the window into the cannon chamber to admit the laser energy [160]. This window must repeatedly withstand a particularly harsh environment of highly reactive high-temperature combustion products from the gun propellant at pressures of up to 440 MPa or 65,000 psi. Failure of the window can be caused by either thermal gradients in the window or mechanical force, or their combination. Previous successes with single-crystal sapphire have sometimes been limited by window deterioration modes suggestive of crystalline behavior. AlON ceramics have been fabricated in the same design as the standard sapphire windows and have been evaluated for gun testing. This process involves a series of experiments in a closed chamber where gun propellant is burned to generate an environment similar to that inside the gun. Windows mounted in two methods have been tested. One of these windows has survived the full pregun test series with no visible damage.

10.6.3 Transparent Armor Design and Dynamic Responses

10.6.3.1 Design and Constructure of Transparent Armors

Construction of transparent armors follows the principles that have been well established for their opaque counterparts. Armor against advanced threats has a typically structure as shown in Fig. 10.19 [165]. The front-face layer should be as hard as possible, in order to damage the projectiles in maximum range. Ideally, it should be harder than the projectile core. Currently, several hardened glasses or glass ceramics are being used. This layer is the weak point of present transparent

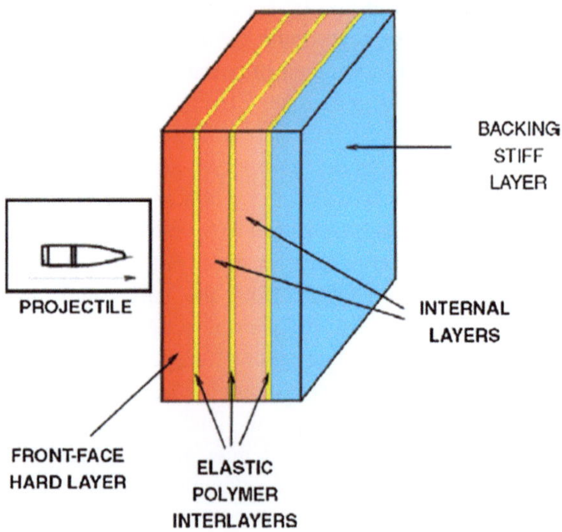

Fig. 10.19 Schematic structure of transparent armor against advanced threats. Reproduced with permission from [165]. Copyright © 2008, Elsevier

armors. The backing layer is used to catch residual projectile fragments and comminuted ceramic particles and hinder the crack propagation together with interlayers. Stiff and tough materials, such as polycarbonate, are generally used for this layer. Glasses or glass ceramics are usually used for the internal layers. Layers are bound together by polymer films, such PVB and PU foils.

10.6.3.2 Modeling of Transparent Armors

The most widely used materials for ballistic transparency protection of armored tactical platforms in commercial and military applications are the low-cost glasses that are backed by polycarbonate [166]. To achieve enhanced protection capability and extended service performance for future-armored windows to protect the soldiers, it is demanding to develop high-strength transparent ceramics as next-generation armor windows. Due to the high cost of testing transparent ceramics, modeling approach is usually carried out, along with ballistic testing to validate armor designs, with transparent $MgAl_2O_4$, striking ply backed with polycarbonate. The influence of defects on the failure of the laminates was specifically characterized, according to both the static and dynamic testing results. In this specific study, finite element modeling (FEM) was used to predict armor designs and reduce number of laminate configurations in experimental testing. A notional ceramic armor system based on spinel/polycarbonate assemblies was used to study the effect of surface and interior, equal area defects on the ballistic behavior of a laminated structure.

The failure probability of ceramic articles with short-term loading has been calculated by taking into account the defects [167, 168]. The effect of various shapes and locations of defects of equal cross-sectional area was simulated. Figure 10.20 shows schematically the various types of defects and the defect locations. The location of the defects at either surface or interior to the sample was considered, while the defect shapes from elliptical to square to circular were included. The Lagrange and Euler solvers were used in the modeling. It was found that the Lagrange solver was applicable to all shapes of defects, due to the erosion of cells of shapes with sharp corners, such as the defects with elliptical and rectangular cross-sectional areas. In contrast, the Euler solver could be applied to all defect shapes. To ensure the consistency of the simulation, the results of the impact simulation of a target without defects were included for comparison. It was found that all the results produced using three solvers were similar.

The effect of surface flaws on ballistic performance of the target was studied by introducing square cuts into the surface of the spinel ceramics, with a width of 1.5 mm and a height of 5 mm. The area density of defects for the spinel ceramics was less than 5 %. Simulation images 11 μs after the impact are shown in Fig. 10.21, demonstrating that the damage progressions for the targets with seven and eleven surface defects were similar [166]. The defects were investigated both at the exposed surface and positioned on the interlayer or directly on PC when the bonding layer was excluded. The interior defects were introduced along the

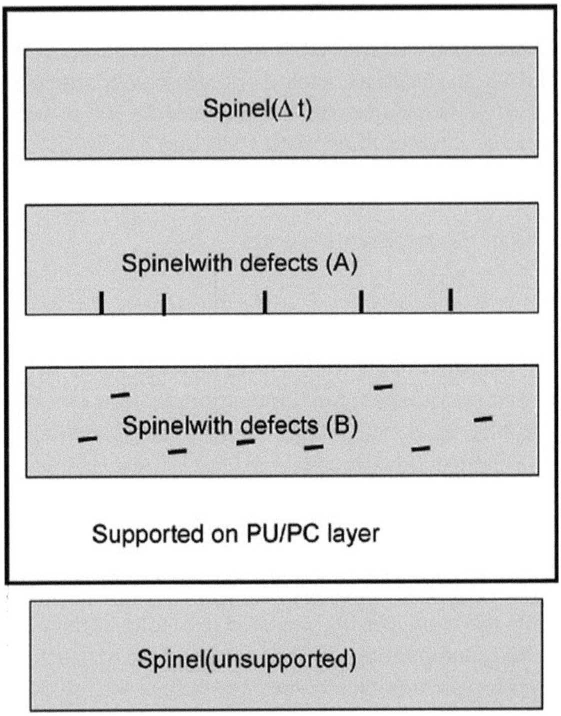

Fig. 10.20 Types and locations of various defects present in spinel ceramics for ANSYS/AUTODYN simulation. Defects were located either at the surface (*A*) or in the interior (*B*) of the spinel layer in the laminated structure. The polymer layers were kept constant. Voids were simulated as ideal gaseous air. Reproduced with permission from [166]. Copyright © 2009, Elsevier

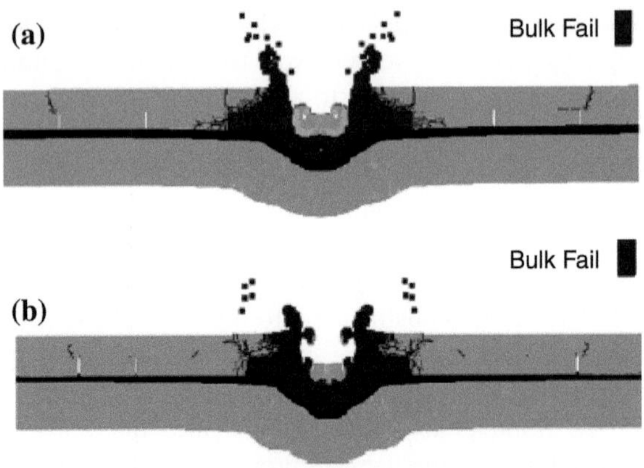

Fig. 10.21 Snapshot images from simulations showing the propagation of the damage into the spinel/polycarbonate stacks after 11 μs with slug moving 1000 m s^{-1} for different numbers of surface defects: **a** 7 and **b** 11. Reproduced with permission from [166]. Copyright © 2009, Elsevier

geometric center-line of the spinel layer, spaced equally from each other. The projectile impacted the target in two ways: (i) perpendicular to the center of a defect and (ii) perpendicular to the target but in between two defects. Experimental tests and simulation results were analyzed by evaluating the extent of damage and residual velocity after the impact.

The simulations also indicated that the internal defects had a reduced effect on the target resistance to penetration, when compared to the surface defects. The average residual velocity of the projectiles into the defect targets showed a three-time increase for the surface defects as compared to the internal defects. Simulations were used to further evaluate the difference in stress concentrations among the targets with different types of defects, including rectangular, elliptical, and circular ones. The simulation results indicated that internal defects with elliptical cross section demonstrated the largest increase in the residual velocity, when compared to the circular and rectangular defects that were located internally in the spinel. In addition, the defect with elliptical cross section with its longer axis parallel to the line of impact led to the highest exit velocity. Figure 10.22 shows time slices from the impact into an ellipse-based defect target, with the progression of damage [166]. The damage progression grew continuously until the midpoint of two neighboring defects. However, at this point, the damage appeared at the next defect, while it continued to grow. This phenomenon was more obvious for the samples with defects having sharp corners, such as those of elliptical and rectangular cross sections. The phenomenon could be attributed to the fact that the

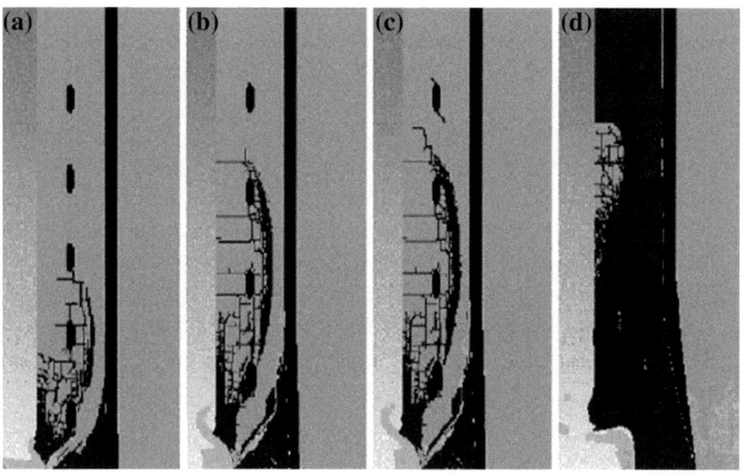

Fig. 10.22 Damage progression in the spinel/polycarbonate laminate with 4 % elliptical defects internally inserted along the center line of the spinel after different time durations: **a** 16 μs, **b** 30.37 μs, **c** 32 μs and **d** 52 μs. The damage progression appeared to grow at the sharp edges of the defects until the ceramic was completely failed. Note that the damage did not penetrate the void areas, but was initiated at the boundaries. Reproduced with permission from [166]. Copyright © 2009, Elsevier

damage wave that arrived at the sharp corners, i.e., points of stress concentrations, was faster than the progressing damage.

First of all, surface defects are the most detrimental to the mechanical performance of the ceramics in resisting the ballistic impact. As a result, preventing the ceramics from damage is essential to improve the ballistic performance of the armors. Also, the inclusion of localized defects to control the spread of damage in the target was not supported by any evidence. In addition, the importance of defects could have been overestimated. Firstly, the defects described in this study are much larger than those that could be present in ceramics processed by using the standard ceramic methods. Secondly, once there are defects, the ceramics would not be transparent. In other words, if a ceramics is transparent, there should not be defects. Furthermore, the presence of defects caused only a small decrease in the ballistic efficiency of the systems and the defects were located just under the impact zone and oriented along the axis of impact. Therefore, it can be concluded that unless the transparent ceramics contains large defects that are visually observable, defects are not a serious problem for transparent ceramics to be used as transparent armors, so that ceramic transparent armors should be ideal candidates of transparent armors.

10.6.3.3 Ballistic Testing and Dynamic Responses of Transparent Armors

Here, an example of the testing of transparent armor is described, which was intended to assess the protective strength of transparent ceramics that are potentially used in transparent armor systems [169]. Not only the ballistic testing method but also the experimental techniques of the visualization of damage and the projectile penetration were discussed in this study. Visualization techniques are essential for understanding the mechanisms of interaction between projectile and target materials. Ballistic testing results, high-speed photographs, and flash radiographs from experiments with transparent armor were discussed in detail.

To evaluate the ballistic resistance of transparent ceramics as part of a transparent armor system, the targets consisted of three main components: (i) a ceramic front layer with thicknesses of 1.3–8.3 mm, (ii) a glass laminate consisting of 1–3 layers of soda-lime float glass, and (iii) a polycarbonate layer with a thickness of 4 mm. The three target components were clamped together by using two steel frames, without using glues. The lateral dimensions of the ceramic tiles were in the range of 85–100 mm^2. The size of the ceramic tiles was selected by considering two factors. The ceramic tiles were sufficiently large to ensure that the ballistic performance had not been affected due to the reduction in size for specific type of projectile, while they are also sufficiently cost-effective for the manufacture of the specimens.

The glass and polycarbonate plates had dimensions of 150 mm × 150 mm. Test series with stacks of glass plates, with lateral dimensions of 400 mm × 400 mm and 150 mm × 150 mm, did not show any difference in the ballistic performance. Therefore, the samples with smaller size were used for the test series, where the

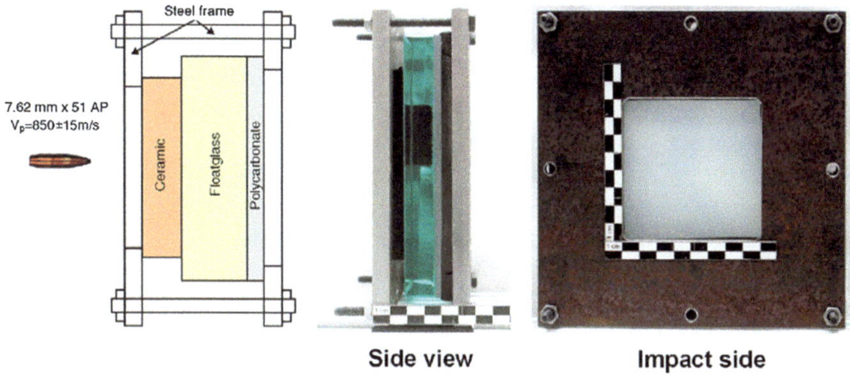

Fig. 10.23 Schematic and photographs of the target assembly. Reproduced with permission from [169]. Copyright © 2009, Elsevier

performance against a single hit was used for the assessment. Another test series were stacks of soda-lime float glass with a total thickness of 59 mm. Two types of stacks were assembled with either 4 layers or 10 layers of the glass. The test results indicated that a lower ballistic resistance was observed when 10 layers of the glass were used. Figure 10.23 shows a schematic of the target assembly and photographs of a typical target, which seems to be translucent because it was not polished [169].

In order to verify the hypothesis of the hierarchical structure in influencing the protective strength of the transparent ceramics, it is necessary to identify the mode of fragmentation of different materials and target designs and record the progresses of both the dwell and penetration. In this case, flash X-ray imaging was used to observe of the projectile–target interaction phases [170, 171]. A schematic of the measurement setup of flash X-ray cinematography is shown in Fig. 10.24 [169]. Instead of several separate X-ray tubes, one multi-anode tube was utilized in this study. In the multi-anode tube, eight anodes were arranged on a circle with a diameter of ~12 cm. This configuration caused only a relatively small parallax for the projections from the different anodes. The process under observation could be X-ray observed at eight different times. The radiation transmitted through the target

Fig. 10.24 Schematic of typical setup of flash X-ray cinematography. Reproduced with permission from [169]. Copyright © 2009, Elsevier

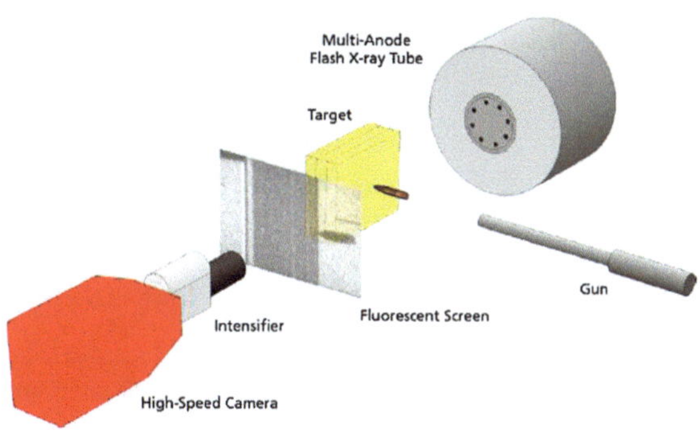

was detected by using a fluorescent screen. The position of the target was between the multi-anode tube and the fluorescent screen, but relatively closer to the fluorescent screen. The fluorescent screen converted the radiograph into an image in the visible wavelength range, which was photographed by using an intensified digital high-speed camera. The maximum frame rate that could be achieved with such the system was dependent on the decay time of the fluorescent screen, the time characteristics of the intensifier and the camera. Frame rates of 100,000 fps have been achieved with a fast-decaying fluorescent screen, which were used in this study.

The first material tested was high-purity, fully dense sintered α-Al_2O_3 ceramics with sub-μm grain sizes. Targets were assembled by using ceramic front layers with the thicknesses of 1.3, 1.5, 2, and 4 mm. For each thickness, the minimum glass backing thickness was determined, according to the projectile that could be stopped. In all tests, yaw and pitch of the projectiles were measured by using the flash radiographs. Residual velocities of the projectiles were estimated from two time exposure flash radiographs. Figure 10.25 shows a representative flash X-ray picture of a test in which the projectile was stopped (top) and a case with perforation of the target (bottom) [169]. The film was exposed three times in both cases. The first exposure showed the projectile shortly before impact, whereas the second and third

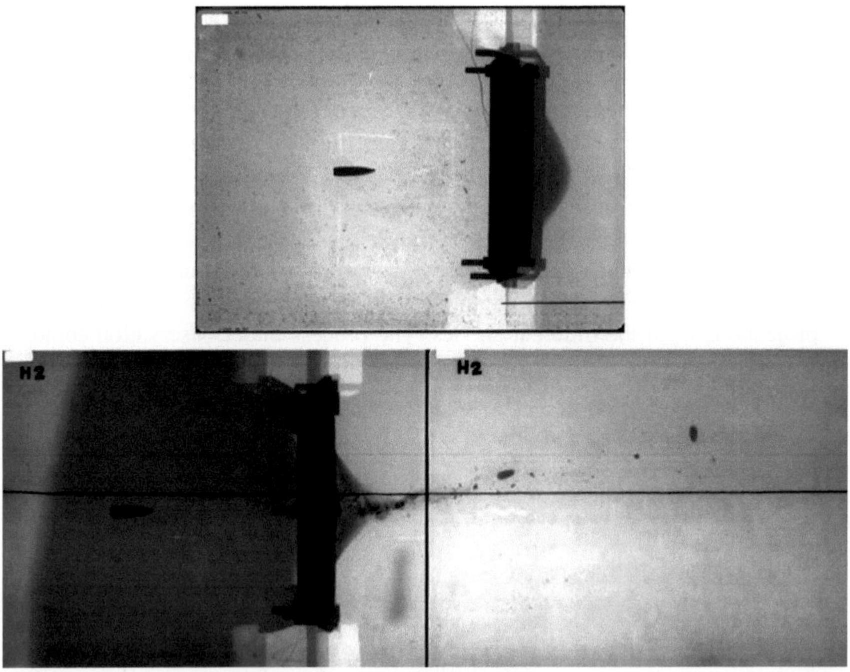

Fig. 10.25 Flash X-ray pictures of the projectile defeat (*top*) and perforation (*bottom*) by the Al_2O_3/glass/polycarbonate targets. Reproduced with permission from [169]. Copyright © 2009, Elsevier

exposures visualized the target deformation and the residual projectile at two different times after perforation.

It has been demonstrated that large weight reductions can be achieved by using a hard front layer of transparent ceramic, when compared to the conventional transparent armor, which consists of multiple glass–plastics–laminates. A target setup has been developed which was able to identify the difference in ballistic performances between different ceramics used as striking ply for transparent armors. The protective strengths of four transparent ceramics have been tested by using the ballistic testing setup, in which armor piercing projectiles with a steel core and targets of ceramic–glass–polycarbonate were used. Very important conclusions have been drawn from these ballistic experimental results, which can be used as guidance for the design of transparent armors by using transparent ceramics.

Dynamic responses of selected transparent ceramics have been evaluated for armor applications, which are critical for many defence and civilian facilities [172]. The transparent armor systems consisted of a wide variety of glass laminate assemblies with polymeric bonding interfaces and baking, as well as polycrystalline ceramic (AlON and spinel) and single crystals (sapphire) as front facing materials. Over the last 20 years, as the threats have escalated and become more varied, the development of specific transparent armor packages with optimized threat resistance has become more and more challenging. The ultimate failure of structural ceramics in impact events is a function of the temporal and spatial interactions of the macro-stresses at the macro-, micro-, and even nanostructural scales, including elastic and inelastic (plastic) deformations, crack nucleation, damage evolution, and the resulting failures at the macro-scale (top down) and/or the nanoscale (bottom up). In order to accelerate the development of validated design and predictive performance models, a systematic series of experimental investigations have been conducted on various glasses, single-crystals (e.g., sapphire), and polycrystalline ceramics (AlON and spinel).

The Edge-on Impact (EOI) test coupled with a high-speed Cranz-Schardin film camera has been extensively used to visualize and quantify stress wave, crack, and damage propagation in a variety of materials, including monolithic and laminated glasses, AlON, and sapphire single crystals. A modified Kolsky bar technique instrumented with a high-speed digital camera has been employed to examine the dynamic deformation and failure of AlON ceramics that underwent uniaxial high strain rate compression, when the samples were tested with either unconfined or confined modes. Real-time photography has clearly identify the critical influence of defects, while postmortem characterization of the fragments after the testing have uncovered micro-scale deformational twining and nanoscale cleavage. These experimental results can be used to validate the theoretical models that have been used to simulate the impact performance of ceramic materials.

The main differences between conventional quasi-static mechanical testing and a dynamic impact event on large samples have been well demonstrated in the open literature. Generally, the strain rate for quasi-static mechanical testing is on the order of 10^{-3}–10^{-2} s^{-1}, while dynamic testing is in the range of 10^{2}–10^{4} s^{-1}. Therefore, the stressed volume in an impact event is much larger than that in the conventional quasi-static testing, which thus leads to a large variety of defects, microstructural inhomogeneities, and various stress concentrated regions. Shock waves are generated in sequences (i) surface wave (Rayleigh wave), (ii) compression/longitudinal wave (dilatational, push or *p* wave), and (iii) shear wave (transverse, shake, or *s* wave). For *n* harmonic normal load in the impact event, the total energy in the shock waves consists of 67 % in the Rayleigh (R) wave, 26 % in the shear (S) wave, and 7 % in the compression (C) wave, as shown schematically in Fig. 10.26 [172]. The relative velocities are in the order of $V_C > V_S > V_R$. The compression and shear waves are more critical for bulk materials. The compression wave can densify the materials in both a reversible and irreversible ways, during which defects are activated and microstructures are altered. Under the impact of shear wave, which has significantly high energy, cracks start to grow, propagate, and coalesce into a damage front. The impact stresses or pressures produced by the impact are significantly higher than the compressive strength of the materials, so that the materials are under an overstressed state.

An EOI test method, equipped with a high-speed Cranz-Schardin camera, with frame rates of up to 10^7 fps, has been used to visualize damage propagation and dynamic fracture of glass and opaque structural ceramics with the back reflection technique [173, 174]. Transmission and shadowgraph mode can be used to observe the behaviors of transparent ceramics. Two different optical configurations have been studied. A regular transmitted light shadowgraph setup was used to observe damage nucleation and propagation, while another one was a modified configuration, in which the sample plates were placed between crossed polarizers and the photoelastic effect, or known as polariscope technique, was employed to visualize the stress waves. Two impact tests at almost same velocities were carried out in transmitted unpolarized (shadowgraphs) and crossed polarized lights, respectively. Figure 10.27 shows an exploded view of the impactor–sample interaction [172]. It is also possible to take both the horizontal (perpendicular to plate) and vertical (perpendicular to long edge) photographs. As a result, various phenomena, such as

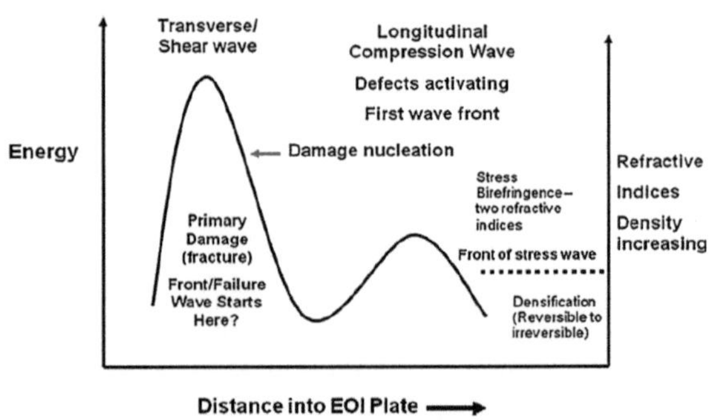

Fig. 10.26 Hypothetical dynamic shock stress waves/energy profiles in an edge-on impact (EOI) test. Reproduced with permission from [172]. Copyright © 2013, Springer Science + Business Media

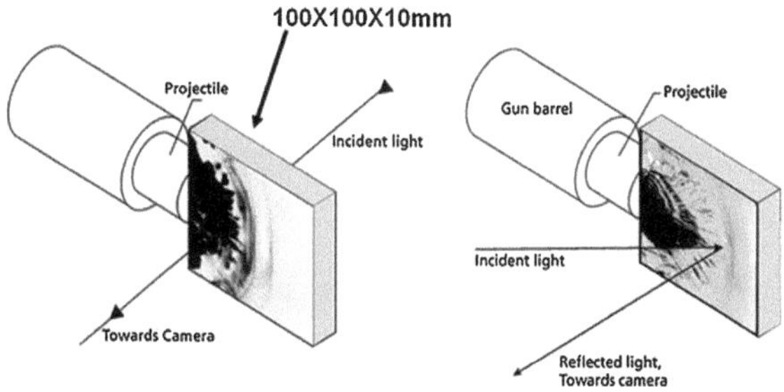

Fig. 10.27 EOI transmitted light (shadowgraph) and reflected light configurations. Reproduced with permission from [172]. Copyright © 2013, Springer Science + Business Media

nucleation of cracks, number, density, and influence of defects, can be observed in real time.

Figure 10.28a shows typical damage patterns observed by using the shadowgraph technique with soda-lime (float) glass at an impact velocity of ~200 m s^{-1}, while Fig. 10.28b demonstrates the typical damage pattern of Starphire™, a high-purity soda-lime glass, at an impact velocity of ~400 m s^{-1} [172]. Figure 10.28c shows the damage pattern of AlON transparent polycrystalline ceramics at the same impact velocity. Based on the progression of the shock waves in the EOI plate, as shown in Fig. 10.26, it has been concluded that the crack centers ahead of the primary fracture front in Fig. 10.28a are nucleated either by the interaction of the compression wave or the shear wave with the defects in the materials, which were dependent on the amplitude of the waves.

In the examples shown in Fig. 10.28a, b, crack nucleation and growth were initiated mainly by the shear waves, whereas in that of Fig. 10.28c, the nucleation caused by the longitudinal wave was dominant. These isolated crack centers had a

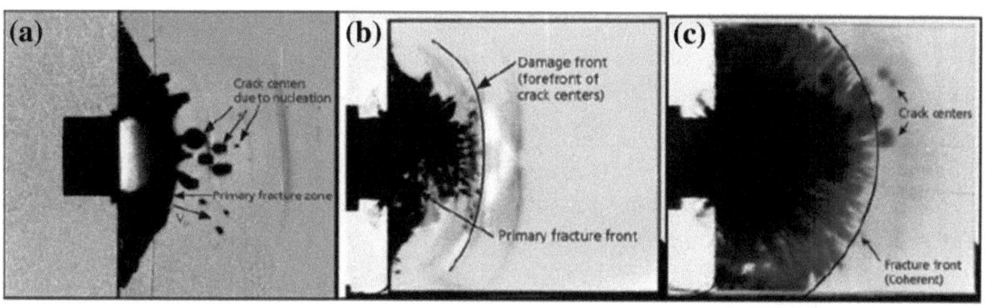

Fig. 10.28 Damage patterns obtained with the shadowgraph technique without using the crossed polarizers. **a** Typical damage pattern of float glass at v_P of ~200 m s^{-1} and **b** typical damage pattern of Starphire™ glass at v_P of ~400 m s^{-1}, for the crossed polarizers image of the stress wave, and **c** typical damage pattern of the AlON ceramics at v_P of ~400 m s^{-1}, in which the coherent fracture front is a circular approximation of the cracks/crack fingers at the head of the damage zone. Reproduced with permission from [172]. Copyright © 2013, Springer Science + Business Media

significant influence on morphology and progression of the primary fracture fronts. In these figures, the primary fracture zone/front was the best approximation of the coherent area of undifferentiated massive cracking. The damage front was a similar circular approximation of the cracks/crack fingers ahead of the primary fracture front.

If there are no isolated crack centers that can be distinguished, the damage velocity refers to the velocity of the primary coherent fracture zone. If the time and position of the appearance of isolated crack centers can be identified, the damage velocity is defined as the slope of the regression straight line through the leading nucleation points. The distance–time plot for the test shown in Fig. 10.28c has been established [172]. Crack velocity refers to the velocity of the expanding isolated cracks. The coherent fracture front shown in Fig. 10.28c is a circular approximation of the cracks/crack fingers at the head of the damage zone. In Fig. 10.28b, the damage front is a circular approximation of the tips of the cracks from the primary fracture front. In most cases, these fracture/damage zones/fronts cannot be clearly differentiated, especially when the number of the isolated cracks initiated in front of the main damage zone is too large.

For convenient discussion, damage velocity refers to the velocity of the primary fracture zone, which crack velocity means the velocity of the expanding isolated cracks. The two optical techniques showed different sensitivities with respect to the stress level that can be visualized. In the shadowgraph images, the light intensity is dependent on the second spatial derivative $\partial^2 n/\partial x^2$ of the refractive index, while for the crossed polarizer setup, the intensity of the transmitted light is dependent on the photo-elastic properties of the materials. As a result, the first visible wave front in the shadowgraph configuration could appear at a different position from the forefront of the stress wave, which is visible by the crossed polarizer setup. The two techniques can be used to visualize different parts of the same stress wave, as shown in Fig. 10.29, with the shadowgraph and the corresponding crossed polarizers photograph of a Starphire™ sample at 8.7 μs after the impact [172].

In the EOI test, transparent polycrystalline AlON specimens with a dimension of $100 \times 100 \times 10$ mm^3 were impacted with steel solid cylinder and saboted sphere impactors at velocities in the range of 270–925 m s^{-1}. Impact tests at nearly same velocities were conducted in the transmitted unpolarized (shadowgraphs) and crossed polarized light configurations. Representative shadowgraphs and crossed polarized light photographs are shown in Fig. 10.30, in which the impactor is visible [172]. The images shown in Fig. 10.30a were those conducted at about 380 m s^{-1}, while those in Fig. 10.30b are the two groups conducted at 820 and 925 m s^{-1}. The EOI tests at velocities of ≤400 m s^{-1} were carried out by using a more controllable gas gun arrangement, while velocities >400 m s^{-1} were realized by using a powder gun. Stress wave and damage velocities, as well as the nucleation of crack centers ahead of the apparent fracture front, could be identified from the experimental results [175, 176]. The photographs revealed several useful features. In the shadowgraphs, distinct growing darkened regions can be observed in rows 1 and 3, which mainly corresponded to a massively fractured damage zone that could be referred to as a failure wave.

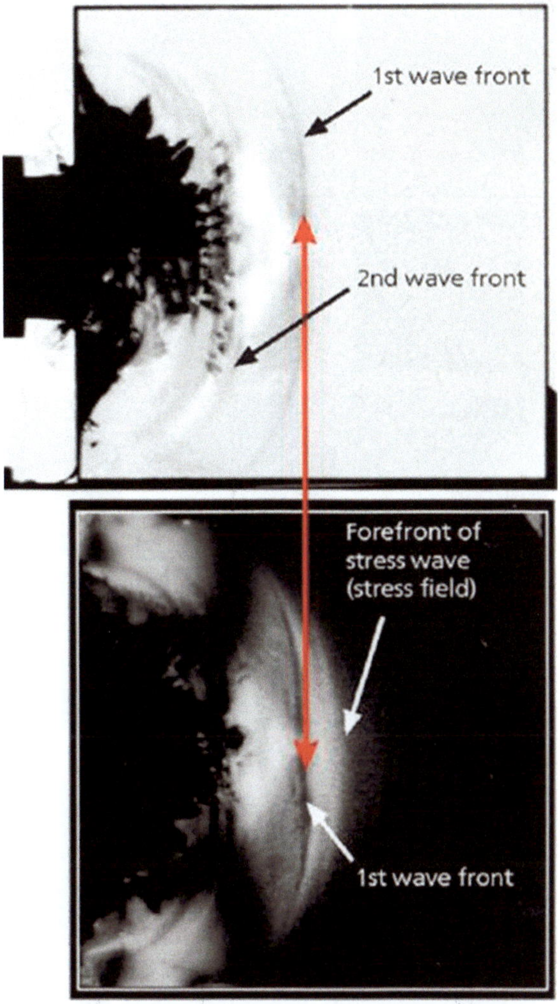

Fig. 10.29 Shadowgraph (*top*) and corresponding crossed polarizer photograph (*bottom*) of Starphire™, 8.7 μs after the impact at ~ 400 m s^{-1}. Reproduced with permission from [172]. Copyright © 2013, Springer Science + Business Media

In addition, the shadowgraphs at 6.7, 7.7, and 8.7 μs revealed the nucleation of the clearly visible crack centers ahead of the damage front, which was probably from bulk defects. This observation was similar to that of the glass shown in Fig. 10.28a, which means that the ceramics might have bulk defects, although they are transparent. Defects, such as stress concentrated regions, can be nucleated to facilitate crack grow. In the crossed polarized light photographs, in rows 2 and 4, a band of visible light was clearly observed, which has been attributed to the presence of stress-induced birefringence or photoelasticity, due to the passing of the compression stress wave. The velocity of this wave should be very close to the longitudinal acoustic wave velocity (c_L) in the AlON ceramics. Figure 10.31 shows the

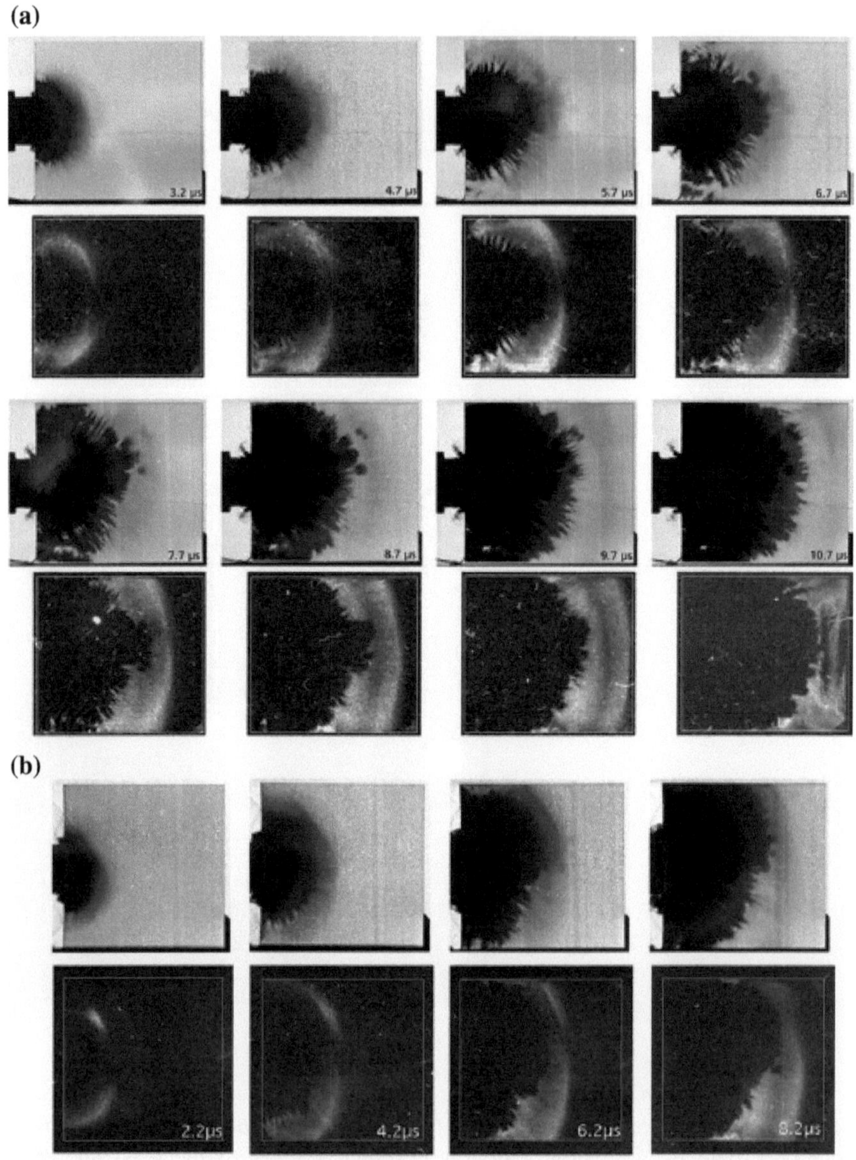

Fig. 10.30 **a** Shadowgraphs (row 1 (*top*) and 3) and crossed polarized light photographs (rows 2 and 4) of the AlON ceramics impacted with a solid steel cylinder at a velocity of ~ 380 m s^{-1}. **b** Shadowgraphs (*top row*) and crossed polarized light photographs of the AlON impacted at 820 m s^{-1} (*top*) and 925 m s^{-1} (*bottom*). Reproduced with permission from [172]. Copyright © 2013, Springer Science + Business Media

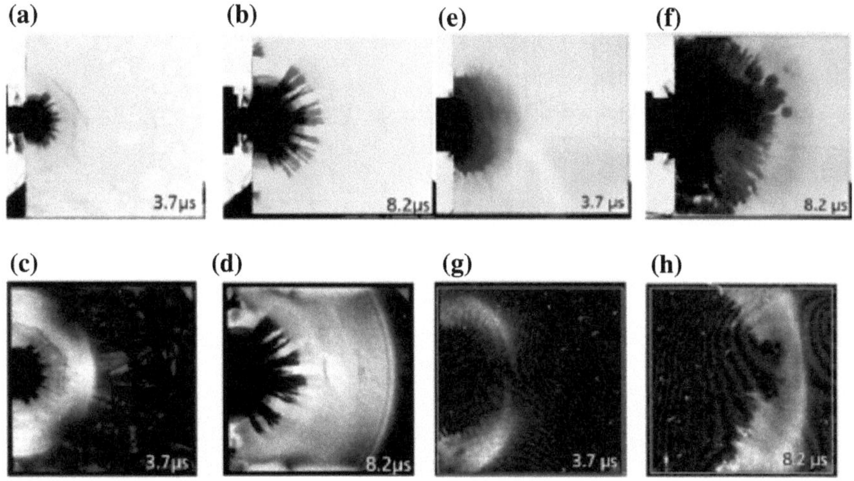

Fig. 10.31 Photographs of the AlON ceramics impacted with spheres and cylinders: **a**, **b** sphere impactor in shadowgraphs, **c**, **d** in crossed polarized light photographs, 3.7 and 8.2 μs, **e**, **f** cylinder impactor in shadowgraphs, **g**, **h** crossed polarized light photographs, 3.7 and 8.2 μs. Reproduced with permission from [172]. Copyright © 2013, Springer Science + Business Media

difference in morphology between the damage fronts due to the impacting with a spherical and solid cylinder impactors [172]. This difference could be related to the difference in weight between the two impactors, i.e., the sphere was 39.1 g while the cylinder was 54 g.

Figure 10.32b–d demonstrates fracture surfaces of the fragment, shown in Fig. 10.32a, that were collected from the samples after the confined EOI test [172]. It is observed that besides intergranular fracture, there is a significant intragranular microcleavage down to the nanoscale, as shown in Fig. 10.32d. Although it is still not clear whether the cleavage is preceded by micro-deformational twinning, this observation can be considered as a plasticity mechanism prior to the occurrence of catastrophic macro-cracking or failure. Similar inelastic deformation mechanisms and damage have been observed in other structural ceramics subjected to high-velocity impact [177].

Similar Edge-on Impact test has also been carried out for coarse-grained polycrystalline spinel rod in order to observe fracture morphology caused by the impact, so as to evaluate contributions of energy dissipation of various deformation and fracture modes [178]. The formation of two distinct damage zones was observed in the target beneath the impact surface. Zone 1 extended from the impact plane into the target with a distance of 18 mm, which consisted of mixed mode fractures that were completely separated from the bar. The second zone (Zone 2) was presented by an additional 18 mm into the bar, which was found to contain intergranular fractures that remained intact. The transition in fracture mode was also evidenced by a change in damage velocity during the damage penetration. Upon impact, the damage initiated at a velocity close to the shear wave velocity of the target

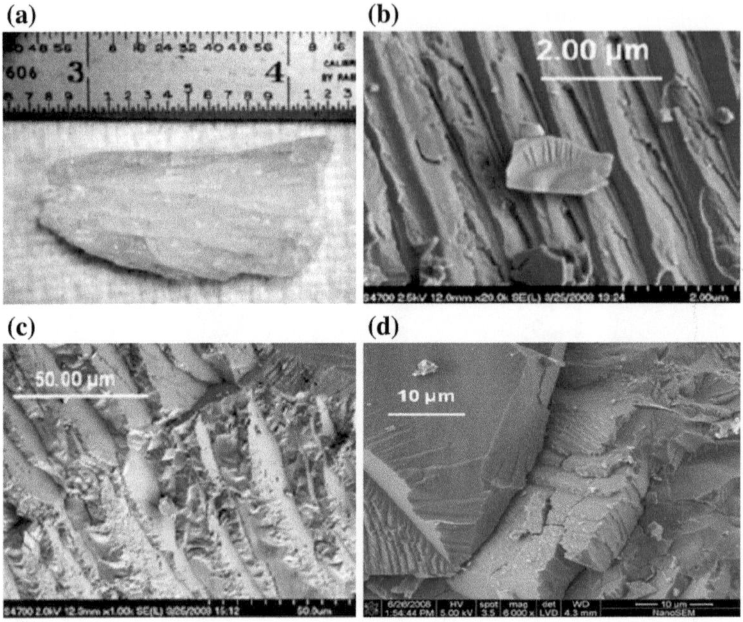

Fig. 10.32 **a** AlON fragment from an EOI test, **b** and **c** SEM photographs of fragment in **a** and **d** SEM photograph of an AlON fragment showing microcleavage down to the nanoscale. Reproduced with permission from [172]. Copyright © 2013, Springer Science + Business Media

materials, but it quickly dropped to about 60 % of the shear wave velocity, indicating a shift to longer fracture path, which was consistent with the transition from transgranular to intergranular fracture. The energies consumed by different deformation and fracture mechanisms have estimated, with which the contributions of each observed mechanism to the dissipation of total impact energy was thus quantified. Elastic wave propagation, fracture surface creation, kinetic energy of ejected fragments, and frictional effects were identified to be the most dominant mechanisms of energy dissipation. Creation of comminuted material had a marginal direct effect on the dissipation of impact energy through fracture energy, while the kinetic energy associated with ejected fragments and friction between the fragments and the ball were also found to be significant dissipation mechanisms.

10.7 Other Applications

Spinel and AlON® transparent ceramics have been evaluated in terms of in vitro cytotoxicity, cell–materials interactions, and tribological performance, for potential wear resistant load bearing implant applications [179]. These transparent ceramics with high hardness in the range of 1334–1543 HV showed in vitro wear rate in the range of 5.3–10 × 10^{-6} mm^3 Nm^{-1} against Al_2O_3 ball at a normal load of 20 N. The in vitro biocompatibility study using hFOB cells showed nontoxicity, while cell–materials interactions of spinel and AlON® transparent ceramics were superior over CP–Ti. Furthermore, the spinel exhibited relatively higher wear resistance and superior cell–materials interactions than the AlON®. The enhanced cellular adhesion and proliferation on spinel was attributed to the positive influence of the surface energy of the polar component.

Figure 10.33 shows sample surfaces imaged after 3 days of culture to investigate the bone cell attachment and spreading on different surfaces [179]. On both spinel (Fig. 10.33a, b) and AlON® (Fig. 10.33c, d) surfaces, hFOB cells were found to show a flat spread shape morphology and well attached to the samples. It has been accepted that osteoblasts are attachment-dependent cells, i.e., they must attach first on the surface and then spread and finally mineralize their extracellular matrix. Dense growth of cells on both the surfaces with numerous cell–cell contacts suggested that both the spinel and AlON® ceramic surfaces are biocompatible and no inhibition to bone cells (hFOB) adhesion and growth were caused. Moreover, it was found that small white granular apatite nodules were formed on the cell surface, as

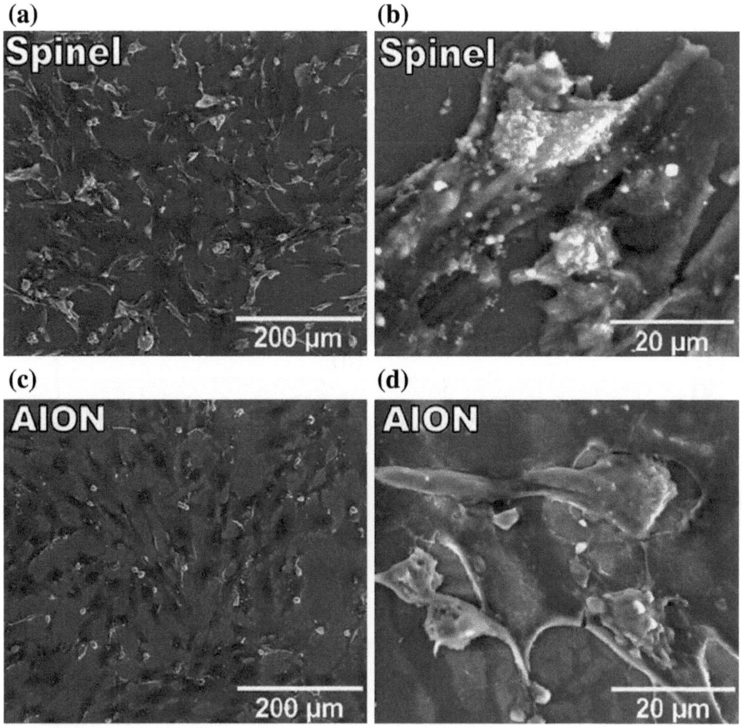

Fig. 10.33 SEM micrographs showing the cell adhesion after 3 days of culture on spinel (**a**, **b**) and AlON® (**c**, **d**) sample surfaces. Reproduced with permission from [179]. Copyright © 2011, Springer Science + Business Media

shown in high-magnification SEM image in Fig. 10.33b, which implied that cells became matured on spinel surface.

Figure 10.34 shows SEM images of the samples cultured for 5 days, demonstrating the cell proliferation and early stage cell differentiation [179]. On spinel surfaces, cells formed a confluent layer covering the entire surface, as shown in Fig. 10.34a, b. Cells were observed to adhere to each other with cellular micro-extensions and connected to substrate in addition to the neighboring cells, as demonstrated in Fig. 10.34b. On AlON® surfaces, cell growth was observed to be less pronounced. From a low-magnification SEM image, as shown in Fig. 10.34c, it was observed that cells were still in the process of confluent layer formation on AlON®, which was unlike spinel surface shown in Fig. 10.34b, where cells had already started to mineralize their extracellular matrix with the abundance deposition of apatite minerals [180, 181].

Adult orthodontics is recently gaining popularity due to its importance in esthetics, oral, and general health [182]. However, none of the currently available alumina- or zirconia-based ceramic orthodontic brackets meet the esthetic

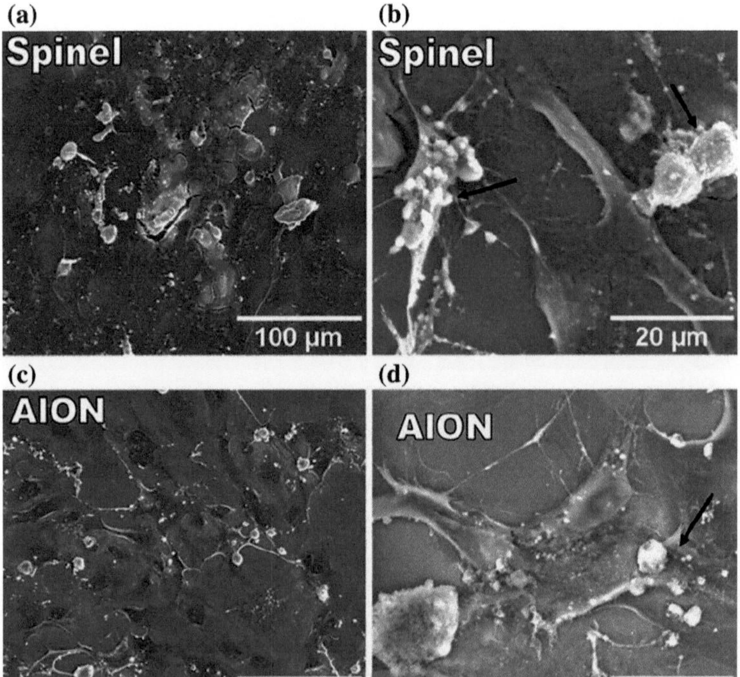

Fig. 10.34 SEM micrographs illustrating the cell adhesion and proliferation after 5 days of culture on spinel (**a**, **b**) and AlON® (**c**, **d**) sample surfaces. Arrows in high magnification SEM images indicate apatite mineral formation. Reproduced with permission from [179]. Copyright © 2011, Springer Science + Business Media

requirements of adult patients. Inherent hexagonal lattice structure and associated birefringence limits the visible light transmission in normal PCA ceramics, which and make them appear white but not transparent. Therefore, transparent magnesium aluminate ($MgAl_2O_4$) spinel was assessed for potential applications as esthetic orthodontic brackets. The transparent spinel ceramics were prepared from commercially available spinel powder, which as compacted through colloidal processing, followed by pressureless sintering and hot isostatic pressing at optimized temperatures and pressures.

The transparent samples were characterized in terms of chemical composition, phase composition, density, hardness, flexural strength, fracture toughness, and optical transmission. Biocompatibility was evaluated with in vitro cell line experiments for cytotoxicity, apoptosis, and genotoxicity. Experimental results indicated that the transparent spinel samples exhibited promising physico-chemical, mechanical, optical properties, and excellent biocompatibility for applications as orthodontic brackets. A piece of the transparent ceramics is shown in Fig. 10.35, together with some tooth samples [182].

It is well known that biomaterial–oral mucosa interface is crucial in determining the biocompatibility of a new material that is intended for dental applications. The materials should be able to withstand the challenges posed by the biological, thermal, and chemical stresses in vivo, without inducing adverse effects [183]. Guidelines have been well defined for evaluating the cytotoxic, apoptotic, and genotoxic effects of dental materials, for developing biocompatible products [184, 185]. Noncompatible materials can affect the cell in various ways. For example, it can directly inhibit the proliferation or growth of the cells. It is also possible that the materials attack the cell membrane, thus leading to cell death. They could also be toxic to the gene, resulting in mutation or impairment of genetic information. Altering the functions of cells, such metabolism and signaling, can be happen in some cases [186].

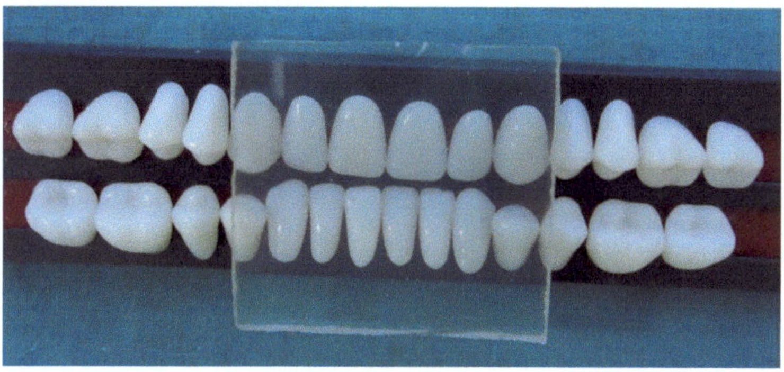

Fig. 10.35 Photographs of a piece of transparent spinel ceramics, together with tooth samples. Reproduced with permission from [182]. Copyright © 2014, Springer Science + Business Media

These mechanisms have been evaluated by using HEK 293 cell line, derived from human embryonic kidney cells, which are widely used in transfection and in vitro cytotoxicity assays [187]. Moreover, HEK is an easily available cell line with high proliferation rate. As a result, it has been widely used for cell culturing experiments to evaluate medical and dental biomaterials [188]. The MTT assay showed no considerable variation in the viability of cells, when treated with different concentrations of spinel and alumina. The average percentage viability of cells at different concentrations of spinel and alumina has been studied. No major difference was observed between the treated and untreated cells. In addition, they did not show any growth inhibition even in high doses, e.g., 0.01 kg m^{-2}. The materials neither inhibited cell proliferation nor affected its morphology, after incubation for 48 h, as shown in Fig. 10.36. Similar results were observed for alumina, which is generally used for making orthodontic brackets, as shown in Fig. 10.37 [182].

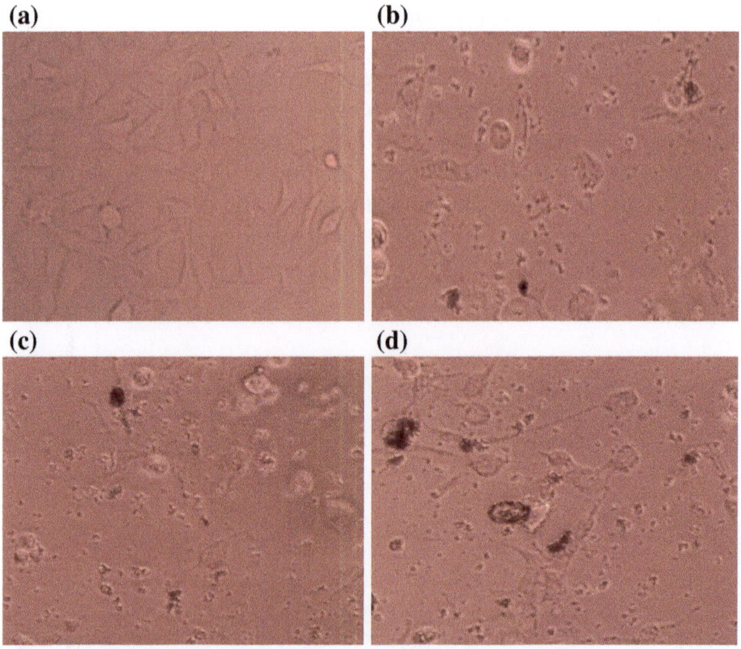

Fig. 10.36 Morphologies of HEK 293 cells with different concentrations of spinel: **a** without treatment, **b** 0.0025, **c** 0.005, and **d** 0.01 kg m^{-2}. Reproduced with permission from [182]. Copyright © 2014, Springer Science + Business Media

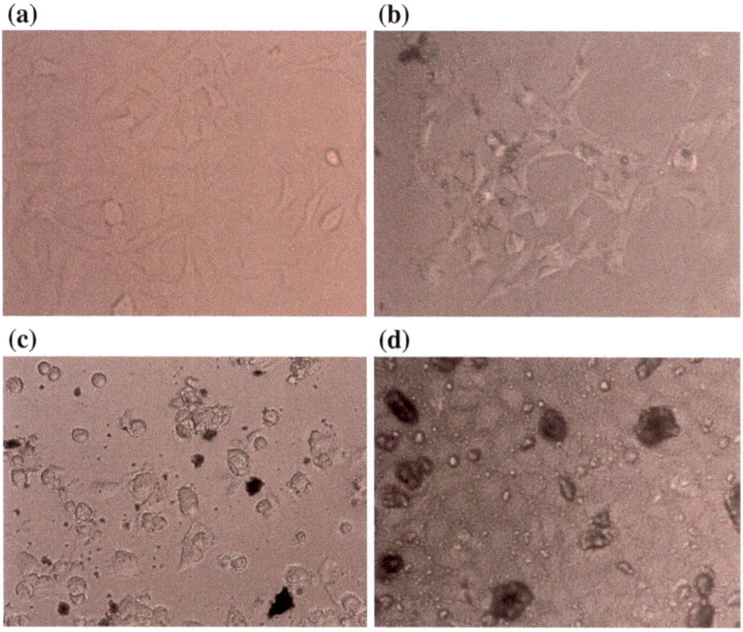

Fig. 10.37 Morphologies of HEK 293 cells with different concentrations of alumina: **a** without treatment, **b** 0.0025, **c** 0.005, and **d** 0.01 kg m^{-2}. Reproduced with permission from [182]. Copyright © 2014, Springer Science + Business Media

10.8 Conclusions

Due to their special thermal and mechanical properties, transparent ceramics can withstand higher temperatures and mechanical stresses than other optical materials, thus making them potential candidates for such applications, which has been demonstrated by several transparent ceramics. Transparent ceramics have also shown their advantages for scintillator applications. Transparent armors based on transparent ceramics could be fabricated with much less thickness and weight. Applications of transparent ceramics used as biomaterials in some specific fields have exhibited very strong potential. However, as compared to those in solid-state lasers, the applications of transparent ceramics discussed in this chapter are much less explored, which thus means that more achievements could be made in this regard. It is also expected that more and more applications will be identified for transparent ceramics, indicating a bright future of this special group of advanced ceramic materials.

References

1. de Groot JJ, van Vliet J (eds) (1986) The high-pressure sodium lamp. Macmillan, London
2. Brown KE, Chalmers AG, Wharmby DO (1982) Tin sodium-halide lamps in ceramic envelopes. J Illum Eng Soc 11:106–114
3. Carleton S, Seinen PA, Stoffels J (1997) Metal halide lamps with ceramic envelopes: a breakthrough in color control. J Illum Eng Soc 26:139–145
4. Jungst S, Lang D, Galvez M (2004) Improved arc tubes for ceramic metal halide lamps. In: Zissis G (ed) Light sources, p 115–124
5. Krell A, Hutzler T, Klimke J (2007) Transparent ceramics for structural applications: part 2: fields of applications. CFI-Ceram Forum Int 84:E50–E56
6. Krell A, Hutzler T, Klimke J (2007) Transparent ceramics for structural applications: part 1: physics of light transmission and technological consequences. CFI-Ceram Forum Int 84: E41–E50
7. Krell A, Strassburger E (2007) Ballistic strength of opaque and transparent armor. Am Ceram Soc Bull 86:9201–9207
8. Hayashi K, Kobayashi O, Toyoda S, Morinaga K (1991) Transmission optical properties of polycrystallline alumina with submicron grains. Mater Trans JIM 32:1024–1029
9. Krell A, Blank P, Ma HW, Hutzler T, van Bruggen MPB, Apetz R (2003) Transparent sintered corundum with high hardness and strength. J Am Ceram Soc 86:12–18
10. Wei GC (2005) Transparent ceramic lamp envelope materials. J Phys D Appl Phys 38:3057–3065
11. Wei GC (2009) Transparent ceramics for lighting. J Eur Ceram Soc 29:237–244
12. Wei GC (2004) Transparent ceramic lamp envelope materials. In: Zissis G (ed) Light sources 2004. IOP Publishing Ltd, Bristol, pp 169–170
13. Wei GC, Lapatovich WP, Browne J, Snellgrove R (2008) Dysprosium oxide ceramic arc tube for HID lamps. J Phys D Appl Phys 41:144014
14. Guenther K, Hartmann T, Sarroukh H (2004) Hg free ceramic automotive headlight lamps. In: Zissis G (ed) Light sources 2004, pp 219–220
15. Monahan RD, Halloran JW (1979) Single-crystal boundary migration in hot-pressed aluminum-oxide. J Am Ceram Soc 62:564–567
16. Scott C, Kaliszewski M, Greskovich C, Levinson L (2002) Conversion of polycrystalline Al_2O_3 into single-crystal sapphire by abnormal grain growth. J Am Ceram Soc 85:1275–1280
17. Thompson GS, Henderson PA, Harmer MP, Wei GC, Rhodes WH (2004) Conversion of polycrystalline alumina to single-crystal sapphire by localized codoping with silica. J Am Ceram Soc 87:1879–1882
18. Lapatovich WP (2004) Recent advances in lighting science. In: Cohen JS, Mazevet S, Kilcrease DP (eds) Atomic processes in plasmas, pp 255–264
19. Hayashi K, Kobayashi O, Toyoda S, Morinaga K (1991) Trnasmission optical-properties of polycrystalline alumina with submicro grains. Mater Trans JIM 32:1024–1029
20. Wei GC (2002) Characterization of translucent polycrystalline alumina (PCA) ceramics. In: Matsui M, Jahanmir S, Mostaghaci H, Naito M, Uematsu K, Wasche R et al (eds) Improved ceramics through new measurements, processing, and standards, pp 135–144
21. Goldman LM, Foti R, Smith M, Kashalikar U, Sastri S (2010) AlON (R) transparent armor. In: Swab JJ (eds) Advances in ceramic armor V, pp 225–232
22. Goldman LM, Hartnett TM, Wahl JM, Ondercin RJ, Olson K (2001) Recent advances in aluminum oxynitride (ALON (TM)) optical ceramic. In: Tustison RW (ed) Window and dome technologies and materials VII. SPIE-Int Soc Optical Engineering, Bellingham, pp 71–78
23. Goldman LM, Twedt R, Balasubramanian S, Sastri S (2011) AlON (R) optical ceramic transparencies for window, dome and transparent armor applications. Window Dome Technol Mater XII 8016:801608

24. Hartnett TM, Bernstein SD, Maguire EA, Tustison RW (1998) Optical properties of ALON (aluminum oxynitride). Infrared Phys Technol 39:203–211
25. Jiang HW, Du HB, Tian TY, Wu H (2010) Influence of Y_2O_3 additive on transparent of AlON ceramics. In: Pan W, Gong JH (eds) Chinese ceramics communications, pp 580–581
26. Qi JQ, Wang YZ, Lu TC, Yu YG, Pan L, Wei N et al (2011) Preparation and light transmission properties of AlON ceramics by the two-step method with nanosized Al_2O_3 and AlN. Metall Mater Trans A Phys Metall Mater Sci 42A:4075–4079
27. Qi JQ, Zhou JC, Pang W, He JF, Su YY, Liao ZJ et al (2007) Study on the preparation of ALON powder by solid state reaction method. Rare Metal Mater Eng 36:88–91
28. Yuan XY, Zhang F, Liu XJ, Zhang Z, Wang SW (2011) Fabrication of transparent AlON ceramics by solid-state reaction sintering. J Inorg Mater 26:499–502
29. McCauley JW, Corbin ND (1979) Phase relations and reaction sintering of transparent cubic aluminum oxynitride spinel (AlON). J Am Ceram Soc 62:476–479
30. Willems HX, Hendrix MMRM, de With G, Metselaar R (1992) Thermodynamics of AlON I: stability at lower temperatures. J Eur Ceram Soc 10:327–337
31. Willems HX, Hendrix MMRM, de With G, Metselaar R (1992) Thermodynamics of AlON II: phase relations. J Eur Ceram Soc 10:339–346
32. Willems HX, Dewith G, Metselaar R (1993) Thermodynamics of AlON 3. Stabilization of AlON with MgO. J Eur Ceram Soc 12:43–49
33. Autrata R (1989) Backscattered electron imaging using single-crystal sintillator detectors. Scan Microsc 3:739–763
34. Blasse G (1994) Scintillator materials. Chem Mater 6:1465–1475
35. Greskovich C, Duclos S (1997) Ceramic scintillators. Annu Rev Mater Sci 27:69–88
36. Heath RL, Hofstadter R, Hughes EB (1979) Inorganic scintillators—review of techniques and applications. Nucl Instrum Methods 162:431–476
37. Knapitsch A, Lecoq P (2014) Review on photonic crystal coatings for scintillators. Int J Mod Phys A 29
38. Nadaraia L, Jalabadze N, Chedia R, Khundadze L (2013) Production of nanopowder and bulk aluminate ceramic scintillators. Ceram Int 39:2207–2214
39. Nikl M, Laguta VV, Vedda A (2008) Complex oxide scintillators: material defects and scintillation performance. Physica Status Solidi B Basic Solid State Phys 245:1701–1722
40. Nikl M, Yoshikawa A, Kamada K, Nejezchleb K, Stanek CR, Mares JA et al (2013) Development of LuAG-based scintillator crystals—a review. Prog Cryst Growth Charact Mater 59:47–72
41. Rodnyi PA, Dorenbos P, Vaneijk CWE (1995) Energy-loss in inorganic scitillators. Physica Status Solidi B Basic Res 187:15–29
42. Ryzhkov VD, Stadnik PE, Yakovlev YA (1984) Development prospects of scintillator—photodiode systems (review). Instrum Exp Tech 27:1045–1056
43. van Eijk CWE (2002) Inorganic scintillators in medical imaging. Phys Med Biol 47: R85–R106
44. Graichen J, Maier K, Schuth J, Siepe A, von Witsch W (2002) Efficiency and directional effects in the detection of low-energy recoil nuclei in a NaI(Tl) single crystal. Nucl Instrum Methods Phys Res Sect A Accelerators Spectrometers Detectors Assoc Equip 485:774–779
45. Picchio M, Savi A, Lecchi M, Landoni C, Gianolli L, Brioschi M et al (2003) Evaluation of the clinical performances of a large NaI(Tl) crystal 3D PET scanner. Q J Nucl Med 47:90–100
46. Nagarkar VV, Gaysinskiy V, Ovechkina EE, Thacker SC, Miller SR, Brecher C et al (2007) Suppression of afterglow in CsI(Tl) by codoping with Eu^{2+}: fabrication of microcolumnar films for high-resolution high-speed imaging. IEEE Trans Nucl Sci 54:1378–1382
47. Valais I, Nikolopoulos D, Kalivas N, Gaitanis A, Loudos G, Sianoudis I et al (2007) A systematic study of the performance of the CsI: Tl single-crystal scintillator under X-ray excitation. Nucl Instrum Methods Phys Res Sect A Accelerators Spectrometers Detectors Assoc Equip 571:343–345

48. Gironi L, Arnaboldi C, Capelli S, Cremonesi O, Pessina G, Pirro S et al (2009) $CdWO_4$ bolometers for double beta decay search. Opt Mater 31:1388–1392
49. Nagornaya L, Onyshchenko G, Pirogov E, Starzhinskiy N, Tupitsyna I, Ryzhikov V et al (2005) Production of the high-quality $CdWO_4$ single crystals for application in CT and radiometric monitoring. Nucl Instrum Methods Phys Res Sect A Accelerators Spectrometers Detectors Assoc Equip 537:163–167
50. Zaldo C, Moya E (1993) Low-temperaure X-ray-induced optical-properties of Bi4Ge3O12 scintillators. J Phy Condens Matter 5:4935–4944
51. Greskovich CD, Cusano D, Hoffman D, Riedner RJ (1992) Ceramic scintillators for advanced, medical X-ray-detectors. Am Ceram Soc Bull 71:1120–1130
52. Martin T, Douissard PA, Seeley ZM, Cherepy N, Payne S, Mathieu E et al (2012) New high stopping power thin scintillators based on Lu_2O_3 and $Lu_3Ga_{5-x}In_xO_{12}$ for high resolution X-ray imaging. IEEE Trans Nucl Sci 59:2269–2274
53. Placious RC, Polansky D, Berger H, Bueno C, Vosberg CL, Betz RA et al (1991) High-density glass scintillator for real-time X-ray inspection. Mater Eval 49:1419–1421
54. Bueno C, Barker MD, Betz RA, Barry RC, Buchanan RA (1995) Nondestructive evaluation of aircraft structures using high-resolution real-time radiography
55. Chen YP, Luo DL (2012) Development of containing Li-6 glass scintillators for neutron detection. J Inorg Mater 27:1121–1128
56. Sun XY, Zhang M, Yu XG (2011) Luminescence properties of Tb^{3+}-activated silicate glass scintillator. Int J Mater Res 102:104–108
57. Sangway PC, Stapleton RE (1977) X-ray-emission spectra of LaOBr: Tm and its impact on screen-film system-design. J Electrochem Soc 124:C110
58. Somaiah K, Narayana MV, Brixner LH (1990) Thermally stimulated luminescence of LaOBr:Tm. Physica Status Solidi A Appl Res 117:K81–K84
59. Shanker V, Ohmi K, Tanaka S, Kobayashi H (1998) Gd_2O_2S: Tb phosphor thin films grown by electron beam evaporation and their photoluminescent and electroluminescent characteristics. IEICE Trans Electron E81C:1721–1724
60. Tao S, Gu ZH, Nathan A (2002) Fabrication of Gd2O2S: Tb based phosphor films coupled with photodetectors for X-ray imaging applications. J Vac Sci Technol A Vac Surf Films 20:1091–1094
61. Timmer JH, van Vuure TL, Bom V, van Eijk CW, de Haas J, Schippers JM (2002) A scintillating GEM for 2D-dosimetry in radiation therapy. Nucl Instrum Methods Phys Res Sect A Accelerators Spectrometers Detectors Assoc Equip 478:98–103
62. Schippers JM, Boon SN, van Luijk P (2002) Applications in radiation therapy of a scintillating screen viewed by a CCD camera. Nucl Instrum Methods Phys Res Sect A Accelerators Spectrometers Detectors Assoc Equip 477:480–485
63. Klyachko AV, Moskvin V, Nichiporov DF, Solberg KA (2012) A GEM-based dose imaging detector with optical readout for proton radiotherapy. Nucl Instrum Methods Phys Res Sect A Accelerators Spectrometers Detectors Assoc Equip 694:271–279
64. Bertrand GHV, Hamel M, Normand S, Sguerra F (2015) Pulse shape discrimination between (fast or thermal) neutrons and gamma rays with plastic scintillators: state of the art. Nucl Instrum Methods Phys Res Sect A Accelerators Spectrometers Detectors Assoc Equip 776:114–128
65. Bertrand GHV, Hamel M, Sguerra F (2014) Current status on plastic scintillators modifications. Chem-A Eur J 20:15660–15685
66. Cherepy NJ, Sanner RD, Beck PR, Swanberg EL, Tillotson TM, Payne SA et al (2015) Bismuth- and lithium-loaded plastic scintillators for gamma and neutron detection. Nucl Instrum Methods Phys Res Sect A Accelerators Spectrometers Detectors Assoc Equip 778:126–132
67. Boyer K, Thompson BD, McPherson A, Rhodes CK (1994) Evidence for coherent electron motions in multiphoton X-ray-production from Kr and Xe clusters. J Phys B Atomic Mol Opt Phys 27:4373–4389

68. Kubo H, Sasaki A, Moribayashi K, Higashijima S, Takenaga H, Shirnizu K et al (2007) Study of highly ionized Xe spectra with 3s-3p and 3p-3d transitions in JT-60U reversed shear plasmas. J Nucl Mater 363:1441–1445
69. Takahashi T, Nakagawa M, Yoshida M, Takeuchi H (1992) Highly stable solid-state X-ray-detector array. Med Phys 19:1161–1166
70. Yoshida M, Nakagawa M, Fujii H, Kawaguchi F, Yamada H, Ito Y et al (1988) Application of Gd_2O_2S ceramic scintillator for X-ray solid-state detectro in X-ray CT. Jpn J Appl Phys Part 2 Lett 27:L1572–L1575
71. Duclos SJ, Greskovich CD, Lyons RJ, Vartuli JS, Hoffman DM, Riedner RJ et al (2003) Development of the HiLight (TM) scintillator for computed tomography medical imaging. Nucl Instrum Methods Phys Res Sect A Accelerators Spectrometers Detectors Assoc Equip 505:68–71
72. Greskovich C, Chernoch JP (1973) Polycrystalline ceramic lasers. J Appl Phys 44:4599–4606
73. Kostler W, Winnacker A, Rossner W, Grabmaier BC (1995) Effect of Pr-codoping on the X-ray-induced afterglow of $(Y,Gd)_2O_3$:Eu. J Phys Chem Solids 56:907–913
74. Godmanis I, Hohenau W (1989) On the nature of afterglow of the X-ray-induced luminescence in crystalline and glassy SiO_2. Physica Status Solidi A Appl Res 111:335–343
75. Yanagida T, Fujimoto Y, Ito T, Uchiyama K, Mori K (2014) Development of X-ray-induced afterglow characterization system. Appl Phys Expr 7
76. Yanagida T, Sato H (2014) Optical and scintillation properties of Nd-doped complex garnet. Opt Mater 38:174–178
77. Knupfer W, Hell E, Mattern D (1999) Novel X-ray detectors for medical imaging. Nucl Phys B Proc Suppl 78:610–615
78. Yamada H, Suzuki A, Uchida Y, Yoshida M, Yamamoto H, Tsukuda Y (1989) A scintillator Gd_2O_2S:Pr, Ce, F for X-ray comuted-tomography. J Electrochem Soc 136:2713–2716
79. Shimizu J, Yamada H, Murata S, Tomita A, Kitamura M, Suzuki A (1991) Optical-confinement-factor dependencies of the K-factor, differential gain, and nonlinear gain coefficient for 1.55 μm InGaAs/InGaAsP MQW and strained-MQW lasers. IEEE Photonics Technol Lett 3:773–776
80. Rodnyi PA, Dorenbos P, Vaneijk CWE (1994) Creation of elecron-hole pairs in inorganic scintillators. In: Weber MJ, Lecoq P, Ruchti RC, Woody C, Yen WM, Zhu RY (eds) Scintillator Phosphor Mater pp 379–385
81. Rodnyi PA, Dorenbos P, Vaneijk CWE (1995) Energy-loss in inorganic scintillators. Physica Status Solidi B Basic Res 187:15–29
82. Devanathan R, Corrales LR, Gao F, Weber WJ (2006) Signal variance in γ-ray detectors—a review. Nucl Instrum Methods Phys Res Sect A Accelerators Spectrometers Detectors Assoc Equip 565:637–649
83. Gruner SM, Tate MW, Eikenberry EF (2002) Charge-coupled device area X-ray detectors. Rev Sci Instrum 73:2815–2842
84. Morgan GP, Glynn TJ, Imbusch GF, Remeika JP (1978) Luminescence from $Al_{2x}Ga_{2(1-x)}O_3$: Cr^{3+}. J Chem Phys 69:4859–4866
85. Morgan GP, Treacy JF, Glynn TJ, Imbusch GF, Remeika JP (1979) Luminescence from $Al_{2x}Ga_{2(1-x)}O_3$:Cr^{3+}. J Lumin 18–9:175–178
86. Donnelly CJ, Glynn TJ, Morgan GP, Imbusch GF (1991) Level-crossing in chromium-doped materials. J Lumin 48–9:283–287
87. Henderson B, Imbusch GF (1988) Optical processes in tunable transition-metal-ion lasers. Contemp Phys 29:235–272
88. Imbusch GF (1992) Energy transfer in heavily doped ruby. J Lumin 53:465–467
89. Maccraith BD, Glynn TJ, Imbusch GF, McDonagh C (1980) Optical excitation transfer in MgO:Cr^{3+}. J Phys C Solid State Phys 13:4211–4216
90. Ilmer M, Schneider J, Ostertag M, Berthold T, Bohaty L, Grabmaier BC (1994) Studies of structural and luminescent properties of $Gd_3Ga_5O_{12}$:Cr and corresponding solid solutions. In:

Delhez R, Mittemeijer EJ (eds) Epdic 3, Pts 1 and 2: proceedings of the 3rd European powder diffraction conference, pp 801–806

91. Yamaji A, Kochurikhin VV, Kurosawa S, Suzuki A, Fujimoto Y, Yokota Y et al (2014) Luminescence properties of $Gd_3Ga_5O_{12}$: Cr single crystals. IEEE Trans Nucl Sci 61:320–322
92. Blasse G, Grabmaier BC, Ostertag M (1993) The afterglow mechanism of chromium-doped gadolinium gallium garnet. J Alloy Compd 200:17–18
93. Forest H, Ban G (1969) Evidence for Eu^{3+}-emission from 2 symmetry sites in $Y_2O_3:Eu^{3+}$. J Electrochem Soc 116:474–478
94. Forest H, Ban G (1971) Random substitution of Eu^{3+} for Y^{3+} in $Y_2O_3:Eu^{3+}$. J Electrochem Soc 118:1999–2001
95. Duclos SJ, Greskovich CD, Oclair CR (1994) Electronic trap defects in Y_2O_3:Eu and (Y, Gd)$_2$O$_3$:Eu X-ray scintillators. In: Weber MJ, Lecoq P, Ruchti RC, Woody C, Yen WM, Zhu RY (eds) Scintillator and phosphor materials, pp 503–509
96. Hsieh J, Williams E, Shaughnessy C, Tang XY, Grekowicz B, Nilsen R et al (2004) Optimization of system design for 64-slice cone beam computed tomography. In: Bonse U (ed) Developments in X-ray tomography Iv, pp 694–700
97. Grassmann H, Moser HG, Dietl H, Eigen G, Fonseca V, Lorenz E et al (1985) Improvements in photodiode readout for small CsI(Tl) crystals. Nucl Instrum Methods Phys Res Sect A Accelerators Spectrometers Detectors Assoc Equip 234:122–124
98. Kostler W, Bayer E, Rossner W, Grabmaier BC (1993) Correlation of time and temperature-dependence of the luminescence of rare-earth activated (Y,Gd)$_2$O$_3$. Nucl Tracks Radiat Meas 21:135–138
99. Cherepy NJ, Kuntz JD, Seeley ZM, Fisher SE, Drury OB, Sturm BW et al (2010) Transparent ceramic scintillators for gamma spectroscopy and radiography. In: Burger A, Franks LA, James RB (eds) Hard X-ray, gamma-ray, and neutron detector physics XII
100. Cherepy NJ, Kuntz JD, Roberts JJ, Hurst TA, Drury OB, Sanner RD et al (2008) Transparent ceramic scintillator fabrication, properties and applications. In: Burger A, Franks LA, James RB (eds) Hard X-ray, gamma-ray, and neutron detector physics X. Spie-Int Soc Optical Engineering, Bellingham
101. Lempicki A, Brecher C, Szupryczynski P, Lingertat H, Nagarkar VV, Tipnis SV et al (2002) A new lutetia-based ceramic scintillator for X-ray imaging. Nucl Instrum Methods Phys Res Sect A Accelerators Spectrometers Detectors Assoc Equip 488:579–590
102. Brecher C, Bartram RH, Lempicki A (2004) Hole traps in Lu_2O_3: Eu ceramic scintillators. I. Persistent afterglow. J Lumin 106:159–168
103. Bartram RH, Lempicki A, Kappers LA, Hamilton DS (2004) Hole traps in Lu_2O_3: Eu ceramic scintillators. II. Radioluminescence and thermoluminescence. J Lumin 106:169–176
104. Yanagida T, Fujimoto Y, Yagi H, Yanagitani T (2014) Optical and scintillation properties of transparent ceramic Yb:Lu_2O_3 with different Yb concentrations. Opt Mater 36:1044–1048
105. Futami Y, Yanagida T, Fujimoto Y, Pejchal J, Sugiyama M, Kurosawa S et al (2013) Optical and scintillation properties of Sc_2O_3, Y_2O_3 and Lu_2O_3 transparent ceramics synthesized by SPS method. Radiat Meas 55:136–140
106. Hui Y, Sun X, Chen J, Li X, Huo D, Liu S et al (2014) The fabrication of monoclinic Gd_2O_3 transparent microspheres and scintillator array via laser heating. IEEE Trans Nucl Sci 61:367–372
107. Yanagida T, Roh T, Takahashi H, Hirakuri S, Kokubun M, Makishima K et al (2007) Improvement of ceramic YAG(Ce) scintillators to (YGd)$_3$Al$_5$O$_{12}$(Ce) for gamma-ray detectors. Nucl Instrum Methods Phys Res Sect A Accelerators Spectrometers Detectors Assoc Equip 579:23–26
108. Cherepy NJ, Kuntz JD, Tillotson TM, Speaks DT, Payne SA, Chai BHT et al (2007) Cerium-doped single crystal and transparent ceramic lutetium aluminum garnet scintillators. Nucl Instrum Methods Phys Res Sect A Accelerators Spectrometers Detectors Assoc Equip 579:38–41

109. Liu S, Feng X, Shi Y, Wu L, Luo J, Wang W et al (2014) Fabrication, microstructure and properties of highly transparent Ce^{3+}:$Lu_3Al_5O_{12}$ scintillator ceramics. Opt Mater 36:1973–1977
110. Hupke R, Doubrava C (1999) The new UFC-detector for CT-imaging. Physica Med 15:315–318
111. Nagarkar VV, Miller SR, Tipnis SV, Lempicki A, Brecher C, Lingertat H (2004) A new large area scintillator screen for X-ray imaging. Nucl Instrum Methods Phys Res Sect B Beam Interact Mater Atoms 213:250–254
112. Bartram RH, Hamilton DS, Kappers LA, Lempicki A (1997) Electron traps and transfer efficiency of cerium-doped aluminate scintillators. J Lumin 75:183–192
113. Bartram RH, Hamilton DS, Kappers LA, Lempicki A, Glodo J, Schweitzer JS et al (1999) Electron traps and transfer efficiency in cerium-doped lutetium oxyorthosilicate scintillators. Radiat Eff Defects Solids 150:11–14
114. Zych E, Brecher C, Wojtowicz AJ, Lingertat H (1997) Luminescence properties of Ce-activated YAG optical ceramic scintillator materials. J Lumin 75:193–203
115. Mihokova E, Nikl M, Mares JA, Beitlerova A, Vedda A, Nejezchleb K et al (2007) Luminescence and scintillation properties of YAG: Ce single crystal and optical ceramics. J Lumin 126:77–80
116. Zorenko Y, Gorbenko V, Konstankevych I, Voloshinovskii A, Stryganyuk G, Mikhallin V et al (2005) Single-crystalline films of Ce-doped YAG and LuAG phosphors: advantages over bulk crystals analogues. J Lumin 114:85–94
117. Lu JR, Ueda K, Yagi H, Yanagitani T, Akiyama Y, Kaminskii AA (2002) Neodymium doped yttrium aluminum garnet ($Y_3Al_5O_{12}$) nanocrystalline ceramics—a new generation of solid state laser and optical materials. J Alloy Compd 341:220–225
118. Li HL, Liu XJ, Huang LP (2005) Fabrication of transparent cerium-doped lutetium aluminum garnet (LuAG:Ce) ceramics by a solid-state reaction method. J Am Ceram Soc 88: 3226–3228
119. Li HL, Liu XJ, Huang LP (2006) Fabrication of transparent Ce:LuAG ceramics by a solid-state reaction method. J Inorg Mater 21:1161–1166
120. Li HL, Liu XJ, Xie RJ, Zeng Y, Huang LP (2006) Fabrication of transparent cerium-doped lutetium aluminum garnet ceramics by co-precipitation routes. J Am Ceram Soc 89: 2356–2358
121. Li HL, Liu XJ, Xie RJ, Zhou GH, Hirosaki N, Pu XP et al (2008) Cerium-doped lutetium aluminum garnet phosphors and optically transparent ceramics prepared from powder precursors by a urea homogeneous precipitation method. Jpn J Appl Phys 47:1657–1661
122. Nikl M, Mares JA, Solovieva N, Li HL, Liu XJ, Huang LP et al (2007) Scintillation characteristics of $Lu_3Al_5O_{12}$: Ce optical ceramics. J Appl Phys 101:033515
123. Yanagida T, Fujimoto Y, Yokota Y, Yoshikawa A, Kuretake S, Kintaka Y et al (2011) Evaluations of pure and ytterbium doped transparent ceramic complex perovskite scintillators. Opt Mater 34:414–418
124. Chipaux R, Cribier M, Dujardin C, Garnier N, Guerassimova N, Mallet J et al (2002) Ytterbium-based scintillators, a new class of inorganic scintillators for solar neutrino spectroscopy. Nucl Instrum Methods Phys Res Sect A Accelerators Spectrometers Detectors Assoc Equip 486:228–233
125. Guerassimova N, Dujardin C, Garnier N, Pedrini C, Petrosyan AG, Kamenskikh IA et al (2002) Charge-transfer luminescence and spectroscopic properties of Yb^{3+} in aluminium and gallium garnets. Nucl Instrum Methods Phys Res Sect A Accelerators Spectrometers Detectors Assoc Equip 486:278–282
126. Antonini P, Belogurov S, Bressi G, Carugno G, Santilli P (2002) Scintillation properties of Yb-doped yttrium-aluminum garnets. Nucl Instrum Methods Phys Res Sect A Accelerators Spectrometers Detectors Assoc Equip 488:591–603
127. Belogurov S, Bressi G, Carugno G, Grishkin Y (2004) Properties of Yb-doped scintillators: YAG, YAP, LuAG. Nucl Instrum Methods Phys Res Sect A Accelerators Spectrometers Detectors Assoc Equip 516:58–67

128. Belogurov S, Bressi G, Carugno G, Moszynski M, Czarnacki W, Kapusta M et al (2003) Characterization of Yb:YAG and Yb:YAP scintillators by means of LAAPD at temperature around 100K. Nucl Instrum Methods Phys Res Sect A Accelerators Spectrometers Detectors Assoc Equip 496:385–389
129. Jiang H, Zou YK, Chen Q, Li KK, Zhang R, Wang Y et al (2005) Transparent electro-optic ceramics and devices. In: Ming H, Zhang XP, Chen MY (eds) Optoelectronic devices and integration, Pts 1 and 2. SPIE-Int Soc Optical Engineering, Bellingham, pp 380–394
130. Song QW, Wang XM, Bussjager R, Osman J (1996) Electro-optic beam-steering device based on a lanthanum-modified lead zirconate titanate ceramic wafer. Appl Opt 35:3155–3162
131. Haertling GH (1999) Ferroelectric ceramics: history and technology. J Am Ceram Soc 82:797–818
132. Zeng X, He XY, Cheng WX, Qiu PS, Xia B (2014) Effect of Dy substitution on ferroelectric, optical and electro-optic properties of transparent $Pb_{0.90}La_{0.10}(Zr_{0.65}Ti_{0.35})O_3$ ceramics. Ceram Int 40:6197–6202
133. Qiao L, Ye Q, Gan JL, Cai HW, Qu RH (2011) Optical characteristics of transparent PMNT ceramic and its application at high speed electro-optic switch. Opt Commun 284:3886–3890
134. Dong ZR, Ye Q, Qu RH, Fang ZJ (2007) Characteristics of a PLZT electro-optical deflector. Chin Optics Lett 5:540–542
135. Peters BR, Reardon PJ, Wong JK (2001) Preliminary investigation of an active PLZT lens. In: Righini GC, Honkanen S (eds) Integrated optics devices V, pp 242–248
136. Tatebayashi T, Yamamoto T, Sato H (1991) Electrooptic variable focal-length lens using PLZT ceramic. Appl Opt 30:5049–5055
137. Tatebayashi T, Yamamoto T, Sato H (1992) Dual focal point electrooptic lens with a fresnel-zone plate on a PLZT ceramic. Appl Opt 31:2770–2775
138. Staudigl G, Benien H, Suchentrunk R (2000) Transparent scratch resistant coatings on plastics. Materialwiss Werkstofftech 31:360–364
139. Peuchert U, Okano Y, Menke Y, Reichel S, Ikesue A (2009) Transparent cubic-ZrO_2 ceramics for application as optical lenses. J Eur Ceram Soc 29:283–291
140. Grujicic M, Bell WC, Pandurangan B (2012) Design and material selection guidelines and strategies for transparent armor systems. Mater Des 34:808–819
141. Salem JA (2013) Transparent armor ceramics as spacecraft windows. J Am Ceram Soc 96:281–289
142. Subhash G (2013) Transparent armor materials. Exp Mech 53:1–2
143. Klein CA (2005) Oxyfluoride glass for high-energy laser windows: thermal lensing issue. Appl Phys Lett 87:231117
144. Klein CA (2006) Calcium fluoride windows for high-energy chemical lasers. J Appl Phys 100:083101
145. Pazol B, McGuire P, Gentilman R, Locher J, Askinazi J (2000) Large area flat and curved sapphire window blanks. In: Marker AJ, Arthurs EG (eds) Inorganic optical materials II, pp 52–58
146. Askinazi J, Wientzen RV, Khattak CP (2001) Development of large aperture, monolithic sapphire optical windows. In: Tustison RW (ed) Window and dome technologies and materials VII, pp 1–11
147. Askinazi J (1991) Technology trends for high-performance windows
148. Askinazi J (1993) Functional challenges for optical window materials
149. Askinazi J (1997) Large aperture, broadband sapphire windows for common aperture, target acquisition, tracking and surveillance systems
150. Borden MR, Askinazi J (1997) Improving sapphire window strength
151. Gentilman R, McGuire P, Fiore D, Ostreicher K, Askinazi J (2003) Low-cost sapphire windows. In: Tustison RW (ed) Window and dome technologies VIII, pp 12–17
152. Gentilman R, McGuire P, Fiore D, Ostreicher K, Askinazi J (2003) Large-area sapphire windows. In: Tustison RW (ed) Window and dome technologies VIII, pp 54–60

153. Gentilman R, McGuire P, Pazol B, Askinazi J, Steindl R, Locher J (1999) High strength edge-bonded sapphire windows. In: Tustison RW (ed) Window and dome technologies and materials VI, pp 282–287
154. McGuire P, Pazol B, Gentilman R, Askinazi J, Locher J (2001) Large area edge-bonded flat and curved sapphire windows. In: Tustison RW (ed) Window and dome technologies and materials VII, pp 12–19
155. Patterson MCL, DiGiovanni AA, Fehrenbacher L, Roy DW (2003) Spinel: gaining momentum in optical applications. In: Tustison RW (ed) Window and dome technologies VIII. SPIE-Int Soc Optical Engineering, Bellingham, pp 71–79
156. Borrero-Lopez O, Ortiz AL, Gledhill AD, Guiberteau F, Mroz T, Goldman LM et al (2012) Microstructural effects on the sliding wear of transparent magnesium-aluminate spinel. J Eur Ceram Soc 32:3143–3149
157. Mroz T, Goldman LM, Gledhill AD, Li DS, Padture NP (2012) Nanostructured, infrared-transparent magnesium-aluminate spinel with superior mechanical properties. Int J Appl Ceram Technol 9:83–90
158. Mroz TJ, Hartnett TM, Wahl JM, Goldman LM, Kirsch J, Lindberg WR (2005) Recent advances in spinel optical ceramic. In: Tustison RW (ed) Window and dome technologies and materials IX. SPIE-Int Soc Optical Engineering, Bellingham, pp 64–70
159. Ramisetty M, Sastri S, Kashalikar U, Goldman LM, Nag N (2013) Transparent polycrystalline cubic spinels protect and defend. Am Ceram Soc Bull 92:20–25
160. Beyer RA, Kerwien H (1999) Evaluation of ALON for cannon window application. In: Tustison RW (ed) Window and dome technologies and materials VI, pp 113–118
161. Krell A, Bales A (2011) Grain size-dependent hardness of transparent magnesium aluminate spinel. Int J Appl Ceram Technol 8:1108–1114
162. Krell A, Hutzler T, Klimke J, Potthoff A (2010) Fine-grained transparent spinel windows by the processing of different manopowders. J Am Ceram Soc 93:2656–2666
163. Krell A, Klimke J, Hutzler T (2009) Advanced spinel and sub-μm Al_2O_3 for transparent armour applications. J Eur Ceram Soc 29:275–281
164. Bayya S, Villalobos G, Kim W, Sanghera J, Chin G, Hunt M et al (2013) Recent developments in transparent spinel ceramic and composite windows. Mater Technol Appl Opt Struct Compon Sub-Syst 8837
165. Klement R, Rolc S, Mikulikova R, Krestan J (2008) Transparent armour materials. J Eur Ceram Soc 28:1091–1095
166. Sands JM, Fountzoulas CG, Gilde GA, Patel PJ (2009) Modelling transparent ceramics to improve military armour. J Eur Ceram Soc 29:261–266
167. Vysotsky MS, Pochtenny EK, Gorbatsevich MI (1992) Summartion of fatigure damages under 2-frequency loading. Dokl Akad Nauk BSSR 36:511–514
168. Vysotsky MS, Pochtenny EK, Gorbatsevich MI, Kurban AV (1991) Fatigure resistance analysis at 2-frequency loading. Dokl Akad Nauk BSSR 35:334–337
169. Strassburger E (2009) Ballistic testing of transparent armour ceramics. J Eur Ceram Soc 29:267–273
170. Frungel F, Alberti H, Thorwart W (1962) High-speed X-ray flash cinematography of small objects. J SMPTE-Soc Motion Picture Telev Eng 71:90–92
171. Frungel F, Thorwart W, Alberti H (1960) High-speed X-ray flash cinematography of small objects. J SMPTE-Soc Motion Picture Telev Eng 69:624
172. McCauley JW, Strassburger E, Patel P, Paliwal B, Ramesh KT (2013) Experimental observations on dynamic response of selected transparent armor materials. Exp Mech 53:3–29
173. Strassburger E (2004) Visualization of impact damage in ceramics using the edge-on impact technique. Int J Appl Ceram Technol 1:235–242
174. Senf H, Strassburger E, Rothenhausler H (1994) Stress wave induced damage and fracture in impacted glasses. J Phys IV 4:741–746
175. Strassburger E, Hunzinger M, Patel Pl, McCauley JW (2013) Analysis of the fragmentation of AlON and spinel under ballistic impact. J Appl Mech Trans ASME 80

176. Strassburger E, Patel P, McCauley JW, Templeton DW (2006) High-speed photographic study of wave propagation and impact damage in fused silica and AlON using the edge-on impact (EOI) method. In: Furnish MD, Elert M, Russell TP, White CT (eds) Shock compression of condensed matter—2005, Pts 1 and 2. American Institute of Physical, Melville, pp 892–895
177. LaSalvia JC, McCauley JW (2010) Inelastic deformation mechanisms and damage in structural ceramics subjected to high-velocity impact. Int J Appl Ceram Technol 7:595–605
178. Haney EJ, Subhash G (2012) Edge-on-impact response of a coarse-grained magnesium aluminate spinel rod. Int J Impact Eng 40–41:26–34
179. Bodhak S, Balla VK, Bose S, Bandyopadhyay A, Kashalikar U, Jha SK et al (2011) In vitro biological and tribological properties of transparent magnesium aluminate (Spinel) and aluminum oxynitride (AlON (R)). J Mater Sci Mater Med 22:1511–1519
180. Bodhak S, Bose S, Bandyopadhyay A (2009) Role of surface charge and wettability on early stage mineralization and bone cell-materials interactions of polarized hydroxyapatite. Acta Biomater 5:2178–2188
181. Bodhak S, Bose S, Bandyopadhyay A (2010) Electrically polarized HAp-coated Ti: in vitro bone cell-material interactions. Acta Biomater 6:641–651
182. Krishnan M, Tiwari B, Seema S, Kalra N, Biswas P, Rajeswari K et al (2014) Transparent magnesium aluminate spinel: a prospective biomaterial for esthetic orthodontic brackets. J Mater Sci Mater Med 25:2591–2599
183. Williams DF (2008) On the mechanisms of biocompatibility. Biomaterials 29:2941–2953
184. Hanks CT, Wataha JC, Sun ZL (1996) In vitro models of biocompatibility: a review. Dent Mater 12:186–193
185. Klein CL, Kohler H, Reclaru L, Susz C, Kirkpatrick CJ (1995) Biocompatibility testing of alloys in an in-vitro cell-culture model. J Dent Res 74:921
186. Orrenius S, Nicotera P, Zhivotovsky B (2011) Cell death mechanisms and their implications in toxicology. Toxicol Sci 119:3–19
187. Cho MH, Niles A, Huang RL, Inglese J, Austin CP, Riss T et al (2008) A bioluminescent cytotoxicity assay for assessment of membrane integrity using a proteolytic biomarker. Toxicol In Vitro 22:1099–1106
188. Negroiu G, Piticescu RM, Chitanu GC, Mihailescu IN, Zdrentu L, Miroiu M (2008) Biocompatibility evaluation of a novel hydroxyapatite-polymer coating for medical implants (in vitro tests). J Mater Sci Mater Med 19:1537–1544

FSC
www.fsc.org
MIX
Papier aus verantwortungsvollen Quellen
Paper from responsible sources
FSC® C105338

If you have any concerns about our products,
you can contact us on
ProductSafety@springernature.com

In case Publisher is established outside the EU,
the EU authorized representative is:
Springer Nature Customer Service Center GmbH
Europaplatz 3, 69115 Heidelberg, Germany

Printed by Libri Plureos GmbH
in Hamburg, Germany